Gerhard Ringel

Rainer Bodendiek · Rudolf Henn (Eds.)

Topics in Combinatorics and Graph Theory

Essays in Honour of Gerhard Ringel

With 178 Figures

Physica-Verlag Heidelberg

Professor Dr. Rainer Bodendiek
Institut für Mathematik und ihre Didaktik
Pädagogische Hochschule Kiel
Olshausenstraße 75
2300 Kiel, FRG

Professor Dr. Rudolf Henn (died 1989)
Institut für Statistik und
Mathematische Wirtschaftstheorie
Universität Karlsruhe
Rechenzentrum, Zirkel 2
Postfach 6980
7500 Karlsruhe, FRG

ISBN-13: 978-3-642-46910-7 e-ISBN-13: 978-3-642-46908-4
DOI: 10.1007/978-3-642-46908-4

CIP-Titelaufnahme der Deutschen Bibliothek
Topics in combinatorics and graph theory: essays in honour of
Gerhard Ringel / Rainer Bodendiek; Rudolf Henn (eds.). -
Heidelberg: Physica-Verl., 1990
ISBN-13: 978-3-642-46910-7
NE: Bodendiek, Rainer [Hrsg.]; Ringel, Gerhard: Festschrift

Softcover reprint of the hardcover 1st edition 1990

7120/7130-543210

Preface

Graph theory and combinatorics are two different areas of discrete mathematics. Although graph theory is intimately related to combinatorics and it is not always easy to decide whether certain contributions belong to graph theory or combinatorics it is obvious that graph theory is not a part of combinatorics. This means that these two areas are two independent members of discrete mathematics. Furthermore, it is nearly superfluous to mention that there are a lot of applications of graph theory to some areas of physics, chemistry, commuciation science, computer technology, operational research, psychology, linguistics, economics and a great number of branches of mathematics including group theory, matrix theory, numerical analysis, probability, topology and combinatorics.

Therefore, it is not very surprising that the graph theory has got a fantastic development during the last thirty years. The number of graph theorists and of papers dealing with graph theoretical topics increases enormously. It is allowed to mention the graph theorists Klaus Wagner, Gerhard Ringel and Horst Sachs as pioneers of graph theory in Germany. Since one of them, Gerhard Ringel, has become 70 last year (1989) , we chose the Thursday during the graph theory meeting, held at Oberwolfach from 3 - 9 June 1990, to honour the work of Gerhard Ringel. About 50 mathematicians from 15 countries participated in lectures and discussions on graph theory related to Ringel's work.

Gerhard Ringel is among the most quoted graph theorists. He has developed concepts and methods of great originality, made a lot of fundamental discoveries and conjectures, and has seen the necessity to introduce graph theory in economics.
In order to manage a fruitful cooperation between economics and graph theory Gerhard Ringel worked together with the economist Rudolf Henn, the co-editor of this volume. Rudolf Henn died in November 1989. Although he was not a mathematician he became a great friend of the graph theory and introduced various graph theoretical methods in economics and wrote several books dealing with a lot of topics of graph theory in economics.

Although he was seriously ill,he managed to settle the arising problems in a wonderful manner such that it was a great honour to me to work together with Rudolf Henn.

This volume contains 86 contributions in honour of Gerhard Ringel given by 128 authors from 22 countries. I wish to thank all the contributors. I also thank Gerda Schmidt and Dr. Claus Thies who have helped preparing and typing the manuscripts.
Professor Heiko Harborth is thanked for giving the appreciation of Gerhard Ringel. Finally, we are especially grateful to Dr. Werner A. Müller from Physica-Verlag for his cooperation in all aspects of the production of this book.

Rainer Bodendiek

Contents

Gerhard Ringel - Seventy

October 28, 1989, was the seventieth return of Gerhard Ringel's birthday. On this occasion around seventy mathematicians have submitted articles for this volume to honor Gerhard Ringel who himself has published around seventy mathematical papers. So there are reasons enough to celebrate seventy.

In the mathematical world of Graph Theory and Combinatorics Gerhard Ringel is a well-known mathematician. Since nearly fourty years he has solved some old conjectures, mainly in topological Graph Theory, and he created several nice new problems which stimulated many other people working in Graph Theory.

Gerhard Ringel was born in Großdorf near Braunau in Austria. After his attendance at the gymnasium of Braunau he had to serve in the German army since 1940. Because of a heart-defect he soon could interrupt his military service for two years, and he studied at the Karls University of Prague. Then, however, more than seven years followed where he was soldier in the German army, and then a prisoner-of-war in the Soviet Union. At the end of 1949 he could continue to study mathematics, now at the University of Bonn. He received his doctorate in 1951, two years later the Habilitation, and he continued at Bonn as a junior faculty member.

In 1960 he moved to the Free University of Berlin as Professor of Mathematics. Also in Berlin, in 1966 he became an Ordinarius and Director of the Mathematical Institute. Within the next four years visiting professorships in Rome (1966), Santa Cruz (1967/68), and Calgary (1969) were included. In 1970 he left Berlin and followed the offered permanent position at the University of California, Santa Cruz, to be the successor of his friend and coauthor J.W.T. Youngs.

In Santa Cruz Gerhard Ringel lives together with his wife, Isolde, in a nice house just at the shore of the Pacific Ocean. From his first marriage he has three children, Gerhard, Ingrid, and Renate.

Gerhard Ringel has several sporting hobbies. In his Institute at Berlin regularly soccer was played. In California he became a surfer. He plays chess, and still today he nearly every day plays tennis. Moreover, he collects butterflies, and for this reason the Ringels travel to exotic tropical countries, such as Costa Rica, Ecuador, or Indonesia.

Still today Gerhard Ringel serves the University of Santa Cruz as Professor. He was Head of Department for more than twelve years. His main research interests are map coloring and embeddings of graphs into surfaces. These topics each year are offered by him in a combinatorics class. So far he had six doctoral students, and three of them have published several common papers with him.

In 1983 Gerhard Ringel was awarded an Honorary Doctorate by the University Fridericiana at Karlsruhe. He has written three books, one of which has been translated into Russian. The title of the forthcoming text book "Pearls in Graph Theory" gives an impression of Ringel's mental attitude to Graph Theory. Gerhard Ringel is member of the editorial boards of Archiv der Mathematik, Journal of Graph Theory, and Journal of Combinatorial Theory B (up to 1983). Since the sixtieth he was chairman, always assisted by his wife Isolde, of all Oberwolfach conferences in Graph Theory scheduled every two or three years.

When Gerhard Ringel started in the fiftieth to work in Graph Theory there were only very few other graph theorists among the German speaking mathematicians, and now there is a large Graph Theory community.

The main mathematical contributions of Gerhard Ringel are the proof of Heawood's map color formula, the method of current graphs, the solution of Heawood's empire conjecture, a characterization of self-complementary graphs, some non-strechable arrangements of pseudolines, minimal triangulations and quadrangulations of 2-manifolds, and results on several measurements of closeness of graphs to planarity, like genus, thickness, coarseness, crossing number, or splitting number. As an example, the Figure gives a perfect splitting of the complete graph K_{12}, and also a map of 12 mutually adjacent empires, each of two countries.

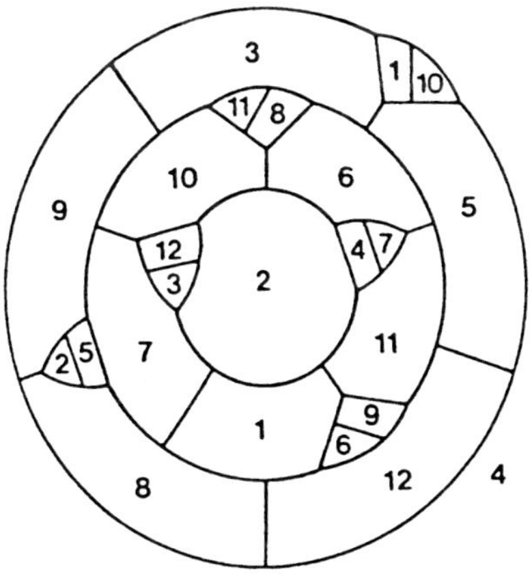

Several nice problems and conjectures created by Gerhard Ringel are research objects for many mathematicians, for example, the conjecture of graceful trees, a six-color problem, recently solved by a Russian mathematician, the Oberwolfach problem, or the recent conjecture, that every connected graph, excluded K_2, is antimagic, that means, there exists a labelling of the q edges with numbers 1 to q such that all sums of the labels at any vertex are different.

Gerhard Ringel's work is part of the blooming epoch of Graph Theory during the last decades. Gerhard Ringel is not an abstract mathematician, he loves Graph Theory, he likes to see things also intuitively, he prefers nice theorems and nice conjectures, he is an aesthet. If sometimes also a book "Pearls of Graph Theorists" should be written, I think, Gerhard Ringel should be one of the pearls.

Heiko Harborth, Braunschweig

List of Publications

G. Ringel

1. Farbensatz für nichtorientierbare Flächen beliebigen Geschlechts. J. Reine Angew. Math. 190 (1952), 129-147.
2. Bestimmung der Maximalzahl der Nachbargebiete auf nichtorientierbaren Flächen. Arch. Math. 4 (1953), 137-142.
3. Über polyedrische Zerlegungen geschlossener Flächen. Habilitationsschrift, Rheinische Friedrich-Wilhelms-Universität Bonn, 1953.
4. Farbensatz für orientierbare Flächen vom Geschlecht p>0. J. Reine Angew. Math. 193 (1954), 11-38.
5. Bestimmung der Maximalzahl der Nachbargebiete auf nichtorientierbaren Flächen. Math. Ann. 127 (1954), 181-214.
6. Lokal-reguläre Zerlegungen geschlossener orientierbarer Flächen. Math. Z. 59 (1954), 484-495.
7. Über drei kombinatorische Probleme am n-dimensionalen Würfel und Würfelgitter. Abh. Math. Sem. Univ. Hamburg 20 (1955), 10-19.
8. Wie man die geschlossenen nichtorientierbaren Flächen in möglichst wenige Dreiecke zerlegen kann. Math. Ann. 130 (1955), 317-326.
9. Teilungen der Ebene durch Geraden oder topologische Geraden. (E. Sperner zum 50. Geburtstag am 9. Dezember 1955 gewidmet.) Math. Z. 64 (1956), 79-102.
10. Über Geraden in allgemeiner Lage. Elem. Math. 12 (1957), 75-82.
11. Färbungsprobleme auf Flächen und Graphen. Math. Monographien, 2. VEB Deutscher Verlag der Wiss., Berlin, 1959, 132 pages.
12. Über das Problem der Nachbargebiete auf orientierbaren Flächen. (Geheimrat Prof. Dr. Lothar Heffter zum 99. Geburtstag gewidmet.) Math. Sem. Univ. Hamburg 25 (1961), 105-127.
13. Konfigurationen und Schließungssätze (Zusammenfassung eines Vortrags). Wiss. Z. Martin-Luther-Univ. Halle-Wittenberg, Math.-Nat. X (1961), 116-117.
14. Das Färbungsproblem auf orientierbaren Flächen (Zusammenfassung eines Vortrags). Wiss. Z. Martin-Luther-Univ. Halle-Wittenberg, Math.-Nat. X (1961), 117.
15. Über eine Klasseneinteilung der zweiecklosen Graphen. Arch. Math. 12 (1961), 231-237.
16. Selbstkomplementäre Graphen. (Hans Petersson zum 60. Geburtstag gewidmet.) Arch. Math. 14 (1963), 354-358.

17. Extremal problems in the Theory of Graphs. Theory of Graphs and its Applications (Proc. Sympos. Smolenice, 1963). Publ. House Czechoslovak Acad. Sci., Prague 1964, 85-90.

18. Das Geschlecht des vollständigen paaren Graphen. Abh. Math. Sem. Univ. Hamburg 28 (1965), 139-150.

19. On the complete bipartite graph. Celebrazioni archimedee del secolo XX Simposio di didattica della matematica, Syracuse (1964). Tipografia "oderisi", Editrice Gubbio, Italia, 141-145.

20. Die toroidale Dicke des vollständigen Graphen. (Ruth Moufang zum 60. Geburtstag am 10. 1. 1965 gewidmet.) Math. Z. 87 (1965), 19-26.

21. Ein Sechsfarbenproblem auf der Kugel. (Emanuel Sperner zum 60. Geburtstag gewidmet.) Abh. Math. Sem. Univ. Hamburg 29 (1965), 107-117.

22. Der vollständige paare Graph auf nichtorientierbaren Flächen. (Wolfgang Franz zum 60. Geburtstag gewidmet.) J. Reine Angew. Math. 220 (1965), 88-93.

23. Über das Geschlecht des vollständigen Graphen. Beiträge zur Graphentheorie. Intern. Koll. Manebach (DDR), 9.-12. Mai 1967. Teubner, Leipzig 1968, 107-112.

24. A six-color problem on the sphere. Theory of Graphs (Proc. Colloq., Tihany, 1966). Academic Press, New York 1968, 265-269.

25. (with J.W.T. Youngs) Solution of the Heawood map-coloring problem. Proc. Nat. Acad. Sci. U.S.A. 60 (1968), 438-445.

26. (with J.W.T. Youngs) Lösung des Problems der Nachbargebiete. Arch. Math. (Basel) 20 (1969), 190-201.

27. (with J.W.T. Youngs) Solution of the Heawood map-coloring problem - Case 11. J. Combinatorial Theory 7 (1969), 71-93.

28. (with J.W.T. Youngs) Solution of the Heawood map-coloring problem - Case 2. J. Combinatorial Theory 7 (1969), 342-352.

29. (with J.W.T. Youngs) Solution of the Heawood map-coloring problem - Case 8. J. Combinatorial Theory 7 (1969), 353-363.

30. (with J.W.T. Youngs) Remarks on the Heawood conjecture. Proof Techniques in Graph Theory (Proc. Second Ann Arbor Graph Theory Conf., Ann Arbor, Mich., 1968) Academic Press, New York, 1969, 133-138.

31. (with J.W.T. Youngs) Das Geschlecht des vollständigen dreifärbbaren Graphen. Comment. Math. Helv. 45 (1970), 152-158.

32. Genus of graphs. Combinatorial Structures and their Applications (Proc. Calgary Internat. Conf., Calgary, Alta., 1969). Gordon and Breach, New York, 1970, 361-366.

33. Das Kartenfärbungsproblem. Selecta Mathematica III. Hei-

delberger Taschenbücher, 86. Springer, Berlin 1971, 27-55.

34. Kleine Verschärfung des Fünf-Farben-Satzes. Symposia Mathematica, Vol. V (INDAM, 1969/70). Academic Press, London 1971, 315-324.

35. Triangular embeddings of graphs. (Dedicated to the memory of J.W.T. Youngs.) Graph Theory and Applications (Proc. Conf. Western Michigan Univ., Kalamazoo, Mich., 1972. Lecture Notes in Math., Vol. 303, Springer, Berlin 1972, 269-281.

36. J.W.T. Youngs (1910-1970). J. Combinatorial Theory Ser. B 13 (1972), 91-93.

37. Map Color Theorem. Die Grundlehren der mathematischen Wissenschaften, Band 209. Springer, New York-Heidelberg 1974, 191 pages.

38. (with Richard K. Guy) Triangular imbedding of $K_n - K_6$. J. Combinatorial Theory Ser. B 21 (1976), 140-145.

39. The combinatorial map color theorem. J. Graph Theory 1 (1977), 141-155.

40. (with Mark Jungerman) The genus of the n-octahedron: regular cases. J. Graph Theory 2 (1978), 69-75.

41. On the genus of the graph $K_n x K_2$ or the n-prism. Discrete Math. 20 (1977/78), 287-294.

42. Nonexistence of graph embeddings. Theory and Applications of Graphs (Proc. Internat. Conf., Western Mich. Univ., Kalamazoo, Mich. 1976). Lecture Notes in Math. 642, Springer, Berlin 1978, 465-476.

43. (with Mark Jungerman) Minimal triangulations on orientable surfaces. Acta Math. 145 (1980), 121-154.

44. A nine color theorem for the torus and the Klein bottle. The Theory and Applications of Graphs (Kalamazoo, Mich. 1980), Wiley, New York 1981, 507-515.

45. (with Brad Jackson) Maps of m-pires on the projective plane. Discrete Math. 46 (1983), 15-20.

46. (with Brad Jackson) Colorings of circles. Amer. Math. Monthly 91 (1984), 42-49.

47. (with Brad Jackson) Heawood's empire problem. J. Combin. Theory Ser. B 38 (1985), 168-178.

48. (with Brad Jackson) The splitting number of complete bipartite graphs. Arch. Math. (Basel) 42 (1984), 178-184.

49. (with Brad Jackson) Empire maps. Graphs and Applications (Boulder, Colo., 1982). Wiley-Intersci. Publ., Wiley New York, 1985, 195- 201.

50. (with Brad Jackson) Splittings of graphs on surfaces. Graphs and APPlications (Boulder, Colo., 1982). Wiley-Intersci. Publ., Wiley New York, 1985, 203-219.

51. (with Brad Jackson) Plane constructions for graphs, networks, and maps: measurements of planarity. Lecture Notes

in Econom. and Math Systems, 226, Springer, Berlin-New York, 1984, 315-324.

52. (with Brad Jackson) Coloring island maps. Bull. Austral. Math. Soc. 29 (1984), 151-165.

53. (with Brad Jackson) Solution of Heawood's empire problem in the plane. J. Reine Angew. Math. 347 (1984), 146-153.

54. (with Nora Hartsfield and Brad Jackson) The splitting number of the complete graph. Graphs Combin. 1 (1985), 311 -329.

55. 250 Jahre Graphentheorie. (Klaus Wagner zum 75. Geburtstag gewidmet.) Graphen in Forschung und Unterricht, Festschrift Kiel, 1985. Franzbecker, Bad Salzdetfurth 1985, 136-152.

56. Vermutungen über numerierbare Graphen. Elem. Math. 41 (1986), 68-74.

57. (with Nora Hartsfield) Minimal quadrangulations of nonorientable surfaces. J. Combin. Theory Ser. A 50 (1989), 186-195.

58. (with Nora Hartsfield) Hamilton surfaces for the complete symmetric tripartite graph. Arch. Math. (Basel) 50 (1988), 470-473.

59. (with Nora Hartsfield) Minimal quadrangulations of orientable surfaces. J. Combin. Theory Ser. B 46 (1989), 84-95.

60. (with Nora Hartsfield) Quadrangular embeddings of the complete even k.partite graph. Discrete Math., to appear.

61. (with Nora Hartsfield) Clean triangulations. Combinatorica, to appear.

62. (with Nora Hartsfield and Brad Jackson) Hamilton surfaces for the complete even symmetric bipartite graph. Discrete Math. 78 (1989), 89-94.

63. (with Nora Hartsfield) Supermagic and antimagic graphs. J. Recreational Math. 21 (1989), 107-115.

64. (with Nora Hartsfield and Dragan Marusic) Nonorientable self-dual embeddings of Cayley graphs. Graphs Combin., submitted.

65. (with Nora Hartsfield) Pearls in Graph Theory. Academic Press, 350 pages, to appear.

66. (with Hanfried Lenz) A brief review on Egmont Köhler's mathematical work. Discrete Math., to appear.

On the Problem of Relative Components of Minimal Graphs

R. Bodendiek, Kiel, FRG
K. Wagner, Köln, FRG

1 Introduction

1935, D.König asked the question on page 199 in his famous book with the title "Theorie der endlichen und unendlichen Graphen" whether there exists a Kuratowski-type theorem for every closed orientable surface F_p, $p \in \mathbb{N}_0$. The modern version of Kuratowski's theorem is:

The minimalbasis $M_1(F_0)$ consists of exactly two graphs, the so-called Kuratowski graphs $K_{3,3}$ and K_5 with $M_1(F_0) = \{K_{3,3}, K_5\}$, where the minimalbasis $M_1(F_0)$ is defined as the set of all $>_1$-minimal graphs of the set $\Gamma(F_0) \subseteq \Gamma$ of all nonplanar graphs. Γ stands for the set of all finite undirected graphs without loops and multiple edges and $>_1$ is the well-known subdivision relation on Γ or $\Gamma(F_0)$, respectively.

About 10 years ago, R.Bodendiek and K.Wagner introduced a new partial ordering relation $>_4$ on Γ, or $\Gamma(F_0)$, respectively. $>_4$ is finer than the subdivision relation $>_1$; i.e. $>_4$ and $>_1$ satisfy the following assertion:

$$\bigwedge_{G,G' \in \Gamma} \left[(G,G') \in >_1 \longrightarrow (G,G') \in >_4 \right].$$

By means of $>_4$, R.Bodendiek and K.Wagner need only one graph to characterize the set of all graphs embeddable (or not embeddable) in the plane F_0. Their short version of the Kuratowski theorem is:

$$M_4(F_0) = \{K_5\},$$

where $M_4(F_0)$ stands for the $>_4$-minimalbasis of F_0 in relation to $>_4$.

Using these modern denotations and concepts, we can express König's question in the following form:

R. Bodendiek, R. Henn (Eds.)
Topics in Combinatorics and Graph Theory

How do the minimalbases $M_1(F_p)$, or $M_4(F_p)$, respectively look like for $p \in N$?

Simplest investigations concerning König's question have already shown that the number of Kuratowski graphs in $M_1(F_p)$, and even in $M_4(F_p)$, grows quite rapidly for increasing $p \in \mathbb{N}$. P.Huneke and H.Glover claim that the list for the minimalbasis $M_1(F_1)$, where F_1 stands for the torus, has over 800 members. Therefore, it is not clear that the minimalbasis $M_i(F_p)$, $i = 1,4$, must be finite for every $p \in \mathbb{N}$.

In [1] and [2] the authors and H.Schumacher showed that the minimalbasis $M_4(N_1)$, where N_1 denotes the projective plane, consists of exactly 12 $>_4$-minimal graphs. At the same time, C.S.Wang et alii proved that $M_1(N_1)$ has exactly 103 $>_1$-minimal elements. D.Archdeacon, H.Glover and P.Huneke investigated the problem to determine all the Kuratowski graphs of N_q, $q \in \mathbb{N}$, ([14]).

It is very interesting that there exists a further-class of closed surfaces, the so-called spindle-surfaces, that differ from the closed orientable surfaces and from the closed nonorientable surfaces by the fact that they have certain singular points. In order to explain the concept of a spindle-surface, we consider a connected graph $G^* = (V^*, E^*)$ without vertices of degree 1 which contains at least one edge and which is allowed either to have multiple edges or to consist of precisely one loop. If G^* is different from a loop, the spindle-surface $S(G^*)$ arises from G^* by substituting each, edge $e = \{v,v'\} \in E^*$, $v \neq v'$, by a sphere $S(e)$ in such a way that two distinct points u and u' are to be identified with $v = u$ and $v' = u'$, and such that the intersection $S(e) \cap S(e')$ of each two spheres $S(e)$, $S(e')$ for each two edges $e,e' \in E^*$ with $e \neq e'$ satisfies:

$$S(e) \cap S(e') = \begin{cases} \emptyset, & \text{if } e \cap e' = \emptyset \\ \{v\}, & \text{if } e \cap e' = \{v\} \\ \{v,v'\}, & \text{if } e,e' \text{ are multiple edges with the two endvertices } v,v' \text{ and if } \{v,v'\} \text{ denotes a set of the two vertices } v,v'. \end{cases}$$

If G^* is a loop with exactly one vertex, then we get the corresponding spindle-surface $S(G^*)$ from a sphere by identifying two distinct points of the sphere.

In both cases, we can give for $S(G^*)$ the following representation:

$$S(G^*) = \bigcup_{e \in E^*} S(e).$$

The spindle-surfaces $S(G^*)$ are topological spaces which fail to be 2-manifolds at a finite number of points. These points are the vertices of G^* and they are said to be singular points of $S(G^*)$.

More graphically, we can say that a spindle-surface $S(G^*)$ arises from G^* by blowing up each edges $e \in E^*$ to a spindle. If $G^* = C_n$, where C_1 denotes the loop with exactly one vertex, C_2 is a double edge with precisely two endvertices v, v', and finally C_n, $n \geq 3$, stands for the cycle of length $n \geq 3$, the corresponding spindle-surface $S(C_n)$ ist particularly interesting because it is very simple, has $n \geq 1$ singular points and arises from a torus F_1 by contracting $n \geq 1$ mutually distinct meridians of F_1 to a single point, each. For sake of brevity, we denote $S(C_n)$ by S_n (Figure 1).

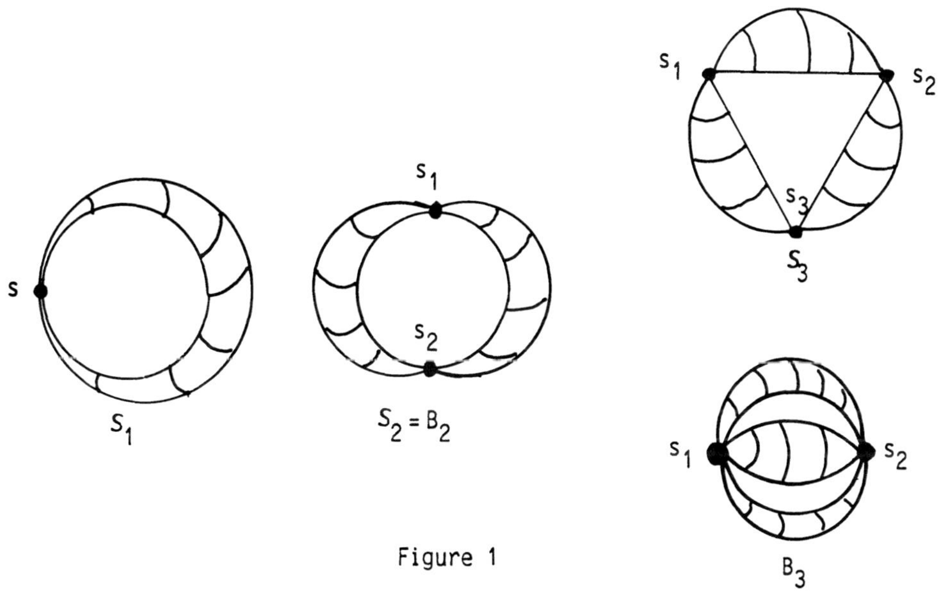

Figure 1

Another interesting class of spindle-surfaces $S(G^*)$ arises from a multiple edge $G^* = M_n, n \geq 2$, with two endvertices v, v' and n edges joining v and v'. The corresponding spindle-surface $S(M_n)$ is called banana surface and denoted by $B_n = S(M_n)$. Figure 1 shows B_2 and B_3.

Because of the relationship between the torus F_1 and the spindle-surfaces S_n, $n \geq 1$, it turns out that every closed spindle-surface S_n, $n \geq 1$ can be represented as a closed chain whose links are leaves with two singular points (not necessarily distinct) and where the two opposite edges of each leaf must be identified. Figure 2 represents plane models of S_1, S_2, S_3, and S_n, $n \geq 4$.

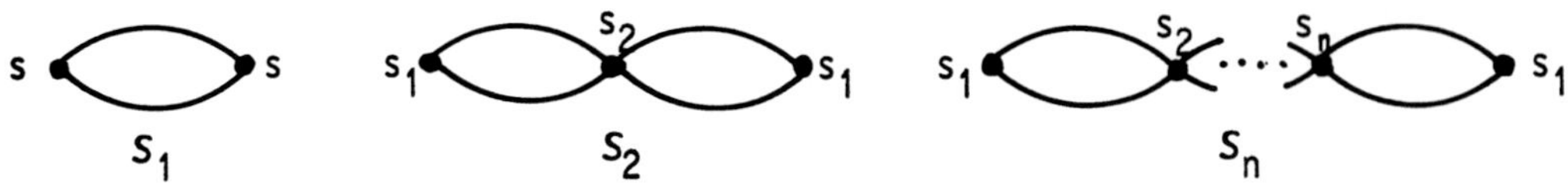

Figure 2

With respect to König's problem, we now ask the question whether K_5 is a S_n-graph or whether K_5 is embeddable in S_n, $n \geq 1$. Figure 3 gives the answer.

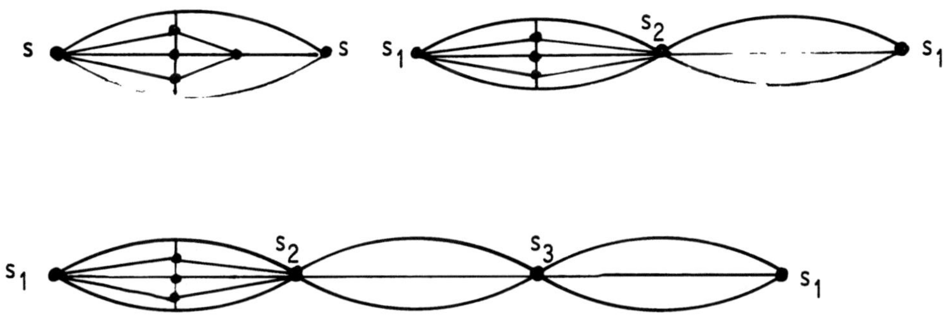

Figure 3

We notice that for $n \geq 3$ K_5 can be embedded in S_n only in such a way that exactly one edge e of K_5 passes through $n - 2$ singular points of S_n with the property that these $n - 2$ singular points are inner points of e and do not belong to the vertex set of K_5.

With respect to a lot of applications it is useful to require that if any edge of a graph $G = (V,E)$ embedded in S_n, $n \geq 1$, or more generally embedded in an arbitrary spindle-surface $S(G^*)$ meets a singular point s of S_n (or $S(G^*)$, respectively), then s must necessarily belong to the vertex set V of G. It is obvious that this additional requirement means no restriction for the embedding of graphs in S_1 and S_2. Therefore, we are able to enlarge König's question in the following way:

Let F denote a closed orientable surface F_p, $p \in \mathbb{N}_0$, or a closed nonorientable surface N_q, $q \in \mathbb{N}$, or a spindle-surface S_n, $n = 1,2$, then König's question can be expressed in the following way:

How do the minimalbases $M_1(F)$ and $M_4(F)$ look like?

Up to this time, R.Bodendiek and K.Wagner could prove the following

Theorem 1:

(a) $M_4(N_1)$ is finite.

(b) $M_4(N_1) = \{G_1,\dots,G_{12}\}$, where the twelve graphs $G_1,\dots,G_{12}$ are given in [1] or [2].

(c) $M_4(S_1)$ is finite.

(d) $|M_4(S_1)| \geq 12$

(e) $\bigwedge_{p \in \mathbb{N}_0} \bigwedge_{i \in \{1,4\}} M_i(F_p)$ is finite

(f) $|M_4(F_1)| \geq 23$.

2 Background, concepts and facts

The essential part of the proof of Theorem 1 is the so-called 'relative component method'. In order to describe this method we consider an arbitrary nonempty connected graph $G = (V,E) \in \Gamma$ and an arbitrary nonempty subgraph $G' = (V',E') \subseteq G$ with $G \neq G'$ and then we investigate the question whether it is possible to construct the graph G by means of G'.

Before we are able to give a sufficient answer we have to introduce some concepts. First we call an edge $e = \{v,v'\}$ of G that does not belong to G' a relative component of order 0 or more graphically a chord of G in relation to G' if, and only if, the two endvertices v and v' belong to G'.
If G" denotes the induced subgraph spanned by all vertices of $V - V'$, then it is obvious that G' and G" are disjoint. If $G'' \neq \emptyset$, then we denote the components of G" by $G''_1,\dots,G''_n$, $n \geq 1$. Since G is connected it is clear that there are edges (so-called bridges) joining vertices of G' and G''_ν, $\nu = 1,\dots,n$.

Let B_ν, $\nu = 1,\dots,n$, denote the graph consisting of all bridges between G' and G''_ν. Then it is obvious that B_ν, $\nu = 1,\dots,n$, is a nonempty connected bipartite graph.
Finally the graph $Q_\nu = G''_\nu \cup B_\nu$, $\nu = 1,\dots,n$, is said to be a relative component of G in relation to G' of order $|G''_\nu|$, where $|G''_\nu|$ denotes the order of G''_ν (number of vertices of G''_ν).

It is easy to see that each relative component Q_ν, $\nu = 1,\dots,n$, of G in relation to G' has got the order ≥ 1. If Q_ν is a relative component of G in relation to G' of order 1, we call Q_ν a star. If the order of Q_ν is 2 we call Q_ν a double star.
For sake of brevity we call the graph G''_ν of $Q_\nu = G''_\nu \cup B_\nu$, $\nu = 1,\dots,n$, the center of Q_ν, and the vertices of $B_\nu \cap G'$ the basis points of Q_ν.

Now we are able to formulate the answer of the question mentioned above in the following way:

Theorem 2: If $G' = (V',E')$ is a nonempty subgraph of a nonempty connected graph $G = (V,E)$ with $G \neq G'$, then G has the shape

$$G = G' \cup \{e_1,\dots,e_m\} \cup Q_1 \cup \dots \cup Q_n,$$

where $m,n \in \mathbb{N}_0$, $m+n \geq 1$, $e_1,\dots,e_m$ are chords of G in relation to G' and $Q_1,\dots,Q_n$ represent the relative components of G in relation to G' of order ≥ 1.

Since the relative component method needs further properties of graphs we consider the set Σ of all closed orientable surfaces F_p, $p \in \mathbb{N}_0$, all closed nonorientable surfaces N_q, $q \in \mathbb{N}$, and all closed spindle-surfaces $S(G^*)$, and define a so-called embedding relation $<_e$ on Σ with $<_e \subseteq \Sigma \times \Sigma$ in the following way:

Let $F, F' \in \Sigma$ be two surfaces. Then we define:

$$F' <_e F \Longleftrightarrow \bigwedge_{G \in \Gamma} \left[G \text{ is embeddable in } F' \longrightarrow G \text{ is embeddable in } F \right].$$

It is quite clear that $<_e$ is a partial ordering relation on Σ and that the following theorem holds:

Theorem 3:

(a) $F_0 <_e F_1 <_e \cdots <_e F_p <_e F_{p+1} <_e \cdots$

(b) $N_1 <_e N_2 <_e \cdots <_e N_q <_e N_{q+1} <_e \cdots$

(c) $\bigwedge_{p \in \mathbb{N}_0} F_p <_e N_{2p+1}$.

It is easy to see that we can formulate a similar assertion for spindle-surfaces. It holds:

Theorem 4: Let $S(G^*)$ be an arbitrary spindle-surface and let $U(G^*)$ denote a subdivision of G^*. Then holds:

$$S(U(G^*)) <_e S(G^*).$$

Proof: Let $G = (V,E) \in \Gamma$ be a graph embeddable in $S(U(G^*))$ and let $e = \{v,v'\}$ denote an edge of G^*. The corresponding path $P(e) \subseteq U(G^*)$ is a path with the same endvertices v and v'. In the same way we have a corresponding surface $S(P) \subseteq S(U(G^*))$ on $S(U(G^*))$. Now we consider the subgraph of G lying on $S(P)$. Since this subgraph $G \cap S(P)$ is planar, it is obvious that $G \cap S(P)$ is embeddable in the spindle $S(e) \subseteq S(G^*)$. As this holds for each edge e of G^* the graph G is embeddable in $S(G^*)$. Hence $S(U(G^*)) <_e S(G^*)$.

An immediate consequence from Theorem 4 is

Theorem 5:

$$\cdots <_e S_n <_e S_{n-1} <_e \cdots <_e S_2 <_e S_1.$$

In the same way we obtain the following chain for the banana spindle-surfaces:

Theorem 6:

$$B_2 <_e B_3 <_e \cdots <_e B_n <_e B_{n+1} <_e \cdots$$

Observing the last three theorems the question arises whether it is possible to construct the $>_i$-minimal graphs of $M_i(F)$, $i = 1,4$, by means of the graphs of $M_i(F')$, where F and F' are two surfaces of Σ with $F' <_e F$. In the following we are going to describe a general method of constructing graphs in $M_1(F)$ by means of graphs in $M_1(F')$ with $F' <_e F$.

Let G be a $>_1$-minimal graph in $M_1(F)$. Because of the definition of the minimalbasis $M_1(F)$ we know that G is not embeddable in F. Since $F' <_e F$ G is not embeddable in F'. Therefore, there is a graph $G' \in M_1(F')$ with $G >_1 G'$. In other words: G contains a subdivision graph U(G') of G'. According to Theorem 2 we can represent G in the way given in (a) with

(a) $G = U(G') \cup Q_1 \cup \ldots \cup Q_n \cup \{e_1,\ldots,e_m\}$,

where $m,n \in \mathbb{N}_0$, $m+n \geq 1$, $e_1,\ldots,e_m$ are chords of G in relation to U(G'), and $Q_1,\ldots,Q_n$ are relative components of G in relation to U(G') of order ≥ 1.

When we now assume that we know all $>_1$-minimal graphs G' of $M_1(F')$ then it is possible to give the $>_1$-minimal graphs of $M_1(F)$ by determining all chords and relative components of order ≥ 1 in relation to U(G'). In order to simplify the following investigations we shall restrict ourselves on $F = N_1$ and $F = S_i$, $i = 1,2$, with

(b) $F_0 <_e N_1$ and $F_0 <_e S_2 <_e S_1 <_e F_1$.

The starting point in the two cases of (b) is F_0. The endpoints are the projective plane N_1 and the torus F_1 while the two spindle-surfaces S_1 and S_2 are certain intermediate stations. According to (a) and $M_1(F_0) = \{K_5, K_{3,3}\}$ we have to consider the subdivision graphs $U(K_5)$ and $U(K_{3,3})$ of the two Kuratowski graphs K_5 and $K_{3,3}$, and to determine the chords and relative components in relation to $U(K_5)$ and $U(K_{3,3})$.

As mentioned above R.Bodendiek and K.Wagner proved that the minimalbasis $M_4(N_1)$ consists of exactly 12 graphs $G_1,\ldots,G_{12}$, where the graphs $G_1 = K_5 \dot\cup K_5$, $G_2 = K_4 * 1 * K_4$ and $G_3 = K_3 * 2 * K_3$ consist of two K_5 which are disjoint, stitched together in 0,1 or 2 vertices. It is clear that these three graphs do not contain a subdivision graph $U(K_{3,3})$. All the other 9 graphs of $M_4(N_1)$ contain a $U(K_{3,3})$. Furthermore it is easy to show that the 10 graphs $G_3,\ldots,G_{12}$ are embeddable in the torus and that G_1, G_2 belong to $M_4(F_1)$.
This fact implies:

(1) $M_4(N_1) \cap M_4(F_1) = \{G_1, G_2\}$.

In the same way we can prove the following three statements:

(2) $M_4(\mathcal{N}_1) \cap M_4(\mathcal{S}_1) = \{G_1, G_2, G_5, G_7\}$,

(3) $M_4(\mathcal{N}_1) \cap M_4(\mathcal{N}_2) = \emptyset$

(4) $K_3 * K_2 * K_3 \in M_4(\mathcal{S}_1) \cap M_4(\mathcal{S}_2)$.

The graph G_5 of $M_4(\mathcal{N}_1)$ is very interesting. Therefore, we discuss its structure. It consists of a $K_{3,3} = \{1,3,5\} * \{2,4,6\}$ and two stars $Q_1 = \{v\} * \{1,3,5\}$, $Q_2 = \{u\} * \{1,3,5\}$, $u \neq v$, with the same basis points 1,3,5. In the same way we can describe the structure of $G_7 \in M_4(\mathcal{N}_1)$. G_7 consists of a $U(K_{3,3})$, arising from a $K_{3,3} = \{1,3,5\} * \{2,4,6\}$ by inserting a new vertex u_1 on the edge $\{2,3\}$ and a further new vertex u_2 on the edge $\{3,4\}$, and two stars $Q_1 = \{v_1\} * \{2,3,4\}$, $Q_2 = \{v_2\} * \{2,3,4\}$, $v_1 \neq v_2$, with the same basis points 2,3,4. Finally, it is also easy to show that G_7 is characterized by the property that it does not contain a $U(K_{3,3})$ with less than 8 vertices.

It turns out that the set of all $>_4$-minimal graphs in $M_4(\mathcal{S}_1)$ or $M_4(\mathcal{F}_1)$, which do not contain a subdivision graph $U(K_{3,3})$, consists of exactly three or exactly five members, respectively. More precisely, we can say:

(5) G_1, G_2 and $G_3' = K_3 * K_2 * K_3$ are the only graphs in $M_4(\mathcal{S}_1)$ which do not contain a subdivision graph $U(K_{3,3})$.

(6) G_1, G_2, G_3'', G_4'' and G_5'' are the only graphs in $M_4(\mathcal{F}_1)$, which do not contain a subdivision graph $U(K_{3,3})$. Figure 4 illustrates the graphs G_3'', G_4'' and G_5''.

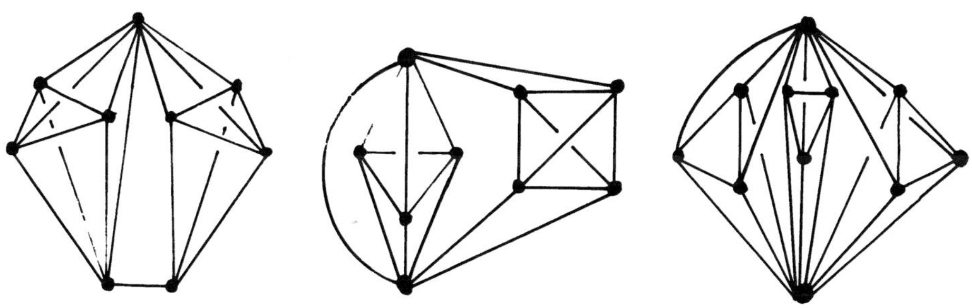

G_3'' G_4'' G_5''

Figure 4

Observing (5) and (6) we obtain the very interesting and important

Theorem 7: (a) If $G \in M_4(S_1) - \{G_1, G_2, G_3\}$, then G contains a subdivision graph $U(K_{3,3})$ with

$G = U(K_{3,3}) \cup Q_1 \cup \ldots \cup Q_n \cup \{e_1, \ldots, e_m\}$, where $m,n \in \mathbf{N}_0, m+n \geq 1$, $e_1, \ldots, e_m$ are chords of G in relation to $U(K_{3,3})$ and where $Q_1, \ldots, Q_n$ are relative components of G in relation to $U(K_{3,3})$ of order ≥ 1.

(b) If $G \in M_4(F_1) - \{G_1, G_2, G_3'', G_4'', G_5''\}$ then G contains a subdivision graph $U(K_{3,3})$ with

$G = U(K_{3,3}) \cup Q_1 \cup \ldots \cup Q_n \cup \{e_1, \ldots, e_m\}$, where $m,n \in \mathbf{N}_0$, $m+n \geq 1$, $e_1, \ldots, e_m$ are chords of G in relation to $U(K_{3,3})$ and where $Q_1, \ldots, Q_n$ are relative components of G in relation to $U(K_{3,3})$ of order ≥ 1.

When we now want to look for $>_4$-minimal graphs G in $M_4(S_1)$ or $M_4(F_1)$ we can assume that G contains a $U(K_{3,3})$. It turns out that it would be useful to choose a graph $\hat{U} = U(K_{3,3})$ instead of $U(K_{3,3})$, where $\hat{U} = \hat{U}(K_{3,3})$ is a subdivision graph $U(K_{3,3}) \subseteq G$ whose number of vertices is minimal. It may happen that $\hat{U} = K_{3,3}$. The graph $G_5 \in M_4(\mathbf{N}_1)$ is a $>_4$-minimal graph whose $\hat{U}$ is equal to $K_{3,3}$. The graph $G_7 \in M_4(\mathbf{N}_1)$ shows that $\hat{U} \neq K_{3,3}$ is possible for $\hat{U} \subseteq G_7$ has got the order 8.

In this section we want to go on by investigating the class of minimal graphs G containing only chords in relation to a $\hat{U} \subseteq G$ with

(7) $G = \hat{U} \cup \{e_1, \ldots, e_m\}$, where $m \geq 1$ and $e_1, \ldots, e_m$ are the chords (= relative components of order 0) of G in relation to $\hat{U}$.

The advantage of using $\hat{U}$ instead of an arbitrary $U(K_{3,3})$ arises from the fact th it may happen that we obtain a new subdivision graph $U'(K_{3,3})$ from $\hat{U}$ by adding a new chord e with the property that $U'(K_{3,3})$ has got less vertices than $\hat{U}$. It is easy to see that we need not consider this case. In other words: In (7), there does not exist a subdivision graph $U'(K_{3,3}) \subseteq G$ with less vertices than $\hat{U}$.

It is quite clear that (7) gives us the great chance of determining all $>_4$-minimal graphs in $M_4(S_1)$ or $M_4(F_1)$ arising from a $\hat{U}$ by adding chords. According to (7) we only need to answer the following two questions:

(i) Which subdivision graph $U(K_{3,3})$ can become a minimal subgraph $\hat{U}$ of a $>_4$-minimal graph G in $M_4(S_1)$ or $M_4(F_1)$ with $\hat{U} \subseteq G$ satisfying (7)?

(ii) Which chords can be added to an arbitrary $\hat{U}$ such that the graph G arising from $\hat{U}$ by adding these chords belongs to $M_4(S_1)$ or $M_4(F_1)$ and satisfies (7)?

A first simple result is: If $\hat{U} = U(\{1,3,5\} * \{2,4,6\})$ denotes a minimal subgraph of a $>_4$-minimal graph G,then it is impossible to add a chord in relation to $\hat{U}$ joining two vertices of the same main path (i...j) of $\hat{U}$ with $i \in \{1,3,5\}$ and $j \in \{2,4,6\}$, for otherwise $\hat{U}$ would not be a minimal subdivision graph in G.

In order to get further results concerning with (i) and (ii) it is useful to divide the set of all chords of G in relation to $\hat{U}$ into four nonempty classes in the following way:

(A) The chord e joins a main vertex v to an inserted vertex v' of $\hat{U}$, where v' does not lie on a main path with the endvertex v.

(B) The chord e joins two inserted vertices of two neighboured main paths of $\hat{U}$.

(C) The chord e joins two inserted vertices of two non-neighboured main paths of $\hat{U}$.

(D) The chord e joins two main vertices i,j of $\hat{U} = U(\{1,3,5\} * \{2,4,6\})$, $i \neq j$, either $i,j \in \{2,4,6\}$ or $i,j \in \{1,3,5\}$ (Figure 5).

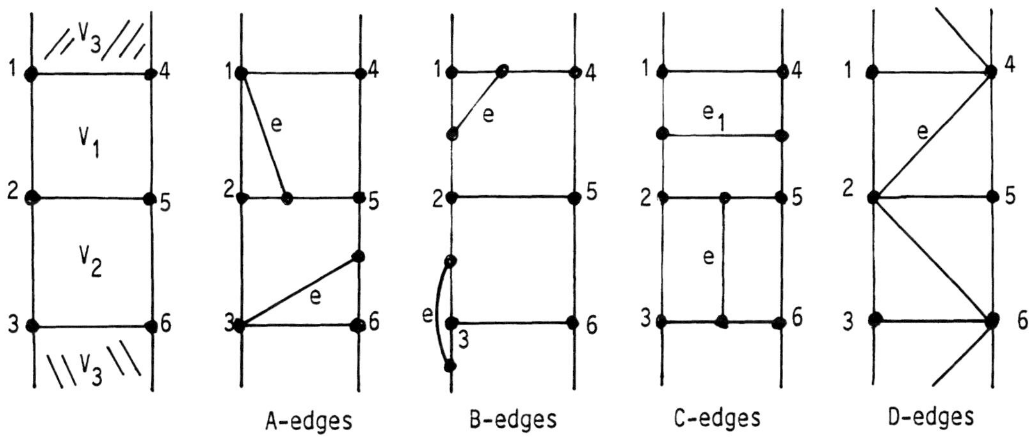

Figure 5

In Figure 5, we represent the embedding of the corresponding $K_{3,3} = \{1,3,5\} * \{2,4,6\}$ in the projective plane $\boldsymbol{N}_1$. $K_{3,3}$ divides $\boldsymbol{N}_1$ into a hexagon (1,2,3,4,5, and three quadrilaterals $V_1 = (1,4,5,2,1)$, $V_2 = (2,5,6,3,2)$ and $V_3 = (3,6,1,4,3)$

If we now consider the case that the minimal graph G of (7) contains a C-edge e_1 (Figure 5), then G contains the subgraph $W = \hat{U} \cup e_1$, where W (Figure 6) is the so-called Wagner graph. Figure 6 illustrates that $W = \hat{U} \cup e_1$ with

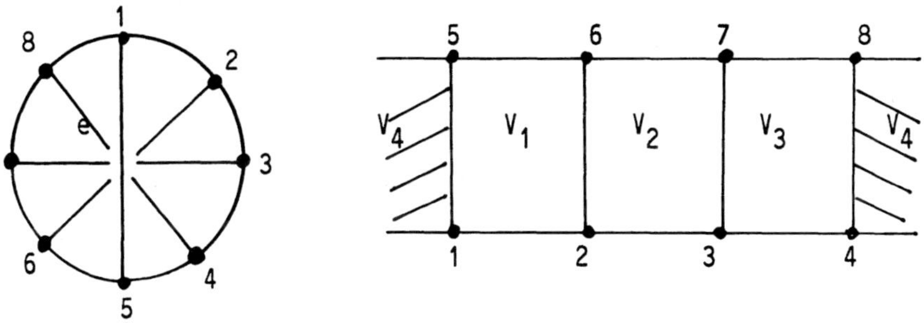

$W = \hat{U} \cup e$

Figure 6

$e_1 = \{4,8\}$ is embeddable in $\boldsymbol{N}_1$ and that $\boldsymbol{N}_1$ is divided into one octagon (1,2,3,...,8,1) and four quadrilaterals V_1, V_2, V_3, V_4. It is quite clear that $\hat{U}$ is minimal and that each of the three main paths 1...5, 2...6 and 3...7 of $\hat{U}$ has got the length 1, each, and that, finally, the octagon (1,...,8,1) of $W = \hat{U} \cup e_1$ necessarily contains all vertices of G. In the same way, we can prove that the two endvertices of a chord $e \neq e_1$ of G must lie on the boundary of the same quadrilateral.

By means of $\hat{U}$ it is easy to show that if two edges e, e_1, $e \neq e_1$, of G are crossing chords in V_i, $i \in \{1,2,3,4\}$, they must be equal to the two diagonals of V_i (Figure 7).

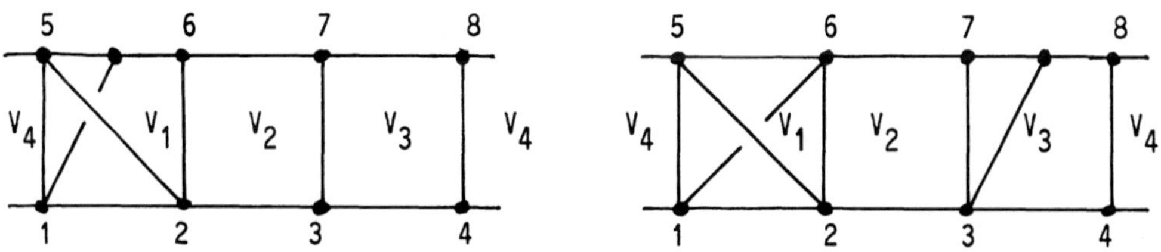

Figure 7

$V_i, i \in \{1,2,3,4\}$, is said to be a quadrilaterial of diagonal type if V_i contains two crossing chords.

It follows from Figure 7 the following assertion:

If V_1 is a quadrilateral of diagonal type, then holds:

(1) The two neighboured quadrilaterals V_2 and V_4 do not have chords.

(2) V_3 does not contain chords or V_3 is at most a quadrilateral of diagonal type.

The proof is not very difficult for it is a simple consequence from the fact that G contains a minimal $\hat{U}$. According to these results we obtain

Theorem 8: If a minimal graph $G = \hat{U} \cup \{e_1,\ldots,e_m\}$, $m \geq 1$, contains only chords and at least one C-edge in relation to $\hat{U}$, then either G is equal to $G_{12} \in M_4(\boldsymbol{N}_1)$ (Figure 8) or none of the four quadrilaterals V_1, V_2, V_3, V_4 contains two crossing diagonals or $G \subseteq G_{12}$ (i.e. G arises from G_{12} only by removing diagonals).

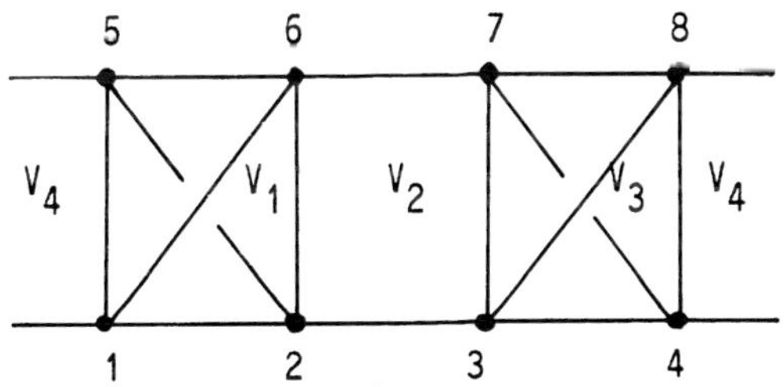

G_{12}

Figure 8

It remains to check which graphs of Theorem 8 belong to $M_4(S_1)$ or $M_4(F_1)$. Since all graphs mentioned in Theorem 8 are embeddable in the torus each graph $G \in M_4(F_1)$ does not contain C-edges. In a similar way we can prove that each graph $G \in M_4(S_1)$ does not contain C-edges in relation to $\hat{U} \subseteq G$.

Therefore, we can restrict ourselves to consider minimal graphs containing at most A-, B- or D-edges.

Since the investigations concerning the question which A-, B- or D-edges are possible in $G \in M_4(F)$, $F = \boldsymbol{N}_1, S_1, F_1$, are long and complicated we can only refer to Figure 9. It shows some examples for $\hat{U}$ with one or two chords.

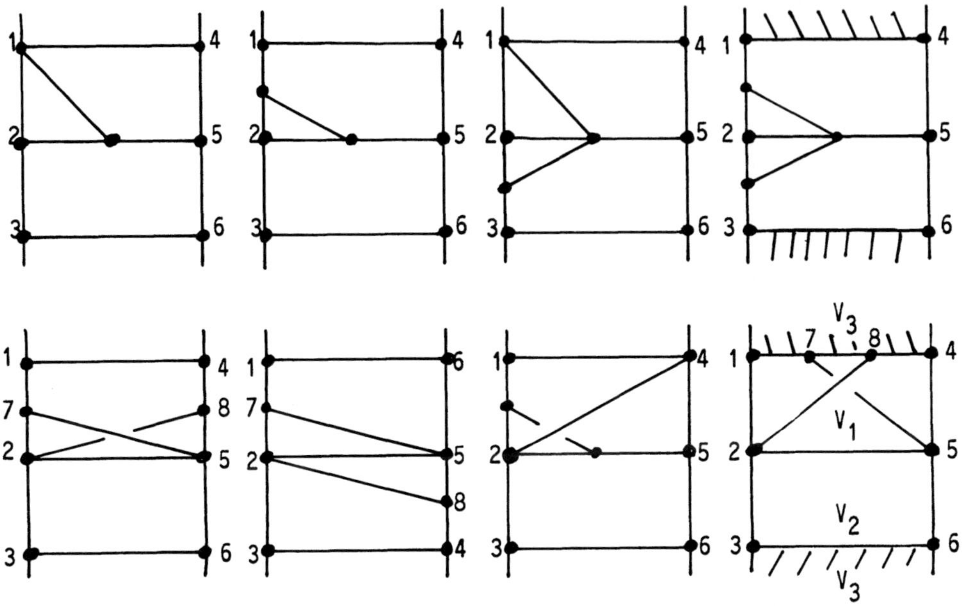

Figure 9

Furthermore, it proves that certain vertices in $\hat{U}$ must be neighbours. When we consider the last case of Figure 9 we immediately recognize that both the vertices 2 and 3 and the vertices 5 and 6 are neighbours in $\hat{U}$, and that V_3 does not contain any chord.

According to Theorem 8 we continue in formulating

Theorem 9:
(a) $\{G \in M_4(S_1) \mid G = \hat{U} \cup \{e_1,\dots,e_m\}\} = \{K_6\}$.
In words: K_6 is the only graph in $M_4(S_1)$ with the property that it contains only chords in relation to a subdivision graph $\hat{U}$.
(b) $\{G \in M_4(N_1) \mid G = \hat{U} \cup \{e_1,\dots,e_m\}\} = \{G_{12}\}$.
In words: G_{12} is the only graph in $M_4(N_1)$ with the property that it contains only chords in relation to a minimal subdivision graph $\hat{U}$.

Now we want to settle the problem of determining the set of all $>_4$-minimal graphs of $M_4(F)$ arising from a minimal subdivision graph $\hat{U}$ by adding chords or relative components of order 1 in relation to $\hat{U}$. It turns out that this problem is much more difficult such that we could solve it only for $F = N_1$.

The corresponding problem for $F = S_1$ or $F = F_1$ is partially open up to this time. A third problem arises from Theorem 7 when we look for the set of $>_4$-minimal graphs in $M_4(F)$ containing a relative component of order ≥ 2 in relation to a $\hat{U}$. In order to give a solution we develope a method which we are going to describe in the following section.

Let G be a graph in $M_4(F)$ containing a relative component Q of order ≥ 2 in relation to a minimal subdivision graph $\hat{U}$. Let Q* denote the star arising from Q by contracting the center of Q to a single point. It is obvious that Q* and Q have got the same basis points in $\hat{U}$. Finally let G* denote the graph arising from G by substituting Q by Q*. Because of $G \in M_4(F)$ the graph G* is embeddable in F. The essential idea of our method is that it is impossible to obtain an embeddable graph by substituting Q* by Q for every embedding of G* in F. In [12] K.Wagner succeeded in giving four distinct graphs (1) - (4) characterizing the relative component Q of order ≥ 2 in relation to a $\hat{U}$.

Before we shall apply these results to S_1 we mention the well-known embedding criterion for S_1.

Theorem 10: A graph $G \in \Gamma$ is embeddable in S_1 if, and only if, there is a vertex v in G such that the graph G - v is embeddable in F_0 in such a way that there exist two countries C_1, C_2 of G - v with the property that each neighbour of v in G lies on the union of the boundaries of C_1 and C_2.

1) We **remark** that a graph $G \in \Gamma$ is planar if, and only if, there exist a vertex v in G and an embedding of G - v in F_0 such that all the neighbours of v lie on the boundary of exactly one country of G - v.

2) The proof of Theorem 10 is very easy so that we can omit it.

Theorem 11: $\{G_1, G_2, G_3', \ldots, G_{12}'\} \subseteq M_4(S_1)$, where G_1, G_2 belong to $M_4(\boldsymbol{N}_1)$, $G_3' = K_3 * K_2 * K_3$, $G_4' = 1 * K_{3,3}$, $G_5' = 5 * 3$, $G_6' = K_6$ and where Figure 9a illustrates the graphs $G_7', \ldots, G_{12}'$.

In the first step of the proof to Theorem 11 we have to show that $G_1, G_2, G_3', \ldots, G_{12}'$ are not embeddable in S_1. In the second step of the proof we must verify that each graph $R_i(G_1)$ or $R_i(G_2)$ or $R_i(G_j')$, $i = 0,1,2,3,4$ and $j \in \{3, \ldots, 12\}$ is embeddable in S_1. In the following we shall restrict ourselves

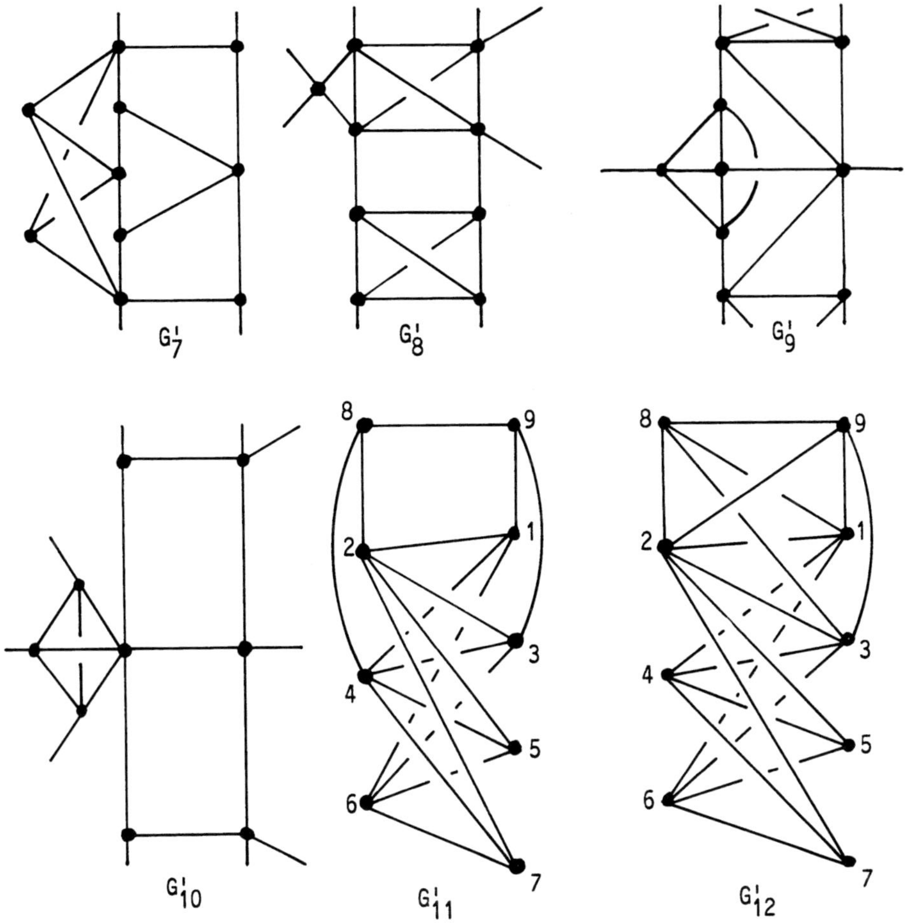

Figure 9a

on G'_6 and G'_{12}. According to the embedding criterion the graph $G'_6 = K_6$ is not embeddable in S_1 because of $K_6 - v = K_5$ for each vertex v of K_6. If we remove an edge e of K_6 then we get $G'_6 - e = 1 * K_4 * 1$. In order to apply Theorem 10 to $G'_6 - e = 1 * K_4 * 1$ we delete a vertex v of $K_4 \subseteq G'_6 - e$. Then it is easy to see that the graph $(G'_6 - e) - v = 1 * K_3 * 1$ is a planar triangular graph satisfying the conditions of Theorem 10. Subsequently $G'_6 - e$ is embeddable in S_1. If we contract an edge of G'_6, then we get the K_5 embeddable in S_1. Since $G'_6 = K_6$ does not contain vertices of degree 3 we cannot apply R_3 and R_4 to G'_6. Therefore, $G'_6 = K_6 \in M_4(S_1)$.

In the same way we consider G'_{12} (Figure 9a). Observing the symmetry of G'_{12} we notice that we can permutate the vertices 4 and 6, or 5 and 7, or 8 and 9 without changing the graph G'_{12}. In order to apply R_0 to G'_{12} we have to consider the following cases for the edge e: {1,2}, {1,4}, {1,8}, {2,3}, {2,5}, {2,8}, {3,4}, {3,8}, {4,5}, {5,6}, {6,7} and {8,9}. Due to the embedding criterion for S_1 we have to look for a vertex v in each of the 12 possibilities and to check whether the graph $R_0(G'_{12} - v) = (G'_{12} - e) - v$ is embeddable in the plane. It is easy to see that we can simplify the investigations because the vertex v = 2 is suitable for all twelve possibilities for e (Figure 10):

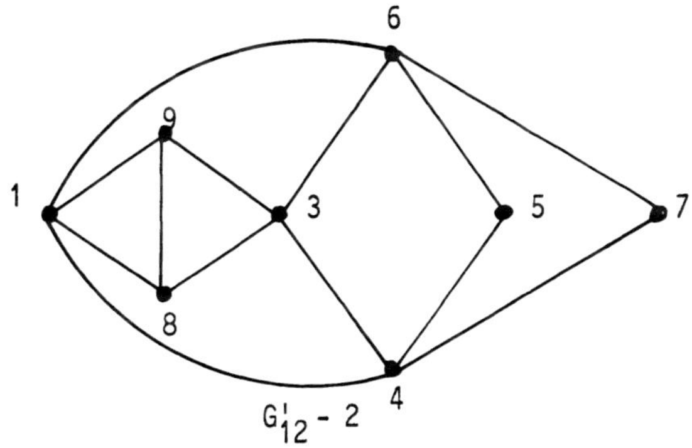

Figure 10

If e denotes the edge {1,2} for example, then all vertices of G'_{12} with exception of the vertices 4 and 6 are neighbours of the vertex 2. In $G'_{12} - e = G'_{12} - \{1,2\}$ are all vertices of G'_{12} with exception of 1,4 and 6 are neighbours of 2. According to Theorem 10 and Figure 10 it follows immediately that $G'_{12} - \{1,2\}$ is embeddable in S_1. In the same simple way we can prove that $G'_{12} - e$ is embeddable in S_1, where e denotes an edge of the remaining 11 cases mentioned above. Observing Figure 10 we can immediately recognize that the graph arising from G'_{12} by contracting e is embeddable in S_1.

Therefore it remains to show that the graph $R_3(G'_{12})$ arising from G' by applying the relation R_3 to the vertex 7 is embeddable in S_1. This follows from Figure 11, since each vertex of $R_3(G'_{12})$ different from the vertex 2 lies either on the boundary of the outer quadrilateral (1,4,5,6,1) or of the triangle (3,8,9,3) (Figure 11). This proves $G'_{12} \in M_4(S_1)$.

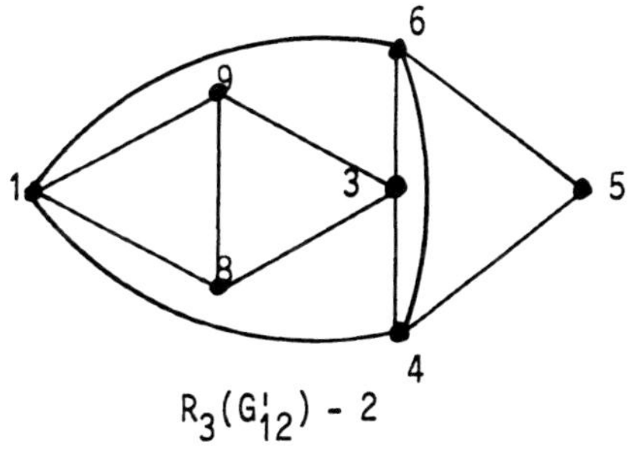

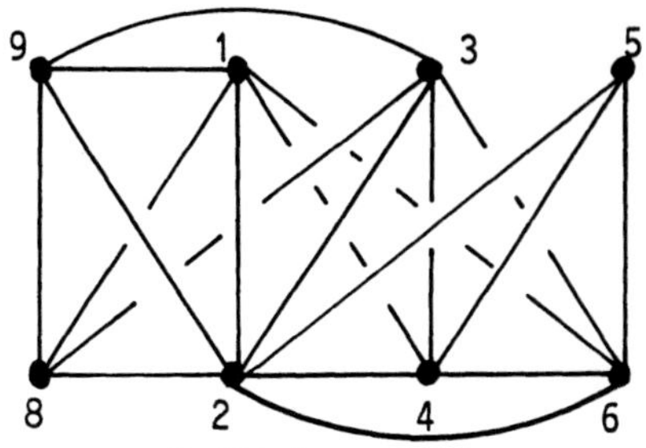

$R_3(G'_{12})$

Figure 11

3 Relative components of order ≥ 2

Now we want to come back to the problem of the relative components of order ≥ 2 of $>_4$-minimal graphs in $M_4(S_1)$. The first aim is to determine all the topologically different embeddings of $K_{3,3}$. It is easy to check because of Theorem 10 that there exist exactly two topologically different embeddings of $K_{3,3}$ in S_1 (Figure 12). The two drawings in the third row of Figure 12 illustrate the corresponding maps of these two embeddings.

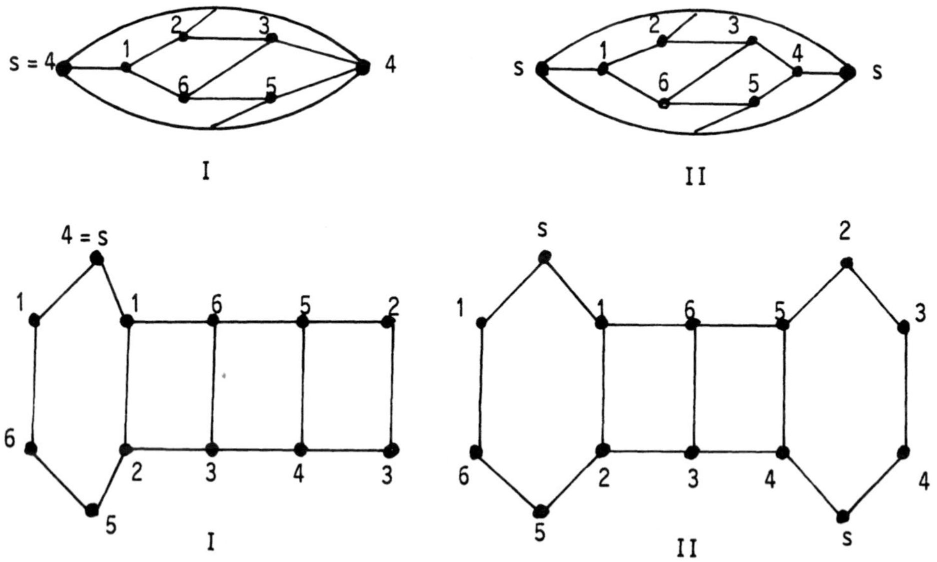

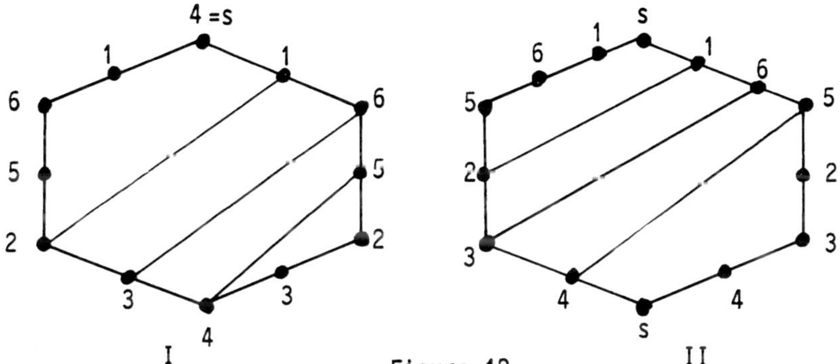

Figure 12

It is well-known that $K_{3,3}$ and also every subdivision graph $U(K_{3,3})$divides the projecitve plane into three quadrilaterals and one hexagon. If we have a look at the two embeddings of $K_{3,3}$ or $U(K_{3,3})$ in Figure 12 we can recognize that the embedding problem on S_1 is more complicated than on $\mathbf{N}_1$, for both the map (I) on S_1 and the map (II) on S_1 contain two countries whose boundaries have vertices and edges that appear twice. We think that this difference between $\mathbf{N}_1$ and S_1 (or between $\mathbf{N}_1$ and F_1) is responsible for the fact that we could determine $M_4(\mathbf{N}_1)$ but that we could not point out $M_4(S_1)$ (or $M_4(F_1)$).

In the next section, we shall show that the bipartite graph $K_{3,4}$ plays the similar role for S_1 as $K_{3,3}$ for $\mathbf{N}_1$ by determining all the $>_4$-minimal graphs in $M_4(S_1)$ containing a subdivision graph $U(K_{3,4})$. In order to do this we first apply the embedding criterion to $U(K_{3,4})$. If we denote the vertices of $K_{3,4}$ by 1,...,7 such that $K_{3,4} = \{2,4,6\} * \{1,3,5,7\}$, then each even vertex of $K_{3,4}$ is joined to an odd vertex by an edge. By means of Theorem 10 we see that $K_{3,4}$ is embeddable in S_1 because there exists a vertex v (for example v = 2,4 or 6) with $K_{3,4} - v = K_{2,4}$ embeddable in the plane (or sphere). It is well-known that $K_{2,4}$ is a quadrilateral graph on the sphere for each of the four countries of $K_{2,4}$ is a quadrilateral on the sphere (Figure 13).

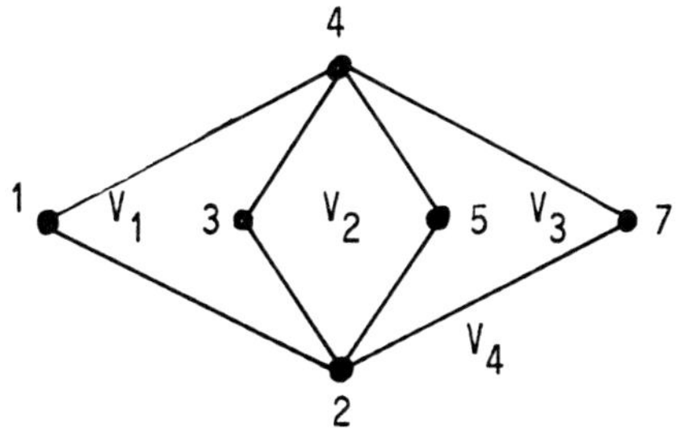

Figure 13

To get an embedding of $K_{3,4}$ in S_1 by means of Figure 13 we have to split up the vertex v = 6 of $K_{3,4}$ into two vertices v',v" (v' ≠ v") and then to put them into two quadrilaterals of $K_{2,4}$ in such a way that each odd vertex of $K_{3,4}$ lies on the boundary of one of the two quadrilaterals. This procedure is obviously unique. So we have to put v' into V_1 and v" into V_3 such that we can join v' to 1 and 3 by two edges in V_1 and, analogically, v" to 5 and 7 by two further edges in V_3. This proves

Theorem 12: $U(K_{3,4})$ has got exactly one embedding in S_1. More precisely: $U(K_{3,4})$ is a quadrilateral graph (Figure 14), where the singular point s of

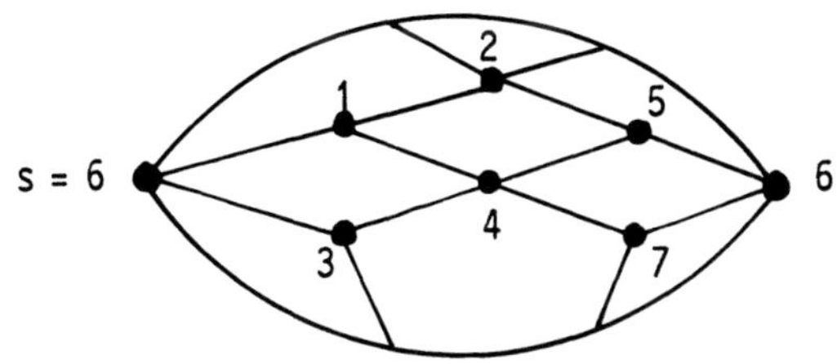

$U(K_{3,4})$
Figure 14

S_1 necessarily equals to one of the even main vertices of $U(K_{3,4})$.

Let $G \in M_4(S_1)$ be a graph containing a $U(K_{3,4})$, and let Q' denote a relative component of G in relation to $U(K_{3,4})$. Q' is said to be a relative component of quadrilateral type if, and only if, there exists a subdivided quadrilateral of $U(K_{3,4})$ containing all basis points of Q'. Then we are able to prove the two following statements:

Theorem 13: If a graph $G \in M_4(S_1)$ contains a relative component Q' in relation to $U(K_{3,4}) \subseteq G$ that is not a quadrilateral type relative component, then G is necessarily equal to the graph $G = G_5' = 3 * 5$ (Theorem 11).

Theorem 14: If a graph $G \in M_4(S_1)$ contains a relative component Q' of order ≥ 2 in relation to a subdivision graph $U(K_{3,4}) \subseteq G$, then $G = G_{11}'$ or $G = G_{12}'$ (Figure 9a).

Proof: Obviously the relative component Q' in relation to $U(K_{3,4})$ is not a quadrilateral type component if either the three even or three of the odd vertices of $U(K_{3,4})$ are basis points of Q' (Figure 15). It is similar to see that Q' is not a quadrilateral type relative component if there are three basis points of Q' which turn to the vertices 2,4,6 or to three of the four

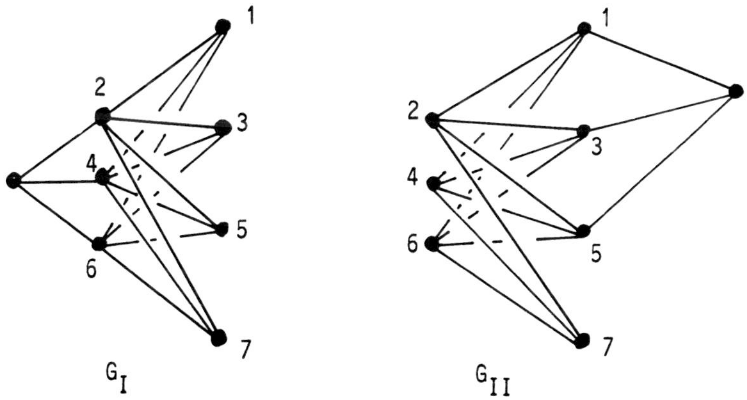

Figure 15

odd vertices of $U(K_{3,4})$ by contracting main paths of length ≥ 2 into paths of length 1. It follows from the hypothesis of Theorem 13 that the graph $U(K_{3,4}) \cup Q' \subseteq G$ is contractible to G_I or G_{II} given in Figure 15. It is clear that $G_I = 5 \star 3$. This implies $G = G_I = 5 \star 3$.
In the second case we obtain $R_3^2(G_{II}) = K_6$. Because of $G >_2 G_{II}$ and $K_6 \in M_4(S_1)$ a contradiction arises. So we must exclude the second case. This proves Theorem 13.

In order to prove Theorem 14 we at first contract the center of $Q' \subseteq G$ to a single point so that we get a star Q* with the same basis points as Q'. Then we consider the graph G* arising from G by substituting Q' by Q*. As mentioned above G* is embeddable in S_1 and Q' cannot be embedded in a country of G* without crossing of edges. Because of $U(K_{3,4}) \subseteq G^*$ the graph $V \cup Q'$, where V is a quadrilateral of $U(K_{3,4})$ in S_1, is not planar. Therefore we can apply Theorem 1 in [12].
Since G is also a $>_2$-minimal graph it is easy to see that the graph $V \cup Q'$ can only satisfy the cases (2) and (3) in Theorem 1 in [12] , where in (2) the graph $Q' = K_2 \star 3$ has three basis points and the center K_2, while in (3) Q' is equal to a so-called H-graph with four basis points and two crossing edges in V (Figure 16).

$Q' = K_2 * 3$ $Q' = H$

Figure 16

Since according to Theorem 12 and Figure 16 the graph $U(K_{3,4}) \cup Q'$ is not embeddable in S_1, we obtain $G = U(K_{3,4}) \cup Q'$. Because of the minimality of G we finally get $G = K_{3,4} \cup (K_2 * 3)$ or $G = K_{3,4} \cup H$. By means of Theorem 10 we come to $G = K_{3,4} \cup (K_2 * 3) = G'_{12}$ and $G = K_{3,4} \cup H = G'_{11}$. This proves Theorem 14.

The summary of the last two theorems is

Theorem 15: If a graph $G \in M_4(S_1)$ contains a $U(K_{3,4})$ and if $G \neq G'_5, G'_{11}, G'_{12}$, then G has only relative components of order 1 or chords in relation to every $U(K_{3,4}) \subseteq G$.

That means that the relative component problem for $>_4$-minimal graphs in $M_4(S_1)$ containing a $U(K_{3,4})$ is totally solved. A comparison between the two Figures 12 and 14 suggest the study of the embeddings of graphs in an arbitrary surface F. If G is a (nonempty and connected) graph embedded in F, then G divides F into certain disjoint regions of F called regions or countries of G in F. A region of G in F is said to be a 2-cell of G in F if, and only if, it is homeomorphic to the open unit disc.
If every region for an embedding of a graph G is a 2-cell, the embedding is said to be a 2-cell embedding.
If G has a 2-cell embedding then G and the union of all 2-cells are said to be a map of F.
Figure 14 shows that $K_{3,4}$ (and also $U(K_{3,4})$) has a 2-cell embedding where the boundary of each region is a quadrilateral. Hence we can say that the embedding of $K_{3,4}$ in S_1 is characterized by the fact that the union $R \cup Bd(R)$ is homeomorphic to the closed disc for each region R of $K_{3,4}$ in S_1, where Bd(R) denotes the boundary of the region R. So it is obvious to define:

A graph G is said to be normally embeddable in F if, and only if, the following two conditions hold:

(1) G is embeddable in F,

(2) Every embedding of G in F is a map with the property that $R \cup Bd(R)$ is homeomorphic to a closed disc for each region R of G in F, where Bd(R) denotes the boundary of R.

Therefore, we can say that $K_{3,4}$ is normally embeddable in S_1 and that every subdivision graph $U(K_{3,4})$ is normally embeddable in S_1. Furthermore, $K_{3,3}$ and every $U(K_{3,3})$ are normally embeddable in N_1. Otherwise it is clear that $U(K_{3,3})$ is not normally embeddable in S_1. Finally, Figure 17 gives an embedding E of $K_{3,3}$ in Klein's bottle N_2.

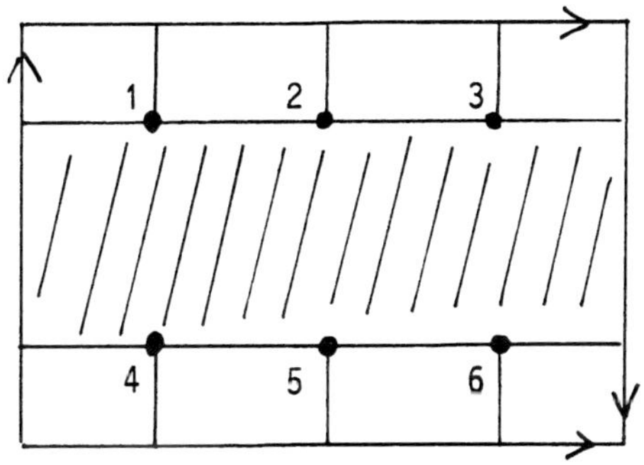

Figure 17

Although the shaded region R of $K_{3,3}$ in N_2 has a cycle as boundary, the embedding E is not a normal embedding of $K_{3,3}$ in N_2, for R is a Möbius strip. Since a Möbius strip is not homeomorphic to an open disc, R is not a 2-cell (or in other words: the embedding E is not a map of $K_{3,3}$ in N_2).

It is not very hard to see that we can generalize the method used in the proof of Theorem 14 for all closed orientable surfaces F_p, $p \in \mathbb{N}$, and for all closed nonorientable surfaces N_q, $q \in \mathbb{N}$. Let F be such a surface. Then we denote the set of all $>_4$-minimal graphs in $M_4(F)$ containing a subdivision graph of an arbitrary graph G' by $M_4(F \mid G >_1 G')$.

Theorem 16: If G' is normally embeddable in F, then at least one of the following three statements holds for **each** $G \in M_4(F \mid G >_1 G')$:

1. G contains a relative component Q of order ≥ 2 in relation to a $U(G') \subseteq G$ such that $G = U(G') \cup Q$ and $Q = K_2 * 3$;
2. G contains a relative component Q of order ≥ 2 in relation to a $U(G') \subseteq G$ such that Q contains a subdivision U(H) of a crossing H-graph (Figure 16) for every embedding of G* in S_1 ;
3. G contains only relative components of order ≤ 1 in relation to every $U(G') \subseteq G$.

Since the proof is similar to the one above mentioned we omit it.

If we investigate statement 2. of Theorem 16 a little bit more precisely, we at first call crossing H-graph shortly cross caps. It is very important to point out that it may happen that a cross cap H of a relative component Q can be drawn without crossing in a suitable embedding of the graph G* arising from G by substituting Q by a star Q* with the same basis points. In order to illustrate this fact we consider the cross cap H (Figure 16) with the basis points 1,4,5,2 in relation to the quadrilateral $V_1 = (1,4,5,2,1)$, where the one vertex of K_2 is joined to the vertices 1,5 and where the other vertex of K_2 is a neighbour of the vertices 4 and 2. It is now quite clear that it is possible to embed H in the hexagon (1,2,3,4,5,6,1) without crossings of edges (Figur 5). In other words we can say that the cross cap H dissolves.
A class of cross caps $H_1,\dots,H_n$, $n \geq 1$, of a relative component Q is said to be complete if, and only if, at least one of the n cross caps does not dissolve for every embedding of G* in *F*, where G* is defined above.
In addition to 2. of Theorem 16 we obtain

Theorem 17: Let $H_1,\dots,H_n$, $n \geq 1$, be a complete class of a relative component Q satisfying 2. of Theorem 16. If we delete the bridges of the graph $H_1 \cup \dots \cup H_n$ then we obtain the center of Q. Furthermore, for each of the basis points $b_1,\dots,b_m$ of Q, $m \in \mathbb{N}_0$, that does not belong to Q holds: There is only one bridg e_μ, $\mu = 1,\dots,m$, such that

$$Q = H_1 \cup \dots \cup H_n \cup \{e_1,\dots,e_m\},$$

where it may happen, that $\{e_1,\dots,e_m\}$ is empty.

Proof: It is quite clear that $H_1 \cup \dots \cup H_n$ is a relative component with a connected center. Assume, $H_1 \cup \dots \cup H_n$ has got a disconnected center. Then we would obtain from Q a new relative component Q' by contracting certain edges of the center of Q that do not belong to $H_1 \cup \dots \cup H_n$ such that the graph $(G-Q) \cup Q'$ is not embeddable in *F*. This is a contradiction to $G \in M_4(F)$. In the same way,

we can show that G is embeddable in *F* if Q contains at least 2 bridges incident to the same b_μ. Therefore, it follows the equality $Q = \bar{Q}'$.
Now we want to take up the cross caps. Each cross cap H of a relative component Q consists of two disjoint paths the so-called side paths of H, where each of the two side paths joins two basis points of Q, and of a third path - the so-called median path - joining the two side paths. Therefore, there exist exactly three distinct cross caps H_1, H_2, H_3 with the same bysis points (Figure 18)

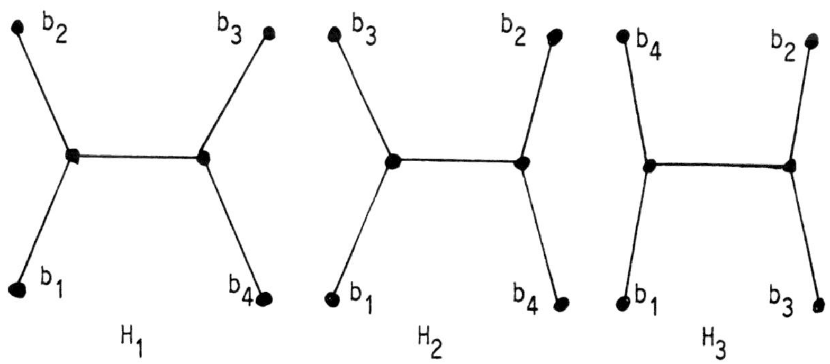

Figure 18

with the property that exactly one of $\{H_1, H_2, H_3\}$ does not dissolve if we draw them into the same country.

Another way for describing relative components Q of Theorem 17 uses normal embeddings of graphs in *F*.

Theorem 18: If a 2-connected graph G contains a subgraph G' normally embeddable in *F*, then holds: Either G is normally embeddable in *F* or G is not embeddable in *F*.

Proof: Without loss of generality we can assume that at least one subgraph $G' \subseteq G$ is normally embeddable in *F* and that G is embeddable in *F*. G' divides *F* in regions such that for each region R of G' holds: $R \cup Bd(R)$ is homeomorphic to the closed unit disc. According to Theorem 7 we consider the relative components of G in relation to G' for each country R. Since the graph $Bd(R) \cup Q_R$, where Q_R denotes the relative component in R, is a planar 2-connected graph, it divides R into regions whose boundary is equal to a cycle. Hence G is normally embeddable in *F*.

We conclude this paper by mentioning two open problems. It is well-known that every graph $G \in M_4(F_p)$, $p \in \mathbb{N}_0$, has got the orientable genus $p+1$. For the graphs $G \in M_4(N_q)$, $q \in \mathbb{N}$, we only know that the nonorientable genus of G is $q+1$ or $q+2$. The open question is: Is the nonorientable genus $q+1$?

The second open problem is a generalization of theorem 1 in [12]. It is well-known that the main hypothesis in Theorem 1 is the normal embedding of graphs in *F*. Therefore, we ask the open question: What is the analogon of Theorem 1 in [12], if we substitute the normal embedding by an arbitrary 2-cell embedding

4 References

[1] Bodendiek,R./Schumacher,H./Wagner,K.: Zur Minimalstruktur der nicht in die projektive Ebene einbettbaren Graphen. J. reine angew. Mathematik 321 (1981), pp 99-112.

[2] Bodendiek,R./Schumacher,H./Wagner,K.: Die Minimalbasis der Menge aller nicht in die projektive Ebene einbettbaren Graphen. J. reine angew. Mathematik 327 (1981), pp 119-142.

[3] Bodendiek,R./Schumacher,H./Wagner,K.: Über Relationen auf Graphenmengen. Abh. Math. Sem. Univ. Hamburg 51 (1981), pp 232-243.

[4] Bodendiek,R./Schumacher,H./Wagner,K.: Über Graphen auf Flächen und Spindelflächen. In: Graphen in Forschung und Unterricht, Herausgeber R.Bodendiek et alii. Franzbecker-Verlag (1985), pp 18-47.

[5] Bodendiek,R./Wagner,K.: Über den Kuratowskischen Satz für beliebige Flächen. Math. Nachr. 124 (1985), pp 234-251.

[6] Bodendiek,R./Wagner,K.: Solution to König's Graph Embedding Problem. Math. Nachr. 140 (1989), pp 251-272.

[7] Glover,H./Huneke,P.: The set of Irreducible Graphs for the Projective Plane. Discrete Math. 22 (1978), pp 234-256.

[8] König,D.: Theorie der endlichen und unendlichen Graphen. Leipzig 1936.

[9] Ringel,G.: Färbungsprobleme auf Flächen und Graphen. Berlin 1959.

[10] Ringel,G.: Map Color Theory. Berlin 1974.

[11] Wagner,K.: Graphentheorie. Mannheim 1970.

[12] Wagner,K.: Eine Klasse minimaler, nichtplättbarer Graphen. Math. Ann. 187 (1970), pp 104-113.

[13] Wagner,K./Bodendiek,R.: Graphentheorie I. Mannheim 1989.

[14] Wang,C.S.: Embedding graphs in the projective plane. Dissertation, Ohio 1975.

Irregular Assignments and Two Problems à la Ringel

M. Aigner, Berlin, FRG
E. Triesch, Aachen, FRG

1. Introduction

One of the best-known facts in graph theory states that any (simple) graph with at least two vertices contains two vertices of the same degree. Hence the following problem, initiated in [5], arises: Consider a weighting $w : E(G) \longrightarrow \{1, \dots, m\}$ and denote by $w(v) = \sum_{v \in e} w(e)$ the weighted degree of the vertex $v \in V$. We call the weighting w *admissible* or an *irregular assignment* if all weighted degrees are distinct. What is the minimum number m for which an admissible weighting exists? This number $s(G)$ is called the *irregularity strength* of G. Our observation above shows $s(G) \geq 2$ for any graph G, and it is an easy exercise to show that $s(G) < \infty$ iff G has no isolated edge and at most one isolated vertex.

The number $s(G)$ has been studied in a variety of papers, see [10] for a survey. As shown in [1], the investigation of $s(G)$ bears a strong number-theoretic flavor. By means of decompositions of the additive group $\mathbf{Z}_n$, it was shown that $s(G) \leq n+1$, n=number of vertices, for any graph G with $s(G) < \infty$. For connected graphs on $n \geq 3$ vertices this can be strengthened to $s(G) \leq n-1$ [8], with the single exception $s(K_3) = 3$. In particular, we have $s(T) \leq n-1$ for any tree on at least three vertices, and in fact it was shown in [1] that $s(T) \leq n-2$ unless T is the star St_n on n vertices for which clearly $s(St_n) = n-1$ holds.

A natural variant of our problem asks for the smallest *size* $i(G)$ of the image of an irregular assignment. That is, $i(G)$ is the smallest number r such that an admissible weighting $w : E(G) \longrightarrow \mathbb{N}$ exists with $|im(w)| = r$. Clearly, $i(G) \leq s(G)$ and $i(G) < \infty$ iff $s(G) < \infty$. As we shall see in section 3,

R. Bodendiek, R. Henn (Eds.)
Topics in Combinatorics and Graph Theory

the determination of $i(G)$ reduces to a purely combinatorial question with interesting connections to packing problems.

In the following two sections, several results on $s(G)$ and $i(G)$ are derived with emphasis on two problems, inspired by Ringel.

2. Another "graceful" problem for trees

Some 25 years ago Ringel posed the following problem: Consider a labelling τ of the vertices of an n-order tree T with the numbers $1, 2, \ldots, n$. τ induces a labelling $\overline{\tau}$ on the edge-set $E(T)$ by setting $\overline{\tau}(uv) = |\tau(u) - \tau(v)|$, i.e. $1 \leq \overline{\tau}(e) \leq n-1$ for all $e \in E(T)$. T is called *graceful* if T admits a labelling τ such that $im(\overline{\tau}) = \{1, \ldots, n-1\}$, i.e. $\overline{\tau} : E(T) \longrightarrow \{1, \ldots, n-1\}$ is an injection. Ringel's famous conjecture asserts that *every* tree is graceful. A great many papers have appeared on the subject (see [6] for a discussion), but a solution seems not in sight.

The notion of irregular assignments suggests a "dual gracefulness" of trees via Kirchhoff's laws. We have noted above that $s(T) \leq n-1$ for any T on at least $n \geq 3$ vertices. Let us call an admissible assignment $w : E(T) \longrightarrow \{1, \ldots, n-1\}$ again *graceful* if w is an injection, i.e. *all* weights $1, \ldots, n-1$ appear on the edges.

Conjecture: *Every tree on $n \geq 3$ vertices admits a graceful assignment.*

We have very little evidence to support this conjecture, but a proof (or disproof) might be somewhat simpler to obtain than for Ringel's original problem.

Proposition 1. *Every path P_n on $n \geq 3$ vertices has a graceful assignment.*

Proof. Number the edges from end-vertex to end-vertex $e_1, e_2, \ldots, e_{n-1}$, and set

$$w(e_i) = \begin{cases} 2i - 1 & (i = 1, \ldots, \lfloor \frac{n}{2} \rfloor) \\ 2(n - i) & (i = \lfloor \frac{n}{2} \rfloor + 1, \ldots, n - 1) \end{cases} .$$

The weighting w is an injection into $\{1, \ldots, n-1\}$, and is clearly admissible. □

Several other classes of trees can be shown to be graceful in our sense, e.g., when the maximal degree is 3, or, on the other hand, when the tree is a star or almost a star, meaning that there is at most one vertex with degree > 2.

3. The general irregularity strength $i(G)$

Suppose $w : E(G) \longrightarrow \mathbb{N}$ is an admissible weighting of the graph G. Let us denote by $St(u)$ the edges incident with u and by $M(u)$ the multiset of the images of $St(u)$. Clearly, $M(u) \neq M(v)$ for all $u \neq v$. Suppose, on the other hand, that we have a coloring of the edges by any set $\{a, b, c, \dots\}$ of, say, size r where $M(u)$ is the multiset of colors assigned to $St(u)$. If $M(u) \neq M(v)$ for all $u \neq v$, then we can plainly achieve an admissible weighting by replacing $a, b, \dots$ by integers far enough apart. E.g. $a = 1$, $b > n, c > n^2$, etc. will do. Hence the determination of $i(G)$ reduces to the following coloring problem: *What is the minimum number $i(G)$ of colors for the edges such that the multisets of colors $M(u)$ are different for any two vertices $u \neq v$?*

Let us look at a few examples.

Note first that $M(u) \cap M(v) \neq \emptyset$ if u, v are adjacent. Furthermore $i(G) \geq 2$ by our initial remark on degree sequences.

(1) $\quad i(K_n) = 3$ for $n \geq 3$.

Proof. Since $|St(u)| = n - 1$ for all u, we must have n different multisets of size $n - 1$. For two colors a, b there exist precisely n such multisets, namely $aa \dots a, aa \dots ab, \dots, bb \dots b$, but the first and last are incompatible since $M(u) \cap M(v) = \emptyset$ cannot occur by our remark above. Hence $i(K_n) > 2$. From $i(K_n) \leq s(K_n)$ and $s(K_n) = 3$ (see [8]) we infer $i(K_n) = 3$. □

(2) $\quad i(K_{m,n}) = \min\{k : \binom{k+m-1}{m} \geq n\}$ for $m < n$,

$\quad i(K_{n,n}) = 3 \quad (n \geq 2)$.

Proof. Arrange the colors in an $m \times n$-matrix. The columns constitute n different multisets, and since there are $\binom{k+m-1}{m}$ multisets with k entries and size m, we must have $\binom{k+m-1}{m} \geq n$. On the other hand, by arranging $n \leq \binom{k+m-1}{m}$ such multisets in lexicographic order, all rows are distinct if $m < n$. To prove the second assertion we consider $K_{n-1,n}$, $n \geq 2$. By what we just proved $i(K_{n-1,n}) = 2$, and by adding an n-th row consisting of a third color, we conclude $i(K_{n,n}) \leq 3$, and thus $i(K_{n,n}) = 3$ since $s(K_{n,n}) \geq 3$ (see [10]). □

(3) Let C_n be the cycle of length n, then $i(C_n) = \min(\min\{r : r \text{ odd}, \binom{r+1}{2} \geq n\}, \min\{r : r \text{ even}, \frac{r^2}{2} \geq n\}) = \min(2.\lceil\frac{1}{4}(-3+\sqrt{8n+1}\rceil+1,\ 2.\lceil\frac{1}{2}\sqrt{2n}\rceil)$.

Proof. Let an admissible coloring of C_n with r colors, say $\{1,2,\ldots r\}$, be given. Consider the multigraph M_r consisting of the complete graph K_r and all r loops at the vertices, and let $f : V(C_n) \longrightarrow E(M_r)$ be the map which associates to $u \in V(C_n)$ the edge $ij \in E(M_r)$ where i,j are the colors of the edges incident with u. If we run through $V(C_n)$ in some orientation, then the images $f(V(C_n))$ form a closed trail in M_r with all edges different. Conversely, any such closed trail in M_r of length n yields an admissible coloring of C_n with at most r colors. If r is odd, then M_r is an Eulerian graph, and the length of a longest closed trail is $\binom{r+1}{2}$. If r is even, then M_r minus a matching with $\frac{r}{2}$ edges is Eulerian, whence the length of a longest closed trail is $\binom{r+1}{2} - \frac{r}{2} = \frac{r^2}{2}$. Solution of the quadratics $\binom{r+1}{2} \geq n$ resp. $\frac{r^2}{2} \geq n$ yields the last formula. □

By the same argument we can prove:

(4) Let P_n be the path with n vertices, then $i(P_n) = \min(2\lceil\frac{1}{4}(-3+\sqrt{8n-7})\rceil+1,\ 2\lceil\frac{1}{2}\sqrt{2n-6}\rceil)$.

(5) Let $G = tC_3$ consist of t disjoint cycles of length 3, then $i(G) = \min(\min\{r : \lfloor\frac{r}{3}\lfloor\frac{r-1}{2}\rfloor\rfloor \geq t,\ r \not\equiv 5 \pmod 6\}, \min\{r : \lfloor\frac{r}{3}\lfloor\frac{r-1}{2}\rfloor\rfloor - 1 \geq t, r \equiv 5 \pmod 6\})$.

Proof. By the same reasoning as in (3), one sees that $i(tC_3)$ is the smallest number r such that t edge-disjoint triangles can be packed in K_r. The maximal number of edge-disjoint triangles in K_r has been determined in [9] with the result

$$\begin{array}{ll} \lfloor \frac{r}{3} \lfloor \frac{r-1}{2} \rfloor \rfloor & r \not\equiv 5 (\mathrm{mod}\ 6) \\ \lfloor \frac{r}{3} \lfloor \frac{r-1}{2} \rfloor \rfloor - 1 & r \equiv 5 (\mathrm{mod}\ 6). \end{array} \qquad \square$$

(6) Let $G = tP_k$ consist of t disjoint paths of size k, then $i(G) \geq 2t$, since G has $2t$ end-vertices. It is clear that $i(G) = 2t$ for t large enough. Hence the question arises as to what the smallest $t = t(k)$ is for given k with $i(G) = 2t$. Let an admissible coloring of G with r colors be given, and consider again M_r. By the same method as in the proof of (3) we see that we have to find t edges-disjoint *open* trails of length $k - 2$ in M_r with distinct end-vertices. If r is odd, then all degrees in M_r are even. Since the end-vertices of the t trails must have odd degree, we conclude

$$t(k-2) \leq \binom{r+1}{2} - t \quad \text{or} \quad t(k-1) \leq \binom{r+1}{2}, \quad r \ \text{odd}.$$

If r is even, then all degrees in M_r are odd. Since there are $r - 2t$ vertices which have an even degree in the combined trails, we infer

$$t(k-2) \leq \binom{r+1}{2} - (\frac{r}{2} - t) \quad \text{or} \quad t(k-3) \leq \frac{r^2}{2}, \quad r \ \text{even}.$$

For $r = 2t$, this implies $t \geq \lceil \frac{k-3}{2} \rceil = \lfloor \frac{k}{2} \rfloor - 1$.

Let $G(t,k) = tP_k$ consist of t disjoint paths of size k. Then $i(G(t,k)) = 2t$ iff $t \geq \lfloor \frac{k}{2} \rfloor - 1$, $k \geq 4$, and $G(t,3) = 2t$ iff $t \geq 1$.

Proof. The case $k = 3$ is obvious, so assume $k \geq 4$. We have already seen that $i(G(t,k)) = 2t$ implies $t \geq \lfloor \frac{k}{2} \rfloor - 1$. Suppose $i(G(t,k)) = 2t$ for some $t \geq \lfloor \frac{k}{2} \rfloor - 1$. Then $i(G(t+1,k)) = 2(t+1)$ by coloring the $(t+1)$-st path alternately. E.g., when $k \equiv 1 \pmod 4$, we may choose the coloring $t+1, 1, t+2, 2, t+1, 3, \ldots, \frac{k-3}{2}, t+2, t+2$. The other cases are treated similarly. Hence the only interesting case is $t = \lfloor \frac{k}{2} \rfloor - 1$. Here we have to decompose M_{2t} into t open trails of length $2t+1$ (k odd) and length $2t$ (k even), respectively. Consider first the case when k is odd. Adding loops at the beginning and end of the t trails, we have to decompose the complete graph K_{2t} into t Hamiltonian paths, and the existence of such decompositions is a well-known result in graph theory (see [4]). The case when k is even reduces to the same question by adding just one loop to the Hamiltonian paths. □

For t in the range $1 \leq t < \lfloor \frac{k}{2} \rfloor - 1$, the bounds derived above may well be exact.

After these examples let us note a result on general graphs.

Proposition 2. *We have $i(G) \leq n-1$ for any n-order graph G with $i(G) < \infty$, except $i(K_3) = 3$.*

Proof. We use induction on n. For $n = 3$ and 4, the result can be seen directly. Suppose then $n \geq 5$. If there exists a vertex v such that $G - v$ has no isolated edge and at most one isolated vertex, then by induction $i(G - v) \leq n - 2$ and hence $i(G) \leq n - 1$ by coloring the star $St(v)$ with a new color. If there is no such vertex, then G is a disjoint union of P_3's and C_3's, and the result follows easily from (5) and (6). □

Another result which can be readily proved is the following.

Proposition 3. *Let T be a tree on at least three vertices without vertices of degree 2. Then $i(T)$ = number of vertices of degree 1.*

It is conjectured (see [10]) that, in fact, $s(T)$ equals the number of vertices of degree 1 for such trees.

The packing result (5) suggests a natural generalization for arbitrary cycles which closely resembles Ringel's famous Oberwolfach problem. Let us first state Ringel's problem: *Given K_r, r odd, which sets of cycles C_i can be packed edge-disjoint such that every edge of K_r is in some cycle?* For example, K_r can be fully decomposed into C_3's iff $r \equiv 1$ or 3 (mod 6). (For references on the Oberwolfach problem see e.g. [2,3,6,7].)

In our context, the general cycle-packing problem reads as follows:

Let $C(i_1, \ldots, i_t)$ be the graph consisting of t disjoint cycles of length $i_1, \ldots, i_t$, respectively. Compute $i(C(i_1, \ldots, i_t))$.

By the same argument as in (5), the problem can be reformulated as follows: What is the smallest r such that the multigraph M_r contains t edge-disjoint closed trails of lengths $i_1, \ldots, i_t$?

To conclude let us consider the example $C(\underbrace{4,4,\ldots,4}_{t})$. We have to pack s C_3's in M_r (together with loops), hence $s \leq r$, and $t-s$ C_4's. Suppose $t \geq r$. The number of vertex-cycle incidences in K_r is $3s+4(t-s)=4t-s$. On the other hand, any vertex is in at most $\frac{r-1}{2}$ cycles if r is odd and $\frac{r-2}{2}$ cycles if r is even. Hence

$$4t-r \leq 4t-s \leq \begin{cases} \frac{r(r-1)}{2} & r \text{ odd} \\ \frac{r(r-2)}{2} & r \text{ even} \end{cases},$$

and we conclude

$$i(C(4,\ldots,4)) \geq \min(2\lceil -\frac{3}{4}+\sqrt{2t+\frac{1}{16}}\rceil + 1,\ 2\lceil\sqrt{2t}\rceil).$$

The following table shows that the right-hand bound is the correct number for $7 \leq t \leq 17$; the small values up to $t=6$ have $t < i(C(4,\ldots,4))$.

t	1	2	3	4	5	6	7	8	9	10	11	12	13	14	15	16	17
$i(C(4,\ldots,4))$	3	5	5	6	7	7	7	8	9	9	9	10	11	11	11	11	12

As an example, $t=7$ is realized by the Fano plane. It is quite plausible that the lower bound is correct for all values of t.

References

1. M. Aigner, E. Triesch, Irregular assignments of trees and forests. Submitted to SIAM J. Disc. Alg. Methods.

2. B. Alspach, R. Häggkvist, Some observations on the Oberwolfach problem. J. Graph Theory 9(1985), 177-187.

3. B. Alspach, P.J. Schellenberg, D.R. Stinson, D. Wagner, The Oberwolfach problem and factors of uniform odd length cycles. J. Combinatorial Theory (A) 52(1989), 20-43.

4. J.-C. Bermond, Hamiltonian graphs. In: Selected Topics in Graph Theory (Beineke-Wilson, eds.), Academic Press 1978, 127-167.

5. G. Chartrand, M. Jacobson, J. Lehel, O. Oellerman, S. Ruiz, F. Saba, Irregular networks. Proc. of the 250th Anniversary Conf. on Graph Theory. Fort Wayne, Indiana, 1986.

6. R. Guy, Unsolved combinatorial problems. Combinatorial Mathematics and its Applications. (D. Welsh ed.), Academic Press 1971.

7. R. Häggkvist, Decompositions of complete bipartite graphs. Surveys in Combinatorics (J. Siemons ed.), Cambridge University Press 1989, 115-147.

8. M. Jacobson, J. Lehel, A bound for the strength of an irregular network. To appear.

9. J.G. Kalbfleisch, R.G. Stanton, Maximal and minimal coverings of $(k-1)$-tuples by k-tuples. Pac. J. Math. 26(1968), 131-140.

10. J. Lehel, Facts and quests on degree irregular assignments. To appear.

11. G. Ringel, Problem No. 25, Theory of graphs and its applications. Proc. of the Symposium, Smolenice 1963 (M. Fiedler ed.), Publ. House of the Czechoslovak Acad. Sciences, 1964.

A Recursive Bound for the Number of Complete K-Subgraphs of a Graph

R. Ahlswede, Bielefeld, FRG
N. Cai, Bielefeld, FRG
Z. Zhang, Bielefeld, FRG

Abstract

The following inequality was conceived as a tool in determining coloring numbers in the sense of Ahlswede, Cai, Zhang ([1]), but developed into something of a seemingly basic nature.

Theorem For any graph $G = (\Omega_n, \mathcal{E})$ with n vertices let T_k be the number of complete k-subgraphs of G. Then for $k \geq 2$

$$T_k \geq \frac{T_{k-1}}{k \cdot n}[2(k-1)|\mathcal{E}| - (k-2)n^2] \,. \tag{1}$$

Proof of the Theorem stated in the abstract. By its definition we have $T_1 = n$. We show (1) by induction on k.

For $k = 2$ $T_2 = |\mathcal{E}| = \frac{n}{2 \cdot n}[2|\mathcal{E}|] = |\mathcal{E}|$, so (1) holds even with equality. For the induction step from $k-1$ to k we need some notation.

$V(m)$ denotes a set with m vertices and $\mathcal{T}_m$ stands for the set of all those sets, which are the vertex set of a complete m-subgraph. We also set

$$\mathcal{E}_v = \{v' : (v, v') \in \mathcal{E}\} \,, \tag{2}$$

$$\mathcal{E}_{V(m)} = \cap_{v \in V(m)} \mathcal{E}_v \,, \tag{3}$$

$$\mathcal{T}_m(V(m-1)) = \{V(m) \in \mathcal{T}_m : V(m) \supset V(m-1)\} \,, \tag{4}$$

and start now with

$$T_k = \frac{1}{k} \sum_{V(k-1) \in \mathcal{T}_{k-1}} |\mathcal{E}_{V(k-1)}| \,. \tag{5}$$

Next we bound $|\mathcal{E}_{V(k-1)}|$ from below with the help of the identity

$$\left| \bigcup_{V(k-2) \subset V(k-1)} \mathcal{E}_{V(k-2)} \right| = \sum_{V(k-2) \subset V(k-1)} |\mathcal{E}_{V(k-2)}| - (k-2)|\mathcal{E}_{V(k-1)}| \,, \tag{6}$$

which holds, because vertices from the union which are counted more than once in the sum are actually counted $k-1$ times and they are exactly the vertices in $\mathcal{E}_{V(k-1)}$.

Since the union has a cardinality not exceeding n, we get

$$|\mathcal{E}_{V(k-1)}| \geq \frac{1}{k-2} \left(\sum_{V(k-2) \subset V(k-1)} |\mathcal{E}_{V(k-2)}| - n \right) \,. \tag{7}$$

R. Bodendiek, R. Henn (Eds.)
Topics in Combinatorics and Graph Theory

Substituting this in (5) yields

$$\begin{aligned}
k(k-2)T_k &\geq \sum_{V(k-1)\in\mathcal{T}_{k-1}} \left(\sum_{V(k-2)\subset V(k-1)} |\mathcal{E}_{V(k-2)}| - n \right) \\
&= \sum_{V(k-2)\in\mathcal{T}_{k-2}} \; \sum_{V(k-1)\in\mathcal{T}_{k-1}(V(k-2))} |\mathcal{E}_{V(k-2)}| - n|\mathcal{T}_{k-1}| \\
&= \sum_{V(k-2)\in\mathcal{T}_{k-2}} |\mathcal{T}_{k-1}(V(k-2))|^2 - n\, T_{k-1} \\
&\geq T_{k-2}\left(\frac{(k-1)T_{k-1}}{T_{k-2}}\right)^2 - n\, T_{k-1} \quad \text{(by convexity of } x^2\text{)} \\
&= \frac{T_{k-1}}{T_{k-2}}((k-1)^2 T_{k-1} - n\, T_{k-2}) \\
&\geq \frac{T_{k-1}}{T_{k-2}}\left((k-1)\frac{T_{k-2}}{n}(2(k-2)|\mathcal{E}| - (k-3)n^2) - n\, T_{k-2}\right) \\
&= \frac{T_{k-1}}{n}(2(k-1)(k-2)|\mathcal{E}| - ((k-1)(k-3)+1)n^2)
\end{aligned}$$

and therefore (1).

The following consequence is useful.

<u>Corollary</u> If for some $\alpha > 0$ $|\mathcal{E}| \geq \frac{k-1}{2k}n^2 + \alpha n^2$, then

$$T_{k+1} \geq \alpha^k\, n^{k+1}\ . \tag{8}$$

<u>Proof:</u> Since $\frac{k-1}{2k} \geq \frac{\ell-1}{2\ell}$ for $\ell = 1, 2, \ldots, k$, the assumption implies

$$2\ell|\mathcal{E}| - (\ell-1)n^2 \geq 2\ell \cdot \alpha n^2$$

and therefore by (1) and since $T_1 = n$

$$T_{\ell+1} \geq \frac{1}{\ell+1}\frac{T_\ell}{n}2\ell\, \alpha n^2 \geq \alpha n T_\ell\ ,$$

which implies (8).

<u>Remark</u> Our result falls into the context of paragraph VI.1 of [2]. A well-known result by Turan ([3]) concerns the determination of the maximal number $t_k(n)$ of edges in an n–graph such that $T_{k+1} = 0$.

The optimal graphs have the following structure:

For $n = km + r, r < k$, partition Ω_n into r sets with $m+1$ vertices and $k-r$ sets with m vertices and include exactly all edges connecting vertices of different sets.

Therefore one has for Turan's function

$$t_k(n) = \binom{r}{2}(m+1)^2 + \binom{k-r}{2}m^2 + r(k-r)(m+1)m \; . \tag{9}$$

It is remarkable that our quite general inequality almost implies this identity. In fact, in an optimal graph clearly $T_k \geq 1$, because otherwise an edge could be added. Therefore from the inequality we conclude

$$|\mathcal{E}| \leq \frac{n^2(k-1)}{2 \cdot k} \tag{10}$$

and if n is a multiple of k , that is, $n = m \cdot k$, then (10) takes the form $|\mathcal{E}| \leq m^2 \binom{k}{2}$ and thus the bound in (9) follows.

For general $n = km + r$ an easy calculation shows that the bound in (10) is tight, if $\frac{(k-r)r}{2k} < 1$. This is for instance always the case also for $r = 1, 2$.

[1] R. Ahlswede, N. Cai and Z. Zhang, "Rich colorings with local constraints", Preprint in SFB Diskrete Strukturen, Bielefeld 1989.

[2] B. Bollobás, "Extremal Graph Theory", Acad. Press, 1978.

[3] P. Turan, "An extremal problem in graph theory (Hungarian)", Mat.Fiz. Lapok, 48, 436–452, 1941.

One-Factorizations of Tensor Products of Graphs

B. Alspach, Burnaby, Canada
J. C. George, Urbana, USA

Introduction And Definitions

A very natural question raised about products of graphs is the following. Do there exist necessary and sufficient conditions on a pair (G,H) of graphs for their product to have some specified property? In particular, for which graphs is the product one-factorizable? Sufficient conditions have been investigated for the cartesian, lexicographic, and tensor products in [**3**], [**4**], and [**5**]. However, the conditions for the tensor product to be one-factorizable are significantly more scanty than those for the other products. The purpose of this paper is to correct this situation somewhat.

We assume the reader is familiar with the standard definitions of graph theory, such as those in [**2**]; other definitions that might be unfamiliar follow. The **tensor product** (Kronecker product, weak product, conjunction, alpha-product) $G \otimes H$ of G and H is the graph whose vertex-set is the cartesian product $V(G) \times V(H)$ of the vertex-sets of G and H, and the edge-set is any pair $[(u,v),(u',v')]$ in which (u,u') is an edge of G, and (v,v') is an edge of H. The edge $[(u,v),(u',v')]$ is called an **appearance** of the edge (v,v') of H in $G \otimes H$. A set of edges of $G \otimes H$ in which each edge of H has at most one appearance is said to be **special**.

A **Frobenius group** is a transitive permutation group F on n letters with the properties that

(1) no non-identity element has two or more fixed points; and
(2) the stabilizer subgroups F_i, $1 \le i \le n$, are non-trivial.

The first author gratefully acknowledges partial suport from the Natural Sciences and Engineering Council under Grant A-4792. The second author expresses his appreciation to the Symbolic Computation Lab in the UIUC department of mathematics for its support. This paper will comprise a portion of the thesis of the second author.

R. Bodendiek, R. Henn (Eds.)
Topics in Combinatorics and Graph Theory

It is known (for example, see **[1]**) that a Frobenius group of degree n can have order at most $n^2 - n$, and that this maximum order is attained if and only if n is a prime power.

THE MAIN RESULT

THEOREM 1. *Let $F(n)$ be the order of the largest group of degree n which satisfies property (1) above, and suppose that H is any graph. If $K_n \otimes H$ has a special one-factor T, then $K_n \otimes H$ has at least $|F(n)|$ edge-disjoint one-factors.*

PROOF: Let Φ and Ψ be automorphisms of $F(n)$. By identifying the vertices $\{u_i\}$ of K_n with the integers $1, 2, \ldots, n$ we define an action of $F(n)$ on the vertices of K_n. Denote by $\Phi(T)$ the set of edges

$$\{\, [(\Phi(u), v), (\Phi(u'), v')] : [(u, v), (u', v')] \in T \,\}.$$

We first note that $\Phi(T)$ and $\Psi(T)$ are one-factors of the tensor product since T is a one-factor and elements of $F(n)$ are bijective. We must show that $\Phi(T)$ and $\Psi(T)$ are edge-disjoint. If the edge $[(\Phi(u), v), (\Phi(u'), v')]$ is the same as $[(\Psi(u), v), (\Psi(u'), v')]$, then we must have $\Phi(u) = \Psi(u)$ and $\Phi(u') = \Psi(u')$, because the edge $[v, v']$ makes only one appearance in T. But $F(n)$ satisfies property (1) above, and the element $\Phi\Psi^{-1}$ has two fixed points, u and u'. Thus $\Phi\Psi^{-1}$ is the identity. This means that two different permutations of $F(n)$ give two edge-disjoint one-factors. ∎

ONE-FACTORIZATIONS

If $K_n \otimes H$ is one-factorizable, then H must be regular, and the product must have an even number of vertices. For Theorem 1 to apply, there must be the required special one-factor T. We have the following result.

THEOREM 2. *$K_n \otimes H$ has a special one-factor T if and only if H has an n-factor.*

PROOF: First assume $K_n \otimes H$ has a special one-factor, and consider the n vertices (u_i, v), $1 \le i \le n$. Each of these vertices must appear in exactly one edge of T. These n edges of T are appearances of distinct edges of H by hypothesis. Since v was arbitrary, this means each vertex of H has degree n or greater. Thus, the edges of H that appear in T form a regular spanning subgraph of degree n.

To prove the converse, without loss of generality assume H is n-regular. Suppose first n is even; take an arbitrary two-factorization of H, say $F_1, F_2, \ldots, F_{n/2}$. Now we orient the edges of H in such a way that each cycle of each two-factor is oriented cyclically.

We will represent the vertices of $K_n \otimes H$ as a rectangular array of vertices with columns indexed by the vertices of K_n and rows indexed by $V(H)$. Project the oriented

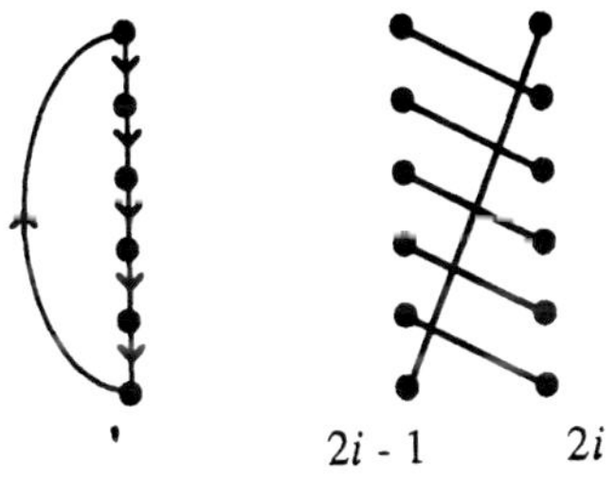

Figure 1

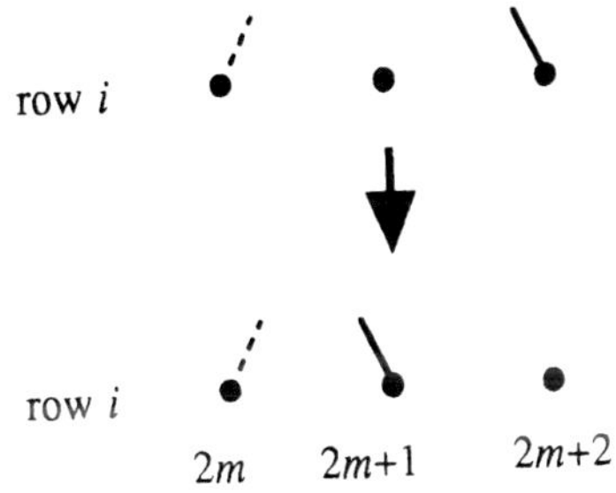

Figure 2

two-factor F_i onto columns $2i-1$ and $2i$ for $1 \leq i \leq \frac{n}{2}$, as shown in Figure 1. The resulting edges are the required special one-factor.

Suppose that $n = 2m+1$ is odd. By Vizing's theorem, H can be edge-colored with $2m+2$ colors. Let F_i be the set of all edges with colors $2i-1$ or $2i$ for $1 \leq i \leq m+1$. Introduce a "dummy" column $2m+2$ at the right of the rectangular array. Now, each subgraph F_i is a spanning subgraph whose components are paths and cycles. Orient the edges of H so that each path or cycle of each F_i is an oriented path or cycle. Now, as above, project each F_i onto columns $2i-1$ and $2i$. We want to be able to eliminate the dummy column. Each row of this rectangular array has precisely one isolated vertex. If, in a given row, this isolated vertex occurs in column $2m+2$, leave that row alone. Suppose that the isolated vertex occurs in column j, $1 \leq j \leq 2m$. Then, shift the edge whose endvertex is at $(i, 2m+2)$ to (i, j), keeping the other end fixed. If the isolated vertex occurs in column $2m+1$, then we must shift the edge with endvertex $(i, 2m+2)$ to $(i, 2)$ and shift the edge with endvertex originally at $(i, 2)$ to $(i, 2m+1)$, keeping the other ends of both edges the same. See Figure 2. ∎

COROLLARY. *Let n be a prime power, and H be a graph with an n-factor. Then $K_n \otimes H$ has $n^2 - n$ edge-disjoint one-factors.*

PROOF: By Theorem 2, $K_n \otimes H$ has a special one-factor; we know that S_n contains a Frobenius group of order $n^2 - n$ since n is a prime power. By Theorem 1, we are done. ∎

COROLLARY. *If H is n-regular, n a prime power, then $K_n \otimes H$ is one-factorizable.*

PROOF: $K_n \otimes H$ is regular of degree $n^2 - n$; hence the one-factors of Theorem 2 form a one-factorization. ∎

FURTHER RESULTS

Theorem 1 has applications to graphs other than the complete graph. Let C_n be the n-cycle; Aut(C_n) is the dihedral group D_n. It is easy to show that if n is odd, D_n is a Frobenius group. It follows that for H regular of degree n, if $C_n \otimes H$ has a special one-factor, then $C_n \otimes H$ is one-factorizable. It is therefore reasonable to prove the following.

THEOREM 3. *If G is n-regular then $C_n \otimes G$ has a special one-factor.*

PROOF: We shall construct a special one-factor of $C_n \otimes G$ by embedding each edge of G into an edge of $C_n \otimes G$ in turn. Assume that we have embedded some edges of G and have come to an edge $[u, v]$ of G. It is enough to show that we can embed this edge so that previously embedded edges are not disturbed.

Since $[u, v]$ has not been embedded, there are vertices (i, u) and (j, v) of $C_n \otimes G$ which are not yet in any embedded edge; let us call such vertices **vacant**. We will place the vertices in a rectangular array with columns corresponding to the vertices of C_n and rows to vertices of G. Since n is odd, the distance from column i to column j is odd in some direction. Without loss of generality, assume that $j - i$ is odd, and $j > i$.

We will now show that we can rearrange the edges already embedded so that any edge already embedded remains embedded, and so that the vertices $(i + 2, u)$ and (j, v) are vacant (we take $i + 2$ modulo n). When this is shown, we will be done. For we need only move the vacant vertex in row u two spaces at a time until its column is $j - 1$; then we can embed the edge $[u, v]$ in the two vacant vertices $(j - 1, u)$ and (j, v). Before we give a general proof, we will give two examples of such rearrangements.

Figure 3

The general proof will employ a bipartite digraph B. Each part will have $|V(G)|$ vertices indexed by $|V(G)|$; the parts will be L and R.

We have the vertex (i, u) vacant, and we want to make the vertex $(i + 2, u)$ vacant. If it is already vacant, we are done; so assume there's an edge incident with it. If the other endvertex of this edge is in column $i + 1$, then we are in the situation of Figure 3 above. Thus we can force $(i + 2, u)$ to be vacant. Otherwise, the other vertex is in

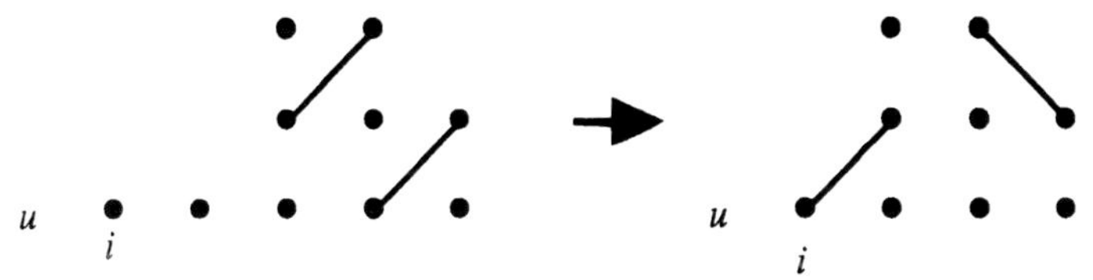

Figure 4

column $i+3$ and in some row, say u_1. To indicate this, place the directed edge from u of L to u_1 of R in B.

If the vertex $(i+1, u_1)$ is vacant, we can replace the edge from $(i+2, u)$ to $(i+3, u_1)$ with the edge from (i, u) to $(i+1, u_1)$, and $(i+2, u)$ will be vacant. Then we are done.

If $(i+1, u_1)$ is not vacant, perhaps the other vertex of its edge is in column $i+2$. Then we are in the position of Figure 4, and we can shift edges to make $(i+2, u)$ vacant. Then we are done.

The only remaining possibility is that $(i+1, u_1)$ is incident to an edge whose other endvertex is, say, (i, u_2). To indicate this, place the directed edge from u_1 of R to u_2 of L in B.

If the vertex $(i+2, u_2)$ is vacant, or if it is incident to an edge whose other vertex is in column $i+1$, then as above, we can shift edges so that $(i+2, u)$ becomes vacant. So, we may assume that we have some edge from $(i+2, u_2)$ to $(i+3, u_3)$, as shown in Figure 5. Possibly $u_3 = u$, but this will not cause any difficulty. We will put a directed edge from u_2 of L to u_3 of R in B.

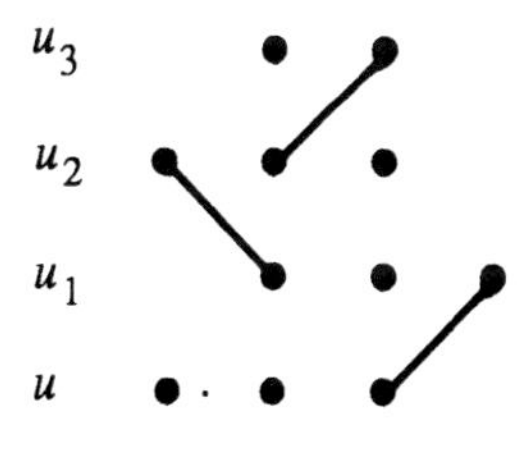

Figure 5

We continue this process. There will be a directed edge from u' of L to u'' of R when there is an edge embedded between $(i+2, u')$ and $(i+3, u'')$. There will be a directed edge from u' of R to u'' of L when there is an edge embedded between $(i+1, u')$ and (i, u''). Each vertex of B has indegree and outdegree at most 1; the indegree of vertex u of L is 0 because (i, u) is vacant. Thus the directed path beginning at u of L ends somewhere in B; following the path backwards enables one to rearrange the edges embedded in $C_n \otimes G$ in such a way that $(i+2, u)$ becomes vacant.

Now we observe that in the same way, we could move the vacancy from (i, u) to $(i-2, u)$; this justifies our assumption above that $j-i$ was odd. Observe also that this process cannot disturb the vacancy at (j, v). The edge rearrangements only disturb vacancies with column distance 1 or 2 away from the current vacancy, and column j is removed from column i by an odd distance greater than 1. ∎

COROLLARY. *If G has order n and can be decomposed into Hamilton cycles, then $G \otimes H$ has a one-factorization whenever H is n-regular.*

PROOF: This follows immediately from the fact that the tensor product distributes over factors. ∎

REFERENCES

1. N. L. Biggs and A. T. White, "Permutation Groups and Combinatorial Structures," London Mathematical Society Lecture Note Series #33, Cambridge University Press, London, 1979.
2. J. A. Bondy and U. S. R. Murty, "Graph Theory with Applications," Elsevier Publishing Company, Inc., New York, 1976.
3. A. Kotzig, *1-Factorizations of Cartesian Products of Regular Graphs*, J. Graph Theory **3** (1979), 23–34.
4. T. Pisanski, J. Shawe-Taylor, and B. Mohar, *1-Factorization of the Composition of Regular Graphs*, Publ. Inst. Math. (Beograd) (N. S.) **33(47)** (1983), 193–196.
5. W. D. Wallis and W. Zhi-jian, *On One-Factorizations of Cartesian Products*, Congr. Numer. **49** (1985), 237–245.

Non-Commutative Geometry and Graphs

J. André, Saarbrücken, FRG

Introduction

Objects of this paper are complete digraphs (directed simple graphs). A colouring of such a digraph is a mapping from the vertices and the arcs onto a set F of colours. We assume that vertices and arcs are differently coloured but do provisionally no further restrictions. Following J. Pfalzgraf [8] we denote this mapping by $\langle , \rangle$. Thus $\langle x,x' \rangle$ is the colour of the arc (x,x') if $x \neq x'$ and of the vertex x if $x = x'$. Such a coloured complete digraph is also called a P-space. We interpret the vertices and colours as proper and improper points resp. and define the line (Linie) $x \square x'$ by (2.1), (2.2). Generally $x \square x' \neq x' \square x$, i.e. the space need not be commutative. A simple example of such a (noncommutative) space is the circle-space: Here $X := \mathbb{R}^2$ and the line $x \square x'$ is the circle through x' with the center x, and its improper point is its radius. Two lines with the same improper point are parallel. Two different parallel lines, however, need not have an empty intersection. The first two sections give a brief introduction to the P-spaces, for more details (with further references) see [2]. Subspaces are point sets being closed with respect to drawing of lines and parallels. A straight line (Gerade) is a line which is also a subspace (sect.3). Arbitrary P-spaces are too general for establishing a reasonable theory. Therefore we additionally assume geometrical conditions which are generalizations of the well known Desargues theorem in affine spaces: We define a sequence Sim_q $(q \in \mathbb{N})$ of geometrical conditions. The simplest non trivial case $q = 2$ leads to the skewaffine spaces. A skewaffine space with commutative $\square$ is affine. An application of the theory of Ramsey-numbers leads to a theorem that a finite selfadjoint skewaffine space in which the number

R. Bodendiek, R. Henn (Eds.)
Topics in Combinatorics and Graph Theory

of proper points is large to that of improper points possesses a staight line (Theorem 6.1). The following Theorem 6.2 specializes this property to permutation groups. In the two concluding sections we consider finite P-spaces and their connections to adjaceny-matrizes and hence to algebraic graph theory. A geometric characterization of distance regular graphs is stated in Theorem 8.2. It might be fruitful to enlarge this theory in order to obtain further relations between graph theory and geometry.

1. The coloured complete digraph

Let $\mathcal{D} = (X,Y)$ be a digraph, X its set of *vertices* and $Y \subseteq X^2\setminus i_X$ its set of *arcs*. It is *complete* if $Y = X^2\setminus i_X$. A *colouring* of such a complete digraph is a surjective mapping $\langle\,,\rangle\colon X^2 \twoheadrightarrow F$, $(x,x') \mapsto \langle x,x'\rangle$. The elements of F are called *colours*. The vertex x is by definition coloured by $\langle x,x\rangle$ and the arc (x,x') $[x,x'\in X,\ x\neq x']$ by $\langle x,x'\rangle$. A coloured complete digraph $\mathcal{R} = (X,\langle\,,\rangle) = (X,\langle\,,\rangle,F)$ is a *P-space* or more briefly a *space* iff

$$\langle x,x\rangle = \langle x',x''\rangle \quad \text{implies} \quad x' = x'' \tag{1.1}$$

That means the sets $E := \{\langle x,x\rangle \mid x\in X\}$ and $R := \{\langle x,x'\rangle \mid x,x'\in X,\ x\neq x'\}$ are disjoint. Exactly in this case the vertices and arcs are coloured by different colours. The elements of R are called *improper points* or *directions*, the vertices (i.e. elements of X) sometimes also *proper points*. In the following we often identify $(X,\langle\,,\rangle)$ with X if no misunderstanding is possible.

A space is *adjoint* iff

$$\langle x,x'\rangle = \langle u,u'\rangle \quad \underline{\text{implies}} \quad \langle x',x\rangle = \langle u',u\rangle \tag{1.2}$$

it is *selfadjoint* iff

$$\langle x,x'\rangle = \langle x',x\rangle \tag{1.2$'$}$$

holds for all $x,x'\in X$. Any selfadjoint space is adjoint. In an adjoint space with $\langle x,x'\rangle = r$ define $r^{\vee} := \langle x',x\rangle$ as the improper point *adjoint* to r. By (1.2) it is well defined. Obviously $r^{\vee\vee} = r$. The space is selfadjoint iff $r^{\vee} = r$ for all $r\in R$.

A *tournament* (cf.e.g. [10],7,§23) is a digraph $\mathcal{D} = (X,Y)$ in which for any $x, x \in X$ with $x \neq x'$ exactly one of the arcs (x,x'), (x',x) belongs to Y. By completing $\mathfrak{D}$ to a complete digraph such that the arcs of $\mathfrak{D}$ and of the complement $\bar{\mathfrak{D}}$ are differently coloured we obtain a tournament:

PROPOSITION 1.1. _A tournament is characterized as an adjoint P-space with exactly two improper points adjoint to each other._ □

Let now be $\mathfrak{R} = (X, \langle , \rangle)$ an arbitrary P-space and $r \in R$. Define

$$Y_r := \{(x,x') \mid \langle x,x' \rangle = r\}. \tag{1.3}$$

Then $\mathfrak{R}_r := (X, Y_r)$ becomes a digraph, the _digraph belonging to the improper point_ r.

2. Lines and parallelity

Let $\mathfrak{R}$ be a P-space. We shall describe a geometric structure on $\mathfrak{R}$. Define a _line_ (Linie) by

$$x \square x' := (x \sqcup x') \cup \{\langle x,x' \rangle\} \tag{2.1}$$

whose _proper part_ is given by

$$x \sqcup x' := \{x\} \cup \{z \in X \mid \langle x,x' \rangle = \langle x,z \rangle\}. \tag{2.1'}$$

The elements x and $\langle x,x' \rangle$ are the _basepoint_ (Aufpunkt) and the _improper point_ of $x \square x'$ resp. Two lines L, L' are _parallel_, denoted by $L \parallel L'$, iff they have the same improper point. Obviously this is an equivalence relation.

A proper part $x \sqcup x'$ of a line is _properly determined_ (_p.d._) iff

$$x \sqcup x' = u \sqcup u' \quad \underline{\text{implies}} \quad \langle x,x' \rangle = \langle u,u' \rangle. \tag{2.2}$$

In this case the improper point of a line is uniquely determined by its proper part. A space is _p.d._ iff all its lines are p.d. It is easy to see that any selfadjoint space is p.d. [2].

Let be $x \in X$ and $r \in R$ and define

$$x \square r := \{x, r\} \cup \{z \in X \mid \langle x,z \rangle = r\}. \tag{2.3}$$

Obviously $|x \square r| \geq 3$ iff the equation $\langle x,z \rangle = r$ has a solution $z \in X$. The _star_ of x is defined by

$$St(x) := \{ f \in F \mid |x \square f| \geq 3 \text{ or } f = \langle x,x \rangle\}. \tag{2.4}$$

Two proper points x, x' are _equivalent_ iff they have the same star (cf.[2],p.24). The equivalence classes are called _orbits_. A space possessing only one orbit is called _transitive_ [2]. In [4] transitive and adjoint spaces are called _rainbows_.

EXAMPLE. Let $(X,\langle,\rangle) = V(G,X)$ be a *group space* to a group G operating on X (cf.e.g. [2]), i.e.

$$\langle x,x'\rangle := G(x,x') := \{(gx,gx')|\ g\in G\}. \tag{2.5}$$

3. Subspaces and straight lines

Let X be a space. A subset $U\subseteq X$ is called a *subspace* (denoted by $U\leq X$) iff

$$x,x',x''\in U,\ z\in X \ \underline{\text{with}}\ \langle x,x'\rangle = \langle x'',z\rangle \ \underline{\text{implies}}\ z\in U \tag{3.1}$$

The sets ϕ, $\{x\}$ $(x\in X)$ and X are always subspaces of a P-space, the *trivial* subspaces. A space only possessing these trivial subspaces is called *primitive*, otherwise *imprimitive*. In a group space $V(G,X)$ (cf.2.) the subspaces coincide with the blocks ([1],[2]).

DEFINITION 3.1. A *straight line* (Gerade) is a line with at least two points whose proper points form a subspace.

We shall characterize straight lines by colourings. A set $C\subseteq X$ with $|C|\geq 2$ is called an *r-clique* $(r\in R)$ iff $\langle x,x'\rangle = r$ for all $x,x'\in C$ with $x\neq x'$. An r-clique is called *closed* iff $x\in C$, $x'\in X\setminus C$ imply $\langle x,x'\rangle \neq r$.

PROPOSITION 3.2. *Any closed r-clique is a straight line.*

PROOF. Let C be a closed r-clique.

(i) C is a subspace, $C\leq X$: Let be $u,v,w\in C$ with $u\neq v$ and $x\in X$ with $r = \langle u,v\rangle = \langle w,x\rangle$. Then $x\in C$ because C is closed

(ii) C is a line: Let $x,x'\in C$ with $x\neq x'$. Then, of course, $x\sqcup x'\subseteq C$. On the other hand let be $x''\in C\setminus\{x\}$. Then $\langle x,x'\rangle = \langle x,x''\rangle = r$ hence $x''\in x\sqcup x'$ by (2.1′), i.e. $C\subseteq x\sqcup x'$. □

We are looking for spaces in which also the converse hold.

THEOREM 3.3. *Let* $\mathcal{R} = (X,\langle,\rangle)$ *be a transitive space. Then* $G\subseteq X$ *is a straight line iff it is a closed r-clique for a suitable* $r\in R$.

PROOF. "$\Leftarrow$": Cf. proposition 3.2.

"$\Rightarrow$": Let $G = x\sqcup x'$ be a straight line with $\langle x,x'\rangle =: r\in R$. Let be $x_1,x_2\in G$, $x_1\neq x_2$. Due to the transitivity of X there exists an $x^*\in X$ with $\langle x_1,x_2\rangle = \langle x,x^*\rangle$. We have $x^*\in G$ since G is a subspace, hence $\langle x_1,x_2\rangle = r$ and G is an r-clique. Now let be $x_1\in G$, $x_2\in X\setminus G$. Assume $\langle x_1,x_2\rangle = r$. Then there exists an $x^*\in X$ with $\langle x,x^*\rangle = \langle x_1,x_2\rangle = r$ and $x^*\in x\sqcup x' = G$. Then $x_2\in G$ since $G\leq X$ contradicting our assumption. Hence G is a closed r-clique. □

4. Simplex-configurations

Let X be a P-space and $q\in\mathbb{N}$. Moreover $x_0,x_1,\ldots x_q,x_0',x_1'\in X$ with $\langle x_0,x_1\rangle = \langle x_0',x_1'\rangle$.
The q-simplex-condition

$$\mathrm{Sim}_q(x_0,x_1,\ldots,x_q,x_0',x_1') \tag{4.1}$$

([2],[3]) means the existence of $x_2',\ldots,x_q'\in X$ such that

$$\langle x_i,x_j\rangle = \langle x_i',x_j'\rangle \tag{4.1'}$$

hold for $0\leq i<j\leq q$.
The condition Sim_q holds in a space iff (4.1), (4.1′) is valid for all possible $x_0,x_1,\ldots,x_q,x_0',x_1'\in X$. Obviously Sim_1 holds in any space.
DEFINITION 4.1. A spaces is called quasiaffine iff Sim_2 holds; a transitive quasiaffine space is skewaffine [2].
REMARK. A quasiaffine space in which additionally $x\square x' = x'\square x$ holds for all points x,x' is an affine space in the usual sense [2].
PROPOSITION 4.2. *In any group space $V(G,X)$ the conditions Sim_q hold for $q\in\mathbb{N}$. Especially such a space is quasiaffine. It is skewaffine iff G acts transitively on X. If X is finite also the converse holds.*
PROOF. [2], p.45(5a), p.48(3). □
The Desargues condition is essentially Sim_3.

5. Skewaffine spaces

Let X be skewaffine. Applying $\mathrm{Sim}_2(x,y,x,x',y')$ we obtain (1.2), hence *any skewaffine space is adjoint*. Also due to (1.2) any skewaffine space is also inverse-skewaffine, i.e. $\breve{\mathrm{Sim}}_2(x,y,z,x',y')$ holds for any $x,y,z,x',y'\in X$ with $\langle x,y\rangle = \langle x',y'\rangle$. This means the existence of $z'\in X$ with $\langle x,z\rangle = \langle x',z'\rangle$ and $\langle z,y\rangle = \langle z',y'\rangle$. Summarizing this we obtain
PROPOSITION 5.1. *Sim_2 is equivalent to $\breve{\mathrm{Sim}}_2$ and (1.2).* □
REMARK. There exist, however, inverse skewaffine spaces not

being skewaffine [3].

From [2], III, §2 we obtain the following results.

PROPOSITION 5.2. Any line in a skewaffine space containing at least three proper points is p.d. □

PROPOSITION 5.3. A p.d. line of a skewaffine space has either exactly one basepoint or all of its points are basepoints; exactly in this case the line is straight. Any parallel to a straight line is also straight. □

DEFINITION 5.4. An r-claw (x,S) $(x \notin S \subseteq X,\ |S| \geq 2,\ r \in R)$ is given by $\langle x,x' \rangle = r$ for all $x' \in S$ but $\langle x_1,x_2 \rangle \neq r$ for $x_1,x_2 \in S$. A maximal r-clique is an r-clique not properly contained in another r-clique. Any closed r-clique (cf. sect.3) is maximal.

Translating the PROPOSITION 5.2 and 5.3 into the language of graph-theory we obtain

THEOREM 5.5. A p.d. line with the improper point r in a skewaffine space is either an r-claw or a closed r-clique (exactly in the last case the line is straight). □

REMARK. Let $\mathcal{R} = (X,\langle\ ,\ \rangle,F)$ be a selfadjoint skewaffine space with at least three proper points on any line, $r \in R$. Then the connected components of the graph $\mathcal{R}_r = (X,Y_r)$ (cf.(1.3)) are either cliques with at least three points or unions of claws not containing any clique with at least three points.

6. Applications to Ramsey-Theory

In this section $\mathcal{R} = (X,\langle\ ,\ \rangle,F)$ is always a finite selfadjoint skewaffine space in which all lines have at least three proper points. We shall give a sufficient condition on the existence of straight lines on $\mathcal{R}$. Let be $|R| = \rho$, hence $|F| = \rho+1$ the rank of the space and $v = |X|$ be its degree. We shall prove that if v is large to ρ then always a straight line does exist. For the following let be (without restriction of generality) $R = \{1,\ldots,\rho\}$. Assume $a_1,a_2,\ldots,a_\rho \in \mathbb{N}\setminus\{1\}$ and let be $r(a_1,\ldots,a_\rho)$ the corresponding Ramsey-number. That means that for any complete graph with at least $r(a_1,\ldots,a_\rho)$ vertices there is at least one $s \in R$ such that the graph contains an s-clique with a_s vertices. It is well known that such a Ramsey-number always exists (cf.e.g.[7]) but no methods are known up to now for computing $r(a_1,\dots,a_\rho)$ exactly. There exist, however, estimations for r, especially for $a_1 = a_2 = \dots = a_\rho = 3$:

$$r(a_1,\dots,a_\rho) \le [\rho!e] + 1 \tag{6.1}$$

(cf.e.g.[7], Theorem 6). Summarizing (6.1) together with the outcomes of the preceeding section we obtain

THEOREM 6.1. <u>A selfadjoint skewaffine space with ρ improper points and at least $[\rho!e]+1$ proper points possesses a straight line with at least three points</u>. □

A group space $V(G,X) = (X,\langle,\rangle)$ (cf. sect.4) is selfadjoint iff the group G is generously transitive, i.e. for any $x,x'\in X$ there are $g\in G$ exchanging x and x' ([2], p.29(4)). Since a straight line is a subspace and a subspace of $V(G,X)$ is a block, Theorem 6.1 gives

THEOREM 6.2. <u>A finite generously transitive group with rank $\rho+1$ and degree $\ge [\rho!e]+1$ is either twofold transitive or imprimitive</u>. □

7. Algebraic Theory

In this and the following section $\mathfrak{R} = (X,\langle,\rangle,F)$ is always a <u>finite</u> P-space with $|X| =: v$ and $E = \{e\}$. Let be $f\in F$. Define the (v,v)-matrix $N_f = (n_{f,xx'})_{x,x'\in X}$ by

$$n_{f,xx'} := \begin{cases} 1 & \langle x,x'\rangle = f, \\ 0 & \langle x,x'\rangle \ne f. \end{cases} \tag{7.1}$$

Hence N_f is an <u>adjacency matrix</u> of the digraph $\mathfrak{R}_f = (X,Y_f)$, c.f. [6], for graphs also [5]. Especially $N_e = I$, the unity matrix. Eigenvalues, included their multiplicities, characteristic and minimal polynomial of N_f are independent of the special order in which $x\in X$ occur in N_f; they only depend on the space $\mathfrak{R}$. Obviously

$$\sum_{f\in F} N_f = J \tag{7.2}$$

where J is the (v,v)-matrix having all 1 as its elements. The <u>adjacency space</u> $\mathbb{V}$ is defined by

$$\mathbb{V} := \left\{ \sum_{f\in F} \alpha_f N_f \middle| \alpha_f \in \mathbb{C} \right\} \tag{7.3}$$

It is a $\mathbb{C}$-vectorspace of dimension v.

PROPOSITION 7.1. <u>The space</u> $\mathfrak{R}$ <u>is adjoint iff</u> $A\in\mathbb{V}$ <u>implies</u> $A^T\in\mathbb{V}$ (A^T is the transpose of A).

PROOF. "$\Longrightarrow$": If $\mathfrak{R}$ is adjoint then $N_{f^\smile} = N_f^T$.

"$\Longleftarrow$": It suffices to assume $N_f^T\in\mathbb{V}$ for all $f\in F$, i.e.

$$N_f^T = \sum_{g\in F} \alpha_g N_g \ .$$

Since all elements of N_f are 0 or 1 we have $\alpha_g \in \{0,1\}$ for all $g\in F$, hence

$$N_f^T = \sum_{g\in A} N_g \tag{7.4}$$

for a suitable $A\subseteq F$. If $|A|>1$ then

$$N_f = N_f^{TT} = \sum_{g\in A} N_g^T = \sum_{h\in B} N_h$$

for a suitable $B\subseteq X$ with $|B|\geq 2$ by (7.4). But this is impossible since the N_f are linearly independent. Hence $|A| = 1$, i.e. $A = \{g\} = \{f^\smile\}$. □

The space $\mathbb{V}$ is closed with respect to the matrix multiplication iff

$$N_r N_s = \sum_{f\in F} a_{rsf} N_f$$

hold for all $r,s\in F$ and suitable $a_{rsf}\in\mathbb{C}$ (then even $a_{rsf}\in\mathbb{N}_0$). In order to find a geometrical equivalent, consider the following sharpening of the condition $\mathrm{Sim}_2^\smile$ (cf.sect.5), the <u>strongly inverse skewaffine condition</u> $\mathrm{Sim}_2^{\smile *}(x,y,z,x',y')$: <u>Whenever</u> $\langle x,y\rangle = \langle x',y'\rangle =: f$ <u>then</u>

$$|\{ z \mid \langle x,z\rangle = r,\ \langle z,y\rangle = s\}| = |\{ z' | \ \langle x',z'\rangle = r, \langle z',y'\rangle = s| \tag{7.6}$$

for all $r,s\in F$. This common value (only dependent on $r,s,f\in F$) is denoted by a_{rsf} and called <u>intersection-number</u>.

A space with (7.6) is called <u>strongly inverse skewaffine</u>. An adjoint strongly inverse skewaffine space is <u>strongly skewaffine</u>. Obviously any strongly (inverse) skewaffine space is also (inverse) skewaffine.

THEOREM 7.2. <u>A space is strongly inverse skewaffine iff its</u>

<u>adjacency space $\mathbb{V}$ is closed with respect to the matrix multiplication. It is strongly skewaffine iff $\mathbb{V}$ is closed with respect to the multiplication and the transpose.</u>

PROOF. [2], p.71, Satz (2) in combination with Proposition 7.1. □

8. Distance-regular graphs

We apply the results of the foregoing section to distance-regular graphs (cf.e.g. [5], sect.20). Let $\mathcal{G}$ be a finite (undirected) connected graph and let $d:X^2 \longrightarrow \mathbb{N}_0$ be the distance function of $\mathcal{G}$ defined in the usual way (cf.e.g. [10], p.30). Define $\langle x,x'\rangle := d(x,x')$. Then $(X,\langle\,,\rangle) =: \mathcal{R}(\mathcal{G})$ is a P-space, the <u>graph-space</u> with respect to $\mathcal{G}$. Here $F = \{0,1,\ldots,D\}$ where D is the <u>diameter</u> (largest possible distance) of $\mathcal{G}$. The adjacency space $\mathbb{V}$ of $\mathcal{R}(\mathcal{G})$ (or simply of $\mathcal{G}$) is then given by

$$\mathbb{V} = \left\{ \sum_{k=0}^{D} \alpha_k N_k \;\middle|\; \alpha_k \in \mathbb{C} \right\} \tag{8.1}$$

(cf.(7.3)) where $N_0 = I$ and $N_1 =: N$ is the <u>adjacency matrix</u> of $\mathcal{G}$ ([5], p.9). Moreover $N_k = 0$ for $k>D$, and by definition $N_{-1} := 0$. Obviously $\mathrm{Dim}\,\mathbb{V} = D+1$. Denote $N_k = (n_{k,xx'})_{x,x'\in X}$ and $NN_k = (n_{xx'}^{(k)})$. Due to the triangle-inequality of d we obtain

$$n_{xx'}^{(k)} = a_{k-1,xx'}n_{k-1,xx'} + a_{k,xx'}n_{k,xx'} + a_{k+1,xx'}n_{k+1,xx'} \tag{8.2}$$

The graph $\mathcal{G}$ is called <u>distance-regular</u> iff the coefficients $a_{k,xx'} \in \mathbb{N}_0$ $(k\in\{0,1,\ldots,D\})$ do not depend on $x,x'\in X$, hence $a_{k,xx'} =: a_k$. For distance-regular graphs (8.2) thus becomes

$$NN_k = a_{k-1}N_{k-1} + a_kN_k + a_{k+1}N_{k+1} \tag{8.2'}$$

$(k\in\{0,1,\ldots,D\})$. A distance-regular graph is characterized by (8.2').

We are now able to state some properties about the structure of the adjacency space $\mathbb{V}$ of a distance-regular graph.

LEMMA 8.1. $N^s \in \mathbb{V}$ <u>for all</u> $s \in \mathbb{N}_0$.

PROOF. Trivially $N^0 = I$ and $N^1 = N \in \mathbb{V}$. Assume N^s is a linear combination of the N_k. Then by (8.2′) N^{s+1} is also a linear combination of the N^k, hence in $\mathbb{V}$. □

COROLLARY. _All polynomials on_ N _are contained in_ $\mathbb{V}$, i.e. $\mathbb{C}[N] \subseteq \mathbb{V}$. □

The minimal polynomial $m_N(X)$ of N has a degree $\geq$ D+1 ([5], p.12, Prop.2.6). Hence the dimension of $\mathbb{C}[N]$ considered as a $\mathbb{C}$-vector-space is also $\geq$ D+1. On the other hand Dim $\mathbb{V}$ = D+1, whence

$$\mathbb{V} = \mathbb{C}[N] \tag{8.3}$$

and $\mathbb{V}$ is closed with respect to the matrix multiplication. All matrices of $\mathbb{V}$ are symmetric for $\mathcal{G}$ is undirected, hence $\mathbb{V}$ is also closed with respect to the transpose. Summarizing these results we obtain

THEOREM 8.2. _A finite connected graph_ $\mathcal{G}$ _is distance-regular iff its graph-space_ $\mathcal{R}(\mathcal{G})$ _is strongly skewaffine_.

PROOF. Consequence of Theorem 7.2. □

COROLLARY. For graph-spaces the condition (7.5) is a consequence of the (seemingly weeker) condition (8.2′). □

PROPOSITION 8.3. _Any distance-regular graph is regular_, _i.e. any vertex is adjacent to the same number of vertices_.

PROOF. We have $J \in \mathbb{V} = \mathbb{C}[N]$. Proposition 3.2 of [5] (p.15) yields this result. □

References

[1] André, J.: Über geometrische Strukturen, die zu Permutationsgruppen gehören. Abh. math. Sem. Univ. Hamburg 44 (1976), 203-221.

[2] André, J.: Endliche nichtkommutative Geometrie. Vorlesungsausarbeitung, ausgearbeitet von D. Brühl Annales Universitatis Saraviensis, Ser.Math. 2 (1988), 1-136.

[3] André, J.: Noncommutative spaces with transitive translation groups. Submitted to Geom.ded.

[4] Aschbacher, M.: Chromatic geometry. Quart.J.Math.Oxford 38 (1987), 277-296.

[5] Biggs, N.: Algebraic graph theory. Cambridge, 1974.

[6] Cvetkovic,D., Doob,M., Sachs,H.: Spectra of graphs. New York, San Francisco, London, 1980.

[7] Greenwood,R.E., Gleason,A.M.: Chromatic graphs, Can.J.Math. 7 (1965), 1-7.

[8] Pfalzgraf, J.: On a model for non commutative geometric spaces. J.Geometry 25 (1986), 147-163.

[9] Pfalzgraf, J.: A note on simlices as geometric configurations. Arch.Math. 49 (1987), 134-140.

[10] Wilson, R.J.: Introduction to graph theory, 2nd ed. London, 1979.

The Complexity of the Graph Embedding Problem

D. Archdeacon, Burlington, USA

Abstract. We investigate the computational complexity of determining if a graph G on v vertices embeds in a surface S. Robertson and Seymour have given an $O(v^3)$ decision algorithm for this embedding problem. We show here how the use the yes/no output from their algorithm to construct the embedding, that is, we self-reduce the search algorithm to the decision algorithm. We conclude that for each fixed surface S there exists an $O(v^{10})$ algorithm for constructing an embedding or answering that no embedding exists.

§1 INTRODUCTION

In this paper graphs may have loops and multiple edges. We consider embeddings of graphs into *surfaces*, compact orientable 2-dimensional manifolds. By a Theorem of Brahana [**B**] these surfaces are homeomorphic to a sphere with g handles attached. This number g is called the *genus* of the surface. The *genus of the graph* G, $\gamma(G)$, is the minimum g such that the graph embeds on the sphere with g handles.

Determining the genus of a graph is one of the fundamental problems in topological graph theory. Until recently it was not known if this was hard. Specifically, it was unknown if the question *"Given a graph G and a natural number k, is $\gamma(G) \leq k$?* " could be answered using a polynomial-time algorithm. Nor was it known if the problem was NP-complete in the sense of Garey and Johnson [**GJ**]. However, in a recent paper Thomassen announced [**T1**] that the graph genus problem is NP-complete. Thomassen's proof has refocused attention on graph embeddings and exactly what can be determined about the genus of a graph in polynomial time.

One modification to the graph genus problem is to restrict the class of graphs under consideration in hopes of finding a good algorithm. One result along these lines is due to Thomassen [**T2**]. He defines a *long-edge-width embedding*, or *LEW* embedding, as one in which every non-contractible cycle is longer than all facial walks. He gives a polynomially bounded algorithm for deciding if an embedding has this property, and also one for deciding if a graph has an embedding of this type. By showing that an *LEW* embedding must be a genus embedding, he obtains a polynomial-time algorithm for the genus of graphs with *LEW* embeddings.

A second modification to the graph genus question is to fix a particular surface S (equivalently, to fix k, the number of handles) and ask the graph embedding problem, *"Does a graph G embed in S?"* The special case *"Is the graph G planar?"* can be solved by a polynomial-time algorithm [**HT**], in fact, the algorithm is linear. The algorithm either outputs that the graph is nonplanar or describes an embedding of the graph, in this sense it is *constructive*. In the same spirit, Filotti et al. [**FMR**] showed that the graph embedding problem was polynomial for each surface. Again

R. Bodendiek, R. Henn (Eds.)
Topics in Combinatorics and Graph Theory

their algorithm is constructive, either responding that the graph did not embed in a surface or producing such an embedding. The degree of their polynomial-time algorithm depended on the surface.

The next major advance followed from the work of Robertson and Seymour on graph minors. They showed that for each surface S there exists a $O(v^3)$ decision algorithm for the graph embedding algorithm (here v is the number of vertices in the graph). It is important to note that Robertson and Seymour's algorithm is nonconstructive. The output of the algorithm is a yes/no answer to the question of whether the graph embeds on S. In the case that the graph does embed, no embedding is produced.

The Robertson-Seymour algorithm does not contradict Thomassen's proof of NP-completeness for the graph genus problem. The order of the Robertson-Seymour algorithms is more accurately described as $c(S)v^3$, where $c(S)$ is a constant depending on the surface S. These constants grow very rapidly with the surface (in fact, super-exponentially).

The purpose of this paper is to prove the following theorem.

THEOREM. *For each surface there exists a polynomial-time algorithm for constructing a graph embedding in that surface or to respond that no such embedding exists.*

Loosely speaking, we turn the Robertson-Seymour *RS* decision algorithm into a constructive (or search) algorithm. This process is generally called *self-reduction* [**MP**] (for additional references see [**BFL**]). The order of the algorithm is v^{10}. The order is determined by making a number of calls to the *RS*-algorithm. Each call asks if a modified version of the original graph embeds in the surface. When done, the algorithm outputs an embedding. Because of the use of the *RS*-algorithm, the order may be more accurately described as $c(S)v^{10}$ where the constant $c(S)$ grows rapidly with the genus of S.

The paper is organized as follows. In Section 2 we show that problem reduces to cellular embeddings of 2-connected cubic graphs. In Section 3 we construct the desired embedding of these 3-regular graphs. A key intermediate step is the assumption that a graph with such an embedding has each facial walk a simple cycle. We close in Section 4 with some concluding remarks.

§2 Reduction to the Cubic Case

In this section we reduce the general graph embedding problem to the special case that the graph is 2-connected, 3-regular, and has only cellular embeddings in the surface. Specifically, suppose that we are given a graph G and a surface S. We will construct a 2-connected 3-regular graph G' and a surface S' such that G embeds in S if and only if G' embeds in S'. Moreover, the embedding of G can easily be reconstructed from the embedding of G'.

Our first reduction is to the case that the graph is 2-connected. This is easily accomplished by the Theorem of Battle et al. [**BHKY**] that the genus of a graph is equal to the sum of the genera of its 2-connected components. Thus we can inductively apply our desired algorithm to the (smaller) 2-connected components. Implicit in [**BHKY**] is a method for piecing together the embeddings of the components to produce an embedding of the whole graph. Thus we may assume that our graph G is 2-connected.

We next note that without loss of generality the graph embeds on the surface in question. If not, then the RS-algorithm will tell us so, and constructing such an embedding is moot. We also note that we may assume the embedding is *cellular*, that is, each component of the complement $S - G$ is homeomorphic to an open 2-cell. If not, then by the work of Youngs [**Y**] the graph has a cellular embedding on some surface S' with fewer handles (that is, with larger Euler characteristic). We can then run our algorithm on the smallest such surface (from among a finite number), and modify the embedding produced by adding in free handles in a face. Alternatively, if one wishes to restrict attention to only cellular embeddings, we can use the result that an embedding into a surface of maximum genus can be found in polynomial time [**FGM**] together with Duke's interpolation theorem [**D**] in constructing the desired embedding.

Our next step will allow us to assume the graph is cubic. We first note that G has no vertices of degree 1 since it is 2-connected. Secondly we note that a vertex of degree 2 may be suppressed without changing the genus. If this operation creates parallel edges, we may delete the vertex and incident edges instead of suppressing it; this leaves the genus unchanged. Hence without loss of generality G has minimum degree at least three.

Suppose that G has a vertex u of degree exceeding 3, and select e_1 and e_2 incident with u. Form a graph G' by *splitting* e_1, e_2 *at* u. Specifically, we add a new vertex u' incident with e_1 and e_2 in place of u, and a new edge e joining u to u'. Note that contracting the edge e in G' recovers the original graph G. Also note that G' embeds in S if and only if G embeds in S with e_1 and e_2 consecutive edges in some face boundary. We consult the RS-algorithm to see if G' embeds in S. If so, we replace G with G' and iterate this procedure. If G' does not embed in S, then we select a different edge e_2 and repeat. This procedure terminates in a cubic graph G' which is 2-connected and embeds in S. From this embedding we can easily reconstruct the embedding of the original G.

What is the order of the above reduction? There are at most v ways to choose the vertex in question. There are $deg(v) \leq v - 1$ ways to choose the edge e_1, and at most $deg(v) - 1 \leq v - 2$ ways to choose e_2. Thus we need at most $O(v^3)$ calls to the RS-algorithm in the reduction procedure. Each call involves a graph which is a split of the original graph. These splits may have up to $O(v^2)$ vertices. Hence the complexity of each call is bounded by $O(v^6)$ (the cube of the number of vertices in the split graph). It follows that the reduction is of order v^9.

We have shown that there exists a cubic graph G' which embeds in S, and from this embedding we can easily reconstruct the embedding of G. We will next show that we can construct this embedding of G' using a number of calls to the RS-algorithm which is at most quadratic in

$|V(G')|$. Since this order is $O(v^2)$, we will make $O(v^4)$ calls to the *RS*-algorithm, each of complexity $O(v^6)$. It follows that the overall algorithm is of order $O(v^{10})$ as desired.

§3 The Cubic Case

In this section we suppose that we start with a cubic graph G on v vertices (and hence with $3v/2$ edges) which embeds on a surface S. We will show how to construct such an embedding using at most $O(v^2)$ calls to the *RS*-algorithm. Following the last paragraph of §2, this will complete the proof of our main theorem.

A cellular embedding of a graph is determined by the set of *facial walks*, those closed walks on the boundary of the 2-cells. (From these walks it is easy to quickly reconstruct the rotation scheme, or some other desired characterization of embeddings.) We will describe a method for finding the edge-sets of these facial walks. However, since the facial walks need not be simple cycles, the edge-sets do not uniquely determine the these walks. To avoid this difficulty we describe a modification to the cubic graph G and surface S with the property that each facial walk of the new graph is a simple cycle.

Let e be an edge of G. Form a new graph G_e as follows. Replace e by a path with four edges, so that there are now three vertices u_1, u_2, u_3 of degree two. Now add two new vertices u_4, u_5, adjacent to each other and each adjacent to the three degree two vertices. The graph $G_e/(G-e)$, where $G-e$ is identified with a point, is nonplanar (in fact, it is isomorphic to K_5). Hence in any embedding, $G_e-(G-e)$ must contain a noncontractible cycle. It follows that G_e embeds in S if and only if there exists an embedding of G in S where e occurs twice in a boundary cycle. We now call the *RS*-algorithm to determine if G_e embeds. If not, then e cannot occur twice in the facial walk of an embedding. If so, then we replace G with $G-e$ (suppressing the degree two vertices), replace S with the surface S' having one fewer handle, and repeat the process for the next edge. This procedure terminates (in at most $3v/2$ steps) with a graph G' which embeds on a surface S' such that no edge is repeated twice in a facial walk. Because G' is cubic, it follows that no vertex is repeated twice in a facial walk; that is, the facial walks are all simple cycles. Observe that it is easy to reconstruct the embedding of G in S from the embedding of G' in S'.

We now have a simple graph for which any embedding on a surface has each facial walk is a simple cycle. Rename if necessary, so that the graph and surface are again G and S. We now describe how to find the edge-set of the facial walks.

Form a graph G' by subdividing each edge e of G and adding two new nonadjacent vertices v_e, v'_e, each of which are adjacent to the two ends and to the midpoint of e. Observe that G' embeds in S, and that for each edge e the two new vertices v_e and v'_e lie in the two different faces (of the restriction to G) incident with e. Note that the nontriangular facial walks of an embedding of G' alternate between degree 3 vertices (the new vertices) and degree 9 vertices (the vertices of G); the remaining vertices (midpoints of edges in G) are of degree 4.

We now augment G' with edges joining the degree three vertices until we obtain a triangulation of S. Specifically, for each pair from the $3v$ degree three vertices of G', we call the RS-algorithm to find if G' together with an edge joining that pair of vertices embeds in S. If not, we move on to the next pair. If so, we add that edge to G' and check the next pair. The procedure terminates with a triangulation G'' of S after at most $3v(3v-1)/2$ calls to the RS-algorithm.

We conclude the algorithm by examining the graph formed by the edges of $G'' - G'$. Each component of this graph has vertices $v_{e_1}, v_{e_2}, \ldots$ corresponding to the edge-set $e_1, e_2, \ldots$ of a facial walk of an embedding of G. From this edge-set we can reconstruct the simple cycle which is the facial walk of the modified graph, and hence the facial walks of the original cubic graph. By the results of Section 2, the Main Theorem is shown.

§4 Conclusion

We have shown that a modification to the Robertson-Seymour decision algorithm can produce a polynomial-time constructive algorithm for the graph embedding problem. The algorithm given here is relatively straightforward. Can a more complicated self-reduction strategy achieve better than $O(v^{10})$? What is the smallest order of a constructive graph embedding algorithm?

The work of Robertson and Seymour produces many such decision algorithms. Is there a general procedure which takes an decision algorithm and produces a constructive algorithm? What is the cost of such a procedure?

We close by noting that the techniques of this paper extend to give polynomial-time algorithms for constructing embeddings of graphs into nonorientable surfaces.

ACKNOWLEDGMENT. The author thanks Mike Fellows and Mike Langston for helpful discussions concerning self-reduction and decision vs. search algorithms. He especially thanks them for uncovering an error in the original complexity analysis.

REFERENCES

[B] H.R. Brahana, *Systems of circuits of two-dimensional manifolds*, Ann. of Math. **30** (1923), 234-243.

[BFL] D.J. Brown, M.R. Fellows and M.A. Langston, *Polynomial-time self-reducibility: theoretical motivations and practical results*, Int. J. Computer Mathematics (to appear).

[BHKY] J. Battle, F. Harary, Y. Kodama and J.W.T. Youngs, *Additivity of the genus of a graph*, Bull. Amer. Math. Soc. **68**, 565-571.

[D] R.A. Duke, *The genus, regional number, and Betti number of a graph*, Canad. J. Math. **18** (1966), 817-822.

[FGM] M.L. Furst, L.L. Gross, and L.A. McGeoch, *Finding a maximum-genus graph imbedding*, preprint (1987).

[FMR] I.S. Filotti, G.L. Miller and J. Reif, *On determining the genus of a graph in $O(v^{O(G)})$ steps.*, Proc. 11th Annual ACM Symp. Theory of Computing (1979), 27-37.

[GJ] M.R. Garey and D.S. Johnson, "Computers and Intractability, a Guide to the Theory of NP-Completeness," W.H. Freeman and Co., San Francisco, 1979.

[GT] J.L. Gross and T.W. Tucker, "Topological Graph Theory," John Wiley & Sons, New York, 1987.

[HT] J.E. Hopcroft and R.E. Tarjan, *Efficient planarity testing*, J. Assoc. Comput. Math. **21** (1974), 549-568.

[MP] A. Meyer and M. Paterson, *With what frequency are apparently intractable problems difficult*, Technical Report, MIT (1979).

[RS] N. Robertson and P.D. Seymour, *Graph minors XIII: The disjoint paths problem*, to appear.

[T1] Carsten Thomassen, *The graph genus problem is NP-complete*, preprint.

[T2] Carsten Thomassen, *Embeddings of graphs with no short noncontractible cycles*, preprint.

[Y] J.W.T. Youngs, *Minimal imbeddings and the genus of a graph*, J. Math. Mech. **12** (1963), 303-315.

Keywords. genus, embedding, computational complexity, NP-complete
1980 *Mathematics subject classifications*: 05C10

Helly Theorems for Dismantlable Graphs and Pseudo-Modular Graphs

H.-J. Bandelt, Maastricht, The Netherlands
H. M. Mulder, Rotterdam, The Netherlands

ABSTRACT. The geodesic convexity of a graph consists of all those subsets of the vertex-set which contain all geodesics (i.e., shortest paths) joining any two of its elements. The Helly number of this convexity is trivially bounded from below by the clique number (i.e., the size of a largest clique). We show that equality between the two numbers hold for graphs which are dismantlable (alias cop-win) or pseudo-modular. This generalizes previously known results for chordal graphs and distance-hereditary graphs, due to Čepoj, Duchet, Jamison and Nowakowski, respectively.

By a *convexity* C on a graph G one usually means a certain system of connected subsets of G closed under intersections; cf. Duchet and Meyniel [7]. Several kinds of graph convexities have been investigated so far; see, for instance, Bandelt [2], Čepoj [5], Farber and Jamison [8]. It is convenient to assume that at least all complete subgraphs belong to C. Then the *Helly number* of C is the least number n such that any finite subsystem of C intersecting n-wise has a non-empty intersection. If G contains an *n-simplex*, that is a complete subgraph on n vertices, then the Helly number of C is at least n. Not much more can be said, of course, except for some general inequalities involving related convexity parameters (such as Carathéodory and Radon numbers) - all belonging to the realm of abstract convexity, cf. Soltan [14] and Van de Vel [15]. To be more specific, let C consist of those sets C for which every induced path in G between any two members of C is contained in C. In this case, the Helly number is just the *clique number* $\omega(G)$ of G, that is, the size of a largest simplex in G; see Duchet [6] or Jamison and Nowakowski [10]. Now, what about the

R. Bodendiek, R. Henn (Eds.)
Topics in Combinatorics and Graph Theory

geodesic convexity ? Its members are sets C that contain every geodesic (i.e., shortest path) of G joining two vertices of C ; henceforth we reserve the name "convex" for such sets C. In contrast to the former convexity, the Helly number $h(G)$ of G with respect to the geodesic convexity can exceed the clique number $\omega(G)$ by any number. Consider, for instance, the subdivision G of the n-simplex ($n \geqq 3$), where each edge of the simplex is replaced by a 2-path; see Jamison and Nowakowski [10]. This bipartite graph evidently inherits its Helly number $h(G)$ from the n-simplex, hence $h(G)=n$. On the other hand, $h(G)$ equals $\omega(G)$ for every distance-hereditary graph G (i.e., graph G in which all induced paths are geodesics; cf. [3]), as is confirmed by the Duchet-Jamison-Nowakowski theorem. This naturally leads to the following question.

For which graphs G is $h(G)$ equal to $\omega(G)$?

Čepoj [5] has shown that this equality actually holds for chordal graphs. His proof relies upon the simplicial elimination scheme, characteristic for chordal graphs. We shall demonstrate below that a more general elimination scheme, defining dismantlable graphs, will still do the job. The key argument that works in the case of distance-hereditary graphs is implicitly based upon a "nearest simplex" property of convex sets in such graphs. This method also extends much further, viz., to the class of pseudo-modular graphs. We cannot offer, though, an approach that would cover both major cases.

Some necessary terminology is provided next. A vertex x is said to be *dominated* by another vertex y if x and all vertices adjacent to x are adjacent to y, in symbols: $N(x)-y \subseteq N(y)-x$ for an edge xy, where the operator N assigns to each vertex all its neighbours. (Vertex-deleted subgraphs or subsets are briefly written as $G-x$, $A-x$, and the like.) *Dismantlable graphs* are recursively defined as follows: the one-vertex graph is dismantlable, and a finite graph with more than one vertex is dismantlable when there is a dominated vertex x whose removal results in a dismantlable graph. Dismantlable graphs, also known as cop-win graphs have been investigated by Nowakowski and Winkler [11], Poston [12], Quilliot [13], and others. The *interval* $I(u,v)$ between two vertices u and v consists of all vertices on geodesics between u and v, in symbols:

$$I(u,v) = \{w \mid d(u,v) = d(u,w) + d(v,w)\},$$

where d denotes the distance function of G. Then, a set C of vertices is convex if and only if $I(u,v) \subseteq C$ for all $u,v \in C$. In particular, every convex set (subgraph) H of G is *isometric* in G, that is, a metric subspace of G with respect to d. A *bridged graph* is a graph not containing any cycle of length greater than three as an isometric subgraph. The main result of Anstee and Farber [1] states that all (finite) bridged graphs are dismantlable.

Following [4] we call a (possibly infinite) graph G *pseudo-modular* if for each triple u,v,w of vertices there exists either a (*median*) vertex x in the intersection of $I(u,v)$, $I(v,w)$, and $I(u,w)$, or a (*pseudo-median*) triangle $\{x,y,z\}$ such that its three edges lie in the following intervals:

$$\{x,y\} \subseteq I(u,v), \qquad \{y,z\} \subseteq I(v,w), \qquad \{x,z\} \subseteq I(u,w).$$

The *convex closure* $co(A)$ of a subset A of the vertex-set is the least convex set containing A. A routine minimality assumption yields that the Helly number of a graph G is at least n if and only if there exists a set A with n vertices of G such that the intersection of all $co(A-x)$ with $x \in A$ is empty.

Theorem 1. Let G be a dismantlable graph, and let $A_1,\ldots,A_n$ $(n>3)$ be convex sets in G such that $\bigcap_{i=1}^{n} A_i = \emptyset$, while $\bigcap_{i=1, i\neq j}^{n} A_i \neq \emptyset$, for $j=1,\ldots,n$. Then there exists an n-simplex K in G with $|A_i \cap K| = n-1$, for $i=1,\ldots,n$.
In particular, $h(G) = \omega(G)$.

Proof. Let us call any n-set X of vertices in G a critical set whenever the convex closures of the $(n-1)$-subsets of X have an empty intersection. For $j=1,\ldots,n$, we choose a vertex a_j in $\bigcap_{i=1, i\neq j}^{n} A_i$. Note that $a_j \notin A_j$, for $j=1,\ldots,n$. The set $A=\{a_1,\ldots,a_n\}$ clearly is a critical set.

By induction on the number of vertices of G, we prove that there exists an n-simplex in G. First, if A equals the vertex-set of G, then we have $A_i = A - a_i$. This implies that $N(a_i)$ is complete, for each i, whence G is an n-simplex by connectivity.

Let G have more than n vertices, and let x be a dominated vertex of G, that is, there is a neighbour y of x with $N(x)-y \subseteq N(y)-x$. If x is not in A, then put $G' = G-x$ and $A_i' = A_i - x$, for each i. Since G' is an isometric subgraph of G, each set A_i' is the convex closure of $A-a_i$ in G'. Clearly we have $\bigcap_{i=1}^{n} A_i' = \emptyset$, whence A is a critical set in G' as well, and by induction, G' contains an n-simplex.

Now assume that x is in A, say $x=a_n$. For $i=1,\ldots,n-1$, let b_i be a neighbour of x on a geodesic between x and a_i. Note that b_i is in A_j, for $i \neq j$ and $1 \leq i,j \leq n-1$. If $b_1,\ldots,b_{n-1}$ are not all distinct, say $b_1 = b_2$, then b_1 is in A_i, for $i = 1,\ldots,n-1$, whence b_1 cannot be in A_n. Now let $A' = (A-x) \cup \{b_1\}$, let $A'_n = A_n$, and $A'_i = co(A'-a_i)$, for $i<n$. Then we have $A'_i \subseteq A_i$, for all i, so that $\bigcap_{i=1}^{n} A'_i = \emptyset$. Thus we have found a critical set A' missing a dominated vertex of G, and by the above argument we have an n-simplex. So we may assume that the vertices $b_1,\ldots,b_{n-1}$ are all distinct. We distinguish two cases.

Case 1. y is among $b_1,\ldots,b_{n-1}$, say $y = b_1$.

Observe that y must be adjacent to $b_2,\ldots,b_{n-1}$ since x is. Assume that the vertices $x,y,b_2,\ldots,b_{n-1}$ do not form an n-simplex. Then, say b_2 and b_3 are not adjacent. As y is common neighbour of b_2 and b_3 it is on a geodesic from b_2 to b_3, whence y is in A_1. Since $y = b_1$ is also in A_i, for $i = 2,\ldots,n-1$, it follows that y is not in A_n and hence not in A either. So we can replace $a_n = x$ by y, that is: put $A' = (A-x) \cup \{y\}$, $A'_n = A_n$, and $A'_i = co(A'-a_i)$, for $i<n$. Then we have $A'_i \subseteq A_i$, for all i, so that $\bigcap_{i=1}^{n} A'_i = \emptyset$. Again we have produced a critical set A' not containing the dominated vertex x, by which Case 1 is settled.

Case 2. y is not among $b_1,\ldots,b_{n-1}$.

Then y is adjacent to $b_1,\ldots,b_{n-1}$ (because x is). We may assume that there are non-adjacent vertices among $b_1,\ldots,b_{n-1}$ (otherwise $x,y,b_1,\ldots,b_{n-1}$ form an $(n+1)$-simplex), say b_1 and b_2 are not adjacent. Then y is on a geodesic between b_1 and b_2, and therefore belongs to A_i, for $i = 3,\ldots,n-1$. If there were another non-adjacent pair among $b_3,\ldots,b_{n-1}$, then y would also be in A_1 and A_2, and so could not be in A_n; but then $A' = (A-x) \cup \{y\}$ were a critical set not containing the dominated vertex x. Hence we may assume that $b_3,\ldots,b_{n-1}$ induce a simplex.

If some vertex b_j ($j \geq 3$) is not adjacent to either b_1 or b_2, then y is on a geodesic between b_1 and b_j, so that y is in A_2, and y is on a geodesic between b_2 and b_j, so that y is in A_1. Therefore y is not in A_n, and as above we have produced a critical set $A' = (A-x) \cup \{y\}$ not containing x. Therefore we may assume that each vertex b_j ($j \geq 3$) is adjacent to at least one of b_1 and b_2. Now assume that some b_j with $j \geq 3$ is not adjacent to b_2. Then b_j must be adjacent to b_1. Suppose that b_1 is not-adjacent to some b_k with $k \geq 3$. Necessarily, $k \neq j$. Thus we may interchange the role of b_k and b_2 and conclude as above that there is a critical set avoiding x. We can therefore assume that b_1 is adjacent to

$b_3,\dots,b_{n-1}$. Now, the vertices $x,y,b_1,b_3,\dots,b_{n-1}$ induce an n-simplex. This settles Case 2, completing the proof. □ □ □

For the next theorem we first need some information on convex sets in pseudo-modular graphs.

Lemma. Let G be a pseudo-modular graph, and let C be a convex set in G. For any vertex u of G not in C, there is a simplex S_u in C such that $I(u,v) \cap S_u \neq \emptyset$, for each v in C.

Proof. Let S be the set of vertices in C having minimal distance, k say, to u. We will show that S is the required simplex. Let v be any vertex in C and let w be any vertex in S. First, if u,v,w have a median y, then y is on a (v,w)-geodesic, and hence y is in C. Moreover, y is on a (u,w)-geodesic, so by minimality of $d(u,w)=k$, we have $y=w$. Thus y is a vertex in S on a (u,v)-geodesic. Second, let x,y,z be a pseudo-median of u,v,w. Then y and z are on a (v,w)-geodesic, and hence they are in C. Clearly $d(u,y)=d(u,z)\leqq d(u,w)$, and so by minimality of k we conclude that $d(u,y)=k$, that is, y is in S. Moreover, y is on a (u,v)-geodesic.

Finally, assume that there are non-adjacent vertices v and w in S. Let x,y,z be a pseudo-median of u,v,w. Since $d(v,w)\geqq 2$, we may assume that, say, z and w are distinct. Now z, being a vertex on a (v,w)-geodesic, is a vertex in C with $d(u,z)\leqq d(u,w)-1$, thus conflicting with the minimality of $d(u,w)$. Hence S is a simplex, and the proof is complete. □ □ □.

Theorem 2. Let G be a pseudo-modular graph, and let $A_1,\dots,A_n$ $(n\geqq 3)$ be convex sets in G such that $\bigcap_{i=1}^{n} A_i=\emptyset$, while $\bigcap_{i=1, i\neq j}^{n} A_i\neq\emptyset$, for $j=1,\dots,n$. Then there exists an n-simplex K in G with $|A_i \cap K| = n-1$, for $i=1,\dots,n$.
In particular, $h(G)=\omega(G)$ (provided that G has no infinite simplices).

Proof. For $j=1,\dots,n$, we choose a vertex a_j in $\bigcap_{i=1, i\neq j}^{n} A_i$, such that $\sum_{k<\ell} d(a_k,a_\ell)$ is as small as possible. Note that a_j is not in A_j, for all j, so that the vertices $a_1,\dots,a_n$ are mutually distinct. Assume that these vertices do not form an n-simplex, that is, say a_1 and a_2 are not adjacent. Let $C \subseteq A_n$ be the convex closure of $\{a_1,\dots,a_{n-1}\}$, and let S be the simplex of nearest vertices in C with respect to a_n. Now, for $j=1,\dots,n-1$, let b_j be a vertex of S on a geodesic between a_n and a_j. Then b_j is in $\bigcap_{\substack{i=1 \\ i \neq j}}^{n} A_j$, whence b_j is not in A_j, and so we can infer that $b_1, \ldots, b_{n-1}$ are all distinct. Since $b_1, \ldots, b_{n-1}$ all lie in the simplex S of nearest vertices in $C \subseteq A_n$ with respect to a_n, the distance sum $\sum_{k<\ell} d(b_k, b_\ell)$, where $b_n = a_n$, is strictly smaller than $\sum_{k<\ell} d(a_k, a_\ell)$. This is in conflict with our minimality assumption. Therefore we conclude that $a_1, \ldots, a_n$ form an n-simplex, which evidently satisfies the required intersection condition. □ □ □

Note that the lemma parallels a result on the induced path convexity, as stated explicitly in Halin's book [9, page 154]. Thus, the theorem of Duchet [6] and Jamison-Nowakowsky [10] is based on the latter result similarly as our Theorem 2 is based on the lemma above.

References

1. R.P. Anstee and M. Farber, On bridged graphs and cop-win graphs, J. Combinatorial Theory Ser. B 44 (1988) 22-28.
2. H.-J. Bandelt, Graphs with intrinsic S_3 convexities, J. Graph Theory 13 (1989) 215-228.
3. H.-J. Bandelt and H.M. Mulder, Distance-hereditary graphs, J. Combinatorial Theory Ser. B 41 (1986) 182-208.
4. H.-J. Bandelt and H.M. Mulder, Pseudo-modular graphs, Discrete Math. 62 (1986) 245-260.
5. V.C. Čepoj, Some properties of d-convexity in triangulated graphs (in Russian), Math. Issledovanija 87 (1986) 164-177.
6. P. Duchet, Convex sets in graphs II: minimal path convexity, J. Combinatorial Theory Ser. B 44 (1988) 307-316.
7. P. Duchet and H. Meyniel, Ensembles convexes dans les graphes I: théorèmes de Helly et de Radon pour graphes et surfaces, European J. Combinatorics 4 (1983) 127-132.
8. M. Farber and R.E. Jamison, Convexity in graphs and hypergraphs, SIAM J. Alg. Disc. Meth. 7 (1986) 433-444.
9. R. Halin, Graphentheorie II, Wiss.Buchgesellschaft Darmstadt, 1981.
10. R.E. Jamison and R. Nowakowski, A Helly theorem for convexity in graphs, Discrete Math. 51 (1984) 35-39.
11. R. Nowakowski and P. Winkler, Vertex-to-vertex pursuit in a graph, Discrete Math. 43 (1983) 235-239.

12. T. Poston, Fuzzy geometry, Ph.D. thesis, University of Warwick, 1971.

13. A. Quilliot, Homomorphismes, points fixes, rétractions et jeux de poursuite dans les graphes, les ensembles ordonnés et les espaces métriques, Thèse, Paris, 1983.

14. V.P. Soltan, Introduction to the axiomatic theory of convexity (in Russian) Štiinca, Kišinev, 1984.

15. M. van de Vel, Abstract, topological, and uniform convex structures, Rapport nr. 353, Vrije Universiteit Amsterdam (Fac. W. en I.), 1989.

On the Level-Oriented Two-Dimensional Packing With Rotation of the Rectangles

G. Bär, Greifswald, GDR

Coffman, Garey, Johnson and Tarjan /1/ have considered the following two-dimensional packing problem: Given a collection of rectangles and a bin with fixed width and unbounded height, pack the rectangles into the bin so that no two rectangles overlap and so that the height to which the bin is filled is as small as possible. We assume that the given rectangles are oriented, each having a specified side that must be parallel to the bottom of the bin. We also assume, with no loss of generality, that the bin width has been normalized to 1 and that width and height of all rectangles are generally less than or equal to 1.

This problem is known to be NP-complete.

For the worst-case behaviour of different approximation algorithms in /1/ asymptotic performance bounds of the form

$$A(L) \leq a \cdot OPT(L) + b$$

have been derived, where

L arbitrary list of rectangles,

A(L) minimal height used by algorithm A when applied to L

OPT(L) minimum possible bin height in which the rectangles in L can be packed.

Coffman et. al. have considered the level-oriented packing algorithms NFDH and FFDH.

NFDH (Next-Fit Decreasing-Height): Order the rectangles of L by decreasing height. After that pack the pieces successively with their width horizontally. The rectangles $R_1, \ldots, R_i$ are placed at the bottom of the bin, each left-justified, until some piece R_{i+1} has a width $w(R_{i+1})$ exceeding $1 - \sum_{j=1}^{i} w(R_j)$. Draw a horizontal line g_1 at height $h(R_1)$. This defines the second level. Start level two with R_{i+1} and proceed in the same manner, and so on.

R. Bodendiek, R. Henn (Eds.)
Topics in Combinatorics and Graph Theory

FFDH (First-Fit Decreasing-Height): Analogously to NFDH, but each rectangle is left-justified placed on the lowest possible level. If none of the existing levels will acommodate this rectangle, a new level is started and the rectangle is packed here.

Coffman et. al. have proved the following assertions.

Theorem 1. For any list L of rectangles

$NFDH(L) \leq 2 \cdot OPT(L) + 1$, $FFDH(L) \leq 1.7 \cdot OPT(L) + 1$,

and the constants 2 and 1.7 are the smallest possible ones.

Theorem 2. For any list L of squares

$FFDH(L) \leq 1.5 \cdot OPT(L) + 1$,

and the constant 1.5 is the smallest possible one.

Let us now consider <u>non-oriented</u> packings; i.e. rotations of 90° are allowed. For this we consider the following approximation algorithms.

RHNFDH (Rotatable - to max. Height - Next-Fit Decreasing-Height):

1. Rotate all rectangles R_i of L so that width(R_i) $\leq$ height(R_i).
2. Order the rectangles of L by decreasing height.
3. Pack according to NFDH.

RHFFDH (Rotatable - to max. Height - First-Fit Decreasing-Height):

1. and 2. as by RHNFDH
3. Pack according to FFDH.

RWNFDH (Rotatable - to max. Width - Next-Fit Decreasing-Height):

1. Rotate all rectangles R_i of L so that width(R_i) $\geq$ height(R_i).
2. Order the rectangles of L by decreasing height.
3. Pack according to NFDH.

RWFFDH (Rotatable - to max. Width - First-Fit Decreasing-Height):

1. and 2. as by RWNFDH.
3. Pack according to FFDH.

These approximation algorithms possess the following performance bounds.

Theorem. Let A be an approximation algorithm with

$A \in \{$ RHNFDH, RHFFDH, RWNFDH, RWFFDH $\}$.

For any list L of rectangles

$A(L) \leq 2 \cdot OPT(L) + 1$,

and the constant 2 is the smallest possible one.

Proof.

1. (See also /1/.) The approximation algorithm A generates for the list L in all l levels with the level heights $s_1, s_2, \dots, s_l$ and $s_1 \geq s_2 \geq \dots \geq s_l$.

Let us denote by w_i the width of the first rectangle of the level i, y_i the total width of all rectangles of the level i and A_i the area of all rectangles of the level i.

For each $i < l$ the first rectangle in the level i+1 does not fit in level i, therefore $y_i + w_{i+1} > 1$.

Since each rectangle in level i has height at least s_{i+1}, it follows

$$A_i + A_{i+1} \geq s_{i+1} \cdot y_i + s_{i+1} \cdot w_{i+1} > s_{i+1} \text{, and}$$

$$A(L) = \sum_{i=1}^{l} s_i = s_1 + \sum_{i=1}^{l-1} s_{i+1} < s_1 + \sum_{i=1}^{l-1} (A_i + A_{i+1})$$

$$< s_1 + 2 \cdot \sum_{i=1}^{l} A_i \leq 1 + 2 \cdot OPT(L) .$$

2. Consider the following example 1. Let a list L of n = 2k rectangles be given with

$$\left.\begin{aligned} w(R_i) &= 1/2 + e \\ h(R_i) &= 1 \end{aligned}\right\} \quad \text{for } i = 1, \dots, n$$

($e > 0$ suitably small.)

a) If rotations are not allowed, then we get
   NFDH(L) = FFDH(L) = OPT(L) = 2k (see fig. 1a).

b) If rotations are allowed, then the algorithms RHNFDH and RHFFDH also generate the quantity 2k. But RWNFDH and RWFFDH yield the value (k + 2ke) (see fig. 1b). This value simultaneously represents the optimal height OPT(L) of the bin packing, if rotation of the rectangles is allowed.
   From this we conclude
   RHNFDH(L)/OPT(L) = RHFFDH(L)/OPT(L) = 2/(1 + 2e) .
   The quotient can be made arbitrarily close to 2 by choosing n suitably large and e suitably small.

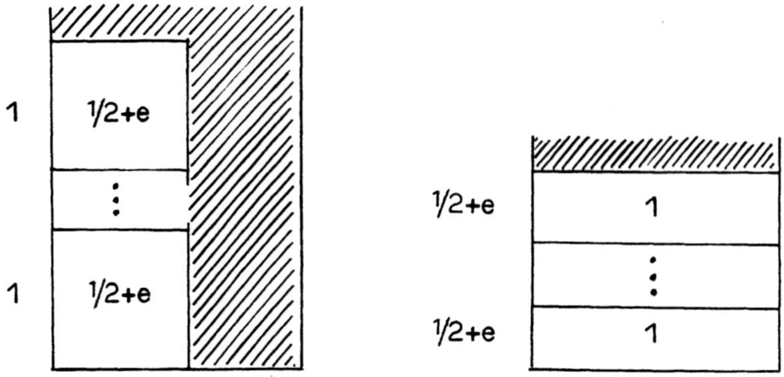

Fig. 1. NFDH- and RWNFDH-packing for example 1.

3. Consider the example 2. Let a list L of $n = 2k$ rectangles be given with

$$\left.\begin{aligned} w(R_i) &= 1/2 + e \\ h(R_i) &= 1/2 \end{aligned}\right\} \quad \text{for } i = 1,\ldots,n$$

($e > 0$ suitably small.)

a) Without rotations we get
   $NFDH(L) = FFDH(L) = OPT(L) = k$ (see fig. 2a).

b) If rotations are allowed then the algorithms RWNFDH and RWFFDH also generate the value k. But
   $RHNFDH(L) = RHFFDH(L) = k(1/2 + e) = OPT(L)$ (see fig. 2b).
   It follows
   $RWNFDH(L)/OPT(L) = RWFFDH(L)/OPT(L) = 2/(1 + 2e)$. □

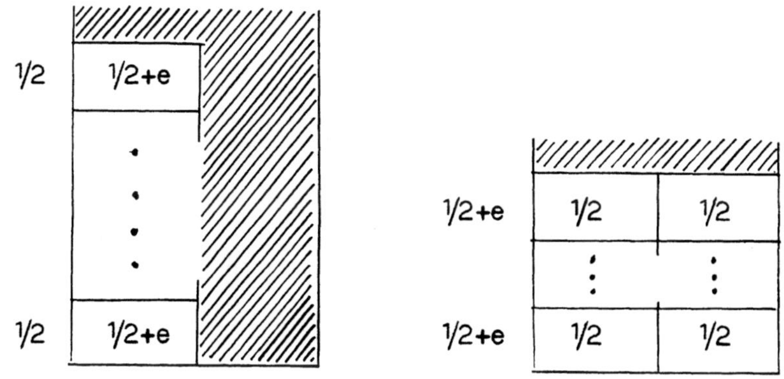

Fig. 2. NFDH- and RHNFDH-packing for example 2.

The theorem shows, that additionally allowed rotations of the rectangles does not yield - as conjectured - a better worst-case behaviour. For First-Fit-algorithms the behaviour even worsens by the transition from nonallowed into allowed rotations. The reason for this lies above all in the fact, that the optimal packing by allowed rotations yields possibly a smaller total packing height.

Above-mentioned examples suggest the introduction of the algorithms RNFDH and RFFDH where

$$\text{RNFDH} = \min\{\text{RHNFDH}, \text{RWNFDH}\}$$
$$\text{RFFDH} = \min\{\text{RHFFDH}, \text{RWFFDH}\}.$$

The performance bounds of the algorithms lie obviously between 1.5 and 2.0 by reasons of the last theorem and the case, that all rectangles are squares.

The following example shows, that the performance bound for RNFDH even cannot be better than 5/3; thus it lies between 5/3 and 2.0 .

Example 3. Let a list L of n = 5k rectangles be given where

$(w(R_i), h(R_i))$: $(1, 1/2+e)$ for $i = 1,\ldots\ldots,k$
$(1/2+e, 1/2+e)$ for $i = k+1,\ldots\ldots,3k$
$(1/2+e, 1/2-e)$ for $i = 3k+1,\ldots\ldots,5k$

We get $\text{RNFDH}(L) = \min\{k(5/2 + 3e), k(5/2 + e)\} = k(5/2 + e)$ and $\text{OPT}(L) = k(3/2 + 3e)$, and it follows

$$\text{RNFDH}(L)/\text{OPT}(L) = (5+ 2e)/(3 + 6e),$$

and the quotient can be chosen arbitrarily close to 5/3 .

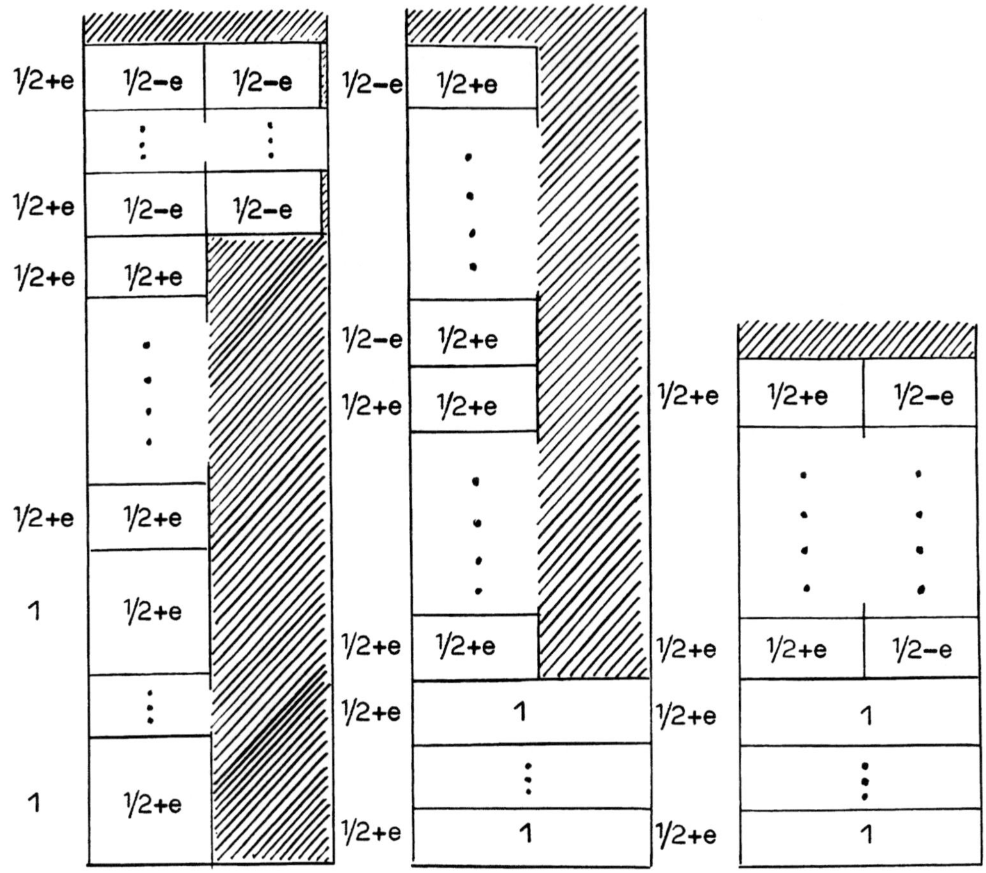

Fig. 3. RHNFDH-, RWNFDH- and OPT-packing for example 3.

Reference.

E. G. Coffman, M. R. Garey, D. S. Johnson, R. E. Tarjan: Performance bounds for level-oriented two-dimensional packing algorithms. SIAM J. Comput. 9 (1980), 808-826

On Planar Tilings With Finitely Many Sorts of Tiles

K. Bezdek, Budapest, Hungary

1. Introduction

The area of mathematical tilings has been of interest for a long time. It is not only an exciting collection of a great deal of artistic tilings but it contains deep theorems and a lot of interesting open problems. Although quite complicated new tilings and theorems are being found there is still much to be discovered.

The purpose of this paper is to prove **Theorem 1** on planar tilings with finitely many sorts of tiles and then to use it in the proofs of some statements (**Theorem 2, 3** and **4**) about n-neighbour tilings.

A collection of convex polygons is said to form a tiling of the Euclidean plane $\mathbf{E}^2$ if the polygons (tiles) cover $\mathbf{E}^2$ such that the individual tiles have disjoint interiors. Two tiles having a boundary point in common are called neighbours. A point in common of at least three tiles is called a vertex of the tiling. For any tiling $\mathbf{T}$ of $\mathbf{E}^2$ let $N(\mathbf{T})=\{x \mid$ there is a tile of $\mathbf{T}$ which possesses x neighbours$\}$. We are going to investigate the tilings given by the following definition. Let $\mathbf{T}_{n_1 n_2 \dots n_k}$ denote the tiling of $\mathbf{E}^2$ which has finitely many sorts of tiles and whose each tile possesses either n_1 or $n_2 \dots$ or n_k vertices, $3 \leq n_1 \leq n_2 = \dots \leq n_k$, $1 \leq k < +\infty$. (Since we have finitely many convex polygons in $\mathbf{E}^2$ such that any tile of $\mathbf{T}_{n_1 n_2 \dots n_k}$ is congruent to one of them therefore $N(\mathbf{T}_{n_1 n_2 \dots n_k})$ is a finite set.) Take a tile T of $\mathbf{T}_{n_1 n_2 \dots n_k}$ having $n_i (1 \leq i \leq k)$ vertices. Let $R(T)=(1/v_1)+\dots+(1/v_{n_i})$ where $v_1, v_2, \dots, v_{n_i}$ denote the degrees of the vertices of T i.e.

R. Bodendiek, R. Henn (Eds.)
Topics in Combinatorics and Graph Theory

the numbers of the tiles which meet at the corresponding vertices of T . It is easy to see that the set $\{R(T)\mid T$ is a tile of $\mathbf{T}_{n_1 n_2 \ldots n_k}\}$ is a finite set. Let $R_s(\mathbf{T}_{n_1 n_2 \ldots n_k})=\min\{R(T)\mid T$ is a tile of $\mathbf{T}_{n_1 n_2 \ldots n_k}\}$ and let $R_m(\mathbf{T}_{n_1 n_2 \ldots n_k})=\max\{R(T)\mid T$ is a tile of $\mathbf{T}_{n_1 n_2 \ldots n_k}\}$. We can now state our first result.

<u>**Theorem 1.**</u> $\{x\mid R_s(\mathbf{T}_{n_1 n_2 \ldots n_k})\leqq x\leqq R_m(\mathbf{T}_{n_1 n_2 \ldots n_k})\}\cap\{y\mid (n_1/2)-1\leqq y\leqq (n_k/2)-1\}\neq\emptyset$ for any tiling $\mathbf{T}_{n_1 n_2 \ldots n_k}$ of $\mathbf{E}^2$, $3\leqq n_1\leqq\ldots\leqq n_k$, $1\leqq k<+\infty$.

Then we get the following immediate

<u>**Corollary.**</u> $((n/2)-1)\in\{x\mid R_s(\mathbf{T}_n)\leqq x\leqq R_m(\mathbf{T}_n)\}$ for any tiling $\mathbf{T}_n$ of $\mathbf{E}^2$, $3\leqq n$.

It is obvious that $\sup_{\mathbf{T}_n}(\max(N(\mathbf{T}_n)))=+\infty$, $3\leqq n$.(One may observe that $n\leqq 6$ for any tiling $\mathbf{T}_n$, $3\leqq n$ of $\mathbf{E}^2$.) Now let us consider further tilings which are slightly different from $\mathbf{T}_n$. Namely, let $\mathbf{T}_{P_n}$ ($\mathbf{T}_{Q_n}$, resp.) be the tiling of $\mathbf{E}^2$ whose each tile is congruent to the convex n-sided polygon P_n (the convex equilateral n-gon Q_n , resp.) of $\mathbf{E}^2$. (As before, it is obvious that $3\leqq n\leqq 6$.) The problem of determining $\sup_{\mathbf{T}_{P_n}}(\max(N(\mathbf{T}_{P_n})))$($\sup_{\mathbf{T}_{Q_n}}(\max(N(\mathbf{T}_{Q_n})))$, resp.), $3\leqq n\leqq 6$ seems to be a non-trivial question. Now we can phrase the following more special problem of L.Fejes Tóth [3] . Let $\mathbf{T}^c_{P_n}$ be the tiling of $\mathbf{E}^2$ whose each tile is congruent to the convex n-sided polygon P_n and whose each tile possesses exactly the same number of neighbours say, $N(\mathbf{T}^c_{P_n})$. (In [3] $\mathbf{T}^c_{P_n}$ is called an N-neighbour tiling.) Prove or disprove that $N(\mathbf{T}^c_{P_n})\in\{6,7,8,9,10,12,14,16,21\}$. A slightly weaker version of the Corollary yields already the following theorem (for more details see [1]).

<u>**Theorem 2.**</u> Let $\mathbf{T}^c_{P_n}$ be the edge-to-edge tiling of $\mathbf{E}^2$ whose each tile is congruent to the convex n-sided polygon P_n and whose each tile possesses the same number $N(\mathbf{T}^c_{P_n})$ of neighbours. Then $N(\mathbf{T}^c_{P_n})\in\{6,7,8,9,10,12,14,16,21\}$.

At this point it is worth mentioning that we have the spherical analogue of the above theorem. We recall it here just for the sake of completeness. Further details can be found in [2].

Theorem 3. Let $\mathbf{T}^c_{P_n}$ be the edge-to-edge tiling on the 2-dimensional unit sphere whose each tile is congruent to the spherical convex n-sided polygon P_n of diameter $< (\pi/2)$ and whose each tile possesses the same number $N(\mathbf{T}^c_{P_n})$ of neighbours. Then $N(\mathbf{T}^c_{P_n}) \in \{5,6,7,8,9,10,11,12,14,17\}$.

Finally, we have arrived at the last theorem, which besides Theorem 1 we are going to prove in this paper. In a sense it completes the investigation based on Theorem 1.

Theorem 4. Let $\mathbf{T}^c_{Q_n}$ be the tiling of $\mathbf{E}^2$ whose each tile is congruent to the convex equilateral n-gon Q_n and whose each tile possesses the same number $N(\mathbf{T}^c_{Q_n})$ of neighbours. Then $N(\mathbf{T}^c_{Q_n}) \in \{6,7,8,10,12\}$.

2. Proof of Theorem 1

(Indirect) Assume that we have a tiling $\mathbf{T}_{n_1 n_2 \ldots n_k}$ such that $(n_1/2)-1 \leqq (n_2/2)-1 \leqq \ldots \leqq (n_k/2)-1 < R_1 \leqq R_2 \leqq \ldots \leqq R_\ell$ (or $R_1 \leqq R_2 \leqq \ldots \leqq R_\ell < (n_1/2)-1 \leqq (n_2/2)-1 \leqq \ldots \leqq (n_k/2)-1$) where $\{R_1, R_2, \ldots, R_\ell\} = \{R(T) \mid T$ is a tile of $\mathbf{T}_{n_1 n_2 \ldots n_k}\}$. Let T be a tile of $\mathbf{T}_{n_1 n_2 \ldots n_k}$ and let $S_0 \subset S_1 \subset \ldots \subset S_{i-1} \subset S_i \subset \ldots$ be the sequence of the neighbourhoods of T defined by $S_0 = T$ and $S'_i = \cup\{T' \mid T' \not\subset S_0 \cup S_1 \cup \ldots \cup S_{i-1}$ is a tile of $\mathbf{T}_{n_1 n_2 \ldots n_k}$ such that $T' \cap (S_0 \cup S_1 \cup \ldots \cup S_{i-1}) \neq \emptyset\}$ and $S_i = S'_i \cup (\cup\{T' \mid T'$ is the tile of $\mathbf{T}_{n_1 n_2 \ldots n_k}$ which lies in the closure of a bounded connected component of $\mathbf{E}^2 \setminus (S'_i \cup (S_0 \cup S_1 \cup \ldots \cup S_{i-1}))\})$, $i=1,2,\ldots$. Then let $U_j = \bigcup_{i=0}^{j} S_i$, $j=0,1,2\ldots$. If $d > 0$ denotes the largest diameter of the tiles of $\mathbf{T}_{n_1 n_2 \ldots n_k}$, then it is clear that U_j is contained in a disk centered at a point of T having radius $(j+1)d$. Now let $v(j)$, $e(j)$ and $f(j)$ denote the number of vertices, edges and faces (tiles) of $\mathbf{T}_{n_1 n_2 \ldots n_k}$ in U_j . Euler's formula says that $f(j)+v(j)=e(j)+1$ i.e. (1) $v(j)/f(j)=(e(j)/f(j))-1+(1/f(j))$ If $f(j,n_1),f(j,n_2),\dots,f(j,n_k)$ denote the number of tiles in U_j having $n_1,n_2,\dots,n_k$ vertices and $f(S_j)$ is the number of tiles in S_j, then it is easy to see that

(2) $2\cdot e(j)=n_1\cdot f(j,n_1)+n_2\cdot f(j,n_2)+\dots+n_k\cdot f(j,n_k)+O(1,j)$ with $f(j)=f(j,n_1)+f(j,n_2)+\dots+f(j,n_k)$ and $o<O(1,j)\leqq n_k\cdot f(S_j)$. Also, if $r(j,1),r(j,2),\dots,r(j,\ell)$ denote the number of tiles T in U_j for which $R(T)$ equals $R_1,R_2,\dots,R_\ell$, then one can observe that

(3) $v(j)=R_1\cdot r(j,1)+R_2\cdot r(j,2)+\dots+R_\ell\cdot r(j,\ell)+O(2,j)$ with $f(j)=r(j,1)+r(j,2)+\dots+r(j,\ell)$ and $o<O(2,j)\leqq n_k\cdot f(S_j)$.

Proposition 1. $\inf_j(f(S_j)/f(j))=o$

Proof: (Indirect) Suppose that $\inf_j(f(S_j)/f(j))>o$. It implies the existence of $\varepsilon>o$ such that $(f(S_i)/f(i))\geqq\varepsilon>o$ for any $i\in\{1,2,\dots\}$. Since $f(S_i)=f(i)-f(i-1)\geqq\varepsilon\cdot f(i)>\varepsilon\cdot f(i-1)$ therefore $f(i)>(1+\varepsilon)\cdot f(i-1)$ consequently, $f(i)>(1+\varepsilon)^i\cdot f(0)=$ $=(1+\varepsilon)^i$ for any $i\in\{1,2,\dots\}$. Thus $\min\{\text{area}(T')\mid T'$ is a tile of $\mathbf{T}_{n_1 n_2\dots n_k}\}<((i+1)^2\cdot d^2\cdot\pi)/(1+\varepsilon)^i$ for any $i\in\{1,2,\dots\}$. But then $\lim_{i\to+\infty}((i+1)^2\cdot d^2\cdot\pi)/(1+\varepsilon)^i)=o$ yields $\min\{\text{area}(T')\mid T'$ is a tile of $\mathbf{T}_{n_1 n_2\dots n_k}\}=o$, a contradiction. □

Without loss of generality we may suppose that $\lim_{j\to+\infty}(f(S_j)/f(j))=o$. Combining (1), (2) and (3) we get

(4) $(R_1\cdot r(j,1)+R_2 r(j,2)+\dots+R_\ell\cdot r(j,\ell))/(r(j,1)+r(j,2)+\dots$
$\dots+r(j,\ell))+O(2,j)/f(j)=(n_1\cdot f(j,n_1))/(2f(j))+\dots$
$\dots+(n_k\cdot f(j,n_k))/(2f(j))-1+(1/f(j))$

Now let $j\to+\infty$. Since $o<O(2,j)/f(j)\leqq(n_k\cdot f(S_j))/f(j)$ and $o<O(1,j)/(2f(j))\leqq(n_k\cdot f(S_j))/(2f(j))$ therefore $\lim_{j\to+\infty}O(2,j)/f(j)=o$, $\lim_{j\to+\infty}O(1,j)/(2f(j))=o$ and $\lim_{j\to+\infty}1/f(j)=o$. Hence lim inf(left side of (4)) $>(n_k/2)-1$ and lim inf(right side of (4)) $\leqq(n_k/2)-1$, a contradiction (or lim inf(left side of (4)) $<(n_1/2)-1$ and lim inf(right side of (4)) $\geqq(n_1/2)-1$, a contradiction).

3. Proof of Theorem 4

It is known that $3\leqq n\leqq 6$. If n=3, then a trivial argument gives that $N(\mathbf{T}^c_{Q_3})\in\{8,10,12\}$. Then suppose that n=4. We would like to prove that $N(\mathbf{T}^c_{Q_4}) \in \{6,7,8,10\}$. Take a tiling $\mathbf{T}^c_{Q_4}$. It might have "fault-lines" . It is obvious that such fault-lines are necessarily parallel to one another. So the tiling can be slipped along the fault-lines to become an edge-to-edge tiling $\mathbf{T}_{Q_4}$ with $N(\mathbf{T}^c_{Q_4}) \leqq \min(N(\mathbf{T}_{Q_4}))$. Since the tiling $\mathbf{T}_{Q_4}$ is a $\mathbf{T}_4$ tiling as well therefore the Corollary implies that $1 \in \{x \mid R_s(\mathbf{T}_{Q_4}) \leqq x \leqq R_m(\mathbf{T}_{Q_4})\}$. Hence there is a tile T of $\mathbf{T}_{Q_4}$ such that $R(T)=(1/v_1)+\ldots+(1/v_4) \geqq 1$. Let N be the number of neighbours of T in $\mathbf{T}_{Q_4}$.

<u>Proposition 2.</u> $4 \leqq N \leqq 10$.

<u>Proof.</u> On the one hand the inequality $4 \leqq N$ is trivial. On the other hand assume that $11 \leqq N$. Since $v_1+v_2+v_3+v_4 = N+8$, $3 \leqq v_1$, $3 \leqq v_2$, $3 \leqq v_3$, $3 \leqq v_4$ and $(1/v_1)+\ldots+(1/v_4) \geqq 1$ therefore either $\{v_1,v_2,v_3,v_4\} = \{N-1,3,3,3\}$ or $\{v_1,v_2,v_3,v_4\} = \{N-2,4,3,3\}$. Now recall that T is a rhombus. Since in both cases T possesses vertices of degree 3 therefore the angles of T are $\pi/3$ and $(2\cdot\pi)/3$. But then either $N-1 \leqq 6$ or $N-2 \leqq 6$, a contradiction. □

As a corollary we get that $N(\mathbf{T}^c_{Q_4}) \leqq 10$. In order to complete the investigation of the case n=4 it is enough to prove

<u>Proposition 3.</u> $N(\mathbf{T}^c_{Q_4}) \notin \{4,5,9\}$.

<u>Proof.</u> If $N(\mathbf{T}^c_{Q_4})=4$ ($N(\mathbf{T}^c_{Q_4})=5$, resp.) , then it is easy to see that $\mathbf{T}^c_{Q_4}$ has to be an edge-to-edge tiling. So if T is an arbitrary tile of $\mathbf{T}^c_{Q_4}$ with vertex-degrees v_1, v_2, v_3 and v_4, then $v_1=v_2=v_3=v_4=3$ ($\{v_1,v_2,v_3,v_4\}=\{4,3,3,3\}$, resp.) . Hence $1 \notin \{x \mid R_s(\mathbf{T}^c_{Q_4}) \leqq x \leqq R_m(\mathbf{T}^c_{Q_4})\} = \{4/3\}$ ($1 \notin \{5/4\}$, resp.) , a contradiction (see the Corollary) . Finally, <u>assume</u> the existence of the tiling $\mathbf{T}^c_{Q_4}$ with <u>$N(\mathbf{T}^c_{Q_4})=9$</u> . First we prove that <u>Q_4 has to be a rhombus with angles $\pi/3$ and $(2\cdot\pi)/3$.</u> Namely, either

(5) $\mathbf{T}^c_{Q_4}$ is an edge-to-edge tiling in which case let T be the tile of $\mathbf{T}^c_{Q_4}$ for which $R(T)=(1/v_1)+(1/v_2)+(1/v_3)+(1/v_4) \geqq 1$ (see Corollary) or

(6) by slipping along the fault-lines we get an edge-to-edge tiling $\mathbf{T}_{Q_4}$ whose tile T possesses 9 or 10 neighbours with $R(T)=(1/v_1)+(1/v_2)+(1/v_3)+(1/v_4) \geqq 1$ (see Corollary and the proof of Proposition 2) . An easy computation yields that in case (5) we have $\{v_1,v_2,v_3,v_4\} \in \{\{8,3,3,3\}, \{7,4,3,3\}, \{6,5,3,3\}, \{6,4,4,3\}\}$ and in case (6) we have $\{v_1,v_2,v_3,v_4\} \in \{\{8,3,3,3\}$,

$\{7,4,3,3\}$, $\{6,5,3,3\}$, $\{6,4,4,3\}$, $\{9,3,3,3\}$, $\{8,4,3,3\}$, $\{7,5,3,3\}$, $\{6,6,3,3\}\}$. However, in both cases T possesses at least one vertex of degree 3 yielding that T is a rhombus of angles $\pi/3$ and $(2\pi)/3$. Next, we show that $\mathbf{T}^c_{Q_4}$ has to be an edge-to-edge tiling. It is a corollary of the trivial statement that if $\mathbf{T}_{Q_4}$ is not an edge-to-edge tiling of $\mathbf{E}^2$ whose tiles are congruent to the rhombus Q_4 of angles $\pi/3$ and $(2\cdot\pi)/3$, then there exists a tile of $\mathbf{T}_{Q_4}$ the number of neighbours of which is $\leqq 8$. Finally, we prove that the edge-to-edge tiling $\mathbf{T}^c_{Q}$ with $N(\mathbf{T}^c_{Q_4}) = 9$ where Q_4 is a rhombus of angles $\pi/3$ and $(2\cdot\pi)/3$ does not exist , a contradiction. Namely, the Corollary yields that $\mathbf{T}^c_{Q_4}$ has to have a tile T with $R(T)=(1/v_1)+(1/v_2)+(1/v_3)+(1/v_4)\geqq 1$. Thus $\{v_1,v_2,v_3,v_4\}\in\{\{6,5,3,3\}, \{6,4,4,3\}\}$. First we assume that $\mathbf{T}^c_{Q_4}$ possesses a tile T with vertex degrees $\{v_1,v_2,v_3,v_4\} = \{6,5,3,3\}$. We know that for any other tile T' of $\mathbf{T}^c_{Q_4}$ with vertex degrees v'_1, v'_2, v'_3 and v'_4 we have $\{v'_1,v'_2,v'_3,v'_4\}\in \{\{6,5,3,3\}, \{6,4,4,3\}, \{5,5,4,3\}, \{5,4,4,4\}\}$. As it is shown in Fig. 1 the information obtained above determines a neighbourhood of T with vertex V of degree 5 whose given neighbourhood however makes it impossible to have that degree.

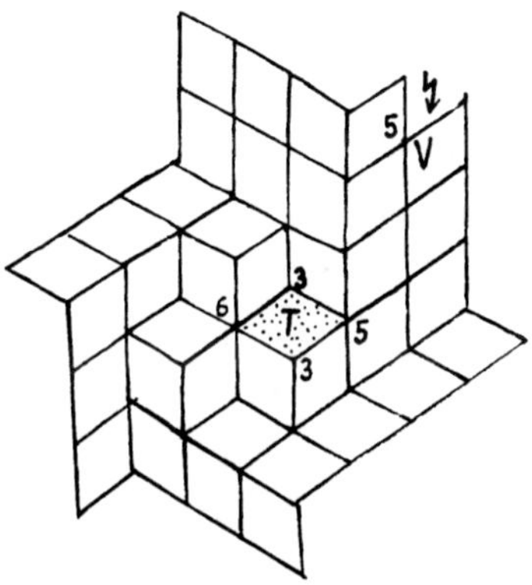

Fig. 1

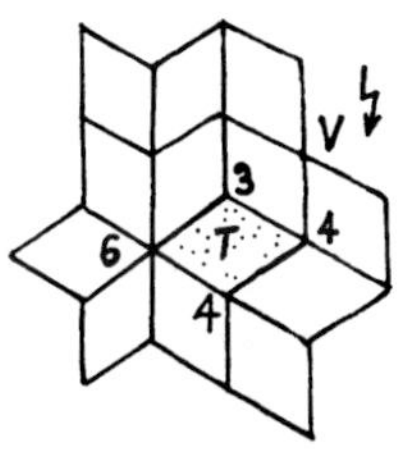

Fig. 2

Second we may assume that $\mathbf{T}^c_{Q_4}$ possesses a tile T with vertex degrees $\{v_1,v_2,v_3,v_4\} = \{6,4,4,3\}$ and tiles T' with vertex degrees

$\{v'_1, v'_2, v'_3, v'_4\} \in \{\{6,4,4,3\}, \{5,5,4,3\}, \{5,4,4,4\}\}$. Fig. 2 shows the uniquely determined neighbourhood of T and the vertex V whose degree then cannot be 6 , a contradiction. □

We are left with the cases n=5 and n=6 . Suppose that n=5 . Take a tiling $\mathbf{T}^c_{Q_5}$. If it is necessary the tiling can be slipped along the fault-lines to become an edge-to-edge tiling $\mathbf{T}_{Q_5}$ with $N(\mathbf{T}^c_{Q_5}) \leq \min(N(T_{Q_5}))$. Then the Corollary implies that $(3/2) \in \{x \mid R_s(\mathbf{T}_{Q_5}) \leq x \leq R_m(\mathbf{T}_{Q_5})\}$ and so $N(\mathbf{T}^c_{Q_5}) \leq 8$. Also it is obvious that $6 \leq N(\mathbf{T}^c_{Q_5})$ hence $N(\mathbf{T}^c_{Q_5}) \in \{6,7,8\}$. Finally, let n=6 and take a tiling $\mathbf{T}^c_{Q_6}$. A simple application of the Corollary gives $N(\mathbf{T}^c_{Q_6}) = 6$ finishing the proof of the theorem.

References

[1] K. Bezdek, Mosaics with each face having the same number of neighbours, Mat. Lapok 28, 1980, 297-303.

[2] K. Bezdek, Optimal arrangements, Thesis of Kandidátus Degree, Budapest, 1984, 92-99.

[3] L. Fejes Tóth, Tessallation of the plane with convex polygons having a constant number of neighbours, Amer. Math. Monthly 82, 1975, 273-276.

Examples of Space-Tiling Polyhedra Related to Hilbert's Problem 18, Question 2

A. Bezdek*, Budapest, Hungary
W. Kuperberg, Auburn, USA

1. Definitions and preliminaries. Let P be a connected polyhedron in the d-dimensional Euclidean space E^d. A system $\{P_i\}$ of congruent copies of P, whose union is E^d and whose interiors are mutually disjoint, is called a *tiling*. If each P_i is a translate of P, then $\{P_i\}$ is a *tiling by translates*. If there exists a subgroup G of the group of isometries of E^d, such that $\{P_i\} = \{ m(P): m \in G\}$, then $\{P_i\}$ is a *tiling by a group of motions*, also called (see [5]) a *tile-transitive tiling* or an *isohedral tiling*.

A *lattice* in E^d is the set of linear combinations with integer coefficients of a set of d linearly independent vectors in E^d. A tiling is a *lattice tiling* if all of its members are translates of each other and if the corresponding translation vectors form a lattice. If P admits a tiling, then P is called a *tile*, and if P admits a tiling by translates, a tiling by a group of motions, or a lattice tiling, then P is called a *translation tile*, a *group tile* (also called a *fundamental region of a group of motions*), or a *lattice tile*, respectively.

Obviously, every lattice tile is a translation tile. We do not know whether the converse is true in the plane E^2. The higher dimensional case is discussed in the following sections, among other, related questions.

---

* Partially supported by the Hungarian National Foundation for Scientific Research, grant number 1238.

R. Bodendiek, R. Henn (Eds.)
Topics in Combinatorics and Graph Theory

2. Introduction. Hilbert's Problem 18 (see [8]), posed in 1900, consists of three questions, second of which reads: "Whether polyhedra also exist which do not appear as fundamental regions of groups of motions, by means of which nevertheless by a suitable juxtaposition of congruent copies a complete filling up the space is possible?" The first example answering this question in the affirmative was produced by K. Reinhardt [13] in 1928. He constructed a (non-convex) polyhedron in E^3 which admits a tiling but not by a group of motions. In the same paper, Reinhardt expressed his conviction, apparently shared by his contemporaries including Hilbert himself, that in E^2 no such example was possible. Reinhardt's polyhedron is obtained by cutting the solid shown on Fig. 1a along the dotted lines and by modifying the congruent pieces as shown on Fig. 1b. It should be noted that Reinhardt's tile is not a translation tile.

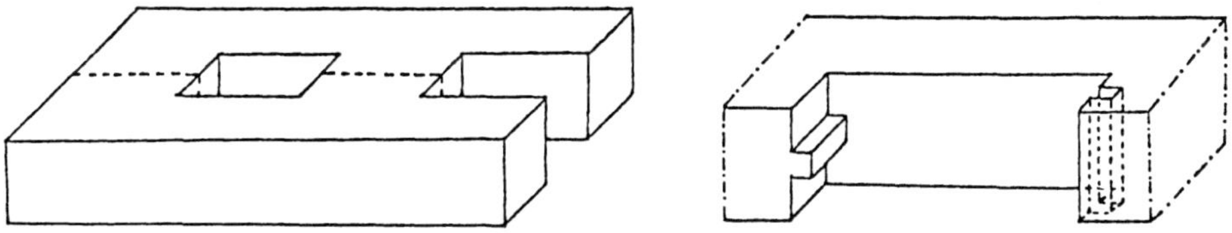

(a) (b)

Figure 1.

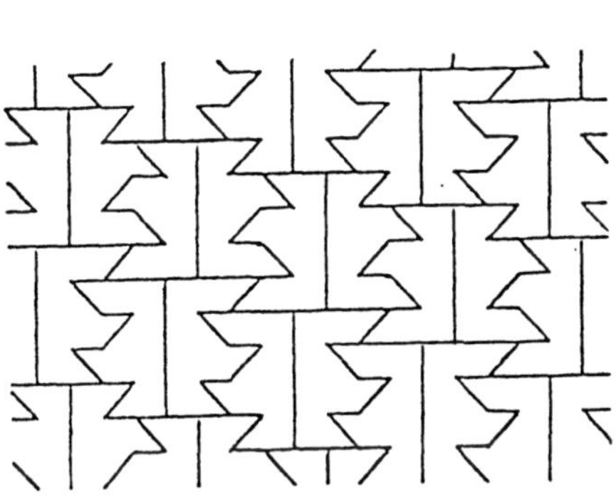

Figure 2.

Figure 3.

The first planar (non-convex) example was given by H. Heesch [6] in 1965 (see Fig. 2). Heesch's example was somewhat simplified (made star shaped but not convex) by J. Milnor [12] in 1976 (see Fig. 3), although already in 1968 R. B. Kershner [9] had produced examples of convex pentagons which tile the plane but not by a group of motions (see Fig. 4a,b,c). This was a remarkable accomplishment of Kershner, even though he did not quite succeed in his attempt of classifying all plane convex tiles. Again, it should be noted that neither of the planar examples mentioned here is a translation tile (Kershner's pentagons cannot be translation tiles due to their convexity, otherwise they would be lattice tiles).

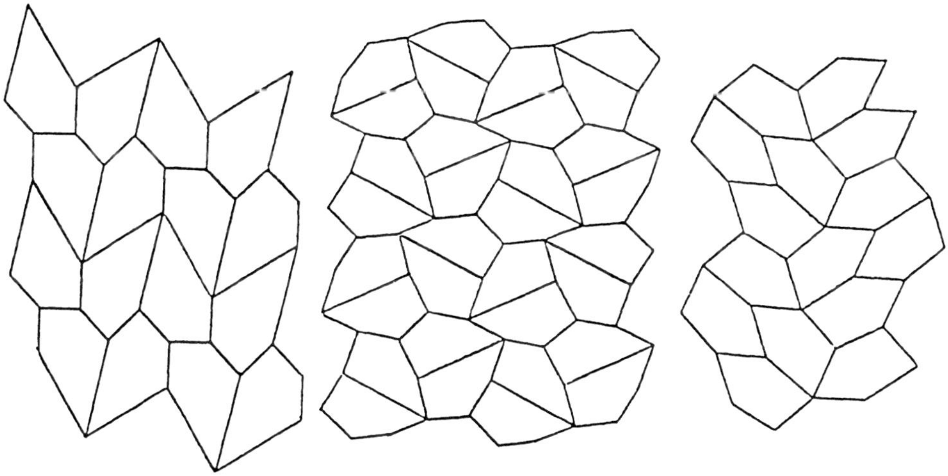

(a) (b) (c)

Figure 4.

In 1972, S. K. Stein [14] gave an example of a centro-symmetric star shaped 10-dimensional polyhedron which is a translation tile but not a lattice tile. Refining Stein's idea, S. Szabó [15] found in 1983 a similar example in dimension 5 and shortly thereafter (in [16], 1985) a 3-dimensional one (see Fig. 5), which is not even a group tile. The examples of Stein and Szabó cannot be improved in the direction of achieving convexity in place of star shape. A theorem proved by B. A . Venkov [17] and, independently, by P. McMullen [10], characterizing all convex translation tiles in E^d (for all d), implies that every convex translation tile is a lattice tile.

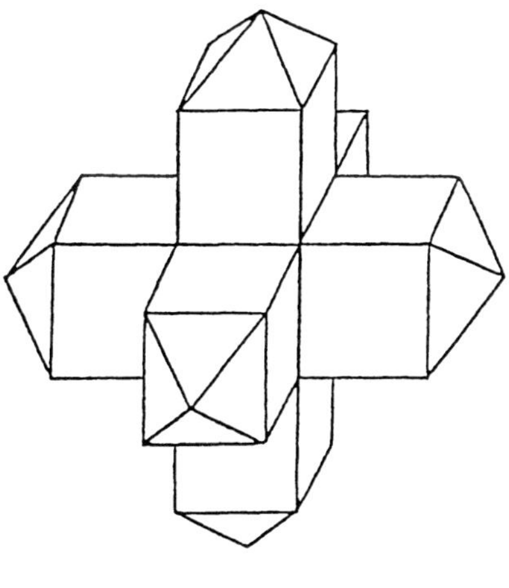

Figure 5.

The ideas of Stein and Szabó are rather non-trivial and their proofs utilize some algebra. The example described in the next section, although not star-shaped, is quite simple, is accompanied by a very short proof, and is based on an idea essentially different from those of Stein and Szabó.

3. The example. Let C_1 denote the unit cube $\{(x_1,x_2,x_3): 0 \le x_i \le 1,\ i=1,2,3\}$ of the Cartesian coordinate space, and let $\hat{C}_1$ be obtained from C_1 by attaching a "protrusion" to the face $x_1 = 1$ and by forming a "cavity" on the face $x_1 = 0$ as shown on Fig. 6a. The protrusion is a tetrahedron whose base lying on the face $x_1 = 1$ is a non-isosceles triangle, whose all three angles at the tip are smaller than 45° and whose height is smaller than 1. The cavity's shape is designed so that it is filled precisely by the translation of the protrusion by the vector [-1,0,0]. Let $T: E^3 \to E^3$ be the translation by the vector [2,0,0] and let $C_2 = T(C_1)$, $\hat{C}_2 = T(\hat{C}_1)$. Let v_1 and w_1 denote the tip of the protrusion on $\hat{C}_1$ and the deepest point of the cavity in $\hat{C}_1$, respectively, and let $v_2 = T(v_1)$, $w_2 = T(w_1)$. Let B the prism of height 1/8 whose base is the parallelogram with vertices (0,-1/2,1), (1,-1/2,1), (3,3/2,1) and (2,3/2,1), and let $P = \hat{C}_1 \cup \hat{C}_2 \cup B$ (see Fig. 6b).

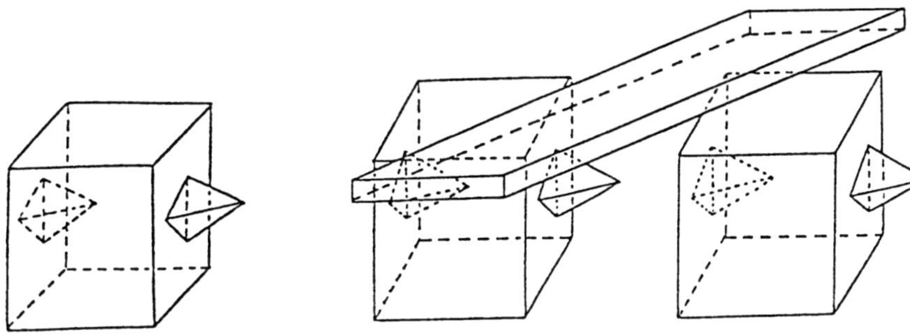

(a) (b)

Figure 6.

As we show below, the polyhedon P is a translation tile, but does not admit a tiling by any group of motions. The first part of this statement follows from the fact that the polyhedron $P \cup (P + [1,0,0])$ admits a lattice tiling of E^3 under the lattice generated by vectors [4,0,0], [-1/2,1,0] and [0,0,9/8]. In order to prove the second part, assume to the contrary, that P is a tile in a tiling by a group of motions G. Then the vertex w_2 of P is a vertex of m(P) for some non-identity element m of G. The only vertices of P that can fit into this deepest point of the cavity by a rigid motion are v_1 and v_2, and the only way to put them there is by a translation by [1,0,0] or by [-1,0,0], respectively. Thus m is a translation of E^3 by one of these two vectors. Then $m^2 \in G$ is a translation of E^3 by [2,0,0] or by [-2,0,0], and in either case $m^2(P)$ overlaps with P, which is not allowed in a tiling.

Remark. The modified cubes $\hat{C}_i$ were used in the construction instead of the cubes C_i only in order to allow a simple proof. The simpler-looking polyhedron $C_1 \cup C_2 \cup B$, without the cavities and protrusions, would serve as an example as well, but with a somewhat more elaborate proof.

4. Problems

1. In the plane, is every translation tile a lattice tile? (Of course, we focus here on non-convex tiles.)

Comment. There are known examples (see [1] and [7]) of plane polygons whose translation packing density is greater than lattice packing density (see [3], [4] and [11]) for the definitions of these concepts).

2. Does there exist a convex body in E^d (for some d) whose translation packing density is greater than lattice packing density?

Comment. One of the most outstanding problems in packing theory can be stated as follows: Is the packing density of a ball in E^d equal to its lattice packing density? The answer is not known for any $d \geq 3$ and is conjectured to be negative for all sufficiently large d (see [11], p.15).

3. Does there exist a convex polyhedron P in E^d which is not a tile, yet a right prism (with some height) with base P is a tile in E^{d+1}?

Comment. The prism constructed by Reinhardt and shown here on Fig. 1a (disregarding the dotted lines) constitutes a non-convex (even non-simply-connected) example related to Problem 3 with $d = 2$. For other examples of tilings of E^3 with non-simply-connected tiles of various topological types, see [2].

References

1. Bezdek, A. and Kertész, G., Counter-examples to a packing problem of L. Fejes Tóth, in: Intuitive Geometry, Coll. Math. Janos Bolyai Soc. 48(1985), 29-36.

2. Debrunner, H. E., Tiling three-space with handlebodies, Studia Sci. Math. Hungar., 21(1986), 201-202.

3. Fejes Tóth, G., New results in the theory of packing and covering, in: Convexity and its applications, Basel, Birkhäuser 1983, 318-359.

4. Fejes Tóth, L., Lagerungen in der Ebene, auf der Kugel und im Raum, Grundl. Math. Wiss. 65, Springer, Berlin 1953.

5. Grünbaum, B. and Shephard, G., Tilings and patterns, Freman, New York 1987.

6. Heesh, H., Aufbau der Ebene aus kongruenten Bereichen, Nachr. Ges. Wiss. Göttingen, New Ser. 1(1935), 115-117.

7. Heppes, A., On the density of translates of a domain, Studia Sci. Math. Hungar., to appear.

8. Hilbert, D., Mathematische Probleme, Göttinger Nachrichten 1900, 253-297.

9. Kershner, R. B., On paving the plane, Amer. Math. Monthly 75(1968), 839-844.

10. McMullen, P., Convex bodies which tile space by translation, Mathematika 27(1980), no. 1, 113-121.

11. Rogers, C. A., Packing and covering, Cambridge Tracts 54, Cambridge Univ. Press, Cambridge 1964.

12. Milnor, J., Hilbert's Problem 18: on crystallographic groups, fundamental domains, and on sphere packing, Proc. Sympos. Pure Math., Vol. 28(1976), Amer. Math. Soc., Providence, RI, 591-506.

13. Reinhardt, K., Zur Zerlung der euklidischen Räume in kongruente Polytope, S.-ber. Preuss. Akad. Wiss. 1928, 150-155.

14. Stein, S. K., A symmetric star body that tiles but not as a lattice, Proc. Amer. Math. Soc. 36(1972), 543-548.

15. Szabó S., A symmetric star polyhedron that tiles but not as a fundamental domain, Proc. Amer. Math. Soc. 89(1983), 563-566.

16. Szabó, S., A star polyhedron that tiles but not as fundamental domain, in: Intuitive Geometry, Coll. Math. Janos Bolyai Soc. 48(1985), 531-544.

17. Venkov, B. A., On a class of Euclidean polyhedra, Vestnik Leningrad. Univ. Ser. Mat. Fiz. Him. 9(1954), no. 2, 11-31.

The Historical Background to Gerhard Ringel's Work

N. L. Biggs, London, Great Britain
E. K. Lloyd, Southampton, Great Britain
R. J. Wilson, Milton Keynes, Great Britain

1. Euler's polyhedral formula

The ancestry of Ringel's work on the Heawood formula can be traced back to the Ancient Greeks. The Greeks were interested in the mathematical properties of regular polyhedra (where all the faces are congruent regular polygons), and found that there are just five such solids. A full account was written by Theaetetus around 400 BC, and it has been argued that Euclid's *Elements* (c. 320 BC) was primarily an introduction to the study of these regular polyhedra. But there is no evidence that the Greeks knew the simple formula relating the numbers v, e and f of vertices, edges and faces of a polyhedron — namely, that $v - e + f = 2$.

This result, often wrongly ascribed to Descartes, was discovered by Leonhard Euler. For many years he corresponded with a fellow mathematician, Christian Goldbach, and in 1750 wrote to him saying that he had found the formula, but could not prove it. Shortly afterwards he wrote two papers on the polyhedral formula. In the first one, he verified it for several families of solids, but admitted his inability to find a general proof. In the second one, he set forth a proof by dissection; it consists of slicing out tetrahedral pieces from the solid in such a way that the number $v - e + f$ is unchanged, until eventually all that remains is a single tetrahedron, for which $v - e + f = 4 - 6 + 4 = 2$. It is an ingenious method — and an apt one, since it uses no metrical concepts. However, it is not obvious that

R. Bodendiek, R. Henn (Eds.)
Topics in Combinatorics and Graph Theory

this slicing procedure can always be carried out, since it may give rise to 'degenerate' polyhedra for which the meaning from the formula is ambiguous. The first correct proof of Euler's formula was not obtained until 1794; it was given by Legendre, using metrical properties of spherical polygons.

Another type of proof, involving planar graphs, was given by Cauchy in 1813. Cauchy's idea was to project the polyhedron onto a plane — that is, to flatten the polyhedron so that it becomes a plane network of polygons; he then gave a direct proof of the relation $v - e + f = 1$ for this network. His proof really does belong to graph theory, for it uses a process of 'triangulation' rather than any metrical concepts. So this procedure, together with the process of projection, can be used to give a topological proof of the original polyhedral formula.

2. Exceptions to Euler's formula

The proofs of Euler, Legendre and Cauchy all require certain assumptions about the polyhedron or graph under consideration. Although these assumptions may seem innocuous at first, the difficulties are actually very profound. Indeed, Imre Lakatos used this topic as the scenario for his philosophical study '*Proofs and Refutations – the Logic of Mathematical Discovery*' [4]. That book contains a series of imaginary attempts to solve the difficulties presented by counter-examples to Euler's formula, employing successive refinements of the conditions and alternative proof techniques; it also contains a wealth of historical material.

In the present context, the most significant modifications of Euler's formula are those which lead to the idea of the genus of a surface. Such modifications were first studied by Lhuilier, who published a paper on them in 1811. Nowadays one thinks of the formulae of Euler, Cauchy and Lhuilier as statements about maps on various surfaces. If the surface is connected and every face is simply-connected, then they assert that the *Euler characteristic*

$$\eta = v - e + f$$

depends only on the surface on which the map is drawn, and not on the map itself. For example, Euler's convex polyhedra correspond to maps on a sphere ($\eta = 2$), whereas Cauchy's plane networks correspond to maps on a plane disc ($\eta = 1$).

The most important class of surfaces considered by Lhuilier are *closed*, in the sense that (unlike the disc) they have no boundary curves and are of finite extent.

The simplest closed surface is the sphere, and the next simplest is the *torus*, or anchor ring. It is easy to check that $\eta = 0$ for the latter surface. In fact, infinitely many closed surfaces can be constructed in the following way. First take a sphere and cut out two circular pieces from its surface; then take a cylindrical 'handle' and glue its ends to the sides of the holes in the sphere; finally, repeat this process g times, to give a *sphere with g handles*. (The torus is the case $g = 1$ and the sphere itself is the case $g = 0$.) The resulting surface is also called the *orientable surface of genus g*, denoted by S_g.

The version of Euler's formula obtained by Lhuilier may be regarded as a formula for the Euler characteristic of a closed orientable surface of genus g — namely, $\eta = 2 - 2g$. This formula was the starting point for an extensive investigation (1861) by Listing, which was to be influential in the development of topology as a branch of modern mathematics. In his work '*Der Census räumlicher Complexe*' (The census of spatial complexes) he defined certain objects, called 'complexes' because they were built up from simpler pieces, and investigated their topological properties — especially the question of how such properties affect the generalization of Euler's formula. The properties were given names such as '*periphraxis*' and '*cyclosis*', borrowed from the biological sciences. Listing's ideas were taken up by other mathematicians — notably, Poincaré in his fundamental researches on algebraic topology at the turn of the century. What Poincaré possessed, and Listing lacked, was the mathematical apparatus needed to formulate topological ideas within the framework of algebra.

Lhuilier's work overlooked the possibility that a surface can be 'one-sided'. This possibility was not discovered until around 1858 when Listing and Möbius independently described the Möbius strip, a one-sided surface with a boundary. Closed one-sided surfaces, such as the *Klein bottle*, were later described by Tietze and others. Their construction is similar to that described above for a sphere with g handles. First, construct a Möbius strip from a rectangle by giving it a half-twist and joining a pair of opposite edges; then take a sphere, cut out a circular piece from its surface, and glue the boundary edge of the Möbius strip to the side of the hole in the sphere (this cannot be done in three-dimensional space); finally, repeat this process g times to give a *sphere with g crosscaps*. (The projective plane is the case $g = 1$, and the Klein bottle is the case $g = 2$.) The resulting surface is also called the *non-orientable surface of genus g*, denoted by T_g.

There is a version of Euler's formula for non-orientable surfaces, analogous to Lhuilier's formula for orientable surfaces. This states that the Euler characteristic of a closed non-orientable surface of genus g depends only on the surface, and is equal to $\eta = 2 - g$.

3. The Heawood conjecture

In 1890 Percy Heawood produced one of the most important papers in the history of graph colourings. He started by pointing out a major error in Kempe's previously-accepted proof of the four colour theorem for maps, and he continued by showing how enough can be salvaged from Kempe's argument to prove that every map on a sphere can be coloured with *five* colours.

Heawood continued by considering the colouring of maps on orientable surfaces other than the sphere. He proved, in particular, that every map on a torus can be coloured with just seven colours, and he gave an example of a map which requires all seven colours. In modern terminology, this amounts to saying that the *chromatic number* of the torus is seven. He also gave an upper bound for the chromatic number of S_g — the minimum number of colours needed to colour any map drawn on a sphere with g handles — namely $\lfloor \frac{1}{2}(7+\sqrt{(1+48g)})\rfloor$, wherever $g \geq 1$. He also asserted, but with insufficient proof, that there are maps on such a surface which need this number of colours. The conjecture that this is the case became known as the *Heawood conjecture*, and it took more than seventy years to prove it.

Thus, in modern terminology and notation, Heawood claimed that the chromatic number $\chi(S_g)$ is equal to

$$h(S_g) = \lfloor \tfrac{1}{2}(7 + \sqrt{(1 + 48g)})\rfloor, \text{ for } g \geq 1.$$

This number $h(S_g)$ is now termed the *Heawood number*, but in fact Heawood showed only that $\chi(S_1) = h(S_1)$, and $\chi(S_g) \leq h(S_g)$, for $g \geq 2$.

In 1891, Heffter established equality for $g \leq 6$. His approach was to consider the *problem of neighbouring regions*, which asks for the minimum value of g which permits n regions to be drawn on a surface of genus g in such a way that each pair of regions is adjacent. Stated in dual form, the problem is to find the minimum genus g such that the complete graph K_n can be embedded on the surface S_g ; this number g is now called the *(orientable) genus* of K_n. It is easy to establish that

$$g(K_n) \geq \lceil \tfrac{1}{12}(n-3)(n-4) \rceil, \text{ for } n \geq 3,$$

and that if equality holds for all $n \geq 3$, then $\chi(S_g) = h(S_g)$ for all $g \geq 1$. Thus the emphasis shifted to trying to show that

$$g(K_n) = \lceil \tfrac{1}{12}(n-3)(n-4) \rceil,$$

and this came to be known as the *complete graph conjecture.*

In 1910, Tietze considered the map colouring problem for the non-orientable surface T_g of genus g. Again it is easy to establish that $\chi(T_g) \leq h\,(T_g)$, for $g \geq 1$, where in this case

$$h\,(T_g) = \lfloor \tfrac{1}{2}(7 + \sqrt{(1 + 24\,g)}) \rfloor.$$

Tietze was able to show that $\chi(T_1) = h\,(T_1)$, but only that $6 \leq \chi(T_2) \leq h\,(T_2) = 7$. It was some years before there was any further progress, but in 1934 Franklin proved that $\chi(T_2) = 6$, a result which turned out to be the solitary exception to the rule. It was later confirmed that $\chi(T_g) = h\,(T_g)$ for $g = 3$, 4 and 6 by Kagno in 1935, for $g = 7$ by Bose in 1939, and for $g = 5$ by Coxeter in 1943.

The non-orientable case may also be considered in terms of neighbouring regions, and dualized as above. Thus the non-orientable genus $\bar{g}$ is the minimum genus such that K_n can be embedded on the surface $T_{\bar{g}}$. Again, an inequality is easy to establish:

$$\bar{g}\,(K_n) \geq \lceil \tfrac{1}{6}(n-3)(n-4) \rceil, \text{ for } n \geq 3.$$

The *complete graph conjecture for non-orientable surfaces* is the assertion that

$$\bar{g}\,(K_n) = \lceil \tfrac{1}{6}(n-3)(n-4) \rceil,$$

noting that this result is not true in general, since Franklin's result implies that $\bar{g}\,(K_7) = 3$.

It follows from Euler's formula that small values of the genus $\bar{g}$ imply large numbers of faces, and this in turn implies that the average face-degree is small.

The best possible cases arise when all faces are triangles, and it turns out that there can exist a triangular embedding of K_n in a non-orientable surface if and only if $\frac{1}{6}(n-3)(n-4)$ is an integer — the so-called *regular cases,* which correspond to non-orientable surfaces for which $n \equiv 0$ or $1 \pmod 3$. The problem of determining the non-orientable genus was eventually solved for both regular and irregular cases by Ringel in 1954.

However, little progress was made on proving the complete graph conjecture for orientable surfaces until 1952, when Ringel settled it for $n = 13$. This was the prelude to what was to become a popular area of research in the 1950s and 1960s. The proof of the complete graph conjecture splits naturally into twelve separate cases, numbered 0, 1, 2, . . ., 11, where case k corresponds to $n \equiv k \pmod{12}$. As before, the best possible cases arise when all faces are triangles, and it turns out that there can be a triangular embedding of K_n in an orientable surface if and only if $\frac{1}{12}(n-3)(n-4)$ is an integer — the so-called *regular cases* 0, 3, 4 and 7. But the first case to be settled (by Ringel) was case 5, as a by-product of his solution for the non-orientable case mentioned above.

In 1961, Ringel settled the problem for cases 7, 10 and 3. Independently, Gustin worked on the problem and published a research announcement that he had settled cases 3, 4 and 7, but he failed to publish full details. He was unaware that Ringel had already done two of these cases, and unfortunately his treatment of case 4 contained a misprint. Nevertheless, Gustin had introduced a new technique — the *current graph* — which was to prove a powerful tool in the work. Case 0 was settled by Terry, Welch and Youngs in 1963, cases 1 and 9 by Gustin and Youngs during the period 1963-65, case 6 by Youngs in 1966, and cases 2, 8 and 11 by Ringel and Youngs late in 1967.

Although this dealt with the main cases, this was not the end of the story since the general proofs did not hold for certain small values of n, and so separate proofs had to be given in such cases. Odd values up to 23 were done by Mayer, a professor of French literature at Montpellier, and Ringel recalls that early in 1968 the only values still outstanding were 30, 35, 47 and 59 which were settled in February 1968 in the following order: 59 by Guy, 35 and 47 by Ringel and Youngs, and 30 by Mayer. (Youngs also did 30 independently in March 1968.) Thus, after nearly 80 years, Heawood's famous conjecture was indeed known to be a theorem. Since the non-orientable analogue had also been settled, the two may be combined into the following theorem.

Map Colour Theorem *Let S be the orientable surface S_g of genus g ($g \geq 1$) or the non-orientable surface T_g of genus g ($g \neq 2$); then the chromatic number $\chi(S)$ is given by*

$$\chi(S) = \lfloor \tfrac{1}{2}(7 + \sqrt{(49 - 24\eta)}) \rfloor,$$

where $\eta = 2 - 2g$, or $2 - g$, is the Euler characteristic of S. The exceptional non-orientable case is the Klein bottle T_2 for which $\chi = 6$, not 7.

The original proof of this theorem consisted of many journal articles totalling some 300 pages, but later improvements in the proofs have been made and certain of the exceptional values have been incorporated into revised general solutions. The orientable case $g = 0$ is the four colour theorem, which was settled by Appel, Haken and Koch in 1976.

References

Sources for papers published prior to 1936 may be found in:

1. N.L. Biggs, E.K. Lloyd and R.J. Wilson, *Graph Theory 1736-1936* (paperback edition), Clarendon Press, Oxford, 1986.

Gerhard Ringel's own extended account of the solution to the Heawood conjecture, and references to most of the results mentioned above, are given in:

2. G. Ringel, *Map Color Theorem*, Springer-Verlag, Berlin, 1974.

Other useful sources are:

3. J. Gross and A.W. Tucker, *Topological Graph Theory*, Wiley, New York, 1987.
4. I. Lakatos, *Proof and Refutations – the Logic of Mathematical Discovery*, Cambridge University Press, Cambridge, 1976.
5. A.T. White, Chapters 2 and 3 of *Selected Topics in Graph Theory* (ed. L.W. Beineke and R.J. Wilson), Academic Press, London, 1978.

Around Three Lemmas in Hamiltonian Graph Theory

D. Bauer, Hoboken, USA
H. J. Broersma, Enschede, The Netherlands
H. J. Veldman, Enschede, The Netherlands

Abstract. Three recent lemmas have led to results in hamiltonian graph theory which generalize and unify many previous results. Several applications of the lemmas, especially to 1-tough graphs, occur in a survey by Bauer, Schmeichel and Veldman. Recent additional applications are surveyed here. In particular, the lemmas are used to obtain an alternative proof of a recent result of Flandrin, Jung and Li.

1. INTRODUCTION

We use [8] for terminology and notation not defined here and consider simple graphs only.

Let G be a graph of order n. The number of vertices in a maximum independent set of G is denoted by $\alpha(G)$, the length of a longest cycle in G by c(G), and the set of vertices adjacent to a vertex v by N(v). For $k \le \alpha(G)$ we define

$$\sigma_k(G) = \min\{\sum_{v \in S} d(v) \mid S \text{ is an independent set of k vertices}\}$$

*) This research was supported in part by the National Security Agency under grant number MDA 904-89-H-2008.

R. Bodendiek, R. Henn (Eds.)
Topics in Combinatorics and Graph Theory

and

$$NC_k(G) = \min\{|\bigcup_{v\in S} N(v)| \ |S \text{ is an independent set of k vertices}\}.$$

For $k > \alpha(G)$ we set $\sigma_k(G) = k(n-\alpha(G))$ and $NC_k(G) = n-\alpha(G)$. Instead of $\sigma_1(G)$ and $NC_1(G)$ we use the more common notation $\delta(G)$. The graph G is *1-tough* if $\omega(G-S) \le |S|$ for every subset S of V(G) with $\omega(G-S) > 1$, where $\omega(G-S)$ stands for the number of components of G–S.

Let C be a cycle of a graph G. If V(G)–V(C) is an independent set, then C is called a *dominating cycle* of G. By $\vec{C}$ we denote the cycle C with a given orientation. If $u,v \in V(C)$, then $u\vec{C}v$ denotes the consecutive vertices on C from u to v in the direction specified by $\vec{C}$. The same vertices, in reverse order, are given by $v\overleftarrow{C}u$. We will consider $u\vec{C}v$ and $v\overleftarrow{C}u$ both as paths and as vertex sets. We use u^+ to denote the successor of u on $\vec{C}$ and u^- to denote its predecessor. If $A \subseteq V(C)$, then $A^+ = \{v^+|v \in A\}$ and $A^- = \{v^-|v \in A\}$.

The three lemmas referred to in the title can be found in [2].

Lemma 1 [2]. Let G be a graph of order n with $\delta(G) \ge 2$ and $\sigma_3(G) \ge n$. Assume G contains a longest cycle $\vec{C}$ which is a dominating cycle. If $v \in V(G)-V(C)$, then $(V(G)-V(C)) \cup (N(v))^+$ is an independent set of vertices.

Lemma 2 [2]. Let G be a 2-connected graph of order n with $\sigma_3(G) \ge n+2$. Then every longest cycle of G is a dominating cycle. Moreover, if G is non-hamiltonian, G contains a longest cycle C such that $\max\{d(v)|v \in V(G)-V(C)\} \ge \frac{1}{3}\sigma_3(G)$.

Lemma 3 [2]. Let G be a 1-tough graph of order $n \ge 3$ with $\sigma_3(G) \ge n$. Then every longest cycle of G is a dominating cycle. Moreover, if G is non-hamiltonian, G contains a longest cycle C such that $\max\{d(v)|v \in V(G)-V(C)\} \ge \frac{1}{3}\sigma_3(G)$.

The first part of Lemma 2 is a result of Bondy [7] and the second part is implicit in the proof of [2, Theorem 10]. The first part of Lemma 3 is [2, Theorem 5] and the second part is implicit in the proof of [2, Theorem 9].

To see that (the first part of) Lemma 2 is best possible, define the graph G_n $(n \ge 8)$ as the join of K_2 and the graph of order n–2 consisting of three disjoint complete graphs, the orders of which pairwise differ by at most one. Then G_n is 2-connected and $\sigma_3(G_n) = n+1$, while G_n contains no dominating cycle. Lemma 3 is also best possible. Consider the graph H_n ($n \geq 10$), where H_n is obtained from the join of K_1 and the graph of order $n-1$ consisting of three disjoint complete graphs, the orders of which pairwise differ by at most one, by choosing a vertex in each of the three complete graphs and adding the edges of a triangle between the three vertices.

It should be noted that Lemmas 1 and 2 admit relatively short proofs, whereas the proof of Lemma 3 in [2] is much more involved.

Lemmas 1, 2 and 3 have recently been used to generalize and unify a number of earlier results. Several examples of this appear in [4], where the focus is on 1-tough graphs. In Sections 2 and 3 we update [4] by surveying recent additional applications of the lemmas. In particular, we show in Section 3 how Lemmas 1 and 3 can be combined to provide a new proof of a recent result of Flandrin, Jung and Li [13]. In Section 4 we conclude by indicating how the lemmas are currently being used to obtain new results.

2. THE FIRST APPLICATIONS OF THE LEMMAS

The main results in [2], obtained by combining Lemma 1 with Lemma 2 and Lemma 3, respectively, are the following.

Theorem 4 [2]. If G is a 2-connected graph of order n with $\sigma_3(G) \geq n+2$, then $c(G) \geq \min\{n, n+\frac{1}{3}\sigma_3(G)-\alpha(G)\}$.

Theorem 5 [2]. If G is a 1-tough graph of order $n \geq 3$ with $\sigma_3(G) \geq n$, then $c(G) \geq \min\{n, n+\frac{1}{3}\sigma_3(G)-\alpha(G)\}$.

We first focus on Theorem 4. As noted in [2], Theorem 4 is best possible in two senses. For $3 \leq p \leq q \leq 2p-2$ we have $\sigma_3(K_{p,q}) = 3p \geq n+2$, $\alpha(K_{p,q}) = q$ and $c(K_{p,q}) = 2p = |V(K_{p,q})| + \frac{1}{3}\sigma_3(K_{p,q}) - \alpha(K_{p,q})$, so the inequality in the conclusion of Theorem 4 is sharp. Furthermore, the bound on $\sigma_3(G)$ imposed in Theorem 4 cannot be lowered: the graph G_n, defined in Section 1, satisfies $\sigma_3(G_n) = n+1$ and $c(G_n) = \lceil\frac{2}{3}(n+1)\rceil$, whereas $\min\{n, n+\frac{1}{3}\sigma_3(G_n)-\alpha(G_n)\} = \min\{n, n+\frac{1}{3}(n+1)-3\} = n$.

Theorem 4 is a common generalization of several earlier results. Since $\sigma_3(G) \geq 3\delta(G)$ for any graph G, Theorem 4 directly implies the following theorem of

Nash-Williams, as observed in [2]. In fact, Theorem 4 is a kind of "long cycle version" of this result.

Corollary 6 [17]. If G is a 2-connected graph of order n with $\delta(G) \geq \max\{\frac{1}{3}(n+2),\alpha\}$, then G is hamiltonian.

Less directly, Theorem 4 contains a number of other previous results. The following three lemmas will be of use in showing this.

Lemma 7 [1]. $\sigma_3(G) \geq \frac{3}{2}\sigma_2(G)$ for any graph G.

Lemma 8 [1]. $\alpha(G) \leq n-NC_k(G)$ for any graph G of order n and any positive integer k.

Lemma 9 [1]. $\sigma_3(G) \geq 3NC_2(G)-n+3$ for any graph G of order $n \geq 3$.

The proofs of Lemmas 7 and 8 are short and easy, the proof of Lemma 9 is slightly more involved.

Using Lemmas 7 and 8 and the obvious fact that $NC_2(G) \geq \frac{1}{2}\sigma_2(G)$ for any graph G, it was shown in [1] that Theorem 4 contains the following result of Ore.

Corollary 10 [18]. If G is a graph of order $n \geq 3$ with $\sigma_2(G) \geq n$, then G is hamiltonian.

Via Lemmas 8 and 9 it was observed in [1] that Theorem 4 also contains the following results in terms of $NC_2(G)$.

Corollary 11 [12]. If G is a 2-connected graph of order n with $NC_2(G) \geq \frac{1}{3}(2n-1)$, then G is hamiltonian.

Corollary 12 [11]. If G is a 2-connected graph of order n with $NC_2(G) \geq n-\delta(G)$, then G is hamiltonian.

If G is a graph of order n, then $\alpha(G) \leq n - \frac{1}{3}\sigma_3(G)$. Hence, as observed in [2], Theorem 4 also contains the following result of Bondy.

Corollary 13 [7]. If G is a 2-connected graph of order n with $\sigma_3(G) \geq n+2$, then $c(G) \geq \min\{n,\frac{2}{3}\sigma_3(G)\}$.

By Lemma 7, $\frac{2}{3}\sigma_3(G) \geq \sigma_2(G)$ for any graph G, so Corollary 13 generalizes Corollary 10 and is a partial generalization of the following long cycle version of Corollary 10, independently established by Bermond and Linial.

Theorem 14 [5,16]. If G is a 2-connected graph of order n, then $c(G) \geq \min\{n, \sigma_2(G)\}$.

Fournier and Fraisse [14] have shown that in Corollary 13 the requirement $\sigma_3(G) \geq n+2$ can be dropped, thus obtaining a full generalization of Theorem 14. By contrast, the requirement $\sigma_3(G) \geq n+2$ is crucial in Theorem 4.

Results related to Theorem 4 occur in [15]. We mention the one which is most closely related.

Theorem 15 [15]. If G is a 2-connected nonhamiltonian graph of order n and every longest cycle of G is a dominating cycle, then

(i) $c(G) \geq 4\delta(G)-3$ or

(ii) there is an independent set S such that $|S| = n-c(G)+\delta(G)$ and $c(G) \geq d(u)+d(v)$ for any two distinct vertices u and v of S.

Corollary 16. If G is a 2-connected graph of order n with $\delta(G) \geq \frac{1}{4}(n+3)$ and every longest cycle of G is a dominating cycle, then $c(G) \geq \min\{n, n+\delta(G)-\alpha(G)\}$.

We now turn our attention to Theorem 5. As Theorem 4, Theorem 5 is best possible in the sense that the bound on $\sigma_3(G)$ cannot be lowered. This is illustrated by the graph H_n defined in Section 1. Furthermore, the quantity $n+\frac{1}{3}\sigma_3(G)-\alpha(G)$ in Theorem 5 cannot be increased by more than one. To see this, construct, for odd $n \geq 15$, the graph I_n from $\overline{K_{\frac{1}{2}(n-1)}} \cup K_{\frac{1}{2}(n-5)} \cup K_3$ by joining every vertex in $K_{\frac{1}{2}(n-5)}$ to all other vertices and by adding a matching between the vertices in K_3 and three vertices in $\overline{K_{\frac{1}{2}(n-1)}}$. Then I_n is 1-tough, $\sigma_3(I_n) = \frac{3}{2}(n-5) \geq n$, $\alpha(I_n) = \frac{1}{2}(n-1)$ and $c(I_n) = n-1 = n+\frac{1}{3}\sigma_3(I_n)-\alpha(I_n)+1$. As shown in [3], the quantity $n+\frac{1}{3}\sigma_3(G)-\alpha(G)$ in Theorem 5 can indeed be increased by one if $\sigma_3(G) = 3\delta(G)$.

Theorem 17 [3]. If G is a 1-tough graph of order $n \geq 3$ with $\delta(G) \geq \frac{1}{3}n$, then $c(G) \geq \min\{n, n+\delta(G)-\alpha(G)+1\}$.

Theorem 17 is a kind of long cycle version of a result of Bigalke and Jung.

Corollary 18 [6]. If G is a 1-tough graph of order $n \geq 3$ with $\delta(G) \geq \max\{\frac{1}{3}n, \alpha(G)-1\}$, then G is hamiltonian.

In [1], Theorem 5 is shown to imply a best possible improvement of Corollary 11.

Corollary 19 [1]. If G is a 2-connected graph of order n with $NC_2(G) \geq \frac{1}{3}(2n-3)$, then G is hamiltonian.

The proof of Corollary 19 in [1] consists of first showing that any graph satisfying its hypothesis must be 1-tough. After noting that $NC_2(G) \geq \frac{1}{3}(2n-3)$ implies $\sigma_3(G) \geq n$ (Lemma 9), the proof proceeds by applying Theorem 5. To see that Corollary 19 is best possible, observe that the graph G_n is nonhamiltonian and $NC_2(G_n) = \lceil\frac{1}{3}(2n-3)\rceil-1$.

For other implications of Theorem 5 we refer to [4].

3. ON A RESULT OF FLANDRIN, JUNG AND LI

For a graph G of order n we define

$$\bar{\sigma}_3(G) = \begin{cases} \min\{d(u)+d(v)+d(w)-|N(u)\cap N(v)\cap N(w)| \mid \{u,v,w\} \text{ is an} \\ \text{independent set of three vertices}\} \text{ if } \alpha(G) \geq 3, \\ 2(n-\alpha(G)) \text{ otherwise.} \end{cases}$$

Flandrin, Jung and Li recently proved the following result.

Theorem 20 [13]. If G is a 2-connected graph of order n with $\bar{\sigma}_3(G) \geq n$, then G is hamiltonian.

The graphs $K_m \vee \overline{K}_{m+1}$ with $m \geq 2$ show that the bound imposed on $\bar{\sigma}_3(G)$ in Theorem 20 cannot be lowered [13]. As shown in [13], Theorem 20 contains Corollaries 10, 12 and 19 (and hence Corollary 11). Theorem 20 does not imply Corollary 6. To see this, consider, for $n \equiv 1 \pmod 3$ and $n \geq 10$, the graph $J_n = K_{\frac{1}{3}(n+2)} \vee (K_{\frac{1}{3}(2n-11)} \cup \overline{K}_3)$. The graph J_n is 2-connected (even 1-tough), $\delta(J_n) = \frac{1}{3}(n+2)$, $\bar{\sigma}_3(J_n) = \frac{2}{3}(n+2)$ and $\alpha(J_n) = 4$, so that $n+\delta(J_n)-\alpha(J_n) \geq n$.

Hence J_n is hamiltonian by Corollary 6, but not by Theorem 20. A fortiori there exist graphs which are hamiltonian by Theorem 4 (and Theorem 5), but not by Theorem 20. On the other hand there exist graphs which are hamiltonian by Theorem 20, but not by either Theorem 4 or 5. For $n \equiv 0 \pmod 6$ and $n > 6$, construct the graph L_n from $K_{\frac{1}{2}n} \vee \overline{K_{\frac{1}{2}n-3}}$ by adding three new vertices and joining each of them to exactly $\frac{1}{3}n$ vertices of $K_{\frac{1}{2}n}$ in such a way that every vertex of $K_{\frac{1}{2}n}$ becomes adjacent to exactly two of them. Then L_n is 2-connected, $\sigma_3(L_n) = \bar{\sigma}_3(L_n) = n$ and $\alpha(L_n) = \frac{1}{2}n$, so that $n+\frac{1}{3}\sigma_3(L_n)-\alpha(L_n) = \frac{5}{6}n$. Hence L_n is hamiltonian by Theorem 20, but not by either Theorem 4 or 5.

Here we use Lemmas 1 and 3 to give an alternative proof of Theorem 20.

Proof of Theorem 20. Let G satisfy the hypothesis of Theorem 20.

We first show that G is 1-tough. Assuming the contrary, let S be a subset of V(G) such that $\omega(G-S) > |S|$. Since G is 2-connected, $|S| \geq 2$. Let u, v and w be three vertices in distinct components of G–S. Every component of G–S contains a vertex adjacent to none of the vertices u, v, w. Furthermore, all vertices adjacent to more than one of the vertices u, v, w are in S. Hence

$$\begin{aligned} d(u)+d(v)+d(w) \leq\ & n-\omega(G-S)+2|N(u)\cap N(v)\cap N(w)| \\ & +(|S|-|N(u)\cap N(v)\cap N(w)|), \end{aligned}$$

implying that

$$d(u)+d(v)+d(w)-|N(u)\cap N(v)\cap N(w)| \leq n-\omega(G-S)+|S| < n.$$

This contradiction shows that, indeed, G is 1-tough.

Assuming G is nonhamiltonian, let $\vec{C}$ be a longest cycle of G and u a vertex of V(G)–V(C) having maximum degree among all vertices of V(G)–V(C). Since G is 1-tough and $\sigma_3(G) \geq \bar{\sigma}_3(G) \geq n$, Lemma 3 implies that C is a dominating cycle and $d(u) \geq \frac{1}{3}n$. Set $A = N(u)$. By Lemma 1, $(V(G)-V(C)) \cup A^+$ and, similarly, $(V(G)-V(C)) \cup A^-$ are independent sets. Since $|A| > \frac{1}{3}|V(C)|$, there exist two vertices v_1 and v_2 of A such that $v_1^+ = v_2^-$. Set $v = v_1^+$, $w = v_2^+$ and, for $z \in V(C)$, $f(z) = |N(z) \cap \{u,v,w\}|$.

Let $x\vec{C}y$ be a subpath of C such that $x \in (A^+)^+$, $y \in A^-$ and $x\vec{C}y \cap A = \emptyset$. Since A^- is an independent set, $vy \notin E(G)$ and hence

$$f(y) \leq 1. \tag{1}$$

We now show that

$$\text{if } z \in x\vec{C}y \text{ and } f(z) = 2, \text{ then } f(z^+) = 0. \tag{2}$$

Suppose $f(z) = 2$ and $f(z^+) \geq 1$. Then z is adjacent to both v and w and, by (1), $z \neq y$. Hence z^+ is adjacent to v or w. If $vz^+ \in E(G)$, then the cycle $v_1uv_2\vec{C}zvz^+\vec{C}v_1$ is longer than C, a contradiction. If $wz^+ \in E(G)$, then the cycle $v_1uv_2vz\overleftarrow{C}wz^+\vec{C}v_1$ contradicts the choice of C. Hence (2) holds.

From (1) and (2) we conclude that

$$\sum_{z \in x\vec{C}y} f(z) \leq |x\vec{C}y|. \tag{3}$$

Now by (3) and the fact that $(V(G)-V(C)) \cup A^+$ is an independent set,

$$d(u)+d(v)+d(w) = \sum_{z \in V(C)} f(z) = \sum_{z \in A} f(z) + \sum_{z \in A^+} f(z) + \sum_{z \notin A \cup A^+} f(z)$$
$$\leq (3|N(u)\cap N(v)\cap N(w)| + 2(|A| - |N(u)\cap N(v)\cap N(w)|)) + 0 + (|V(C)| - 2|A|),$$

implying that

$$n-1 \geq |V(C)| \geq d(u)+d(v)+d(w) - |N(u)\cap N(v)\cap N(w)| \geq \bar{\sigma}_3(G) \geq n.$$

This contradiction completes the proof. □

4. NEW RESULTS

Very recently, Lemmas 1, 2 and 3 were used to prove the following results.

Theorem 21 [9]. If G is a 2-connected graph of order n with $\sigma_3(G) \geq n+2$, then $c(G) \geq \min\{n, n+NC_3(G)-\alpha(G)\}$.

Theorem 22 [9]. If G is a 1-tough graph of order $n \geq 3$ with $\sigma_3(G) \geq n$, then $c(G) \geq \min\{n, n+NC_3(G)-\alpha(G)\}$.

Let G be a graph of order n. Since $NC_3(G) \geq \frac{1}{3}\sigma_3(G)$, Theorems 21 and 22 generalize Theorems 4 and 5, respectively. In the proof of Theorem 20 given in Section 3 it was shown that G is 1-tough if G satisfies the hypothesis of Theorem 20. By Lemma 8, $n+NC_3(G)-\alpha(G) \geq 2NC_3(G)$. Furthermore, it is easily shown that $2NC_3(G) \geq \bar{\sigma}_3(G)$ [9]. Hence, in fact, Theorem 22 is a common generalization of Theorems 5 and 20. Since $n+NC_3(G)-\alpha(G) \geq 2NC_3(G) \geq 2NC_2(G)$, Theorems 21 and 22, respectively, also generalize the following two results. Stronger versions of these results appear in [1].

Corollary 23 [1]. If G is a 2-connected graph of order n with $\sigma_3(G) \geq n+2$, then $c(G) \geq \min\{n, 2NC_2(G)\}$.

Corollary 24 [1]. If G is a 1-tough graph of order $n \geq 3$ with $\sigma_3(G) \geq n$, then $c(G) \geq \min\{n, 2NC_2(G)\}$.

As in Theorems 4 and 5, the bounds imposed on $\sigma_3(G)$ in Corollaries 23 and 24 cannot be lowered. The observation that the hypotheses of Corollaries 23 and 24 constitute sharp sufficient conditions for the existence of a dominating cycle led to the question whether the existence of a dominating cycle alone is sufficient to draw the conclusion of Corollaries 23 and 24. This question is answered by the following result.

Theorem 25 [10]. If G is a 2-connected graph and G contains a dominating cycle, then $c(G) \geq \min\{n, 2NC_2(G)\}$ unless G is the Petersen graph.

Corollaries 23 and 24 follow from Theorem 25 in combination with Lemma 2 and 3, respectively. By combining Theorem 25 with other sufficient conditions for the existence of a dominating cycle, improvements of some known results not mentioned here can be obtained. For details we refer to [10].

We close with the remark that, in fact, stronger results than Theorems 21 and 22 will appear in [9]. The subscript 3 of NC in Theorems 21 and 22 can be replaced by a higher number. In [9] it will also be shown that the subscript 2 of NC in Corollaries 23 and 24 can be replaced by a linear function of n. We give no details, since we are not yet sure of the final shape the results in [9] will take.

REFERENCES

[1] D. Bauer, G. Fan and H.J. Veldman, Hamiltonian properties of graphs with large neighborhood unions, preprint (1989).

[2] D. Bauer, A. Morgana, E.F. Schmeichel and H.J. Veldman, Long cycles in graphs with large degree sums, to appear in Discrete Math.

[3] D. Bauer, E.F. Schmeichel and H.J. Veldman, A generalization of a theorem of Bigalke and Jung, Ars Combinatoria 26 (1988) 53-58.

[4] D. Bauer, E.F. Schmeichel and H.J. Veldman, Some recent results on long cycles in tough graphs, to appear in Proc. 6th Int. Conf. on the Theory and Applications of Graphs, Kalamazoo, 1988.

[5] J.-C. Bermond, On Hamiltonian walks, in Proc. 5th British Combinatorial Conf., Aberdeen, 1975, Utilitas Math. (1976) 41-51.

[6] A. Bigalke and H.A. Jung, Über Hamiltonsche Kreise und unabhängige Ecken in Graphen, Monatsh. Math. 88 (1979) 195-210.

[7] J.A. Bondy, Longest paths and cycles in graphs of high degree, Research Report CORR80-16, Univ. of Waterloo, Waterloo, Ontario (1980).

[8] J.A. Bondy and U.S.R. Murty, Graph Theory with Applications (Macmillan, London and Elsevier, New York, 1976).

[9] H.J. Broersma, J. van den Heuvel and H.J. Veldman, Long cycles, degree-sums and neighborhood unions, in preparation.

[10] H.J. Broersma and H.J. Veldman, Long dominating cycles and paths in graphs with large neighborhood unions, preprint (1989).

[11] R.J. Faudree, R.J. Gould, M.S. Jacobson and L. Lesniak, Neighborhood unions and highly hamiltonian graphs, preprint (1988).

[12] R.J. Faudree, R.J. Gould, M.S. Jacobson and R.H. Schelp, Neighborhood unions and hamiltonian properties in graphs, to appear in J. Combin. Theory (B).

[13] E. Flandrin, H.A. Jung and H. Li, Hamiltonism, degree sum and neighborhood intersections, to appear in Discrete Math.

[14] I. Fournier and P. Fraisse, On a conjecture of Bondy, J. Combin. Theory (B) 39 (1985) 17-26.

[15] P. Fraisse and H.A. Jung, Longest cycles and independent sets in k-connected graphs, in: V.R. Kulli, ed., Recent Studies in Graph Theory (Vishwa International Publications, Gulbarga, India, 1989) 114-139.

[16] N. Linial, A lower bound on the circumference of a graph, Discrete Math. 15 (1976) 297-300.

[17] C.St.J.A. Nash-Williams, Edge-disjoint hamiltonian circuits in graphs with vertices of large valency, in: L. Mirsky, ed., Studies in Pure Mathematics (Academic Press, London, 1971) 157-183.

[18] O. Ore, Note on hamiltonian circuits, Amer. Math. Monthly 67 (1960) 55.

A Note on Metric Properties of Infinite Graphs

T. Böhme, Ilmenau, GDR

1. Introduction

1.1. Definitions

Throughout this note by a graph G we mean an undirected connected graph without loops or multiple edges, defined on a countably infinite vertex set $V(G)$. The edge set of G is denoted by $E(G)$ and the valency of a vertex x of G by $v(x:G)$. A graph G is said to have universally bounded valencies if there is positive number L such that $v(x:G) = L$ for all $x \in V(G)$.

As usual we denote by R the set of all real numbers and by Z the set of all integers. R^2 is the Cartesian plane, with the usual linear structure and the euclidean distance function d. $Z^2 = Z \times Z$ is considered to be a subset of R^2. With respect to the vectoraddition Z^2 forms an abelian group $(Z^2,+)$ isomorphic to the direct sum of two infinite cyclic groups.

A metric space is a pair (M,d_M), where M is a nonempty set and d_M is a function from M M into R, subject to the following conditions:

(M1) $d_M(x,y) \geq 0$ always.

(M2) $d_M(x,y) = 0$ iff $x=y$.

(M3) $d_M(x,y) = d_M(y,x)$ always.

(M4) $d_M(x,y) + d_M(y,z) \geq d_M(x,z)$ always.

Under this conditions, d_M is called a distance function for M. By abuse of language, we may refer to the set M as a metric space. A nonempty subset M′ of a metric space M is always considered a metric space endowed with the induced distance function $d_{M'}$, i.e. $d_{M'}(x,y) = d_M(x,y)$ for all $x,y \in M'$.

R. Bodendiek, R. Henn (Eds.)
Topics in Combinatorics and Graph Theory

Let (M,d_M) and $(M',d_{M'})$ be metric spaces, then an injection β from M into M′ is said to be isometric iff

$$d_M(x,y) = d_{M'}(\beta(x),\beta(y)) \text{ for all } x,y \in M.$$

Let G be a graph, then there is a natural way to define a distance function d_G for V(G): For $x,y \in V(G)$, $d_G(x,y)$ is the least number of edges a path joining x and y can have. An isometric bijection from V(G) onto itself is said to be an automorphism of G. The set of all automorphisms of G forms a group. We denote it by aut G.

Let (M,d_M) be a metric space and β be an injection from m into R^2. Define

$$\text{dis } \beta = \sup \{|d_M(x,y) - d(\beta(x),\beta(y))| \,;\, x,y \in M\} .$$

Obviously, β is isometric iff dis β = o. Furthermore define dis M = inf dis β where the infimum is taken over all injections β from M into R^2.

Let (M,d_M) be a metric space and q be a positive real number. Then we denote by qM the metric space (M,qd_M).

1.2. The problem

During a stay in Budapest in the beginning of 1988 the author heard about the following

Conjecture: For no graph G there is a bijection β from V(G) onto Z^2 such that dis $\beta < \infty$.

Regrettably, the author knows almost nothing about the background of this conjecture. The only facts he knows are that the conjecture arose from investigations in the field of VLSI - design and that it was communicated by Paul Erdös to some graph theorists in Budapest. Note that the conjecture becomes false if the condition for β to be bijective is dropped. A simple example for that is an infinite path mapped onto the integral points of the x-axis such that consecutive vertices are mapped onto consecutive integral points.

Let us assume for a moment the conjecture to be false. Then there are a graph G, a positive number K and a bijection β from V(G) onto Z^2 such that $|d_G(x,y) - d(\beta(x),\beta(y))| \leq K$ for all $x,y \in V(G)$.

It is not hard to see that the images of all neighbours of a vertex $x \in V(G)$ are contained in a circular disk of radius K+1 centered at $\beta(x)$. Consequently, there is a positive number L such that $v(x:G) \leq L$ for all $x \in V(G)$. Therefore we shall restrict ourselves to the consideration of graphs with universally bounded valencies.

Now we denote by A the group of all isometric bijections of Z^2 onto itself. If $a \in A$ then $b = \beta^{-1} \circ a \circ \beta$ is a bijection of V(G) onto itself. A simple computation yields

$$|d_G(x,y) - d_G(b(x),b(y))| \leq 2K$$

for all $x,y \in V(G)$.

Therefore, from the authors point of view, it seems interesting to consider the set of all graphs G with $V(G) = Z^2$ and the property that A is a subgroup of aut G. The main result of the present note is the following

Theorem. Let G be a graph with vertex set Z^2 and the property that A is a subgroup of aut G. Then dis $G = \infty$.

2. Proof

We need some auxiliary considerations. The simple proofs of the following two propositions are left to the reader.

Proposition 1. Let M be a finite metric space. Then dis M = o iff there is an isometric injection from M into R^2. □

Proposition 2. Let M be a metric space. Then

(a) for every nonempty subset M′ of M, dis M $\geq$ dis M′;

(b) for every positive q, dis qM = qdis M. □

The main idea of the proof of the theorem is the following. In order to prove dis $G = \infty$ for a graph G, construct an infinite sequence $(V_i)_{i=1,2,...}$ of subsets of V(G) and an infinite sequence $(q_i)_{i=1,2,...}$ of positive integers such that

(S1) dis $V_1 \neq o$,

(S2) $V_i = q_i V_1$ for all positive integers i, and,

(S3) $\lim_{i \to \infty} q_i = \infty$.

Then, by Proposition 2, $\lim_{i\to\infty} \text{dis } V_i = \infty$ and, therefore, dis $G = \infty$.

<u>Proposition 3.</u> Let (M,d_M) be a metric space and h, k positive integers such that
$M = \{x_1,x_2,x_3,x_4\}$ and
$d_M(x_1,x_2) = d_M(x_2,x_3) = d_M(x_3,x_4) = d_M(x_4,x_1) = h$,
$d_M(x_1,x_3) = d_M(x_2,x_4) = k$.
Then dis $M \neq o$.

<u>Proof.</u> Suppose there were an isometric bijection ß from M into R^2. Then $ß(x_1),\ldots,ß(x_4)$ would be the vertices of a square with side length h. Thus $k = \sqrt{2}\,h$. This contradicts the condition for h, k to be integers. □

We need the concept of a Cayley graph. Let B be a group and F be a generating set of B, subject to the following conditions:
(C1) $1 \in F$.
(C2) $x \in F$ implies $x^{-1} \in F$.
Here 1 denotes the identity element of B and x^{-1} is the inverse of x.
The Cayley graph H for B with respect to F is defined by $V(H) = B$ and $(x,y) \in E(H)$ iff $x^{-1}y \in F$. Obviously, $v(x:H) = |F|$ for all $x \in V(H)$.
It is not too hard to see that for every $b \in B$ the function f_b, defined by $f_b(x) = bx$ for $x \in V(H)$, is an automorphism of H. Furthermore, the automorphism F_b $(b \in B)$ form a subgroup B' of aut H isomorphic to B that acts regularly on V(H), i.e. for every ordered pair (x,y) there is an uniquely determined $f \in B'$ such that $f(x) = y$. Conversely, if H is a graph and B' is a subgroup of aut H that acts regularly on V(H) then H is isomorphic to a Cayley graph for B' with respect to some generating set (see [1]).
Let G be a graph with universally bounded valencies such that $V(G) = Z^2$ and A is a subgroup of aut G. We notice that A is generated by $\{a_1,a_2,a_3,a_4\}$ where $a_1((x_1,x_2)) = (x_1+1,x_2)$, $a_2((x_1,x_2)) = (x_2,x_1)$, $a_3((x_1,x_2)) = (-x_1,x_2)$, and

$a_4((x_1,x_2)) = (-x_1+1,x_2)$ for all $\underline{x} = (x_1,x_2) \in Z^2$.
Obviously, the subgroup of A spanned by a_{1_2} and $a_2 \circ a_1 \circ a_2$ is isomorphic to $(Z^2,+)$ and acts regularly on $Z^2 = V(G)$. Therefore we consider G a Cayley graph for $(Z^2,+)$ with respect to a finite generating set F of $(Z^2,+)$ having the additional property that $a_i \in \text{aut } G$ for $i = 2,3,4$.
For the sake of brevity we omit the proof of the following proposition.

<u>Proposition 4.</u> Let G be a Cayley graph for $(Z^2,+)$ with respect to a finite generating set F and $\underline{x} = (x_1,x_2)$ an element of Z^2. Then there is a positive integer $m = m(\underline{x})$ such that $d_G(\underline{o},nm\underline{x}) = nd_G(\underline{o},m\underline{x})$ for all positive integers n. (Here $\underline{o} = (o,o)$ and $n\underline{x} = (nx_1,nx_2)$.)

Let $p = m((1,o))$ and $p' = m((o,1))$. The fact that $a_2((1,o)) = (o,1)$ where $a_2 \in \text{aut } G$ implies $p = p'$.
Put

$$V_j = \{\underline{o},\ jp(1,o), jp(o,1),\ jp(1,1)\}, \quad j = 1,2,\ldots .$$

Note that $V_j \subset V(G)$. G being a Cayley graph, $d_G(\underline{o},p(1,o)) = d_G(p(o,1),p(1,1))$ and $d_G(\underline{o},p(0,1)) = d_G(p(1,o),p(1,1))$. Applying the automorphisms a_2 and a_3 or a_4, $d_G(\underline{o},p(1,o)) = d_G(\underline{o},p(0,1))$ and $d_G(p(1,o),p(o,1)) = d_G(\underline{o},p(1,1))$. Hence, by Proposition 3, dis $V_1 \neq o$. Furthermore, Proposition 4 yields $V_j = jV_1$ for $j = 1,2,\ldots$.
Therefore, $\lim_{j\to\infty} \text{dis } V_j = \infty$ and, consequently, dis $G = \infty$.

References

[1] Biggs, N.L.: Algebraic Graph Theory. Cambridge University Press, Cambridge 1974.

Automorphism Groups of Directed Cayley Graphs

U. Baumann, Dresden, GDR
M. Lesch, Dresden, GDR
I. Schmeichel, Dresden, GDR

1. Introduction

A famous result of R. Frucht (1938) initiated a widespread interest for investigating automorphism groups of graphs. Cayley graphs are essential to the construction of graphs whose automorphism groups are isomorphic with a given abstract group. Many authors realized the proposal of L. Babai /2/ to investigate properties of Cayley graphs.
In this field it is obvious to concern with automorphism groups of these graphs, N. Biggs, W. Imrich, M. Watkins et al. obtained interesting results for.
This paper deals with problems of algebraic graph theory. The graphs considered are isomorphic with directed Cayley graphs. Regular subgroups of the automorphism group of such graphs determine partitions of their arc sets. There are interesting relations between these partitions and properties of subgroups of the automorphism group. Especially graph automorphisms, permutations of the arc set and relations to group automorphisms are considered.

2. Basic properties of Cayley graphs

Let H be a finite group and ω a generating set of H not

R. Bodendiek, R. Henn (Eds.)
Topics in Combinatorics and Graph Theory

containing the identity of this group. H and ω determine a directed Cayley graph $G(H,\omega)$ in the following way: H is the vertex set of $G(H,\omega)$ and $\{(h_i,h_j): h_i,h_j \in H,\ h_i^{-1}h_j \in \omega\}$ is the arc set of this digraph. Directed Cayley graphs are finite simple regular connected digraphs.

An automorphism a of a digraph is a permutation of its vertex set which maps each arc (u,v) onto an arc (a(u),a(v)) of this digraph. A(G) denotes the group of all automorphisms of the digraph G.

In addition, a(u,v) is defined by a(u,v) = (a(u),a(v)) for every permutation a of the vertex set and all pairs (u,v) of vertices.

A permutation group S on X is regular iff for any two elements u,v of X there is exactly one element of S mapping u onto v. The number n of the elements of X is the order of S.

It is well known that a finite simple connected digraph is isomorphic with a (directed) Cayley graph iff its automorphism group contains a regular subgroup /3/.

If S is a regular subgroup of the automorphism group of such a graph G with the vertex set X, x_1 a fixed element of X and $\{(x_1,y_i): i=1,\dots,r\}$ the set of all arcs with start vertex x_1, then the elements s_i of S determined by $s_i(x_1)=y_i$ constitute a generating set ω_{x_1} of S, G is isomorphic with the directed Cayley graph $G(S,\omega_{x_1})$, and $\bigcup_{i=1}^{r}\{s(x_1,s_i(x_1)): s \in S\}$ is the arc set of G.

The notations and notions mentioned above shall be used.

This paper deals with automorphism groups of digraphs, which are isomorphic with directed Cayley graphs. All graphs considered can be described by $G=(X, \bigcup_{i=1}^{r}\{s(x_1,s_i(x_1)): s \in S\})$, where S is a regular subgroup of A(G). In all following considerations X is the vertex set and $\bigcup_{i=1}^{r}\{s(x_1,s_i(x_1)): s \in S\}$ is the arc set of G.

Each subset $\{s(x_1,s_i(x_1)): s \in S\}$ of the arc set of G contains exactly $|S|$ elements and determines a set of cycles $((s(x_1),s_is(x_1)),(s_is(x_1),s_i^2s(x_1)),\dots,(s_i^{m-1}s(x_1),s_i^ms(x_1)))$ $(s \in S)$ of equal length in the arc set of G, where m denotes the order of s_i. Any two of these cycles are either equal or vertex disjoint. If $\{s(x_1,s_i(x_1)): s\in S\}\cap\{s(x_1,s_j(x_1)): s\in S\}$ contains an element $(s(x_1),s_is(x_1)) = (s'(x_1),s_js'(x_1))$, then $s=s'$ and $s_is=s_js'$ by regularity of S. Thus distinct subsets $\{s(x_1,s_i(x_1)): s\in S\}$, $\{s(x_1,s_j(x_1)): s\in S\}$ are arc disjoint.

It is useful for further investigations to describe the arc set of G using the elements of the centralizer Z of S in the symmetric group on X. This centralizer consists of those permutations z on X satisfying $zs=sz$ for all $s\in S$.

<u>Lemma 1.</u> Let $s_1,\dots,s_n$ denote the elements of a regular permutation group on X.
The permutations $\begin{pmatrix} s_1(u)\dots\ s_n(u) \\ ss_1(v)\dots ss_n(v)\end{pmatrix}$ $(s\in S)$ constitute the centralizer Z of S in the symmetric group on X for arbitrary fixed elements u,v of X. Z is a regular permutation group on X. /5/

Let x_1 be a fixed element of X. The elements $z_1,\dots,z_n$ of Z shall be enumerated by $z_i = \begin{pmatrix} s_1(x_1)\dots\ s_n(x_1) \\ s_is_1(x_1)\dots s_is_n(x_1)\end{pmatrix}$ for $i=1,\dots,n$. Thus $s_i(x_1)=z_i(x_1)$ holds for $i=1,\dots,n$, and the permutation $\begin{pmatrix} z_1(x_1)\dots\ z_n(x_1) \\ z_iz_1(x_1)\dots z_iz_n(x_1)\end{pmatrix}$ maps each element $z_j(x_1)$ to $z_iz_j(x_1)=z_j(z_i(x_1))=z_j(s_i(x_1))=s_js_i(x_1)=s_i(s_j(x_1))=s_i(z_j(x_1))$.
Hence $s_i = \begin{pmatrix} z_1(x_1)\dots\ z_n(x_1) \\ z_iz_1(x_1)\dots z_iz_n(x_1)\end{pmatrix}$ for $i=1,\dots,n$. Now from Lemma 1 follows that S is the centralizer of Z in the symmetric group on X.
By Lemma 1 $\{s(x_1,s_i(x_1)): s\in S\} = \{(s(x_1),s_is(x_1)): s\in S\} = \{(s(x_1),z_i(s(x_1))): s\in S\} = \{(x,z_i(x)): x\in X\}$ for $i=1,\dots,n$.
Now the arc set of G can be given using the elements of Z:
$G=(X, \dot{\bigcup}_{i=1}\{(x,z_i(x)): x\in X\})$.
There are two further possibilities to describe G if Z is a subgroup of A(G).

<u>**Theorem 1.**</u> $G=(X,\dot{\bigcup}_{i=1}\{z(x_1,z_i(x_1)): z\in Z\})=(X,\dot{\bigcup}_{i=1}\{(x,s_i(x)): x\in X\})$ iff Z is a subgroup of A(G).

Proof: Let Z be a subgroup of A(G). If $i=1,\dots,r$, then $\{z(x_1,z_i(x_1)): z\in Z\} = \{z(x_1,s_i(x_1)): z\in Z\}$ is a subset of the arc set of G. Every arc of G is an element of one of these sets. Thus $G=(X,\bigcup_{i=1}^{r}\{z(x_1,z_i(x_1)): z\in Z\})$.

Let $G=(X,\bigcup_{i=1}^{r}\{z(x_1,z_i(x_1)): z\in Z\})$. An arbitrary element z_j of Z maps an arbitrary arc $z(x_1,z_i(x_1))$ onto the arc $zz_j(x_1,z_i(x_1))$. Hence Z is a subgroup of A(G).

Since S is the centralizer of Z, by Lemma 1 $\{z(x_1,z_i(x_1)): z\in Z\}=$ $=\{(z(x_1),z_iz(x_1)): z\in Z\}=\{(z(x_1),s_i(z(x_1))): z\in Z\}=\{(x,s_i(x)): x\in X\}$

Now the arc set of G can be given using the elements of S: $G=(X,\bigcup_{i=1}^{r}\{(x,s_i(x)): x\in X\})$. □

Later on the equalities $\{z(x_1,z_i(x_1)): z\in Z\}=\{(x,s_i(x)): x\in X\}$ $(i=1,\dots,n)$ stated in the proof will be used.

3. The generation of graph automorphisms

It is obvious that the order of the automorphism group of a directed Cayley graph is not smaller than the cardinality of its vertex set. Some classes of graphs with regular automorphism groups are known /1/, /6/, /7/, /8/.

There is an interesting connection between graph automorphisms of the graphs considered, permutations of the elements of S and permutations of the elements of Z. For each element a of A(G) there are permutations $a' = \begin{pmatrix} s_1\dots s_n \\ s_1'\dots s_n' \end{pmatrix}$, $a'' = \begin{pmatrix} z_1\dots z_n \\ z_1'\dots z_n' \end{pmatrix}$ and vertices x,y,u,v such that $a = \begin{pmatrix} s_1(x)\dots s_n(x) \\ s_1'(y)\dots s_n'(y) \end{pmatrix} =$ $= \begin{pmatrix} z_1(u)\dots z_n(u) \\ z_1'(v)\dots z_n'(v) \end{pmatrix}$. The automorphism a is said to be induced by a' and a''.

In addition, a is called induced by a' and a'', respectively, if a is an arbitrary permutation of the vertex set of G, too. By means of induced automorphisms a tight upper bound of the order of the automorphism group of directed Cayley graphs can be given.

<u>Theorem 2.</u> The order of A(G) is not greater than $r!\cdot(|X| - 1 - r)!\cdot|X|$.

Proof: Let $a \in A(G)$, $a(x_1)=y$, and let s_n be the identity. Then there is a permutation $\begin{pmatrix} s_1 \dots s_n \\ s_1' \dots s_n' \end{pmatrix}$ such that $a = \begin{pmatrix} s_1(x_1) \dots s_n(x_1) \\ s_1'(y) \dots s_n'(y) \end{pmatrix}$ and $s_n = s_n'$. This permutation is uniquely determined.
To find a necessary condition satisfied by the permutations of the elements of S which induce automorphisms of G, an arbitrary automorphism $\bar{a}$ of G, mapping x_1 onto w, is considered:
$\bar{a} = \begin{pmatrix} s_1(x_1) \dots s_n(x_1) \\ \bar{s}_1(w) \dots \bar{s}_n(w) \end{pmatrix}$, $s_n = \bar{s}_n$.
Let $w = s_1(x_1)$. Then $\bar{a}$ maps each arc $(x_1, s_i(x_1))$ onto $(w, \bar{s}_i(w)) = (s_1(x_1), s_1\bar{s}_i(x_1)) = s_1(x_1, s_1\bar{s}_i s_1^{-1}(x_1))$. Hence $s_1\bar{s}_i s_1^{-1} \in \{s_1, \dots, s_r\}$ and $\bar{s}_i \in \{s_1^{-1}s_1s_1, \dots, s_1^{-1}s_rs_1\}$ for each $s_i \in \{s_1, \dots, s_r\}$, and there are at most $r!\cdot(|X| - 1 - r)!$ permutations $\begin{pmatrix} s_1 \dots s_n \\ \bar{s}_1 \dots \bar{s}_n \end{pmatrix}$ of the elements of S satisfying $s_n = \bar{s}_n$ and inducing automorphisms of G which map x_1 to w.
Since w is an arbitrary vertex, $|A(G)| \leq r!\cdot(|X| - 1 - r)!\cdot|X|$ holds. □

Equality holds, for example, if G is a complete symmetric digraph. The estimation given in Theorem 2 holds for a larger class of graphs and digraphs, too.
Now special subgroups of A(G) are investigated. Properties of permutations of the elements of S determine properties of induced automorphisms of G.

<u>Lemma 2.</u> The permutations on X induced by the identical permutation of the elements of S constitute the group Z.

Proof: Let $a = \begin{pmatrix} s_1(x) \dots s_n(x) \\ s_1(y) \dots s_n(y) \end{pmatrix}$. Then $a \in Z$ by Lemma 1. Further an arbitrary element z_i of Z is induced by $\begin{pmatrix} s_1 \dots s_n \\ s_1 \dots s_n \end{pmatrix}$ since

$$z_i = \begin{pmatrix} s_1(x_1)\dots & s_n(x_1) \\ s_is_1(x_1)\dots & s_is_n(x_1) \end{pmatrix} = \begin{pmatrix} s_1(x_1) & \dots s_n(x_1) \\ s_1(s_i(x_1))\dots & s_n(s_i(x_1)) \end{pmatrix}. \square$$

The group Z can also be characterized in a different way.

<u>Theorem 3.</u> Z is the group of all permutations on X which map each set $\{(x,s_i(x)): x\in X\}$ $(i=1,\dots,n)$ onto itself.

Proof: Let $(y,s_i(y))$ be an arbitrary pair, $y=s_k(x_1)$ and $z_j\in Z$. Then $z_j(y,s_i(y)) = z_j(s_k(x_1),s_ks_i(x_1)) = (s_js_k(x_1),s_js_ks_i(x_1)) =$
$= (s_js_k(x_1),s_i(s_js_k(x_1)))\in\{(x,s_i(x)): x\in X\}$.
Let t be a permutation on X which maps every set $\{(x,s_i(x)$: $x\in X\}$ $(i=1,\dots,n)$ onto itself. Then $\{(x,s_i(x)): x\in X\} =$
$= t(\{(x,s_i(x)): x\in X\}) = \{(t(x),s_it(x)): x\in X\} =$
$= \{(t(x),t^{-1}s_it(t(x))): x\in X\}$. Hence $s_i(y) = t^{-1}s_it(y)$ for all vertices y and $s_i = t^{-1}s_it$ $(i=1,\dots,n)$. Thus t is an element of the centralizer Z of S. □

<u>Corollary 1.</u> If a permutation t on X maps each set $\{(x,s_i(x)):$ $x\in X\}$ $(i=1,\dots,r)$ onto itself, then t is contained in Z.

Proof: The equality $\{(x,s_i(x)): x\in X\} = \{z(x_1,z_i(x_1)): z\in Z\}$ holds for $i=1,\dots,r$. The set $\{s_1,\dots,s_r\}$ generates S. Hence,
by $z_iz_j = \begin{pmatrix} s_1(x_1)\dots & s_n(x_1) \\ s_js_is_1(x_1)\dots & s_js_is_n(x_1) \end{pmatrix}$ for all $i,j\in\{1,\dots,n\}$,
the set $\{z_1,\dots,z_r\}$ is a generating set of Z. An element $\bar{z}$ of Z satisfies $t(x_1) = \bar{z}(x_1)$. Now $t(x) = \bar{z}(x)$ $(x\in X)$ shall be proved.
Let x be an arbitrary element of X. Then there is an element z of Z such that $z(x_1)=x$. This element z is a product $z_1''\cdot\dots\cdot z_k''$ of elements of $\{z_1,\dots,z_r\}$. The permutation t maps the sequence $((x_1,z_k'(x_1)),(z_k''(x_1),z_{k-1}'z_k''(x_1)),\dots$
$\dots,(z_2'\cdot\dots\cdot z_k'(x_1),z(x_1)))$ onto the sequence $((\bar{z}(x_1),\bar{z}(z_k'(x_1))),$ $(\bar{z}(z_k''(x_1)),\bar{z}(z_{k-1}'z_k'(x_1))),\dots,(\bar{z}(z_2'\cdot\dots\cdot z_k'(x_1)),\bar{z}(z(x_1))))$.
Thus $t(z(x_1)) = \bar{z}(z(x_1))$ and $t=\bar{z}$. □

As it was mentioned above, $G=(X,\bigcup_{i=1}^{r}\{s(x_1,s_i(x_1)): s\in S\})=$

$=(X, \bigcup_{i=1}^{r} \{(x,z_i(x)): x \in X\})$. Now it is possible to characterize the permutations mapping each set $\{(x,z_i(x)): x \in X\}$ onto itself.

Theorem 3'. S is the group of all permutations on X which map each set $\{(x,z_i(x)): x \in X\}$ $(i=1,\dots,n)$ onto itself. □

In a similar way one can find

Corollary 1'. If a permutation t on X maps each set $\{(x,z_i(x)): x \in X\}$ $(i=1,\dots,r)$ onto itself, then t is contained in S. □

Now the question arises which relations exist between properties of nonidentical permutations of the elements of S, induced graph automorphisms and permutations of the arc set induced by them. In this section the considerations are restricted to group automorphisms of S.
If Z is a subgroup of the automorphism group of the graph G, then the set of permutations on X induced by the inner automorphisms of S can be characterized in the following way.

Theorem 4. Let Z be a subgroup of A(G). Then the group of all inner automorphisms of the group S induces the subgroup SZ of A(G).

Proof: Since Z is the centralizer of S, $\{sz: s \in S, z \in Z\}$ constitutes a group generated by $S \cup Z$. This group SZ is a subgroup of A(G).
Let $sz \in SZ$. Then there are vertices x,y such that
$z = \begin{pmatrix} s_1(x) & \dots s_n(x) \\ s_1(s^{-1}(y)) & \dots s_n(s^{-1}(y)) \end{pmatrix}$ by Lemma 2. Therefore $sz = zs =$
$= \begin{pmatrix} s_1(x) \dots & s_n(x) \\ s^{-1}s_1s(y) \dots & s^{-1}s_ns(y) \end{pmatrix}$. Thus sz is induced by the inner automorphism $\begin{pmatrix} s_1 & \dots & s_n \\ s^{-1}s_1s & \dots & s^{-1}s_ns \end{pmatrix}$ of S.
Let $a' = \begin{pmatrix} s_1 & \dots & s_n \\ s^{-1}s_1s & \dots & s^{-1}s_ns \end{pmatrix}$ be an arbitrary inner automorphism of the group S and let $a = \begin{pmatrix} s_1(x) \dots & s_n(x) \\ s^{-1}s_1s(y) \dots & s^{-1}s_ns(y) \end{pmatrix}$

be induced by a'.

Now $a = \begin{pmatrix} s_1(x) \ldots & s_n(x) \\ s_1(s^{-1}(y)) \ldots & s_n(s^{-1}(y)) \end{pmatrix} \cdot \begin{pmatrix} s_1(s^{-1}(y)) \ldots & s_n(s^{-1}(y)) \\ s(s_1(s^{-1}(y))) \ldots & s(s_n(s^{-1}(y))) \end{pmatrix}$

The first factor is an element of Z by Lemma 2, and the second factor is an element of S. Thus $a \in SZ$. □

<u>Corollary 2.</u> The subgroup of A(G) mentioned in Theorem 4 has the order $\dfrac{|X|^2}{|S \cap Z|}$.

Proof: The cosets s_iZ, s_jZ are equal iff $s_j \in s_iZ$, i.e. $s_j \in s_i(S \cap Z)$. Thus $\dfrac{|X|^2}{|S \cap Z|}$ is the order of the subgroup considered. □

Each element of SZ maps each set $\{s(x_1, s_i(x_1)) \colon s \in S\}$ onto a set $\{s(x_1, s_j(x_1)) \colon s \in S\}$ and each set $\{(x, s_i(x)) \colon x \in X\}$ onto a set $\{(x, s_k(x)) \colon x \in X\}$. This assertion follows from Theorem 6. In general there are further permutations on X which permute these sets.

Let $A_p(G)$ denote the group of all automorphisms of G generating permutations of $\{\{s(x_1, s_i(x_1)) \colon s \in S\} \colon s_i \in S\}$. This group $A_p(G)$ has a remarkable property:

<u>Theorem 5.</u> $A_p(G)$ is the group of those automorphisms of G which are induced by group automorphisms of S.

Proof: Let $a \in A_p(G)$. Then $a = \begin{pmatrix} s_1(x_1) \ldots & s_n(x_1) \\ a^{-1}s_1a(a(x_1)) \ldots & a^{-1}s_na(a(x_1)) \end{pmatrix}$.

Now $a^{-1}s_ka(\{s(x_1, s_i(x_1)) \colon s \in S\}) = s_ka(\{s(x_1, s_j(x_1)) \colon s \in S\}) =$
$= a(\{s(x_1, s_j(x_1)) \colon s \in S\}) = \{s(x_1, s_i(x_1)) \colon s \in S\}$. By Theorem 3' the permutations $a^{-1}s_ka$ $(s_k \in S)$ are elements of S.

The permutation $\begin{pmatrix} s_1 & \ldots & s_n \\ a^{-1}s_1a & \ldots & a^{-1}s_na \end{pmatrix}$ induces a. For arbitrary elements s_i, s_j the equality $a^{-1}s_ia \cdot a^{-1}s_ja = a^{-1}s_is_ja$ holds. Therefore a is induced by a group automorphism of S.

Let $a' = \begin{pmatrix} s_1 \ldots s_n \\ s_1' \ldots s_n' \end{pmatrix}$ be a group automorphism of S and

$a = \begin{pmatrix} s_1(x)\ldots s_n(x) \\ s_1'(y)\ldots s_n'(y) \end{pmatrix}$ an automorphism of the graph G.

With $x_1 = s_k(x) = s_l(y)$ the equations $a(s_j(x_1,s_i(x_1))) =$
$= a(s_j(x_1),s_is_j(x_1)) = a(s_ks_j(x),s_ks_is_j(x)) =$
$= (s_k's_j'(y),s_k''s_i''s_j'(y)) = (s_l^{-1}s_k''s_j'(x_1),s_l^{-1}s_k's_i's_j'(x_1)) =$
$= s_l^{-1}s_k''s_j'(x_1,s_l^{-1}s_k''s_i'(s_k')^{-1}s_l(x_1))$ hold. Thus a maps each set $\{s(x_1,s_i(x_1)): s \in S\}$ onto $\{s(x_1,s_l^{-1}s_k''s_i'(s_k'')^{-1}s_l(x_1)): s \in S\}$, and therefore $a \in A_p(G)$. □

Furthermore, the group $A_p(G)$ has a property concerning the generated permutations of the arc set.

<u>Theorem 6.</u> All elements of $A_p(G)$ generate permutations of $\{\{(x,s_i(x)): x \in X\}: s_i \in S\}$.

Proof: Let $a \in A_p(G)$. Then there is a group automorphism $\begin{pmatrix} s_1\ldots s_n \\ s_1'\ldots s_n' \end{pmatrix}$ of S satisfying $a = \begin{pmatrix} s_1(x)\ldots s_n(x) \\ s_1'(y)\ldots s_n'(y) \end{pmatrix}$.
With $x_1 = s_k(x) = s_l(y)$ the equations $a(s_j(x_1),s_i(s_j(x_1))) =$
$= a(s_ks_j(x),s_ks_js_i(x)) = (s_k's_j'(y),s_k's_j's_i'(y))$ are valid.
Hence a maps each set $\{(x,s_i(x)): x \in X\}$ onto $\{(x,s_i'(x)): x \in X\}$. □

Furthermore one can prove that all automorphisms of G generating permutations of $\{\{(x,s_i(x)): x \in X\}: s_i \in S\}$ are elements of $A_p(G)$.
To characterize $A_p(G)$ it is sufficient to require $A_p(G)$ to be the group of all automorphisms of G permuting the sets $\{s(x_1,s_i(x_1)): s \in S\}$ $(i=1,\ldots,r)$. Since the elements of $A_p(G)$ are automorphisms of G, it is obvious that every element of $A_p(G)$ which permutes $\{\{s(x_1,s_i(x_1)): s \in S\}: s_i \in S\}$ also permutes $\{\{s(x_1,s_i(x_1)): s \in S\}: i=1,\ldots,r\}$. Now assume that an automorphism a of G permutes the sets $\{s(x_1,s_i(x_1)): s \in S\}$

(i=1,...,r). Then $a = \begin{pmatrix} s_1(x_1)\ldots & s_n(x_1) \\ a^{-1}s_1a(a(x_1))\ldots & a^{-1}s_na(a(x_1)) \end{pmatrix}$ is

induced by the group automorphism $\begin{pmatrix} s_1 & \ldots & s_n \\ a^{-1}s_1a & \ldots & a^{-1}s_na \end{pmatrix}$ of S.

Note that $a^{-1}s_ja$ maps each set $\{s(x_1,s_i(x_1)): s \in S\}$ (i=1,...,r) onto itself.

Therefore a is an element of $A_p(G)$.

If Z is a subgroup of A(G), then $A_p(G)$ is the group of all automorphisms of G which permute the sets $\{(x,s_i(x)): x \in X\}$ (i=1,...,r).

It is easy to see that each element of $A_p(G)$ is induced by various permutations of the elements of S and by exactly one group automorphism of S.

The group $A_p(G)$ can also be characterized by other algebraic means. The normalizer of S in the symmetric group on X is denoted by S^N. All permutations p on X satisfying $p^{-1}Sp = S$ constitute the group S^N.

<u>Theorem 7.</u> $A_p(G) = A(G) \cap S^N$.

Proof: Every element a of $A_p(G)$ is induced by a group automorphism $\begin{pmatrix} s_1\ldots s_n \\ s_1'\ldots s_n' \end{pmatrix}$. There are vertices x,y satisfying

$a = \begin{pmatrix} s_1(x)\ldots s_n(x) \\ s_1'(y)\ldots s_n'(y) \end{pmatrix}$. The permutation $a^{-1}s_ia =$

$= a^{-1}\begin{pmatrix} s_1(x)\ldots & s_n(x) \\ s_1s_i(x)\ldots s_ns_i(x) \end{pmatrix} a = \begin{pmatrix} s_1'(y)\ldots & s_n'(y) \\ s_1's_i'(y)\ldots s_n's_i'(y) \end{pmatrix} = s_i'$

is an element of S for each $s_i \in S$. Therefore $A_p(G) \subseteq A(G) \cap S^N$.

Let $a \in A(G) \cap S^N$. Then $a(\{s(x_1,s_i(x_1)): s \in S\}) =$

$= \{a^{-1}sa(a(x_1),a(s_i(x_1))): s \in S\}$ for each $s_i \in S$. The permutations $a^{-1}sa$ $(s \in S)$ are elements of S, and $(a(x_1),a(s_i(x_1)))$ is an element of a set $\{s(x_1,s_j(x_1)): s \in S\}$. Thus a maps $\{s(x_1,s_i(x_1)): s \in S\}$ onto $\{s(x_1,s_j(x_1)): s \in S\}$.

By $a \in A(G)$, a is an element of $A_p(G)$. □

Now one can ask if S^N is a subgroup of A(G). To answer the question a property of regular permutation groups and their centralizers is needed. Since S is a regular permutation group on X, the normalizers S^N, Z^N of S and Z, respectively, in the symmetric group on X coincide. Thus Z being a subgroup of A(G) is a necessary condition for the equality of $A_p(G)$ and S^N.

<u>Theorem 8.</u> Let Z be a subgroup of A(G). Then $A_p(G) = S^N$ iff for all $t \in S^N$ and all $s_i \in \{s_1,\dots,s_r\}$ the product $t^{-1}s_it$ is an element of $\{s_1,\dots,s_r\}$.

Proof: The arc set of G is $\bigcup_{i=1}^{r}\{(x,s_i(x)): x \in X\}$ by Theorem 1.
Let $t \in A_p(G) = S^N$, $s_i \in \{s_1,\dots,s_r\}$. Then $(x_1,s_i(x_1))$ and $t(x_1,s_i(x_1)) = (t(x_1),t^{-1}s_it(t(x_1)))$ are arcs of G.
Now $t^{-1}s_it \in S$ by $t \in S^N$ and therefore $t^{-1}s_it \in \{s_1,\dots,s_r\}$.
Let $t \in S^N$ and $s_i \in \{s_1,\dots,s_r\}$ imply $t^{-1}s_it \in \{s_1,\dots,s_r\}$.
Then t maps an arbitrary arc $(x,s_i(x))$ onto an arc $(t(x),s_it(x)) = (t(x),t^{-1}s_it(t(x)))$. Thus t is an automorphism of G and $A_p(G) = S^N$ by Theorem 7. □

Above a connection between properties of permutations of the elements of S and properties of graph automorphisms was stated. There is a similar connection between permutations of the elements of Z and graph automorphisms. The proofs of the following results are analogous to previous ones. They will be omitted.
First it shall be stated that the permutations on X induced by the identical permutation of the elements of Z constitute the group S. If Z is a subgroup of A(G), then not only the group of all inner automorphisms of the group S but also the group of all inner automorphisms of the group Z induces the subgroup SZ of A(G). The group $A_p(G)$ is the group of those automorphisms of G which are induced by group automorphisms of Z, too.
From this one can deduce that the order of $A_p(G)$ is the

product of $|X|$ and the number of automorphisms of the group Z mapping $\{z_1,\dots,z_r\}$ onto itself.

With this knowledge the permutations of the elements of Z inducing graph automorphisms can be characterized.

Theorem 9. A permutation $\begin{pmatrix} z_1(x)\dots z_n(x) \\ z_1'(y)\dots z_n'(y) \end{pmatrix}$ is an element of A(G) iff for each $z_k \in Z$ and each $z_i \in \{z_1,\dots,z_r\}$ there is an element z_j' of $\{z_1,\dots,z_r\}$ satisfying $(z_k z_i)' = z_k' \cdot z_j'$.

Proof: Let $a = \begin{pmatrix} z_1(x)\dots z_n(x) \\ z_1'(y)\dots z_n'(y) \end{pmatrix} \in A(G)$, $z_k \in Z$ and $z_i \in \{z_1,\dots,z_r\}$.

Then $(z_k(x), z_i(z_k(x)))$ and $a(z_k(x), z_i(z_k(x))) =$ $= (z_k'(y), (z_k z_i)'(y))$ are arcs of G. Therefore the permutation z_j' determined by $(z_k z_i)' = z_k' \cdot z_j'$ is an element of $\{z_1,\dots,z_r\}$.

Now let the permutation $a = \begin{pmatrix} z_1(x)\dots z_n(x) \\ z_1'(y)\dots z_n'(y) \end{pmatrix}$ be induced by a permutation $\begin{pmatrix} z_1\dots z_n \\ z_1'\dots z_n' \end{pmatrix}$ with the properties mentioned in Theorem 9. Then a maps an arbitrary arc $(z_k(x), z_i(z_k(x)))$ onto $(z_k'(y), z_k' z_j'(y))$, and this pair is an arc of G. Hence $a \in A(G)$. □

References:

/1/ Babai, L.: On a conjecture of M.E. Watkins on graphical regular representations of finite groups. Compositio Math., Vol. 37, Fasc. 3, 1978, 291-296

/2/ Babai, L.: On the abstract group of automorphisms. London Math. Soc. Lecture Note Series 52, 1981, 1-40

/3/ Biggs, N.L.: Algebraic graph theory. Cambridge, 1974

/4/ Frucht, R.: Herstellung von Graphen mit vorgegebener abstrakter Gruppe. Compositio Math., Vol. 6, 1938, 239-250

/5/ Huppert, B.: Endliche Gruppen. Berlin, Heidelberg, New York, 1967

/6/ Imrich, W.: Graphs with transitive abelian automorphism groups. Comb. Theory and its Appl. II, Amsterdam, 1970, 651-656

/7/ Nowitz, L.A.; Watkins, M.E.: Graphical regular representation of non-abelian groups. Canad. J. Math., Vol. 24, No. 6, 1972, 993-1018

/8/ Watkins, M.E.: On the action of non-abelian groups on graphs. J. Comb. Theory B, Vol. 11, 1971, 95-104

Triangular Embeddings of Tensor Products of Graphs

A. Bouchet, Cedex † France
B. Mohar, Ljubljana, Yugoslavia

ABSTRACT

Given triangular embeddings of graphs G and H and an assignment of signs ± 1 to the angles of the embedding of G, with the property that the sum of the signs around each vertex of G is ± 1 and the product of the four signs at each edge is $+1$, triangular embeddings of the tensor (also called categorical) product $G \otimes H$ of G and H are constructed. This generalizes previously known results about embeddings of tensor products of graphs.

1. INTRODUCTION

Our graphs will be finite and simple. By a surface we mean a compact surface without boundary. We assume that the reader is familiar with basic notions of topological graph theory, and refer to the book [GT].

Let G and H be graphs and let $G \cdot H$ denote some graph product of G and H. It is an important problem to determine any genus embeddings of $G \cdot H$, i.e., embeddings of the graph with the minimal possible genus, knowing some genus embeddings of G and

† Supported in part by P.R.C. Mathématiques et Informatique.
‡ Supported in part by the Research Council of Slovenia, Yugoslavia.

R. Bodendiek, R. Henn (Eds.)
Topics in Combinatorics and Graph Theory

H. In the case of the Cartesian product $\times$ of graphs, several results about embeddings of the graph $G \times H$ are known [P1, P2, W]. Cf. also [GT]. There are also some works about the embeddings of the lexicographic [B3] and the tensor products [B4, Ž] of graphs. In this paper we consider the tensor (categorical) product.

If G and H are graphs then the *tensor product* of G and H is the graph $G \otimes H$ having vertex set $V(G \otimes H) = V(G) \times V(H)$ and with vertices (u, v) and (u', v') being adjacent if and only if u is adjacent to u' in G and v is adjacent to v' in H. By some authors this product is also called conjunction, or the categorical product.

We shall restrict ourselves to a particular, simpler but still difficult problem: Let G and H be given graphs together with triangular embeddings $i : G \to \Sigma_1$ and $j : H \to \Sigma_2$. Recall that "triangular" means that all the faces of the embeddings are triangles. The problem is to find a triangular embedding of the tensor product $G \otimes H$. Notice that triangular embeddings of simple graphs are always minimal genus embeddings (either orientable, or nonorientable). It will be assumed, moreover, that all the vertices of the graph G have odd degrees. For our problem this is a very natural condition. Namely, in order for the graph $G \otimes H$ to have triangular embeddings, the neighbors of each vertex must span a cycle. The neighborhood graph of a vertex (u, v) in $G \otimes H$ is equal to the tensor product of neighborhood graphs of v and u in G and H, respectively. If these are both even cycles, for example, their product is disconnected, and hence $G \otimes H$ trivially admits no triangular embeddings.

We construct triangular embeddings of $G \otimes H$, $\kappa : G \otimes H \to \Sigma$, with the property that there are (simplicial) maps $\pi_1 : \Sigma \to \Sigma_1$ and $\pi_2 : \Sigma \to \Sigma_2$ such that each face in Σ is projected bijectively onto a face in Σ_i, $i = 1, 2$. Mappings π_i with this property are called *coverings with folds* (cf. [B1, B2] for more details about such mappings). Combinatorially they define simplicial maps between simplicial complexes corresponding to the triangular embeddings. In our construction, the mapping π_1 will have no folds, so topologically it will be a branched covering of surfaces with branch points at vertices of G only. Our construction uses certain valuation of the angles of the embedding of the graph G. We believe that such valuations always exist. Our main result, Theorem 1, generalizes previous results [B4] on triangular embeddings and the genus of tensor products of graphs.

2. VALUATIONS OF MAPS

Let $i : G \to \Sigma$ be a 2-cell embedding. The vertices, edges, and faces together with the incidency relations determine a combinatorial structure, called a *map*, or a *cell complex*. We will denote such a map as a triple, $M = (G, \Sigma, i)$. For a map M denote by $V(M) = V(G)$, $E(M) = E(G)$, and $F(M)$ the sets of vertices, edges, and faces of M, respectively. Each face F of M is bounded by a sequence of edges of M, and this determines the *angles* of F, looking at F as being a polygon. We shall denote by $A(M)$ the set of angles of all faces of M. It is clear what we shall mean by the incidency of angles and vertices, edges, or faces of M. Notice that $|A(M)| = 2|E(M)|$. Recall that a map is *triangular* if all its faces are triangles. Triangular maps with a simple underlying graph G can also be viewed as simplicial complexes.

A (homology) *cycle* of G, or of M, is any subgraph of G with all vertices of even degree. An *angle valuation* of a map M is an assignment $\nu : A(M) \to \{-1, +1\}$ with the property that for each edge e, the product of signs of the four angles incident with e is equal to $+1$. There is an obvious correspondence between the cycles and angle valuations of maps. Under this correspondence, the signs of two angles incident with the same vertex and the same edge e are different if and only if e lies in the corresponding cycle. Therefore, each angle valuation determines a unique cycle, and a cycle determines angle valuations around each vertex of G up to the sign.

Cycles and/or angle valuations will be used in the construction of triangular embeddings of tensor products in the next section. They will have to satisfy an additional condition:

(P) For each vertex $v \in V(M)$, the sum of the signs of angles around v is equal to ± 1.

Notice that the property (P) holds, or does not hold, simultaneously for all angle valuations determined by a given cycle of M. So we will speak about cycles with, or without, the property (P). It is easily seen that each vertex of G must be of odd degree if there is some cycle with property (P). We believe that the converse is also true.

CONJECTURE. *An orientable triangular map (G, Σ, i) admits an angle valuation with property (P) if and only if the graph G has only odd degrees.*

For example, if the graph G has a perfect matching Q and all vertices have odd degree then the set $E(G) \setminus Q$ is a homology cycle with property (P).

3. CONSTRUCTION

Let (G, Σ_1, i) and (H, Σ_2, j) be triangular maps. We want to construct a triangular map for the tensor product $G \otimes H$ of the two graphs. In our construction, which depends on the existence of an angle valuation of G with property (P), the obtained map will be a (branched) covering without folds above (G, Σ_1, i), and a covering with folds above (H, Σ_2, j). Let us recall that a covering with folds is nothing else as a simplicial mapping between the corresponding simplicial complexes. By a *fold* we mean that two triangles with a common edge e in the map above are mapped to the same triangle below. The edge e of the above map where the folding appears is said to be *singular*.

THEOREM 1. *Let (G, Σ_1, i) and (H, Σ_2, j) be triangular maps. If the graph G admits a homology cycle C with property (P) then there is a triangular embedding of the tensor product $G \otimes H$ into some surface Σ. The projections of $G \otimes H$ onto the factors G and H determine the corresponding map of $G \otimes H$ to be a covering without folds above (G, Σ_1, i) and a covering with folds above (H, Σ_2, j). For the latter covering with folds, the singular edges of $G \otimes H$ are exactly those edges which are mapped by the first covering onto the edges of C.*

Proof. Assume that C is a cycle of G with property (P), and let $\nu : A(M) \to \{-1, +1\}$ be a corresponding angle valuation of the map $M = (G, \Sigma_1, i)$. Let Q denote the set of all triangles of $G \otimes H$ lying simultaneously above the faces of M and of $M' = (H, \Sigma_2, j)$, with respect to the canonical projections of the product $G \otimes H$ onto its factors, i.e.

$$Q = \left\{ \{(u_1, v_1), (u_2, v_2), (u_3, v_3)\} \;\middle|\; \{u_1, u_2, u_3\} \in F(M), \{v_1, v_2, v_3\} \in F(M') \right\} .$$

Pick now any two triangles T_1 and T_2 in Q which have a common edge, say $T_k = \{(u, v), (u', v'), (u_k, v_k)\}$, $k = 1, 2$. We call T_1 and T_2 *adjacent* if

(a) $u_1 \neq u_2$, and

(b) $v_1 = v_2$ if the edge uu' of G is in C, and $v_1 \neq v_2$ if $uu' \notin C$.

This relation determines a graph with vertex set Q. Pick a connected component $Q' \subseteq Q$ of this graph. We claim that the triangles of Q' determine a triangular map with the properties stated in the theorem. Notice that the choice of Q' was arbitrary, hence other triangular embeddings of $G \otimes H$ can be obtained by taking other connected parts of the graph of Q. In fact, it can be shown that Q has exactly two components, so it factors into two triangular maps of $G \otimes H$. Note also that the graph of Q' is just the dual graph of the corresponding triangular embeddnig.

First of all, it is clear by definition of the adjacency that in the simplicial complex determined by the triangles of Q' each edge belongs to exactly two triangles. Let T be an arbitrary triangle of Q', and (u,v) a vertex of T. By the above, the link of (u,v) in Q' is a 2-regular graph. We shall see that it must be connected and that it contains all the vertices of $G \otimes H$ adjacent to (u,v). Let $u_1, u_2, \ldots, u_k$ be the neighbors of u in G, enumerated in the same order as they appear around u on the surface Σ_1. Enumerate similarly the neighbors $v_1, v_2, \ldots, v_q$ of v in H. We may assume that $(u_1, v_1) \in T$. Now, consecutive vertices on the link of (u,v) in Q' are $(u_1, v_{s(1)})$, $(u_2, v_{s(2)})$, $\ldots$, $(u_k, v_{s(k)})$, $(u_1, v_{s(k+1)})$, $\ldots$, $(u_k, v_{s(2k)})$, $(u_1, v_{s(2k+1)}), \ldots$. The indices at the vertices u_l are consecutive numbers modulo k, and $s(1) = 1$, while the other values of $s(t)$ depend on C. Without loss of generality we may assume that the sum of angle signs around u induced by C is $+1$. Then it is clear that $s(k+1) = s(1) + 1$, $s(2k+1) = s(k+1) + 1$, etc. This implies that for each t, $s(t+k) = s(t) + 1 \pmod{l}$, and we are done. Similarly, if the sum is -1 then $s(t+k) \equiv s(t) - 1 \pmod{l}$ for each t.

Since G and H are connected, so is $G \otimes H$, since by the property (P) of C, the degrees of G are odd, and so G is not bipartite. By the above, Q' contains $G \otimes H$, and the proof is finished. □

NOTE. It is easily seen from the above proof that the property (P) could be weakened to the following condition:

(P') For each vertex $u \in V(G)$, the sum of the angle values around u is relatively prime with all degrees of vertices of H.

Notice that (P') does not imply G to have odd degrees, but if a vertex of G has even degree then H must have only odd degrees.

So far we did not say a word about orientability of the constructed triangular embeddings of $G \otimes H$. This is important to know since the existence of triangular embeddings of $G \otimes H$ into Σ determines either the orientable, or the nonorientable genus of this graph, depending on the orientability of Σ.

Let us mention only a very basic result about orientability - two simple sufficient conditions for the orientability of Σ. Recall that a homology cycle C in G is a *boundary cycle* with respect to the given map (G, Σ_1, i) if there is a set $\mathbf{F}$ of faces of the map such that C contains precisely those edges of G which are lying in exactly one face of $\mathbf{F}$.

PROPOSITION 1. *Let (G, Σ_1, i) and (H, Σ_2, j) be triangular maps, and C a homology cycle of G having property (P). In each of the following two cases:*

(a) Σ_1 *is orientable, or*

(b) Σ_2 *is orientable and C is a boundary cycle in (G, Σ_1, i),*

the triangular map $(G \otimes H, \Sigma, \kappa)$ constructed in Theorem 1 is orientable.

Proof. It is well-known that branched covers over orientable surfaces must be orientable. This implies that (a) is sufficient to guarantee the orientability of Σ.

To verify the sufficiency of (b) we recall a result of [B1] which states that Σ is orientable if Σ_2 is orientable and the set of singular edges of $G \otimes H$ with respect to the projection to (H, Σ_2, j) is a boundary cycle. Let π be the branched covering of the map $(G \otimes H, \Sigma, \kappa)$ onto (G, Σ_1, i). If C is a boundary cycle and $\mathbf{F}$ a corresponding set of faces whose boundary is C, then $\pi^{-1}(\mathbf{F})$ is a set of faces in Σ whose boundary are precisely those edges which project to C by π. By Theorem 1, these are exactly the singular edges of the covering of Σ to Σ_2, and so we are done.

REFERENCES

[B1] A.Bouchet, *Covering triangulations with folds*, in "Theory of applications of graphs," Ed. G. Chartrand, Wiley, 1981, pp. 127–147.

[B2] A. Bouchet, *Constructions of covering triangulations with folds*, J. Graph Theory 6 (1982) 57–74.

[B3] A. Bouchet, *Triangular imbeddings into surfaces of a join of equicardinal independent sets following an Eulerian graph*, in "Theory and applications of graphs, Proc. Michigan 1976," Springer-Verlag, 1978, pp. 86–115.

[B4] A. Bouchet, *Produit tensoriel de rotations*, Journées de Combinatoire et Informatique de Bordeaux, Juin 1975, Editions Univ. Bordeaux 1, pp. 53–59.

[GT] J. Gross, T. Tucker, *Topological graph theory*, Wiley, 1987.

[P1] T. Pisanski, *Genus of cartesian products of regular bipartite graphs*, J. Graph Theory 4 (1980) 31–42.

[P2] T. Pisanski, *Nonorientable genus of cartesian products of regular graphs*, J. Graph Theory 6 (1982) 391–402.

[W] A.T. White, *The genus of repeated cartesian products of bipartite graphs*, Trans. Amer. Math. Soc. 151 (1970) 393–404.

[Ž] V. Železník, *Quadrilateral embeddings of the conjunction of graphs*, Math. Slovaca 38 (1988) 89–98.

Computing Light Edges in Planar Graphs

O. V. Borodin, Novosibirsk, USSR

1. Survey of results.

We consider planar graphs and pseudographs with the sets of vertices V, edges E, and faces F. Remind that unlike pseudographs, graphs do not contain loops and multiple edges. The weight $w(e)$ of an edge $e=(a,b)$ is defined to be the degree sum $s(a)+s(b)$ of its end vertices. Kotzig proved [5] that each 3-connected planar graph contains an edge of the weight at most 13, the bound being the best possible. This result was further developed in various directions by Kotzig and his followers [1-6,8,10]. The interest to these investigations increased recently due to discovering its applications in coloring problems [1]. Thus, Kronk and Mitchem's conjecture $x_{vef}(G) \leq \Delta(G)+4$ [7] on entire coloring the vertices, edges, and faces of planar graphs with the maximal degree $\Delta(G)$ was proved [1] for $\Delta(G) \geq 12$. (The case $\Delta(G)=3$ was solved in [7].) Moreover, for $\Delta(G) \geq 18$ we have the precise bound $x_{vef}(G) \leq \Delta(G)+2$ [1].

As follows from [2-6,8,10], if in this or that class of graphs one can guarantee the existence of an edge whose weight does not exceed some constant, then the number of such edges, called light in this context, is great enough. For example, Kotzig proved [6] that if the minimal degree, δ, of a planar bipartite graph is at least 3, then $10e_{3,3}+5e_{3,4}+2e_{3,5} \geq 120$, where $e_{i,j}$ is the number of edges joining i-vertices with j-vertices. Moreover, all the coefficients of this inequality are the best possible. Similarly, if all the faces have the length 5, then $e_{3,3} \geq 30$ [6].

R. Bodendiek, R. Henn (Eds.)
Topics in Combinatorics and Graph Theory

These results may be extended as follows (it's trivial that the minimal face size, ρ , in each planar pseudograph with $\delta \geq 3$ does not exceed 5):

<u>Theorem 1.</u> Let a planar pseudograph has $\delta \geq 3$. Then
(i) if $\rho \geq 4$, then $10e_{3,3}+5e_{3,4}+2e_{3,5} \geq 120$;
(ii) if $\rho=5$, then $e_{3,3} \geq 30$.

Thus, for $\rho>3$ our problem is apparently solved (when $\rho<3$, the minimal weight of an edge is clearly not restricted from above). The situation is more complicated when 3-faces are allowed. Even for the special case of triangulations, the problem of obtaining the best, in this or that sense, bounds of the type $\sum_{i,j} a_{i,j}e_{i,j} \geq M$ remains open. Grünbaum conjectured [8] that

$$\begin{aligned}20e_{3,3}&+15e_{3,4}+12e_{3,5}+10e_{3,6}+6\tfrac{2}{3}e_{3,7}+5e_{3,8}+3\tfrac{1}{3}e_{3,9}+2e_{3,10}+\\&+12e_{4,4}+7e_{4,5}+5e_{4,6}+4e_{4,7}+2\tfrac{2}{3}e_{4,8}+\tfrac{2}{3}e_{4,9}+\\&+4e_{5,5}+2e_{5,6}+\tfrac{1}{3}e_{5,7}+\\&+12e_{6,6} \geq 120 .\end{aligned}$$

Soon after, also for triangulations without loops and multiple edges, Jucovič proved [4] that

$$\begin{aligned}20e_{3,3}&+25e_{3,4}+16e_{3,5}+10e_{3,6}+6\tfrac{2}{3}e_{3,7}+5e_{3,8}+2\tfrac{1}{2}e_{3,9}+2e_{3,10}+\\&+20e_{4,4}+11e_{4,5}+5e_{4,6}+5e_{4,7}+5e_{4,8}+3e_{4,9}+\\&+8e_{5,5}+2e_{5,6}+2e_{5,7}+2e_{5,8} \geq 120 .\end{aligned} \qquad (*)$$

One can see that the coefficients $a_{i,j}$ in (*) for $i \leq j \leq 5$ are worse than those conjectured by Grünbaum [8]. However, Jucovič gives [4] examples of graphs which imply $a_{3,4} \geq 22$, $a_{3,5} \geq 15$, $a_{4,4} \geq 12$, and $a_{4,5} \geq 7$.

As a corollary of our main Theorem 4, we shall obtain

<u>Theorem 2.</u> For each planar triangulation without loops and multiple edges,

$$\begin{aligned}20e_{3,3}&+25e_{3,4}+16e_{3,5}+10e_{3,6}+6\tfrac{2}{3}e_{3,7}+5e_{3,8}+2\tfrac{1}{2}e_{3,9}+2e_{3,10}+\\&+20e_{4,4}+11e_{4,5}+5e_{4,6}+1\tfrac{2}{3}e_{4,7}+\\&+8e_{5,5}+2e_{5,6} \geq 120 .\end{aligned}$$

As was supposed by Jucovič [4], not all the coefficients in (*) were the best possible; this is confirmed also by Theorem 2. But attempts to improve them may encounter resistence, even in the simplest cases. For example, Grünbaum announced [2] for triangulations with $\delta=5$ that $2e_{5,5}+e_{5,6}\geqslant 60$, and proved a weaker assertion: if $e_{5,5}=0$ then $e_{5,6}\geqslant 60$. On the other hand, due to the attempts to solve the Four Color Problem, it was established long ago [9] that $e_{5,5}+e_{5,6}>0$ for $\delta=5$. However, Grünbaum represents in [3] an example of a triangulation with $\delta=5$ and $2e_{5,5}+e_{5,6}=56$ which was communicated to him by Fisk. In [3], there is also given without proof the bound $4e_{5,5}+e_{5,6}\geqslant 60$ for $\delta=5$ (clearly, $\delta\leq 5$ for each planar graph). We have

<u>Theorem 3.</u> For each planar triangulation with $\delta=5$, the following inequalities are valid:

(i) $2\frac{4}{7}e_{5,5}+e_{5,6}\geqslant 60$;

(ii) $2e_{5,5}+e_{5,6}+\frac{2}{7}e_{5,7}\geqslant 60$.

The second inequality in Theorem 3 shows that with the additional condition $e_{5,7}=0$ upon a triangulation, Grünbaum's conjecture from [2] becomes true. In fact, it is valid under a weaker assumption of the absence of such 7-vertices which are adjacent to six or seven 5-vertices.

Also for triangulations, Grünbaum proved [2] that if there are no edges of the weight 12 or less, then $e_{3,10}\geqslant 60$.

Now turn our attention to the problem of estimating the number of light edges in those planar graphs and pseudographs which contain both triangular and non-triangular faces. Jucovič remarks in [4] that the proof of (*) does not work for non-triangular graphs; one may assert only the existence of three edges of the weight at most 13. Subsequent papers [3,10] also concern triangulations only (of the plane and other surfaces).

Our main result is

<u>Theorem 4.</u> For each planar pseudograph with $\delta\geqslant 3$, $\rho\geqslant 3$, the following bounds are true:

(i) $30e_{3,3}+25e_{3,4}+16e_{3,5}+10e_{3,6}+6\frac{2}{3}e_{3,7}+5e_{3,8}+3e_{3,9}+2e_{3,10}+$
$+\frac{5}{6}e_{3,11}+20e_{4,4}+11e_{4,5}+5e_{4,6}+1\frac{2}{3}e_{4,7}+$
$+8e_{5,5}+2e_{5,6}\geqslant 120$;

(ii) $a_{3,3}=40$, ... , $a_{3,9}=2\frac{1}{2}$, ... , $a_{3,11}=0$; all the other coefficients are the same as in (i).

The coefficient $a_{3,3}$ in (i) is the best possible as shown by the planar pseudograph obtained by identifying four copies of K_4 along two vertices, which results in an edge of the multiplicity 4. Similar constructions show the unimprovability of $a_{3,t}$ for all $t>3$ in (i). It can be shown also that $a_{3,3}=40$ in (ii).

The following two assertions, in a sense mutually complementary, describe two classes of pseudographs for which the coefficients $a_{i,j}$ do not exceed the level declared in [8]. Moreover, all the coefficients in Theorems 5 and 6 are the best possible.

<u>Theorem 5.</u> If a planar pseudograph has $\delta \geqslant 3$, $\rho \geqslant 3$, and $e_{3,3}=e_{3,4}=e_{3,5}=e_{4,4}=e_{4,5}=e_{5,5}=0$, then

$$\begin{aligned}&10e_{3,6}+6\tfrac{2}{3}e_{3,7}+5e_{3,8}+2\tfrac{1}{2}e_{3,9}+2e_{3,10}+\\&+5e_{4,6}+1\tfrac{2}{3}e_{4,7}+\\&+2e_{5,6}\geqslant 120 .\end{aligned}$$

<u>Theorem 6.</u> If a planar pseudograph has $\delta \geqslant 3$, $\rho \geqslant 3$, and $\Delta \leqslant 6$, then

$$\begin{aligned}&20e_{3,3}+15e_{3,4}+12e_{3,5}+10e_{3,6}+\\&+10e_{4,4}+7e_{4,5}+5e_{4,6}+\\&+4e_{5,5}+2e_{5,6}\geqslant 120 .\end{aligned}$$

It follows from above that the profile of $a_{i,j}$ depends essentially of restrictions posed on the class of (pseudo)graphs under consideration.

We precede the proof of our main result - Theorem 4 - by a few remarks:

(R1) Theorem 2 follows immediately from the part (i) of Theorem 4 and the fact that $e_{3,3}=0$ for all triangulations without loops and multiple edges having 5 or more vertices, while $e_{3,3}=20$ for K_4. The requirement of triangulicity in Theorem 2 is essential as shown by identifying five copies of K_4 along K_2 (the resulting graph has $a_{3,3}=24$).

(R2) The proof of Theorem 1 repeats Kotzig's argument [6], while those of Theorems 3 and 6 use the techniques demonstrated below. Theorem 3 is in fact valid for all planar pseudographs with $\rho \geqslant 3$

(R3) Our approach is suitable for arbitrary surfaces as well.

2. Proof of Theorem 4.

Let P be a counter-example to the part (i) of Theorem 4. We want first to construct a pseudograph P^* which is a triangulation and also a counter-example to (i).

Suppose, P is non-triangular. A diagonal is defined to be an edge whose adding leaves the pseudograph planar and does not create 1- or 2-faces. A diagonal $d=(x,y)$ is checked if $s(x)+s(y)\leq 9$ (in P). We henceforth denote by x,y the end vertices of a diagonal at hand; moreover, we assume that $s(x)\leqslant s(y)$. If a non-checked diagonal d may be pointed out for P, then due to non-increasing $a_{i,j}$ over each coordinate, $P+d$ is also a counter-example to (i), and we are done.

Let each diagonal of P is checked. For a diagonal d of a face f, denote by d' a diagonal which cuts from f the triangular face with x in its boundary. Similar diagonal defined by y is denoted by d''.

We want to find such a diagonal which being added in, the sum $\Sigma = \sum_{i,j} a_{i,j}e_{i,j}$ in (i) does not increase; that is, the pseudograph obtained would also be a counter-example to (i).

Suppose first that there exists a diagonal d with $s(x)=4$. When d is added in, Σ is increased by at most $a_{5,5}=8$. On the other hand, one of end vertices of d' has in P the degree 3 or 4 since d' is checked. Hence, with the addition of d either some (3,4)-edge becomes (3,5)-edge, or some (4,4)-edge becomes (4,5)-edge. This eitherwise decreases Σ by 9. Thus, adding d results in total decreasing Σ.

We may now assume that each diagonal is incident with a 3-vertex; i.e., $s(x)=3$. Let d from now on be the heaviest diagonal.

If $s(y)=3$, then the addition of our d increases Σ by $a_{4,4}=20$ due to increasing $e_{4,4}$ by 1, and decreases Σ by $4(a_{3,3}-a_{3,4})=20$ due to the fact that the four (3,3)-edges adjacent to d' and d'' and incident with x,y become (3,4)-edges. Thus, Σ decreases totally.

If $s(y)=4$, then, similarly, Σ increases by $a_{4,5}=11$, while decreases by 5 due to turning a (3,3)-edge at x into a (3,4)-edge, and by 9 due to turning a (3,4)-edge at y into a (3,5)-edge; that is, decreases totally.

If, at last, $5 \leq s(y) \leq 6$, then turning a (3,3)-edge at x into a (3,4)-edge already compensate appearing d .

Thus, soon or later we discover a diagonal whose addition leaves our pseudograph a counter-example. Hence, after a finite number of steps we arrive at a counter-example P^* which is a triangulation, thereby completing the first part of our proof.

Euler formula $|V|-|E|+|F|=2$ for P^* may be written as $\sum_{v \in V} (s(v)-6)=-12$.

We shall say that each vertex v has <u>a charge</u> which is equal to $s(v)-6$. The negative charge of each <u>minor</u> vertex (i.e., of degree 3,4, or 5) is distributed equally among the incident edges; that is, by -1, -1/2, and -1/5, respectively. Consider (i,j)-edges, where $i \leq j \leq 5$, and call a vertex z <u>dual</u> to the edge (a,b) if there exists a face [abz] . We transfer the following charge from every such (i,j)-edge to each of the two vertices dual to it: 1/2 if $j \leq 4$, and 1/5 if $j=5$. Similarly, for $i<6$, $j>6$, every (i,j)-edge recieves $\min\{(j-6)/\lfloor j/2 \rfloor, (6-i)/i\}$ from its j-vertex. This general rule has the two exceptions: each (3,9)-edge recieves $7/10<3/4$, while each (3,11)-edge recieves $11/12<1$. Remark that the sum of resulting charges of edges and vertices remains to be equal -12.

Let us verify that the resulting charge is non-negative for each vertex. Clearly, only the case of j-vertex v with $j>6$ needs consideration.

We label some of the edges which are either adjacent to v , or incident with v ; each step of the labelling procedure consist in the following. Let v be incident with a 3-face [vab] which is bounded by the edges $e_1=(v,a)$, $e_2=(v,b)$, and $e_3=(a,b)$, where the edge e_3 is dual to v and has not been labelled yet. Let also $s(a) \leq s(b)$ and $4 \leq s(b) \leq 5$. Then we label the edges e_3 and e_2 (the last of them may be labelled twice).

The following property of the resulting labelling is of importance for our purposes: Let the edges incident with v be denoted clockwise by $e^1, e^2, \dots, e^{s(v)}$, with $z^1, z^2, \dots, z^{s(v)}$ being their second end vertices, respectively. Then for each two consecutive unlabelled edges e^r, e^{r+1} (addition mod s(v)) with both z^r and z^{r+1} minor, we have

$s(z^r)=s(z^{r+1})=3$.

This implies in particular that if v is not dual to any (3,3)-edges, then at most $\lfloor j/2 \rfloor$ unlabelled edges join v with minor vertices. Since v transfers at most $(j-6)/\lfloor j/2 \rfloor$ to each of them, v transfers to unlabelled edges at most $j-6$ totally, which is equal to the initial charge of v. As for the labelled edges joining v with minor vertices, they recieve from v not more totally than v recieves from its dual edges; one may therefore assume that every edge e_3 defined above "feeds" from e_2 directly, passing v aside. Thus, the resulting charge is now seen to be non-negative in this case.

Let from now on v be dual to $k \geqslant 1$ edges of the type (3,3) and incident with l non-labelled edges which end in minor vertices. It follows from the definition of labelling that if the vertices z^r, z^{r+1} defined above both have degree 3, then each of the edges e^{r-1}, e^{r+2} is either labelled, or ends in a minor vertex (observe that $e^{r-1} \neq e^{r+2}$, since $P^* \neq K_4$). This yields $k \leqslant \lfloor s(v)/3 \rfloor$. We must verify that the initial charge $s(v)-6$ together with $k/2$ obtained by v from its dual (3,3)-edges, suffice to saturate all unlabelled edges incident with v.

Let $7 \leqslant s(v) \leqslant 11$. We tabulate below some easy upper bounds for l based on the properties of our labelling, depending on $s(v)$ and k :

	7	8	9	10	11
1:	2	2	3	3	4
2:	0	1	1	2	2
3:	-	-	0	0	1

Remind that each 7-, 8-, ... , 11-vertex transfers to each incident unlabelled edge the charge $q(v)$ which does not exceed 1/3, 1/2, 7/10, 4/5, and 11/12, respectively. It can be verified directly that in each of 3x5=15 cases the required inequality $s(v)-6+k/2 \geqslant (2k+l)q(v)$ holds. (Remark that one should distinguish between the lower integer function and the upper one; for example, $\lfloor 3.14 \rfloor = 3 < \lceil 3.14 \rceil = 4$.)

Now let $s(v) \geqslant 12$. Then $q(v)=1$ and $l \leqslant \lceil (s(v)-3k-1)/2 \rceil$. We must prove that $s(v)-6+k/2 \geqslant 2k+l$. Let us prove moreover that $s(v)-6+k/2 \geqslant 2k+\lceil (s(v)-3k-1)/2 \rceil$, or $s(v)-6-k/2 \geqslant \lceil (s(v)-k-1)/2 \rceil$.

Since $s(v) \geqslant 12$, we have $s(v)-6 \geqslant \lceil s(v)/2 \rceil$, but $\lceil s(v)/2 \rceil -k/2 \geqslant \lceil \lceil s(v)/2 \rceil -k/2-1/2 \rceil \geqslant \lceil (s(v)-k-1)/2 \rceil$. Hence, the resulting charge of each vertex is non-negative.

As for the edges, precisely the following (i,j)-edges, where $i \leq j$, have negative resulting charges: $i=3$, $j \leqslant 11$; $i=4$, $j \leqslant 7$ $i=5$, $j \leqslant 6$. In fact these charges are easily verified to be equal to $-a_{i,j}/10$. This implies $-\sum_{i,j} a_{i,j}e_{i,j}/10 \leqslant -12$, which completes the proof of (i).

The assertion (ii) is proved along the same lines.

References

1. O.V.Borodin, Simultaneous colorings of graphs on the plane, Met. diskret. anal. 45 (Novosibirsk, 1987) 21-27 (in Russian).
2. B.Grünbaum, Acyclic colorings of planar graphs, Isr. J. Math. 14 (1973) 390-408.
3. B.Grünbaum and G.C.Shephard, Analogs for tillings of Kotzig's theorem on minimal weight of edges, Ann. discret. math. 12 (1982) 129-140.
4. E.Jucovič, Strengthening of a theorem about 3-polytopes, Geom. dedic. 13 (1974) 233-237.
5. A.Kotzig, Contribution to the theory of Eulerian polyhedra, Mat. čas. 5 (1955) 101-103 (in Russian).
6. A.Kotzig, From the theory of Euler's polyhedra, Mat. čas. 13 (1963) 20-34 (in Russian).
7. H.Kronk and J.Mitchem, A seven color theorem on the sphere, Discrete math. 5 (1973) 253-260.
8. Theorie combinatorie, Proc. Int. Colloq. Rome, 1973 (Accad. nacion. dei lincei, Rome, 1976).
9. P.Wernicke, Über den Kartografischen Vierfarbensatz, Math. Ann. 58 (1904) 412-426.
10. J.Zaks, Extending Kotzig's theorem, Isr. J. Math. 45 (1983) 281-296.

On the Domination Problem for Bipartite Graphs

A. Brandstädt, Jena, GDR

1. Introduction

The Domination problem on graphs is one of the basic algorithmic problems on graphs and can be formulated as follows:
Let $G=(V,E)$ be a graph which is finite, undirected and simple (i.e. without self-loops and multiple edges).
A subset V' of V <u>dominates</u> G if for all vertices u of $V-V'$ there is a vertex $v \in V'$ which is adjacent to u in G i.e.

$$\bigwedge_{u \in V-V'} \bigvee_{v \in V'} uv \in E.$$

Let $\gamma(G)$=minimum size of a subset $V' \subseteq V$, which dominates G.
Because of its many applications and also theoretical importance the Domination problem plays a central role in algorithmic graph theory.
Let DOMINATING SET (DS) be the following decision problem:

$$DS = \{(G,k): \gamma(G) \leq k\}.$$

It is well-known that DS and also variants of this problem (where the dominating set is connected, independent, a clique etc.) are NP-complete on graphs and also on special graph classes (cf. /8/,/13/,/14).
It was shown in /7/ that DS is NP-complete even on chordal bipartite graphs - a special class of bipartite graphs which will be defined below.
We will study here another variant of Domination which is natural for bipartite graphs. The problem is mentioned already in /12/:

R. Bodendiek, R. Henn (Eds.)
Topics in Combinatorics and Graph Theory

Let $G=(X,Y,E)$ be a bipartite graph. A subset $X' \subsetneq X$ <u>dominates</u> <u>Y</u> if $\bigwedge_{y \in Y} \bigvee_{x \in X'} xy \in E$.

Analogously define that a subset $Y' \subsetneq Y$ dominates X.

Let

$$\gamma_X(G) = \begin{cases} \text{minimum size of a subset } X' \subsetneq X \text{ which dominates} \\ Y \text{ if there is such a dominating set} \\ \\ \infty \quad \text{otherwise} \end{cases}$$

and anlogously define $\gamma_Y(G)$.

Let X-DS= $\{(G,k): \gamma_X(G) \leq k\}$ and analogously define Y-DS.

It is easy to see that X-DS is NP-complete on bipartite graphs and for a class $\mathcal{G}$ of bipartite graphs with the property that with $G=(X,Y,E) \in \mathcal{G}$ also $G'=(Y,X,E)$ is in $\mathcal{G}$ the problems X-DS and Y-DS coincide on $\mathcal{G}$.

There are, however, classes of bipartite graphs which are not so symmetric and on such classes the computational complexity of X-DS and Y-DS may differ (an example for this case will be given).

The aim of this paper is to study the computational complexity of X-DS and Y-DS on some special classes of bipartite graphs which are closely connected to chordal, strongly chordal and split graphs.

2. Special graph classes

Let $C=(v_1,\dots,v_k)$ be a cycle in G. The edge $v_iv_j \in E$ is a <u>chord of C</u> if v_i and v_j are not neighbours in C.

If $C=(v_1,\dots,v_{2k})$ is a cycle of even length 2k in G then the chord $v_iv_j \in E$ of C is an <u>odd chord of C</u> if the distance in C between v_i and v_j is odd.

A graph $G=(V,E)$ is <u>chordal</u> (cf. /5/,/10/) if each cycle in G of length at least 4 has a chord.

A graph $G=(V,E)$ is <u>strongly chordal</u> (cf. /6/) if G is chordal and each cycle in G of even length at least 6 has an odd chord.

A graph G=(V,E) is a split graph (cf. /7/,/10/) if G and $\bar{G}=(V,\bar{E})$ with $vw \in \bar{E}$ iff $vw \notin E$ are chordal graphs. Equivalently, a split graph is a graph which can be partitioned into an independent set I and a clique C: $V = C \cup I$, $C \cap I = \emptyset$.

There are connections between chordal graphs and certain bipartite graphs with chordality properties (cf. e.g. /10/, /6/,/1/):

"Chordality" for bipartite graphs can be defined similarly as for graphs:

A bipartite graph G=(X,Y,E) is chordal bipartite if each cycle in G of length at least 6 has a chord (cf. /10/).

Let $N(x)=\{y:\ y \in V \text{ and } xy \in E\}$, $V = X \cup Y$.

An edge $uv \in E$ is bisimplicial in G if $N(u) \cup N(v)$ induce a complete bipartite subgraph in G.

A sequence of edges $u_1v_1,\ldots,u_kv_k$ is a perfect edge elimination scheme (p.e.e.s.) of G(cf. /10/) if for i=1,...,k u_iv_i is bisimplicial in the subgraph induced by $V - \bigcup_{j=1}^{i-1} \{u_j,v_j\}$ and the subgraph of G induced by $V - \bigcup_{j=1}^{k} \{u_j,v_j\}$ has no edges.

A bipartite graph G is perfect elimination bipartite if G has a p.e.e.s. (cf. /10/).

Chordal bipartite graphs are also perfect elimination bipartite (cf. /10/) but not vice versa.

We introduce another extension of chordal bipartite graphs which we call semichordal graphs (cf. /1/):

Let G=(X,Y,E) be a bipartite graph and $C=(x_1,y_1,\ldots,x_k,y_k)$ be a cycle in G of length $2k \geq 6$, $x_i \in X$, $y_i \in Y$, i=1,...,k.

C has an X-star if there is an $x \in X$ such that $xy_{i_j} \in E$, j=1,2,3.

Analogously define a Y-star.

A graph is semichordal (X-semichordal, Y-semichordal) if each chordless cycle of length at least 6 contains an X-star or a Y-star (an X-star, a Y-star).

Obviously, each chordal bipartite graph is semichordal.

The following examples show the incomparability of the classes of semichordal graphs and perfect elimination bipartite graphs w.r.t. set inclusion:

G_1:

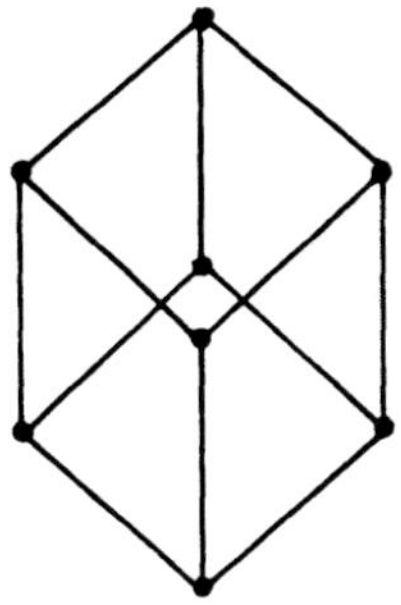

is semichordal but not perfect elimination bipartite

G_2:

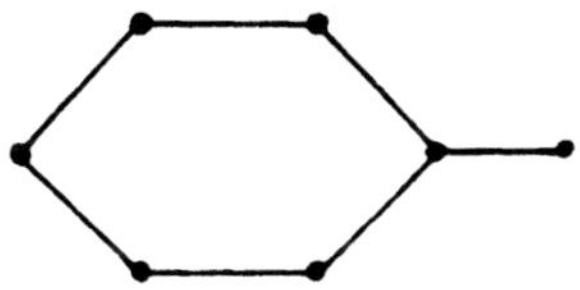

is perfect elimination but not semichordal

3. Some standard constructions of bipartite graphs

For a graph $G=(V,E)$ let T_G denote the set of its triangles and let C_G denote the set of its maximal cliques i.e. cliques which are not contained in a larger clique of G. Then the vertex-clique graph $B_C(G)$ of G is defined as the bipartite graph

$B_C(G) = (V,C_G,E')$ with $xc \in E'$ if $x \in c$, $x \in V$, $c \in C_G$.

The vertex-triangle graph $B_T(G)$ of G is defined analogously as the bipartite graph

$B_T(G) = (V,T_G,E')$ with $xt \in E'$ if $x \in t$, $x \in V$, $t \in T_G$.

For a bipartite graph $G=(X,Y,E)$ let

$\text{split}_X(G) = (X,Y,E')$ with $E' = E \cup \{xx' : x,x' \in X, x \neq x'\}$,

i.e. the corresponding split graph if X becomes a clique. The following properties are known:

Theorem 1:

i) (/3/)

For a bipartite graph $G=(X,Y,E)$ $\text{split}_X(G)$ is strongly chordal iff G is chordal bipartite,

ii) (/6/)
G is strongly chordal iff $B_C(G)$ is chordal bipartite,

iii) (/1/)
G is chordal iff $B_C(G)$ is Y-semichordal.

iv) (/1/)
Suppose that each edge of G is contained in a triangle. Then G is chordal iff $B_T(G)$ is Y-semichordal.

v) (/1/)
Let G be a split graph. Then $B_C(G)$ is perfect elimination bipartite.

Now let

$B_\alpha(\text{CHORD}) = \{B_\alpha(G)\text{: G a chordal graph}\}$,
$B_\alpha(\text{SPLIT}) = \{B_\alpha(G)\text{: G a split graph}\}$,
$B_\alpha(\text{STRONGCH}) = \{B_\alpha(G)\text{: G a strongly chordal graph}\}$,
$\alpha \in \{T, C\}$.

Let CBG denote the class of chordal bipartite graphs,
SEMICH denote the class of semichordal bipartite graphs,
PEB denote the class of perfect elimination bipartite graphs.

Figure 1 shows the inclusions between the classes mentioned above:

Figure 1:

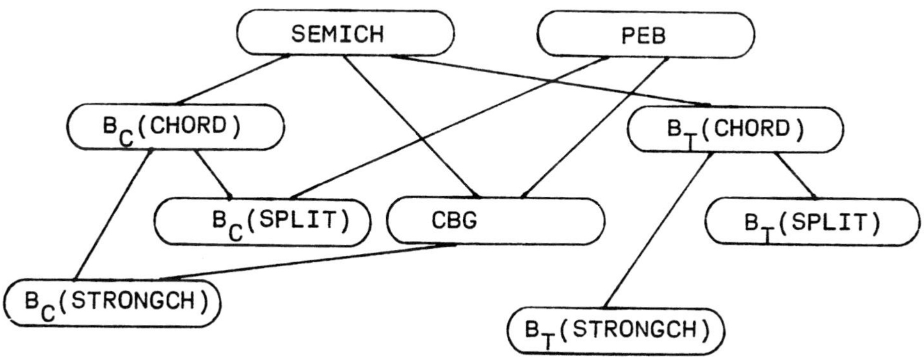

4. X-Domination and Y-Domination for special bipartite graphs

The special graph classes which will be considered here are the classes defined above.
Note that the classes CBG, SEMICH, and PEB are closed under the exchange of X and Y in the graph G=(X,Y,E). Thus the problems X-DS and Y-DS coincide on these classes.

Theorem 2:
The problem X-DS is solvable in polynomial time on chordal bipartite graphs.

This bases on the polynomial time solvability of the minimum dominating clique problem for strongly chordal graphs (cf. /15/) and i) of Theorem 1.
Theorem 2 is interesting in comparison with the NP-completeness of DS on CBG (cf. /16/).

Theorem 3:
The problem X-DS is NP-complete on the class B_C(SPLIT).

The result can be obtained by a reduction from the set cover problem MINIMUM COVER ([SP 5] of /8/) or by a reduction from the VERTEX COVER problem ([GT 1] of /8/). The second reduction yields also degree bounds, and it turns out that also the Steiner tree problem and the Connected Dominating Set problem on the class B_C(SPLIT) are NP-complete.
On the other hand it is a simple observation that Y-DS is solvable in polynomial time on the class B_C(CHORD) (and subclasses of it).
For $G \in B_T$(CHORD) the X-DS problem asks for a minimum set of vertices of G which contains at least one vertex of each triangle of G.
For chordal graphs this is equivalent to the problem of finding a minimum set of vertices of G which contains at least one vertex of each cycle of G which is known as the feedback vertex set problem on undirected graphs.

In /9/ and independently in /2/ and /11/ it was shown that X-DS is solvable in polynomial time on vertex-triangle graphs of chordal graphs i.e. X-DS(B_T(CHORD)) is solvable in polynomial time.
For $G \in B_T$(CHORD) the Y-DS problem asks for a minimum set of triangles of G which covers all vertices of G.
The complexity of this problem is not known.
For split graphs instead of chordal graphs, however, the problem can be reduced to a maximum flow problem:

Theorem 4 (/16/):
The problem Y-DS is solvable in polynomial time for vertex-triangle graphs of split graphs i.e. for $G \in B_T$(SPLIT).

Figure 2: Survey of the computational complexity of X-DS and Y-DS:

P - solvable in polynomial time
NP-c - NP-complete
? - open

	X-DS	Y-DS
CBG	P	
B_C(SPLIT)	NP-c	P
B_C(CHORD)	NP-c	P
B_T(CHORD)	P	?
B_T(SPLIT)	P	P
SEMICH	NP-c	
PEB	NP-c	

References:

/1/ A. Brandstädt, Classes of bipartite graphs related to chordal graphs, Fo.-Ergebn. d. FSU Jena N/88/10, 1988, to appear in Discr. Appl. Math.

/2/ D.G. Corneil, J. Fonlupt, The complexity of generalized clique covering, manuscript 1987.

/3/ E. Dahlhaus, manuscript Habilitationsschrift, Bonn, 1988.

/4/ P. Damaschke, H. Müller, Domination in convex and chordal bipartite graphs, manuscript 1989.

/5/ G. Dirac, On rigit circuit graphs, Abhandlungen Math. Seminar d. Univ. Hamburg 25, 1961, 71-76.

/6/ M. Farber, Characterizations of strongly chordal graphs, Discr. Math. 43 (1983), 173-189.

/7/ S. Földes, P.L. Hammer, Split graphs, Proc. 8th South-eastern Conf. Combinatorics, Graph Theory and Computing, 1977, 311-315.

/8/ M.R. Garey, D.S. Johnson, Computers and Intractability, Freeman, San Francisco, 1979.

/9/ F. Gavril, M. Yannakakis, The maximum k-colorable sub-graph problem for chordal graphs, Inf. Proc. Letters 24 (1987), 133-137.

/10/ M.C. Golumbic, Algorithmic Graph Theory and Perfect Graphs, Academic Press, 1980.

/11/ Th. Gundermann, Feedback vertex set for chordal graphs, manuscript, 1987.

/12/ S.T. Hedetniemi, R. Laskar, A bipartite Theory of Graphs: I, Congressus Numerantium 55 (1986), 5-14.

/13/ D.S. Johnson, The NP-completeness Column: An Ongoing Guide (10th edition), J. Algor. 5, 147-160 (1984).

/14/ D.S. Johnson, The NP-completeness Column: An Ongoing Guide (16th edition), J. Algor. 6, 434-451 (1985).

/15/ D. Kratsch, A linear time algorithm for the minimum dominating clique problem on strongly chordal graphs, Fo.-Ergebn. FSU Jena, N/87/29, 1987.

/16/ D. Kratsch, personal communication, 1987.

/17/ H. Müller, A. Brandstädt, The NP-completeness of STEINER TREE and DOMINATING SET for chordal bi-partite graphs, Theor. Comp. Science 53 (1987), 257-265.

Polyhedral Maps With Few Edges

U. Brehm, Berlin, FRG

ABSTRACT. We give a lower bound for the number of edges which a polyhedral map of given topological type can have. For several manifolds we find the exact minimal value for the number of edges.

INTRODUCTION

DEFINITION. A *polyhedral map (p. map)* is a finite tesselation of a compact 2-manifold without boundary by closed topological discs *(facets)* such that the intersection of any two or three tiles is empty, a point *(vertex)* or a topological interval *(edge)*. A polyhedral map is called *weakly neighborly (w.n.p. map)* if any two vertices lie in a common facet.

Weakly neighborly polyhedral maps in general have been studied in [5], where also results on p. maps are included. The w.n.p. maps of various manifolds have been classified in [6], [2], [3], [4] and [7].

Polyhedral maps arise as a canonical generalization of triangulated manifolds. They arise also as the combinatorial structure of polyhedra (which are 2-manifolds) in $\mathbb{R}^n$ with convex facets. We regard p. maps as the underlying combinatorial structure of the tesselation.

Note that polyhedral maps can be defined purely combinatorially by the incidences between vertices and facets. Exchanging the rôles of the vertices and of the facets we get again a polyhedral map, the *dual* polyhedral map.

Now a very natural question is to ask for the smallest possible number of vertices, edges and facets, respectively, which a polyhedral map of given topological type can have.

For the case of the number of vertices we can restrict ourselves to triangulated manifolds because each facet can be triangulated without adding new vertices and the definition of a p. map ensures that for any such triangulation no edge can appear twice. The minimal number of

R. Bodendiek, R. Henn (Eds.)
Topics in Combinatorics and Graph Theory

vertices of a triangulation has been determined for each 2-manifold in [9] and [8] by G. Ringel and G. Ringel together with M. Jungerman, respectively. This result is closely related to the famous Map Color Theorem of Ringel and Youngs (cf. [10] and [11]).
By duality one gets the same number for the minimal number of facets of a polyhedral map of given topological type. So the only of our three problems which remains open is the question for the minimal number of edges of a polyhedral map, where even not a good lower bound seems to be known. In the present paper we give such a lower bound and determine for some 2-manifolds the exact minimal number of edges of p. maps.

LOWER BOUNDS FOR THE NUMBER OF EDGES

Let V, E, F denote the number of vertices, edges, facets, respectively, of a p. map. Define

$Y := V + F$

$\chi := V - E + F = Y - E$ the Euler characteristic.

Now we can state the main results of this section.

THEOREM 1. For every polyhedral map M holds

$$E \leq \frac{Y}{8}(\sqrt{2Y} + 2) \qquad (1)$$

$$-\chi \leq \frac{Y}{8}(\sqrt{2Y} - 6) \qquad (2)$$

Equality holds if and only if M is a w.n.p. map and V = F and all facets of M have the same number of vertices.

REMARK. The conditions for equality imply that there exists a number k such that k k-gons meet in each vertex and $V = F = (k-1)^2$, $E = \frac{1}{2}k(k-1)^2$, $\chi = -\frac{1}{2}(k-4)(k-1)^2$. Furthermore M^{dual} is also a w.n.p. map.

COROLLARY 1. For each polyhedral map of Euler characteristic χ holds

$$E \geq G(\chi) - \chi, \text{ where} \qquad (3)$$

$$G(\chi) := \min\{y \in \mathbb{N} \mid y(\sqrt{2y} - 6) \geq -8\chi \text{ and } y \geq 8\}$$

and $\mathbb{N}$ denotes the set of positive integers.

We can give also a more explicit lower bound for E which is for large $|\chi|$ almost as good as that one of Corollary 1 and which gives for $\chi \to -\infty$ asymptotically the same result.

THEOREM 2. For each polyhedral map with $\chi \leq 0$ holds

$$E > |\chi| + 2|2\chi|^{2/3} + 4|2\chi|^{1/3} + 6. \tag{4}$$

PROOF of Theorem 1. Let M be a polyhedral map. We assume w.l.o.g. $V \leq F$ (otherwise regard the dual map). Let $k_1,\ldots,k_F$ denote the numbers of vertices of the facets $f_1,\ldots,f_F$. Then

$$\sum_{i=1}^{F} k_i = 2E \tag{5}$$

$$V(V-1) \overset{(a)}{\geq} 2E + \sum_{i=1}^{F} k_i(k_i-3) \overset{(5)}{=} \sum_{i=1}^{F} k_i^2 - 4E \overset{(b)}{\geq} \frac{(2E)^2}{F} - 4E \tag{6}$$

(a) because the number of diagonals of f_i is $\frac{1}{2}k_i(k_i-3)$ and because all diagonals and edges have different pairs of endpoints because M is a p.map. Equality holds if and only if M is weakly neighborly.

(b) using (5) the inequality follows from the Cauchy-Schwarz inequality with equality if and only if $k_1 = k_2 = \ldots = k_F$; (regard the vectors $(1,\ldots,1)$, $(k_1,\ldots,k_F) \in \mathbb{R}^F$).

(6) implies

$$4E^2 - 4FE - V(V-1)F \leq 0$$

thus with $t := \frac{V}{F+V}$, recalling $Y = V + F$

$$E \leq \frac{1}{2}\left(F + \sqrt{F(F+V^2-V)}\right) = \frac{Y}{2}\left(1 - t + \sqrt{(1-t)(1-2t+Yt^2)}\right).$$

Now define

$$f(Y,t) := -t + \sqrt{(1-t)(1-2t+Yt^2)}.$$

We shall show

$$f(Y,t) < f\left(Y, \frac{1}{2}\right) \quad \text{for } Y \geq 18, \quad 0 < t < \frac{1}{2}. \tag{7}$$

Proof of (7): For Y = 18 holds

$$f\left(18,\frac{1}{2}\right) - f(18,t) = 1 - f(18,t) = 1 + t - \sqrt{(1-t)(1-2t+18t^2)} > 0,$$

because $(1+t)^2 - (1-t)(1-2t+18t^2) = t(1-2t)(5-9t) > 0$ for $0 < t < \frac{1}{2}$.

On the other hand we have for all Y and for $0 \leq t \leq \frac{1}{2}$

$$\frac{\partial f}{\partial Y}(Y,t) = \frac{1}{2} t^2 (1-t)^{1/2} (1-2t+Yt^2)^{-1/2} \quad \le \frac{1}{2} t ((1-t)/Y)^{1/2} \le \frac{1}{4} (2Y)^{-1/2} = \frac{\partial f}{\partial Y}(Y, \frac{1}{2}) .$$

This implies (7).

Thus we get for $Y \ge 18$

$$E \le \frac{Y}{8}(2 + \sqrt{2Y})$$

$$-\chi \le \frac{Y}{8}(\sqrt{2Y} - 6)$$

with equality if and only if M is a w.n.p. map, $k_1 = ... = k_F$ and $V = F$, (see (a), (b), (7), noting $t = \frac{1}{2} \Leftrightarrow V = F$).

It remains to show Theorem 1 for $Y \le 17$.
For a p. map clearly holds $Y = F + V \ge 8$.
For $8 \le Y \le 15$ holds

$$-2 \le \frac{Y}{8}(\sqrt{2Y} - 6) .$$

We show that a p. map with $Y \le 15$ is a sphere. Note that in a p. map a "diagonal" of a facet can be added as an edge yielding a new p. map with one additonal edge. Dually one can "split" a vertex of valence ≥ 4 by adding an edge in the dual map and then dualizing again. Because every p. map except the tetrahedron contains a facet which is not a triangle, or a vertex of valence > 3 one can get a p. map with exactly one additonal edge (except for the tetrahedron). Thus we can assume w.l.o.g. $Y = 15$ and $V \le F$ as usual. We assume $\chi \le +1$, thus $E \le 14$ and show that this leads to a contradiction.
$4 \le V \le 6, F \ge 9$ gives a contradiction to (6), thus $V = 7, F = 8$. Here we have $V(V-1) = \frac{(2E)^2}{F} - 4E$ for $E = 14$, thus equality in (a) and (b), thus $k_1 = ... = k_8 = 7/2$ which is impossible because the k_i's must be integers.

Finally we have to consider the case $16 \le Y \le 17$.
For $Y \le 17$ holds

$$-1 \le \frac{Y}{8}(\sqrt{2Y} - 6)$$

We show that a p. map with $Y \le 17$ is either a sphere or a projective plane.
We can assume w.l.o.g. $Y = 17, V \le F$.
If $\chi \le 0$, then $E \le 17$ and we get for every V with $4 \le V \le 8$ and $F = 17 - V$ a contradiction to (6). □

PROOF of the Remark. Let k be the number of vertices of each facet. Let v be some vertex and h be the number of facets meeting in that vertex. Because M is a w.n.p. map each vertex $w \neq v$ is either contained in exactly one facet meeting at v or (v,w) is an edge. Thus $V-1 = h(k-2)$, which shows that the number h does not depend on the vertex v. Now $V = F$ and $E = \frac{1}{2} F k = \frac{1}{2} V h$ imply $k = h$. Thus $V = F = (k-1)^2$, $E = \frac{1}{2}k(k-1)^2$,
$\chi = -\frac{1}{2}(k-4)(k-1)^2$. ❑

PROOF of Corollary 1. (2) and $Y \geq 8$ imply

$$E + \chi = Y \geq \min\left\{y \in \mathbb{N} \;\middle|\; y(\sqrt{2y} - 6) \geq -8\chi \text{ and } y \geq 8\right\}.$$ ❑

PROOF of Theorem 2. Corollary 1 implies for $\chi \leq 0$ that $Y \geq 18$. We shall show for all $Y \geq 18$

$$Y > 2\left(\frac{Y}{4}(\sqrt{2Y} - 6)\right)^{2/3} + 4\left(\frac{Y}{4}(\sqrt{2Y} - 6)\right)^{1/3} + 6. \tag{8}$$

From (8) follows Theorem 2 using (2), $-\chi = |\chi|$ and $E = |\chi| + Y$.

(8) is equivalent to

$$1 > \left(1 - 3\sqrt{\frac{2}{Y}}\right)^{2/3} + 2\left(1 - 3\sqrt{\frac{2}{Y}}\right)^{1/3}\sqrt{\frac{2}{Y}} + \frac{6}{Y}. \tag{9}$$

Substituting $z^3 := 1 - 3\sqrt{\frac{2}{Y}}$ we get

$$0 > -1 + z^2 + 2z\left(\frac{1-z^3}{3}\right) + 3\left(\frac{1-z^3}{3}\right)^2 = \frac{1}{3}(z-1)^3(z^3 + 3z^2 + 4z + 2) \tag{10}$$

(10) is obviously true for all $z \in [0,1)$ which implies that (9) and thus (8) are true for all $Y \geq 18$. ❑

MINIMAL NUMBERS OF EDGES

We determine for some 2-manifolds the exact value of the minimal number of edges of a polyhedral map.

First we define for even numbers $\chi \leq +2$ $E_+(\chi)$ as the smallest number E for which a p. map with E edges exists on the orientable 2-manifold of Euler characteristic χ,
and for integers $\chi \leq +1$ $E_-(\chi)$ as the smallest number E for which a p. map with E edges exists on the non-orientable 2-manifold of Euler characteristic χ.

THEOREM 3.

$E_+(2) = 6$

$E_-(1) = 15$

$E_+(0) = E_-(0) = 18$

$E_-(-1) = 23$
$E_-(-2) = 26$
$E_+(-8) = 40$
$E_+(-26) = 78$.

PROOF. $\chi = +2$: The p. map on the 2-sphere ($\chi = +2$) with minimal number of edges is just the tetrahedron (E = 6, Y = 8), for which equality holds in (1) and (2).

$\chi = +1$: In the proof of Theorem 1 we showed that $E \geq 15$. It turns out that there are exactly seven p. maps on the projective plane $\mathbb{R}P^2$ with 15 edges. This follows from [1]. One of these is the "hemi-icosahedron" (see Fig. 1a) which is the unique 6-vertex triangulation of $\mathbb{R}P^2$. In Figure 1 opposite vertices (and edges, respectively,) on the boundary are identified. It turns out that these seven p. maps are exactly the seven w.n.p. maps on $\mathbb{R}P^2$ (see [7]). They occur in pairs of dual maps a-b, c-d, e-f and one self-dual map g.

$\chi = 0$: Here (3) says $E \geq 18$. There are exactly two p. maps on the torus and one p. map on the Klein-bottle with 18 edges (see Fig. 2) (see [6] and [7]).Note that equality holds in (1) and (2) so that the Remark can be applied.

$\chi = -1$: We show that E = 22 leads to a contradiction . If E = 22, then Y = 21. W.l.o.g. $V \leq F$. $4 \leq V \leq 9$, $F = 21 - V$ leads to a contradiction to (6). Thus V = 10, F = 11. Let p_i denote the number of i-gons. (6) implies either $p_4 = 11$ or $p_3 = p_5 = 1$, $p_4 = 9$.
$p_4 = 11$ is impossible because in each vertex at most 4 quadrangles can meet (a vertex can "see" at most $V - 1 = 9$ other vertices via an edge or a diagonal), which would imply $V \geq 11$.
$p_3 = p_5 = 1$, $p_4 = 9$ imply equality in (a), thus our map would be a w.n.p. map. In [2] we have determined all w.n.p. maps with $\chi = -1$. One of these has 24 edges, the other three have 25 edges.
In Figure 3 a we show a map with $\chi = -1$ and 23 edges. Its automorphism group is isomorphic to $Z_2 \times Z_2$.

$\chi = -2$: We show that E = 25 leads to a contradiction. If E = 25, then Y = 23. W.l.o.g. $V \leq F$.
$4 \leq V \leq 10$, $F = 22 - V$ leads to a contradiction to (6) . Thus V = 11, F = 12. Here we get eqality in (6) which implies that our map is a w.n.p. map. In [3] and [4] we have determined all w.n.p. maps with $\chi = -2$. All of these have 30 edges.

In Figure 3 b we show a non-orientable p. map with $\chi = -2$ and 26 edges. Its automorphism group is isomorphic to Z_2.

$\chi = -8$: Here (3) says $E \geq 40$. We construct an orientable p. map with $\chi = -8$, $E = 40$, $V = F = 16$ as follows: Let the set of vertices be the cyclic group Z_{16}. Then the pentagon (0,1,2,5,8) (cyclic order of the vertices) and its images under Z_{16} give our map. Note that it is self-dual and has Z_{16} as a vertex- and facet-transitive automorphism group . In Figure 4 we show this p. map. The numbers in the centers of the facets denote the names of the facets showing the self-duality .The Petrie polygons 0 1 2 3 ... and 0 3 6 9 ... of our map are Hamiltonian cycles.

$\chi = -26$: If $E = 77$, then $Y = 51$. W.l.o.g. $V \leq F$, but $4 \leq V \leq 25$, $F = 51 - V$ leads to a contradiction to (6). Thus $E \geq 78$.
We construct an orientable p. map with $\chi = -26$, $E = 78$, $V = F = 26$ as follows: Let the set of vertices be the cyclic group Z_{26}. Then the hexagon (0, 1, 2, 5, 8, 17) (cyclic order of the vertices) and its images under Z_{26} give our map. Note that it is self-dual and has Z_{26} as a vertex- and facet-transitive automorphism group. The Petrie polygons 0 1 2 3..., 0 3 6 9...,.0 9 18 27... are edge disjoint Hamilton cycles which cover all edges of our map.

❑

REMARKS. 1) By a similar construction as for $\chi = -8$ or $\chi = -26$ one can get easily infinite sequences of orientable Hamiltonian self-dual p. maps with vertex- and facet-transitive automorphism groups.
2) The problem of determining $E_+(\chi)$ for all even $\chi \leq 2$ and $E_-(\chi)$ for all $\chi \leq 1$ seems to me to be completely untractible. On the other hand it should not be too difficult to determine these numbers for some further values of χ .
3) Note that for $\chi = +2, 0, -8$ we can get equality in (1), (2), (3). For $\chi = +1, -1, -2, -26$ the minimal edge number exceeds the lower bound in (3) by 1.

CONJECTURES

1) There exist infinitely many w.n.p. maps for which hold equality in (1).
2) There exists a universal constant K such that

$$E(\chi) \leq |\chi| + 2|2\chi|^{2/3} + K|\chi|^{1/3}$$

for all $\chi \leq -1$, where $E(\chi) := \max(E_+(\chi), E_-(\chi))$ for χ even and $E(\chi) := E_-(\chi)$ for χ odd.

Conjecture 2 states that our lower bound in Theorem 2 is not too far from the exact value for the minimal number of edges.

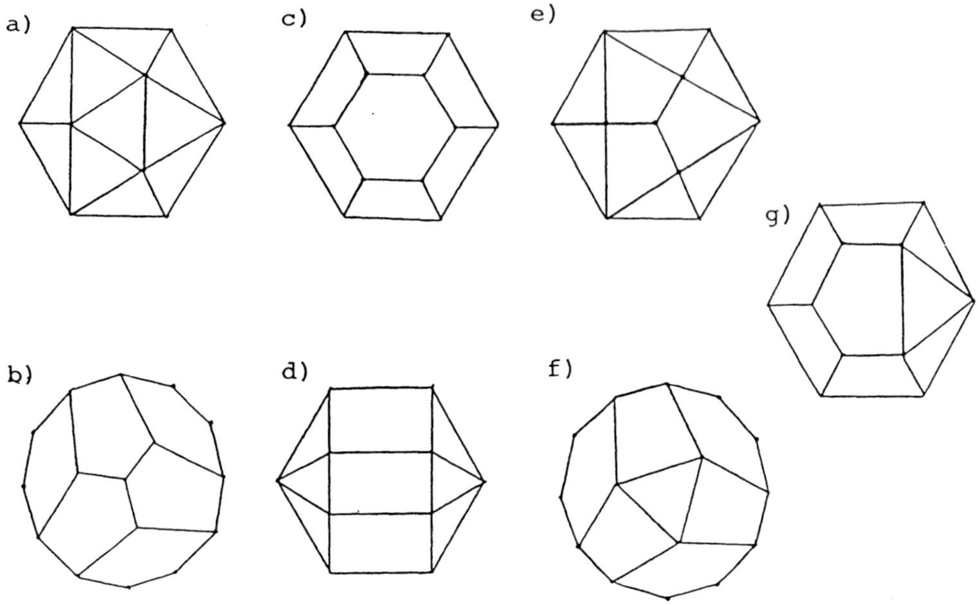

FIGURE 1

The seven polyhedral maps on $\mathbb{R}P^2$ with minimal number of edges 15 (opposite points on the boundary are identified)

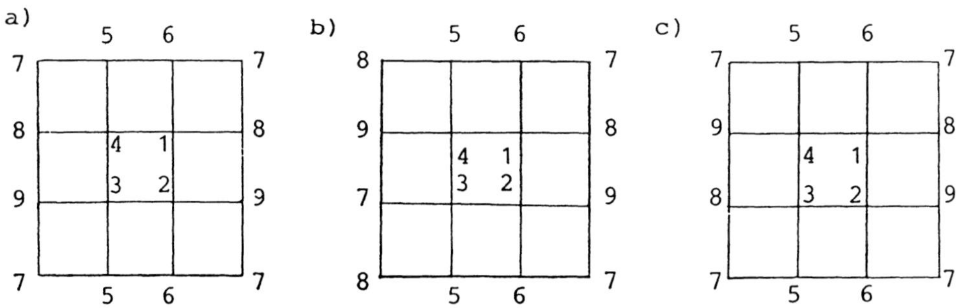

FIGURE 2

The three polyhedral maps with $\chi = 0$ (a,b torus, c Klein-bottle) with minimal number of edges 18.

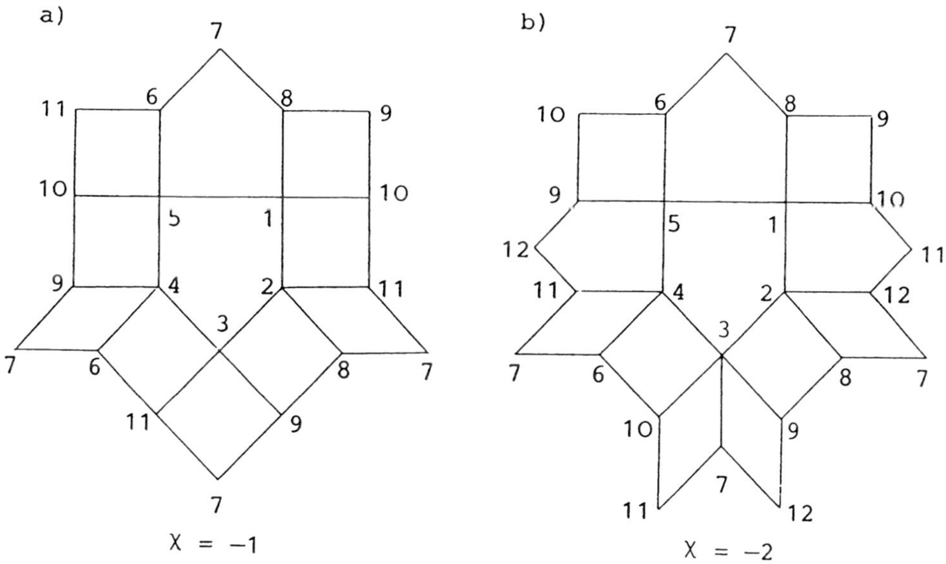

FIGURE 3

a) A polyhedral map with $\chi = -1$ with minimal number of edges 23

b) A polyhedral map with $\chi = -2$ with minimal number of edges 26

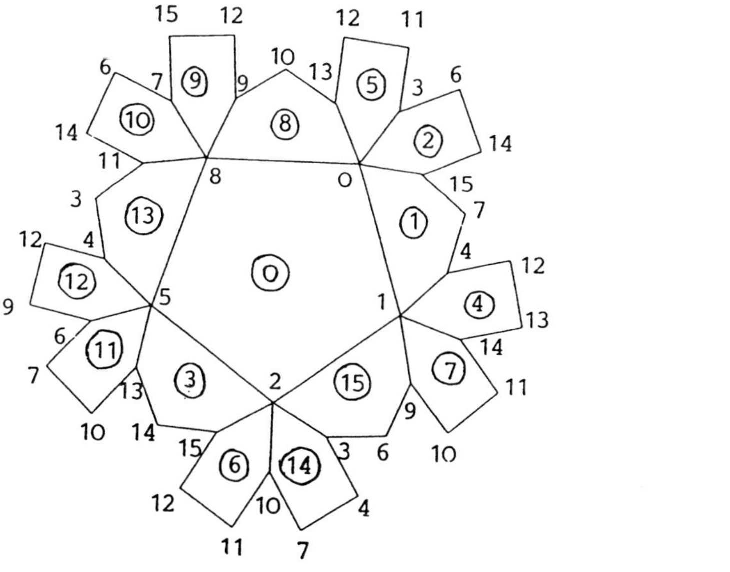

FIGURE 4

An orientable map of genus 5 ($\chi = -8$) with minimal number of edges 40 and Z_{16} as automorphism group

REFERENCES

[1] D. Barnette, Generating Projective Plane Polyhedral Maps, to appear, J. Comb. Theory, Ser. B.

[2] A. Altshuler and U. Brehm, The Weakly Neighborly Polyhedral Maps on the 2-Manifold with Euler Characteristic -1, Discrete & Comput. Geom. 1 (1986), 355-369.

[3] A. Altshuler and U. Brehm, Nonexistence of Weakly Neighborly Polyhedral Maps on the Orientable 2-Manifold of Genus 2, J. Comb. Theory 42, Ser. A (1986), 87-103.

[4] A. Altshuler and U. Brehm, The Weakly Neighborly Polyhedral Maps on the Non-orientable 2-Manifold with Euler Characteristic -2, J. Comb. Theory 45, Ser. A (1987), 104-124.

[5] U. Brehm and A. Altshuler, On Weakly Neighborly Polyhedral Maps of Arbitratry Genus, Israel J. Math. 53 (1986), 137-157.

[6] U. Brehm and A. Altshuler, The Weakly Neighborly Polyhedral Maps on the Torus, Geom. Dedicata 18 (1985), 227-238.

[7] U. Brehm, The Weakly Neighborly Polyhedral Maps on the Projective Plane, the Möbius Strip and the Klein Bottle (in preparation).

[8] M. Jungermann and G. Ringel, Minimal Triangulations on Orientable Surfaces, Acta Math. 145 (1980), 121-154.

[9] G. Ringel, Wie man die geschlossenen nichtorientierbaren Flächen in möglichst wenig Dreiecke zerlegen kann, Math. Ann. 130 (1955), 317-326.

[10] G. Ringel, Map Color Theorem, Springer Verlag Berlin, 1974.

[11] G. Ringel, J. W. T. Youngs, Solution of the Heawood Map-Coloring Problem, Proc. Nat. Acad. Sci. U.S.A. 60 (1968), 438-445.

Aut $G_{m,n}$ for the Hasse Graph $G_{m,n}$ of the Subword Poset $B_{m,n}$ of an m-Ary Cyclic Word of Length n

G. Burosch, Rostock, GDR
J.-M. Laborde, Grenoble, France

I) Introduction

There exists a rich literature on automorphism groups for (undirected) graphs. This subject is specially developed for distance transitive graphs (one of them is the n-cube). We consider here a non distance transitive graph $G_{m,n}$ obtained from the hypercube by identifying some of its vertices and we charcaterise its automorphism group in terms of S_p the symmetric group on p elements. Starting point for this study has been when investigating the poset $B_{m,n}$ of all sub-words from a given word of length n:

$$u_{m,n} = 01....(m\text{-}1)01....(m\text{-}1)............01....(m\text{-}1)01...(r\text{-}1),$$

where $n=qm+r$ with $0 \leq r < m$. $G_{m,n}$ is just the Hasse graph of $B_{m,n}$. $G_{m,n}$ for $n \leq m$ is an n-dimensional cube Q_n. The automorphism group of $B_{m,n}$ (as a poset) were characterised for m=2 in [Bu Fr Rö] and in general in [Bu Gr La].

The main result in this paper is the following

Theorem 1 :

$$\mathrm{Aut}\, G_{m,n} = \begin{cases} S_2{}^n \otimes S_n & \text{for} \quad 1 \leq n \leq m \\ S_2{}^2 \otimes S_{m-1} & \text{for} \quad n = m+1 \\ S_2 \otimes S_{2m-n} & \text{for} \quad m+2 \leq n \leq 2m-1 . \\ S_2 & \text{for} \quad 2m \leq n \end{cases}$$

R. Bodendiek, R. Henn (Eds.)
Topics in Combinatorics and Graph Theory

This is to be compared with

Theorem 1' [Bu Gr La] :

$$\operatorname{Aut} \boldsymbol{B}_{m,n} = \begin{cases} S_n & \text{for } \quad 1 \leq \; n \leq m \\ S_2 \otimes S_{2m-n} & \text{for } m+1 \leq \; n \leq 2m-1 \\ S_2 & \text{for } \quad 2m \leq \; n\,. \end{cases}$$

In paragraph **II** we introduce some concepts and properties of $\boldsymbol{B}_{m,n}$ as taken from earlier studies on $\boldsymbol{B}_{m,n}$. In paragraph **III** we give some lemmas concerning isomorphy. Paragraph **IV** contains a proof for Theorem 1 and the last paragraph gives some hints for two open problems.

II) Basic facts about $\boldsymbol{B}_{m,n}$

Let $\boldsymbol{P}=(X,\leq)$ be a poset. If for $x\in X$ and $y \in X$, $x<y$ and it does not exist any $z \in X$ $x<z<y$, we note as usually $x <\!\!\cdot\, y$. If X has a smallest element a, we call a function r: X→ IN a rank function if it fulfills $r(a)=0$ and $r(y)=r(x)+1$ for all $x <\!\!\cdot\, y$. If $\boldsymbol{P}$ has a smallest element and a rank function then $\boldsymbol{P}$ is called ranked. In a ranked poset $\boldsymbol{P}$ the kth level, for k=0,1,... is defined as $\{x\in X,\ r(x)=k)\}$. Every poset $\boldsymbol{P}$ can be associated with its Hasse diagram $\mathcal{H}(\boldsymbol{P})$, the digraph (X,E) with vertex set the underlying set of $\boldsymbol{P}$ and

$$(x,y) \in E \;\Leftrightarrow\; x<\!\!\cdot\, y\ .$$

Removing the orientation we obtain the Hasse graph $\mathcal{G}(\boldsymbol{P})$, i.e. the graph (X,E') such that

$$\{x,y\} \in E' \;\Leftrightarrow\; x<\!\!\cdot\, y \text{ or } y<\!\!\cdot\, x\ .$$

We denote by Aut $\boldsymbol{P}$, Aut $\mathcal{H}(\boldsymbol{P})$, Aut $\mathcal{G}(\boldsymbol{P})$ the corresponding automorphism groups. Consider $\boldsymbol{B}_n$ the Boolean lattice defined as the set of all $(x_1,x_2,....,x_n)\in \{0,1\}^n$ with the order

$(x_1,x_2,....,x_n) \leq (y_1,y_2,....,y_n) \quad \Leftrightarrow \quad x_i \leq y_i \quad i=1,2, ...n.$ $\boldsymbol{Q}_n$ denotes the Hasse graph of $\boldsymbol{B}_n$, commonly called hypercube, n-dimensionnal cube or n-cube.

We have Aut $\boldsymbol{B}_n$=Aut $\mathcal{H}(\boldsymbol{B}_n)$ but $|\operatorname{Aut} \boldsymbol{Q}_n| = 2^n |\operatorname{Aut} \boldsymbol{B}_n|$. Indeed the symmetry of $\boldsymbol{Q}_n$ allows much more automorphisms than just in B_n. Comparing Theorems 1 and 1' we see that the symmetry of Q_n after identification of the vertices leading to $G_{m,n}$ is so destroyed that Aut $G_{m,n}$ = Aut $B_{m,n}$ for $n \geq 2m-1$.

Given an undirected graph $G(V,E)$ and one of its vertices $a \in V$ a so called canonical poset $\mathcal{P}_a(G)$ of G in respect to a is defined taking

$x \leq y$ iff it exists in G a shortest path from a to y through x.

Lemma 1 ([Bu 2],[Mu],[Sa]) An undirected connected graph is bipartite iff it is the Hasse graph of all its canonical posets.

The word $\mathbf{a}=a_1a_2....a_t$ is a subword of $\mathbf{b}=b_1b_2....b_s$ iff $a_i=b_{j_i}$ for $i=1,2,....,t$ and $j_1<j_2<....<j_t$. We define $W_{m,n}$ as the set of all subwords (including the empty word Ø) of

$$u_{m,n} = u_1u_2....u_n \qquad u_1=0, \quad u_{i+1} = u_i +1 \ (\mathrm{mod}\ m).$$

The length of a word (its number of letters) has evidently the properties to be a rank function. So if we denote "**a** is a subword of **b**" as $\mathbf{a} \leq \mathbf{b}$, $(W_{m,n}, \leq)$ turns to be a ranked poset we will denote $B_{m,n}$. Ø is the smallest element of $B_{m,n}$. Corresponding Hasse graph (resp. diagram) will be denoted $G_{m,n}$ (resp. $H_{m,n}$).

[Bu Fra Rö], [Bu Gr La Wa] and [Bu Gr La] give some properties of $B_{m,n}$.

As in the present paper we are interested in $G_{m,n}$ in contrast to $B_{m,n}$ we just present the following description of $B_{m,n}$, which allows a simple construction of $G_{m,n}$.

We say a graph $G_1 =(V_1, E_1)$ is obtained from $G =(V,E)$ by an elementary identification of both vertices $x, y \in V$ if we get V_1 from V by deleting x and y and adding a new vertex x' which will be connected to exactly those remaining vertices in V being previously connected to x or y. More generally we say **G'** is obtained from **G** by vertices identification in the case we have a sequence of graphs starting with **G** ending with **G'** such that each of the graphs in-between is obtained from the preceding by an elementary identification.

For two n-tuples

$\mathbf{x} = (x_1,x_2,..,x_s,....,x_t,...,x_n) \in \{0, 1\}^n$

$\mathbf{y} = (y_1,y_2,..,y_s,....,y_t,...,y_n) \in \{0, 1\}^n$

we call **y** an m-admissible transposition of **x** iff

- $t = s + m$
- $x_{s+1} = ... = x_{t-1} = 0$
- $x_s = y_t = 1$ and $x_t = y_s = 0$ or $x_s = y_t = 0$ and $x_t = y_s = 1$.

The set $[\mathbf{x}]_m$ consists of **x** plus all elements which can be obtained from **x** by successive m-admissible transpositions. As an example for **x** = (11000) and m=3, we have $[\mathbf{x}]_m = \{(11000), (10001), (00011)\}$. The partition of $\{0, 1\}^n$ in the sets $[\mathbf{x}]_m$ is for a given m≥2 a partition in equivalent classes.

Lemma 2 [Bu Gr La Wa] : For given integers m≥2, n≥1 the graph $G_{m,n}$ derives from the hypercube Q_n by identification of all vertices in each equivalent class as defined above. (For n≤m $G_{m,n} = Q_n$.)

Key for the proof is the interpretation of an n-tuple $\mathbf{x} \in \{0,1\}^n$ as a choice vector w(**a**) for a subword **a** of $u_{m,n} = u_1u_2....u_n$, where $\mathbf{a} = u_{j_1}, u_{j_2}, ..., u_{j_k}$ with $j_1<j_2<....<j_k$ and $\{j_1, j_2, ..., j_k\} = \{i,\ 1 \le i \le n \text{ and } x_i = 1\}$. Figure 1 shows $G_{3,5}$ as well as the vertices of the original Q_5 from which it has been constructed.

In an undirected graph G , denote $d_G(x,y)$ or $d(x,y)$ (resp. $d_G(x)$ or $d(x)$) the distance between two vertices x and y (resp. the degree of a vertex x). In the digraph H , denote $d^+_H(x)$ (resp. d^-_H) the outdegree (resp. indegree) of a vertex x.

Lemma 3 [Bu Gr La Wa] :
For given integers m≥2, n≥1

- $d_{G_{m,n}}(\emptyset) = m$,
- $d_{G_{m,n}}(\mathbf{u}_{m,n-m}) = m(n-m+1) > d_{G_{m,n}}(\mathbf{a}_{m,n-m})$ for all $\mathbf{a} \in W_{m,n}$, $\mathbf{a} \neq \mathbf{u}_{m,n-m}$.

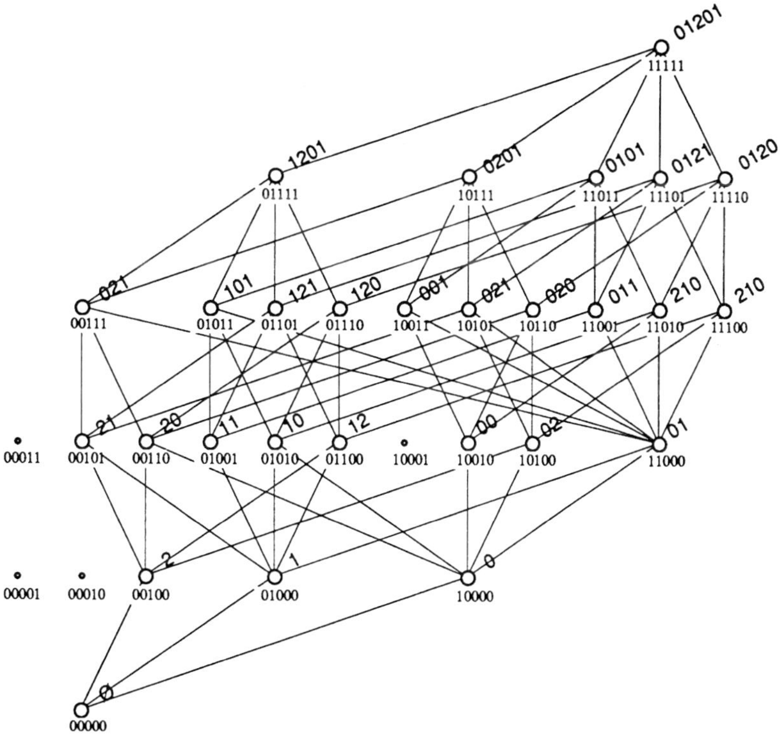

Figure 1

Lemma 4 :

For given integers m≥2 and n>m the minimal degree δ of the graph is given by

$$\delta(G_{m,n}) = \begin{cases} d_{G_{m,n}}(\emptyset) = m < d_{G_{m,n}}(\mathbf{a}) & \forall\, \mathbf{a} \neq \emptyset \text{ and } n>m+1 \quad (1) \\ d_{G_{m,n}}(\emptyset) = d_{G_{m,n}}(00)=m<d_{G_{m,n}}(\mathbf{a}) & \forall\, \mathbf{a} \neq \emptyset,\ \mathbf{a} \neq 00 \text{ and } n=m+1 \quad (2) \end{cases}$$

Proof : The neighbours of Ø are exactly the vertices 0,1,....., m-1. Thereby $d_{G_{m,n}}(\emptyset)=m$.

Consider now **a** s.t. $|\mathbf{a}|=k\neq 0$. After [Bu Gr La Wa] Lemma 4 gives

$$d_{G_{m,n}}(\mathbf{a}) \geq d^{+}_{H_{m,n}}(\mathbf{a}) + 1 = \begin{cases} n-k+1 \text{ for } \mathbf{a} \nleq \mathbf{u}_{m,n-m}, \\ (k+1).m - k + 1 \text{ for } \mathbf{a} \leq \mathbf{u}_{m,n-m}. \end{cases}$$

Case 1 : $\mathbf{a} \nleq \mathbf{u}_{m,n-m}$.

For k ≤n-m we have n-k+1 ≥ n-(n-m)+1=m+1 > m and then d(**a**) ≥ m+1 > d(Ø).

For $k > n-m+1$ we have after [Bu Gr La Wa] exactly one choice vector $w(\mathbf{a}) \in \{0,1\}^n$ associated to the word $\mathbf{a}$. This $w(\mathbf{a})$ has exactly k 1's and then

$$d_{G_{m,n}}(\mathbf{a}) = d_{Q_n}(w(\mathbf{a})) = n \geq m+1 > d(\emptyset).$$

For $k=n-m+1$ and $d^-_{H_{m,n}}(\mathbf{a}) \geq 2$ is again $d_{G_{m,n}}(\mathbf{a}) \geq (n-(n-m+1)+1)+1 = m+1 > d(\emptyset)$

For $k=n-m+1$ and $d^-_{H_{m,n}}(\mathbf{a}) = 1$ $\mathbf{a} = a_1a_2....a_k$ with $a_1=a_2=....=a_k$.

Those equalities lead to $n-m+1= k \geq n/m +1$ and so we have

$$n \leq n/m + m \quad . \tag{3}$$

We will show that (3) leads to a contradiction for $n/m \geq 2$.

Assume $n/m \geq 2$ and $n=qm+r$ with $0 \leq r < m$.

We write $n= qm+r = m + (q-1)m + r \leq q+m$ and get

$$(q-1)m + r \leq q \quad . \tag{4}$$

If $q \leq m$ it follows from (4) $m \geq q \geq (q-1)m + r \geq m+r$; so we must have $r=0$, $q=2$ and $n = 2m$.

Thereby $k = n-m+1 = m+1$. But because every letter occurs at most twice $2 \geq k = m+1 \geq 3$, a contradiction.

If $q>m$ it follows from (4)

$$q+r \geq q(m-1) + r = qm - q + r < qm - m + r = (q-1)m + r \leq q$$

and $q+r < q$ once more a contradiction.

So must be $q \leq 1$, hence, from $n>m$, $q=1$. Then by (3) $n=m+1$ and $n-m+1=k$ gives $k=2$.

Finally because of $\mathbf{u}_{m,m+1} = 01....(m-1)0$, we obtain $\mathbf{a}=00$. Indeed $d^-_{G_{m,n}}(00) = 1$ and $d^+_{G_{m,n}}(00)=n-2=m-1$, and so $d_{G_{m,n}}(00)=m=d(\emptyset)$.

For all $\mathbf{a} \in W_{m,n}$, $\mathbf{a} \neq \emptyset$, $\mathbf{a} \neq 00$, $\mathbf{a} \nleq \mathbf{u}_{m,n-m}$ holds $d_{G_{m,n}}(\mathbf{a}) > m=d_{G_{m,n}}(\emptyset)$.

Case 2 : $\mathbf{a} \leq \mathbf{u}_{m,n-m}$.

W.L.O.G. let $\mathbf{a} \neq \emptyset$, i.e. $k \geq 1$. Then $k(m-1) \geq 1$, thereby

$d_{G_{m,n}}(\mathbf{a}) \geq d^+_{H_{m,n}}(\mathbf{a}) = (k+1)m - k \geq m+1 > d(\emptyset)$. ■

III) Some lemmas about automorphisms

Lemma 5 : For each poset $\boldsymbol{P}$ $\quad \text{Aut}\,\boldsymbol{P} = \text{Aut}\,\mathcal{H}(\boldsymbol{P})$. (5)

The proof is obvious . ■

For an undirected graph $\boldsymbol{G}$ let $\text{Aut}_a\boldsymbol{G}$ be the subgroup of those automorphisms in Aut$\boldsymbol{G}$ for which a is a fixed point.

Lemma 6 : For each poset $\boldsymbol{P}$ with a smallest element a

$$\text{Aut}\,\mathcal{H}(\boldsymbol{P}) \subseteq \text{Aut}_a\boldsymbol{G}(\boldsymbol{P}) .$$

Proof : Denote $\mathcal{H}(\boldsymbol{P})$=(X,E) , $\mathcal{G}(\boldsymbol{P})$=(X,E') and $\varphi \in \text{Aut}\,\mathcal{H}(\boldsymbol{P})$. φ is a bijection $X \rightarrow X$ with $\varphi(a)=a$ and it remains to prove

$$\{x,y\} \in E' \quad \Leftrightarrow \{\varphi(x),\varphi(y)\} \in E' .$$

Because of (5) we get

$$\{x,y\} \in E' \quad \Leftrightarrow x <\!\!\cdot\, y \text{ or } y <\!\!\cdot\, x \quad \Leftrightarrow \varphi(x) <\!\!\cdot\, \varphi(y) \text{ or } \varphi(y) <\!\!\cdot\, \varphi(x)$$
$$\Leftrightarrow \{\varphi(x), \varphi(y)\} \in E' . \ \blacksquare$$

Lemma 7 : For a ranked poset $\boldsymbol{P}$=(X,≤) with smallest element a

$$\text{Aut}_a\,\mathcal{G}(\boldsymbol{P}) \subseteq \text{Aut}\,\mathcal{H}(\boldsymbol{P}) . \tag{6}$$

Proof : We show easily

$$\forall\, x \in X \ \ r(x) = d_{\mathcal{G}(\boldsymbol{P})}(a,x) \quad . \tag{7}$$

If $\varphi(x) \in \text{Aut}_a\,\mathcal{G}(\boldsymbol{P})$

$$d_{\mathcal{G}(\boldsymbol{P})}(a,x) = d_{\mathcal{G}(\boldsymbol{P})}(a,\varphi(x)). \tag{8}$$

It remains just (6) to be proved. Consider $(x,y) \in E$. So $\{x,y\} \in E'$ and then $\{\varphi(x),\varphi(y)\} \in E'$. The hypothesis $(\varphi(y),\varphi(x)) \in E$ leads by (7) and (8) to the following contradiction

$$r(\varphi(x))-1 \quad = r(\varphi(y)) = d_{\mathcal{G}(\boldsymbol{P})}(a,\varphi(y)) = d_{\mathcal{G}(\boldsymbol{P})}(a,y) = r(y) = r(x) + 1$$
$$= d_{\mathcal{G}(\boldsymbol{P})}(a,x) + 1 = d_{\mathcal{G}(\boldsymbol{P})}(a,\varphi(x)) + 1 = r(\varphi(x)) + 1 .$$

Then we get $\{\varphi(x),\varphi(y)\} \in E$. On a similar way, considering φ^{-1}, we obtain $(x,y) \in E$ from $(\varphi(x),\varphi(y)) \in E$. ■

Lemma 7 does not hold for any poset with a smallest element as shown on Figure 2 :

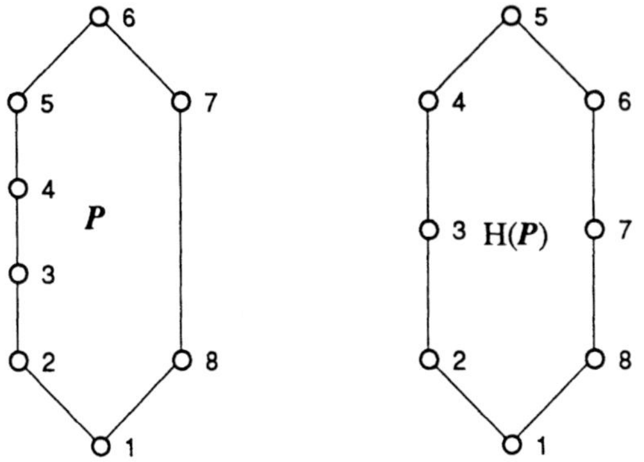

Aut $\boldsymbol{P}$ = {id} but $\begin{pmatrix} 12345678 \\ 18765432 \end{pmatrix} \in \mathrm{Aut}_1\, \mathcal{G}(\boldsymbol{P})$

Figure 2.

Lemmas 5 and 6 show that in order to determine Aut $\boldsymbol{G}_{m,n}$ we have to take care of all elements from $\boldsymbol{B}_{m,n}$. We start showing that in general

$$\mathrm{Aut}\, \boldsymbol{B}_{m,n} \neq \mathrm{Aut} \boldsymbol{G}_{m,n}.$$

Consider the case n=m+1 and let us define on $\boldsymbol{W}_{m,m+1}$ a poset isomorphic to $\boldsymbol{B}_{m,n}$ with smallest element 00 . We need on $\boldsymbol{W}_{m,m+1}$ a poset $\boldsymbol{Q}_{m,m+1}$ isomorphic to $\boldsymbol{B}_{m,m+1}$ and with smallest element 00 (cf. Figures 2 and 3).

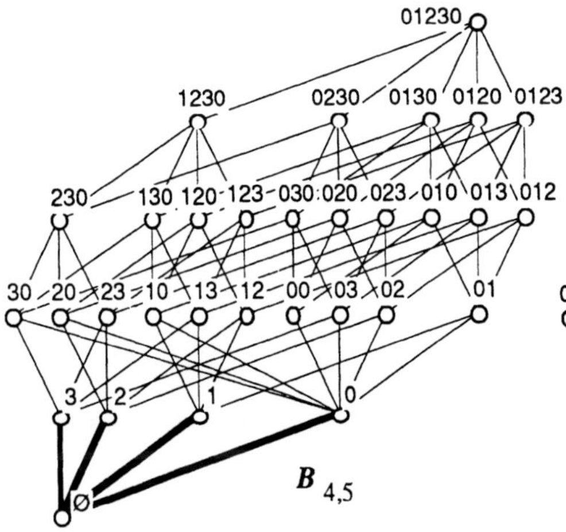

Figure 3.

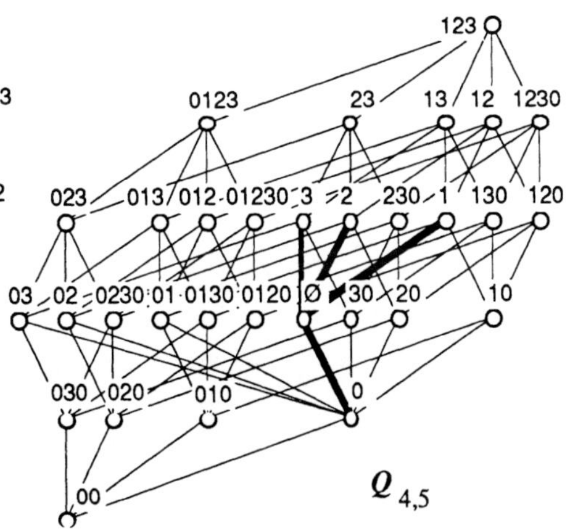

Figure 4.

For $\mathbf{a} \in W_{m,m+1}$, $\mathbf{b} \in W_{m,m+1}$ we define

$\mathbf{a} <\!\!\bullet' \mathbf{b}$ iff $\mathbf{b}$ can be obtained from $\mathbf{a}$ by deleting a 0 or adding of one letter $i \neq 0$.

$\mathbf{a} <' \mathbf{b}$ iff $\exists\ \mathbf{a}_0, \mathbf{a}_1, \ldots, \mathbf{a}_t \in W_{m,m+1}$ satisfying $\mathbf{a}=\mathbf{a}_0 <\!\!\bullet' \mathbf{a}_1 <\!\!\bullet' \ldots <\!\!\bullet' \mathbf{a}_t = \mathbf{b}$.

$\mathbf{a} \leq' \mathbf{b}$ iff $\mathbf{a} = \mathbf{b}$ or $\mathbf{a} <' \mathbf{b}$.

It is easy to show that precedings definitions give on $W_{m,m+1}$ a poset $Q_{m,m+1}$. (Cf. Figures 3 and 4.)

Lemma 8 : In $Q_{m,m+1}$ are the words 00 the smallest element and 12....(m-1) the greatest. The following mapping $\tau : W_{m,m+1} \rightarrow W_{m,m+1}$ is an isomorphism from $B_{m,m+1}$ on $Q_{m,m+1}$ with fixed point 0 :

$$\tau(\mathbf{a}) = \begin{cases} \mathbf{a}' & \text{if } \mathbf{a} = 0\mathbf{a}'0, \\ 0\mathbf{a}' & \text{if } \mathbf{a} = \mathbf{a}'0 \text{ and } 0 \not\leq \mathbf{a}', \\ \mathbf{a}'0 & \text{if } \mathbf{a} = 0\mathbf{a}' \text{ and } 0 \not\leq \mathbf{a}', \\ 0\mathbf{a}0 & \text{if } 0 \not\leq \mathbf{a}. \end{cases}$$

Proof : τ is easily shown as a bijection. From $\mathbf{a} <\!\!\bullet \mathbf{b}$ follows $\tau(\mathbf{a}) <\!\!\bullet \tau(\mathbf{b})$. Thereby we consider the four cases (where $0 \not\leq \mathbf{b}'$): $\mathbf{b} = 0\mathbf{b}'0$, $\mathbf{b} = 0\mathbf{b}'$, $\mathbf{b} = \mathbf{b}'0$, $\mathbf{b} = \mathbf{b}'$. Each of those cases is subdivided in subcases according the letter j added to $\mathbf{a}$ for obtaining $\mathbf{b}$, is 0 or not.

Thus in the first case

$\mathbf{a} = 0\mathbf{a}'$ and $\mathbf{b} = 0\mathbf{a}'0$

or $\mathbf{a} = \mathbf{a}'0$ and $\mathbf{b} = 0\mathbf{a}'0$

or $\mathbf{a} = 0\mathbf{a}'0$ and $\mathbf{b} = 0\mathbf{b}'0$ with $0 \not\leq \mathbf{b}'$ and $\mathbf{b}'$ is obtained by adding of one letter to $\mathbf{a}'$.

It follows

$\tau(\mathbf{a}) = \mathbf{a}'0$ and $\tau(\mathbf{b}) = \mathbf{a}'$

or $\tau(\mathbf{a}) = 0\mathbf{a}'$ and $\tau(\mathbf{b}) = \mathbf{a}'$

or $\tau(\mathbf{a}) = \mathbf{a}'$ and $\tau(\mathbf{b}) = \mathbf{b}'$.

In each case we have $\tau(\mathbf{a}) <\!\!\bullet' \tau(\mathbf{b})$. The other cases can be handled the same way and conversely we conclude from $\tau(\mathbf{a})$ <•' $\tau(\mathbf{b})$ to **a <• b** .

Corollary : $\boldsymbol{H}_{m,m+1}$ is the Hasse diagram and $\boldsymbol{G}_{m,m+1}$ is the Hasse graph of $\boldsymbol{B}_{m,m+1}$.

IV) Proof of Theorem 1

Lemma 9 : For any integer $m\geq 2$ $\quad \mathrm{Aut}\, \boldsymbol{G}_{m,m+1} = S_2{}^2 \otimes S_{m-1}$.

Proof: By [Bu Gr La] $\mathrm{Aut}\, \boldsymbol{B}_{m,m+1} = S_2 \otimes S_{m-1}$. Each of those $2(m-1)!$ poset automorphisms is also a graph automorphism of $\boldsymbol{G}_{m,m+1}$. Composing each of them with the graph automorphism τ (as defined in Lemma 8) gives again an automorphism of $\boldsymbol{G}_{m,m+1}$. It is easy to see that all those $2^2(m-1)!$ automorphisms are different. Moreover $\tau^2 = \mathrm{id}$, so $S_2{}^2 \otimes S_{m-1}$ is a subgroup of $\mathrm{Aut}\, \boldsymbol{G}_{m,m+1}$. By (2) we get

$$\mathrm{Aut}\, \boldsymbol{G}_{m,m+1} = \Omega_1 \cup \Omega_2 ,$$

where $\Omega_1 = \mathrm{Aut}_{\emptyset}\, \boldsymbol{G}_{m,m+1}$ and $\Omega_2 = \{ \varphi \in \mathrm{Aut}\, \boldsymbol{G}_{m,m+1} ,\ \varphi(\emptyset) = 00 \}$.

By Lemmas 5-7 $\Omega_1 = \mathrm{Aut}\, \boldsymbol{B}_{m,m+1}$ and then $|\Omega_1| = 2(m-1)!$ follows from [Bu Gr La]. For each automorphism $\varphi \in \Omega_2$ $\tau\circ\varphi \in \Omega_1$. For $\varphi \in \Omega_2$, $\varphi' \in \Omega_2$, $\varphi \neq \varphi'$ we have $\tau\circ\varphi \neq \tau\circ\varphi'$. This means $|\Omega_2| \leq |\Omega_1|$ and $|\mathrm{Aut}\, \boldsymbol{G}_{m,m+1}| \leq 2^2(m-1)!$. ■

Lemma 10 : For integers n and $m\geq 2$

$$\mathrm{Aut}\, \boldsymbol{G}_{m,n} = \begin{cases} S_2 \otimes S_{2m-n} & \text{for } m+1 < n \leq 2m-1 , \\ S_2 & \text{for } n \geq 2m . \end{cases}$$

Proof: By (1) the vertex $\emptyset$ is a fixed point for every automorphism from $\boldsymbol{G}_{m,n}$. This means $\mathrm{Aut}\, \boldsymbol{G}_{m,n} = \mathrm{Aut}_{\emptyset} \boldsymbol{G}_{m,n}$. Therefore the assertion follows from Lemmas 5-7 together with Theorem about $\mathrm{Aut}\, \boldsymbol{G}_{m,n}$ from [Bu Gr La] . ■

We make use of the well known results about $\mathrm{Aut}\, \boldsymbol{Q}_n = S_2{}^n \otimes S_n$ and obtain the

Lemma 11 : For integers n, m with $1 \leq n \leq m \geq 2$

$$\mathrm{Aut}\, \mathcal{Q}_n = \mathrm{Aut}\, \mathcal{G}_{m,n} = S_2{}^n \otimes S_n .$$

($\mathrm{Aut}\, \mathcal{Q}_n = \mathrm{Aut}\, \mathcal{G}_{m,n}$ because here $\mathcal{Q}_n = \mathcal{G}_{m,n}$).

Remark :
A possible way to show $\mathrm{Aut}\, \mathcal{Q}_n = S_2{}^n \otimes S_n$ consists in proving the following assertions :
- if v is fixed there are 2^n possibilities to choose its image v' and n! possibilities to choose the images of its neighbors in the neighborhood of v';
- using the property of $\mathcal{Q}_n$ to be a (0, 2) ([La],[Mu]) graph it is not difficult to see that each automorphism of $\mathcal{Q}_n$ is completely determined in the level $N_i = \{x \in \mathcal{Q}_n,\ d_{\mathcal{Q}_n}(v, x)=i\}$ for $i \geq 2$, by its restriction to preceding levels;
- starting with any permutation of the first level N_1 we actually build out level by level an automorphism of $\mathcal{Q}_n$.

V) Two open problems

A set $M=\{a_1,......a_t\} \subseteq V$ is called a *recognising set* of $\mathcal{G}$ iff every vertex $u \in V$ is uniquely determined by its distance vector $(d_{\mathcal{G}}(u,a_1),....,d_{\mathcal{G}}(u,a_t))$ [Ga Sa]. By the last remark the neighbourhood of any vertex in $\mathcal{Q}_n$ is a recognising set .

Problem 1: Determine the asymptotic value of the minimal cardinality of a recognising set for $\mathcal{G}_{m,n}$.

(We do not know this value even for $\mathcal{Q}_n$)

We made use above the assertion that $\mathcal{P}_{\emptyset}(\mathcal{G}_{m,n}) = \boldsymbol{B}_{m,n}$ and $\mathcal{P}_{00}(\mathcal{G}_{m,n})$ are isomorphic for n=m+1. Obviously they are not isomorphic to $\mathcal{P}_0(\mathcal{G}_{m,n})$. It would be interesting to know all canonical posets of $\mathcal{G}_{m,n}$.

Problem 2: Determine the number of non isomorphic canonical posets $\mathcal{P}_a(\mathcal{G}_{m,n})$ for $a \in \boldsymbol{W}_{m,n}$.

References

[Bu Fra Rö] G. Burosch, U. Franke, S. Röhl: Über Ordnungen von Binärworten, *Rostocker Math. Kolloquium* (to appear)

[Bu Gr La] G. Burosch, H.-D, O.F. Gronau, J.-M. Laborde: Automorphism Group of the Subword Poset of m-ary Words (submitted to *Discrete Mathematics*)

[Bu Gr La Wa] G. Burosch, H.-D, O.F. Gronau, J.-M. Laborde, I. Warnke: On Posets of m-ary Words (submitted to *Order*)

[Bu 1] G. Burosch: Erkennende Teilmengen der Ordnung der Teilworte eines m-ären Wortes, *Rostocker Math. Kolloquium* (to appear)

[Bu 2] G. Burosch: Hasse-Graphen spezieller Ordnungen, in Wagner, Bodendiek : Graphentheorie I, 157-235, BI Verlag Mannheim 1989

[La] J.-M.Laborde: Caractérisation locale du graphe du n-cube, *Actes du colloque Algèbre Appliquée et Combinatoire* Grenoble (1978) 198-200

[Mu] H.M.Mulder: The Interval Function of a Graph, Amsterdam 1980

[Ga Sa] Г.П. Гаврилов, А.А. Сапоженко (Gavrilov Sapojenko): Учебник дискретной математики, Moscou 1977

Status of Graceful Tree Conjecture in 1989

I. Cahit, Mersin, Turkey

1. Introduction

The root of graph labellings go back to a problem stated by Gerhard Ringel. Let T be a given tree with n vertices; then the edges of K_{2n-1} can be partitioned into (2n-1) trees isomorphic to T [1] . A.Rosa in his well-known 1966 paper mentioned stronger version of this conjecture which was attributed to A. Kotzig. Let T be a given tree with n vertices, then the edges of K_{2n-1} can be partitioned cyclically into (2n-1) trees isomorphic to T [2]. Rosa was first to give the connections of these conjectures with certain valuations (labellings) of the vertices of a tree [2], but the name "graceful" was coined by Golomb [3] which popularizes the β-labelling (in terms of Rosa's terminolgy) problem of trees. As today well over 100 papers have appeared in the literature around these conjectures but none of them have triggered any single hope towards its resolution (e.g., see surveys [4], [5], [10]).

In the literature there exits basically a single unsuccessful general method for tree labellings. For this the base tree T_B of a given tree T has been defined as a tree obtained from T by deleting all its end-vertices. For example any path is a base tree of a caterpillar while any caterpillar is a base tree of a lobster. Paths and caterpillars are graceful but graceful labelling of a lobster is a conjecture due to Bermond [5]. Another similar and yet unsuccessful method is to classify trees by considering their diameters. It is known that trees with diameters 2 and 3 [2], [4] and with diameter 4 [6], [7] are graceful but the problem of graceful labelling of trees of diameter 5 is still open.

Recently the author has taken the graceful tree labelling problem in a quite different way as follows: Let G=(V,E) be a graph with vertex set V and edge set E. A labelling of G

R. Bodendiek, R. Henn (Eds.)
Topics in Combinatorics and Graph Theory

is a mapping $f : V \to \{0,1,\dots,k\}$, $k \leq e$ and for each $v \in V$, $f(v)$ is called the label of v under f and for each $x=uv \in E$ the label of x under $\bar{f}$ is given by $|f(u)-f(v)|$. Denote by $v_f(i)$ (resp. $e_{\bar{f}}(i)$) the number of vertices (resp. edges) with label $i \in \{0,1,\dots,k\}$ under the labelling f of G and the labelling f is called (k+1)-equitable (or simply equitable) if we have $|v_f(i)-v_f(j)| \leq 1$ and $|e_{\bar{f}}(i)-e_{\bar{f}}(j)| \leq 1$, $i \neq j$, $i,j = 0,1,\dots,k$. A graph G is (k+1)-equitable if it admits a (k+1)-equitable labelling. For k=1, 2-equitable labelling has been called a special name "cordial labelling" [8] . For k=e, the (e+1)-equitable is exactly same as graceful labelling of G.

In this paper we give a graceful labelling technique for the trees which we call it canonic spiral labelling (CSL). Then we show that all trees of diameter four and certain trees of diameter five can be labelled with this labelling technique gracefully. The last section summarizes the results so far obtained on cordial and 3-equitable labellings

2. Canonic Spiral Labellings

A labelling f of a tree T is a bijection between the vertex set V(T) and the set $N=\{1,2,\dots,n\}$. The vertex b of T for which f(b)=1 is called the base of the labelling. For a vertex v in T, let V(v) be the (base) set of vertices u of T (including v) for which the distance d(u,v) (from v to u) is even and let s(v) be the number of vertices in V(v). If T is a tree having a graceful labelling f with base b then the size s of T under f is s=s(b). A labelling f of T with base b is a parity labelling if it induce, by restriction, a bijection between V(v) and the set $\{1,2,\dots,s\}$. Let $\bar{f}$ be the induce edge labelling which is a mapping from the edge-set E(T) onto a set $\bar{N}$ of nonnegative integers given by $\bar{f}\{(u,v)\} = |f(u)-f(v)|$ then f is called graceful labelling of T if $\bar{N} = \{1,2,\dots,n-1\}$ [2], [3] . T is a graceful tree if it has a graceful labelling. Note that if f is a graceful labelling so is $f'(v)=n+1-f(v)$. An labelling f_α of T is called

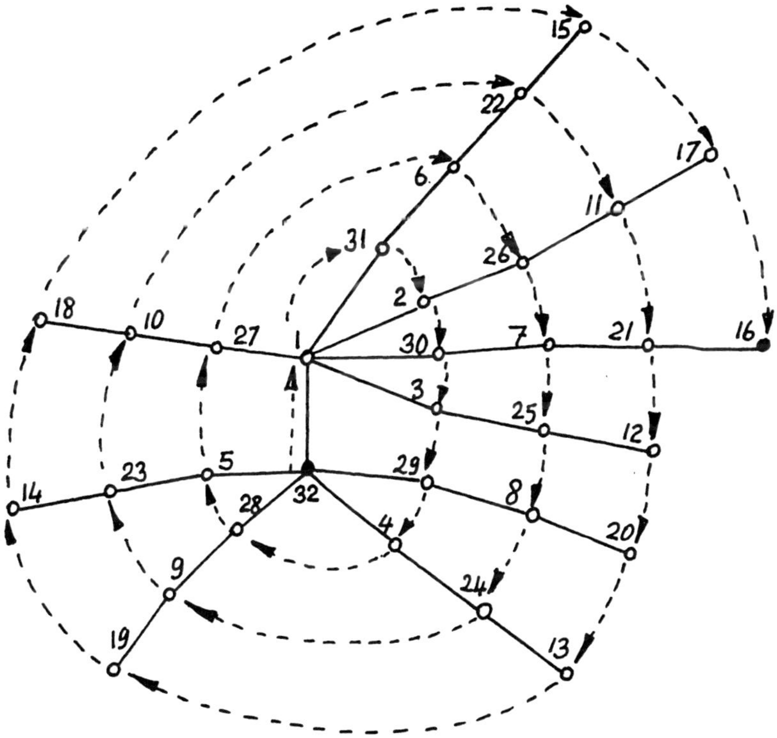
15
22
17
6
11
31
26
2
18
10
27
1
30
7
21
16
3
25
12
23
5
29
14
28
32
8
20
4
9
24
19
13

(c)

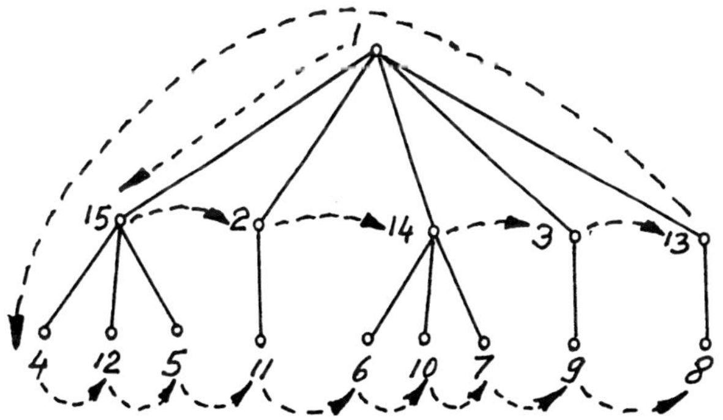
1
15
2
14
3
13
4
12
5
11
6
10
7
9
8

(b)

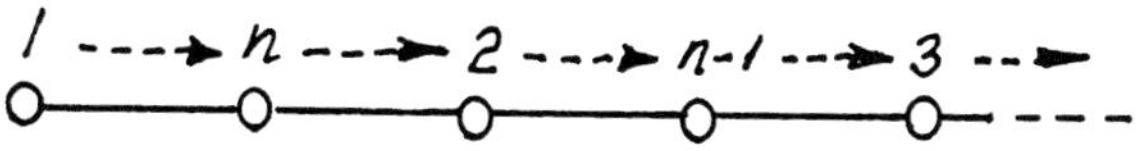
1
n
2
n-1
3

(a)

Fig. 1.

α-labelling [2] (or balanced [11], interlaced [10]) if and only if there is ainteger r such that whenever u an v are adjacent in T either $f(u) \leq r < f(v)$ or $f(v) \leq r < f(u)$ is satisfied.

Let P_n be an path on n vertices. Label the vertices of P_n starting with the leftmost end-vertex of P_n with label (in that order) 1,n,2,n-1,3,... . Such an α-labelling f_α has a several properties: (a) The numbers in $N=\{1,2,...,n\}$ are assigned to the vertices consecutively in a min-max (or a max-min for f'_α) fashion. (b) Once the label $x \in N$ is assigned to a vertex v then the next label $y \in N$ of the vertex u adjacent to v is uniquely determined. We call such an <u>ordering</u> of labels 1,2,...,n as <u>canonic</u> ordering. In the case of the path P_n, canonic ordering of vertex labels match with the edges of P_n. That is it constitutes a virtual labelled directed path (see Fig. 1) which traverses all of the vertices of P_n once and only once.

<u>Definition:</u> We call that vertices of a tree T has been labelled with the canonic spiral labelling (CSL) f_s technique if
(a) the resulting labelling f_s of T is graceful and
(b) if there is a predetermined <u>rule</u> of assignment of vertex labels to the vertices of T <u>canonically</u>.

Let f_α be an α-labelling of T with n vertices. Since f_α is also a parity labelling we can write

$$f_\alpha : V(b) \rightarrow \{1,2,...,s\}, \quad \overline{V}(b) \rightarrow \{s+1,s+2,...,n\}$$
$$\overline{f}_\alpha : E(T) \rightarrow \{1,2,...,n-1\},$$

where $V(T) = V(b) \cup \overline{V}(b)$, b is the base vertex of T for which $f_\alpha(b)=1$. Now add a positive integer k to each element of $\overline{V}(b) = \{s+1,s+2,...,n\}$ to obtain $\overline{V}'(b) = \{k+s+1,...,k+n\}$. The induce edge label-set under this new vertex labelling f'_α becomes $E'(T) \rightarrow \{k+1,k+2,...,k+n-1\}$. Clearly f'_α so obtained is not a graceful labelling of T any longer since the edge labels of T from 1 to k are now missing.

<u>Question</u>: How one can glue the k edges to the vertices of the labelled tree (under f_α) gracefully without relabelling vertices of T or how many ways to glue the k edges to the vertices of T gracefully ?

Canonic spiral labelling technique gives a (partial) answer

to these questions. Unfortunately it does not cover all possible derived trees of T. Due to the limitation of space we give only an example to show an application of the CSL. It can be seen from the Fig. 2 that the virtual directed path of CSL traverses new vertices of T' around all (see Fig.2) of or the subset of vertex labels $\{1,2,\ldots,s\}\cup\{s+k+1,\ldots,n+k\}$ of T. We have the following theorem which is a outcome of the CSL construction of graceful trees.

Theorem 1. Let T_b be an any α-valuable tree. Assume $|V(b)| = |\bar{V}(b)|$. Let f_α be an α-labelling of T_b. Derive the new tree T from T_b either by

(1) connecting odd number of edges to each vertex of T_b except to the vertex x, $x \in V(T_b)$ with $|f_\alpha(x)-f_\alpha(z)|=1$, where $z \notin V(T_b)$, $z \in V(T)$ or

(2) connecting even number of edges to each vertex of T_b or

(3) connecting even number of edges to each vertex of T_b which have labels $\{1,2,\ldots,m\}$ and $\{n-m+1,\ldots,n\}$ and odd number of edges which have labels $\{m+1,m+2,\ldots,n-m\}$.

Then the trees T obtained by (1),(2) and (3) are graceful.

When the given α-valuable tree is a isolated vertex or an edge the CSL around these enable us to study of the labelling trees of diameter four and five.

3. Labelling trees of diameter four and five

The class S(4) of trees of diameter 4 can be partitioned into three subclasses $S(4)=S(0,4)\cup S(1,4)\cup S(2,4)$ where S(0,4) contains only the path P_4, S(1,4) contains caterpillars of diameter 4, and S(2,4) contains all trees of diameter 4 which are not caterpillar. The difficult case was S(2,4) [12] and the general tree T(4) of this subclass has been settled independently by the author [6] and Z. Shi-Lin [7]. The general tree T(4) is shown in Fig.3 together with the CSL of the vertices.

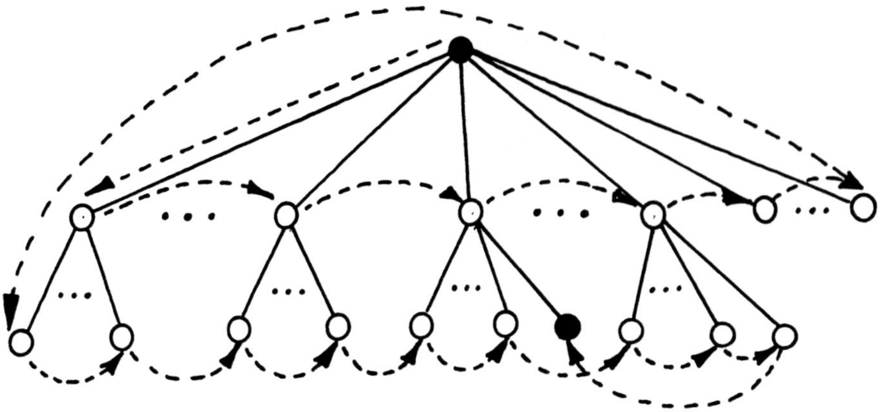

Fig. 3

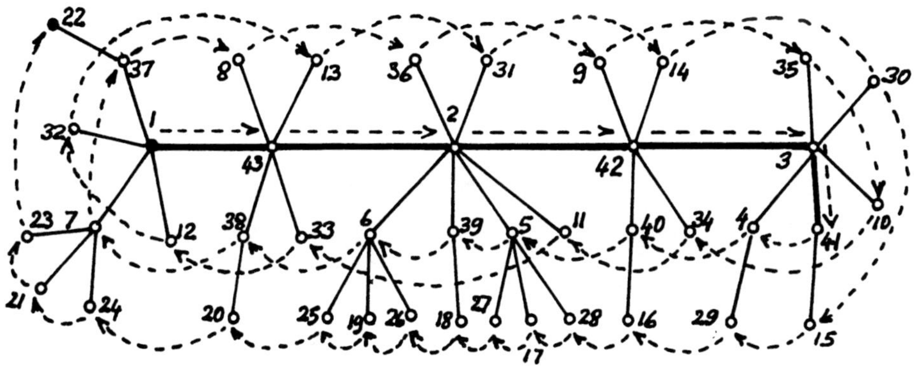

Fig. 2

Lemma 1 [6] . Let T be a graceful tree with the labelling f. Let v be vertex of T for which f(v)=1. Then the new tree $T' = T \cup S_k$ is graceful, where S_k is a k-star with k edges and is connected from one of its end-vertex to the vertex v of T.

With Lemma 1 above and the CSL of the T(4) of Fig.3 one can state,

Theorem 2. All trees of diameter four are graceful.

Let $T_{r,k}$ be denote a rooted tree in which all vertices except the root and the end-vertices are of even degrees and the lengths of all paths from the root to end-vertices are of length k. Then we have,

Theorem 3. The tree $T_{r,k}$ has a graceful labelling.

The class S(5) of trees of diameter five can be partitioned into three subclasses $S(5)=S(0,5) \cup S(1,5) \cup S(2,5)$ where S(0,5) contains all paths and caterpillars of diameter five, S(1,5) contains exactly a single subtree of diameter four from the subclass S(2,4) and S(2,5) contains exactly two verex disjoint trees from the subclass S(2,4). The trees in the subclass S(0,5) admit graceful labellings [12] . The CSL technique for labelling trees in S(1,5) and S(2,5) can not cover all trees of diameter five. The general diameter 5 tree is consisted of an edge with end-points x_1 and x_2 and two disjoint subtrees from S(2,4); attachment of the roots of these subtrees are at the vertices x_1 and x_2. The vertices x_1 and x_2 are respectively incident at p_1 and p_2 edges (the first level edges). The other ends of p_1 (p_2) edges are further, respectively incident at p_{11} (p_{21}) odd bunches of leaf edges and p_{12} (p_{22}) even bunches of leaf edges (the second level edges). That is $p_1=p_{11}+p_{12}$ and $p_2=p_{21}+p_{22}$ (see Fig.4).

We will consider only the case p_1=odd, p_2=even. The cases p_1,p_2 =even and p_1,p_2 =odd can be handled with the aid of Lemma 1, therefore is omitted.

Theorem 4. Let T(5) be the general tree of diameter five shown in Fig.4 . Then there exists the CSL f of T(5) except for

(1) p_{11}= odd , p_{12}=even , p_{21}=odd , p_{22}=odd and
(2) p_{11}= odd , $p_{12}=0$, $p_{21}=0$, p_{22}=even.

We omit details of the proof. In Fig.4 we have shown the CSLs of T(5) for

(a) p_{11}= even , p_{12}=odd , p_{21}=even , p_{22}=even
(b) p_{11}= even , p_{12}=odd , p_{21}=odd , p_{22}=odd
(c) p_{11}= odd , p_{12}=even , p_{21}=even , p_{22}=even,
($p_{12}\neq 0$, $p_{21}\neq 0$).

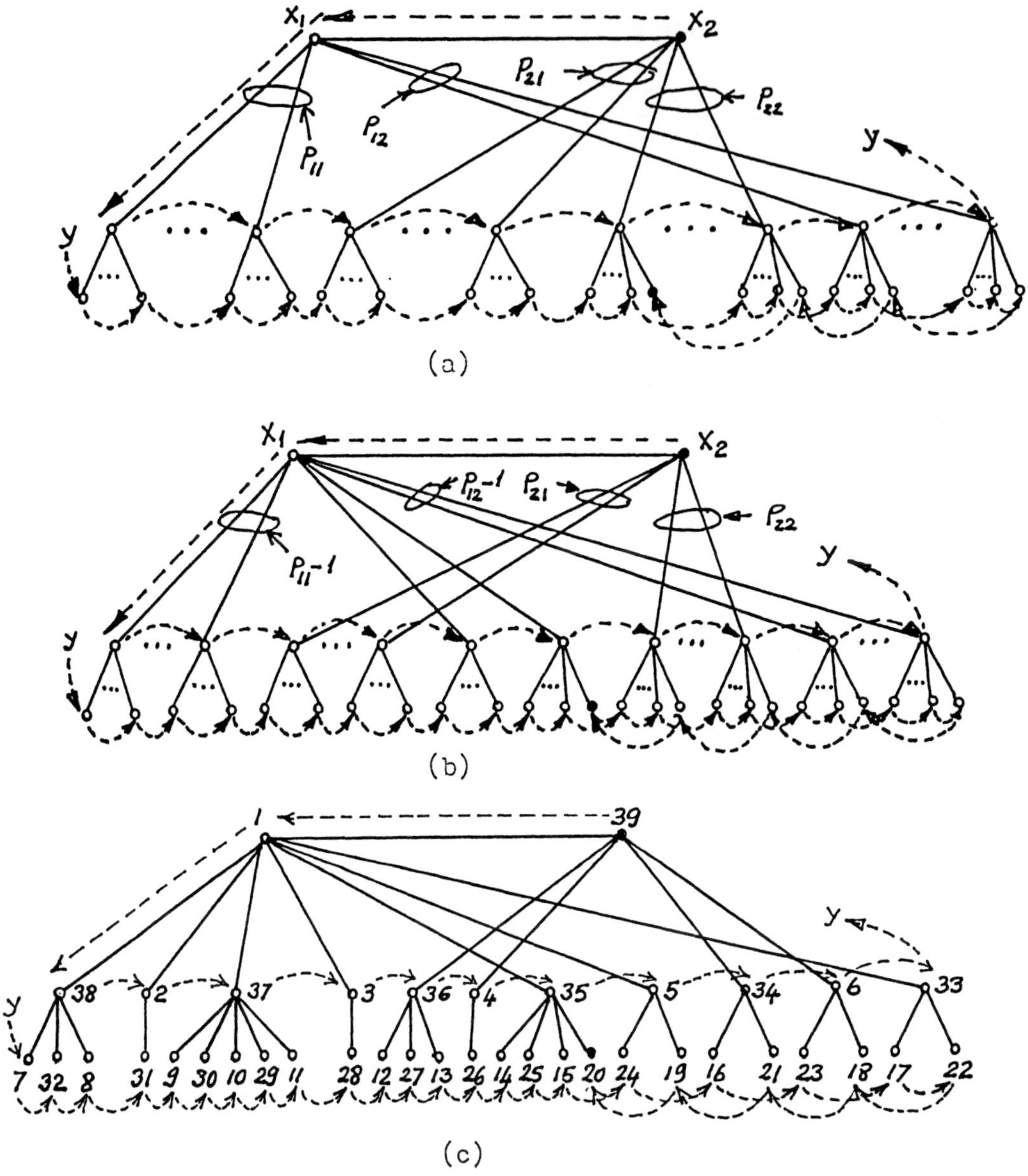

Fig. 4 Graceful labellings of T(5).

(a) p_{11}=even, p_{12}=odd, p_{21}=even, p_{22}=even

p_{11}=even, p_{12}=odd, p_{21}=odd, p_{22}=odd

(b) p_{11}=odd, p_{12}=even, p_{21}=even, p_{22}=even

(c) An example.

4. Cordial and 3-equitable tree labellings

Cordial labellings and its generalization to (k+1)-equitable labellings of trees have opened new avenues to investigate a possible proof of graceful tree conjecture. An attempt to show inductively that whether all trees admit equitable labellings is failed [9] .

Theorem 5 [8] . Every tree is cordial.

In [8] a simple inductive proof on the number of vertices as well as an algorthmic proof of this theorem have been given. The so called "horse racing algorithm" initially chooses an arbitrary vertex of T and label it with 0. A general step selects an unlabelled vertex adjacent to a labelled vertex and label it with 0 or 1 in such a way that the cordiality condition $|v_f(1) - v_f(0)| \leq 1$, $|e_{\bar{f}}(1) - e_{\bar{f}}(0)| \leq 1$ is satisfied at each step, where $v_f(i)$ (resp. $e_f(i)$) is the number of vertices (resp. edges) with labels i=0,1 under the cordial labelling f of T.

When the variety of labels are selected as 0,1, and 2 , i.e. , 3-equitable labelling, we couldn't able to provide either an inductive or an algorithmic proof to show that all trees admit 3-equitable labellings. The results so far obtained is summariezed below :

Theorem 6. For a given tree T with n vertices if the number of internal vertices I_i is at most $\lfloor \frac{1}{3} n \rfloor$ or the number of end-vertices I_e is at least $\lfloor \frac{1}{3} n \rfloor$ then the tree T has 3-equitable labelling .

Theorem 7. All caterpillars are 3-equitable.

Theorem 8. A tree with fewer than five end-vertices has a 3-equitable labelling.

References

[1] G. Ringel, Problem 25 in : Theory of Graphs and its Applications,(Proc. Internat. Sympos. Smolenice 1963), Nakl. CSAV, Praha 1964, p.162.

[2] A. Rosa, On certain valuations of the vertices of a graph, in: Theory of Graphs (Proc. Internat. Sympos. Rome 1966), Gordon and Breach-Dunod, N.Y.-Paris 1967, pp. 349-355.

[3] S.W. Golomb, How to number of a graph, in: Graph Theory and Computing (ed. R.C. Read), Academic Press, N.Y. (1972), pp. 23-37.

[4] G.S. Bloom, A chronolgy of the Ringel-Kotzig conjecture and the continuing quest to call all trees graceful, in : Topics in Graph Theory (Annal of the N.Y. Academy of Sciences, 328,ed. F. Harary), N.Y., 1979,pp.32-51.

[5] J.C.-Bermond, Graceful graphs, radio antennae and French windmills, in: Proc. of Open Univ. Conf., May 1978, (Pitman, London 1979),pp. 13-37.

[6] I. Cahit, Graceful labelling of diameter four trees, preprint.

[7] Z. Shi-Lin, private communication.

[8] I. Cahit, Cordial graphs,: A weaker version of graceful and harmonious graphs, Ars Combinatoria, 23(1987), pp. 201-208.

[9] I. Cahit, Equitable tree labellings, preprint.

[10] K.M. Koh, D.G. Rogers and T. Tan, A geaceful arboretum: a survey of graceful trees, in: Proc. of the First Franco-Southeast Asian Math. Conf.,Singopore, May 1979, (vol 2), pp.278-287.

[11] D.A. Sheppard, The factorial representation of balanced labelled graphs, Discrete Math., 15 (1976),pp.379-388.

[12] C. Huang, A. Kotzig, A. Rosa, Further results on tree labellings, Utilitas Math., 21C (1982),pp.31-48.

Embedded Graphs, Facial Colorings, and Double Cycle Covers

P.A. Catlin, Detroit, USA

We use the notation of Bondy and Murty [4], except that graphs have no loops. Let H be a subgraph of G. The contraction G/H is the graph obtained from G by contracting all edges of H and deleting any resulting loops. Denote

$$O(G) = \{\text{odd-degree vertices of } G\}.$$

The graph G is called an even graph if $O(G) = \emptyset$, and an even graph is called supereulerian if it also connected. Denote

$$S\mathcal{L} = \{\text{supereulerian graphs}\}.$$

Let G be a graph embedded on a surface. A coloring of the faces is called packed if at each vertex of degree at least 3, the incident faces have been assigned at least three different colors, and if each edge is incident with faces of two different colors. If G is 3-regular with no cut edge, then any proper facial coloring is packed. We shall generalize a criterion of Archdeacon in order to give a reduction method to determine whether a graph has a packed facial 3-coloring or 4-coloring on some surface.

R. Bodendiek, R. Henn (Eds.)
Topics in Combinatorics and Graph Theory

The subgraph H of a graph G is said to evenly span G if

(i) each vertex of G is of even degree in H;

(ii) each vertex of degree at least 3 in G has nonzero degree in H; and

(iii) each component of H contains evenly many vertices of $O(G)$.

We say that G is evenly spanned if some subgraph H evenly spans G. Denote

$$\mathcal{ES} = \{\text{evenly spanned graphs}\}.$$

Archdeacon [2] noted that $\mathcal{SL} \subset \mathcal{ES}$, and he proved for graphs with minimum degree at least 3 that membership in $\mathcal{ES}$ is equivalent to the existence of a packed 3-coloring of the faces of some embedding on some surface. The smallest 2-edge-connected graph not evenly spanned is the Petersen graph.

Let G be a graph with no cut edge. A 3-splitting of G is any 3-regular graph H that can be converted to G by a sequence of edge contractions and edge-subdivisions.

The following result was first proved by Tutte [15] for 3-regular graphs, and Archdeacon [2] obtained the present version:

Theorem 1 [2] Let G be a graph with $\delta(G) \geq 3$. The following are equivalent:

(a) $G \in \mathcal{ES}$;

(b) G has a 3-splitting H with $\chi'(H) = 3$;

(c) G embeds on some orientable surface with a packed 4-coloring of the faces;

(d) G embeds on some surface with a packed 4-coloring of the faces;

(e) G embeds on some surface with a packed 3-coloring of the faces. □

Define $\mathcal{S}_3$ to be the family of graphs G such that $G \in \mathcal{S}_3$ whenever there is a partition $E(G) = E_1 \cup E_2 \cup E_3$ such that $O(G[E_k]) = O(G)$ for all $k \in \{1,2,3\}$. In [9] we noted that $\mathcal{SL} \subset \mathcal{S}_3$ properly. A double cycle cover of G is a collection of cycles of G (multiplicities allowed) such that each edge of G is in exactly two cycles in the collection. A double cycle cover of G is called k-colorable, $k \geq 2$, if the collection of cycles can be partitioned into k subcollections, where the cycles in each subcollection are edge-disjoint (and hence induce an even subgraph of G).

Theorem 2 Let G be a graph. Then conditions (a) through (e) of Theorem 1 are equivalent to each other, and they are equivalent to each of these:

(f) $G \in \mathcal{S}_3$;

(g) G has a 3-colorable double cycle cover;

(h) G has a 4-colorable double cycle cover.

Proof: Any graph with a vertex of degree 1 satisfies none of the conditions (a) through (h). Suppose that G is a smallest counterexample with $\delta(G) = 2$. By the minimality of G, no component and no endblock of G is a cycle. Therefore, there is a graph G' with $\delta(G') \geq 3$ such that G can be obtained from G' by subdivisions of edges. For each of the conditions (a) through (h), it is easy to check that G satisfies that condition if and only if G' does. Thus, the first part of Theorem 2 follows by applying Theorem 1 to G'.

The equivalence of (f) and (g) is easy, since $G - E_k$ is an even graph if and only if $O(G[E_k]) = O(G)$, where $k \in \{1,2,3\}$. The equivalence of (g) and (h) was proved by Bermond, Jackson, and Jaeger [3].

To prove that (f) is equivalent to the other conditions, we shall prove (f) $\Leftrightarrow$ (a) by proving $\mathcal{S}_3 = \mathcal{ES}$. First, suppose that $G \in \mathcal{ES}$. Then G has a subgraph H that evenly spans G, and hence that satisfies conditions (i), (ii), and (iii) of the definition. Define $E_3 = E(G) - E(H)$. Then $O(G[E_3]) = O(G)$, by (i). Since H is an even graph, it is an easy consequence of (iii) that there is a partition $E(H) = E_1 \cup E_2$ such that $O(G[E_k]) = O(G)$, $k \in \{1,2\}$. Therefore, $G \in \mathcal{S}_3$. This proves $\mathcal{ES} \subseteq \mathcal{S}_3$.

Before proving $\mathcal{S}_3 \subseteq \mathcal{ES}$, we first present a lemma and some notation. Let H be a graph, let $v \in V(G)$, and let e and e' be distinct edges of H that are incident with v. Denote by $H(v,e,e')$ the graph with edge set $E(H)$ and with vertex set $V(H) \cup \{v'\}$, where $v' \notin V(H)$, where e and e' are incident in $H(v,e,e')$ with v' instead of v, and where all other incidences in $H(v,e,e')$ are the same as in H. This lemma is an easy consequence of the definitions:

Lemma For any graph H, for any $v \in V(H)$, and for any pair $\{e,e'\}$ of distinct edges of H incident with v, H is an even graph if and only if $H(v,e,e')$ is an even graph. $\square$

Let G be a graph, and let $E(G) = E_1 \cup E_2 \cup E_3$ and $E(G) = E_1' \cup E_2' \cup E_3'$ be two partitions of $E(G)$. For any vertex $v \in V(G)$ and for each $k \in \{1,2,3\}$, denote by $d_k(v)$ (respectively, $d_k'(v)$) the number of incidences of v with edges of E_k (respectively, E_k').

Suppose that $G \in \mathcal{S}_3$. Then there is a partition

$$E(G) = E_1 \cup E_2 \cup E_3 \tag{1}$$

with

$$(2) \qquad d_k(v) \equiv d(v) \pmod 2$$

for all $v \in V(G)$ and $k \in \{1,2,3\}$. Define the set $T = T(E_3)$ by

$$T = \{x \in V(G) \mid d_3(x) = d(x) \geq 3\},$$

and suppose that the partition $E_1 \cup E_2 \cup E_3$ satisfying (1) and (2) has been chosen to minimize $|T|$. We claim that T is empty.

By way of contradiction, suppose that $|T| > 0$. Then G has a vertex $x \in T$, where $d_3(x) = d(x) \geq 3$. Define the set

$$S = \{v \in V(G) \mid d_1(v) = 2 \text{ and } d_3(v) = d(v) - 2 > 0\}.$$

If $v \in S$ then v is incident with edges e_1 and e_3 (say), where $e_1 \in E_1$ and $e_3 \in E_3$. By (1) and (2), $G[E_1 \cup E_3]$ is an even subgraph of G, and so the lemma implies that $G[E_1 \cup E_3](v, e_1, e_3)$ is an even subgraph of $G(v, e_1, e_3)$, a graph of order $|V(G)| + 1$. Apply the lemma in this way at each vertex $v \in S$, to convert G into a graph G_0, say, of order $|V(G)| + |S|$. Thus, $G_0[E_1 \cup E_3]$ is an even subgraph of G_0, and each $v \in V(G_0) - V(G)$ is incident in G_0 with exactly two edges, one in E_1 and one in E_3.

Since $G_0[E_1 \cup E_3]$ is an even graph it has no cut edge, and so the vertex x is in a cycle C, say, in $G_0[E_1 \cup E_3]$. We may also regard C as a closed trail in $G[E_1 \cup E_3]$ on the same edge set. Define

$$E_1' = E_1 \triangle E(C); \quad E_2' = E_2; \text{ and } \quad E_3' = E_3 \triangle E(C),$$

where $E_k \triangle E(C)$ denotes the symmetric difference of E_k and $E(C)$. Then $E_1' \cup E_2' \cup E_3'$ is a partition of $E(G)$ satisfying $d_k'(v) \equiv d(v) \pmod 2$ for all $v \in V(G)$ and $k \in \{1,2,3\}$.

Denote $T' = T(E_3')$. We claim $T' \subset T$ properly. If $v \in V(G) - V(C)$ then $d_3'(v) = d_3(v)$ and $v \in T \Leftrightarrow v \in T'$. If $v \in V(C) \cap S$, then by the construction of the even graph $G_0[E_1 \cup E_3]$ containing C, v is incident with at least one edge in $E_1 \triangle E(C)$. Hence $v \notin T'$. If $v \in (V(C) \cap T) - S$ then $d(v) - 2 = d_3(v) - 2 = d_3'(v)$ and $d_1'(v) = 2$, and so $v \notin T'$. If $v \in V(C) - (S \cup T)$ then either $d_3'(v) \neq d(v)$ or $d(v) < 3$, and so $v \notin T'$. Hence $T' \subset T$, and since $x \in T - T'$, this containment is proper, contrary to the minimality of $|T|$. Thus, $T = \emptyset$, as claimed.

Define $H = G[E_1 \cup E_2]$. We claim that H evenly spans G. By (1) and (2), H satisfies (i) of the definition. Since T is empty, H satisfies (ii). If a component H_0 of H contains an odd number of vertices of $O(G)$, then by (2), $O(H_0[E_1])$ is that odd set of vertices in $O(G)$, an impossibility since $|O(H[E_1])|$ is even. Thus, (iii) holds and the claim is proved. Hence $G \in \mathcal{ES}$ and so $\mathcal{S}_3 \subseteq \mathcal{ES}$. □

Corollary [2] If G satisfies any of the conditions (a) through (e), then G has a double cycle cover. □

Define $\mathcal{S}_3^O$ to be the family of graphs H such that for any supergraph G of H, this equivalence holds:

(3) $$G \in \mathcal{S}_3 \Longleftrightarrow G/H \in \mathcal{S}_3.$$

Let $\mathcal{C}_3$ be the family of graphs G such that for any even subsets $X, Y \subseteq V(G)$ there are disjoint subsets $E_X, E_Y \subseteq E(G)$ such that $O(G[E_X]) = X$ and $O(G[E_Y]) = Y$. It was proved [7, Corollary 13A] that $\mathcal{C}_3 \cup \{C_4\} \subseteq \mathcal{S}_3^O$, and so $\mathcal{S}_3^O$ contains all cycles of length at most 4. Also, $\mathcal{C}_3$ contains every graph that has two edge-disjoint spanning trees. By repeated applications of (3), the question of membership in $\mathcal{S}_3$ can be reduced to the consideration of graphs with no nontrivial subgraphs in $\mathcal{C}_3$. An analogous reduction method for determining membership in $\mathcal{SL}$ was introduced in [5]. Both of these reduction methods are special cases of a more general method discussed in [6] and [8].

A lifting theorem of Fleischner [10] (see [11], p. 4) can also be used to investigate membership in $\mathcal{S}_3$. It can be applied either before or after (3) has been used repeatedly. Suppose that v is a vertex of degree at least 4 in G. By Fleischner's Theorem [10], for some pair of distinct edges e and e' incident with v, $G \in \mathcal{S}_3$ if and only if $G(v, e, e') \in \mathcal{S}_3$.

Szekeres [14] and Seymour [13] conjectured that any graph with no cut edge has a double cycle cover. For graphs with no cut edge and having no subgraph that is a subdivision of the Petersen graph, Matthews [12] (generalizing a conjecture of Tutte [16] on 3-regular graphs) conjectured that they are in $\mathcal{S}_3$, and Alspach and Zhang [1] showed that they have a double cycle cover.

REFERENCES

1. B. Alspach and C. Q. Zhang, Cycle coverings of graphs. Submitted.

2. D. Archdeacon, Face colorings of embedded graphs. J. Graph Theory **8** (1984) 387-398.

3. J. C. Bermond, B. Jackson, and F. Jaeger, Shortest coverings of graphs with cycles. J. Combinatorial Theory (B) 35 (1983) 297-308.

4. J. A. Bondy and U. S. R. Murty, "Graph Theory with Applications". American Elsevier, New York (1976).

5. P. A. Catlin, A reduction method to find spanning eulerian subgraphs. J. Graph Theory 12 (1988) 29-45.

6. P. A. Catlin, A reduction method for graphs. Congressus Numerantium **65** (1988) 159-170.

7. P. A. Catlin, Double cycle covers and the Petersen graph. J. Graph Theory. To appear.

8. P. A. Catlin, The reduction of graph families closed under contraction. Submitted.

9. P. A. Catlin and H.-J. Lai, On supereulerian graphs. Submitted.

10. H. Fleischner, Eine gemeinsame Basis für die Theorie der Eulerschen Graphen and den Satz von Petersen. Monatsh. Math. **81** (1976) 267-278.

11. F. Jaeger, A survey of the cycle double cover conjecture. Annals of Discrete Math. **27** (1985) 1-12.

12. K. R. Matthews, On the eulericity of a graph. J. Graph Theory **2** (1978) 143-148.

13. P. D. Seymour, Sums of circuits, in "Graph Theory and Related Topics" (J. A. Bondy and U. S. R. Murty, eds.) Academic Press, New York (1979) 341-355.

14. G. Szekeres, Polyhedral decomposition of cubic graphs. Bull. Austral. Math. Soc. **8** (1973) 367-387.

15. W. T. Tutte, On the imbedding of linear graphs in surfaces. Proc. London Math Soc. **51** (1949) 474-483.

16. W. T. Tutte, A geometrical version of the four color problem. Proc. Chapel Hill Conf., University of N. Carolina Press, Chapel Hill (1969) 553-560.

On Peripheral Vertices in Graphs

G. Chartrand[1], Michigan, USA
G. Johns, Michigan, USA
O. R. Oellermann[1,2], Durban, South Africa

ABSTRACT

The periphery of a graph G is the subgraph of G induced by the vertices in G whose eccentricity equals the diameter of G. Graphs that are peripheries of graphs with specified diameter are characterized. For a graph G, the antipodal graph A(G) of G is the graph having the same vertex set as G and edge set $E(A(G)) = \{uv \mid u, v \in V(G)$ and $d_G(u, v) = \text{diam } G\}$. For a graph G and positive integer n, the nth iterated antipodal graph $A^n(G)$ of G is defined as the graph $A(A^{n-1}(G))$ where $A^1(G)$ is A(G) and $A^0(G)$ denotes G. It is shown that there exist nonnegative integers n_1 and n_2 such that $n_1 < n_2$ and $A^{n_1}(G) \cong A^{n_2}(G)$. The smallest postitive integer k for which there exists an integer ℓ such that $A^{\ell}(G) \cong A^{\ell+k}(G)$ is called the antipodal period of G. It is shown that for every positive integer k there exists a graph with antipodal period k.

1. Introduction

Let G be a graph with vertices u and v. If there exists a u–v path in G, then the *distance* d(u, v) between u and v is the length of a shortest u–v path in G. If there is no u–v path in G, then we define $d(u, v) = \infty$. The *eccentricity* e(v) of a vertex v of G is $e(v) = \max \{d(u, v) \mid u \in V(G)\}$. The *radius* rad G of G is $\min \{e(v) \mid v \in V(G)\}$ while the *diameter* diam G of G is $\max \{e(v) \mid v \in V(G)\}$. The subgraph of G induced by those vertices having eccentricity rad G is called the *center* of G. For other terminology we follow [5].

[1] Research supported in part by Office of Naval Research Contract N0014–88–K–0018.
[2] Research supported in part by the Foundation of Research Development and the Universtiy of Natal.

R. Bodendiek, R. Henn (Eds.)
Topics in Combinatorics and Graph Theory

Much research has dealt with properties of the center of a graph. In particular, Hedetniemi (see [4]) showed the following.

Theorem A Every graph is the center of some graph.

In this paper we investigate concepts that are associated with vertices of maximum eccentricity in a graph.

2. The Periphery of a Graph

The *periphery* P(G) of a graph G is the subgraph of G induced by those vertices having eccentricity diam G. The periphery of a graph has potential applications to chemistry since the peripheral attachments of a molecule are quite often the deciding factor in a chemical reaction.

If $G \cong K_p$, then P(G) = G, that is, every complete graph is the periphery of some graph. Bielak and Sysło [3] completely characterized graphs that are peripheries.

Theorem B A graph G of order p is the periphery of some graph if and only if $G \cong K_p$ or rad G ≥ 2.

If G is a graph of order p with rad G ≥ 2, then the graph H constructed by Bielak and Sysło in the proof of Theorem B such that P(H) $\cong$ G has diameter 2. This fact suggests the following problem. If G is a graph and $n \geq 2$ is an integer, does there exist a graph H with diameter n and having G as its periphery? The next result provides a solution to this problem.

Theorem 1 Let G be a graph and $n \geq 2$ an integer. Then G is the periphery of a graph with diameter n if and only if rad G $\geq$ n.

Proof Assume G is the periphery of a graph H with diameter n. Suppose, to the contrary, that rad G < n. Then there is some vertex v of G whose distance from every vertex in G is less than n. Since v belongs to the periphery of H, there is some vertex u of H such that $d_H(v, u) = \text{diam}_n H = n$. Moreover, $u \in V(G)$. However, this is not possible since $n = d_H(v, u) \leq d_G(v, u) < n$. Therefore rad G $\geq$ n.

Conversely, suppose now that G is a graph whose radius is at least n. We consider two cases.

Case 1 Suppose $n = 2k$ for some positive integer k.

Let H be obtained by joining every vertex of G to a new vertex v and then subdividing each edge of the type vu where $u \in V(G)$ a total of $k - 1$ times. It can now be shown in a straightforward manner that $P(H) = G$ and that diam $H = n$.

Case 2: Suppose $n = 2k + 1$ for some positive integer k.

Suppose $V(G) = \{v_1, v_2, \ldots, v_p\}$. Let F be a copy of K_p with $V(F) = \{u_1, u_2, \ldots, u_p\}$. We obtain H by joining vertex v_i to u_i for $1 \leq i \leq p$ and then subdividing each edge u_iv_i a total of $k - 1$ times. Once again it can be shown in a straightforward manner that $P(H) = G$ and diam $H = n$. ❑

In the next section we consider another graph associated with the vertices of maximum eccentricity in a graph.

3. The Antipodal Period of a Graph

In the preceding section we investigated the structure of the subgraph of a given graph G induced by the vertices of maximum eccentricity in G. We now study properties of a graph having the same vertex set as G and whose edges are determined by pairs of vertices whose distance in G is diam G. More specifically, for a graph G, the *antipodal period* $A(G)$ of G is the graph with vertex set $V(A(G)) = V(G)$ and edge set $E(A(G)) = \{ uv \mid u, v \in V(G)$ and $d_G(u, v) = \text{diam } G\}$. Antipodal graphs were introduced in [8] and studied in [1], [2], [6] and [7]. In particular, the following characterization of antipodal graphs was given in [1].

Theorem C A graph G is an antipodal graph if and only if it is the antipodal graph of its complement.

In [6] an alternate form of Theorem C was stated.

Theorem D A graph G is an antipodal graph if and only if

(1) diam $\overline{G} = 2$ or
(2) $\overline{G}$ is disconnected and the components of $\overline{G}$ are complete graphs

For a graph G of order p and an integer $n \geq 1$, the nth *iterated antipodal graph* $A^n(G)$ of G is defined to be the graph $A(A^{n-1}(G))$ where $A^1(G)$ denotes $A(G)$ and $A^0(G)$ is G. Since there exist only finitely many graphs of order p, there exist nonnegative integers, n_1 and n_2 with $n_1 < n_2$ such that $A^{n_1}(G) \cong A^{n_2}(G)$. Moreover, $A^{n_1+i}(G) \cong A^{n_2+i}(G)$ for all

$i \geq 0$. With these observations in mind we define the *antipodal period* of a graph G as the smallest positive integer k for which there exists a nonnegative integer ℓ such that $A^{\ell}(G) \cong A^{\ell+k}(G)$. We will determine all positive integers k for which there exists a graph with antipodal period k. We begin with a collection of useful lemmas.

Lemma 1 Let d be a positive integer. Then $C_{2d+1} \cong A(C_{2d+1})$. Moreover, if u and v are two vertices of C_{2d+1} with $d_{C_{2d+1}}(u, v) = n$, then there exists a u–v path in $A(C_{2d+1})$ of length 2n and $d_{A(C_{2d+1})}(u, v) = \min\{2n, 2d - 2n + 1\}$.

Proof Suppose that C_{2d+1}: $v_1, v_2, \ldots, v_{2d+1}, v_1$. Then diam $C_{2d+1} = d$ and for each vertex v_i, $1 \leq i \leq 2d+1$, there exist exactly two vertices v_j and v_k with $j \equiv (i + d) \bmod (2d+1)$ and $k \equiv (i+d+1) \bmod (2d+1)$ such that $d(v_i, v_j) = d(v_i, v_k) = d$. Thus, v_i is adjacent only to v_j and v_k in $A(C_{2d+1})$, so that $A(C_{2d+1})$ is 2–regular.

Suppose u and v are distinct vertices of C_{2d+1} and that $d_{C_{2d+1}}(u, v) = n$ with $1 \leq n \leq d$. We may assume that $u = v_{2d+1}$ and $v = v_n$. Then,

$$P:\ u = v_{2d+1}, v_{d+1}, v_1, v_{d+2}, v_2, \ldots, v_{d+n}, v_n = v$$

is a u–v path in $A(C_{2d+1})$. Hence $A(C_{2d+1})$ is connected and thus isomorphic to C_{2d+1}. Since P has length 2n it now follows that $d_{A(C_{2d+1})}(u, v) = \min\{2n, 2d - 2n + 1\}$. ❑

Lemma 2 Let $m \geq 2$ be an integer and let $p_1, p_2, \ldots, p_m$ be a sequence of positive integers with $p_1 > 1$. If $G \cong \bigcup_{i=1}^{m} K_{p_i}$, then G has antipodal period 2.

Proof Since $\overline{G} \cong K_{p_1, p_2, \ldots, p_m}$ has diameter 2, it follows from Theorem C that G is an antipodal graph. Moreover, $A(G) = \overline{G} \not\cong G$. Further, $A^2(G) = A(\overline{G}) \cong G$. Hence G has antipodal period 2. ❑

Before presenting the next result, we need some additional terminology. Let $p \geq 7$ be an integer and let n be an integer such that $2 \leq n \leq \lfloor \frac{p-2}{2} \rfloor$. Then the graph G(n, p) is defined as follows. The graph G(n, p) has vertex set $V(G(n,p)) = \{v_1, v_2, \ldots, v_{p-2}, u_1, u_2\}$ and edge set

$$E(G(n, p) = \{(v_i, v_{i+1}) \mid 1 \leq i \leq p-3\} \cup \{v_{p-2}v_1, v_{p-2}u_1, v_2u_1, v_nu_2, v_{n+2}u_2\}.$$

Now let G(1, p) be the graph $A(G(\lfloor \frac{p-2}{2} \rfloor, p))$. Then the order of G(n, p) is p and $d_{G(n,p)}(u_1, u_2) = n$ for all n, $1 \leq n \leq \lfloor \frac{p-2}{2} \rfloor$. Note that, in each graph G(n, p), the vertices u_1

and v_1 are similar and the vertices u_2 and v_{n+1} are similar. We are now prepared to prove our next result.

Lemma 3 If $p \geq 9$ is an odd integer, then $A^2(G(\lfloor \frac{p-2}{2} \rfloor, p)) \cong G(2, p)$.

Proof Let $G = G(\lfloor \frac{p-2}{2} \rfloor, p)$ and suppose that the vertex set of G is as described prior to this lemma. If u and v are vertices of $G - u_2$, then $d_{G-u_2}(u, v) = d_G(u, v)$. Moreover, diam G = diam $G - u_2$. Therefore, $A(G - u_2) < A(G)$. Observe that $G - u_2$ is obtained from a $(p-2)$–cycle by joining a new vertex to two vertices on the cycle at distance 2 apart. It can be shown in a straight forward manner that $A(G - u_2) \cong G - u_2$. Since G contains exactly three vertices whose distance from u_2 is diam G, namely u_1, v_1 and v_{p-2} it follows that $A(G) = G(1, p)$ is obtained from $A(G - u_2)$ by adding u_2 as well as the edges u_2u_1, u_2v_1 and u_2v_{p-2}.

By a similar argument it can be shown that $A(G(1, p) - u_2) < A(G(1, p))$ and that $A(G(1, p) - u_2) \cong G(1, p) - u_2$. There exist exactly two vertices, say w and x in $G(1, p)$ whose distance from u_2 is diam $(G(1, p))$. Since u_1 and u_2 are adjacent vertices on a $(p - 2)$–cycle in $G(1, p)$, one of the two vertices, say w, satisfies $d(u_1, w) = d(u_2, w)$ = diam $G(1, p)$ in $G(1, p)$. Therefore, $d_{A(G(1, p))}(u_1, u_2) = 2$ and $A^2(G) \cong G(2, p)$. ❑

We now present our last lemma.

Lemma 4 If $p \geq 11$ is an odd integer and $2 \leq n \leq \lfloor \frac{p-2}{2} \rfloor$, then $A(G(n, p)) \cong A(G(2n, p))$.

Proof Let $G = G(n, p)$ have vertex set and edge set as described prior to Lemma 3. As in the proof of Lemma 1 for each vertex v there exist exactly two vertices in the set $\{v_1, v_2, \ldots, v_{p-2}\}$, namely, v_j and v_k where $j \equiv i + (\frac{p-1}{2}) \bmod p$ and $k \equiv i + (\frac{p+1}{2}) \bmod p$ such that $d_G(v_i, v_j) = d_G(v_i, v_k) = \frac{p-1}{2}$ = diam G. Hence the subgraph induced by the vertices $v_1, v_2, \ldots, v_{p-2}$ in $A(G)$ is the cycle

$$C: v_{p-2}, v_{\frac{p-1}{2}}, v_1, v_{\frac{p+1}{2}}, v_2, v_{\frac{p+3}{2}}, \ldots, v_{\frac{p-3}{2}}, v_{p-2}.$$

Let $v'_1 = u_1$, $v'_{n+1} = u_2$, and for $i \neq 1, n + 1$ let $v'_i = v_i$. Then the subgraph induced by the vertices $v'_1, v'_2, \ldots, v'_{p-2}$ in $A(G)$ is the cycle

$$C': v'_{p-2}, v'_{\frac{p-1}{2}}, v'_1, v'_{\frac{p+1}{2}}, v'_2, v'_{\frac{p+3}{2}}, \ldots, v'_{\frac{p-3}{2}}, v'_{p-2}.$$

Further, $A(G)$ has edge set $E(C) \cup E(C')$. Thus, the only vertices having degree 3 in $A(G)$ are $v_{\frac{p-1}{2}}$, $v_{\frac{p+1}{2}}$, $v_{n+\frac{p-1}{2}}$, and $v_{n+\frac{p+1}{2}}$, indices expressed modulo $p - 2$. The remaining vertices have degree 2 in A(G). Observe that both u_1 and v_1 are adjacent with $v_{\frac{p-1}{2}}$ and $v_{\frac{p+1}{2}}$ in A(G) and both u_2 and v_{n+1} are adjacent with $v_{n+\frac{p-1}{2}}$ and $v_{n+\frac{p+1}{2}}$ in A(G). Further,

$$d_C(v_{(p-1)/2}, v_{(p+1)/2}) = d_C(v_{n+(p-1)/2}, v_{n+(p+1)/2}) = 2.$$

By Lemma 1, we know that $d_C(v_1, v_{n+1}) = 2n$. Hence $d_{A(G)}(u_1, u_2) = 2n$. Therefore,

$$A(G) = A(G(n, p)) \cong G(2n, p). \quad \square$$

With the aid of the preceding lemmas we now obtain the following.

Theorem 2 For every positive integer k there is a graph with antipodal period k.

Proof For $k = 1$ or 2 the theorem follows from Lemma 1 and Lemma 2, respectively.

Suppose now that $k \geq 3$. Let $p = 2^k + 3$ and consider G(2, p). From Lemma 4 it follows for $m = 1, 2, \ldots, k-1$ that $A^m(G(2, p)) \cong G(2^{m+1}, p)$. Hence,

$$A^{k-2}(G(2, p)) \cong G(2^{k-1}, p).$$

Using this fact together with Lemma 3, we obtain

$$\begin{aligned} A^k(G(2, p)) &\cong A^2(A^{k-2}(G(2, p))) \\ &\cong A^2(G(2^{k-1}, p)) \\ &\cong A^2(G(\lfloor \tfrac{p-2}{2} \rfloor, p)) \cong G(2, p). \end{aligned}$$

Since $G(2^s, p) \not\cong G(2^t, p)$ for $1 \leq s < t \leq k-1$ it follows that k is the smallest positive integer for which there exists a nonnegative integer ℓ such that

$$A^{\ell}(G(2, p)) \cong A^{\ell+k}(G(2, p)),$$

namely, $\ell = 0$ is such an integer. Therefore G(2, p) has antipodal period k. $\square$

REFERENCES

[1] R. Aravamudhan and B. Rajendran, On antipodal graphs. *Discrete Math.* **49** (1984) 193–195.

[2] R. Aravamudhan and B. Rajendran, A note on "on antipodal graphs". *Discrete Math.* **58** (1986) 303–305.

[3] H. Bielak and M.M. Sysło, Peripheral vertices in graphs. Hungarika **18** (1983) 269–275.

[4] F. Buckley, Z. Miller and P.J. Slater, On graphs containing a given graph as center. *J. Graph Theory* **5** (1981) 427–434.

[5] G. Chartrand and Linda Lesniak, *Graphs & Digraphs (Second Edition),* Wadsworth & Brooks/Cole, Monterey CA (1986).

[6] G. Johns, Generalized distance in graphs. Ph.D. dissertation (1988).

[7] B. Rajendran, Topics in graph theory: Antipodal graphs. Ph.D. thesis (1985).

[8] R. Singleton, There is no irregular Moore graph. *Amer. Math. Monthly* **75** (1968) 42–43.

The Vertex-Degrees of Steiner Minimal Trees in Minkowski Planes

D. Cieslik, Greifswald, GDR

1. Introduction

The Steiner problem of minimal trees has been considered in the plane with Euclidean and rectilinear norm (c.f. for instance /1/ and /2/). We intend to discuss this problem for arbitrary Minkowski planes.

Denote by $M_2(B)$ a affine plane with norm $\|.\|$ and unit ball $B=\{x \mid \|x\| \leq 1\}$. $M_2(B)$ is called Minkowski plane. We consider graphs $G=(V,E)$, which are embedded in the plane in the sense that V is a finite subset of $M_2(B)$ and each edge in E is a line segment $\underline{vv}'=\{tv+(1-t)v' \mid 0 \leq t \leq 1\}$. The length of G in $M_2(B)$ is defined by $l_B(G)=\sum_{\underline{vv}' \in E} \|v-v'\|$.

The <u>Steiner problem of minimal trees</u> is the following: Given a finite set N in $M_2(B)$. Find a connected graph $G=(V,E)$ in $M_2(B)$ with $V \supseteq N$ and with minimal length.

A solution of this problem must be a tree, which is be called Steiner minimal tree (SMT) of N and the points from $V-N$ are called Steinerpoints. From the triangle inequality of the norm it follows that every Steinerpoint has degree greater than two.

It is easy to see that the cardinality of the set $V-N$ of Steinerpoints of a SMT of N is forced to be bounded and a simple calculation yields the bound card $(V-N) \leq$ card $N - 2$.

We are going to get an other restriction for SMT : The degrees of vertices are bounded. These bounds will be calculated.

R. Bodendiek, R. Henn (Eds.)
Topics in Combinatorics and Graph Theory

2. The results

<u>Theorem 1.</u> For every finite set N in $M_2(B)$ there is a SMT of N in which every vertex has degree less or equal than four.

Proof. Let v be a vertex in a SMT $T=(V,E)$. If the degree of v exceed four, then there are points q_1, q_2 and q_3 in the set $\{p \mid p \in \underline{vv'} \cap \{x \mid \|x-v\| = (1/2)\cdot \min_{\underline{vv'}\in E} \|v-v'\|\},\ \underline{vv'}\in E\}$ with $v \notin \underline{q_1q_2}$ and $\underline{q_1q_2} \cap \underline{vq_3} \neq \emptyset$. Let $q \in \underline{q_1q_2} \cap \underline{vq_3}$ be an additional Steinerpoint. We construct T' by $T'=(V \cup \{q\}, E \cup \{\underline{q_jq} \mid j=1,2,3\} \cup \cup \{\underline{vq}\} - \{\underline{q_jv} \mid j=1,2,3\})$. By the triangle inequality $l_B(T') \leq l_B(T)$. If T' contains vertices of degree greater than four, then the procedure will be repeated.

Theorem 1 yields a first result for bounding the degrees of the vertices in a SMT. But we need bounds for all SMT.

<u>Theorem 2.</u> Let N be a finite set in $M_2(B)$.

a) Every vertex of a SMT of N has degree less or equal than six.

b) The degree six is possible if and only if B is an affinregular hexagon.

c) Every Steinerpoint of a SMT of N has degree less or equal than five.

From the proofs of theorem 1 and 2 we obtain

<u>Theorem 3.</u> Let B be a strictly convex or smooth unit ball. Then every vertex of a SMT of a finite set in $M_2(B)$ has degree less or equal than four.

Especially Steinerpoints with degree three can be met in Euclidean plane and with degree four in plane with rectilinear norm (see /1/ and /2/). A Minkowski plane possessing a Steinerpoint in a SMT of degree five is unknown.

3. The proof of Theorem 2

Assume, that $T=(V,E)$ is a SMT of N, v in V, $r=(1/2)\cdot \min_{\underline{vv'}\in E} \|v-v'\|$ and $\{v_1,\dots,v_d\}=\{v' \mid \underline{vv'}\in E\}$ where d is the degree of v. We define additional Steinerpoints: $q_i \in \{x \mid \|x-v\|=r\} \cap \underline{vv_i}$ for

$i=1,\dots,d$ and we consider for $N'=\{q_i \mid i=1,\dots,d\}$ the SMT $T'=(N'\cup\{v\},\{\underline{vq_i} \mid i=1,\dots,d\})$.

By means of translations and homotheties we can assume that $N' \subseteq bdB=\{x \mid \|x\|=1\}$ and $v=o$.

The following concept turns out to be helpful. A finite subset N of bdB is called correct if $\underline{ov} \cap \{x \mid \|x-v'\|<1\}=\emptyset$ for all v,v' in N, $v\neq v'$. It is easy to see that

($\underline{1}$) If $T=(N\cup\{o\},\{\underline{ov} \mid v\in N\})$ is a SMT of N or $N\cup\{o\}$ and $N\subseteq bdB$, then N is correct.

A theorem by Laugwitz (see /3/) implies that the natural number $\max\{\text{card } N \mid N\subseteq bdB, \text{card } N \text{ is finite and } N \text{ is correct}\}$ exists, since $\|v-v'\|\geq 1$ holds in a correct set N for all v,v' $(v\neq v')$.

We are going to determine this number.

Let bdK denote the boundary and let K^o denote the interior of a convex set K. For p in bdK denote by $C_p(K)$ the support cone of K in p. $bdC_p(K)$ is the union of two halflines starting in p. We denote by $a(p)$ the angle between these lines.

($\underline{2}$) Let B be a unit ball and p in bdB. Then $0<a(p)\leq\pi$.

If there is only one support line in $p\in bdK$, then p is called regular, otherwise singular.

($\underline{3}$) Let B be a unit ball and p in bdB. Then p is regular if and only if $a(p)=\pi$.

We define: $bd_sB=\{p\in bdB \mid p \text{ singular}\}$.

($\underline{4}$) The cardinality of bd_sB is at most countable. If it is finite, it is obliged to be even.

(The statements (2),(3) and (4) are proved in /4/.)

($\underline{5}$) Let B be a unit ball. If $N\subseteq bd_sB$ and card N is finite, then $\sum_{p\in N} a(p) \geq (\text{card } N - 2)\cdot\pi$. Equality holds if and only if N equals bd_sB and B is the convex hull of N.

(5) is a consequence of the well-known fact that for N holds $\sum_{p\in N} (\pi-a(p))\leq 2\pi$ (see /4/).

(6) If $N \subseteq bdB$ is correct and v is in N, then

$$bdB \cap (C_o(\{x \mid \|x-v\| \leq 1\}))^o \cap N = \{v\}.$$

Proof. If v lies in bdB, then -v lies in bdB too and a(-v) is equal to a(v).
Defining p_1 and p_2 by $\{p_1, p_2\} = bdB \cap bdC_o(\{x \mid \|x-v\| \leq 1\})$, we obtain that none further point of N lies in the interior of the Jordan-arc running in bdB from p_1 to p_2 through v. This is the assertion.
(6) says that every point in bdB is contained in at most two arcs $bdB \cap C_o(\{x \mid \|x-v\| < 1\})$ where v is a point in a correct set. Following

(7) If $N \subseteq bdB$ is correct, then $\sum_{v \in N} a(v) \leq 4\pi$.

Let $N \subseteq bdB$ be a correct set possessing exactly r regular and s singular points. (2),(3),(5) and (7) say that $4\pi \geq \sum_{v \in N} a(v) \geq (s-2)\pi + r\pi$. This implies that $r+s \leq 6$ and in accordance with (1) this yields the desired result.
Moreover, equality holds if and only if B is the convex hull of the singular points of N. Hence, s=4 or s=6.
For s=4 B is forced to be a parallelogramm spanned by the singular points $v_1, \ldots, v_4$ of N. But this is a contradiction, since for the regular points v_5 and v_6 of N holds $v_5, v_6 \notin B$ due to the fact that $H_j = C_o(\{x \mid \|x-v_j\| \leq 1\})$, j=5,6 are half-planes where $o \in bdH_j$ and $v_i \notin H_j$ for i=1,...,4 and j=5,6 (see (3) and (6)).
Finally let r=0 and s=6 with $B = conv\{v_1, \ldots, v_6\}$ for $v_i \in N \cap bd_sB$, i=1,...,6 . Place $v_1, \ldots, v_6$ on bdB in a zyclic sequence. By $a(v_i, v_j)$ we denote the angle between $\underline{ov_i}$ and $\underline{ov_j}$.
Since N is correct, (5),(6) and (7) implies
$a(v_i, v_{(i+1)\bmod 6}) + a(v_{(i+1)\bmod 6}, v_{(i+2)\bmod 6}) = a(v_{(i+1)\bmod 6})$
for i=1,...,6. Hence $conv\{o, v_i, v_{(i+1)\bmod 6}, v_{(i+2)\bmod 6}\}$ is a parallelogramm for all i=1,...,6. Now B is an affin-regular hexagon.
Conversly we can assume that B=conv N with $N = \{\pm(1,0), \pm(0,1), \pm(1,1)\}$, since all affin-regular hexagons are affin equiva-

lent. Then $T=(N\cup\{o\},\{\underline{ov}\mid v\in N\})$ is an SMT of $N\cup\{o\}$ with degree of o equals six.

c) is a consequence of the last part of the proof. If o is a Steinerpoint of degree six, then we replace $(N\cup\{o\},\{\underline{oq_i}\mid i=1,\ldots,6\})$ by $(N,\{\underline{q_iq_{(i+1)}}\mid i=1,\ldots,5\})$. But this is a contradiction to the minimality.

Thus, theorem 2 is proved.

4. The vertex-degrees of SMT in higher-dimensional Minkowski spaces

Denote by $M_d(B)$ a d-dimensional Minkowski space with unit ball B. We define the Steinerproblem of minimal trees in the same manner as in the two-dimensional case.

It is easy to see that there are natural numbers $z_d(B)$ exist so that the degrees of all vertices in a SMT of a finite set in $M_d(B)$ are bounded by $z_d(B)$.

Especially we obtain in the d-dimensional space with rectilinear norm $z_d=2d$ and in the d-dimensional Euclidean space $z_d=3$.

Moreover, with similar arguments as in the proof of theorem 2 we obtain

Theorem 4. Let B be a smooth unit ball in a d-dimensional Minkowski space. Then $z_d(B)\leq 2d$.

For the general case in /5/ and /6/ is proved

Theorem 5. Let B be a unit ball in a d-dimensional Minkowski space. Then $z_d(B)\leq 3^d-1$.

I conjecture that the last inequality is not sharp.

Conjecture. For every Minkowski space $M_d(B)$ holds

$$z_d(B)\leq 2\,(2^d-1)\ .$$

Equality holds if and only if B is affin equivalent to $\mathrm{conv}(I\cup -I)$ with $I=\{(t_1,\ldots,t_d)\mid 0\leq t_i\leq 1,\ i=1,\ldots,d\}$.

Theorem 2 a) and b) show us that this conjecture is true in the two-dimensional case.

References

/1/ E.N. Gilbert, H.O. Pollak : Steiner minimal trees; SIAM J. Appl. Math., 16(1968), 1-29.
/2/ M. Hanan : On Steiners Problem with rectilinear distance; SIAM J. Appl. Math., 14(1966), 255-265.
/3/ D. Laugwitz: Konvexe Mittelpunktsbereiche und normierte Räume; Math. Zeitschrift, 61(1954), 235-244.
/4/ I.M. Jaglom, W.G. Boltjanski: Konvexe Figuren; Berlin, 1956.
/5/ D. Cieslik: Knotengrade kürzester Bäume in endlich dimensionalen Banachräumen; to appear in Rost. Math. Koll..
/6/ D. Cieslik: Knotengrade kürzester Bäume in Banach-Minkowski Räumen; in preparation.

Unfolding Weighted Consensus Orders into Consistent Numerical Scales

J. Czyzowicz, Québec, Canada
A. Pelc, Ontario, Canada
I. Rival, Ontario, Canada

The aim of this paper is twofold: first, to introduce an apparently new model in the factor analysis of preference data; second, to establish several related theoretical results about ordered sets, which seem to be of independent mathematical interest.

Suppose that a population sample, say of students, is asked to rank a list of alternatives, say of career prospects. Our objective is to find out just how many criteria are used in making career decisions. The experimental data consist of, essentially linearly ordered lists of career choices, yet, many, possibly conflicting criteria may be involved in the actual decisions taken. How should these be "unfolded"?

Thus, it may be all prefer the career of high technology entrepreneur(e) to sanitation inspector (i) and the career of supreme court judge (j) to building superintendent(s). As unanimous preference is transitive such pairs give rise to an order on the underlying list of alternatives. There are, however, several ways to produce an order, especially if there is little unanimity, yet widespread agreement. One approach is simple majority. However, even in quite simple cases the majority relation may not be transitive. A common resolution to this objection is to find the smallest positive integer k such that the relation at least k members of our sample prefer career x to career y, is itself a transitive relation. This order may call a "k-majority order" or, for our purposes, a consensus order. In symbols, we may write x > y to stand for related pairs in this consensus order. Of course, unanimous preferences always belong to the consensus order, too. Thus,

R. Bodendiek, R. Henn (Eds.)
Topics in Combinatorics and Graph Theory

e > i and j > s. And there may be many consensus preferences (which are not unanimous at all), say medical school dean (d) > mathematician (m), and j > e, j > d, d > s. Of course, additional relations are implied by transivity; thus, j > d and d > i imply j > i. For convenience, we may illustrate this consensus order by an "upright" diagram in which only immediate successors and immediate predecessors of a point are joined to it noncomparabilities too, such as d, e or i, s. A member of our sample may well rank all pairs leaving no pair noncomparable at all.

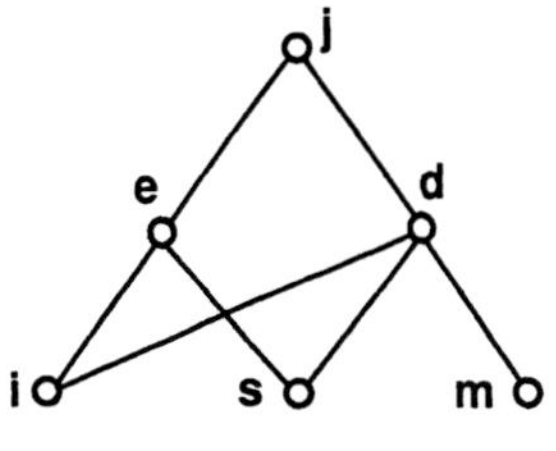

Figure 1

In addition to asking the students to rank the career alternatives we may have asked them to assign numerical values to their preferences too. Thus, Ada may assign to her preferece of dean over superintendent the weight $w_A(d, s) = 7$. On the other hand, Boris prefers entrepreneur over inspector by $w_B(e, i) = 9$. Claude less preoccupied with glitter, $w_C(e,i) = 2$. In contrast, Ada may have actually preferred inspector to dean - although the consensus had it d > i. We associate with each student of our sample a separate copy of the consensus order with weights assigned to those edges of it which correspond to numerical values chosen by that student (see Figure 2). Each student is in this way assigned a <u>weighted consensus order</u>. Thus, not every edge of Boris' order is weighted for he neglected to assign a value, for instance, to $w_B(d,s)$. Also, not all of Ada's weights are included for she actually preferred mathematician (m) to dean (d) by $w_A(m,d) = 3$ yet, the consensus order put d > m.

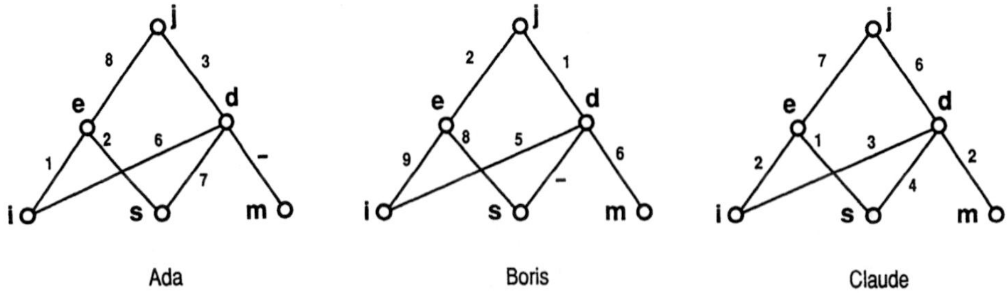

Figure 2

The actual preference data of an individual may be made according to several criteria such as salary, job security, career progress, etc., and the criteria themselves may vary from one individual to another. Thus, Boris values independence and money, while Claude social service and security. Ada, the intellect. But, of course, it is the purpose of our experiment to determine just which criteria (and perhaps which psychological traits) explain the measurements. Or, without labelling the relevant criteria or traits involved we may have in mind to find the least number of numerical scales needed to account for the data "consistent with the consensus". This idea we shall now make more precise.

Let P be a (finite) ordered set. Its underlying elements represent the alternatives and its ordered pairs $a > b$ stand for consensus preferences as determined by the population sample. Because of transitivity it is convenient and economical to focus especially on the covering relation of the order. Thus, a <u>covers</u> b (a is an <u>upper cover</u> of b), written $a \succ b$, if $a > b$ and, for any, c in P, $a \geq c > b$ implies $a = c$. We also speak of <u>covering pairs</u> (a,b), or (b,a). (In Figure 1, e covers i but j does not cover i.) Say that P is <u>weighted</u> if there is an assignment of positive real numbers to some of its covering relations (edges in its diagram); thus, a partial map w of $\{(a,b) : a \succ b \text{ in } P\}$ to R^+. This weighted order is <u>consistent</u> if there is an order-preserving map f from the vertices of P to the real numbers r such that, for any covering, $a \succ b$,

$$w(a, b) = f(a) - f(b)$$

for which w(a,b) is defined (see Figure 3). Otherwise, we say

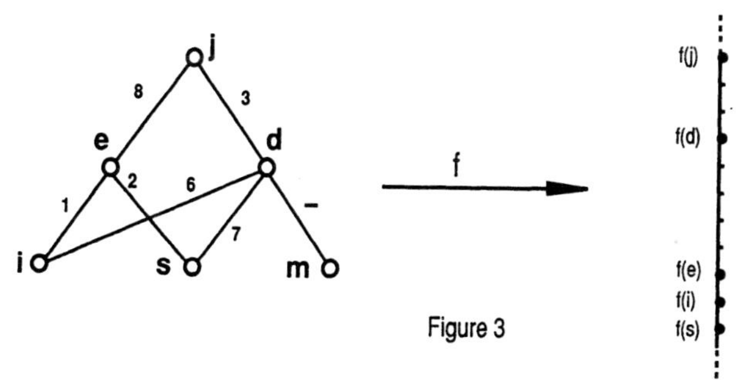

Figure 3

it is _nonconsistent_. Loosely speaking weighted ordered set is consistent if its rankings are "one-dimensional". Although underlying Ada's preferences there may be several contending criteria (salary, progress through the ranks, security, etc.), and, indeed, certain of her preferences may be based on several of them simultaneously, we shall postulate that, in view of her preference data, with respect to her weighted consensus order, one single numerical scale suffices to consistently resolve her preference choices. Actually we are already overlooking a possible "inconsistency" in the preference data. We have considered only the weights of pairs corresponding to covering pairs. There may be further preference data collected concerning comparable pairs with intervening elements. Thus, $w_A(e,i) = 1$ and $w_A(j,e) = 8$ yet perhaps Ada has assigned a weight 7 to the comparable pair $j > i$; thus, $w_A(e, i) + w_A(j, e) \neq w_A(j, i)$. We could well consider such "additivity" constraints too by replacing diagrams with suitably enlarged directed graphs showing such additional edges as are needed; nevertheless, we shall ignore this possibility here and tacitly assume that such additivity prevails whenever it is implied by the consensus order.

A _directed cycle_ C in P is a sequence $(a_1, a_2, \ldots, a_n)$ of its elements such that (a_i, a_{i+1}), for $i \leq n-1$, and (a_n, a_1) are covering pairs. Say that a covering pair (a_i, a_{i+1}) (modulo n) in a directed cycle C is _positive_ if $a_{i+1} \succ a_i$, and _negative_ if $a_i \succ a_{i+1}$. the set of positive covering pairs in C is its _positive part_ C^+, and its set of negative covering pairs its negative part C^-. The positive or negative part is _complete_ if each of its covering pairs is assigned a weight. The sum of the weights of the weighted covering pairs in C^+ is denoted by $\Sigma\, C^+$, and the sum of the weights in C^- by $\Sigma\, C^-$. Here is our first result; it seems to be of independent mathematical interest. Apparently the notion of complete consistency has been studied before (cf. Doignon 1987 and Walther 1984). Nevertheless we are unaware of any corresponding result for the general case.

THEOREM 1. *A weighted ordered set is consistent if and only if, for every directed cycle whose negative part is complete,*

$$\Sigma\, C^{-} \quad \text{is} \quad \begin{cases} \text{equal to } \Sigma\, C+, \text{ if } C+ \text{ is complete} \\ \text{larger than } \Sigma\, C+, \text{ if } C+ \text{ is not complete.} \end{cases}$$

What if the preference data of an individual cannot be resolved along a single scale? Thus, no single psychological dimension, such as salary, career progress, job security, or any other, yet unnamed, linear scale, can account for the measurements recorded for Claude, each of whose covering pairs is weighted (see Figure 2), nor for Boris, whose weight function is not defined for every covering pair. How many "dimensions" are needed to "unfold" the preference data of an individual? Again,many psychological and material factors may contribute to an individual's career preferences. We shall be more precise.

Say that a weighted order P is *chain* n-*consistent* if there are n consistent subsets such that, for any integer m, every sequence c_1 >- c_2 >- ... >- c_m is contained entirely in one of these consistent subsets. See (Figure 4.) We postulate that, for any single covering chain a single scale should suffice,although different scales may be needed for different covering chains. Thus, in Boris' weighted consensus order, the two covering chains j >- e >- i and j >- d >- i cannot both be resolved by the same single linear scale.

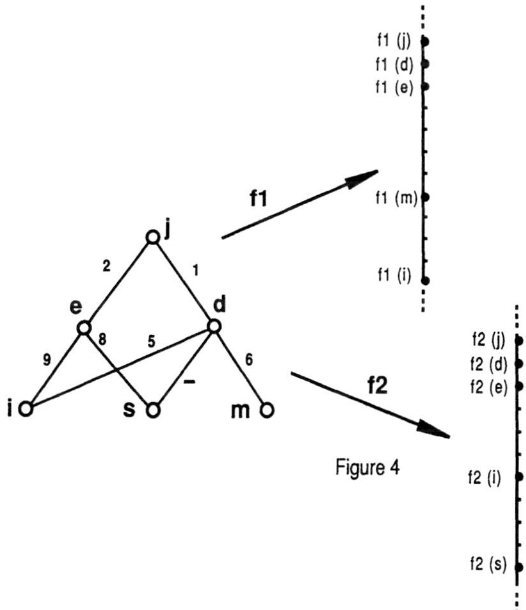

Figure 4

Yet another approach to the study of the multidimensionality of a nonconsistent order is this. Say that a weighted order P is *edge* n-*consistent* if there is a partition of the set of

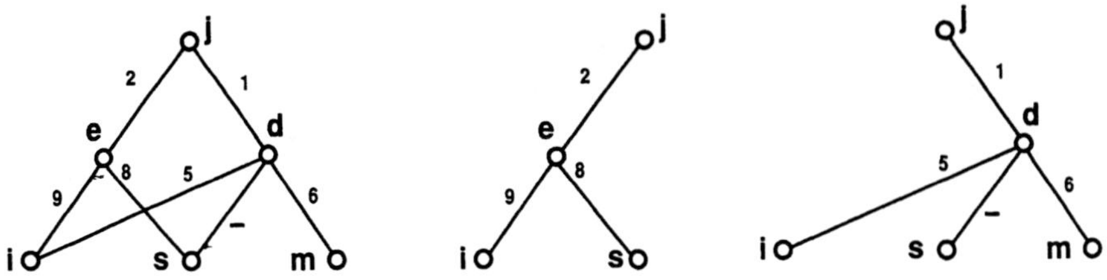

Figure 5

all covering pairs into n consistent subsets(See Figure 5). In this case not every covering chain need be contained entirely in a consistent subset.

THEOREM 2. Let n be any positive integer.

(i) There are weighted orders which are not chain n-consistent.

(ii) There are weighted orders which are not edge n-consistent.

There is perhaps a resemblance between our concept of "multidimensionality" and the concept of "order-dimension". Indeed, it is an interesting question in measurement theory to find the least number of real-valued functions $f_1, f_2, \ldots, f_k$ of the elements of an ordered set P such that, for all a, b in P, a > b in P if and only if, for every $i \le k$, $f_i(a) > f_i(b)$ (cf. K. Baker , P. Fishburn and F. Roberts (1971)). The resemblance is only cursory though: for one thing order-dimension involves no weights on the covering pairs.

Proof of the Theorem 1. To prove the "only if" part let P be a consistent weighted consensus order with corresponding map f. Let C be a directed cycle in P. Let $m = \min \{f(x) : x \in C\}$ and $M = \max \{f(x) : x \in C\}$. If the positive and negative parts of C are complete then

$$M-m = \sum C^+ = \sum C^-.$$

Next, suppose that C^- is complete but C^+ is not.
Then

$$M-m = \sum C^- = \sum_{(x,y) \text{ in } C^+} f(y) - f(x) .$$

As C^+ is not complete and $f(y) > f(x)$, for any (x, y) in C^+,

$$\sum_{(x,y) \text{ in } C^+} f(y) - f(x) > \sum C^+ .$$

To prove the "if" part suppose that an arbitrary directed cycle C of a weighted order satisfies the conditions set out in the statement of the theorem. Let $P_1, P_2, \ldots, P_k$ be the partition of P into equivalence classes defined by $x\theta y$ just if $x = y$ or there is a sequence $x = a_1, a_2, \ldots, a_n = y$ such that each (a_i, a_{i+1}) is a weighted covering pair in P. Our aim now is to show there is an order-preserving, real-valued map for P to establish the consistency of this weighted consensus order. To this end suppose, first, that k=1, that is all elements of P are equivalent. Define $f(a)=0$ for some element a in P. Suppose inductively that an element q in P has value x, assigned previously. Put

$$f(s) = \begin{cases} x + w(s,q), & \text{is } s >\!- q \\ x - w(s,q), & \text{if } q >\!- s \end{cases}$$

for all upper and lower covers of q. If this can be done consistently for all elements q in P, the map f will do. If not, then a different value must be assigned to an element which was already assigned a value previously. This, however, implies the existence of a directed cycle $C = (a_1, a_2, \ldots, a_n)$ for which $\sum C^+ \neq \sum C^-$. Obviously both the negative and positive parts of C are complete, which contradicts the condition of the theorem, thus proving the "if" part in this case.

Next suppose that there are at least two equivalence classes of θ. Without loss of generality we may now assume that each of them (considered as a suborder of P) is consistent. For $j \leq k$ let p_j be any element of P_j. Let $f(p_j) = x_j$. Then, for every p in P_j, we must have $f(p) = x_j + t$, for some real t, where $t = \sum \{w(a_i,a_{i+1}) : a_i >\!- a_{i+1})\} - \sum \{w(a_i,a_{i+1}) : a_{i+1} >\!- a_i\}$, in any sequence $p_j = a_1, \ldots, a_n = p$, from the definition of θ.

What are the connections between the different numbers x_j? For any couple j_1, j_2 identify all covering pairs (p, q) where p belongs to P_{j_1} and q to P_{j_2}. Belonging to different equivalence classes they are all unweighted covering pairs. The existence of any such pair q >- p implies the inequality $f(q) > f(p)$. Since $f(p) = x_{j_1} + t$ and $f(q) = x_{j_2} + s$, there are, among all such inequalities, at most two (one of the form $x_{j_1}+c(j_1,j_2) < x_{j_2}$ corresponding to a pair $q'(j_1,j_2) \succ p'(j_1,j_2)$ and the other of the form $x_{j_1}+d(j_1,j_2) > x_{j_2}$ corresponding to a pair $q''(j_1,j_2) \succ p''(j_1,j_2)$) which are the "sharpest", that is, imply all others for the couple (j_1, j_2) of indices. It is easy to see that, if all those inequalities are satisfied, the map f is as required and hence the consistency of the weighted consensus order P is equivalent to the existence of solutions of the following system of inequalities

$$\begin{cases} x_{j_1} + c(j_1, j_2) < x_{j_2} \\ x_{j_1} + d(j_1, j_2) > x_{j_2} \end{cases}$$

for given constants $c(j_1, j_2)$, $d(j_1, j_2)$, $j_1, j_2 \le k$. (In any particular case some of those inequalities may, of course, be absent).

When does such a system of inequalities have a solution? There is a solution only in the case that there exists a sequence $l_1, l_2, \ldots, l_m$ for which

$$x_{l_i} + c(l_i, l_{i+1}) < x_{l_{i+1}}, \text{ for } i = 1, 2, \ldots, m-1$$

but

$$x_{l_1} + d(l_1, l_m) > x_{l_m}$$

and

$$\sum_{i=1}^{m-1} c(l_i, l_{i+1}) \ge d(l_1, l_m).$$

Let A_{l_i}, for $2 \le i \le m$, be a sequence of elements of P_i which account for the θ-equivalence of $p'(l_i, l_{i+1})$ and $q'(l_{i-1}, l_i)$. In particular, let A_1 be a sequence of elements of P_1 associated with the θ-equivalence of $p'(l_1, l_2)$ and $q'(l_m, l_1)$. Consider the directed cycle C defined by

$$p'(l_1, l_2), q'(l_1, l_2), A_{l_2}, p'(l_2, l_3), q'(l_2, l_3), A_{l_3}, \ldots, q'(l_{m-1}, l_m), A_{l_m}, p'(l_m, l_1), q'(l_m, l_1), A_{l_1}$$

(cf.Figure 6). Clearly, its negative part is complete (negative covering pairs occur only on "vertical channels" and hence are weighted). Its positive part is not: the covering pairs $q'(l_i, l_{i+1}) >\!\!- p'(l_i, l_{i+1})$ are unweighted. On the other hand,

$$\Sigma c^+ - \Sigma c^- = \sum_{i=1}^{m-1} c(l_i, l_{i+1}) - d(l_1, l_m) \geq 0.$$

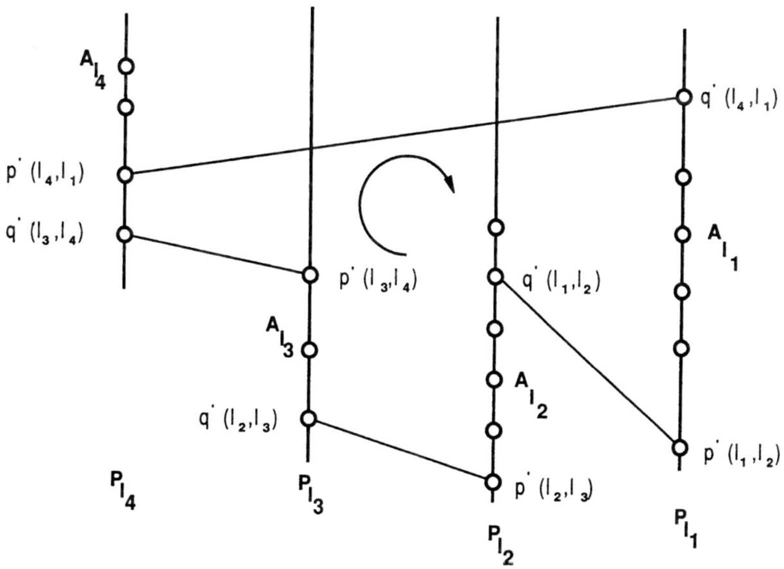

Vertical channels correspond to equivalence classes of θ. The arrow indicates the orientation of the directed cycle C.

Figure 6

Thus, we have shown that if θ has at least two equivalence classes and P is nonconsistent then the condition from the theorem is violated. This concludes the proof.

Proof of the Theorem 2. The proof of (i) is easy to verify using the weighted ordered set illustrated in Figure 7. Part (ii) involves a somewhat more elaborate construction. To this end let P be an ordered set whose corresponding undirected graph G of covering pairs has large chromatic number, say, larger than 2^n. (See either B. Bollobas (1977) or J. Nesetril and V. Rödl (1978), (1979) for the construction of such ordered sets. All of these originate in the probabilistic construction of P. Erdös (1959), (1961) according to which there are graphs of arbitrarily large chromatic number and arbitrarily large girth.) Suppose, moreover, that the covering pairs are all labelled log p_i, using a different prime number for each covering pair. For contradictions, suppose that the edges of G could be partitioned into n subsets corresponding to consistent subsets $S_1, S_2, \ldots, S_n$ of the weighted order P, no pair having a common covering pair. If one such consistent subset S_i contains a directed cycle C then, according to the theorem $\sum C^+ = \sum C^-$ which, in turn, is associated with two distinct subsets of the set of primes used to weight all of the edges. Thus,

$$\sum_I \log p_i = \sum_J \log p_j.$$

so,

$$\pi_I \; p_i = \pi_J \; p_j,$$

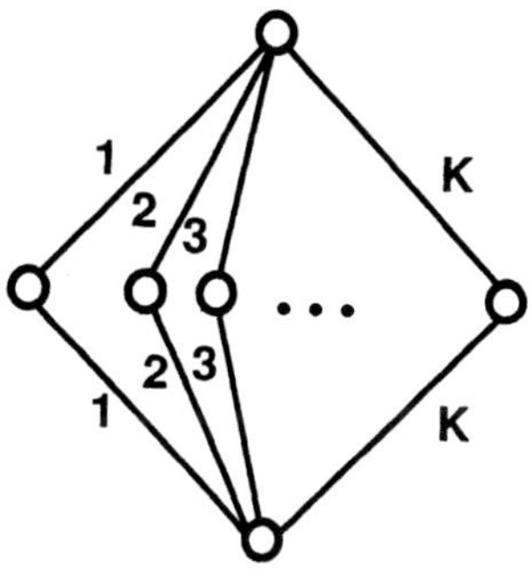

Figure 7

although the prime numbers chosen are all distinct. It follows that none of the subsets S_i contains a directed cycle at all. This, in turn, means that the chromatic number of the subgraph of G corresponding to S_i is at most two, for each $i \le n$. In that case, though, the chromatic number of the whole underlying graph G is at most 2^n, contrary to our hypothesis. This completes the proof.

Two final comments are in order.

As a positive model in the factor analysis of preference data we would propose that it actually be implemented with an available data set to compare the results with existing interpretations. Indeed, it is not at all clear, for instance, just how the numbers of scales needed for the individual weighted consensus orders should themselves be analysed.

From a mathematical viewpoint it would be interesting to have a characterization of, say, chain two-consistent weighted consensus orders. We have no further insight at this writing.

Acknowledgement. We are grateful to anonymous referees especially for pointing out references to the complete case of theorem 1.

REFERENCES

K.A. Baker, P.C. Fishburn, F.S. Roberts (1971) Partial orders of dimension 2, Networks **2**, 11-28.

B. Bollobas (1977) Colouring lattices, Algebra Universalis, 7, 313-314.

J.P. Doignon (1987) Multiple semi-orders SIAM J. of Alg. and Disc. Methods.

P. Erdös (1959) Graph theory and probability, Canad. J. Math. **11**, 34-38.

P. Erdös (1961) Graph theory and probability II, Canad. J. Math. **13**, 346-352.

J. Nesetril and V. Rödl (1978) On a probabilistic graph-theoretical method, Proc. Amer. Math. Soc. **72**, 417-421.

J. Nesetril and V. Rödl (1979) A short proof of the existence of highly chromatic graphs without short cycles, J. Comb. Th. **B 27**, 225-227.

F.S. Roberts (1979) Measurement Theory, Encyclopedia of Mathematics and Its Applications. **7**, Addison-Wesley.

H. Walther (1984) Ten applications of graph theory, Reidel.

Forbidden Ordered Subgraphs

P. Damaschke, Jena, GDR

1. Introduction

There are many special classes of undirected graphs which occur permanently in the literature in several contexts. These classes are interesting for their structural properties, and they are motivated from applications.
A classical paper concerned with 2-dimensional partial orders is [5]. In the sixties, numerous authors studied comparability and interval graphs. In the following years the catalogue of interesting graph classes has further increased. References to original papers can be found in [9]. Other surveys were written by P.Duchet [4] and R.H.Möhring [14]. We can assert that interval graphs are a favourite class [10].
Since the concept of NP-completeness was introduced it has turned out that almost all natural decision problems on graphs are NP-complete, and therefore computationally intractable [8]. So it was an obvious question whether the structural properties of special graph classes help to design efficient algorithms when the problems are restricted to these classes. At present there is an immense list of results in this direction. For a bibliography see e.g. the series by D.S.Johnson [11].
In the present paper, we give a uniform description (so-called FOSG-characterizations) of such well-known graph classes which are usually defined by very different principles (intersection

R. Bodendiek, R. Henn (Eds.)
Topics in Combinatorics and Graph Theory

graphs, graphs related to posets, partitions of the vertex set, chordality properties, forbidden subgraphs, recursive definitions etc.). This shows in some sense the genuineness of this world of graph classes, and it gives a new general view on it. The paper could be the startpoint of a series.

In section 2, we introduce our concept of FOSG-classes and we give some elementary properties. In section 3, we list a lot of FOSG-characterizations which are easy to prove from the known theory. Section 4 is concerned with an FOSG-characterization of chordal bipartite graphs. For proving it we must deeper consider the structure of such graphs, thereby we take advantage of a relationship to strongly chordal graphs. In section 5, the well-known perfect elimination orderings are generalized; we can say something about recognition and forbidden subgraphs of classes generated by elimination orderings. Our approach leads to a number of interesting and, apparently, difficult open problems, mentioned in section 6.

2. Preliminaries

We use the term "*graph*" $G=(V,E)$ only for a finite, undirected graph with neither loops nor multiple edges. V is the vertex set and E is the edge set of G. For $x \in V$ we define $N(x)=\{y: xy \in E\}$.

Directed graphs are called "*digraphs*" here. $H \leq G$ means that the graph H is an *induced subgraph* of G. (We will omit the word "induced".) The relation $\leq$ defines a poset on the class of all graphs.

A ***graph class*** $\mathcal{G}$ is a finite or infinite set of graphs. $\mathcal{G}$ is called ***hereditary*** if $G \in \mathcal{G}$ and $H \leq G$ implies $H \in \mathcal{G}$.

Let $\mathcal{G}$ be a hereditary graph class. A graph F is called a *forbidden subgraph* if $F\notin\mathcal{G}$, and $F-v\in\mathcal{G}$ for all vertices v of F. The set of all forbidden subgraphs is denoted by $\mathcal{F}\mathcal{G}$. Then we have $G\in\mathcal{G}$ iff there is no $F\in\mathcal{F}\mathcal{G}$ with $F\leq G$. Clearly, $\mathcal{F}\mathcal{G}$ is an antichain in the $\leq$-poset of all graphs. Conversely, let $\mathcal{F}$ be an arbitrary antichain of graphs. Then there exists a unique hereditary graph class $\mathcal{G}$ so that $\mathcal{F}=\mathcal{F}\mathcal{G}$.

An *ordered graph* L is a triple (V,E,Q) where G=(V,E) is a graph with n vertices, and Q is an ordering $(v_1,\dots,v_n)$ of the vertices. G is the *underlying graph* of L. We also say that L is a *layout* (of G). In a layout, we define $N_i[v_j]=\{v_j\}\cup\{v_k:\ i\leq k,\ v_jv_k\in E\}$ for $i\leq j$.

We will describe layouts in the form (*,*)...(*,*)(*)...(*), where the symbols "*" represent numbers from {1,...,n}; (i,j) denotes an edge between v_i and v_j, and (k) means that v_k is an isolated vertex. A layout M=(W,F,R) is called a *sublayout* of L=(V,E,Q) if $W\subseteq V$, the underlying graph of M is the subgraph of (V,E) induced by W, and R is the restriction to W of the ordering Q.

The *complement* L^c of a layout L is obtained from L by replacing the underlying graph G of L by its complement G^c. The *dual layout* L^* is obtained from L by reverse the ordering of vertices. For a graph class $\mathcal{G}$ we define $\mathcal{G}^c=\{G:\ G^c\in\mathcal{G}\}$. For a set $\mathcal{L}$ of layouts we define $\mathcal{L}^c=\{L:\ L^c\in\mathcal{L}\}$, and $\mathcal{L}^*=\{L:\ L^*\in\mathcal{L}\}$.

Let $\mathcal{L}$ be a set of layouts, and let G be a graph. A layout L of G is called *$\mathcal{L}$-feasible* (or short: feasible) if it contains no sublayout in $\mathcal{L}$. We denote by $(\mathcal{L})$ the class of all graphs G which have at least one feasible layout. For $\mathcal{L}=\{L_1,\dots,L_k\}$ we write $(L_1,\dots,L_k)$ instead of $(\{L_1,\dots,L_k\})$.

$\mathcal{G}$ is called an *FOSG-class* if $\mathcal{G}=(\mathcal{L})$ for some finite $\mathcal{L}$. A set $\mathcal{M}$ of layouts is called *order invariant* if for each layout $M\in\mathcal{M}$ all layouts obtained from M by changing the ordering of vertices also belong to $\mathcal{M}$. The following properties are trivial:

Proposition 1.

(1) $(\mathcal{L})$ is hereditary.

(2) $(\mathcal{L}^c)=(\mathcal{L})^c$.

(3) $(\mathcal{L}^*)=(\mathcal{L})$.

(4) $(\mathcal{L}\cup\mathcal{M})\subseteq(\mathcal{L})\cap(\mathcal{M})$.

(5) If $\mathcal{M}$ is order invariant then we have equality in (4).

(6) $\mathcal{G}$ with finite $\mathcal{F}\mathcal{G}$ is always an FOSG-class.

(7) If $\mathcal{G}$ is an FOSG-class and $\mathcal{F}\mathcal{H}$ is finite then $\mathcal{G}\cap\mathcal{H}$ is an FOSG-class. ∎

Remarks.

(4): We cannot state equality in general; an example is given in section 3.

(6): From an obvious cardinality argument we obtain the existence of hereditary $\mathcal{G}$ which are not FOSG-classes. On the other hand, we have many examples for FOSG-classes $\mathcal{G}$ with infinite $\mathcal{F}\mathcal{G}$.

(7): We do not know whether $\mathcal{G}\cap\mathcal{H}$ is an FOSG-class whenever $\mathcal{G},\mathcal{H}$ are FOSG-classes.

3. A general observation

Let us introduce some denotations for the 8 layouts of 3 vertices:

ch=(1,2)(1,3),

cp=(1,2)(2,3),

cl=(1,2)(1,3)(2,3).

The symbols ch^c, cp^c, cl^c, ch^*, ch^{c*} are clear then.

Theorem 1. There hold the following FOSG-characterizations:

(ch)	chordal graphs [4]
(cp)	comparability graphs [4]
(ch,cp)	arborescence comparability graphs
(ch,cl)	forests
(ch,ch^{c})	threshold graphs
(ch,cp^{c})	interval graphs [15]
(ch,ch^{*})	proper interval graphs [3]
(ch,ch^{c*})	split graphs
(cp,cl)	bipartite graphs
(cp,cp^{c})	permutation graphs (cf. [5])
(ch,cp,ch^{c*})	threshold graphs
(ch,cl,cp^{c})	unions of caterpillars of hairlength 1 [12]
(cp,cl,cp^{c})	bipartite permutation graphs.

Further, the following classes are FOSG-classes: split interval graphs, split permutation graphs, cographs, chordal permutation graphs, circular arc graphs, unions of caterpillars, perfectly ordered graphs, graphs of bounded degree, graphs of bounded clique size, forests of bounded depth, outerplanar graphs. ■

Remarks.

1. Some of the characterizations given here are aleady known in other terms (cf. references), but they occur occasionally among other ones. For the assertions of Theorem 1 without reference, one can find short (but mostly non-trivial) proofs using the known theory which is presented e.g. in [9]. These proofs are left to the interested reader.

2. We have two incomparable FOSG-characterizations of threshold graphs.

3. In some cases, we can overlook the system of all feasible layouts of a fixed graph G of the class, e.g.:

(ch): perfect elimination orderings of G.

(cp): linear extensions of posets with comparability graph G.

(ch,cl): Fix a root in every connected component T of G and consider T as a rooted tree, i.e. as a digraph. Then a layout is feasible iff every edge $v_i v_j$ (i<j) in the layout is directed from v_j to v_i.

(ch,cp^c): orderings of vertices with respect to the left endpoints of intervals in the interval models of G.

(cp,cp^c): the permutations defining G.

4. We have (ch,cp)⊂(ch)∩(cp), since e.g. P_4 belongs to the difference of the two classes. It is open whether (ch)∩(cp) is an FOSG-class.

4. Special bipartite graphs

In this section, I and J denote the colour classes of a bipartite graph. (Note that I,J are not uniquely determined.)
Let I,J be a partition of V in an arbitrary graph G=(V,E). We say that I,J are *separated* in a layout L of G if all vertices of I occur before all vertices of J.
Further we define:
$\mathcal{B}$= {cp, cl, (1,2)(3,4), (1,2)(3,4)(1,4)}

Lemma 1. ($\mathcal{B}$) is the class of bipartite graphs. The feasible layouts of G are exactly those layouts L for which there exists a 2-colouring I,J of G so that I,J are separated in L. ■
The proof is straightforward.
A bipartite graph is called *convex* if I can be linearly ordered so that N(v) consists of consecutive vertices for every v∈J. If the analogous holds also for J, then the graph is called *biconvex*.

Proposition 2. $(\mathcal{B}\cup\{(1,2)(1,4)(3)\})$ is the class of convex bipartite graphs, and $(\mathcal{B}\cup\{(1,2)(1,4)(3),\ (1,4)(3,4)(2)\})$ is the class of biconvex bipartite graphs. ∎

This is a consequence of Lemma 1.

A chordal graph is called *strongly chordal* if it has a perfect elimination ordering (*PEO*) such that $i\le j\le k$ implies $N_i[v_j]\subseteq N_i[v_k]$. Such a PEO is called a *strong elimination ordering* (*SEO*). An equivalent condition is that $i<j<k<l$, $v_iv_l\in E$, $v_iv_k\in E$, $v_jv_k\in E$ imply $v_jv_l\in E$ [6]. So one can easily see:

Proposition 3. (ch, $(1,3)(1,4)(3,4)(2,3)$) is the class of strongly chordal graphs. ∎

Lemma 2. [2] For a bipartite graph G with colour classes I,J let S be the graph obtained from G by adding edges so that J becomes a clique. Then G is chordal bipartite iff S is strongly chordal. ∎

For the proof we need a third characterization from [6]: A chordal graph is strongly chordal iff each cycle of even length at least 6 has an odd chord.

Note that S in Lemma 2 is a strongly chordal split graph.

Lemma 3. [13] Every strongly chordal split graph has a SEO in which all vertices of I occur before all vertices of J. ∎

This can be proved as follows: Consider a SEO L not satisfying our assertion. Then there must exist vertices $v_m\in J$, $v_{m+1}\in I$. We exchange v_m and v_{m+1} in L and get a new layout L'. Show that L' is a SEO, too. So we can step by step move the J-vertices to the right and the I-vertices to the left of the layout, respectively.

Theorem 2. $(\mathcal{B}\cup\{(1,4),(1,3),(2,3)\})$ is the class of chordal bipartite graphs.

Proof. Let G be chordal bipartite. We define S like in Lemma 2. Then S is strongly chordal, and we consider a fixed SEO of the special kind obtained in Lemma 3. Then we delete all additional edges in J. We denote the layout so obtained by L. Clearly, L is $\mathcal{B}$-feasible. Assume that L has a sublayout (1,4)(1,3)(2,3). Since I,J are separated, the vertices 1,2 belong to I, and the vertices 3,4 belong to J. By adding an edge between vertices 3,4 we get a contradiction to the fact that we started with a SEO. So the assumption is false, and L is feasible.

Conversely, let L be a feasible layout of a graph G. By completing J to a clique, we get a SEO of a strongly chordal graph. Then Lemma 2 yields that G is chordal bipartite. ∎

5. Generalized elimination orderings

In this section, $N_H(v)$ denotes the neighbourhood of the vertex v in the graph $H\leq G$. Instead of $N_G(v)$ we will further write $N(v)$.

A graph G is chordal iff it has a layout such that $N_i[v_i]-\{v_i\}$ is a clique for all i. We want to generalize this concept, replacing the term "clique" by a member of some other fixed hereditary graph class $\mathcal{G}$. The graph class generated by $\mathcal{G}$ in this way is named $\mathcal{EG}$. A layout satisfying the condition $N_i[v_i]-\{v_i\}\in\mathcal{G}$ for all i is called an *elimination ordering* (with respect to $\mathcal{G}$).

Proposition 4. $\mathcal{EG}$ is hereditary.

Proof. It suffices to show: If $G\in\mathcal{EG}$ then $G-v\in\mathcal{EG}$. Let L be an elimination ordering. We have $v=v_k$ for some k. We delete v_k from L. The layout so obtained is an elimination ordering of G-v. (Here we use essentially that $\mathcal{G}$ is hereditary.) ∎

Proposition 5. If $G\in\mathcal{E}\mathcal{G}$ and $N(v)\in\mathcal{G}$ then there exists an elimination ordering with $v_1=v$. In other words: An elimination ordering can be produced greedy-like by choosing always an arbitrary vertex v with $N_R(v)\in\mathcal{G}$ where R is the actual rest graph.

Proof. The greedy-algorithm fails iff we get a rest graph R at any time such that $N_R(v)\notin\mathcal{G}$ for all v. Clearly, $R\notin\mathcal{E}\mathcal{G}$. Due to Proposition 4, this implies $G\notin\mathcal{E}\mathcal{G}$, a contradiction. ∎

Proposition 6. If $\mathcal{G}$ can be recognized in time p(n) then $\mathcal{E}\mathcal{G}$ can be recognized in time $n^2p(n-1)/2$.

Proof. We use the algorithm from Proposition 5. In the k-th step we have to check at most n-k+1 graphs $N_R(v)$ whether they belong to $\mathcal{G}$. This gives the time bound. ∎

For a layout L, we define 1L to be the layout obtained from L by adding a new first (i.e. leftmost) vertex adjacent to all vertices of L. We define $1\mathcal{L}=\{1L: L\in\mathcal{L}\}$. Then we have:

Proposition 7. $(1\mathcal{L})\subseteq\mathcal{E}(\mathcal{L})$. If $\mathcal{L}$ is order invariant then it holds equality. Especially, if $\mathcal{F}\mathcal{G}$ is finite then $\mathcal{E}\mathcal{G}$ is an FOSG-class. ∎

Let $\mathcal{G}$ be an arbitrary graph class. By $\mathcal{N}\mathcal{G}$ we denote the class of all connected graphs G so that $N(v)\in\mathcal{G}$ for each vertex v of G. If $\mathcal{G}$ is hereditary then the class of disjoint unions of $\mathcal{N}\mathcal{G}$-graphs is also hereditary. If $\mathcal{G}$ is an antichain then so is $\mathcal{N}\mathcal{G}$.

Proposition 8. $\mathcal{NFG}\subseteq\mathcal{FEG}$.

Proof. Let be $G\in\mathcal{NFG}$. Then $N(v)\notin\mathcal{G}$ for each vertex v. Consequently, we cannot find a vertex v_1 for an elimination ordering, and hence $G\notin\mathcal{EG}$. Now assume that even $G-v\notin\mathcal{EG}$ for some v. Due to the proof of Proposition 5, there exists a graph $H\leq G-v$ with $N_H(w)\notin\mathcal{G}$ for all H-vertices w. Since $N_H(w)\leq N(w)$ and $G\in\mathcal{NFG}$, we have $N_H(w)\in\mathcal{FG}$, thus $N_H(w)=N(w)$ for each w. This contradicts the fact that G is connected. ■

It is left to the reader to study examples of classes $\mathcal{EG}$.

6. Open problems

1. Is $(\mathcal{L})$ properly contained in the class of all graphs, for every $\mathcal{L}$? We are far from a general solution. It seems that the problem requires methods from higher algebra and combinatorics. Especially, there even remain some sets $\mathcal{L}$ with only one layout of 4 vertices for which our question is open; e.g. $\mathcal{L}=\{(1,2)(2,3)(3,4)\}$.
2. Let $\mathcal{G},\mathcal{H}$ be FOSG-classes. Is $\mathcal{G}\cap\mathcal{H}$ always an FOSG-class? Especially, is (ch)∩(cp) an FOSG-class?
3. Not every hereditary graph class is an FOSG-class. What can one say about restrictions to $\mathcal{FG}$ for FOSG-classes $\mathcal{G}$?
4. Which FOSG-classes have a unique minimal FOSG-characterization?
(cf. Theorem 1, Remark 2)
5. Is the class of circle graphs an FOSG-class?
6. Give sufficient conditions for $(\mathcal{L}\cup\mathcal{M})=(\mathcal{L})\cap(\mathcal{M})$ and for $\mathcal{NFG}=\mathcal{FEG}$.

In a forthcoming paper, we consider graph classes $\mathcal{G}$ such that every hereditary subclass $\mathcal{H}$ of $\mathcal{G}$ has finite $\mathcal{F}\mathcal{H}$.

References

[1] V.Chvatal, Perfectly ordered graphs, Annals of Discr. Math.21 (1984), 63-65

[2] E.Dahlhaus, private communication

[3] P.Duchet, Representations, noyaux en theorie des graphes and hypergraphes, Doct. Diss. Univ. Paris 6 (1979)

[4] P.Duchet, Classical perfect graphs, Annals of Discr. Math. 21 (1984), 67-96

[5] B.Dushnik, E.W.Miller, Partially ordered sets, Amer. J. Math. 63 (1941), 600-610

[6] M.Farber, Characterizations of strongly chordal graphs, Discr. Math. 43 (1983), 173-189

[7] S.Foldes, P.L.Hammer, Split graphs, Proc. 8th Southeastern Conf. on Comb., Graph Th. and Computing (1977), 311-315

[8] M.R.Garey, D.S.Johnson, Computers and intractability, a guide to the theory of NP-completeness, Freeman, San Francisco 1979

[9] M.C.Golumbic, Algorithmic graph theory and perfoct graphs, Acad. Press, New York 1980

[10] M.C.Golumbic, Interval graphs and related topics, Discr. Math. 55 (1985), 113-121

[11] D.S.Johnson, The NP-completeness column: an ongoing guide, in: J. of Algorithms, to be continued

[12] A.Kotzig, Paare Hajös Graphen, Casop. Pest. Mat. 88 (1963), 236-241

[13] D.Kratsch, private communication

[14] R.H.Möhring, Computationally tractable classes of ordered sets, Preprint Reihe Math. 181/1987, Techn. Univ. Berlin(West)

[15] G.Ramalingham, C. Pandu Rangan, A unified approach to domination problems on interval graphs, Info. Proc. Letters 27 (1988), 271-274

On Normal Tournaments With the Least Number of 3-Cycles

D. C. Demaria, Torino, Italy
G. M. Gianella, Torino, Italy

We prove that all but one of the hamiltonian tournaments of order n ($n \geq 4$) with the least number of 3-cycles are normal tournaments and we characterize these tournaments among normal tournaments.

Introduction.

MOON proved in [5] that the minimal number of 3-cycles that a hamiltonian tournament of order n can contain is n-2, and BURZIO and DEMARIA characterized these tournaments (called *B_n tournaments*) by giving their structure and score sequence (see [3]).

In this paper we prove that all but one of the hamiltonian tournaments H_n with n-2 3-cycle are a class of special normal tournaments (see Theorems 3.2 and 3.3).

1. Preliminaries.

Let T_n be a tournament of order n. A cycle C of T_n is *coned* by a vertex v (i.e. v *cones* C) if there exists a vertex $v \in V(T_n) - V(C)$ such that either $v \to V(C)$ or $V(C) \to v$.

Let H_n be a hamiltonian tournament. If no vertex of $H_n - \langle C \rangle$ cones C, C is said to be *non-coned* in H_n and the vertices of H_n - C are called *poles* associated to C.

A non-coned cycle C of H_n is *minimal* if each cycle C' such that $V(C') \subset \subset V(C)$ is coned by at least one vertex of H_n. A minimal cycle is *characteristic* if it possesses the shortest length of the minimal cycles and its length is the *cyclic characteristic* $cc(H_n)$ of H_n. Then the difference n - $cc(H_n)$ is called the *cyclic difference* $cd(H_n)$ of H_n.

A vertex v of H_n is called a *neutral vertex* of H_n if H_n - v is hamiltonian, and the number of neutral vertices of H_n is denoted by $\nu(H_n)$.

We recall that between $\nu(H_n)$ and $cd(H_n)$, the inequality $\nu(H_n) \geq cd(H_n)$

R. Bodendiek, R. Henn (Eds.)
Topics in Combinatorics and Graph Theory

holds (see[2]).

DEFINITION 1.1 - The tournament A_n ($n \geq 4$), with vertex set $V(A_n) = \{a_1, a_2, \dots, a_n\}$ and arc set $E(A_n) = \{(a_i, a_j) \ / \ j < i-1 \text{ or } j = i+1\}$ is called a *bineutral tournament.*

REMARK - It is known that A_n is the unique hamiltonian tournament that has exactly 2 neutral vertices. Moreover it has n-2 3-cycles.

Now we recall the definitions and the structural characterization of normal tournaments (see Nos. 3 and 6 of [4]).

DEFINITION 1.2 - A hamiltonian tournament H_n is said to be *normal* if it contains only one minimal cycle (the characteristic one) or equivalently if $cd(H_n) = \nu(H_n)$.

PROPOSITION 1.1 - Let H_n be a normal tournament, then its characteristic cycle C_k is either the 3-cycle A_3 ($k = 3$) or a bineutral tournament A_k ($k \geq 4$).

PROPOSITION 1.2 - Let H_n be a normal tournament with cyclic characteristic k ($k \geq 3$) and let A_k be its characteristic cycle. A pole z, associated to A_k, must have the following adjacencies with respect to A_k:

1) $(a_{i+1}, a_{i+2}, \dots, a_k) \to z \to (a_1, a_2, \dots, a_i)$ $i = 1,2,\dots,k-1$;
2) $(a_i, a_{i+2}, a_{i+3}, \dots, a_k) \to z \to (a_1, a_2, \dots, a_{i-1}, a_{i+1})$ $i = 1,2,\dots,k-1$.

DEFINITION 1.3 - The pole z is called *pole of kind* i *and class* 1 or *class* 2 and denoted by x_i or y_i if its adjacencies are given by the previous conditions 1) or 2) respectively.

THEOREM 1.3 - A tournament H_n is normal if, and only if, the following conditions are true:

a) H_n allows a bineutral tournament A_k ($k \geq 4$) or the 3-cycle A_3 as a minimal tournament;
b) the poles associated to A_k are of the (k-1) kinds of class 1 ($x_1, x_2, \dots, x_{k-1}$) and of the (k-1) kinds of class 2 ($y_1, y_2, \dots, y_{k-1}$), considered in Definition 1.3;
c) the subtournament P_{n-k} of the poles associated to A_k is not hamiltonian;

d) P_{n-k} allows the following transitive composition $P_{n-k} = Tr_{k+1} *(T^{(0)}, T^{(1)}, \dots, T^{(k)})$ (that is $T^{(i)} \rightarrow T^{(j)} \Longleftrightarrow i > j$) where:

$T^{(0)}$	is formed by poles of type	x_1;
$T^{(1)}$	" " " " " "	x_1, x_2, y_1;
$T^{(2)}$	" " " " " "	x_1, x_2, x_3, y_2;
....	....	....
$T^{(i)}$	" " " " " "	x_{i-1}, x_i, x_{i+1}, y_i;
....	....	....
$T^{(k-2)}$	" " " " " "	x_{k-3}, x_{k-2}, x_{k-1}, y_{k-2};
$T^{(k-1)}$	" " " " " "	x_{k-2}, x_{k-1}, y_{k-1};
$T^{(k)}$	" " " " " "	x_{k-1},

and each component different from $T^{(0)}$ and $T^{(k)}$ can be empty or not.

2. Tournaments B_n.

Following [3], we will denote a hamiltonian tournament of order n with the least number of 3-cycles by B_n and the class of all these hamiltonian tournaments by $\mathfrak{B}$. The following propositions hold:

PROPOSITION 2.1 - Every component $S^{(i)}$ of a B_n tournament is transitive.

PROPOSITION 2.2 - Every hamiltonian subtournament H_m of a tournament B_n, is a tournament B_m.

PROPOSITION 2.3 - Every (non-trivial) quotient R_p of a tournament B_n, is a tournament B_p.

Now, we prove the following propositions:

PROPOSITION 2.4 - If v is a neutral vertex of B_n, then there exists exactly one 3-cycle in B_n passing through the vertex v.

Proof - In fact, $B_n - v \in \mathfrak{B}$ by Proposition 2.2, because $B_n - v$ is a hamiltonian subtournament of B_n. Then $B_n - v$ has n-3 3-cycles. Hence, taking off the neutral vertex v, we take off exactly one 3-cycle. ∎

PROPOSITION 2.5 - Every tournament B_n such that $cc(B_n) = k$ $(k \geq 4)$, is normal.

Proof - a) If C_k is a characteristic cycle of B_n, then C_k is isomorphic to the bineutral tournament A_k.

In fact, if C_k was not bineutral, it would have at least three neutral vertices x, y, z and there would exist three poles called x', y', z' coning $C_k - x$, $C_k - y$, $C_k - z$ (see [2]), respectively. If, for example, $x' \to y'$, $x \to x'$ and $y \to y'$, we obtain $x \to x' \to (C_k - x)$, $y \to y' \to (C_k - y)$. This is a contradiction because x' would be a vertex of two 3-cycles $x' \to y' \to x \to x'$ and $x \to x' \to x^* \to x$, where x^* is the predecessor of x in the cycle C_k (fig 1).

Similarly, if $x' \to y'$, $x' \to x$ and $y' \to y$ we have $(C_k - y) \to y' \to y$, $(C_k - x) \to x' \to x$, and in the same way we obtain a contradiction.

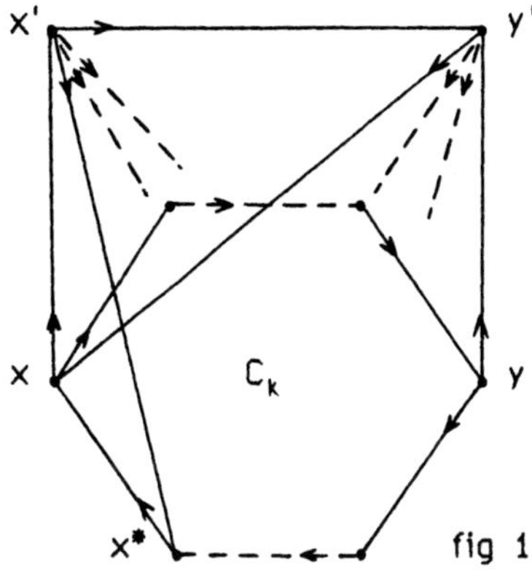

fig 1

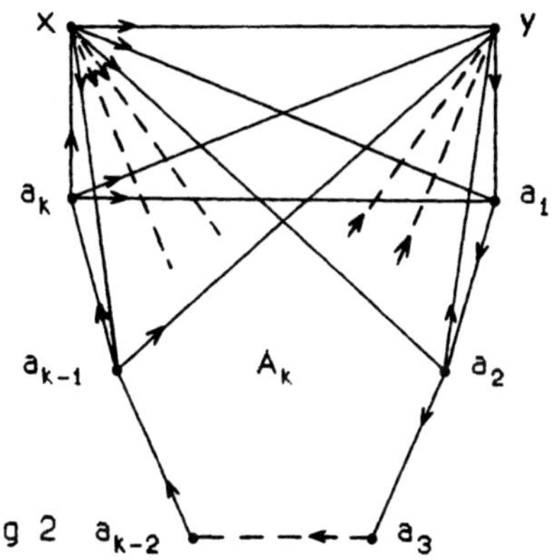

fig 2

b) If A_k, with $V(A_k) = \{a_1, a_2, \ldots, a_k\}$, is a characteristic cycle of B_n, there exists a bineutral tournament A_{k+2} allowing A_k as a characteristic cycle.

In fact, let x and y be two poles of B_n such that x cones $A_k - a_k$ and y cones $A_k - a_1$. If $a_1 \to y \to (A_k - a_1)$ were true, there would exist two 3-cycles of vertex y $y \to a_k \to a_1 \to y$; $y \to a_{k-1} \to a_1 \to y$. But that is a contradiction. Dually, $(A_k - a_k) \to x \to a_k$ is also a contradiction. So it must be $(A_k - a_1) \to y \to a_1$; $a_k \to x \to (A_k - a_k)$.

Moreover, $y \to x$ cannot be true, because there would be two 3-cycles $y \to x \to a_2 \to y$; $y \to x \to a_{k-1} \to y$, which is a contradiction.

Thus $x \to y$ must be true and so x and y are the only neutral vertices of $A_k U\{x,y\}$ that is isomorphic to A_{k+2} (fig 2).

c) B_n is normal.

In fact no vertex $a_i \in A_k$ can be a neutral vertex of B_n because each a_i is a vertex of at least two 3-cycles contained in A_{k+2}, considered above. Therefore the number $\nu(B_n)$ of neutral vertices of B_n is n-k. Hence $\nu(B_n) = = cd(B_n)$. ∎

PROPOSITION 2.6 - The subtournament of the poles of B_n is transitive.

Proof - In fact, in the opposite case, B_n would contain a 3-cycle H_3 and every vertex of H_3 would be a vertex of at least two 3-cycles. ∎

PROPOSITION 2.7 - Every pole of B_n is of class 1 (see Definition 1.3).

Proof - There is no pole y_j of class 2 in B_n because it would be a vertex of two 3-cycles. In fact:

If $2 \leq j \leq k-2$, the pole y_j is a vertex of two 3-cycles $y_j \to a_{j-1} \to a_j \to y_j$ and $y_j \to a_{j+1} \to a_{j+2} \to y_j$ (fig 3).

In the cases $j = 1$ and $j = k-1$, either of the two 3-cycles becomes a 3-cycle of the subtournament A_{k+2} of B_n, not contained in A_k. ∎

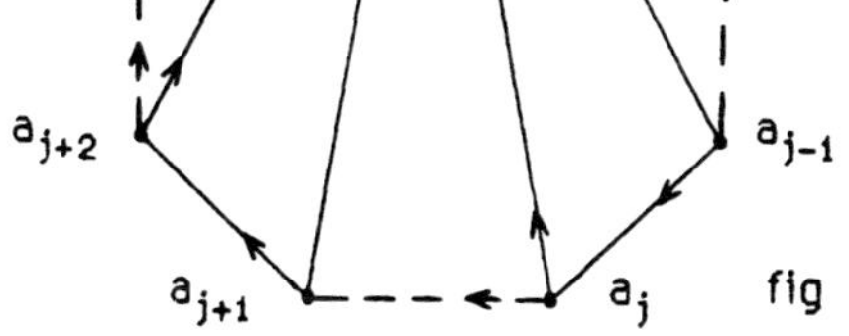

fig 3

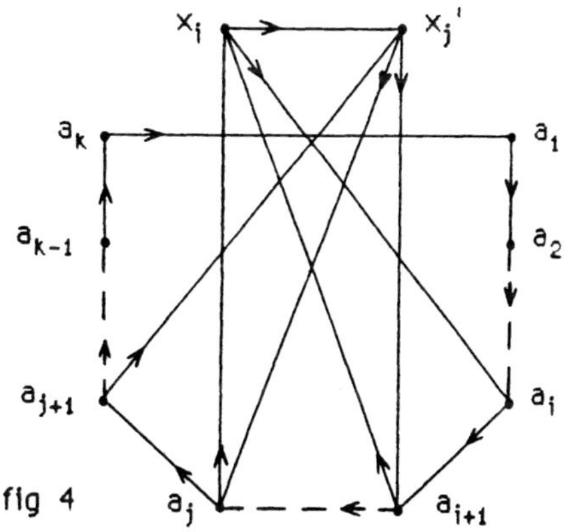

fig 4

PROPOSITION 2.8 - Between two poles of class 1 x_i and x_j' of B_n, the following rules of adjacencies hold $x_i \to x_j' \Rightarrow j \leq i$.

Proof - In fact, let x_i and x_j' be two poles of class 1 such that $x_i \to x_j'$.

If $j > i$, the pole x_i is a vertex of the two 3-cycles $x_i \to a_i \to a_{i+1} \to x_i$, $x_i \to x_j' \to a_{i+1} \to x_i$ and the pole x_j' is a vertex of the two 3-cycles $x_j' \to a_j \to a_{j+1} \to x_j'$, $x_j' \to a_j \to x_i \to x_j'$, and this is a contradiction (fig 4).

Conversely, if $j \leq i$, the pole x_i is a vertex of the only 3-cycle $x_i \to a_i \to a_{i+1} \to x_i$ and the pole x_j' is a vertex of the only 3-cycle $x_j' \to a_j \to a_{j+1} \to x_j'$. ∎

3. Relations between B_n tournaments and normal ones.

PROPOSITION 3.1 - Let be H_n a normal tournament with $cc(H_n) = k$ ($k \geq 3$), such that the subtournament P_{n-k} of its poles is transitive, each of its poles is of class 1 and the previous Proposition 2.8 holds. Thus H_n has exactly n-2 3-cycles.

Proof - In fact, the characteristic cycle A_k contains k-2 3-cycles, but the transitive subtournament P_{n-k} of the poles does not contain 3-cycles. Then there exists exactly one 3-cycle passing through any vertex of P_{n-k}. Hence H_n contains exactly n-2 3-cycles. ∎

In such a way, by Propositions 2.5, 2.6, 2.7, 2.8, we prove the following theorem:

THEOREM 3.2 - A tournament H_n, with $cc(H_n) = k$ ($k \geq 4$), belongs to $\mathfrak{B}$ if, and only if:

a) H_n is normal;

b) its subtournament of the poles is transitive;

c) each of its poles is of class 1;

d) the following rules of adjacencies hold between two poles x_i and x'_j of kind i and j respectively $x_i \to x_j' \Rightarrow j \leq i$. ∎

THEOREM 3.3 - A tournament H_n with $cc(H_n) = 3$ belongs to $\mathfrak{B}$ if, and only if, either of the following two cases are verified:

1) H_n is a normal tournament, such that:

a) the subtournament of the poles is transitive;

b) the poles of H_n are all of class 1;

c) every pole of type x_2 precedes every pole of type x_1.

2) H_n is a non-normal tournament, that is, the composition of a transitive tournament and two singletons with one 3-cycle.

Proof - If the conditions of case 1) or of case 2) hold, it is obvious that $H_n \in \mathfrak{B}$.

Conversely, first of all, from Proposition 2.4, we obtain $\nu(B_n) \leq n-2$. On the other hand we have $\nu(B_n) \geq cd(B_n) = n-3$. So, the number ν of the non-neutral vertices of B_n is either 3 or 2.

1) If $\nu = 3$, the tournament B_n must be normal and satisfy the conditions of case 1), by properties similar to Propositions 2.6, 2.7 and 2.8.

2) If $\nu = 2$, since for any neutral vertex of B_n there exists exactly one 3-cycle, then every 3-cycle must contain the two non-neutral vertices of B_n. Hence case 2) holds. ∎

REMARK - In conclusion, the tournaments B_n ($n \geq 4$) are all normal but one: the composition of 3-cycle (see Theorem 3.3-2)).

From the enumeration of B_n for $n \geq 4$ given in [3], we have that the number of normal tournaments B_n is $2^{n-4} - 1$, whereas the simple tournaments belonging to $\mathfrak{B}$ are all normal and they are F_{n-4}, where F_i is the i-th Fibonacci number.

References

[1] Beineke L.W. and Reid K.B., Tournaments, Selected Topics in Graph Theory. Edited by Beineke L.W. and Wilson R.J., Academic Press, New York (1979).

[2] Burzio M. and Demaria D.C., On a classification of hamiltonian tournaments, Acta Univ. Carol. - Math. Phys., Vol 29. No 2 (1988), 3-14.

[3] Burzio M. and Demaria D.C., Hamiltonian tournaments with the least number of 3-cycles, to appear.

[4] Demaria D.C. and Gianella G.M., On normal tournaments, to appear in Conf. Semin. Mat. Univ. Bari.

[5] Moon J.W., On subtournaments of a tournament, Canad. Math. Bull.,vol. 9 No.3 (1966), 297-301.

[6] Moon J.W., Topics on tournaments, Holt, Rinehart and Winston, New York, (1968).

Two-Irregular Graphs

R. J. Faudree*, Tennessee, USA
R. J. Gould*, Tennessee, USA
M. S. Jacobson**, Tennessee, USA
R. H. Schelp***, Tennessee, USA

1. INTRODUCTION.

One of the most elementary results in graph theory is that a graph on n vertices must have at least two it its vertices with the same degree. Thus, it seems natural to consider those graphs in which no more than two of its vertices have the same degree for each possible degree, and we call such a graph $2-$*irregular*.

In this paper we consider the following question: Which graphs G of order n are embeddable in a $2-$irregular graph of order n? Clearly an arbitrary such graph G must have size at most $3n-7$, since the graph in which 3 vertices have maximum degree cannot be embedded in a $2-$irregular graph of the same order. A question left unanswered is whether each (n,q) graph with $q \leq 3n-7$ is embeddable in a $2-$irregular graph of order n.

In this paper we follow standard notation similar to that of [1] and [2]. In particular, we let $d_G(x)$ denote the degree of x in G and $N_G(x)$ denote the neighborhood of x in G. When the graph G is obvious, the subscript G will be deleted.

2. RESULTS.

There are two principal results in the paper, each of which give sufficient conditions for a graph to be embeddable in a $2-$irregular graph of the same order. We state these results first.

THEOREM 2. *Let G be a graph of order $n \geq \frac{3}{2}d^4$, where $d \geq 8$. If $\Delta(G) \leq d$, then G is embeddable in a $2-$irregular graph of order n.*

THEOREM 3. *For all $n \geq 8^4 3^5/2$ any (n,q) graph G with*

$$q \leq \frac{3}{2}n + \frac{1}{2}\left(\frac{2}{3}\right)^{1/4} n^{1/4}$$

is embeddable in a $2-$irregular graph of order n.

In order to establish the above theorems we need the following three results.

*This research is partially supported by ONR researh grant N000014-88-K-0070
**This research is partially supported by ONR researh grant N000014-85- K-0694
***This research is partially supported by NSA researh grant MDA 904-89-H-2026

R. Bodendiek, R. Henn (Eds.)
Topics in Combinatorics and Graph Theory

LEMMA 1. *Let G be a 2−irregular graph with ℓ vertices of degree $\geq c$. Then, G contains a k−matching where $k = min\{c, \lfloor \frac{\ell}{2} \rfloor\}$.*

THEOREM 1. *Let G be a graph of order n containing two vertices x and y with disjoint neighborhoods, each neighborhood of order at most $d \leq \lfloor \frac{n}{48} \rfloor - 2$. If the graph $G - x - y$ is 2−irregular, then G is embeddable in a 2−irregular graph G^* of order n.*

LEMMA 2. *Let G be a graph of order n and let $V(G)$ be partitioned into sets H_1 and H_2 with $H_1 = \{x_1, x_2, \cdots, x_{\lfloor n/3 \rfloor}\}$. If H_1 is an independent set of vertices and for each ℓ, $1 \leq \ell \leq \lfloor n/3 \rfloor$, the set of vertices $\{x_i \mid 1 \leq i \leq \ell\}$ has at most 2ℓ adjacencies in the set H_2, then G is embeddable in a 2−irregular graph L of order n.*

3. PROOFS.

PROOF OF LEMMA 1: The k−matching will be found by a greedy algorithm that we now describe. Let S be the ℓ−set of vertices of degree $\geq c$.

Choose the edge x_1y_1 such that x_1 is a vertex in S of smallest degree and (when $S \cap N(x_1) \neq \emptyset$) y_1 is a vertex in $S \cap N(x_1)$ of smallest degree. If $S \cap N(x_1) = \emptyset$, then let y_1 be any vertex of $N(x_1)$. Given that j independent edges $x_1y_1, x_2y_2, \cdots, x_jy_j$ have been selected, choose $x_{j+1}y_{j+1}$ such that x_{j+1} is a vertex in $S - \{x_1, y_1, x_2, y_2, \cdots, x_j, y_j\}$ of smallest degree and y_{j+1} is a vertex of smallest degree in $(S - \{x_1, y_1, x_2, y_2, \cdots, x_j, y_j\}) \cap N(x_{j+1})$ (when this set is nonempty). Let y_{j+1} be any unused vertex in $N(x_{j+1})$ when $S - \{x_1, y_1, x_2, y_2, \cdots, x_j, y_j\} \cap N(x_{j+1}) = \emptyset$.

Suppose that this process stops after edges $x_1y_1, x_2y_2, \cdots, x_my_m$ have been found with $m < \min\{c, \lfloor \frac{\ell}{2} \rfloor\}$. Then, $S - \{x_1, y_1, x_2, y_2, \cdots, x_m, y_m\} \neq \emptyset$ and it follows that the vertex x_{m+1} in $S - \{x_1, y_1, x_2, y_2, \cdots, x_m, y_m\}$ of smallest degree has all of its adjacencies in the set $\{x_1, y_1, x_2, y_2, \cdots, x_m, y_m\}$. Let $b = d(x_{m+1})$.

Further, suppose that x_{m+1} is adjacent to t vertices in the set $\{x_1, x_2, \cdots, x_m\}$, so that $b \leq m + t$. By the algorithm, if $x_{m+1}x_r \in E(G)$ for some $1 \leq r \leq m$, then $y_r \in S$ and $d(y_r) \leq d(x_{m+1})$. Also, from the algorithm $c \leq d(x_1) \leq d(x_2) \leq \cdots \leq d(x_m) \leq d(x_{m+1})$, so that ther are at least $m + t + 1$ vertices of S of degree at most b. Therefore, since G is 2−irregular, $b \geq c + (m+t-2)/2 + 1/2$, which implies $m + t \geq b \geq c + (m+t)/2 - 1/2$. Hence, $2m \geq m + t \geq 2c - 1$, from which it follows that $m \geq c$. This contradicts that $m < c$ and completes the proof of Lemma 1. ■

PROOF OF THEOREM 1: For convenience let $G' = G - x - y$ and let z be a vertex in G' of largest degree. Since the neighborhoods of x and y are disjoint, we assume that $xz \notin E(G)$. The method of proof is such that there is no loss of generality in assuming that $d_G(x) \leq d_G(y)$.

We first prove that the theorem is true when there is a matching of order $\lceil n/4 \rceil$ in the complementary graph $\overline{G' - (N_G(x) \cup N_G(y) \cup \{z\})}$. Later we show for $d \leq \lfloor n/48 \rfloor - 2$ that this is an appropriate assumption.

Let M be a matching of order $\lceil n/4 \rceil$ in $\overline{G' - (N_G(x) \cup N_G(y) \cup \{z\})}$ and set

$$D_x = N_G(x) \cap V(G')$$

$$D_y = N_G(y) \cap V(G'), \quad \text{and}$$

$$D = V(G') - (D_x \cup D_y \cup V(M) \cup \{z\}).$$

Then $\lceil n/2 \rceil - 2 \geq |D| \geq \lfloor n/2 \rfloor - 2d - 4$. Set $d^* = |D|$ and partition D into two sets D_1 and D_2 and let $|D_1| = i$. Form the sequence of graphs $G_0, G_1, \cdots, G_{\lfloor d^*/2 \rfloor}$ such that for each i : $V(G_i) = V(G)$ and

$$E(G_i) = E(G) \cup \{xw \mid w \in D_1\} \cup \{yw \mid w \in D_2\} \cup \{yz\} \cup \{uw \mid uw \in E(M)\}.$$

We assume throughout (since it does not effectively alter the proof) that $xz \notin E(G)$, and $yz \in E(G)$. The cases where $xz \in E(G)$ or neither x nor y is adjacent to z can be handled in a similar way.

Clearly G is a subgraph of G_i and G_i is 2−irregular, unless it contains three vertices of degree $d_{G_i}(x)$ (one of which is x) or three vertices of degree $d_{G_i}(y)$ (one of which is y). Observe that $d_{G_i}(x) = d_G(x) + i \leq d_{G_i}(y) = d_G(y) + d^* - i$ for $0 \leq i \leq \lfloor d^*/2 \rfloor$.

Let $V(M) = \{u_1, v_1, u_2, v_2, \cdots, u_t, v_t\}$ and $E(M) = \{u_1v_1, u_2v_2, \cdots, u_tv_t\}$ with $t = \lceil n/4 \rceil$. Suppose for some i, $1 \leq i \leq \lfloor d^*/2 \rfloor$, that G' does not contain two vertices of degree $d_{G_i}(x) - 1$. Then either G_i is 2−irregular or there exist two vertices in G' of degree $d_{G_i}(y) - 1$. Since the proof is complete (under the original matching assumption) when G_i is 2 irregular, suppose the contrary. Thus G' contains two vertices of degree $d_{G_i}(y) - 1$.

Next, we sequentially alter G_i to show that there are more vertices of relatively "low" degree in G'. For each j, $1 \leq j \leq t$, let G_{ij} be the graph with $V(G_{ij}) = V(G_i)$ and

$$E(G_{ij}) = (E(G_i) - \{u_\ell v_\ell \mid 1 \leq \ell \leq j\}) \cup \{u_\ell y \mid 1 \leq \ell \leq j\} \cup \{v_\ell y \mid 1 \leq \ell \leq j\}.$$

But $d_{G_i}(w) = d_{G_{ij}}(w)$ for each $w \in V(G_{ij}) - \{y\}$, so if G_{ij} is not 2−irregular, then G' contains two vertices of degree $d_{G_i}(y) + 2j - 1$. Thus, if G_{ij} is not 2−irregular for each j, then G' contains two vertices of each of the following degrees:

$$d_{G_i}(y) - 1, d_{G_i}(y) + 1, \cdots, d_{G_i}(y) + 2t - 1.$$

Now the above discussion applies as well to the similar graph G_i' obtained from G_i by inserting the edge xz. Since z is a vertex of highest degree in G', G_i' is 2−irregular unless it contains three vertices of degree $d_{G_i'}(x)$ or three vertices of degree $d_{G_i'}(y)$. Also, $d_{G_i'}(x) = d_{G_i}(x) + 1$. Since it has been assumed that G' contains at most one vertex of degree $d_{G_i}(x) - 1 = d_{G_i'}(x) - 2 = d_{G_{i-1}'}(x) - 1$, this means that G_{i-1}' is 2−irregular unless G' contains two vertices of degree $d_{G'_{i-1}}(y) - 1 = d_{G_{i-1}}(y) - 1 = d_{G_i}(y)$. Assuming G'_{i-1} is not 2−irregular, this means, by defining graphs $G'_{i-1,j}$ from G'_{i-1} in the same way that G_{ij} was defined from G_i, that G' contains two vertices of each of the following degrees:

$$d_{G_i}(y), d_{G_i}(y) + 2, \cdots, d_{G_i}(y) + 2t.$$

In summary, we have that G' contains two vertices of each of the degrees

$$d_{G_i}(y) - 1, d_{G_i}(y), \cdots, d_{G_i}(y) + 2t,$$

and so G' contains at least $4t + 4$ vertices. Since $t = \lceil n/4 \rceil$, G' contains at least $n + 4 > n - 2$ vertices, a contradiction. This means that for each i, $1 \leq i \leq \lfloor d^*/2 \rfloor$, G_i contains two vertices of degree $d_{G_i}(x) = d_G(x) + i$. Hence G' contains two vertices of each of the following degrees:

$$d_G(x), d_G(x) + 1, \cdots, d_G(x) + \lfloor d^*/2 \rfloor - 1.$$

By a similar argument, altering the degree of x when forming the graph G_{ij} from G_i rather than the degree of y, it follows that G' contains two vertices of each of the degrees:

$$d_G(y) + d^* - 2, d_G(y) + d^* - 3, \cdots, d_G(y) + \lceil d^*/2 \rceil - 1.$$

Since $d_G(x) \leq d_G(y)$, this implies that G' contains

$$2(d^* - 1) = 2d^* - 2 \geq 2(\lfloor n/2 \rfloor - 2d - 4) - 2 \geq n - 4d - 11$$

vertices of degree at most

$$d_G(y) + d^* - 2 \leq d + \lceil n/2 \rceil - 2 - 2 = d + \lceil n/2 \rceil - 4.$$

It is easy to show that G' cannot have such a large number of vertices of this bounded degree. To see that this is the case, first enlarge both D_x and D_y in $V(G')$ to disjoint subsets D'_x and D'_y, each of order d. Next select two disjoint ℓ−element subsets B_ℓ and C_ℓ of smallest degree in $V(G') - (D'_x \cup D'_y)$, for $1 \leq \ell \leq s$, where $s = \lfloor n/6 \rfloor$. Finally, let $A_\ell = V(G') - (D'_x \cup D'_y \cup B_\ell \cup C_\ell)$, $1 \leq \ell \leq s$. Form the graphs H_ℓ, $1 \leq \ell \leq s$, from G by making x adjacent to $A_\ell \cup D'_x \cup D'_y \cup B_\ell$ and y to $A_\ell \cup D'_x \cup D'_y \cup C_\ell$. Then, $d_{H_\ell}(x) = d_{H_\ell}(y) \geq \lceil 2n/3 \rceil - 2 + \ell$, since $|A_\ell| \geq \lceil 2n/3 \rceil - 2d - 2$. Since H_ℓ is 2−irregular, unless G' contains a vertex of degree $d_{H_\ell}(x) - 2$, the theorem follows unless G' contains vertices of each of the degrees:

$$d_{H_1}(x) - 2 < d_{H_2}(x) - 2 < \cdots < d_{H_s}(x) - 2.$$

But $\lceil 2n/3 \rceil - 3 \leq d_{H_1}(x) - 2$, so that G' contains at least $n - 4d - 11$ vertices of degree at most $d + \lceil n/4 \rceil - 4$ and at least $s = \lfloor n/6 \rfloor$ vertices of degree at least $\lceil 2n/3 \rceil - 3$, a contradiction for $d \leq \lfloor n/48 \rfloor - 2$.

We complete the proof by showing for $d \leq \lfloor n/48 \rfloor - 2$ that there is a matching of order $\lceil n/4 \rceil$ in the graph $\overline{G' - (N_G(x) \cup N_G(y) \cup \{z\})}$. First observe that G' a 2−irregular graph implies that it contains at most $\lfloor (n-2)/2 \rfloor$ vertices of degree $> \lceil 3(n-2)/4 \rceil - 1$. Therefore, $\overline{G'}$ contains at least $\lceil (n-2)/2 \rceil$ vertices of degree $\geq \lfloor (n-2)/4 \rfloor$. By Lemma 1, $\overline{G'}$ contains a $\geq \lfloor (n-2)/4 \rfloor$ matching M.

Set $M' = \overline{G' - (N_G(x) \cup N_G(y) \cup \{z\})} \cap M$. As was done earlier, let $D_x = N_G(x) \cap V(G'), D_y = N_G(y) \cap V(G')$, and $D = V(G') - (D_x \cup D_y \cup V(M') \cup \{z\})$, and let $D_1 \cup D_2$ be a partition of D into two sets with $|D_1| = i$ and $|D| = d^*$. Form the sequence of graphs $G_0, G_1, \cdots, G_{\lfloor d^*/2 \rfloor}$ such that for each i, $V(G_i) = V(G)$ and

$$E(G_i) = E(G) \cup \{xw \mid w \in D_1\} \cup \{yw \mid w \in D_2\} \cup \{yz\} \cup \{uw \mid uw \in E(M')\}.$$

Again, it is clear that G is a subgraph of G_i, and G_i is 2−irregular unless G' contains two vertices of degree $d_{G_i}(x) - 1$ or two vertices of degree $d_{G_i}(y) - 1$. But, for each i,

$$d_{G_i}(x) - 1 = d_{G_i}(x) + i - 1 \leq d_G(y) + d^* - i - 1 \leq d_{G_i}(y) - 1 \leq d + \lceil (n-2)/2 \rceil - i.$$

Therefore, G' contains at least $2(\lfloor d^*/2 \rfloor) + 1 \geq d^* \geq \lceil (n-2)/2 \rceil - 2d - 1$ vertices of degree $\leq \lceil (n-2)/2 \rceil + d$.

We partition $V(G')$ into three sets A, B, and C as follows: Let A be any set of $\lceil (n-2)/2 \rceil - 2d - 1$ vertices of degree $\leq \lceil (n-2)/2 \rceil + d$, let B be those vertices of degree $\geq \lfloor 7(n-2)/8 \rfloor$, and let $C = V(G') - (A \cup B)$. Observe that $|B| \leq \lceil n/4 \rceil + 1$, so

$$|C| \geq n - 2 - (\lceil (n-2)/2 \rceil - 2d - 1) - (\lceil n/4 \rceil + 1) \geq \lfloor n/4 \rfloor + 2d - 3.$$

Also, each vertex of C has degree $\leq \lfloor 7(n-2)/8 \rfloor$, so that in $\overline{G'}$ the vertices of C are of degree $\geq \lceil (n-2)/8 \rceil$. In $\overline{G'}$ the vertices of A are of degree $\geq \lfloor (n-2)/2 \rfloor - d - 1$.

Since $|C| \geq \lfloor n/4 \rfloor + 2d - 3$, with each vertex of degree $\geq \lceil (n-2)/8 \rceil$ in $\overline{G'}$, $\overline{G'}$ has a $\lceil (n-2)/8 \rceil$ matching M_1 with at least $\lceil (n-2)/8 \rceil$ vertices of the matching in the set C. Further, each vertex in $A - V(M_1)$ has degree

$$\geq \lfloor (n-2)/2 \rfloor - d - 2 - \lfloor (n-2)/4 \rfloor$$

in $\overline{G' - M_1}$. Also

$$|A - V(M_1)| \geq \lceil (n-2)/2 \rceil - 2d - 1 - \lceil (n-2)/8 \rceil \geq \lceil 3(n-2)/8 \rceil - 2d - 2.$$

Therefore, by Lemma 1, $\overline{G' - M_1}$ contains a matching M_2 of order $\lfloor 3(n-2)/16 \rfloor - d - 1$. Then $M_1 \cup M_2$ is a matching of order

$$\lfloor 3(n-2)/16 \rfloor - d - 1 + \lceil (n-2)/8 \rceil \geq \lfloor 5(n-2)/16 \rfloor - d - 1$$

in $\overline{G'}$. Since at most $2d+1$ of the vertices of $M_1 \cup M_2$ are in $N_G(x) \cup N_G(y) \cup \{z\}$, if $\lfloor 5(n-2)/16 \rfloor - 3d - 2 \geq \lceil n/4 \rceil$, then $\overline{G' - (N_G(x) \cup N_G(y) \cup \{z\})}$ contains the required matching. But this follows when $d \leq \lfloor n/48 \rfloor - 2$. ■

PROOF OF LEMMA 2: Let $H_1 = \{x_1, x_2, \cdots, x_{\lfloor n/3 \rfloor}\}$ and $H_2 = \{y_1, y_2, \cdots, y_{\lceil 2n/3 \rceil}\}$. We sequentially label the vertices of H_2 with $y_1, y_2, \cdots, y_{\lceil 2n/3 \rceil}$ such that the adjacencies between H_1 and H_2 (in G) satisfy the conditions of the Lemma. It is easy to see that the labelling of H_2 can be done such that these adjacencies will be preserved in the graph L that is described below.

There are two different descriptions for L depending upon whether $2\lfloor n/3 \rfloor < \lceil 2n/3 \rceil$ or $2\lfloor n/3 \rfloor = \lceil 2n/3 \rceil$. In the first case the edge set of L is

$$E(L) = \{y_i y_j \mid 1 \leq i, j \leq \lceil 2n/3 \rceil, i \neq j\} \cup \{x_i y_j \mid 1 \leq j \leq 2i,\ 1 \leq i \leq \lfloor n/3 \rfloor\}.$$

In the second case

$$\begin{aligned} E(L) =& \{y_i y_j \mid 1 \leq i, j \leq \lceil 2n/3 \rceil, i \neq j\} \cup \{x_i y_j \mid 1 \leq j \leq 2i,\ 1 \leq i < \lfloor n/3 \rfloor - 1\} \\ & \cup \{x_{\lfloor n/3 \rfloor - 1} y_j \mid 1 \leq j \leq \lceil 2n/3 \rceil - 1\} \cup \{x_{\lfloor n/3 \rfloor} y_j \mid 1 \leq j \leq \lceil 2n/3 \rceil, j \neq \lceil 2n/3 \rceil - 1\}. \end{aligned}$$

Observe that in the first case the degree sequence of L is

$$d_L(x_i) = 2i \qquad \text{for } 1 \leq i \leq \lfloor n/3 \rfloor,$$

$$d_L(y_j) = d_L(y_{j+1}) = n - (j+1)/2 \qquad \text{for } j = 1, 3, \cdots, \lfloor 2n/3 \rfloor - 1,$$

$$d_L(y_j) = \lceil 2n/3 \rceil - 1 \qquad \text{for } 2\lfloor n/3 \rfloor \leq j \leq \lceil 2n/3 \rceil.$$

In the second case the degree sequence of L is

$$d_L(x_i) = 2i \qquad \text{for } 1 \leq i \leq \lfloor n/3 \rfloor - 2,$$

$$d_L(x_{\lfloor n/3 \rfloor - 1}) = d_L(x_{\lfloor n/3 \rfloor}) = \lceil 2n/3 \rceil - 1,$$

$$d_L(y_j) = d_L(y_{j+1}) = n - (j+1)/2 \qquad \text{for } j = 1, 3, \cdots, \lfloor 2n/3 \rfloor - 5,$$

$$d_L(y_{\lceil 2n/3 \rceil - 3}) = d_L(y_{\lceil 2n/3 \rceil - 1}) = \lceil 2n/3 \rceil + 1,$$

$$d_L(y_{\lceil 2n/3 \rceil}) = d_L(y_{\lceil 2n/3 \rceil - 2}) = \lceil 2n/3 \rceil.$$

In each case the graph L is 2–irregular and contains G as a subgraph. ■

PROOF OF THEOREM 2: The proof uses Theorem 1. Let S be a largest independent set of vertices in G. Since the independence number $\beta(G) \geq n/(d+1)$, the cardinality of the set $|S| \geq n/(d+1)$.

Sequentially remove pairs of vertices in $V(G) - S$ that have disjoint neighborhoods in the present graph obtained from G by the deletions. Continue this removing of pairs of vertices until a subgraph H of G is obtained such that each pair of vertices in $V(H) - S$ have a common neighbor in H. But then $S \subseteq V(H)$ and any vertex x in $V(H) - S$ is at a distance ≤ 2 from any other vertex of $V(H) - S$. Also, since S is maximally independent, each vertex in $V(H) - S$ is adjacent to some vertex of S. Let x be a fixed vertex of $V(H) - S$ and assume x is adjacent to t, $t \leq d$, vertices of S. Since $\Delta(G) \leq d$ and x is at distance ≤ 2 from each vertex of $V(H) - S$, $V(H) - S$ has at most

$$(d-t)(d-2) + t(d-1) + (d-t) + 1 \leq (d-1)d + 1$$

vertices. Further, these $(d-t)(d-2) + t(d-1) + (d-t) + 1$ vertices have at most

$$(d-t)(d-2)(d-1) + t(d-1)(d-1) + d \leq d(d-1)^2 + d$$

adjacencies in S.

Let T denote the vertices of S adjacent to some vertex in $V(H) - S$. By Lemma 2, if $|S - T| \geq \frac{1}{3}|V(H)|$, then H is embeddable in a 2–irregular graph H'. Further, if the pairs of vertices of G removed to form H are replaced a pair at a time, the 2–irregular graph H' can be enlarged, through repeated use of Theorem 1, to obtain a 2–irregular graph G' containing G.

Thus the proof is complete if $|S - T| \geq \frac{1}{3}|V(H)|$ and as long as each graph obtained by replacing the pairs of removed vertices satisfies Theorem 1. But

$$|S - T| \geq |S| - d^3 + 2d^2 - 2d \geq \frac{1}{3}(|S| + d^2 + 1) \geq \frac{1}{3}|V(H)|$$

as long as $2|S| \geq 3d^3 - 5d^2 + 6d + 1$. Also, $2|S| \geq 2n/(d+1)$. Hence the above inequalities hold as long as $n \geq \frac{3}{2}d^4 - d^3 + \frac{1}{2}d^2 + \frac{7}{2}d + 1$. Also, for $n \geq \frac{3}{2}d^4$ and $d \geq 8$, we have that

$$|V(H)| \geq \beta(G) \geq n/(d+1) \geq \frac{3d^4}{2(d+1)} \geq 48d + 144.$$

But $n' \geq 48d + 144$ implies $\lfloor n'/48 \rfloor - 2 \geq d$, so that H satisfies the conditions of Theorem 1, as well as those graphs obtained by replacing pairs of removed vertices. Thus, under the given conditions, Theorem 2 holds. ■

PROOF OF THEOREM 3: There is no loss of generality in assuming G has n vertices and the maximum number of edges allowed. Observe from Theorem 2 that we may assume there exists a vertex z such that $d_G(z) > (\frac{2}{3})^{1/4}n^{1/4}$. Throughout the proof we assume z is the vertex of largest degree. Since $d_G(z) > (\frac{2}{3})^{1/4}n^{1/4}$ and $q \leq \frac{3}{2}n + \frac{1}{2}(\frac{2}{3})^{1/4}n^{1/4}$, G must contain either a vertex of degree $0, 1$ or 2.

The first stage of the proof consists of repeatedly deleting a set of 3 (or 4) vertices incident to a total of at least $5 \geq (\frac{3}{2})3$ (or $6 \geq (\frac{3}{2})4$) edges such that if the deleted graph is 2–irregular, then the graph prior to deletion is embeddable in a 2–irregular graph. We first describe one step in this deletion process and then discuss how ofter the deletion process is repeated. There are several possibilities to consider in the deletion process, each of which dictates the choice of the 3 (or 4) vertex set to be deleted. We consider these possibilities as separate cases. It is assumed that the graph to which the deletion process is applied is a (n_1, q_1) graph G_1 with $n_1 < q_1$.

Case I: ***G_1 contains an isolated vertex u.***

Since G_1 contains an isolated vertex and $n_1 < q_1$, there exists a pair of vertices v and w in G_1 which together are incident to at least 5 edges of G_1. Let $G_2 = G_1 - \{u, v, w\}$, and assume that G_2' is a 2–irregular graph of order $n_1 - 3$ containing G_2. Let G_2^* be the graph with $V(G_2^*) = V(G_1)$ and

$$E(G_2^*) = E(G_2') \cup \{vx \mid x \in V(G_2^*) - \{v\}\} \cup \{wx \mid x \in V(G_2^*) - \{w, u\}\}$$
$$\cup \{wu \mid \text{when } G_2' \text{ has at most one isolated vertex}\}.$$

It is clear that $d_{G_2^*}(u) = 1$ or 2. If $d_{G_2^*}(u) = 1$, then G_2' has at least two isolated vertices, so that for $x \in V(G_2')$,

$$2 \leq d_{G_2^*}(x) \leq |V(G_2^*)| - 4,\ d_{G_2^*}(v) = |V(G_2^*)| - 1, \text{ and } d_{G_2^*}(w) = |V(G_2^*)| - 2.$$

If $d_{G_2^*}(u) = 2$, then G_2^* has at most two vertices of degree 2. Also, then for $x \in V(G_2')$,

$$d_{G_2^*}(x) \leq |V(G_2^*)| - 2, \text{ and } d_{G_2^*}(v) = d_{G_2^*}(w) = |V(G_2^*)| - 1.$$

Thus, for either possiblitity, when G_2' is 2–irregular, then so is G_2^* with G_1 a subgraph of G_2^*. Also, G_2 is a (n_2, q_2) graph with $n_2 = n_1 - 3$ and $q_2 \leq q_1 - 5$.

Case II: $\delta(G_1) = 1$

Let $u \in V(G_1)$ such that $d_{G_1}(u) = 1$, and let $vu \in E(G_1)$. If $d_{G_1}(u) \geq 2$, then since $n_1 < q_1$, there exists a vertex w in G_1 such that v and w are together incident to at least 5 edges of G_1. In this case define G_2, G_2' and G_2^* as was done in Case I. Then G_2^* contains G_1, is 2–irregular when G_2' is 2–irregular, and G_2 is an (n_2, q_2) graph with $n_2 = n_1 - 3$ and $q_2 \leq q_1 - 5$.

If $d_{G_1}(v) = 1$, then there exist vertices w and t in G_1, each of degree at least 3. This means that the set $\{u, v, w, t\}$ is incident to at least 6 edges of G_1. In this case let $G_2 = G_1 - \{u, v, w, t\}$ and assume that G_2' is a 2−irregular graph of order $n_1 - 4$ containing G_2. Define G_2^* by $V(G_2^*) = V(G_1)$ and

$$E(G_2^*) = E(G_2') \cup \{wx \mid x \in V(G_2^*) - \{w, u, v\}\} \cup \{tx \mid x \in V(G_2^*) - \{t, u, v\}\}$$
$$\cup \{wu, vt \mid \text{when } G_2' \text{ has no isolated vertex}\}.$$

We check the degrees of $V(G_2^*)$. If G_2' has no isolated vertices, then $d_{G_2^*}(u) = d_{G_2^*}(v) = 2$, $d_{G_2^*}(w) = d_{G_2^*}(t) = |V(G_2^*)| - 2$, and $d_{G_2^*}(x) \leq |V(G_2^*)| - 3$ for $x \in V(G_2')$. If G_2' has isolated vertices, then $d_{G_2^*}(u) = d_{G_2^*}(v) = 1$, $d_{G_2^*}(w) = d_{G_2^*}(t) = |V(G_2^*)| - 3$, and $2 \leq d_{G_2^*}(x) \leq |V(G_2^*)| - 4$ for $x \in V(G_2')$. Thus, in either case, if G_2' is 2−irregular, then so is G_2^* with G_1 a subgraph of G_2^*. For this subcase G_2 is an (n_2, q_2) graph with $n_2 = n_1 - 4$ and $q_2 \leq q_1 - 6$.

Case III: $\delta(G_1) = 2$

Since $n_1 < q_1$, there exist vertices $u, v, w \in V(G_1)$ such that $uv, uw \in E(G_1)$, $d_{G_1}(u) = 2$, and the set $\{u, v, w\}$ is incident to at least 5 edges of G_1. We assume that this set $\{u, v, w\}$ is chosen so that $d_{G_1}(u) \leq d_{G_1}(v) \leq d_{G_1}(w)$ with the degree of w as large as possible.

Let $G_2 = G_1 - \{u, v, w\}$ and assume that G_2' is a 2−irregular graph of order $n_1 - 3$ containing G_2. Let G_2^* be the graph with $V(G_2^*) = V(G_1)$ and

$$E(G_2^*) = E(G_2') \cup \{vx \mid x \in V(G_2^*) - \{v\}\} \cup \{wx \mid x \in V(G_2^*) - \{w\}\}$$
$$\cup \{ux \mid \text{when } G_2' \text{ has at least two isolated vertices with}$$
$$x \in V(G_2') - \{r\} \text{ and } r \text{ a fixed isolated vertex of } G_2'\}.$$

If G_2' has at most one isolated vertex, then $d_{G_2^*}(u) = 2$, $3 \leq d_{G_2^*}(x) \leq |V(G_2^*)| - 2$ for all but at most one vertex $x \in V(G_2')$, and $d_{G_2^*}(v) = d_{G_2^*}(w) = |V(G_2^*)| - 1$. If G_2^* has at least two isolated vertices, then $d_{G_2^*}(x) \leq |V(G_2^*)| - 3$ for each $x \in V(G_2)$, $d_{G_2^*}(u) = |V(G_2^*)| - 2$ and $d_{G_2^*}(v) = d_{G_2^*}(w) = |V(G_2^*)| - 1$.

It is clear that either possibility gives a graph G_2^* which contains G_1 as a subgraph, is 2−irregular when G_2' is 2−irregular, and G_2 is an (n_2, q_2) graph with $n_2 = n_1 - 3$ and $q_2 \leq q_1 - 5$.

When sequentially applying the deletion process to G it is assumed that the highest degree vertex z will be deleted as soon as possible. This can, in fact, be done the first time the deletion process is applied, unless $\delta(G) = 2$ and z is not adjacent to a vertex of degree 2. After enough deletions, even when $\delta(G_1) = 2$, vertex z will be deleted.

We next see that this deletion process can be stopped when the graph G_0, which results after many deletions, satisfies

$$C' n^{1/4} \leq |V(G_0)| \leq (\frac{2}{3})^{1/4} n^{1/4},$$

where $C' = (\frac{2}{3})^{1/4} - \epsilon$ for some ϵ, $\frac{1}{3}(\frac{2}{3})^{1/4} \leq \epsilon \leq \frac{1}{2}(\frac{2}{3})^{1/4}$. Assume that after repeated deletions, a graph G_0 results with $|V(G_0)| = C' n^{1/4}$. Then $n - C' n^{1/4}$ vertices and at least $\frac{3}{2}(n - C' n^{1/4} -$

4) + $(\frac{2}{3})^{1/4}n^{1/4} + 2$ edges have been deleted. Recall that at least $(\frac{2}{3})^{1/4}n^{1/4} + 2$ edges were deleted when z was deleted. This means that $|V(G_0)| = C'n^{1/4}$ and

$$|E(G_0)| \leq \frac{3}{2}n + \frac{1}{2}(\frac{2}{3})^{1/4}n^{1/4} - \frac{3}{2}(n - C'n^{1/4} - 4) - (\frac{2}{3})^{1/4}n^{1/4} - 2$$
$$\leq (\frac{3}{2}C' - \frac{1}{2}(\frac{2}{3})^{1/4})n^{1/4} + 4 \leq C'n^{1/4}.$$

The last inequality requires $4 \leq \frac{1}{2}\epsilon n^{1/4}$, which holds since $n \geq (8/\epsilon)^4 \geq 8^4 3^5/2$.

Thus, a (n_0, q_0) graph G_0 has been obtained by the deletion process with $q_0 \leq n_0$. We complete the proof by showing that G_0 is embeddable in a 2−irregular graph G_0^* of order n_0. Clearly, G_0^* can be enlarged, replacing the deleted vertices a 3−set (or 4−set) at a time in reverse order to the deletion. At each step in the enlarging process, a 2−irregular graph is obtained with G a subgraph of the enlarged 2−irregular graph.

To embed G_0 in a 2−irregular graph we apply Lemma 2. Sequentially delete $x_1 \cup N^*(x_1), x_2 \cup N^*(x_2), \cdots x_{\lfloor n_0/3 \rfloor} \cup N^*(x_{\lfloor n_0/3 \rfloor})$, from G_0, where x_i is the smallest degree in $G_0 - \cup_{j=1}^{i-1}(x_j \cup N^*(x_j))$ and $N^*(x_i)$ is the neighborhood of x_i in $G_0 - \cup_{j=1}^{i-1}(x_j \cup N^*(x_j))$. Let $H_1 = \{x_1, x_2, \cdots, x_{\lfloor n_0/3 \rfloor}\}$ and $H_2 = \cup_{j=1}^{\lfloor n_0/3 \rfloor} N^*(x_j)$. Clearly, H_1 is independent. Also, since $q_0 \leq n_0$, the set of vertices $\{x_j \mid 1 \leq j \leq i\}$ have at most $2i$ adjacencies in H_2. Thus, by Lemma 2, G_0 is embeddable in a 2−irregular graph G_0^* of order n_0. This completes the proof of Theorem 3. ■

4. CONCLUDING REMARKS.

It is not known whether the sufficient conditions given in Theorems 2 and 3 can be weakened with the same conclusions. The sufficient conditions given in the theorems were determined by the method of proof used. Probably the most interesting question is the one mentioned earlier, namely, is each (n, q) graph with $q \leq 3n - 7$ embeddable in a 2−irregular graph of order n?

Another question of interest is whether there exists a "universal" 2−irregular graph of order n that contains all graphs of order n and of bounded degree d, for fixed d and sufficiently large n. One candidate for such a universal graph is the graph H_n (for n even) obtained from the disjoint union of two complete graphs $K_{n/2}$ with vertices $\{x_1, x_2, \cdots, x_{n/2}\}$ and $\{y_1, y_2, \cdots, y_{n/2}\}$ respectively, by adding the edges $x_i y_j$ for $1 \leq i \leq n/2$ and $1 \leq j \leq i$. Unfortunately H_n is not universal. It can, if fact, be shown by probabilistic methods (for $d \geq 17$) that there exists a graph G_n of order n and maximal degree d such that $K_{n/4,n/4} \not\subseteq \overline{G_n}$. Since $K_{n/4,n/4} \subset \overline{H_n}$, it is not possible for $G_n \subseteq H_n$, and thus H_n is not universal. It is true that H_n is universal for $d \leq 3$.

REFERENCES.

[1] J.A. Bondy and U.S.R Murty, "Graph Theory with Applications," MacMillan, London, 1976.

[2] G. Chartrand and L. Lesniak, "Graphs and Digraphs," Wadsworth and Brooks / Cole, Belmont, CA, 1986.

Cell Complexes and Lower Bounds in Computational Geometry

Th. Fischer, Berlin, GDR

1. Introduction

The characterization of the intrinsic complexity of computation problems by upper and lower bounds is one of the major tasks in computational geometry. In the model of algebraic computation trees almost all known nontrivial lower bounds are of order $\Omega(n \log n)$, where n is a parameter associated with the input size of such problems. The main purpose of this paper is to present a combinatorial technique which allows to derive lower bounds exceeding this barriere.

In the following section we describe the main idea of this method and specify the model of computation. Then we illustrate our technique by deriving a lower bound for a d-dimensional version of the all nearest neighbor problem. Finally, in the last section we derive an improved bound for a fundamental planar problem that can be obtained in a similar way.

2. Preliminaries

In this paper we are interested in determining a bound on the intrinsic complexity of the following computation problem:

All nearest neighbors of S_1 in S_2: Given sets S_1 and S_2 of n points each, for any point p in S_1 find all its nearest neighbors in S_2.

Note that a point z is a nearest neighbor of x, if $d(x,z) = \min \{ d(x,y) \mid y \in S \}$, where $d(x,y)$ denotes the

R. Bodendiek, R. Henn (Eds.)
Topics in Combinatorics and Graph Theory

Euclidean distance between points x and y. This problem is closely related to various other proximity problems that have been investigated in the literature, cf. with [V] and [PS].

Now we shortly recall the fundamental notion of algebraic computation trees (ACT) to which we refer in subsequent sections. Let $W \subseteq R^n$ be any set. Then the membership problem for W requires to decide whether a point $(x_1,...,x_n) \in R^n$ belongs to W. An algebraic computation tree for W is a tree that consists of operational nodes, branching nodes, and leaves. Given any input sequence the program traverses a path in the tree starting in its root. At each operational node an associated arithmetic operation is executed and at each branching node a branching is made according to the outcome of a comparison $x : 0$, where : denotes an arithmetic comparison and x is an input variable or a variable computed in an ancestor of that node. The time complexity C(T) of an algebraic computation tree T is defined as its depth, i.e. the maximum number of nodes traversed until a leaf returning one of the possible answers (yes or no) is reached. Fore a more detailed definition we refer to [PS] and [M].

According to a very general result of Ben-Or the height of any algebraic computation tree that correctly solves this problem must satisfy the bound

$$C(T) \geq \frac{\log \#W - n \log 3}{1 + \log 3} \approx 0.38 \log \#W - O(n) ,$$

where #W is used to denote the number of disjoint connected components of W [B] [M].

In almost all cases that have been studied so far this number is n!, which only implies a bound of order $\Omega(n \log n)$. In the following sections lower bounds are obtained by a method that can be announced as "grouping and mapping". This means that the input sequence is subdivided into two parts: While the first group of input objects is used to construct a cell complex that is considered as being fixed, the second group is used to create a large number of mappings into the faces of the cell complex. Then a crucial point is to show that distinct mappings correspond to disjoint connected components. In the following section we illustrate this idea by a bound derived from the properties of Voronoi diagrams.

3. Voronoi diagrams and nearest neighbor problems

Given any set S of n points in R^d, the Voronoi diagram $Vor(S)$ is a subdivision of R^d into regions such that the points within a fixed region have the same nearest neighbor in S. More precisely, for any $p \in S$, there is a polyhedral region defined by

$$Cell(p) = \{ x \in R^d \mid d(x,p) \leq d(x,p') \text{ for } p' \in S \}.$$

The k-dimensional polyhedra bounding the cells are the k-faces of $Vor(S)$, and the set of all k-faces is denoted by $Vor^k(S)$. Moreover, for any point $q \in R^d \setminus S$, let $C(q)$ denote the subset of points in S that are closest to q, i.e.

$$C(q) = \{ p \mid p \in S \ \& \ d(p,q) = d_{min} \},$$

where $d_{min} = \min\{ d(p,q) \mid p \in S \}$. If $C(q) = C(q')$ for two points q and q', then these points are called equivalent. Apparently, the Voronoi diagram $Vor(S)$ is just the cell complex whose faces are the equivalence classes of points. Hence, for each k-face f of $Vor(S)$, we may define $C_f = C(q)$, for q any point in f. In general, the cardinality of C_f satisfies the inequality $\#C_f \geq d + 1 - k$ (with equality for $k = d$ and $k = d - 1$, cf. with [E]). A Voronoi diagram is called simple, if $\#C_v = d + 1$ for each vertex (0-face) v of $Vor^o(S)$.

The combinatorial structure of d-dimensional Voronoi diagrams has been investigated by V. Klee [K]. A cornerstone for the lower bound arguments applied below is his result on $m(n,d)$, the maximum number of vertices of a Voronoi diagram on n points in d-dimensional space.

$$[d/2]!\, n^{[d/2]} \leq m(n,d) \leq 2([d/2]!\, n^{[d/2]}), \quad d \text{ even},$$

$$(*)$$

$$\frac{([d/2] - 1)!}{e}\, n^{[d/2]} \leq m(n,d) < [d/2]!\, n^{[d/2]}, \quad d \text{ odd},$$

where $[r]$ is used to denote the smallest integer greater than or equal to r, $r \in R$. Note that throughout this paper we assume $n > d$ and n to be sufficiently large.

Klee's result immediately implies that the construction of d-dimensional Voronoi diagrams can be a very hard problem. Besides this trivial consequence it can be used also for deriving lower bounds on the time complexity of some quite different (but certainly easier) computation problems in R^d [F1]. As an example we mention the following result which follows from (*) by a simple information theoretic argument:

Theorem 1 [F1]. Given n points in R^d, in the worst case the search for all nearest neighbors of a query point requires $\lceil d/2\rceil \log n + O(d \log d)$ time.

A similar argument applies in the case of the computation of all nearest neighbors of the points of a set S_1 in another set S_2. The main idea is to associate a lower bound with the number of distinct mappings of the set S_1 into the set of vertices of the Voronoi diagram induced by S_2. In the following we present a somewhat stronger result that is obtained for a counting version of this problem.

Theorem 2. In the worst case counting all nearest neighbors of points of S_1 in another set S_2 requires $\Omega(d\ n \log n)$ time in the ACT model of computation.

Proof. The proof is obtained by deriving a bound on a closely related decision problem. Assume the input be given as a sequence of 2n points $(p_1,\ldots,p_n,q_1,\ldots,q_n)$ such that $S_1 = \{p_1,\ldots,p_n\}$ and $S_2 = \{q_1,\ldots,q_n\}$. For any $p \in S_1$ let $N(p)$ denote the set of all nearest neighbors of p in S_2, i.e. $N(p) \subseteq S_2$. Then consider the problem of deciding whether a given sequence belongs to a set $W \subset R^{2dn}$ that is defined by the property

$$\#N(p_1) = \ldots = \#N(p_n) = d + 1. \tag{1}$$

In order to obtain a good lower bound on $\#W$ we assume a configuration of points $q_1,\ldots,q_n$ such that

$Vor(S_2)$ is simple and satisfies the lower bounds given in (*). (2)

As pointed out in [K] we may assume that $Vor(S_2)$ is simple because the points forming S_2 can be slightly perturbed such that their Voronoi diagram becomes simple without decreasing its number of vertices.

The idea of this assumption is to consider the set S_2 as being fixed and then to use the freedom in the selection of S_1 for creating a possibly large number of distinct nearest neighbor configurations. Hence consider all possible configurations such that S_1 is a subset of $Vor^O(S_2)$. Since $Vor(S_2)$ is simple, the condition (1) is satisfied in this case, i.e. all these configurations belong to W. Hence any injective mapping of S_1 into $Vor^O(S_2)$ induces a configuration that belongs to W. Now it remains to show that distinct mappings create distinct connected components of W.

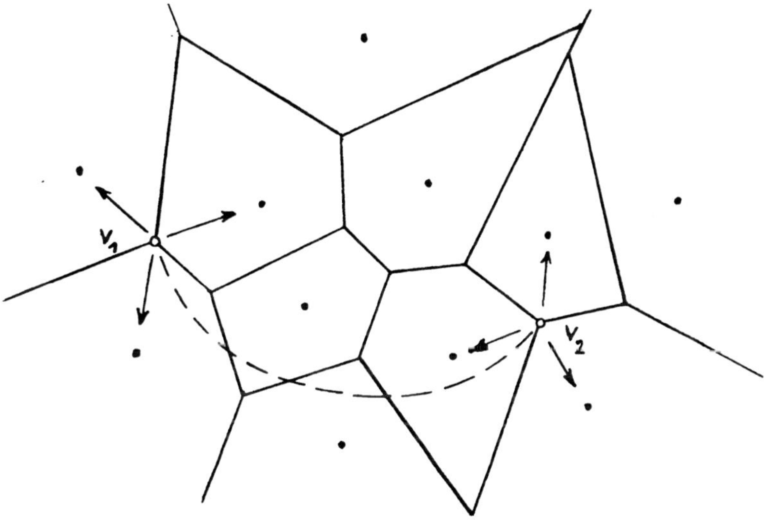

Fig. 1

For any two distinct mappings there must be at least one point in S_1 that is mapped onto distinct vertices in $Vor^O(S_2)$. Let p_i denote this point and consider any movement of p_i starting from a vertex v_1 to another vertex v_2. This corresponds to a curve in R^{2dn} that connects the corresponding input sequences $s(v_1)$ and $s(v_2)$. For $p_i = v_1$ and $p_i = v_2$ we have $N(p_i) = C_{v_1}$ and $N(p_i) = C_{v_2}$, respectively, and the assumed simplicity of $Vor(S_2)$ implies $\#N(p_i) = = \#C_{v_1} = \#C_{v_2} = d + 1$. However, since $v_1 \neq v_2$, any curve connecting v_1 and v_2 (the projection of the curve connecting $s(v_1)$ and $s(v_2)$) must intersect with some k-face of $Vor(S_2)$ (for $k > 0$), and it is not hard to see that the number of nearest neighbors will then become strictly less than $d + 1$, cf. with Fig. 1.

Note that the case that there is a point on the curve connecting $s(v_1)$ and $s(v_2)$ where the two vertices (and thus p_i) become incident can be excluded because this would imply that $\#N(p_i)$ would become greater than $d + 1$ at this point. Thus the components corresponding to any two distinct mappings are disjoint. Therefore it suffices to count the number of injective mappings of S_1 into $Vor^o(S_2)$ that can occur in the worst case. Apparently, this number is given by

$$\#W \geq \frac{m(n,d)!}{(m(n,d)-n)!}$$
$$\geq (m(n,d)-n)^n.$$

By taking the logarithm and assuming n to be sufficiently large we obtain

$$\log \#W \geq n \log m(n,d) - O(n \log n).$$

Then a straightforward evaluation using (*) yields the expression

$$\log \#W \geq [d/2]n \log n + n \log[d/2] - O(n \log n),$$

which for $n > d$ can be subsumed to the lower bound

$$\log \#W \geq [d/2]n \log n - O(n \log n),$$

which yields an $\Omega(d\, n \log n)$ bound on $C(T)$. Finally, we observe that it is immediately apparent that the counting problem mentioned in Theorem 2 is at least as difficult as the decision problem concerning W. Moreover, a similar argument shows that the same bound is also valid for the reporting version of the all nearest neighbor problem.

4. Arrangements of lines and cell complexes defined by boxes

In his book H. Edelsbrunner asks for the existence of an algorithm which decides in time $o(n^2)$ whether or not a set of n points in the Euclidean plane contains three collinear points [E]. By using the method introduced above we can derive a lower bound for this problem which indeed exceeds any other nontrivial lower bound known in planar computational geometry.

Let m be any integer, $m \geq 1$, and let 4m points be distributed on a circle as shown in Fig. 2. Obviously, no three of these points are collinear. Consider the arrangement of lines induced by pairs of points taken from the sets $\{P_1,\ldots,P_m\}\times\{P_{2m+1},\ldots,P_{3m}\}$ and $\{P_{m+1},\ldots,P_{2m}\}\times\{P_{3m+1},\ldots,P_{4m}\}$, resp. Since both sets are of size m^2, it is easy to see that at least m^4 cells are created in this way.

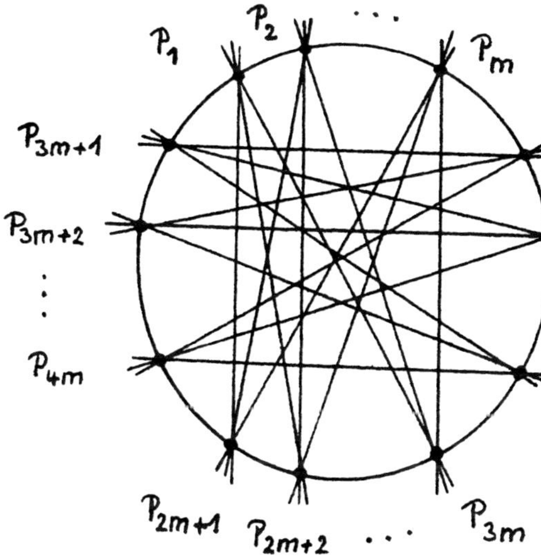

Fig. 2

Now consider the family of configurations that are obtained if the remaining $n - 4m$ points are mapped into these cells such that again no three points become collinear. As in the proof of Theorem 2 it can be seen that distinct mappings correspond to disjoint connected components of the set $R^{2n}\setminus W$ associated with a negative answer of the decision problem. Thus we have $\#(R^{2n}\setminus W) \geq m^{4(n-4m)}$ in this case. Hence, e.g. for $m = n/16$, there are $(n/16)^{3n}$ connected components and Ben-Or's bound yields the following result:

Theorem 3. Deciding the collinearity of any three points within a set of n points requires more than $1.14\, n \log n - O(n)$ steps in the ACT model.

Here the crucial point is that we have found a lower bound of order $\Omega(n \log n)$ with a constant factor greater than one. This shows that collinearity testing is a harder problem than e. g. sorting or element uniqueness which both can be solved with only $n \log n + O(n)$ comparisons.

Finally, we mention that nontrivial $\Omega(d\ n \log n)$ time bounds have been also obtained for a number of familiar computation problems in the geometry of d-dimensional boxes. The perhaps most interesting result concerns the problem of computing the measure of a union of boxes (Klee's rectangle problem extended for a set of boxes), see [F2].

The importance of these results mainly consists in the separation from strictly easier computation problems in R^d such as lexicographic sorting or element uniqueness which both can be solved in $O(n \log n + dn)$ time. Moreover, all these results show that the algorithmic complexity of geometric computation problems essentially depends on the structural complexity of different cell complexes which are associated with these problems in a natural way. Thus the commom opinion that lower bounds are essentially determined by the cardinality of the set of permutations (cf. [PS], p. 186) has to be modified. Theorem 3 shows that even in the planar case a finer classification of geometric problems is possible.

References

[B] M. Ben-Or: Lower bounds for algebraic computation trees, Proc. STOC'83, 80-86

[E] H. Edelsbrunner: Algorithms in Combinatorial Geometry. Berlin, Heidelberg, New York etc. 1987

[F1] Th. Fischer: Lower bounds proofs via Voronoi diagrams, J. Inform. Process. Cybern. 25 (1989), 233-238

[F2] Th. Fischer: Lower bounds in the geometry of boxes (to appear)

[K] V. Klee: On the complexity of d-dimensional Voronoi diagrams, Archiv der Math. 34 (1980), 75-80

[M] K. Mehlhorn: Datenstrukturen und Algorithmen, Teil 1: Sortieren und Suchen. Stuttgart 1986

[PS] F.P. Preparata, M.I. Shamos: Computational Geometry - an Introduction. New York etc., 1985

[V] P.M. Vaidya: An optimal algorithm for the all nearest neighbors problem, Proc. FOCS'86, 117-122

Characterizing Directed Postman Tours

H. Fleischner, Wien, Austria
E. Wenger, Wien, Austria

Abstract

In 1960 Guan Meigu proved a characterization of postman tours in (undirected) graphs, which does not hold for digraphs. We present a characterization of directed postman tours analogous to the undirected case.

Preliminaries.

For notation and terminology, see [2,3]. Let D be a digraph with vertex set $V(D)$ and arc set $A(D)$ and let G_D denote the (undirected) graph induced by D by ignoring the orientation of the arcs. In particular, $V(D)$ and $A(D)$ are always finite.

For a cost function $c : A(D) \to R$ and $B \subseteq A(D)$ define $c(B) := \sum_{a \in B} c(a)$ and for a subdigraph D_1 of D, $c(D_1) := c(A(D_1))$. For a walk W in D and $a \in A(D)$ let $\lambda_W(a)$ denote the number of occurrences of a in W, A_v the set of arcs incident with $v \in V(D)$, A_v^+ the set of arcs incident from v and A_v^- the set of arcs incident to v. Define $c(W) = \sum_{a \in A(D)} \lambda_W(a)c(a)$. Moreover, for $B \subseteq A(D)$ let $\langle B \rangle$ denote the induced subdigraph with $V(\langle B \rangle) = \{v \in V(D) | \ v \textit{ is incident with some arc of } B\}$ and $A(\langle B \rangle) = B$. For a subdigraph D_1 of D define $\mathit{diff}_{D_1}(v) := od_{D_1}(v) - id_{D_1}(v)$, $v \in V(D_1)$.

Definition 1. Let D be a strongly connected digraph and let $c : A(D) \to R^+ \cup \{0\}$ be a cost function. A closed covering walk W in D such that $c(W)$ is minimum, and every cycle K satisfying $c(K) = 0$ contains at least one arc appearing precisely once in W, is called a *directed postman tour* (**DPT** for short). The problem of finding a **DPT** is called the *Directed Postman Problem* (**DPP** for short).

R. Bodendiek, R. Henn (Eds.)
Topics in Combinatorics and Graph Theory

Originally, the problem of finding a closed covering walk of minimal length was formulated only for graphs, at the end of the 1950-ies by Guan Meigu. He studied the question how a mailman can do his job by passing all the streets in a certain district and return to the starting point such that the total length of the walk is as short as possible [4]. Whence the name Postman Tour. A characterization of postman tours in graphs has been proven in [4].

Theorem 1. Let G be a finite connected graph with vertex set $V(G)$ and edge set $E(G)$, and let c be a cost function $c: E(G) \to R^+ \cup \{0\}$. A closed covering walk W is a postman tour if and only if W satisfies the following conditions:

1)$1 \leq \lambda_W(e) \leq 2$ for each $e \in E(G)$

2) for every cycle $C \subseteq G$ the condition $c(H \cap C) \leq \frac{1}{2}c(C)$ is true, where $H = \langle\{e \in E(G) \mid \lambda_W(e) = 2\}\rangle$.

Guan Meigu's paper and especially Theorem 1, although having greatly influenced the development of a whole branch in operations research (i.e. postman tours of various kinds), apparently did not find all that much attention. For, as Guan Meigu noticed himself, Theorem 1 is impractical from an algorithmic point of view. However, in proving that the decision problem for a special type of eulerian trails (called A-trails) in planar 3-connected eulerian graphs is NP-complete, Theorem 1 is used implicitly plus the fact that the (undirected) postman problem can be solved in polynomial time (see the proof of [1], Theorem 2). These facts are part of our motivation for presenting this paper; for, Theorem 2 of the present paper, although impractical from an algorithmic point of view, might be of relevance in theoretical applications.

It is easy to see that Theorem 1 does not hold for digraphs. If $|od(v) - id(v)| > 2min\{od(v), id(v)\}$ for some $v \in V(D)$ of a digraph D, then a directed postman tour W of D cannot satisfy condition **1)** of Theorem 1. To establish a characterization of a DPT analogous to Theorem 1 we proof an auxiliary result about digraphs.

Definition 2. Let D_r and D_b be two arc-disjoint subdigraphs of a digraph D with $A(D) = A(D_r) \cup A(D_b)$; w.l.o.g., D_r and D_b are spanning subdigraphs. We call D_r and D_b a ***diff-pair of*** D if for all $v \in V(D)$, $\mathit{diff}_{D_r}(v) = \mathit{diff}_{D_b}(v)$ (we call this equation (**DIFF**) for short).

Lemma 1. Let D be a digraph and D_r and D_b be a diff-pair as defined in Definition 2. Then either $A(D) = \emptyset$ or $A(D)$ can be decomposed into non-empty classes $K_i \subseteq A(D), 1 \leq i \leq m$, such that for $G_{K_i} := G_{<K_i>}$

1) G_{K_i} is a cycle for $i = 1, \ldots, m$.

2) $D_{r,i} := \langle A(D_r) \cap K_i \rangle \cup V(D)$ and $D_{b,i} := \langle A(D_b) \cap K_i \rangle \cup V(D)$ are a diff-pair of $\langle K_i \rangle \cup V(D)$ for $i = 1, \ldots, m$.

Proof. If $A(D) = \emptyset$ the lemma is true. Whence let us assume that $A(D) \neq \emptyset$. First of all, we conclude that $d(v) \equiv 0 (mod\ 2)$ because D_r and D_b are a diff-pair. Therefore, $A(D)$ has a decomposition $S = \{K_1, \ldots, K_m; m \geq 1\}$ whose elements satisfy **1)**. To prove that there is even a decomposition S whose elements K_i satisfy both **1)** and **2)**, consider two cases according to the value of $\Delta(D)$ (maximum degree of D).

If $\Delta(D) = 2$, then for $K_i := A(C_i)$ where C_i are the weakly connected components of D, it follows that G_{K_i} is a cycle for $i = 1, \ldots, m$. $S = \{K_1, \ldots, K_m\}$ is a partition of $A(D)$ and satisfies **2)**. For, if we introduce an orientation of the (undirected) cycle G_{K_i} then **(DIFF)** implies that *either all arcs of $D_{r,i}$ are directed according to the orientation of G_{K_i} and all arcs of $D_{b,i}$ are directed inverse to this orientation, or vice versa* **(D)**. Note that in a digraph D with $\Delta(D) = 2$ **(DIFF)** and **(D)** are equivalent.

If $\Delta(D) > 2$ we construct a digraph D' with $\Delta(D') = 2$ (we assume that D_r and D_b are transformed together with D, D' a.s.o.). For each vertex v with $d(v) = 2k_v > 2$ introduce k_v 2-valent vertices as follows: first split away all pairs of arcs a^-, a^+ so that $a^- \in A_v^-$, $a^+ \in A_v^+$ and both arcs are either elements of $A(D_r)$ or elements of $A(D_b)$; then split away pairs a_r, a_b with $a_r \in A(D_r)$ and $a_b \in A(D_b)$. Because D_r and D_b are a diff-pair either $\{a_r, a_b\} \subset A_v^+$ or $\{a_r, a_b\} \subset A_v^-$. By this construction we end up with a digraph D' such that $\Delta(D') = 2$. D_r and D_b are also a diff-pair in D' and therefore, by the first part of the proof, $A(D')$ has a decomposition S' whose elements satisfy **1)** and **2)**. If S' corresponds to a decomposition S of $A(D)$ satisfying **1)**, then the lemma holds; if not, there must be some $K' \in S'$ and some $v \in V(D)$ with $|K' \cap A_v| > 2$. K' can be decomposed into two subsets K_1' and K_2' with $K' = K_1' \cup K_2'$, $K_1' \cap K_2' = \emptyset$ and $V(\langle K_1' \rangle) \cap V(\langle K_2' \rangle) = \{v', v''\}$ where v' and v'' are two of the k_v 2-valent vertices in $V(D')$ corresponding to $v \in V(D)$. We obtain D'' from D' by identifying v' and v''.

Define $S'' := (S' - \{K'\}) \cup \{K_1', K_2'\}$, then D'' and S'' also satisfy **(DIFF)** and **(D)**, and S'' is a decomposition in D'' satisfying **1)** and **2)**. Now it is possible to argue as above with D'' and S'' in place of D' and S', a. s. o.. After $j \leq \sum_{v \in V(D)} k_v$ steps we arrive at a digraph $D^{(j)}$ (possibly $D^{(j)} = D$) with a decomposition $S^{(j)}$ of $A(D^{(j)})$ satisfying **1)** and **2)** and corresponding to a decomposition S of $A(D)$. This finishes the proof of the lemma.

With the help of Lemma 1 we are now able to characterize a DPT in a way similar to Theorem 1.

Theorem 2. Let D be a strongly connected digraph with cost function $c : A(D) \to R^+ \cup \{0\}$ and let W be a closed covering walk in D. W is a DPT in D if and only if the following two properties are fulfilled:

1) $D_0 := \langle \{a \in A(D) | \lambda_W(a) > 1\} \rangle$ is acyclic.

2) For every $K \subseteq A(D)$ such that G_K is a cycle it follows that $c(D_{r,K}) \leq \frac{1}{2}c(K)$ whenever the spanning subdigraphs $D_{r,K} \subseteq \langle K \cap A(D_0) \rangle \cup V(D)$ and $D_{b,K} := \langle K - A(D_{r,K}) \rangle \cup V(D)$ are a diff-pair of $\langle K \rangle \cup V(D)$.

Proof. First suppose that W is a DPT. We construct a digraph $D^+ \supseteq D$ with $V(D^+) = V(D)$ by duplicating the arcs of $A(D)$ $\lambda_W(a) - 1$ times and extending c onto $A(D^+)$. D^+ is eulerian and W corresponds to an eulerian trail W^+ in D^+. Moreover, $c(W) = c(D^+)$.

If D_0 contains a cycle C, then $D^+ - A(D)$ contains a cycle C^+ corresponding to C and $D_1^+ := D^+ - A(C)$ is also an eulerian digraph with $D_1^+ \supseteq D$. Each eulerian trail W_1^+ of D_1^+ corresponds to a closed covering walk W_1 of D. It is easy to see that $c(W_1) = c(W_1^+) = c(D_1^+) = c(D^+) - c(C) \leq c(D^+) = c(W)$. Since W is a DPT and W_1 is a closed covering walk in D, we necessarily have $c(W) = c(W_1)$ and therefore $c(C) = 0$, a contradiction to the assumption that W is a DPT. This proves the validity of **1)**.

For proving that also **2)** is valid let $K \subseteq A(D)$ be such that G_K is a cycle and let $D_{r,K}$ and $D_{b,K}$ be a diff-pair as defined in **2)**. We construct D^+ as above and form D_2^+ by changing the arc set of D^+: for every $a \in A(D_{r,K})$ delete one of the (multiple) arcs in D^+ corresponding to a and add one extra parallel arc to every $a' \in A(D_{b,K})$.

As $D_{r,K}$ and $D_{b,K}$ are a diff-pair and therefore fulfill condition **(D)** in the proof of Lemma 1, D_2^+ is also eulerian. For, in the transition from D^+ to D_2^+ either both in- and out-degree of any vertex of K are increased by 1 or decreased by 1 or remain unchanged. Each eulerian trail W_2^+ of D_2^+ corresponds to a closed covering walk W_2 in D. We have $c(W_2) \geq c(W)$ because W is a DPT. From the construction of W_2 it follows that $c(W_2) = c(W) - c(D_{r,K}) + c(K - A(D_{r,K})) = c(W) - 2c(D_{r,K}) + c(K) \geq c(W)$. Therefore, $2c(D_{r,K}) \leq c(K)$ which proves the validity of **2)**.

Now assume that a closed covering walk W satisfies **1)** and **2)**. Let W_o be a DPT and D^+ and D_o^+ be the eulerian superdigraphs of D corresponding to W, W_o respectively (for the construction of D^+, see above). We define $D_r := D^+ - A(D), D_b := D_o^+ - A(D)$ and $\bar{D} := D_r \cup D_b$, regarding D_r and D_b as arc-disjoint digraphs. As W and W_o are closed covering walks and because of the construction of D^+ and D_o^+, the equation $\mathit{diff}_{D^+-A(D)}(v) = \mathit{diff}_{D_o^+-A(D)}(v) = -\mathit{diff}_D(v)$ is true for each $v \in V(D) = V(D^+ - A(D)) = V(D_o^+ - A(D))$. So D_r and D_b are a diff-pair in $\bar{D}$.

If $A(\bar{D}) = \emptyset$, then D is eulerian, $c(W) = c(W_o)$ and W is a DPT as well. If $A(\bar{D}) \neq \emptyset$ Lemma 1 can be applied to $\bar{D}, D_r$ and D_b. Let $S = \{K_1, \ldots, K_m, m \geq 1\}$ be a decomposition of $A(\bar{D})$ satisfying **1)** and **2)** of Lemma 1 and let $D_{r,i}$ and $D_{b,i}$ be as defined in that lemma. Moreover we define $D_0 := \langle\{a \in A(D) | \lambda_W(a) > 1\}\rangle$ and $D_0^o := \langle\{a \in A(D) | \lambda_{W_o}(a) > 1\}\rangle$. For an arbitrary but fixed i, let K be the arc set in D corresponding to K_i and let $D_{r,K}$ and $D_{b,K}$ correspond to $D_{r,i}$ and $D_{b,i}$, respectively.

If $|K| = 1$, then K_i consists of two parallel arcs, one belonging to $A(D_r)$ and one belonging to $A(D_b)$. Therefore, $c(D_{r,K}) = c(D_{b,K}) = \frac{1}{2}c(K_i)$. If $|K| > 1$, then G_K is a cycle, and by definition $A(D_{r,K}) \subseteq K \cap A(D_0)$ and $A(D_{b,K}) \subseteq K \cap A(D_0^o)$. Since $D_{r,i}$ and $D_{b,i}$ are a diff-pair in $\bar{D}$ it follows that $D_{r,K}$ and $D_{b,K}$ are also a diff-pair in D. Having already proved that a DPT satisfies **1)** and **2)** we can conclude that both covering walks W and W_o fulfill **1)** and **2)**. In particular, $c(D_{r,K}) \leq \frac{1}{2}c(K)$ and $c(D_{b,K}) \leq \frac{1}{2}c(K)$ imply $c(K) \geq c(D_{r,K}) + c(D_{b,K}) = c(K_i) = c(K)$. We have thus proved that $c(D_{r,i}) = c(D_{r,K}) = c(D_{b,K}) = c(D_{b,i})$, $1 \leq i \leq m$, and therefore $c(D_r) = \sum_{i=1}^m c(D_{r,i}) = \sum_{i=1}^m c(D_{b,i}) = c(D_b)$. It follows that $c(W) = c(D) + c(D_r) = c(D) + c(D_b) = c(W_o)$; hence W is a DPT. This finishes the proof of the theorem.

References

[1] Andersen, L.D.; Fleischner, H.; The NP-Completeness of Finding A-Trails in Eulerian Graphs and of Finding Spanning Trees in Hypergraphs, to appear.

[2] Beineke, L.W.; Wilson, R.J.; Selected Topics in Graph Theory 2, Academic Press, London, 1983.

[3] Bondy, J.A.; Murty, U.S.R.; Graph Theory with Applications, The Macmillan Press Ltd, London, 1976.

[4] Guan, M.; Graphic Programming Using Odd or Even Points, Chinese Math., Vol.1, No.3 (1962) 273-277.

Some Properties of "Almost All" Functions From p_k

I. D. Giudjenov, Blagoevgrad, Bulgaria

This paper treats some properties of "almost all" k-valued functions connected to the different kinds of dependence of functions on their arguments.

0. INTRODUCTION

During the last 30 years some questions treating separable sets and essential variables of functions are intensivly studied. O.B. Lupanov [5] introduced this theory, than Breithbart [1] treated some therminological issues of from the algebra of logic, and K.N.Chimev [9] treated k-valued and arbitrary functions. He especially concetrates his efforts on the theory of separable sets.

As well - known fact is that the theory of k-valued functions, and in particular the theory of Boolean functions, pays a special attention to the properties of "almost all" functions.

1. DEFFINTIONS AND NOTATION

Used is the terminology from [5],[11-12]

Definition 1. Let p_k (Q) be a set of all k-valued functions of n variables, which have the quality Q. We say, that "almost all" functions of p_k have the quality Q, if

$$\lim_{n\to\infty}\frac{|p_k(Q)|}{k^{k^n}} = 1 \quad \text{or} \quad \lim_{n\to\infty}\frac{|p_k\backslash p_k(Q)|}{k^{k^n}} = 0.$$

Definition 2. We say that the function $f(x_1,x_2,\dots,x_n)$ depends on the variable x_1, $1 \le i \le n$ by first way iff there exist two sequence $\alpha' = (\alpha_1,\alpha_2,\dots,\alpha_{i-1},\alpha_i', \alpha_{i+1},\dots,\alpha_n)$ and $\alpha'' = (\alpha_1,\alpha_2,\dots,\alpha_{i-1}, \alpha_i'',\alpha_{i+1},\dots,\alpha_n)$ where $\alpha_i' \ne \alpha_i''$ such that $f(\alpha') \ne f(\alpha'')$.

This dependence is known as an essential dependence i.e. if the function $f(x_1,x_2,\dots,x_n)$ depends on the variable x_i, $1 \le i \le n$

R. Bodendiek, R. Henn (Eds.)
Topics in Combinatorics and Graph Theory

by first way, so we say that the function f depends essentially on x_i, or x_i is an essential variable for f.

Definition 3. We say that the function $f(x_1,x_2,\ldots,x_n)$ depends on the variable x_i, $1 \le i \le n$, $n \ge 2$ by second way iff for every sequence $(\alpha_1,\alpha_2,\ldots,\alpha_{i-1},\alpha_{i+1},\ldots,\alpha_n)$ the function $f(\alpha_1,\alpha_2,\ldots,\alpha_{i-1},x_i,\alpha_{i+1},\ldots\alpha_n)$ depends on the variable x_i by first way.

Definition 4. We say that the function $f(x_1,x_2,\ldots,x_n)$ depends on the variable x_i, $1 \le i \le n$, $n \ge 2$ by third way iff there exist two constants α_i and α''_i $(\alpha'_i \ne \alpha''_i)$ such that for every sequence $(\alpha_1,\alpha_2,\ldots,\alpha_{i-1},\alpha_{i+1},\ldots,\alpha_n)$ it is hold

$$f(\alpha_1,\alpha_2,\ldots,\alpha_{i-1},\alpha'_i,\alpha_{i+1},\ldots,\alpha_n) \ne$$
$$f(\alpha_1,\alpha_2,\ldots,\alpha_{i-1},\alpha''_1, \alpha_{i+1},\ldots,\alpha_n).$$

Definition 5. We say that the function $f(x_1,x_2,\ldots x_n)$ depends on the variable x_i, $1 \le i \le n$, $n \ge 2$ by fourth way iff for every two costants α'_i and α''_i $(\alpha'_i \ne \alpha''_i)$ there exists a sequence $(\alpha_1,\alpha_2,\ldots,\alpha_{i-1},\alpha_{i+1},\ldots,\alpha_n)$ such that $f(\alpha_1,\ldots,\alpha_{i-1},\alpha'_i,\alpha_{i+1},\ldots,\alpha_n) \ne f(\alpha_1,\ldots,\alpha_{i-1},\alpha''_1, \alpha_{i+1},\ldots,\alpha_n)$.

Definition 6. We say that the function $f(x_1,x_2,\ldots x_n)$ depends on the variable x_i, $1 \le i \le n$, $n \ge 2$ by fifth way iff there exists a sequence $(\alpha_1,\alpha_2,\ldots,\alpha_{i-1},\alpha_{i+1},\ldots,\alpha_n)$ such that for every two costants α'_i and α''_i $(\alpha'_i \ne \alpha''_i)$ it is hold

$$f(\alpha_1,\alpha_2,\ldots,\alpha_{i-1},\alpha'_i,\alpha_{i+1},\ldots,\alpha_n) \ne$$
$$f(\alpha_1,\alpha_2,\ldots,\alpha_{i-1},\alpha''_1,\alpha_{i+1},\ldots,\alpha_n).$$

Definition 7. We say that the function $f(x_1,x_2,\ldots x_n)$ depends on the variable x_i, $1 \le i \le n$, $n \ge 2$ by sixth way iff for every two costants α'_i and α''_i $(\alpha'_i \ne \alpha''_i)$ and every sequence $(\alpha_1,\alpha_2,\ldots,\alpha_{i-1},\alpha_{i+1},\ldots,\alpha_n)$ it is hold

$$f(\alpha_1,\alpha_2,\ldots,\alpha_{i-1},\alpha'_i,\alpha_{i+1},\ldots,\alpha_n) \ne$$
$$f(\alpha_1,\alpha_2,\ldots,\alpha_{i-1},\alpha''_1,\alpha_{i+1},\ldots,\alpha_n).$$

In [7] this way of dependence is treated as dependence on H. We make use of [8] to denote with M_r^k (x^n) the set of all k-valued functions of n variables, which depend on every of these variables by $\bar{r}^{th}$ way, where r = 1,2,...,6. We find in [8] too, that

$$M_6^k (X^n) \subset M_3^k (X^n) \subset M_2^k (X^n) \subset M_1^k (X^n),$$
$$M_6^k (X^n) \subset M_5^k (X^n) \subset M_4^k (X^n) \subset M_1^k (X^n).$$

In [2] for the functions of algebra of logic it is proved that

$|M_1^2(X^n)| = \sum_{j=0}^{n}(-1)^j \binom{n}{j} 2^{2^{n-j}}$, and in [10] for the k-valued functions

$$|M_1^k(X^n)| = \sum_{j=0}^{n}(-1)^j \binom{n}{j} k^{k^{n-j}} .$$

2. RESULTS.

Theorem 1. "Almost all" k-valued functions depend on every of their variables by the first way.

For the two valued functions this theorem is stated as a problem in [2], and for k-valued functions it is proved in [10].

Theorem 2. "Almost all" k-valued functions do not depend on at least one ot their variables by the second way.

Proof: Let $f(x_1,x_2,\dots,x_n) \in P_k$, $n \geq 1$. We take use of Shenon's analog (see [10])and we become that every k-valued function of n variables is completety determined bya ordered k^{n-1}- tuple of functions of a variable.

Then from definition 3 and the aforesaid it follows that the function f depends on the variable x by second way iff f is determined by an ordered k^{n-1}tuple functions of the variable x, $(f_1,f_2,\dots f_{k^{n-1}})$ such that every the function f_i, $1 \leq i \leq k^{n-1}$ depends on the variable x by the first way

But $|M_1^k(X_1^1)| = k^k - k$.

Therefore the number of k-valued functions on n variables which depends on every of their variables by second way do not exceed the number of possible k^{n-1}, tuples of functions resulting from among the functions of $M_1^k(x^1)$.

Then for $|M_2^k(X^n)|$ we obtain

$$|M_2^k(X^n)| \leq (k^k - k)^{k^{n-1}} .$$

For the relation $\dfrac{|M_2^k(X^n)|}{k^{k^n}}$ we have

$$\frac{|M_2^k(X^n)|}{k^{k^n}} = \frac{(k^k - k)^{k^{n-1}}}{k^{k^n}} = \frac{(k^k - k)^{k^{n-1}}}{k^k} .$$

Than

$$\lim_{n\to\infty} \frac{|M_2^k(X^n)|}{k^{k^n}} = \lim_{n\to\infty} \frac{(k^k - k)^{k^{n-1}}}{k^k} = \lim_{n\to\infty} \left(1 - \frac{1}{k^{k-1}}\right).$$

The limit of k^{n-1} is ∞ as when $n \to \infty$, we obtain

$$\lim_{n\to\infty}\frac{|M_2^k(X^n)|}{k^{k^n}} = \lim_{n\to\infty}(1 - \frac{1}{k^{k-1}})^{k^{n-1}} = \lim_{k^{n-1}\to\infty}(1 - \frac{1}{k^{k-1}})^{k^{n-1}}.$$

But $1 - \frac{1}{k^{k-1}} < 1$ Therefore

$$\lim_{n\to\infty}\frac{|M_2^k(X^n)|}{k^{k^n}} \; 0.$$

The proof is completed.

Bearing in mind that $M_6^k(X^n) \subset M_3^k(X^n) \subset M_2^k(X^n)$, so we obtain that next two corollaries are true:

Corollary 1.

$$\lim_{n\to\infty}\frac{|M_3^k(X^n)|}{k^{k^n}} = 0.$$

Corollary 2.

$$\lim_{n\to\infty}\frac{|M_6^k(X^n)|}{k^{k^n}} = 0.$$

Definition 8.[12] We say that the set $\left\{x_{i_1},x_{i_2},\dots,x_{i_m}\right\}$, $1 \le m \le n$; $i_1,i_2,\dots,i_m \in \{1,2,\dots,n\}$ is c-separable for the function $f(x_1,x_2,\dots,x_n) \in p_k$,if for each n-m costants of the variables in $R_f \setminus \{x_{i_1},x_{i_2},\dots,x_{i_m}\}$ we obtain a function which depends essentially on each of the variables $x_{i_1},x_{i_2},\dots,x_{i_m}$.

From definitions 3 and 8 follons that the funiction $f(x_1,x_2,\dots,x_n)$ from p_k depends on x_i, $1 \le i \le 2$, $n \ge 2$ by second way iff when the set $\{x_i\}$ is c-separable for f.

In [12] it is proved that if the functin $f(x_1,x_2,\dots,x_n)$ from p_k, each set of essential variables is c-separable for f when for f each one - element set is c-separable.

From the aforesaid it follows that theorem 2 can be stated the next way: To "almost all" k-valued functions at least one one - element set is not c-separable. Theorem 2 generalites the results from [6] and adds to theorem 1.5.1 from [6].

Definition 9. We say that the function f(x) from p_k is distiuguishing all over if for each two constants α_2 $(\alpha_1 \ne \alpha_2)$ for x it is hold $f(\alpha_1) \ne f(\alpha_2)$.

From definition 9 follows that the function $f(x_1,x_2,..,x_n)$ is distiuguishing all over, if it takes k-values.

L e m m a 3 The number of the functions a a variable from p_k distiuguishing all over, is k !.

T h e o r e m 4. "Almost all" functions from p_k depend on at least one of their variables by fifth way.

P r o o f. Let $f(x_1,x_2,..,x_n) \in p_k$, $n \geq 2$.
As in theorem 2 we obtain that the function will not depend on at least one of its variables by fifth way iff when there exist an ordered k^{n-1}- tuple of functions of a variable $(f_1,f_2,...,f_{k^{n-1}})$ such that no one of the functions is distiuguishing all over.
By $N_5^k(X^4)$ we will denote the set of all k-valued function, which do not depend on any of their variables by fifth way. Bearing in mind that the namber of the k-valued functions of a variable,which are not distiuguishing all over, is $k^k-k!$, so for $|N_5^k(X^n)|$ we obtained

$$| N_5^k(X^n) | \leqslant (k^k- k!)^{k^{n-1}} .$$

For the limit of the relation $\frac{|N_5^k(X^n)|}{k^{k^n}}$ we find the following:

$$\lim_{n\to\infty} \frac{| N_5^k(X^n) |}{k^{k^n}} = \lim_{n\to\infty} \frac{(k^k - k!)^{k^{n-1}}}{k^{k^n}} - \lim_{n\to\infty} (1 - \frac{k!}{k^k})^{k^{n-1}} .$$

But $\lim_{n\to\infty} k^{n-1} = \infty$. Therefore

$$\lim_{n\to\infty} \frac{| N_5^k(X^n) |}{k^{k^n}} = \lim_{k^{n-1}\to\infty} (1 - \frac{k!}{k^k})^{k^{n-1}} .$$

As $(1 - \frac{k!}{k^k}) < 1$, so we obtain that

$$\lim_{n\to\infty} \frac{| N_5^k(x^n) |}{k^{k^n}} = 0$$

The proof is complete.
As $M_5^k(X^n) \subset M_4^k(X^n)$ so, the following corollary is true:

C o r o l l a r y 1. "Allmost all" function from p_k depend on at least of one of their variables by fourth way.

T h e o r e m 5."Almost all" k-valued functions depend on each of their variable by fourth way.

P r o o f. $f(x_1,x_2,..,x_n) \in p_k$, $n \geq 1$. From definition 5

if follows that the function f does not depend on the variable x_i, $1 \le i \le n$, iff when there exist a k-tuple of functions $(f_1, f_2, \ldots, f_k)$ of (n-1) variables in which at least two of the functions are equal.
Therefore the namber of functions from p_k which do not depend on a variable by a fourth way, do not exceed

$$n \binom{k}{2} \left(k^{k^{n-1}} \right)^{k-2} \le nk^2 \left(k^{k^{n-1}} \right)^{k-1} = \frac{k^{k^n} + 2 + \log_k n}{k^{k^{n-1}}} =$$

$$= k^{k^n} - k^{n-1} + \log_2 n + 2$$

Then for $\left| M_4^k (X^n) \right|$ obtain

$$\left| M_4^k (X^n) \right| \ge k^{k^n} - k^{k^n} - k^{n-1} + \log_2 n + 2 .$$

$$\lim_{n\to\infty} \frac{\left| M_4^k (X^n) \right|}{k^{k^n}} = 1 - \lim_{n\to\infty} \frac{1}{k^{k^{n-1}} \log_k n - 2} = 1$$

The proof is complete.
The set of all essential variables for f will be denoted by R_f.
Bearing in mind lemma 2 from [13] and theorem 5, so we obtain that next theorem is true.

T h e o r e m 6. For "almost all" k-valued functions it is hold: If $R_{\varphi_1} \cap R_{\varphi_j} = \emptyset$ for $i \ne j$ and $i, j \in \{, 1, 2, \ldots, n\}$ and $f_1 = f(x_1 = \varphi_1, x_2 = \varphi_2, \ldots, x_n = \varphi_n)$ where

$$R_f \cap R_{\varphi_i} = \emptyset, \text{ for } i = \overline{1n} , \text{ so}$$

$$R_{f_1} = R_{\varphi_1} \cup R_{\varphi_2} \cup \ldots \cup R_{\varphi_n} .$$

R E F E R E N C E S

1. Брейтбарт Ю.Я., о существенных переменных функции алгебры логики. Доклады АН СССР, 172, № 1, 9-10.

2. Гавраилов Г.П., А.А.Сапоженко, Сборник задач по дискретной математике. Москва, 1977л

3. Денев Й.Д., Оценка на броя на подфункциите за почти всички двоични функции. Год. на СУ,Мат.факулт., 1974, т.66, 353-362.

4. Денев Й.Д., Сложность реелизации почти всех функции алгебры логики методом каскадов. Матем.заметки, 1972, 12, 769-780.

5. Лупанов О.Б., Об одном классе схем из функциональных

елементов. Проблемы кибернетики, 7, 1962, 61-114.

6 Мирчев И.А., И.Д.Гюдженов, Свойствата на един клас булеви функции, свързани с понятието с-отделимост на множества от аргументи на функциите. Год. на ВПИ - Благоевград, Математика, т. *IV*, кн. 1 59-64.

7. Чимев К.Н., Върху един начин на зависимост на някои функции от P_k от техните аргументи. Год. на ВТУЗ, Математика, т. *IV* кн.1, 1967, 5-12.

8. Чимев К.Н., Върху различните начини на зависимост на функциите от многозначната логика от аргументите им. Год. на ВТУЗ, Математика, т. *V*. кн.3, 1968/69, 101-110.

9. Чимев К.Н., Отделими множества от аргументи на функциите. Благоевград, 1983, 207 с.

10. Чимев К.Н., И.Д.Гюдженов, Подфункции и мощност на някои класи функции. Благоевград, 1987, 219 с.

11. Cimev K.N., Separable sets of arguments of functions. MTA SzTAKI Tanulmanyok, 180, 1986, 1-174=

12. Shtrakov Sl.Vl., Dominating and annuling sets of variables for functions. Blagoevgrad, 1987, 177

13. Чимев К.Н., Върху съществената зависимост на функциите от техните аргументи. Год. на ВПИ - Благоевград, Математика, т. *III*, кн.3, 1986, 22-29.

Composition of Facets of the Clique Partitioning Polytope

M. Grötschel, Augsburg, FRG
Y. Wakabayashi, Sao Paulo, Brazil

Abstract

In [1] we have introduced the clique partitioning problem and studied the associated polyhedron, the so-called clique partitioning polytope. In this paper we continue these polyhedral investigations; in particular, we present new classes of facets and methods to construct new facet-defining inequalities from given facet-defining inequalities.

0. Introduction and Notation

Let $K_n = (\mathcal{V}_n, \mathcal{E}_n)$ denote the complete graph on n nodes without loops and multiple edges, i. e., every two different nodes of K_n are linked by exactly one edge. An edge set $A \subseteq \mathcal{E}_n$ is called a **clique partitioning** of K_n if there is a **partition** $\{W_1, \ldots, W_k\}$ of $\mathcal{V}_n$ (i. e., each W_i is nonempty, $W_i \cap W_j = \emptyset$ for $i \neq j$, and $\bigcup_{i=1}^{k} W_i = \mathcal{V}_n$) such that A is the union of all those edges in $\mathcal{E}_n$ that have both endnodes in W_i, for some $i \in \{1, \ldots, k\}$. The **clique partitioning problem** (for short: CPP) is the task to find, for a given complete graph $K_n = (\mathcal{V}_n, \mathcal{E}_n)$ with edge weights $c_e \in \mathbf{R}$ for all $e \in \mathcal{E}_n$, a clique partitioning $A^* \subseteq \mathcal{E}_n$ such that $c(A^*) := \sum_{e \in A^*} c_e$ is as small as possible.

The clique partitioning problem is a combinatorial optimization version of a clustering problem in data analysis and has many interesting applications, among others, in zoology, economics, and the political sciences — see, for instance, [2], [3], [4], [5], [6]. This problem is $\mathcal{NP}$-hard. To solve instances coming up in practical applications, we have proposed in [2] an LP-based cutting plane procedure that utilizes our polyhedral investigations [1] of the associated "clique partitioning polytope". This approach works quite well; in particular, we could solve all practical applications we could get hold of to optimality. Although — to date — our method is the only one that is able to solve the larger ones of the problems described in [2], we feel that, to attack really large scale problem instances, more information about the clique partitioning polytope is necessary. Thus this paper is a continuation of our polyhedral work [1] on the clique partitioning problem and aims at enlarging our reservoir of facet-defining inequalities that can be used in a cutting plane procedure.

We use standard graph theory terminology. So a graph is denoted by $G = (V, E)$ where V is the node set and E the edge set of G. For our problems loops and multiple edges are irrelevant, so we assume throughout that all graphs considered are simple. If $H = (W, F)$ and $G = (V, E)$ are graphs with $W \subseteq V$ and $F \subseteq E$ then H is called

R. Bodendiek, R. Henn (Eds.)
Topics in Combinatorics and Graph Theory

a subgraph of G. We will perform many operations with subgraphs of K_n which we distinguish by using subscripts. Therefore we use the symbol $\mathcal{V}_n$ for the node set and the symbol $\mathcal{E}_n$ for the edge set of K_n in order to create no confusion. For $v \in V$, $G - v$ denotes the graph obtained from G by removing v. For $W \subseteq V$, $G[W]$ is the subgraph of G induced by W. It will be convenient to use the following notation, where $S, T, S_1, \ldots, S_k \subseteq V$ and $F \subseteq E$:

$$\begin{aligned} E(S) &:= \{uv \in E \mid u, v \in S\}, \\ E(S_1, \ldots, S_k) &:= \bigcup_{i=1}^{k} E(S_i), \\ [S:T] &:= \{uv \in \mathcal{E}_n \mid u \in S, v \in T\}, \\ V(F) &:= \{v \in V \mid v \text{ is the endnode of some edge in } F\}. \end{aligned}$$

To denote the set of edges in $G = (V, E)$ with one endnode in S and the other in T we write

$$E[S:T] := E \cap [S:T].$$

Using this notation, an edge set $A \subseteq \mathcal{E}_n$ is a clique partitioning of K_n if and only if there is a partition $\{W_1, \ldots, W_k\}$ of $\mathcal{V}_n$ such that $A = \mathcal{E}_n(W_1, \ldots, W_k)$; moreover, for the complete graph $K_n = (\mathcal{V}_n, \mathcal{E}_n)$ and every two disjoint node subsets S, T of $\mathcal{V}_n$, $[S:T] = \mathcal{E}_n[S:T]$ holds.

A **cycle** C of length k is an edge set of the form $\{v_1v_2, v_2v_3, \ldots, v_{k-1}v_k, v_1v_k\}$, where $v_i \neq v_j$ if $i \neq j$. For $k \geq 4$, the set $\overline{C} := \{v_iv_{i+2} \mid i = 1, \ldots, k-2\} \cup \{v_1v_{k-1}, v_2v_k\}$ is called the set of **2-chords** of C. A **triangle** is a cycle of length three. A **wheel** is the union of a cycle and the set of edges that link some node not on the cycle with all nodes of the cycle. A graph $G = (V, E)$ is **bipartite** if its node set can be partitioned into two nonempty subsets V_1, V_2 such that all edges of G have one endnode in V_1 and the other in V_2. Every partition of V with this property is called a **bipartition** of V.

If $G_1 = (V_1, E_1)$ and $G_2 = (V_2, E_2)$ are two graphs then the graph $(V_1 \cup V_2, E_1 \cup E_2)$ is called the **union** of G_1 and G_2 and is denoted by $G_1 \cup G_2$. (We assume that the union operation does not produce multiple edges, so $G_1 \cup G_2$ is a simple graph.)

1. The Clique Partitioning Polytope

To formulate the clique partitioning problem in polyhedral resp. linear programming terms we associate with it a polyhedron in the following way. Let $\mathbf{R}^{\mathcal{E}_n}$ denote the real vector space where every component x_e of a vector $x \in \mathbf{R}^{\mathcal{E}_n}$ is indexed by an edge e of the complete graph $K_n = (\mathcal{V}_n, \mathcal{E}_n)$. To avoid trivialities, we assume throughout the paper that $n \geq 3$. For every edge set $A \subseteq \mathcal{E}_n$, $\chi^A \in \mathbf{R}^{\mathcal{E}_n}$ denotes its incidence vector, i. e., $\chi^A_e = 1$ if $e \in A$ and $\chi^A_e = 0$ if $e \notin A$. The convex hull of all incidence vectors of clique partitionings of K_n is called the **clique partitioning polytope** (of K_n) and is denoted by $\mathcal{P}_n$, i. e.,

$$\mathcal{P}_n = \operatorname{conv}\{\chi^A \in \mathbf{R}^{\mathcal{E}_n} \mid A \text{ is a clique partitioning of } K_n\}.$$

Since the vertices of $\mathcal{P}_n$ are in one-to-one correspondence with the clique partitionings of K_n, it follows immediately that the CPP can be formulated as the problem

$$\begin{array}{ll} \text{minimize} & c^T x \\ \text{subject to} & x \in \mathcal{P}_n. \end{array}$$

This is a linear program in the sense that a linear objective function is to be minimized over a polytope. To apply LP-techniques this formulation is of no use unless $\mathcal{P}_n$ can be represented by a system of linear inequalities. Since the clique partitioning problem is $\mathcal{NP}$-hard, it follows from general results of complexity theory that it is very unlikely that an explicit complete description can ever be obtained; but we were able to determine large classes of valid and facet-defining inequalities for $\mathcal{P}_n$, see [1], and we continue these investigations in this paper.

Recall at this point that an inequality $a^T x \leq \alpha$ is called **valid** for $\mathcal{P}_n$ if $\mathcal{P}_n \subseteq \{x \in \mathbf{R}^{\mathcal{E}_n} \mid a^T x \leq \alpha\}$. A valid inequality $a^T x \leq \alpha$ is said to **define a facet** of $\mathcal{P}_n$ if the **face** $F_a := \{x \in \mathcal{P}_n \mid a^T x = \alpha\}$ of $\mathcal{P}_n$ is a facet, i. e., if F_a is a face of dimension one less than the dimension of $\mathcal{P}_n$ (the dimension of a set S is the cardinality of the largest set of affinely independent points in S minus one). If F is a subset of $\mathcal{E}_n$ then we use the symbol $x(F)$ as a short-hand notation for the sum $\sum_{e \in F} x_e$.

The following theorem is a summary of some of the results presented in [1].

(1.1) Theorem. *Let $K_n = (\mathcal{V}_n, \mathcal{E}_n)$ be a complete graph with $n \geq 3$ nodes, and let $\mathcal{P}_n \subseteq \mathbf{R}^{\mathcal{E}_n}$ be the clique partitioning polytope of K_n.*

(a) The dimension of $\mathcal{P}_n$ is equal to $|\mathcal{E}_n| = n(n-1)/2$.

(b) For every edge $e \in \mathcal{E}_n$, the **trivial inequalities** *$x_e \geq 0$ and $x_e \leq 1$ are valid for $\mathcal{P}_n$. Every inequality $x_e \geq 0$ defines a facet of $\mathcal{P}_n$, but no inequality $x_e \leq 1$ does.*

(c) For every three different nodes $i, j, k \in \mathcal{V}_n$, each of the three associated **triangle inequalities**

$$\begin{aligned} x_{ij} + x_{jk} - x_{ik} &\leq 1 \\ x_{ij} - x_{jk} + x_{ik} &\leq 1 \\ -x_{ij} + x_{jk} + x_{ik} &\leq 1 \end{aligned}$$

defines a facet of $\mathcal{P}_n$.

(d) For every two disjoint nonempty subsets S, T of $\mathcal{V}_n$, the **2-partition inequality induced by S and T** *(for short: $[S,T]$-***inequality***)*

$$x([S:T]) - x(\mathcal{E}_n(S)) - x(\mathcal{E}_n(T)) \leq \min\{|S|, |T|\}$$

is valid for $\mathcal{P}_n$. It defines a facet of $\mathcal{P}_n$ if and only if $|S| \neq |T|$.

(e) For every cycle $C \subseteq \mathcal{E}_n$ of length at least 5 and its set $\overline{C}$ of 2-chords, the **2-chorded cycle inequality**

$$x(C) - x(\overline{C}) \leq \left\lfloor \frac{|C|}{2} \right\rfloor$$

is valid for $\mathcal{P}_n$. It defines a facet of $\mathcal{P}_n$ if and only if $|C|$ is odd.

(f) *For every even cycle $C \subseteq E_n$ of length at least 8, for every node $z \in \mathcal{V}_n$ not in the node set $\mathcal{V}_n(C)$ of C, and for every bipartition $\{V, \overline{V}\}$ of $\mathcal{V}_n(C)$, the* **2-chorded even wheel inequality**

$$x(C \cup R) - x(\overline{C} \cup \overline{R}) \leq \frac{|C|}{2}$$

defines a facet of P_n, *where* $\overline{C}$ *is the set of 2-chords of* C *and* $R := \{zv \mid v \in V\}$, $\overline{R} := \{zv \mid v \in \overline{V}\}$. □

The aim of this paper is to construct further inequalities defining facets of $\mathcal{P}_n$. We will, in particular, generalize the 2-partition inequalities using some "glueing" and "lifting" techniques.

2. G-Induced $[S,T]$-Inequalities and Construction ∇

Let $G = (V, E)$ be a subgraph of $K_n = (\mathcal{V}_n, \mathcal{E}_n)$ and let $\{S, T\}$ be a partition of V, where S or T may possibly be empty. Then the inequality

(2.1) $$x(E[S:T]) - x(E(S)) - x(E(T)) \leq \min\{|S|, |T|\}$$

is called a **general 2-partition inequality induced by G, S, and T**, or for short, a **G-induced $[S,T]$-inequality**. Note that the order of S and T plays no role, so a G-induced $[S,T]$-inequality is also a G-induced $[T,S]$-inequality.

Every $[S,T]$-inequality (introduced in (1.1) (d)) is a $K_{|S \cup T|}$-induced $[S,T]$-inequality where $K_{|S \cup T|}$ is the complete subgraph of K_n induced by the node set $S \cup T$. So every G-induced $[S,T]$-inequality, $S \neq \emptyset \neq T$, can be obtained from the $[S,T]$-inequality by setting some of the positive and negative coefficients to zero. G-induced $[S,T]$-inequalities are not necessarily valid with respect to $\mathcal{P}_n$. They are, however, obviously valid if S or T is empty. This case is trivial and only included for technical reasons.

(2.2) Definition. *Let S, T be two disjoint subsets of $\mathcal{V}_n$ and let $G = (V, E)$ be a subgraph of K_n with $V = S \cup T$. G is called* **$[S,T]$-valid** *(with respect to $\mathcal{P}_n$) if the G-induced $[S,T]$-inequality is valid for $\mathcal{P}_n$. G is called* **strongly $[S,T]$-valid** *(with respect to $\mathcal{P}_n$) if for every node set $W \subseteq V$ the $(G - W)$-induced $[S \setminus W, T \setminus W]$-inequality is valid for $\mathcal{P}_n$.* □

So, for a strongly $[S,T]$-valid graph $G = (V, E)$, $G - W$ is $[S \setminus W, T \setminus W]$-valid for all $W \subseteq V$, in fact, $G - W$ is strongly $[S \setminus W, T \setminus W]$-valid.

(2.3) Remark. (a) If $G = (V, E)$ is a subgraph of K_n with $E = \emptyset$ then G is strongly $[S,T]$-valid for every partition $\{S, T\}$ of V.

(b) It follows immediately from (1.1) (d) that every complete subgraph $G = (V, E)$ of K_n is strongly $[S,T]$-valid for every partition $\{S, T\}$ of V (S or T possibly empty). □

We will now introduce a construction that can be used to combine strongly $[S_i, T_i]$-valid graphs G_i ($i = 1, 2$) into new strongly $[S,T]$-valid graphs. Let us first describe it in terms of an operation on two graphs.

Suppose we have two disjoint graphs $G_1 = (V_1, E_1)$ and $G_2 = (V_2, E_2)$ and, for each $i \in \{1,2\}$, we are given two subsets S_i', T_i' of V_i such that S_i' and T_i' are disjoint, $|T_1'| = |T_2'|$, and the induced subgraph $G[T_i']$ of G_i is complete. Assume furthermore that a bijection $\varphi : T_1' \to T_2'$ is given. Let $G = (V, E)$ be the graph obtained from G_i, S_i', T_i' $(i = 1, 2)$ by identifying each node v of $G_1[T_1']$ with the corresponding node $\varphi(v)$ of $G_2[T_2']$ and adding all edges with one endnode in S_1' and the other in S_2'. We call this operation **Construction** ∇ — see Figure 2.1 for a pictorial description.

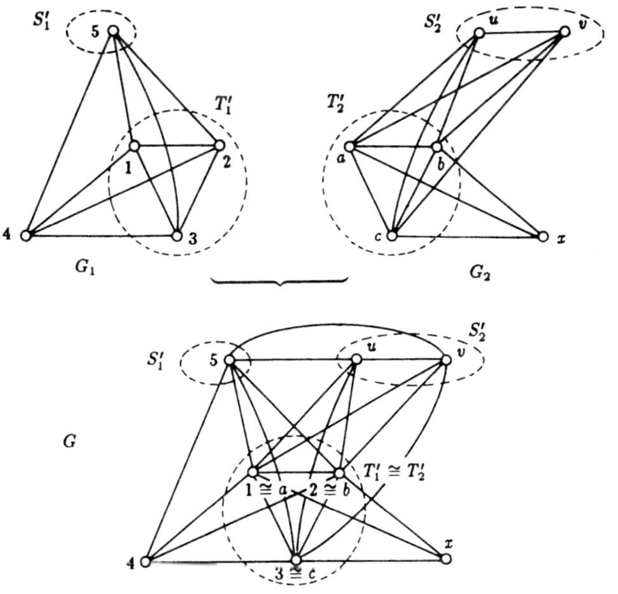

Figure 2.1 Example of Construction ∇

In order to avoid the necessity of specifying the bijection φ it is more convenient for us to work on subgraphs G_1 and G_2 of the complete graph K_n whose intersection is a complete subgraph. So let us redefine Construction ∇ for that case.

(2.4) Definition. *Let $G_1 = (V_1, E_1)$ and $G_2 = (V_2, E_2)$ be two subgraphs of $K_n = (\mathcal{V}_n, \mathcal{E}_n)$ and let $S_1' \subseteq V_1$, $S_2' \subseteq V_2$, and $T' \subseteq \mathcal{V}_n$ be node sets such that*

(A.1) $\quad V_1 \cap V_2 = T'$;

(A.2) $\quad S_1' \subseteq V_1 \setminus T', \;\; S_2' \subseteq V_2 \setminus T'$;

(A.3) $\quad G_i[T']$ *is complete for* $i = 1, 2$.

Let $G = (V, E)$ be the subgraph of K_n obtained from the union $G_1 \cup G_2$ of G_1 and G_2 by adding all edges with one endnode in S_1' and the other in S_2'. We will say that G is obtained from G_1 and G_2 by Construction $\nabla(S_1', S_2'; T')$ and write $G = G_1 \nabla G_2$. □

Note that in (2.4) we replace the identification process (which depends on φ) by assuming that the two subgraphs overlap in T'. This way φ is given implicitly. Also observe that the cases $S_1 = \emptyset$, $S_2 = \emptyset$, or $T' = \emptyset$ are allowed in Construction $\nabla(S_1', S_2'; T')$. It is immediately clear from the above definition that the following holds.

(2.5) Remark. Let $G_1 = (V_1, E_1)$ and $G_2 = (V_2, E_2)$ be subgraphs of $K_n = (\mathcal{V}_n, \mathcal{E}_n)$ satisfying the assumptions of (2.4), and let $G = G_1 \nabla G_2$ be the subgraph of K_n obtained by Construction $\nabla(S_1', S_2'; T')$. Then, for all $W \subseteq \mathcal{V}_n$, the graphs $G_1 - W$ and $G_2 - W$ and the node sets $S_1' \setminus W$, $S_2' \setminus W$, and $T' \setminus W$ satisfy the assumptions of (2.4). So Construction $\nabla(S_1' \setminus W, S_2' \setminus W; T' \setminus W)$ is well-defined, and $G - W = (G_1 - W)\nabla(G_2 - W)$ holds. □

(2.6) Theorem. *Let $G_1 = (V_1, E_1)$ and $G_2 = (V_2, E_2)$ be two subgraphs of $K_n = (\mathcal{V}_n, \mathcal{E}_n)$. For $i = 1, 2$, let $\{S_i, T_i\}$ be a partition of V_i and let $S_i' \subseteq V_i$, $T' \subseteq \mathcal{V}_n$ be node sets such that*

(B.1) $V_1 \cap V_2 = T_1 \cap T_2 = T'$;

(B.2) $S_i' \subseteq S_i$;

(B.3) $G_i[T']$ *is complete;*

(B.4) G_i *is strongly* $[S_i, T_i]$*-valid with respect to* $\mathcal{P}_n$*;*

(B.5) *no node in* $S_i \setminus S_i'$ *is adjacent to a node in* T'.

Let $G = (V, E)$ be the subgraph of K_n obtained from G_1 and G_2 by Construction $\nabla(S_1', S_2'; T')$ and set $T := T_1 \cup T_2$, $S := S_1 \cup S_2$. Then G is strongly $[S, T]$-valid with respect to $\mathcal{P}_n$.

Proof. Note that assumptions (B.1), (B.2), (B.3) imply assumptions (A.1), (A.2), (A.3). So Construction $\nabla(S_1', S_2'; T')$ can be performed.

We prove the theorem by induction on $\nu := |V_1| + |V_2| - |T'| = |V|$. The result is obvious for $\nu \leq 3$. (Actually, the only interesting case is $\nu = 3$ and $|S_i| = |S_i'| = |T'| = 1$ where we obtain a triangle inequality (1.1) (c) from two trivial inequalities of type $x_e \leq 1$. Observe also that, for $|S_i| = |T'| = 1$ and $S_i' = \emptyset$, we have $E = \emptyset$ by (B.5).)

Assume now that the theorem holds for $\nu \geq 3$ and let G_1, G_2, T' be such that $|V_1| + |V_2| - |T| = \nu + 1$. We have to prove that $G - W$ is $[S \setminus W, T \setminus W]$-valid with respect to $\mathcal{P}_n$ for all $W \subseteq V$.

Case 1. $W \neq \emptyset$. By Remark (2.5), $G - W = (G_1 - W)\nabla(G_2 - W)$; by assumption (B.4), $G_i - W$ is strongly $[S_i \setminus W, T_i \setminus W]$-valid; and thus (since $|V_1 \setminus W| + |V_2 \setminus W| - |T' \setminus W| < \nu + 1$) $G - W$ is strongly $[S \setminus W, T \setminus W]$-valid by induction hypothesis.

Case 2. $W = \emptyset$. To prove that G is $[S, T]$-valid we use the fact — proved in Case 1 — that $G - v$ is $[S \setminus \{v\}, T \setminus \{v\}]$-valid for all $v \in V$. By adding the sum of the left-hand sides of the $|S|$ valid inequalities

$$x(E[S \setminus \{v\} : T]) - x(E(S \setminus \{v\})) - x(E(T)) \leq \min\{|S| - 1, |T|\}, \ v \in S$$

to the sum of the left-hand sides of the $|T|$ valid inequalities

$$x(E[S : T \setminus \{v\}]) - x(E(S)) - x(E(T \setminus \{v\})) \leq \min\{|S|, |T| - 1\}, \ v \in T$$

and estimating the sum of the $|S|+|T|$ right-hand sides from above we obtain the valid inequality

$$(\nu-1)(x(E[S:T])-x(E(S))-x(E(T))) \leq (\nu+1)\min\{|S|,|T|\}-\min\{|S|,|T|\}.$$

Dividing by $\nu-1$ we get

$$x(E[S:T])-x(E(S))-x(E(T)) \leq \min\{|S|,|T|\}+\tfrac{1}{\nu-1}\min\{|S|,|T|\}.$$

It follows from $\nu \geq 3$ and $\min\{|S|,|T|\} \leq \lfloor \nu/2 \rfloor$ that $\min\{|S|,|T|\}/(\nu-1) < 1$ which implies that $x(E[S:T])-x(E(S))-x(E(T)) \leq \min\{|S|,|T|\}$ is valid for $\mathcal{P}_n$. □

We will now prove the main (technical) result of our paper that will be used later to derive interesting classes of facet-defining inequalities for $\mathcal{P}_n$. Recall that a **matching** is a subset M of the edges of a graph such that no two edges in M have a common endnode; an s-**matching** is a matching with s elements. A node that is in some edge of a matching M is said to be **covered** by M.

(2.7) Theorem. *Let $G_1=(V_1,E_1)$ and $G_2=(V_2,E_2)$ be two subgraphs of $K_n=(\mathcal{V}_n,\mathcal{E}_n)$. For $i=1,2$, let $\{S_i,T_i\}$ be a partition of V_i and let $S_i' \subseteq V_i$, $T' \subseteq \mathcal{V}_n$ be node sets such that*

(C.1) $V_1 \cap V_2 = T_1 \cap T_2 = T'$, $|T'| \geq 2$;

(C.2) $\emptyset \neq S_i' \subseteq S_i$, $|S_i| \leq |T_i \setminus T'|$;

(C.3) $G_i[S_i' \cup T']$ is complete;

(C.4) G_i is strongly $[S_i,T_i]$-valid with respect to $\mathcal{P}_n$ and the associated $[S_i,T_i]$-inequality defines a facet of $\mathcal{P}_n$;

(C.5) no node in $S_i \setminus S_i'$ is adjacent to a node in T';

(C.6) for every pair of nodes w,z with $w \in T_i \setminus T'$ and $z \in T'$, G_i has an $|S_i|$-matching $M_i(w,z)$ contained in $E_i[S_i:(T_i \setminus T') \cup \{z\}]$ that does not cover w;

(C.7) G_i has an $|S_i|$-matching N_i contained in $E_i[S_i:T_i \setminus T']$.

Let $G=(V,E)$ be the subgraph of K_n obtained from G_1 and G_2 by Construction $\nabla(S_1',S_2';T')$ and let $S:=S_1 \cup S_2$, $T:=T_1 \cup T_2$. Then G is strongly $[S,T]$-valid and the G-induced $[S,T]$-inequality defines a facet of $\mathcal{P}_n$.

Proof. The assumptions (C.1),...,(C.5) obviously imply the assumptions (B.1),..., (B.5) of Theorem (2.6). So, Construction $\nabla(S_1',S_2';T')$ is well-defined and the graph $G=G_1\nabla G_2$ is strongly $[S,T]$-valid with respect to $\mathcal{P}_n$.

Let $a^Tx \leq \alpha$ be the G-induced $[S,T]$-inequality, i. e., $a^Tx = x(E[S:T])-x(E(S)) - x(E(T)) \leq \min\{|S|,|T|\} = \alpha$, and let $F_a := \{x \in \mathcal{P}_n \mid a^Tx = \alpha\}$. Assume that $b^Tx=\beta$ defines a hyperplane such that $F_b := \{x \in \mathcal{P}_n \mid b^Tx=\beta\}$ is a facet of $\mathcal{P}_n$ with $F_a \subseteq F_b$. We will show that there exists a real number $\pi \neq 0$ such that $b=\pi a$. This will prove the theorem.

Let us, for $i=1,2$, denote the G_i-induced $[S_i,T_i]$-inequality by $(a^i)^Tx \leq \alpha_i$. So $a^Tx=(a^1)^Tx+(a^2)^Tx+x(E(T'))-x([S_1':S_2'])$. It follows from (C.2) that $\alpha_i=|S_i|$ $(i=1,2)$ and $\alpha=|S|=|S_1|+|S_2|$.

We prove that $b_e = \pi a_e$ for all $e \in \mathcal{E}_n$ and some $\pi \in \mathbf{R}$.

Case 1. $e \in \mathcal{E}_n(V_1) \cup \mathcal{E}_n(V_2)$.

Since $(a^1)^T x \leq \alpha_1$ defines a facet of $\mathcal{P}_n$ there are $m := |\mathcal{E}_n|$ clique partitionings $A_1, \ldots, A_m$ whose incidence vectors are linearly independent and satisfy $(a^1)^T x \leq \alpha_1$ with equality. By (C.7) there exists an $|S_2|$-matching $N_2 \subseteq [S_2 : T_2 \setminus T']$. Hence the edge sets $B_j := (A_j \cap \mathcal{E}_n(V_1)) \cup N_2$ are clique partitionings of K_n such that $a^T \chi^{B_j} = \alpha$ for $j = 1, \ldots, m$; and therefore

$$b^T(\chi^{B_j} - \chi^{B_m}) = 0 \quad \text{for } j = 1, \ldots, m-1$$

holds. Let X be the $(m-1) \times |\mathcal{E}_n|$ matrix whose rows are the vectors $\chi^{B_j} - \chi^{B_m}$, $j = 1, \ldots, m-1$. All columns of X corresponding to edges $e \in \mathcal{E}_n \setminus \mathcal{E}_n(V_1)$ are zero, and it follows from the fact that the vectors $\chi^{A_1}, \ldots, \chi^{A_m}$ are linearly independent that the $(m-1) \times |\mathcal{E}_n(V_1)|$ submatrix Y of X corresponding to the edges $e \in \mathcal{E}_n(V_1)$ has rank $|\mathcal{E}_n(V_1)| - 1$. So the kernel $\{y \in \mathbf{R}^{\mathcal{E}_n(V_1)} \mid Yy = 0\}$ of Y has dimension 1. Since the vector $\hat{a}^1 \in \mathbf{R}^{\mathcal{E}_n(V_1)}$ obtained from a^1 by deleting all components corresponding to edges in $\mathcal{E}_n \setminus \mathcal{E}_n(V_1)$ and the vector $\hat{b} \in \mathbf{R}^{\mathcal{E}_n(V_1)}$ obtained from b in the same way satisfy $Y\hat{a}^1 = Y\hat{b} = 0$ and since $\hat{a}^1 \neq 0$ we know that there exists a real number π such that $\hat{b} = \pi \hat{a}^1$. This implies $b_e = \pi a_e^1$ for all $e \in \mathcal{E}_n(V_1)$.

By symmetry we obtain that there exists a real number π' such that $b_e = \pi' a_e^2$ for all $e \in \mathcal{E}_n(V_2)$. By (C.1), $T' = V_1 \cap V_2$ and $|T'| \geq 2$, and by (C.3) $G[T']$ is complete. So there is an edge $f \in E(T') = E_1(T') = E_2(T')$. Since $a_f = a_f^1 = a_f^2 = -1$ we can conclude that $\pi = \pi'$, and thus there exists a real number π such that

(1) $$b_e = \pi a_e \quad \text{for all } e \in \mathcal{E}_n(V_1) \cup \mathcal{E}_n(V_2).$$

Case 2. $e = uv$ with $u \in S_1'$ and $v \in S_2'$.

Let z_1 and z_2 be two different nodes in T'. Let N_1, N_2 be the two matchings existing by (C.7) and let $u' \in T_1 \setminus T'$, $v' \in T_2 \setminus T'$ be the nodes such that $uu' \in N_1$ and $vv' \in N_2$. Set

$$A := N_1 \cup N_2 \text{ and}$$
$$B := (A \setminus \{uu', vv'\}) \cup \{uv, uz_1, uz_2, vz_1, vz_2, z_1 z_2\}.$$

Then A and B are clique partitionings; χ^A obviously satisfies $\chi^A \in F_a \subseteq F_b$, while (C.3) yields that $\chi^B \in F_a \subseteq F_b$. Thus (1) implies $0 = b^T\chi^A - b^T\chi^B + b_{uu'} + b_{vv'} - b_{uv} - b_{uz_1} - b_{uz_2} - b_{vz_1} - b_{vz_2} - b_{z_1 z_2} = -b_{uv} - \pi$. From this we obtain

(2) $$b_e = -\pi \quad \text{for all } e \in [S_1' : S_2'].$$

Case 3. $e = uv$ with $u \in T_1 \setminus T'$ and $v \in T_2 \setminus T'$.

Let z_1, z_2 be any two nodes in T' and let $M_1(u, z_1) \subseteq E_1[S_1 : (T_1 \setminus T') \cup \{z_1\}]$, $M_2(v, z_2) \subseteq E_2[S_2 : (T_2 \setminus T') \cup \{z_2\}]$ be $|S_i|$-matchings ($i = 1, 2$) not covering u and v, respectively. Such matchings exist by (C.6). Set

$$A := M_1(u, z_1) \cup M_2(v, z_2), \; B := A \cup \{e\}.$$

Then A and B are clique partitionings with $\chi^A, \chi^B \in F_a \subseteq F_b$, and we can conclude from $0 = \beta - \beta = b^T\chi^B - b^T\chi^A = b_e$ that

$$(3) \qquad b_e = 0 \text{ for all } e \in [T_1 \setminus T' : T_2 \setminus T'].$$

Case 4. $e = uv$ with $u \in T_i \setminus T'$ and $v \in S_j$ for $i, j \in \{1,2\}$, $i \neq j$.

Let $N_j \subseteq E_j[S_j : T_j \setminus T']$ be the $|S_j|$-matching existing by (C.7) and let $v' \in T_j \setminus T'$ be the node with $vv' \in N_j$. Let z be any node in T' and $M_i(u,z)$ be the $|S_i|$-matching existing by (C.6). Set

$$A := M_i(u,z) \cup N_j \text{ and } B := A \cup \{uv, uv'\}.$$

Then A and B are clique partitionings with $\chi^A, \chi^B \in F_a \subseteq F_b$. So $0 = b^T\chi^B - b^T\chi^A = b_{uv} + b_{uv'}$, and (3) implies

$$(4) \qquad b_e = 0 \text{ for all } e \in [T_i \setminus T' : S_j] \text{ with } i, j \in \{1,2\}, i \neq j.$$

Case 5. $e = uv$ with $u \in S_i \setminus S_i'$ and $v \in S_j$ for $i, j \in \{1,2\}$, $i \neq j$.

Let N_i and N_j be the matchings existing by (C.7) and let $u' \in T_i \setminus T'$, $v' \in T_j \setminus T'$ be the nodes with $uu' \in N_i$ and $vv' \in N_j$. Set

$$A := N_i \cup N_j, \; B := A \cup \{uv, u'v', uv', u'v\}.$$

Then A and B are clique partitionings with $\chi^A, \chi^B \in F_a \subseteq F_b$. Therefore, $0 = b_{uv} + b_{u'v'} + b_{uv'} + b_{u'v}$ and (3) and (4) imply

$$(5) \qquad b_e = 0 \text{ for all } e \in [S_i \setminus S' : S_j] \text{ with } i, j \in \{1,2\}, i \neq j.$$

Case 6. $e = uv$ with $u \in \mathcal{V}_n \setminus V$.

This case is trivial and we obtain

$$b_e = 0 \text{ for all } e \notin \mathcal{E}_n(V).$$

Altogether we have now shown that $b = \pi a$, and clearly $\pi \neq 0$. Thus $a^Tx \leq \alpha$ defines a facet of $\mathcal{P}_n$. □

We would like to remark that the statement of Theorem (2.7) holds under slightly more general conditions. These are, however, rather complicated and technical. We have decided to present here the systems (C.1),…,(C.7). These assumptions are relatively easy to understand and are sufficient for the derivation of our main classes of facet-defining inequalities. An immediate consequence of Theorem (2.7) is the following.

(2.8) Theorem. *Let $G_1 = (V_1, E_1)$ and $G_2 = (V_2, E_2)$ be two complete subgraphs of $K_n = (\mathcal{V}_n, \mathcal{E}_n)$. For $i = 1, 2$, let $\{S_i, T_i\}$ be a partition of V_i and let $T' \subseteq \mathcal{V}_n$ be node sets such that*

(D.1) $\quad V_1 \cap V_2 = T_1 \cap T_2 = T'$, $|T'| \geq 2$;

(D.2) $\quad 1 \leq |S_i| \leq |T_i \setminus T'|$.

Let $G = (V, E)$ be the subgraph of K_n obtained from G_1 and G_2 by Construction $\nabla(S_1, S_2; T')$ and let $S := S_1 \cup S_2$, $T := T_1 \cup T_2$. Then G is strongly $[S, T]$-valid and the G-induced $[S, T]$-inequality defines a facet of $\mathcal{P}_n$.

Proof. For $i = 1, 2$, set $S_i' = S_i$ then the assumptions (C.1), (C.2), (C.3), (C.5), (C.6), (C.7) are obviously satisfied; (C.4) is satisfied by Remark (2.3) (b). Thus (2.8) follows from (2.7). □

Figure 2.2 shows two graphs that are obtained by Construction ∇ from two complete subgraphs of K_n. The associated general $[S, T]$-inequalities define facets of $\mathcal{P}_n$ for $n \geq 6$ and $n \geq 9$, respectively, by Theorem (2.8).

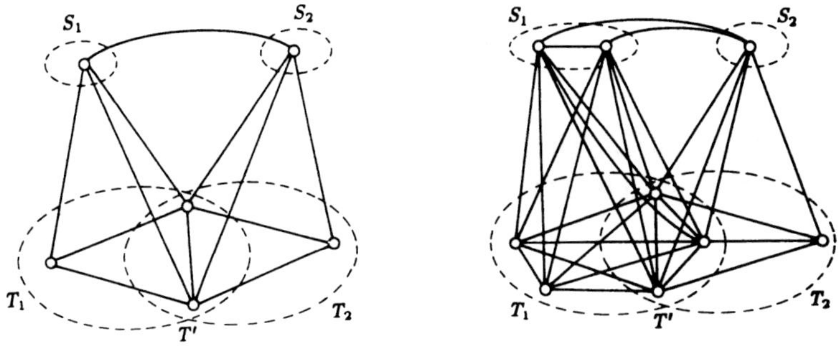

Figure 2.2 Graphs inducing general 2-partition inequalities.

Theorem (2.8) has been cast in a way that Theorem (2.7) is directly applicable. The following version of it is probably easier to remember.

(2.9) Corollary. *Let S_1, S_2, T_1, T_2, T be five mutually disjoint subsets of the node set $\mathcal{V}_n$ of K_n such that $|T| \geq 2$ and $1 \leq |S_i| \leq |T_i|$ for $i = 1, 2$. Then the (general 2-partition) inequality*

$$x([S_1 : T_1 \cup T]) + x([S_2 : T_2 \cup T]) - x([S_1 : S_2]) - x(\mathcal{E}_n(S_1)) - x(\mathcal{E}_n(S_2)) -$$
$$x([T_1 : T]) - x([T_2 : T]) - x(\mathcal{E}_n(T_1)) - x(\mathcal{E}_n(T_2)) - x(\mathcal{E}_n(T)) \leq |S_1| + |S_2|$$

defines a facet of $\mathcal{P}_n$. □

3. Two Further Compositions

We will now describe two ways of applying Construction ∇ iteratively that can be used to produce new facet-defining inequalities for $\mathcal{P}_n$.

(3.1) Definition. *Let $G_1 = (V_1, E_1), \ldots, G_p = (V_p, E_p)$, $p \geq 2$, be complete subgraphs of K_n, let $\{S_i, T_i\}$ be a partition of V_i, $i = 1, \ldots, p$, and let T'_i, $i = 1, \ldots, p-1$, be disjoint subsets of $\mathcal{V}_n$ such that*

(E.1) $V_i \cap V_{i+1} = T_i \cap T_{i+1} = T'_i$ *for* $i = 1, \ldots, p-1$;

(E.2) $V_i \cap V_{i+k} = \emptyset$ *for* $i = 1, \ldots, p-2$ *and* $k = 2, \ldots, p-i$.

Set $\tilde{G}_1 := G_1$, and for $i = 2, \ldots, p$, let $\tilde{G}_i$ be the graph obtained from $\tilde{G}_{i-1}$ and G_i by Construction $\nabla(S_{i-1}, S_i; T'_{i-1})$. Let us denote the graph $\tilde{G}_p$ constructed this way by $G = (V, E)$. We say that G is the graph obtained from complete graphs $G_1, \ldots, G_p$ by **Construction** $\nabla(S_1, \ldots, S_p; T'_1, \ldots, T'_{p-1})$, *or (not specifying details) by a* **repeated nonoverlapping ∇-construction.** □

Figure 3.1 shows the scheme of a repeated nonoverlapping ∇-construction.

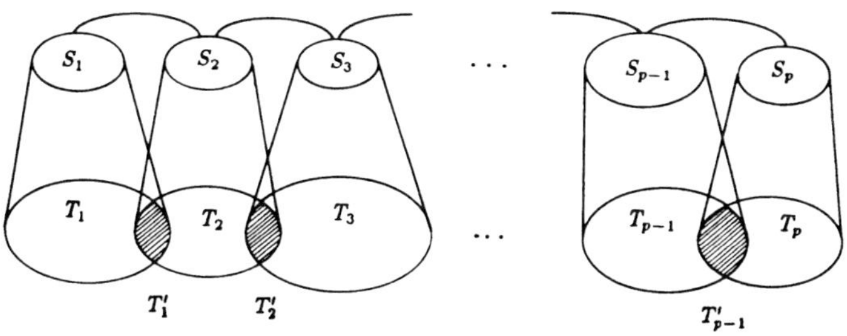

Figure 3.1

(3.2) Theorem. *Let $G_1 = (V_1, E_1), \ldots, G_p = (V_p, E_p)$, $p \geq 2$, be complete subgraphs of $K_n = (\mathcal{V}_n, \mathcal{E}_n)$, let $\{S_i, T_i\}$ be a partition of V_i for $i = 1, \ldots, p$; and let T'_i, $i = 1, \ldots, p-1$, be disjoint subsets of $\mathcal{V}_n$ such that*

(F.1) $V_i \cap V_{i+1} = T_i \cap T_{i+1} = T'_i$ *and* $|T'_i| \geq 2$ *for* $i = 1, \ldots, p-1$;

(F.2) $V_i \cap V_{i+k} = \emptyset$ *for* $i = 1, \ldots, p-2$ *and* $k = 2, \ldots, p-i$;

(F.3) $1 \leq |S_i| \leq |T_i| - \max\{|T'_{i-1}|, |T'_i|\}$ *for* $i = 2, \ldots, p-1$,
$1 \leq |S_1| \leq |T_1| - |T'_1|$, $1 \leq |S_p| \leq |T_p| - |T'_{p-1}|$.

Let $G = (V, E)$ be the subgraph of K_n obtained from $G_1, \ldots, G_p$ by Construction $\nabla(S_1, \ldots, S_p; T'_1, \ldots, T'_{p-1})$ and let $S := \bigcup_{i=1}^p S_i$, $T := \bigcup_{i=1}^p T_i$. Then G is strongly $[S, T]$-valid and the G-induced $[S, T]$-inequality defines a facet of $\mathcal{P}_n$.

Proof. The assumptions imply that the repeated nonoverlapping ∇-construction (3.1) can be performed. Let us denote the graphs constructed in this process by $\tilde{G}_i = (\tilde{V}_i, \tilde{E}_i)$ and set $\tilde{S}_i := \bigcup_{j=1}^i S_i$, $\tilde{T}_i := \bigcup_{j=1}^i T_i$, $\tilde{S}'_i := S_i$, $i = 1, \ldots, p$.

By Theorem (1.1) (d) all G_i-induced $[S_i, T_i]$-inequalities define facets of $\mathcal{P}_n$, and applying Theorem (2.8) to $\tilde{G}_1 = G_1$ and G_2 we get that, for the graph $\tilde{G}_2 = (\tilde{V}_2, \tilde{E}_2)$ obtained by Construction $\nabla(\tilde{S}'_1, S_2; T'_1)$ from $\tilde{G}_1$ and G_2, the $\tilde{G}_2$-induced $[\tilde{S}_2, \tilde{T}_2]$-inequality is strongly valid for $\mathcal{P}_n$ and defines a facet of $\mathcal{P}_n$. It is easy to see that $\tilde{G}_2 = (\tilde{V}_2, \tilde{E}_2)$ and $G_3 = (V_3, E_3)$ with the partitions $\{\tilde{S}_2, \tilde{T}_2\}$ of $\tilde{V}_2$ and $\{S_3, T_3\}$ of V_3 and additional sets

$S_1' := \tilde{S}_2'$, $S_2' := S_3$, $T' := T_2'$ satisfy all assumptions of Theorem (2.7). So, for the graph $\tilde{G}_3 = (\tilde{V}_3, \tilde{E}_3)$ obtained from $\tilde{G}_2$ and G_3 by Construction $\nabla(S_1', S_2'; T') = \nabla(\tilde{S}_2', S_3; T_2')$, the $\tilde{G}_3$-induced $[\tilde{S}_3, \tilde{T}_3]$-inequality is strongly valid for $\mathcal{P}_n$ and defines a facet of $\mathcal{P}_n$.

Repeating this process iteratively we can conclude that for $G = \tilde{G}_p = \tilde{G}_{p-1} \nabla G_p$ the G-induced $[S, T]$-inequality is strongly valid and defines a facet of $\mathcal{P}_n$. □

An easier to read version of the above theorem — which includes Corollary (2.9) as a special case — is the following.

(3.3) Corollary. *Let $S_1, \ldots, S_p, T_1, \ldots, T_p, T_1', \ldots, T_{p-1}'$, $p \geq 2$, be mutually disjoint subsets of the node set $\mathcal{V}_n$ of K_n. Set (for notational convenience) $T_0' := T_p' := \emptyset$ and assume that*

$$|T_i'| \geq 2 \quad \text{for } i = 1, \ldots, p-1;$$
$$1 \leq |S_i| \leq |T_i| + \min\{|T_i'|, |T_{i-1}'|\} \quad \text{for } i = 1, \ldots, p$$

is satisfied. Then the (general 2-partition) inequality

$$\sum_{i=1}^{p} x([S_i : T_i \cup T_i' \cup T_{i-1}']) - \sum_{i=1}^{p-1} x([S_i : S_{i+1}]) - \sum_{i=1}^{p} x([T_i : T_i' \cup T_{i-1}'])$$
$$- \sum_{i=1}^{p} (x(\mathcal{E}_n(S_i)) + x(\mathcal{E}_n(T_i)) + x(\mathcal{E}_n(T_i'))) \leq \sum_{i=1}^{p} |S_i|$$

defines a facet of $\mathcal{P}_n$.

□

Figure 3.2 shows a graph obtained by a repeated nonoverlapping ∇-construction of complete graphs.

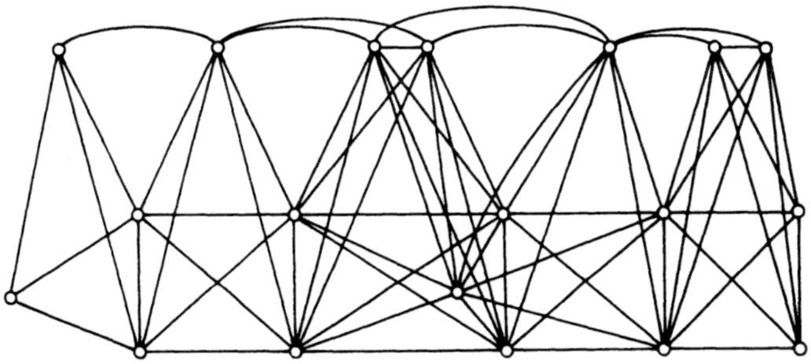

Figure 3.2 Nonoverlapping ∇-composition of complete graphs.

Another way of making iterative use of Construction ∇ is the following.

(3.4) Definition. *Let $G_1 = (V_1, E_1), \ldots, G_p = (V_p, E_p)$, $p \geq 2$, be complete subgraphs of K_n, let $\{S_i, T_i\}$ be a partition of V_i for $i = 1, \ldots, p$ and let T' be a subset of $\mathcal{V}_n$ such that*

(G.1) $\quad V_i \cap V_j = T_i \cap T_j = T'$ *for* $1 \leq i < j \leq p$.

Set $\overline{G}_1 := G_1$, and for $i = 2, \ldots, p$, let $\overline{G}_i$ be the subgraph of K_n obtained from $\overline{G}_{i-1}$ and G_i by Construction $\nabla(S_1 \cup \ldots \cup S_{i-1}, S_i; T')$. Let us denote the graph $\overline{G}_p$ obtained *this way by $G = (V, E)$. We say that G is the graph obtained from complete graphs $G_1, \ldots, G_p$ by* **Construction** $\nabla(S_1, \ldots, S_p; T')$, *or (without specifying details) by a* **repeated totally overlapping ∇-construction**. □

The basic scheme of a repeated totally overlapping ∇-construction is displayed in Figure 3.3.

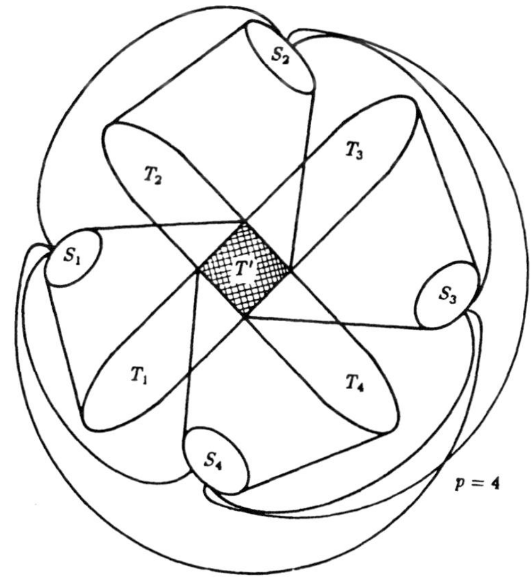

Figure 3.3

Facet-defining inequalities can be obtained with the repeated totally overlapping ∇-construction as follows.

(3.5) Theorem. *Let $G_1 = (V_1, E_1), \ldots, G_p = (V_p, E_p)$, $p \geq 2$, be complete subgraphs of K_n, let $\{S_i, T_i\}$ be a partition of V_i for $i = 1, \ldots, p$ and let $T' \subseteq \mathcal{V}_n$ such that*

(H.1) $V_i \cap V_j = T_i \cap T_j = T'$ *for* $1 \leq i < j \leq p$, $|T'| \geq 2$;

(H.2) $1 \leq |S_i| \leq |T_i| - |T'|$ *for* $i = 1, \ldots, p$.

Let $G = (V, E)$ be the subgraph of K_n obtained from $G_1, \ldots, G_p$ by Construction $\nabla(S_1, \ldots, S_p; T')$ and let $S := \bigcup_{i=1}^p S_i$, $T := \bigcup_{i=1}^p T_i$. Then G is strongly $[S, T]$-valid and the G-induced $[S, T]$-inequality defines a facet of $\mathcal{P}_n$.

Proof. The result follows — as in the proof of (3.2) — by induction from Theorem (2.7). The proof is straightforward and is left to the reader. □

A more digestible form of Theorem (3.5) is the following.

(3.6) Corollary. *Let $S_1, \ldots, S_p, T_1, \ldots, T_p, T'$, $p \geq 2$, be mutually disjoint subsets of the node set $\mathcal{V}_n$ of K_n. Set $S := \bigcup_{i=1}^p S_i$ and assume that*

$$|T'| \geq 2;$$
$$1 \leq |S_i| \leq |T_i| - |T'| \quad \text{for } i = 1, \ldots, p$$

holds. Then the (general 2-partition) inequality

$$\sum_{i=1}^{p} x([S_i : T_i \cup T']) - x(\mathcal{E}_n(S)) - x(\mathcal{E}_n(T')) - \sum_{i=1}^{p}(x(\mathcal{E}_n(T_i)) + x([T_i : T'])) \leq |S|$$

defines a facet of $\mathcal{P}_n$. □

A graph obtained by a repeated totally overlapping ∇-construction from 3 complete graphs is shown in Figure 3.4. Here $|T'| = 2$ and the edge forming $\mathcal{E}_n(T')$ is drawn by a thick line. The sets S_1, S_2, and S_3 have cardinality 1, 1, and 2 respectively.

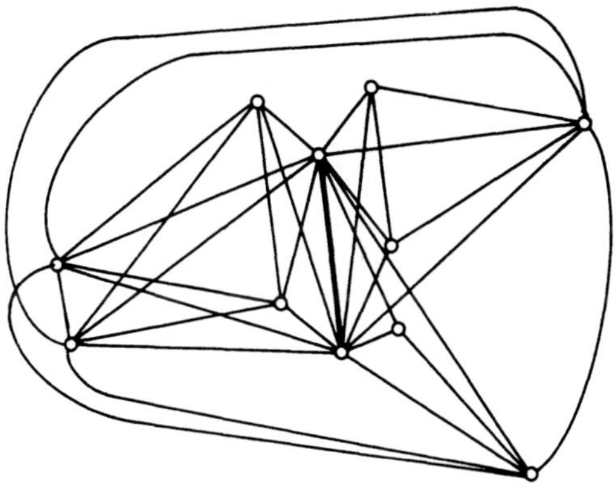

Figure 3.4 Totally overlapping ∇-composition of complete graphs.

References

[1] Grötschel, M. & Wakabayashi, Y. [1987a]: "Facets of the clique partitioning polytope", Report No. 6, Schwerpunktprogramm der Deutschen Forschungsgemeinschaft, Universität Augsburg, Augsburg, West Germany, to appear in *Mathematical Programming 1989.*

[2] Grötschel, M. & Wakabayashi, Y. [1987b]: "A cutting plane algorithm for a clustering problem", Report No. 9, Schwerpunktprogramm der Deutschen Forschungsgemeinschaft, Universität Augsburg, Augsburg, West Germany, to appear in *Mathematical Programming 1989.*

[3] Marcotorchino, J. F. & Michaud, P. [1980]: "Optimisation en analyse des données relationnelles", in: Diday, E. et al. (eds.), *Data Analysis and Informatics,* North-Holland (1980) 655–670.

[4] Opitz, O. & Schrader, M. [1984]: "Analyse qualitativer Daten: Einführung und Übersicht", *Operations Research Spektrum* 6 (1984) 67–83.

[5] Tüshaus, U. [1983]: *Aggregation binärer Relationen in der qualitativen Datenanalyse,* Mathematical Systems in Economics 82, Hain, Königstein, 1983.

[6] Wakabayashi, Y. [1986]: *Aggregation of Binary Relations: Algorithmic and Polyhedral Investigations,* Ph. D. Thesis, Universität Augsburg, West Germany (1986).

Optimal Edge-Numbering of Binary Trees

N. Grünwald, Warnemünde, GDR

Let $G = (V,E)$ be a graph with vertex-set $V(G)$ and edge-set $E(G)$. A 1-1 mapping $f: E(G) \rightarrow \{1,2, \dots, |E(C)|\}$ will be called an edge-numbering of G. $E(v)$ means the set of all edges of G which are incident to the vertex v of G. Define

$$D_f(v) = \max_{e,e' \in E(v)} |f(e) - f(e')| \quad \text{and}$$

$$W_f(G) = \max_{v \in V(G)} D_f(v) .$$

An optimal edge-numbering of G is an edge-numbering of G which minimizes $W_f(G)$. Define

$$W(G) = \min \{ W_f(G): f \text{ is an edge-numbering of } G \} .$$

The following lower bound can be derived in a straightforward way

<u>Theorem 1</u>: Let $D(G)$ denote the diameter of G. Then

$$W(G) \geq \lceil (|E(G)| - 1)/(D(G) - 1)\rceil .$$

Proof: Take an arbitrary edge numbering f. There are two edges u and v numbered 1 and $|E(G)|$. It follows that $1 + (D(G)-1)\, W_f(G) \geq |E(G)|$ or $W_f(G) \geq \lceil (|E(G)|-1)/(D(G)-1)\rceil$.

Let $T_{2,k}$ denote the complete binary tree in which the i-th vertex level consist of 2^{i-1} vertices, each vertex in vertex level $i<k$ has 2 "sons" at vertex level $i+1$, and the i-th edge level consist of t^i edges, each edge in edge level $i<k-1$ has t "daughters" at edge level $i+1$ and $(t-1)$ "sisters" at edge level i. Define $e_{i,j}$ the j-th edge of the i-the edge level from left to right.

<u>Theorem 2:</u> For an optimal edge-numbering of $T_{2,k}$ is

$$W(T_{2,k}) = \lceil (2^k - 3)/(2k - 3)\rceil$$

Proof: By Theorem 1 we have $W(T_{2,k}) \geq \lceil (2^k - 3)/(2k - 3)\rceil$

It suffices to find an edge-numbering f with $W_f(T_{2,k}) = \lceil (2^k - 3)/(2k - 3)\rceil$. We shall prove that for the complete binary tree $T_{2,k}$ the following algorithm pro-

R. Bodendiek, R. Henn (Eds.)
Topics in Combinatorics and Graph Theory

duces an edge-numbering f with $W_f(T_{2,k}$ = $\lceil (2^k - 3)\ (2k - 3) \rceil = x$
We define f by specifying in increasing order the edges chosen from the left half of $T_{2,k}$, whereas the right half is chosen in the reverse order of the left half.
In our algorithm for each s = 1, 2, ... k-2 we number x edges of the left half of $T_{2,k}$. For s = k - 1 we number the remaining edges of the left half of $T_{2,k}$. For all s we number from left to right using first unused integers.

Algorithm:

Step 1: For s = 1 number x edges in the left-most part of the (k - 1) edge level by 1, 2, ..., x.

Step 2: For s = 2, 3, ..., k - 3 perform.

(i) t = s. Number the mothers and their sisters, say y_s, of the edges numbered for s = t - 1.

(ii) Number $x - y_s$ unnumbered edges of the (k - 1) edge level. If in the (k - 1) edge level only $x - y_s - z_s$ edges unnumbered, so number these edges and z_s unnumbered edges in the left most part of the (k - 2) edge level and so on.

Step 3: For s = k - 2 perform:

(i) Number the mothers and their sisters, say y_{k-2}, of the edges numbered for s = k - 3.

(ii) Number $x - y_{k-2} - 1$ unnumbered edges of the (k-1) edge level. If in the (k - 1) edge level only $x - y_{k-2} - 1 - z_{k-2}$ edges unnumbered, so number these edges and z_{k-2} unnumbered edges in the left most part of the (k - 2) edge level and so on.

(iii) Number $e_{3,3}$ (if $e_{3,3}$ is still unnumbered) with $x(k - 2)$.

Step 4: For s = k - 1 perform:

(i) Number $e_{1,1}$.

(ii) Number $e_{3,4}$ (if $e_{3,4}$ is still unnumbered).

(iii) Number the mothers and their sisters of the edges numbered for s = k - 2.

(iv) Number the remaining edges at will.

We show now that the edge-numbering f produced by our algorithm satisfies $W_f(T_{2,k}) = \lceil (2^k - 3)/(2k - 3) \rceil = x$.

For $k \leq 6$, $W_f(T_{2,k})$ can be verified directly.

For y_i, the number of mothers and their sisters of edges numbered for s = i by our algorithm, we get

$$y_i \leq \lceil x/2 \rceil + (i - 1) < x - 2 \quad \text{for } k \geq 7. \qquad (1)$$

(If a sister (no mother) of edge level j is numbered, the next number get an edge of edge level at most j - 1)

Assume we have a vertex v with $D_f(v) = |f(e) - f(e')| > x$. (2)

Clearly by our algorithm and by (1), e and e' are not sisters. Suppose e is the mother of e'. Define s(e) the value of s in our algorithm where the edge e is numbered.

<u>Lemma 1</u>: $|s(e) - s(e')| = r \in \{0,1\}$ (3)

Proof: If the edge e will be numbered after e' is numbered, we are through. So e will be numbered before e' is numbered. Let e" the sister of e and T_e and $T_{e''}$ the subtrees of $T_{2,k}$ induced by e and e".
(see Fig. 1)

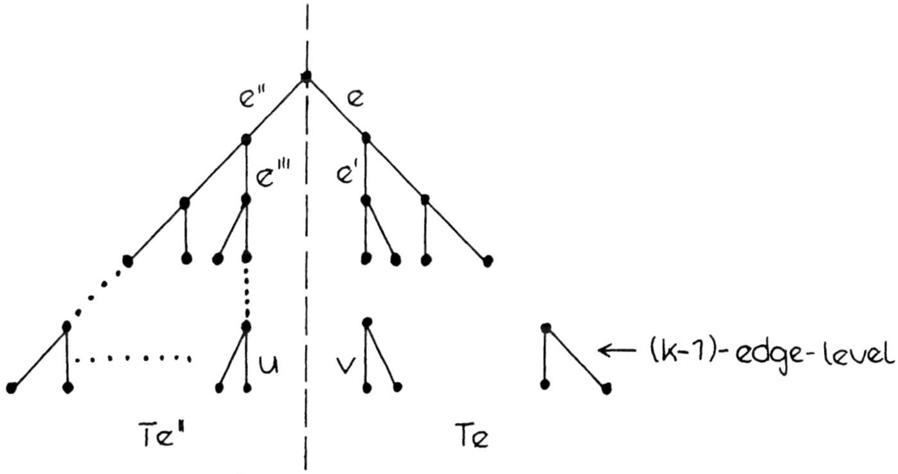

Fig. 1

Let u the right most edge of $T_{e''}$ of edge level k - 1 of $T_{2,k}$ and v the left most edge of T_e of edge level k - 1 of $T_{2,k}$. Clearly by our algorithm and by (1), $s(v) - s(u) = t_1 \in \{0,1\}$ and consequently step by step of our algorithm

$$s(e') - s(e''') = t_2 \in \{0,1\} . \qquad (4)$$

We have also by our algorithm $s(e) = s(e'') = s(e''') + t_3$, $t_3 \in \{0,1\}$. If $t_3 = 0$, then $s(e''') = s(e')$ and it follow (3). If $t_3 = 1$, we derive from (4) $s(e') - s(e) = t_4 \in \{-1,0\}$ and it follows (3).

Remark: Take notice that Lemma 1 is also correct for the special numbered edges $e_{3,3}$ and $e_{3,4}$.

For the incident edges $e_{k-i,1}$ and $e_{k-i-1,1}$, first numbered for s = i and s = i + 1, we have

$$f(e_{k-i-1,1}) - f(e_{k-i,1}) = x \ , \ i = 1,2, \ldots (k-2).$$

Let e be an edge numbered for s = i, e' an edge numbered for s = i + 1, $2 \leq i + 1 < k - 1$. Since for every pair mother and sister of edges numbered for s = i + 1 from left to right we have numbered for s = i at least the both daughters of the mother in the same order. Hence

$$f(e') - f(e) \leq x \text{ in contradiction with (2).}$$

For s = k - 2 we number in the order $e_{2,1}, e_{2,2}, \ldots, e_{3,3}, \ldots$

For s = k - 1 we number in the order $e_{1,1}, e_{3,4}, \ldots$.

Since $e_{3,4}$ is daughter of $e_{2,2}$ and sister of $e_{3,3}$ and the daughters of $e_{3,4}$ are numbered either for s = k - 2 after $e_{2,2}$ or for s = k - 1 the edge $e_{3,4}$ can't be one of the edges e and e' of (2).

Since $e_{3,3}$ is daughter of $e_{2,2}$ and sister of $e_{3,4}$ and the daughters of $e_{3,3}$ are numbered either for s = k - 2 or for s = k - 1 (than the edge $e_{3,3}$ are numbered last for s = k-2) the edge $e_{3,3}$ can't be one of the edges e and e' of (2).

With the same argument as above for the other edges numbered for s = k - 2 and s = k - 1 we get the contradiction to (2).

This concludes the proof of the theorem 2.

References

[1] Grünwald, N., K. Weber: On optimal pair sequences. J. Inf. Process. Cybern. EIK 25 (1989) 3, 75-86

[2] Harper, L.H.: Optimal assignments of numbers to vertices. J. Soc. Indust. Appl. Math. 12 (1964), 131-135.

[3] Hwang, F.K., I.C. Lagarias: Minimum range sequences of all k-subsets of a set. Discrete Math. 19 (1977), 257-264.

[4] Mitchison, G., R. Durbin: Optimal numberings of an N x N array. SIAM Algebraic Discrete Methods 7 (1986), 571 - 582.

On Independent Vertices and Edges of a Graph

I. Gutman, Kragujevac, Yugoslavia

Abstract

Let the numbers of k-element sets of independent vertices and edges of a graph G be denoted by n(G,k) and m(G,k), respectively. Some properties of the numbers n(G,k) and m(G,k) are outlined. In particular, it is shown that the circuits are the only connected graphs for which the equality n(G,k) = m(G,k) is satisfied for all values of k.

In this paper we consider finite graphs without loops and multiple edges. Let G be such a graph; its numbers of vertices and edges will be denoted by n and m, respectively.

The number of ways in which $k \geq 2$ independent (i.e. mutually nonadjacent) vertices can be selected in G is denoted by n(G,k). Hence n(G,k) is the number of distinct k-element sets of independent vertices of the graph G. It is both consistent and convenient to define n(G,0) = 1 and n(G,1) = n.

In an analogous manner the number of distinct k-element sets of independent (i.e. mutually nonincident) edges of G is denoted by m(G,k). Furthermore, m(G,0) = 1 and m(G,1) = m.

In the present paper we are interested in the relations between the numbers n(G,k) and m(G,k). It should be obvious that graphs with few edges have large n(G,k) and small m(G,k) values whereas graphs with many edges have small n(G,k) and large m(G,k) values. In line with this observation are also the following two elementary results.

R. Bodendiek, R. Henn (Eds.)
Topics in Combinatorics and Graph Theory

Lemma 1. Let G be a graph possessing a pair of nonadjacent vertices v_r and v_s. Let $G+e_{rs}$ be the graph obtained from G when the vertices v_r and v_s are connected by a new edge. Then for all $k \geq 0$, $n(G+e_{rs},k) \leq n(G,k)$ and $m(G+e_{rs},k) \geq m(G,k)$. For at least one value of k the above inequalities are strict; e. g. the first inequality is strict for $k = 2$, the second for $k = 1$.

Lemma 2. Denote by K_n the complete graph on n vertices and by $\bar{K}_n$ its complement. Let G be a graph with n vertices. Then for all $k \geq 0$, $n(\bar{K}_n,k) \geq n(G,k) \geq n(K_n,k)$ and $m(\bar{K}_n,k) \leq m(G,k) \leq m(K_n,k)$. Equality for all values of k is achieved only if either $G = K_n$ or $G = \bar{K}_n$.

We note in passing that

$$n(\bar{K}_n,k) = \frac{n!}{k!(n-k)!} \qquad \text{for} \quad k \geq 0$$

$$n(K_n,k) = \begin{cases} n & \text{for} \quad k = 1 \\ 0 & \text{for} \quad k > 1 \end{cases}$$

$$m(\bar{K}_n,k) = 0 \qquad \text{for} \quad k \geq 1$$

$$m(K_n,k) = \frac{n!}{k!2^k(n-2k)!}$$

From Lemma 2 we see that there exist graphs for which $n(G,k) > m(G, k)$ holds for all $k \geq 1$ and that there exist graphs for which $n(G,k) < m(G,k)$ holds for all $k \geq 1$. Somewhere "between" the graphs K_n and $\bar{K}_n$ there may exist graphs for which $n(G,k) = m(G,k)$ would hold for all $k \geq 1$. Indeed, it is very easy to find such graphs:

Theorem 1a. Denote by C_n the circuit with n vertices. Then the equalities $n(C_n,k) = m(C_n,k)$ are satisfied for all $k \geq 0$.

Proof. It is sufficient to observe that $m(G,k) = n(L(G),k)$ where L(G) stands for the line graph of G. Then Theorem 1a is a consequence of $L(C_n) = C_n$. ∎

From the above proof is evident that the relations $n(G,k) = m(G,k)$ are obeyed also if all components of the graph G are circuits.

It is somewhat less immediate to see that the reverse of Theorem 1a is also true, at least for connected graphs. We namely have the following result.

Theorem 1b. Let G be a connected graph. If the equalities $n(G,k) = m(G,k)$ are satisfied for all $k \geq 0$, then G is a circuit.

In order to prove Theorem 1b we need some preparations. The basic properties of the numbers $n(G,k)$ and $m(G,k)$ have been established in several previously published works [1 - 5]. We first quote without proof three of these results.

Lemma 3. Let G be a graph and v_r its vertex. Denote by N_r the set of vertices of G containing v_r and its first neighbors. Then for $k \geq 1$, $n(G,k) = n(G-v_r,k) + n(G-N_r,k-1)$.

Lemma 4. Let G be a graph and v_r and v_s its two adjacent vertices. Denote by e_{rs} the edge connecting v_r and v_s. Then for $k \geq 1$, $m(G,k) = m(G-e_{rs},k) + m(G-v_r-v_s,k-1)$.

Lemma 5 [1,3]. Denote by P_n the path with n vertices. If F is an acyclic graph (= forest) with n vertices then for all $k \geq 0$, $n(F,k) \geq n(P_n,k)$ and $m(F,k) \leq m(P_n,k)$. Equality for all values of k is achieved only if $F = P_n$.

Lemma 6. If G is a connected graph and $n(G,k) = m(G,k)$ holds for all $k \geq 0$, then G is unicyclic.

Proof. Setting $k = 1$ we obtain $n = m$. A connected graph with equal numbers of vertices and edges is unicyclic. ∎

Lemma 7. For $k \geq 1$ an $n \geq 5$,

$$n(C_n,k) = n(C_{n-1},k) + n(C_{n-2},k-1). \qquad (1)$$

Proof. The validity of (1) for $k = 1$ is immediate. It is also easy to check that (1) holds for $n = 5$. Therefore we may restrict ourselfs to the case $k \geq 2$, $n \geq 6$.

Apply Lemma 3 to any vertex of C_n. Then $n(C_n,k) = n(P_{n-1},k) + (P_{n-3},k-1)$. By Lemma 3 we further have $n(P_{n-1},k) = n(P_{n-2},k) + n(P_{n-3},k-1)$ and $n(P_{n-3},k-1) = n(P_{n-4},k-1) + n(P_{n-5},k-2)$. This implies $n(C_n,k) = [n(P_{n-2},k) + n(P_{n-4},k-1)] + [n(P_{n-3},k-1) + n(P_{n-5},k-2)]$ and Eq. (1) follows by means of one more application of Lemma 3. ∎

Lemma 8. If U is a unicyclic graph with n vertices then $m(U,k) \le m(C_n,k)$ holds for all $k \ge 0$. Equality is achieved only if $U = C_n$.

Proof. Every unicyclic graph U possesses an edge e_{rs} belonging to the circuit of U. Then by Lemma 4, $m(U,k) = m(U-e_{rs},k) + m(U-v_r-v_s,k-1)$. Besides $U-e_{rs}$ and $U-v_r-v_s$ are acyclic graphs with n and n-2 vertices, respectively. Because of Lemma 5, $m(U-e_{rs},k) \le m(P_n,k)$ and $m(U-v_r-v_s,k-1) \le m(P_{n-2},k-1)$ implying $m(U,k) \le m(P_n,k) + m(P_{n-2},k-1)$. On the other hand, in accordance with Lemma 4, $m(P_n,k) + m(P_{n-2},k) = m(C_n,k)$.

It is easily seen that C_n is the only unicyclic graph having the property $U-e_{rs} = P_n$ and $U-v_r-v_s = P_{n-2}$. Therefore if $U \ne C_n$ then at least one of the inequalities given in Lemma 8 is strict.

Lemma 9. If U is a unicyclic graph with n vertices then $n(U,k) \ge n(C_n,k)$ holds for all $k \ge 0$. Equality is achieved only if $U = C_n$.

Proof follows by induction on the number of vertices of U. By direct checking we verify Lemma 9 for n = 3 (one unicyclic graph) and n = 4 (three unicyclic graphs).

Assume now that Lemma 9 is true for graphs with n-1 and n-2 vertices. We have to show that then Lemma 9 is true for graphs with n vertices.

We have to distinguish between the following cases.

Case 1: $U = C_n$. Then Lemma 9 holds in a trivial manner.

Case 2: $U \ne C_n$ and U contains an isolated vertex v_r. Then by Lemma 3, $n(U,k) = n(U-v_r,k) + n(U-v_r,k-1)$. By the induction hypothesis, $n(U-v_r,k) \ge n(C_{n-1},k)$ and $n(U-v_r,k-1) \ge n(C_{n-1},k-1)$.
From Lemma 7 we further have $n(C_{n-1},k-1) = n(C_{n-2},k-1) + n(C_{n-3},k-2) \ge n(C_{n-2},k-1)$ and therefore $n(U,k) \ge n(C_{n-1},k) + n(C_{n-2},k-1) = n(C_n,k)$. The latter inequality is strict at least for k = 2.

Thus Lemma 9 holds in the case 2.

Case 3: $U \neq C_n$ and U does not contain isolated vertices. Then U necessarily possesses a vertex v_r of degree one. Let the unique neighbor of v_r be denoted by v_s.

Subcase 3a: The vertex v_s does not belong to the circuit of U. Then by applying Lemma 3 we arrive at

$$n(U,k) = n(U-v_r,k) + n(U-v_r-v_s,k-1). \qquad (2)$$

Note that $U-v_r$ and $U-v_r-v_s$ are unicyclic graphs with n-1 and n-2 vertices, respectively. The induction hypothesis yields then $n(U-v_r,k) \geq n(C_{n-1},k)$ and $n(U-v_r-v_s,k-1) \geq n(C_{n-2},k-1)$ and because of Eq. (1), $n(U,k) \geq n(C_n,k)$. The induction hypothesis implies that for some values of k the latter inequality is strict.

Thus Lemma 9 holds in the subcase 3a.

Subcase 3b: The vertex v_s belongs to the circuit of U. Then (2) holds but now $U-v_r$ is a unicyclic graph with n-1 vertices whereas $U-v_r-v_s$ is an acyclic graph with n-2 vertices. Because of the induction hypothesis, $n(U-v_r,k) \geq n(C_n,k)$ and because of Lemma 5, $n(U-v_r-v_s,k-1) \geq n(P_{n-2},k-1)$. Connecting the two terminal vertices of P_{n-2} by a new edge one obtains C_{n-2} and then from Lemma 1 it follows that $n(P_{n-2},k-1) \geq n(C_{n-2},k-1)$. Consequently, we again have that $n(U,k) \geq n(C_{n-1},k) + n(C_{n-2},k-1) = n(C_n,k)$ with a strict inequality at least for $k = 2$.

Thus Lemma 9 holds also in the subcase 3b.

This completes the proof of Lemma 9. ■

It is worth noting that although Lemmas 8 and 9 have fully analogous forms, their proofs are quite different.

Proof of Theorem 1b. Combining Lemmas 8 and 9 with Theorem 1a we conclude that for any unicyclic graph U with n vertices,

$$m(U,k) \leq m(C_n,k) = n(C_n,k) \leq n(U,k).$$

If $U \neq C_n$ then for some values of k the above inequalities are strict. This means that no unicyclic graph other than C_n has the property $n(U,k) = m(U,k)$ for all $k \geq 0$. On the other hand in Lemma 6 we established that no connected graph other than unicyclic can have the required property. ■

Theorems 1a and 1b can be combined as follows.

Theorem 1. Let G be a connected graph. Then the equalities n(G,k) = = m(G,k) are satisfied for all $k \geq 0$ if and only if G is a circuit.

The extension of Theorem 1 to disconnected graphs is by no means straightforward. We wish to end this paper by formulating a pertinent hypothesis.

Conjecture. The equalities n(G,k) = m(G,k) are satisfied for all k ≥ 0 if and only if every component of G is a circuit.

References

[1] *I.Gutman, Partial ordering of forests according to their characteristic polynomials, in: A.Hajnal and V.T.Sós (Eds.), Combinatorics, North-Holland, Amsterdam 1978, pp. 429-436.*

[2] *I.Gutman, Graphs with greatest number of matchings, Publ.Inst.Math. (Beograd)* **27** *(1980) 67-76; Correction, Ibid.* **32** *(1982) 61-63.*

[3] *I.Gutman and F.Harary, Generalizations of the matching polynomial, Utilitas Math.* **24** *(1983) 97-106.*

[4] *I.Gutman, Graphs with maximum and minimum independence numbers, Publ. Inst. Math (Beograd)* **34** *(1983) 73-79.*

The Outerthickness & Outercoarseness of Graphs I. The Complete Graph & The n-Cube

R. K. Guy, Alberta, Canada
R. J. Nowakowski, Alberta, Canada

Many of the results of this paper were announced without proof in [9]; a parallel paper, discussing the complete bipartite graph, will appear in [11]. Some results were also obtained independently by L.S. Mel'nikov and presented at the 6th Hungarian Combinatorics Colloquium in Eger on 81-07-10.

A graph is **planar** if it can be imbedded in the plane (or sphere). The complement of the graph in such an imbedding is a collection of open cells. If there is an imbedding with all the vertices of the graph on the boundary of a single cell, the graph is said to be **outerplanar**.

Halin [12, and see also 7] has shown that a graph is outerplanar just if it does *not* contain a subgraph homeomorphic to the complete graph, K_4, or to the complete bipartite graph, $K_{2,3}$.

The **thickness** of a graph is the least number of planar graphs having the original graph for their union; i.e. the least number of parts in an edge-partition of the graph, with the edges in each part forming a planar graph.

The **outerthickness** $\theta_o(G)$, of a graph G is defined similarly, but with "outerplanar" in place of "planar".

The outerthickness of the complete graph

It can be seen, from Euler's formula, for example, that an outerplanar graph on n vertices contains at most $2n-3$ edges. The **complete graph**, K_n, has each of its vertices joined to every other, and contains $\frac{1}{2}n(n-1)$ edges. The outerthickness of the complete graph is therefore at least

$$\left\lceil \frac{n(n-1)}{2(2n-3)} \right\rceil = \left\lceil \frac{n}{4} + \frac{n}{4(2n-3)} \right\rceil = \left\lceil \frac{n+1}{4} \right\rceil \text{ for } n > 1.$$

Theorem 1 The outerthickness of K_n is $\theta_o(K_n) = \left\lceil \frac{n+1}{4} \right\rceil$, except that $\theta_o(K_7) = 3$.

Proof: From the preceeding remark, the outerthickness is at least $\lceil (n+1)/4 \rceil$. Figure 1 displays equality for $n \lesssim 6$.

R. Bodendiek, R. Henn (Eds.)
Topics in Combinatorics and Graph Theory

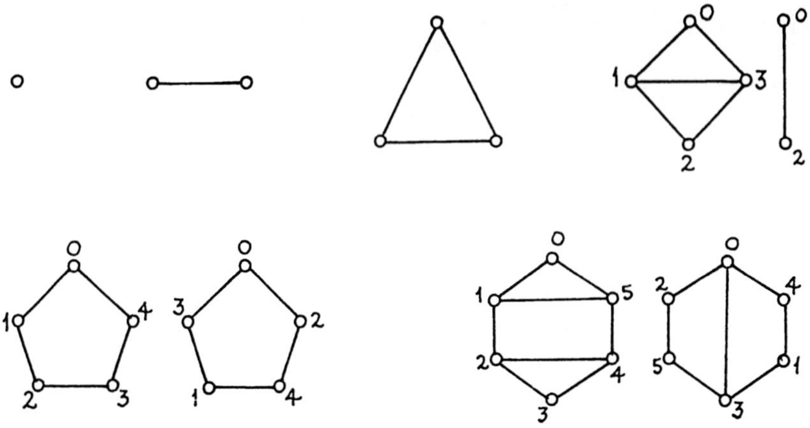

Figure 1. $\theta_0(K_n) = \lceil (n+1)/4 \rceil$, $1 \leq n \leq 6$

To see that $\theta_0(K_7) = 3$ and not 2, note that K_7 can be partitioned into three Hamilton circuits (Figure 2), but it is not the union of two outerplanar graphs. An outerplanar graph on 7 vertices contains at most $2 \times 7 - 3 = 11$ edges, so one of the two would contain 11 edges (with 10 in the other). An outerplanar graph with 7 vertices and 11 edges is isomorphic to one of the four graphs in Figure 3 (compare Fig. 11.9 on p. 108 of [13]). In each case the complementary graph contains the non-outerplanar graph $K_{2,3}$ (vertices 1 & 2 each joined to 4, 5 & 6).

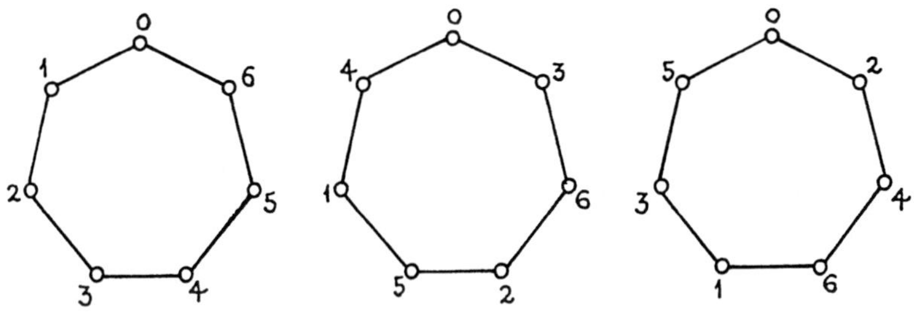

Figure 2. Three hamilton circuits whose union is K_7.

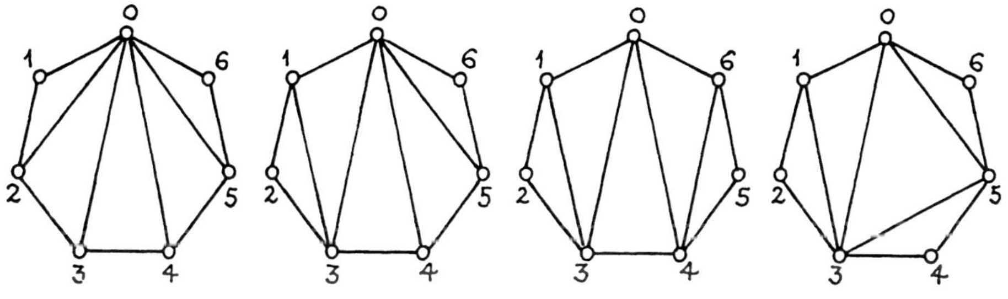

Figure 3. The four maximal outerplanar graphs on 7 vertices.

We first prove the general result when $n = 4r$, exhibiting $r + 1$ outerplanar graphs whose union is K_{4r}. We then successively modify these graphs to accommodate the additional edges of K_n when $n = 4r + 1$, $4r + 2$ or $4r + 3$. Note that $\lceil (n+1)/4 \rceil = r + 1$ in each case.

Label the vertices of K_{4r} with the residue classes modulo $4r$, represented by $0, 1, \ldots, 4r - 1$. First construct the graph of Figure 4, in which the vertex 0 is joined to each of the r vertices $r, r+1, \ldots, 2r-1$, these last forming a path labelled boustrophedon, i.e. label every other vertex from right to left, and then intermediately from left to right. To clarify the alternating nature of the labelling, and for our later convenience, the graph is redrawn in the two cases, $r = 2m$ and $2m + 1$ in Figures 5.0 and 5.1.

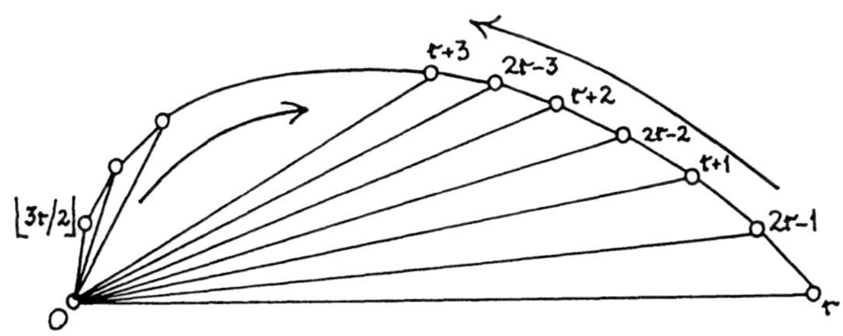

Figure 4. Boustrophedon labelling.

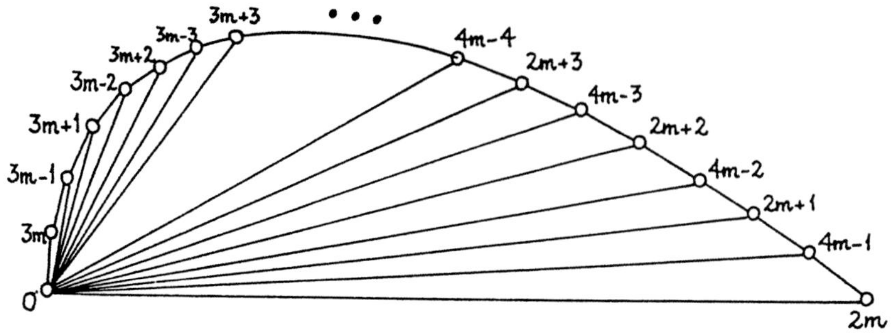

Figure 5.0. $r = 2m$.

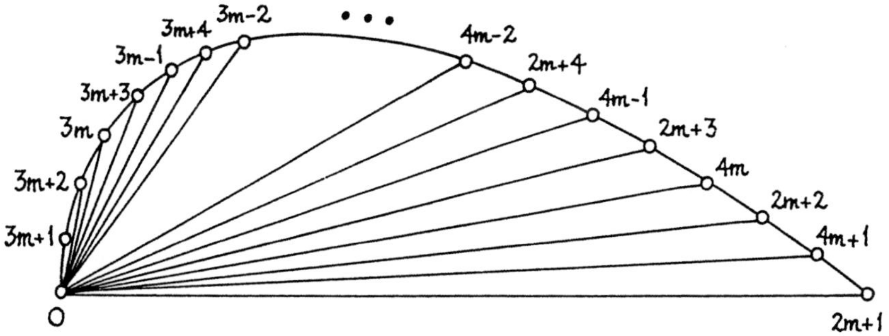

Figure 5.1. $r = 2m + 1$

Next, take three copies of Figure 4, with the labels increased by r, $2r$, and $3r$ modulo $4r$. On identifying vertices having the same label, 0, r, $2r$, and $3r$, these combine with Figure 4 to make **graph zero**, depicted in Figure 6.
Graph g, $0 \leq g \leq r-1$, is a copy of graph zero with each vertex label increased by g mod $4r$. **Graph** r comprises two sets of r disjoint edges $(g, g+2r)$ and $(g+r, g+3r)$. $0 \leq g \leq r-1$, which correspond to the diagonals of the squares in the graphs g. We refer to these sets respectively as the **dexter** and **sinister diameters.**

To check that each of the $2r(4r-1)$ edges of K_{4r} occurs in one of the graphs g, $0 \leq g \leq r$, visualize the edges as chords of a circle on which the $4r$ vertices 0, 1, ..., $4r-1$ form a regular $4r$-gon. Define the **length** of a chord as the angle subtended at the centre of the circle, measured in units of size $\pi/2r$.

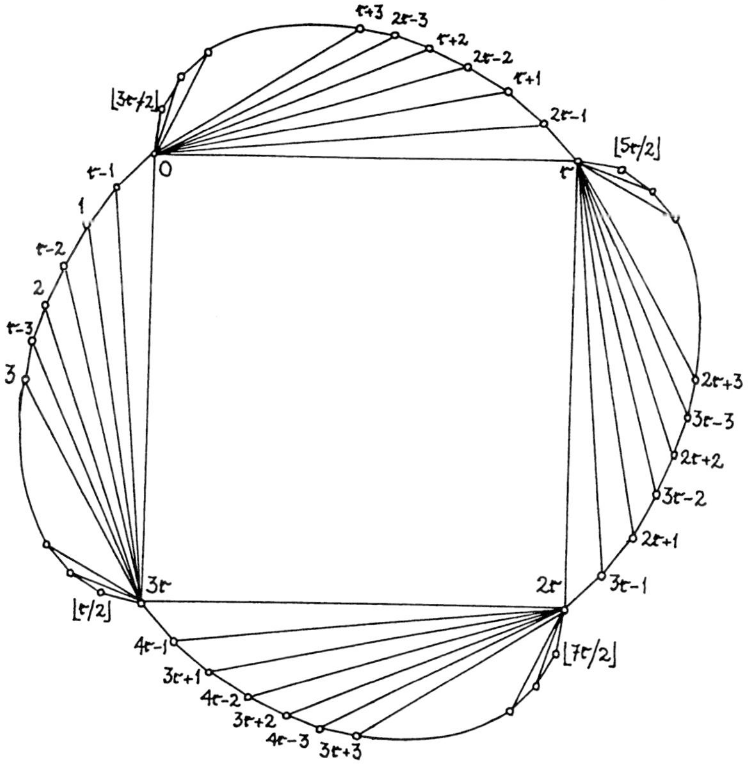

Figure 6. Graph zero.

The **short chords**, of lengths 1, 2, ..., $r-1$, occur in the boustrophedon labelled paths. The **middle chords**, of length r, form the sides of the squares in graphs 0 to $r-1$. The **long chords**, of lengths $r+1$, $r+2$, ..., $2r-1$, are in the "fans", (Fig. 4) and have one end at a corner of a square. The **diameters**, of length $2r$, are all in graph r.

To show that $\theta_0(K_{4r+1}) = r+1$, it is easy to adjoin a single vertex, labelled $4r$, to graph r and connect it to the ends of the $2r$ diameters, forming a "windmill", which has also been called the "friendship" graph. However, to simplify the case $n = 4r+2$, we describe a different procedure.

Two new vertices, A and B, each of valence two, are adjoined to each of the graphs 0, 1, ..., $r-1$, by joining them to pairs of adjacent vertices, as indicated in Table 1. For example, in Figures 5.0 and 5.1, A is adjacent to the two vertices immediately above 0. The remaining $2r$ adjacencies for A and B, as well as the edge AB itself, occur in graph r, exhibited in Figure 7. We consider the cases $r = 2m$ and $r = 2m + 1$ separately.

$r=2m$					$r=2m+1$			
A		B		graph	A		B	
3m–1	3m	5m–1	5m	0	3m+1	3m+2	5m+2	5m+3
7m	7m+1	m	m+1	1	7m+4	7m+5	m+1	m+2
3m+1	3m+2	5m+1	5m+2	2	3m+3	3m+4	5m+4	5m+5
7m+2	7m+3	m+2	m+3	3	7m+6	7m+7	m+3	m+4
.	.	.	.	.	.	.	.	.
.	.	.	.	.	.	.	.	.
m–4	m–3	3m–4	3m–3	2m–3	m–4	m–3	3m–3	3m–2
5m–3	5m–2	7m–3	7m–2	2m–2	5m–1	5m	7m	7m+1
m–2	m–1	3m–2	3m–1	2m–1	m–2	m–1	3m–1	3m
				2m	5m+1	5m+2	7m+2	7m+3

Table 1. Pairs of vertices to which A and B are joined in graphs, 0, 1, ..., $r-1$.

Note that all $2r+1$ diameters are still accommodated in graph r: each of the $4r+2$ vertices is the end of such an edge. Up to r of these edges can be transferred to graphs 0, 1, ..., $r-1$ as either dexter or sinister diagonals.

Indeed, this needs to be done if we are to prove theorem 1 in the case $n = 4r+3$. We have been able to exhibit sets of $r+1$ outerplanar graphs, whose union is K_{4r+3} for $2 \leq r \leq 9$. It is hoped to give a general pattern in the second part of this paper [11].

The outerthickness of the n-cube

Label the 2^n vertices of the n-cube with n-tuples of zeros and ones. Two vertices are connected just if their n-tuples differ in exactly one coordinate. The vertices are n-valent, and the number of edges is $n \cdot 2^{n-1}$. The n-cube is bipartite, the parts being determined by the parity of the number of ones in the vertex labels. To determine the outerthickness we use the following lemma.

Lemma. An outerplanar bipartite graph on $p+q$ vertices, $1 \leq p \leq q$, contains at most $2p+q-2$ edges.

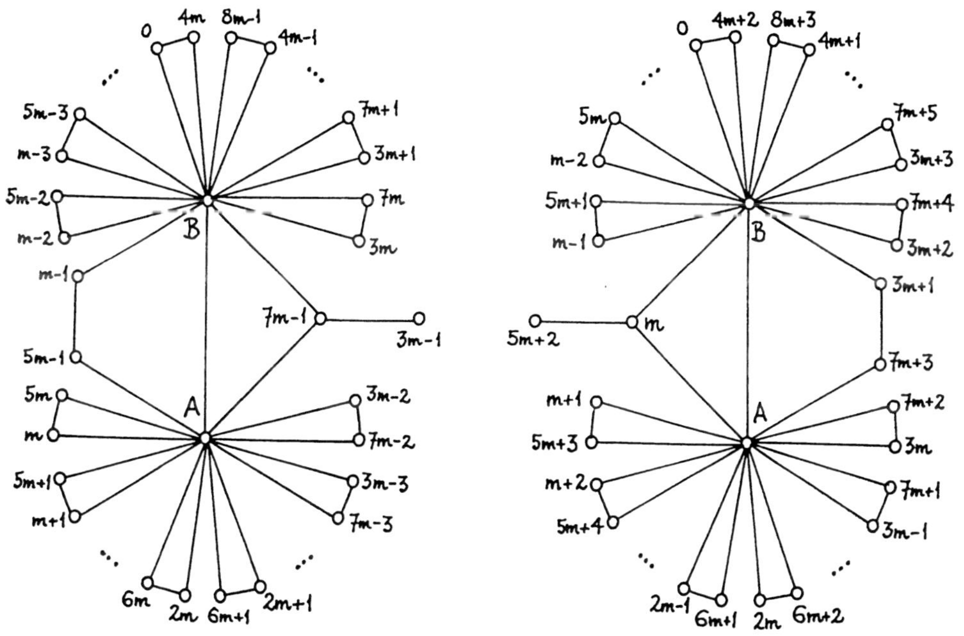

$r = 2m$ $\qquad$ $r = 2m + 1$

Figure 7 Graph r in the two cases $r = 2m$ and $r = 2m + 1$.

Proof: We may assume that the graph is connected, else we could insert an additional edge joining two components, since we are assuming that each part contains at least one vertex. If there is no circuit, the graph is a tree and contains $p + q - 1 \leq 2p + q - 2$ edges, with equality just if $p = 1$ and the graph is the star, $K_{1,q}$. The circuits of a bipartite graph are even. Suppose that the largest circuit contains $2k$ vertices, k of them "odd" vertices and k "even". Since the graph is outerplanar, we may assume that there is no vertex inside the circuit. No vertex outside the circuit is joined to two non-adjacent vertices of the circuit, either directly or via longer paths, since these two paths and the two parts of the circuit would form a homeomorph of $K_{2,3}$. Nor does a path connect two adjacent vertices of the circuit, else a longer circuit would be formed.

So the graph consists of a $2k$-circuit, possibly with some inside chords, and with blocks adjacent to various vertices. If $k = 2$, the 4-circuit contains no chords and just $4 = 2 \times 2 + 2 - 2 = 2k + k - 2$ edges. Consider a $2k$-circuit, $k > 2$, with the maximum number of inside chords. The "shortest" inside chord cuts off one "odd" and one "even" vertex, each of valence two. The subgraph obtained by omitting these two vertices is a $(2k-2)$-circuit, which, by induction, may be assumed to contain at most $(2k-2)+(k-1)-2=3k-5$ edges, so that the original $2k$-circuit contained at most $3k-5+3=2k+k-2$ edges.

Identify the k odd vertices in the circuit and also the k even vertices, reducing it to a single edge. Assume, inductively, that the result holds for this contracted graph on $(p-k+1)+(q-k+1)$ vertices, i.e. that there are at most $2(p-k+1)+(q-k+1)-2$ edges. Then, before contraction, the graph contained at most

$$2(p-k+1)+(q-k+1)-2+(3k-2)-1=2p+q-2 \text{ edges.} \quad \blacksquare$$

Theorem 2. The outerthickness of the n-cube is $\theta_o(Q_n) = \left\lceil \frac{n+1}{3} \right\rceil$.

Proof: From the lemma $\theta_o(Q_n) \geq \dfrac{n\cdot 2^{n-1}}{3\cdot 2^{n-1}-2} = \left\lceil \dfrac{n+1}{3} \right\rceil$, since $\dfrac{n\cdot 2^{n-1}}{3\cdot 2^{n-1}-2} = \dfrac{n}{3} + \epsilon$, where $0<\epsilon<1/3$ for $n\geq 3$, and it's easily verified for $n=0, 1, 2$. We will show that equality holds.

It is convenient to define a 2^k-**ladder** as a (maximal outerplanar) subgraph of Q_k with 2^k vertices. For $k=0, 1, 2$ this is simply Q_k itself. For $k\geq 2$ a 2^k-ladder has 2^k-4 trivalent vertices and two pairs of adjacent bivalent vertices. It consists of 2^{k-1} parallel edges joined by two 2^{k-1}-paths, or a hamilton circuit equipped with a set of parallel chords. It may not be immediately clear that such a ladder always exists, but this can be seen inductively by forming a 2^{k+1}-ladder from two copies of a 2^k-ladder, which has a pair of adjacent bivalent vertices, labelled r and s, say, where r and s are k-tuples of zeros and ones which differ in just one coordinate. Prefix the k-tuple labels of one copy with 0 and the other with 1. Join the two 2^k-ladders by edges joining $0r$ to $1r$, and $0s$ to $1s$.

The truth of the theorem for $0\leq n\leq 5$ is displayed in Fig. 8. Q_0 is a single vertex, labelled with the empty set. Q_1 consists of two copies of Q_0, one labelled by prefixing zero (to the empty set), the other by taking the *complementary* label, and then joining them by an edge. Q_2 comprises two copies of Q_1, labelled by prefixing the labels of one by 0, the other by 1, and then joining these two 2-ladders to form a 4-ladder. So far we have just one graph, which is outerplanar in each case.

Q_3 consists of two copies of Q_2, whose labels we prefix by 0 and by 1, and four edges joining corresponding points. Two of these (shown broken in Fig. 8) are used to form an 8-ladder from the two 4-ladders, and two form a second graph: $\theta_0(Q_3) = 2$.

Q_4 is formed from two copies of Q_3. Prefix the labels of one copy with 0, and take the labels of the other to be the *complements* of those of the first. There are also 2^3 edges joining corresponding vertices of the two copies. The first outerplanar graph now consists of two 8-ladders. The second contains four edges, each of type $0r - 0s$ or $1t - 1u$, where r, s and t, u are pairs of "adjacent" labels of Q_3. This second graph can accommodate the 2^3 joining edges, forming four 4-paths of type

$$1r —— 0r —— 0s —— 1s \quad \text{or} \quad 0t —— 1t —— 1u —— 0u.$$

Notice that the interior, bivalent vertices of the paths in the second graph are the bivalent vertices at the ends of the ladders in the first graph, while the monovalent ends of the paths are the trivalent vertices of the ladders, so that in the union, each vertex has valence four.

Q_5 is formed from two copies of Q_4, with vertex labels prefixed by 0 and by 1, together with 2^4 joining edges. The first outerplanar graph contains four 8-ladders. The second contains eight 4-paths (i.e., paths with 4 vertices), which may be joined by the 2^4 edges to form four 8-ladders also. $\theta_0(K_5) = 2$.

Graphs 1 and 2 for Q_6 each contain 2 copies of the corresponding graphs for Q_5, with vertex labels prefixed by 0 and by 1, that is, eight 2^3-ladders. These are joined by 2^3 edges (2^2 pairs of type $0r$ —— $1r$, $0s$ —— $1s$, where r, s are adjacent vertices of Q_5; exactly analogous to the broken lines in Fig. 8) to form four 2^4-ladders. The remaining $2^{6-1} - 2^3$ $2^3 - 2^1$ joining edges form a new graph 3.

Graphs 1 and 2 for Q_7 each contain two copies of graphs 1 and 2 for Q_6, the vertex labels for one copy obtained by prefixing those of Q_6 with zero, and for the other copy by taking the complementary labels (which will each start with 1). Graph 3 contains two copies of the 2^4 edges of graph 3 for Q_6, labelled in the same way, together with the 2^{7-1} joining edges, forming 2^5 4-paths.

Similarly, graphs 1 and 2 for Q_8 contain sixteen 2^4-ladders, while in graph 3, 2^6 4-paths are joined by 2^{8-1} edges to form 2^5 2^3-ladders.

The construction continues inductively. For Q_{3m} $(m \geq 2)$, graphs 1 and 2 contain 2^{2m-2} 2^{m+2}-ladders; graph 3 contains 2^{2m-1} 2^{m+1}-ladders; graph i, $3 \leq i \leq m$, contains 2^{2m+i-4} 2^{m+4-i}-ladders; and graph $m+1$ consists of 2^{3m-2} separate edges.

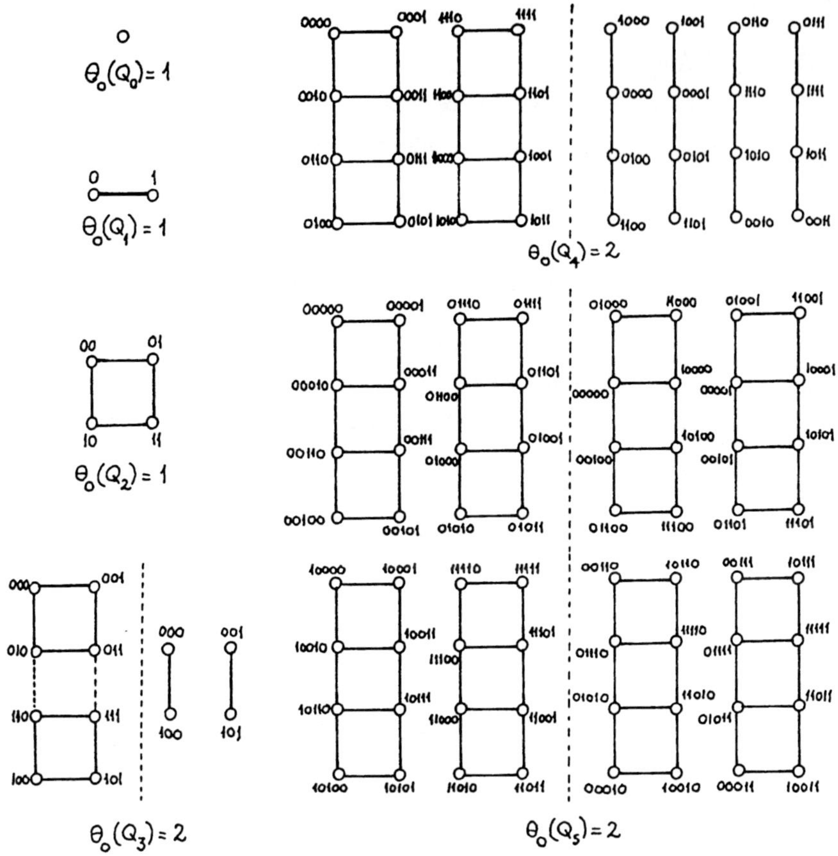

Figure 8. $\theta_0(K_n) = \lceil (n+1)/3 \rceil,\ 0 \leq n \leq 5$.

For $Q_{3m+1} (m \geq 2)$, graphs 1 and 2 comprise 2^{2m-1} 2^{m+2}-ladders; graph i, $3 \leq i \leq m$, contains 2^{2m+i-3} 2^{m+4-i}-ladders; and graph $m+1$ consists of 2^{3m-1} 4-paths. Notice that the two copies of Q_{3m} are labelled by prefixing 0 in the first case and by taking complements in the second. In the other two stages of the induction it suffices to prefix 0 to one copy and 1 to the second.

For $Q_{3m+2} (m \geq 1)$, graphs 1 and 2 comprise 2^{2m} 2^{m+2}-ladders; and graph i, $3 \leq i \leq m+1$ contains 2^{2m+i-2} 2^{m+4-i}-ladders. This completes the induction, the construction, and the proof of Theorem 2. ∎

The outercoarseness of the complete graph.

The outercoarseness of a graph is a dual concept to that of outerthickness:

it is the maximum number of parts in an edge-partition of the graph, each part to be non-outerplanar.

Theorem 3 (Beineke [2, Theorem 13]) The outercoarseness of the complete graph is

$$\xi_o(K_n) = \lfloor n(n-1)/12 \rfloor .$$

Proof: A non outerplanar graph contains at least six edges, and, if exactly six, it is isomorphic to K_4 or $K_{2,3}$. So

$$\xi_o(K_n) \leq \lfloor \tbinom{n}{2}/6 \rfloor = \lfloor n(n-1)/12 \rfloor .$$

We show that equality holds for all n by displaying the edges of K_n in a triangular array (Fig. 9), the upper half of its adjacency matrix. This is partitioned into either 2 by 3 rectangles, representing isomorphs of $K_{2,3}$, or $1+2+3$ triangles, representing K_4. Unused edges are shown as shaded squares, always less than 6 in number. In K_9, the two regions marked A together form a copy of K_4. The 12 by $n-12$ rectangle in the final inductive step can be exactly packed with $2n-24$ 2 by 3 rectangles if $n \geq 14$. Sufficient rows of four 2 by 3 rectangles (as in K_{14}) are used, except that, if n is odd, there is one row of six 3 by 2 rectangles (as in K_{15}). ∎

The outercoarseness of the n-cube.

A non-outerplanar subgraph of the n-cube contains at least eight edges. In Figure 10 three homeomorphs of $K_{2,3}$ are shown, one with eight edges and the others with nine, Fig. 10(c) being essentially four-dimensional. Figure 10(d) is a homeomorph of K_4, which must contain at least nine edges; it contains Fig. 10(a) as a subgraph, so is not of use for our present purpose.

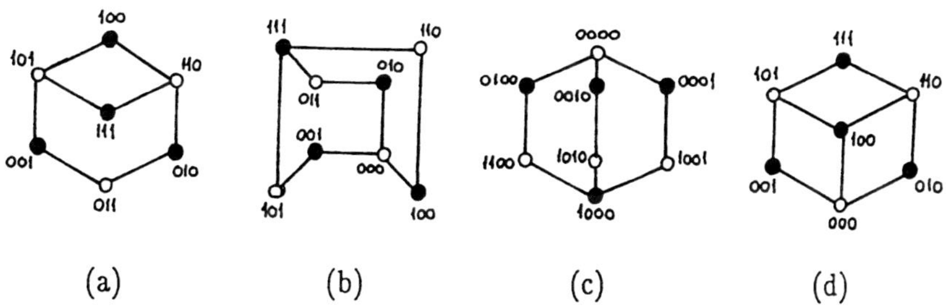

Figure 10. Non-outerplanar subgraphs of the n-cube.

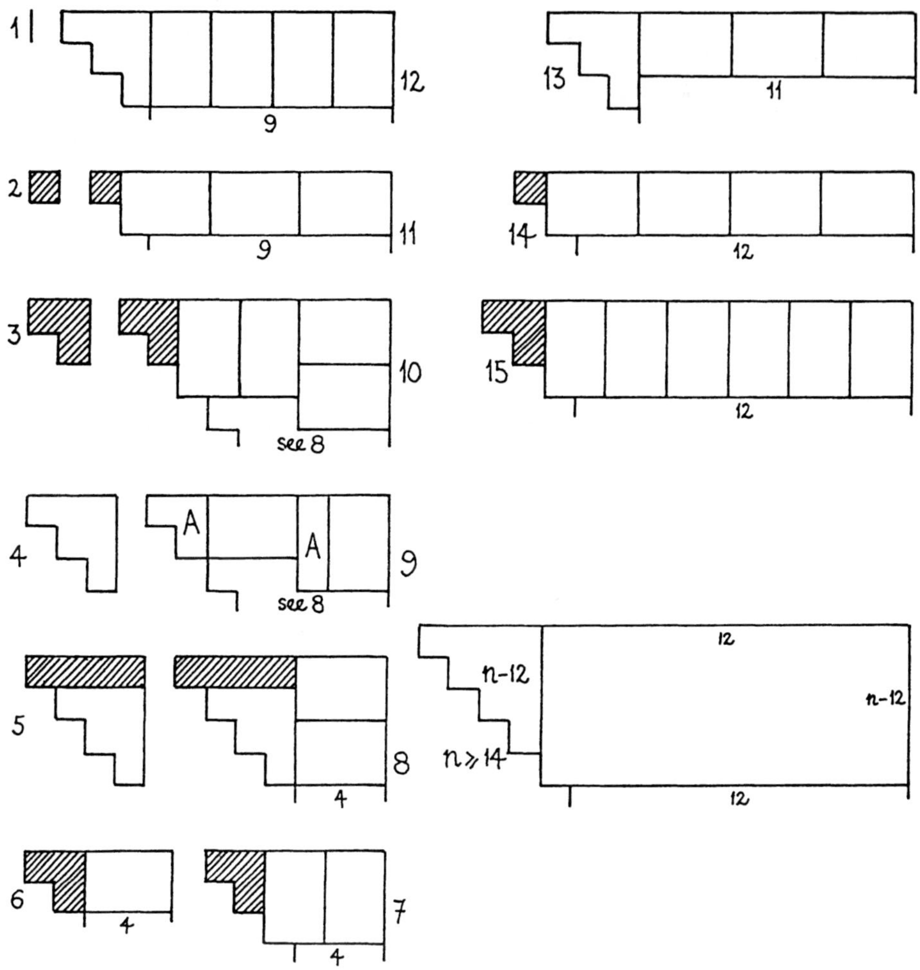

Figure 9. The outercoarseness of the complete graph is $\lfloor n(n-1)/12 \rfloor$

Theorem 4. The outercoarseness of the n-cube, $\xi_o(Q_n)$ is bounded by

$$(0.96n-1.15)2^{n-4} < \xi_o(Q_n) \leq n \cdot 2^{n-4}$$

except that the lower bound may not hold for $n=2$, 5, 6 and 9.

Proof: The upper bound is immediate, since Q_n contains $n \cdot 2^{n-1}$ edges and a non-outerplanar subgraph contains at least eight edges. Clearly $\xi_o(Q_n)=0$ for $n=0$, 1, 2, and $\xi_o(Q_3)=1$. To see that $\xi_o(Q_4)=3$, note that two copies of Fig. 10(a), with vertex labels prefixed by 0 and by 1, are subgraphs of Q_4, and that their complement contains Fig. 10(c). On the other hand, at least two vertices of each non-outerplanar graph are trivalent, whereas those of Q_4 are tetravalent, so that some edges do not belong to non-outerplanar graphs, and $\xi_o(Q_4) < 32/8$. Notice that the unused edges of Q_4 comprise the square whose vertices are $xy11$ and three edges $x101$, $x110$, $x011$.

So Q_5 can be partitioned into 2 copies of the three non-outerplanar subgraphs of Q_4, 1 copy of Q_3, 3 of Q_2 and 6 of Q_1, which contain $2 \times 3 + 1 = 7$ edge-disjoint non-outerplanar graphs. By counting valences, it can be shown that $7 \leq \xi_o(Q_5) \leq 9$.

Similarly, Q_6 can be partitioned into 2^2 copies of the three non-outerplanar subgraphs, one copy of Q_4, 3 of Q_3 and 6 of Q_2, and $\xi_o(Q_6) \geq 2^2 3 + 3 + 3 \times 1 = 18$.

Generally, Q_n can be partitioned into 2^{n-4} copies of the three subgraphs, one copy of Q_{n-2}, 3 of Q_{n-3} and 6 of Q_{n-4}, so that

$$\xi_o(Q_n) \geq 3 \cdot 2^{n-4} + \xi_o(Q_{n-2}) + 3\xi_o(Q_{n-3}) + 6\xi(Q_{n-4}).$$

This gives the lower bounds

	18	46	105	238	543	1213	2655	5806	12624	27193	58260	124477
for $n =$	6	7	8	9	10	11	12	13	14	15	16	17

It can be checked that the inequality of the theorem holds apart from the exceptions noted. Assume inductively that $\xi_o(Q_n) > (0.96n - 1.15)2^{n-4}$ for $n \geq 14$. Then

$$\begin{aligned}
\xi_o(Q_{n+1}) > 3 \cdot 2^{n-3} &+ (0.96(n-1) - 1.15)2^{n-5} + 3(0.96(n-2) - 1.15)2^{n-6} \\
&+ 6(0.96(n-3) - 1.15)2^{n-7} \\
&= (7.68n - 1.52)2^{n-6} \\
&= (7.68(n+1) - 9.2)2^{n-6} \\
&= (0.96(n+1) - 1.15)2^{n-3}
\end{aligned}$$

and the result follows by induction. ∎

REFERENCES

1. J. Battle, F. Harary & Y. Kodama, Every planar graph with nine points has a nonplanar complement, *Bull. Amer. Math. Soc.* 68(1962) 569-571; MR27#5248.

2. Lowell W. Beineke, A survey of packings and coverings in graphs, in G. Chartrand & S.F. Kapoor (eds.) The Many Facets of Graph Theory, (Proc. Conf. Kalamazoo, 1968) Springer Lecture Notes 110(1969) 45-53; MR41#1562.

3. L.W. Beineke & R.K. Guy, The coarseness of $K_{m,n}$, *Canad. J. Math.* 21(1969) 1086-1096; MR41#6727.

4. L.W. Beineke & F. Harary, The thickness of the complete graph, *Canad. J. Math.* 17(1965), 850-859; MR29#1636.

5. L.W. Beineke, F. Harary & J.W. Moon, On the thickness of the complete bipartite graph, *Proc. Cambridge Philos. Soc.*, 60(1964) 1-5; MR28#1611.

6. G. Chartrand, D. Geller & S. Hedetniemi, Graphs with forbidden subgraphs *J. Combin. Theory Ser. B.* 10(1971) 12-41; MR44#2645.

7. G. Chartrand & F. Harary, Planar permutation graphs, *Ann. Inst. H. Poincaré Sec. B(N.S.)* 3(1967) 433-438; MR37#2626.

8. D.P. Geller, Outerplanar graphs [see F. Harary, Graph Theory, Addison-Wesley, 1969, pp. 108 & 245].

9. Richard K. Guy, Outerthickness and outercoarseness of graphs, in T.P. McDonough & V. Mavron, Combinatorics, (Proc. British Combin. Conf., Aberystwyth 1973) CUP 1974 57-60.

10. R.K. Guy & L.W. Beineke, The coarseness of the complete graph, *Canad. J. Math.* 20(1966) 888-894; MR37#2633.

11. Richard K. Guy & Richard J. Nowakowski, The outercoarseness of graphs, in Festschrift for Klaus Wagner's 80th birthday, 1990 (to appear).

12. R. Halin, Über einen graphentheoretischen Basisbegriff und seine Anwendung auf Färbensprobleme, Doctoral thesis, Köln, 1962.

13. Frank Harary, Graph Theory, Addison-Wesley, Reading MA, 1969.

14. M. Kleinert, Die Dicke des n-dimensionalen Würfel-Graphen, *J. Combin. Theory* 3(1967) 10-15; MR35#2776.

15. Jean Mayer, Decomposition de K_{16} en trois graphes planaires, *J. Combin. Theory Ser. B*, 13(1972) 71.

16. John Mitchem, The point-outercoarseness of complete n-partite graphs, *Compositio Math.* 26(1973) 101-110; MR52#2967.

17. John Mitchem, The point-outerthickness of complete n-partite graphs, *Compositio Math.* 29(1974) 55-61; MR50#6928.

18. W.T. Tutte, On the non-biplanar character of the complete 9-graph, Canad. *Math. Bull.* 6(1963) 319-330; MR28#2535.

On Some Graphic Aspects of Addition Theorems

Y. O. Hamidoune, Paris, France

Abstract.

During the last fifty years , Davenport, Mann, Kempermann and many others proved some inequalities relating the cardinality of the sum of two sets in a group to the cardinalities of the original sets. This subject is known as Additive Group Theory.

Almost all these questions have a natural interpretation as statements about the connectivity of Cayley directed graphs.

The Cauchy-Davenport theorem is equivalent to the fact that a Cayley graph with prime order has connectivity equal to the outdegree. Chowla's Theorem says that the same property holds for Cayley graphs on $\mathbb{Z}_n$ defined by subsets of $\mathbb{Z}_n^*$.

We was able to prove a common generalization to several addition theorems with the following graphic interpretation.

Let G be a finite Abelian group and $B \subset G \setminus 0$. There exists a nonnull subgroup H of G such that any cut separating two elements of H has cardinality at least $|B|$. In other words the local connectivities inside H are optimal.

This subgroup exists also when the group is not Abelian and $B=B^{-1}$.

R. Bodendiek, R. Henn (Eds.)
Topics in Combinatorics and Graph Theory

I. The Cauchy-Davenport Theorem

Let G be a group. The subgroup generated by a subset S will be denoted by $\langle S\rangle$. Let A and B be two subsets of G . As usual we write

$AB=\{xy \mid x\in A \text{ and } y\in B\}$.

In the Abelian case we use the notation $A+B$.

The following result was proved by Cauchy [C] and rediscovered by Davenport [Da1] . In [Da2] , Davenport mentioned that this result is in fact due to Cauchy . It is called now the Cauchy-Davenport Theorem:

Theorem I.1 (the Cauchy-Davenport Theorem) .

Let p be a prime number , and let A and B be two proper subsets of $\mathbb{Z}_p$ such that $A+B\neq\mathbb{Z}_p$. Then $|A+B|\geq|A|+|B|-1$.

We explain below the motivations of Cauchy and Davenport .

As a piece of his famous proof that every natural number is a sum of four squares, Lagrange proved the following result.

Theorem I.2 (Lagrange 1770).

For all $x,y\in\mathbb{Z}_p$ such that $y\neq 0$, there are $u,v\in\mathbb{Z}_p$ such that $x=tu^2+v^2$.

We quote from Cauchy (Mémoire sur les nombres):

"Ce théorème [Theorem I.2] a quelque analogie avec un autre plus simple et dont voici l'énoncé [$(\mathbb{Z}_p\backslash 0,.)$ is a group] . Ce dernier théorème ayant été démontré très simplement par M. Legendre dans son Introduction à la théorie des nombres, l'analogie m'a porté à croire qu'il devait exister une démonstration semblable du théorème de M. Lagrange. Le travail que j'avais entrepris dans le dessein de parvenir à cette démonstration m'a conduit, en outre, à la démonstration de quelques autres théorèmes que je vais exposer successivement, et dont plusieurs me paraissent nouveaux."

In order to see that Theorem II.2 is a consequence Of Cauchy Theorem , we observe that Theorem II.2 is equivalent to the relation $(\mathbb{Z}_p)^2+(\mathbb{Z}_p)^2=\mathbb{Z}_p$.

By Cauchy Theorem this is the case unless

$|(\mathbb{Z}_p)^2+(\mathbb{Z}_p)^2| \geq ((p+1)/2)+((p+1)/2)-1=p$.

But in this case also we have $(\mathbb{Z}_p)^2+(\mathbb{Z}_p)^2=\mathbb{Z}_p$. □

In order to explain the motivation of Davenport we need a brief introduction to Additive number theory .

Let $A\subset\mathbb{N}$. The Schnirelman density of A is $\sigma(A)=\inf_n \frac{|A\cap[1,n]|}{n}$.

The following result was proved by Khintchin in 1932.

Theorem I.3 (cf. [H-R]).

If $0\in A\cap B$ **and** $\sigma(A)=\sigma(B)$, **then** $\sigma(A+B)\geq \mathrm{Min}(1,\sigma(A)+\sigma(B))$.

Khintchin conjectured that Theorem I.3 remains true when $\sigma(A)\neq\sigma(B)$. This conjecture was known as the $\alpha+\beta$-hypothesis.

Davenport's result is the p-analogue to the $\alpha+\beta$- conjecture.

Note that the $\alpha+\beta$ - conjecture was solved by Mann :

Theorem I.4 (Mann , 1942, cf. [H-R, M]).

If $0\in A\cap B$, **then** $\sigma(A+B)\geq \mathrm{Min}(1,\sigma(A)+\sigma(B))$.

A well known consequence of Mann's theorem is the fact that a set B such that $0\in B$ and $\sigma(B)>0$ is a base , i.e. $\exists k$ such that $kB=B+B+\ldots+B=\mathbb{N}$

II. Some generalizations to the Cauchy-Davenport Theorem.

Cauchy mentioned that his result is not true for composite numbers.

A first extension to composite numbers was proposed by Chowla : ($\mathbb{Z}_n^*$ denotes the set of units of the ring $\mathbb{Z}_n$) .

Theoerm II.1 (Chowla [Ch]) .

Let $A\subset\mathbb{Z}_n$ and let $B\subset\mathbb{Z}_n^*$. Then either $A\cup(A+B)=\mathbb{Z}_n$ or $|A\cup(A+B)|\geq|A|+|B|$.

It is easy to show that Theorem II.1 implies the Cauchy-Davenport Theorem .

We have a method for proving the above results based on the following property.

Let G be a group and let B be a finite generating proper

subset such that $G \not\subset B \cup \{1\}$. Let $\mathcal{F}$ be the set of finite subsets X of G such that $X \cup XB \neq G$.

We set $\varkappa(G,B)=\mathrm{Min}\{ |XB\setminus X| \; ; \; X\in\mathcal{F} \}$.

Theorem II.2 [H2] .

There exists a finite subgroup M distinct from G such that $\varkappa(G,B)=|MB\setminus M|$ or $\varkappa(G,B)=|BM\setminus M|$.

Theorem II.2 is given in a graphic language in [H2, Theorem 3.1] .

Using the Theorem II.2 , we can give short proofs for Cauchy–Davenport and Chowla theorems.

Let us show how to deduce the Cauchy-Davenport Theorem :

1. The validity of the Cauchy–Davenport theorem for all subsets B containing 0 implies its validity for all B.

Suppose $A+B\neq\mathbb{Z}_p$ and let $x\in-B$, $B'=B-x$. We have $0\in B'$. Also $A+B'\neq\mathbb{Z}_p$.

Therefore $|A+B'|\geq|A|+|B'|-1$, and hence $|A+B|\geq|A|+|B|-1$.

2. We assume $0\in B$ and $A+B\neq\mathbb{Z}_p$. In particular $\mathbb{Z}_p\not\subset B\cup\{0\}$. By Theorem II.2, $\varkappa(\mathbb{Z}_p,B)=|B+\{0\}\setminus\{0\}|=|B|-1$.

By the definition of $\varkappa$, we have $|(A+B)\setminus A|\geq\varkappa(\mathbb{Z}_p,B)=|B|-1$.

Since $0\in B$, we have $|A+B)|-|A|=|(A+B)\setminus A|\geq|B|-1$.□

Shepherdson proved the following nice extension for the Cauchy–Davenport Theorem.

Theorem II.3 (Shepherdson [S]).

Let G be an Abelian group and let A,B be two finite subsets of G such that $B\subset A$. Then either $|A\cup(A+B)|\geq|A|+|B|$ or there is $b\in B$ such that $\langle b\rangle\subset A\cup(A+B)$.

We obtained the following generalization to the above results.

Theorem II.4 [H3] .

Let G be an Abelian group and let B be a finite subset of G . Then there exists $b\in B$ with the following property. For

every $A\subset G$ **such that** $A\cap\langle b\rangle\neq\emptyset$, **either** $|A\cup(A+B)|\geq|A|+|B|$ **or** $\langle b\rangle\subset A\cup(A+B)$.

Remarks.

1. Theorem II.4 implies Chowla's Theorem and hence the Cauchy-Davenport Theorem.

Let $B\subset\mathbb{Z}_n^*$. Hence $\langle b\rangle=\mathbb{Z}_n$, for any $b\in B$. By the above theorem either $A\cup(A+B)\supset\mathbb{Z}_n$ or $|A\cup(A+B)|\geq|A|+|B|$, which is Chowla's theorem. ∎

2. This result implies Shepherdson's Theorem. Suppose $B\subset A$. It follows that $A\cap\langle b\rangle\neq\emptyset$, for every $b\in B$. By Theorem II.4, either $|A\cup(A+B)|\geq|A|+|B|$ or $\langle b\rangle\subset A\cup(A+B)$.

3. The element b in Theorem II.4 does not depend on A , while in the statement of Shepherdson the b depends on both A and B .

4. Shepherdson's Theorem does not imply the Cauchy-Davenport Theorem. We observe also that Shepherdson's Theorem holds when $0\in A$ even if $B\not\subset A$. This remark is not clear from Theorem II.3 .

I do not know if this result is true for nonabelian groups. I can prove it only if B is symmetric .

Theorem II.5 [H3] .

Let G **be a group and let** B **be a finite subset of** G **such that** $B=B^{-1}$**. Then there exists** $b\in B$ **with the following property. For every** $A\subset G$ **such that** $A\cap\langle b\rangle\neq\emptyset$, **either** $|A\cup(AB)|\geq|A|+|B|$ **or** $\langle b\rangle\subset A\cup(AB)$.

III KNESER'S THEOREM .

A very nice addition theorem has been proved by M. Kneser.

Theorem III.1 (Kneser [K,M]).

Let G **be an Abelian group and let** A **and** B **be two finite nonvoid subsets of** G . **There exists a subgroup** H **of** G , **such that** $A+H+B=A+B$ **and** $|A+B|\geq|A|+|B|-|H|$.

The following generalization of Kneser's Theorem is due to Diderrich .

Theorem III.2 (Diderrich [Di]).

Let G be a group and let A and B be two finite nonvoid subsets of G such the elements of B commute . There exists an Abelian subgroup H of G such that AHB=AB and $|AB| \geq |A|+|B|-|H|$.

Diderrich conjectured the validity of Theorem (1.D) for all groups. But Olson found a counter-example to this conjecture [O].

Actually Diderrich Theorem can be obtained as a corollary of Kneser's Theorem , which is apparently weaker [H4] .

I think that Diderrich conjecture is true if $B=B^{-1}$. Probably Theorem II.5 could be used to prove this case.

IV. Cayley graphs .

By a graph we mean a directed graph , loops are allowed .

Let G be a group and $S \subset G$. The Cayley graph of G with respect to S is $Cay(G,S)=(G,U)$, where $U=\{(x,y) : y \in xS\}$.

In this part we we shall consider only the case where the groups and graphs are finite . In this case, some Addition theorems are equivalent to bounds on the number of disjoint paths joining the unity to some other points in a Cayley graph.

The connectivity of a graph X is

$\kappa(X)=\text{Min}\ \{ |T| : X \setminus T \text{ is not strongly connected or } |X\setminus T|=1\}$.

We shall denote by $\tau(x,y)$ the maximum number of openly disjoint paths from x to y

The following statement is equivalent to the Cauchy-Davenport Theorem.

Theorem IV.1 [H1] .

Let $S \subset \mathbb{Z}_p$. then $\kappa(Cay(\mathbb{Z}_p,S))=|S\setminus 0|$.

Actually this statement is equivalent to the Cauchy-Davenport Theorem

A proof of this result based on the Cauchy-Davenport Theorem requires some easy lemmas on the connectivity.

Let us show how to obtain the Cauchy-Davenport Theorem from Theorem IV.1.

We may assume that $0\in B$, since the statement is invariant under translation. Suppose that $A+B\neq \mathbb{Z}_p$. The Cauchy-Davenport Thoorom is equivalent to

$|A+B|-|A|=|(A+B)\setminus A|\geq |B|-1$ (since $0\in B$).

But $\Gamma^+(A)\setminus A$ is a cutset. By Theorem 4.1 and the definition of the connectivity we have $|(A+B)\setminus A|=|\Gamma^+(A)\setminus A|\geq \varkappa = |B\setminus 0|=|B|-1$.

Chowla's Theorem can be formulated in the following equivalent form.

Theorem IV.2 (Chowla Theorem).

Let $S\subset \mathbb{Z}_n^*$. Then $\varkappa(Cay(\mathbb{Z}_n,S))=|S|$.

This result follows easily from a property of atoms proved in [H2].

Theorem II.4 has the following graphic form.

Theorem IV.3.

Let G be a finite Abelian group and $B\subset G\setminus 0$. $\exists b\in B$, such that $\forall x,y\in\langle b\rangle$, $\tau(x,y)=|B|$.

In order to show that Theorem IV.3 implies Theorem II.4.

Suppose $A\cap\langle b\rangle\neq\emptyset$ and $\langle b\rangle\not\subset A\cup(A+B)=A\cup\Gamma^+(A)$. Let $a\in A\cap\langle b\rangle$ and $c\in\langle b\rangle\setminus(A\cup A+B)$. $\Gamma^+(A)\setminus A$ separates a from c. Since $\tau(a,c)=|B|$. We have $|\Gamma^+(A)\setminus A|\geq |B|$.

Bibliography.

[C]. A. Cauchy, Recherches sur les nombres, J. Ecole polytechnique 9(1813), 99-116.

[Ch]. I. Chowla, A theorem on the addition of residue classes: Application to the number $\Gamma(k)$ in Waring's problem. Proc. Indian Acad. Sci. 2(1935), 242-243.

[Cha] L. Chatrowski, A new generalization to the Davenport's Pilai's theorem on the Addition of residue classes, C.R. (Doklady), Acad. Sc. URSS XLV (1944), 315-317.

[Da1]. H. Davenport, On the addition of residue classes, J. London Math. Soc. 10(1935), 30-32.

[Da2]. H. Davenport, A historical note, J. London Math. Soc. 22 (1947), 100-101.

[Di] G.T. Diderrich, On Kneser's addition theorem in groups, Proc. Amer Math. Soc. (1973), 443-451.

[H1]. Y.O. Hamidoune, Sur les atomes d'un graphe orienté, C.R. Acad. Sc. Paris A 284 (1977), 1253-1256.

[H2]. Y.O. Hamidoune, On the connectivity of Cayley digraphs, Europ. J. Combinatorics 5 (1984), 309-312.

[H3]. Y.O. Hamidoune, On a subgroup related to some Addition theorems, Discrete Math. 55(1985),323-326.

[H4] Y.O. Hamidoune, On a subgroup contained in words with bounded length,

[H-R] H. Halberstam and K.F. Roth, "Sequences", Oxford 1966, second edition , Springer, 1983.

[K] M. Kneser, Abschäetzung der asymptotischen dichte von Summenmengen, Math. Z: 58(1953), 459-484.

[M]. H.B. Mann, Addition theorems : The addition theorems of group theory and number theory, Interscience, New York, 1965.

[O] J. E. Olson, On the symmetric difference of two sets in a group, Europ. J. Combinatorics 7 (1986), 43-54.

[S]. J.C. Shepherdson, on the addition of elements of a sequence, J. London Math. Soc. 22(1947), 85-88.

On the Circumference of Regular Polyhedral Graphs

J. Harant, Ilmenau, GDR
H.-J. Walther, Ilmenau, GDR

Introduction

Consider the class Γ of all 3-regular polyhedral graphs, that is, the class of the planar 3-connected graphs in which each vertex has degree 3. Let C be a circuit of a graph $G \in \Gamma$. If one removes from G the vertices of C and all edges incident with them, the resulting graph G-C disintegrates into connected components $K_1, K_2, \ldots, K_k$. If C is a hamiltonian circuit of G (that is a circuit containing all vertices), then G-C is obviously empty, but in all other cases, however, $k > 0$ holds. Let C^i be the set of those vertices of C having at least one neighbour in K_i. Since each vertex has the degree 3 in G, a vertex from C^i is incident with exactly one edge which does not belong to C, that is, this vertex is adjacent to one vertex from K_i. In addition, there exists for any vertex $x \in C$ at most one index i such that x is adjacent to one vertex of K_i. If one adds to K_i the vertices of C^i and also all edges whose one end-vertex belongs to C^i and the other one to K_i, then the resulting graph B_i is called a bridge of G over C. Calling also an edge whose two end-vertices belong to C, but not the edge itself, a bridge, then we see that to each vertex of C there exists exactly one bridge in which this vertex lies. A vertex belonging to both C and B_i is called a touch point of the bridge B_i. Let a bridge over a path be defined accordingly. Let $\Gamma(w)$ be the class of graphs G in in which there exists for any longest circuit C of G a bridge B over C with at least w touch points.

In accordance with [1], we understand by the shortness exponent $\sigma(\Delta)$ of a class Δ of graphs G

R. Bodendiek, R. Henn (Eds.)
Topics in Combinatorics and Graph Theory

$$\sigma(\Delta) := \liminf_{G\in\Delta} \frac{\log c(G)}{\log n(G)}$$

with c(G) being circumference of G, that means the length of a longest circuit of G, and n(G) the number of vertices of G.

In [5] it has been shown that $\sigma(\Gamma(w)) < 1$ for any integer $w > 0$.

In [3] it has been shown that $\sigma < 1$ holds even in the case that additionally only a finite number of types of faces is allowed, but the following problem remained open there:
Let $\Gamma^{p_1,p_2,\dots,p_r}$ (we assume $p_1 < p_2 < \dots < p_r$) be the class of the 3-regular polyhedral graphs for which it holds: To any elementary face f there exists a number p_i ($i \in \{1,2,\dots,r\}$) such that f forms a p_i-gon. The least number r has to be found such that there exist numbers $p_1,p_2,\dots,p_r$ with $\sigma(\Gamma^{p_1,p_2,\dots,p_r} \cap \Gamma(w)) < 1$.
To show that this least number r is equal to 2 we prove the

Theorem

$\sigma(\Gamma^{4,q} \cap \Gamma(w)) \leq \log 26/\log 27 < 1$
for any positive integer w and any odd integer $q \geq 341$.

Proof of the Theorem

We shall prove the theorem only in the case $q = 1 (\mathrm{mod}\ 4)$. The case $q = 3 (\mathrm{mod}\ 4)$ can be shown analogously.
Using suitable structures T_i we shall first construct a sequence $\{H_i\} \subset \Gamma^{4,q}$ such that each longest circuit of H_i has a bridge with at least $3(2^{i+1}-2)$ touch points. Finally, starting with $G_1 = H_i$ for some i with $3(2^{i+1}-2) \geq w$ we shall construct a sequence $\{G_j\} \subset \Gamma^{4,q} \cap \Gamma(w)$ with

$$\lim_{j\to\infty} \frac{\log c(G_j)}{\log n(G_j)} \leq \log 26/\log 27.$$

1. Construction of structures S of type (a,b,c)

Let G be a planar 3-connected cubic graph with the following two properties:

(i) There exists a vertex P incident with an (a+1)-gon, a (b+1)-gon, and a (c+1)-gon (a,b,c need not be pairewise distinct necessarily).

(ii) Any other elementary face is a 4-gon or a q-gon.

Let the structure S of type (a,b,c) be emerging from G by deleting the vertex P and maintaining the three edges originally incident with P as half-edges.
For example, the well-known Tutte-figure (see Fig. 1) is a structure of type (3,4,5) with q = 5, since in the interior we have only 4-gons and 5-gons and the number of boundary vertices is 3,4,5, resp.

Fig. 1

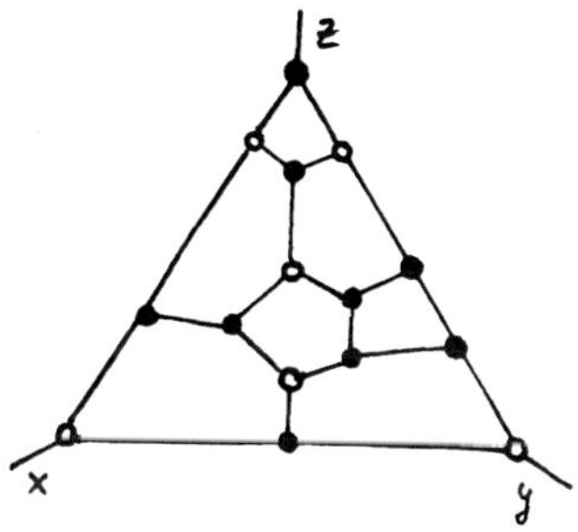

Obviously, a structure S of type (a,b,c) is also a structure of type (a',b',c'), whereas a',b',c' is a permutation of a,b,c.
In [2] , the following was proved:

Lemma 1

For any odd $q \geq 43$ there exists a structure S of type (a,b,c) containing at least one vertex P incident with three 4-gons and $a,b,c \in \{8,9,\ldots,q-8\}$ can be chosen arbitrarily. Furthermore, a longest path through S connecting any two of the three half-edges of S contains all vertices of S. The vertex P of S is called a black vertex, all the other vertices of S are called white ones.

Let us consider an operation Ω which generates a structure S' of type (a,b,c) out of a structure S of type (a,b,c) such that for the numbers $|S|$ and $|S'|$ of vertices of S and S', resp., the relation holds:

$$|S'| = |S| + 6q - 24 \qquad \text{(see Fig. 2).}$$

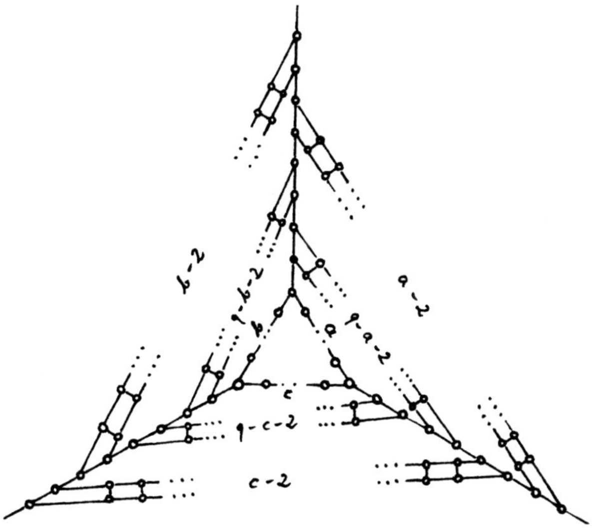

Fig. 2

S' arises from S by "bordering" the structure S of type (a,b,c) with the help of 6 "ladders" with q-a-2, a-2, q-b-2, b-2, q-c-2, c-2, resp., steps.

2. Construction of the sequence $\{H_i\}$

Now let $q \geqslant 431$ with $q \equiv 1 (\mathrm{mod}\ 4)$ be a fixed integer.
We choose a set $M = \{S_1, S_2, S_3, S_4, S_5\}$ of 5 fixed structures S_i of type $(8,8,q-63)$, $(q-63, q-330, \frac{q-81}{2})$, $(8,8,\frac{q+59}{2})$, $(\frac{q-65}{2}, q-70, 8)$, $(15, q-196, 40)$, resp..
Let $\alpha := |S_1| + |S_2| + 4|S_3| + 2|S_4| + S_5$.

We take a further fixed structure F of type (8,8,8) and construct, with the help of the operation Ω, a structure F' of type (8,8,8) with $|F'| > |F| + \alpha + 10$ by repeated "bordering".

Now, we choose a set M' of 8 fixed structures $R_1, \ldots, R_8$ of type (8,8,8), (8,8,q-18), (q-10,q-18,8), (q-18,8,q-24), (8,8,12), (8,q-18,q-15), (8,8,13), $(8,q-18,\frac{q-69}{4})$, resp., such that $|R_i| > |F'|$ (i = 1,2,...,8). Because of the Lemma 1 and the possibility of applying operation Ω such an M' can be found.

Now, we construct a structure W of type (30,18,q-49) by replacing each of the 9 black vertices of the Tutte-figure (see Fig. 1) by a suitable structure R M' according to Fig. 3.

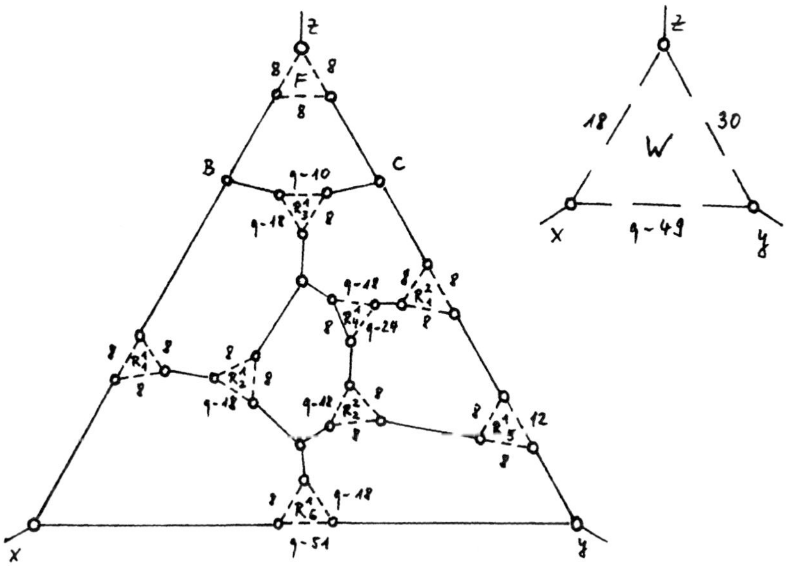

Fig. 3

Further we construct a structure V of type $(31,18,\frac{q-61}{4})$ by replacing each of the 9 black vertices of the Tutte-figure by a suitable structure $R' \in M'$ according to Fig. 4.

It is well-known that any (x,y)-path of the Tutte-figure T^0 (see Fig. 1) omits exactly one black vertex and vice versa: To any black vertex X of T^0 there exists an (x,y)-path of T^0 omitting X and only X (see [3], [4], [5]), (a (u,v)-path of a structure S, $u,v \in \{x,y,z\}$, is a longest u and v connecting path through the structure S).

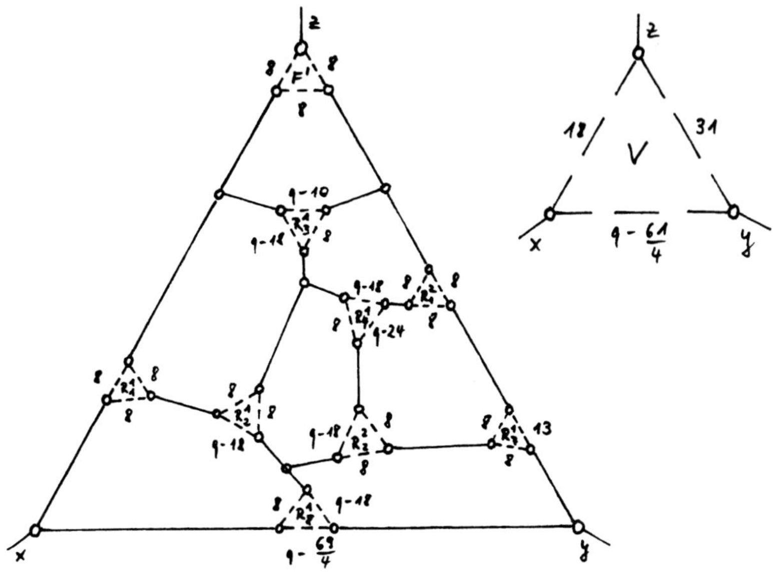

Fig. 4

Because of $|F| < |F'| < |S|$ for any $S \in M'$ it holds that an arbitrary (x,y)-path of W contains no vertex of the structure F but all other vertices of W and, an arbitrary (x,y)-path of V contains no vertex of structure F' but all other vertices of V.

Let V^1 and V^2 be two pieces of the structure V and let W^1 be a piece of the structure W. In accordance with Fig. 5 we combine these three structures to the new structure T_1 of type $(61,36,\frac{q-61}{2})$.

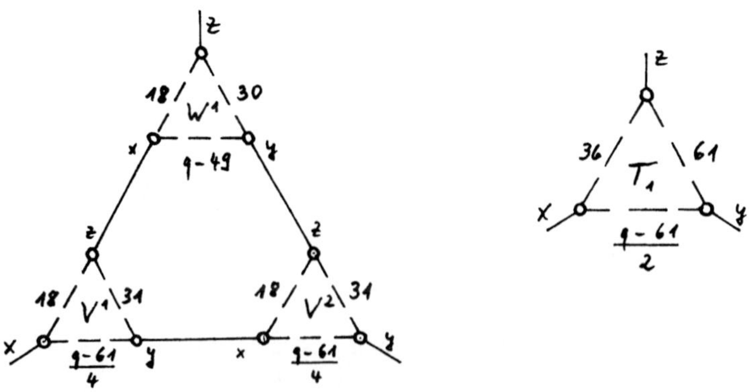

Fig. 5

The structure T_1 has the following properties:

1. A longest path through T_1 is an (x,y)-path of T_1 and each (x,y)-path of T_1 contains all vertices but no vertex of the structure F of W.
2. An (x,z)-path of T_1 contains all vertices but no vertex of the structure F' of V^1, and a (y,z)-path of T_1 contains all vertices but no vertex of the structure F' of V^2.
3. An (x,z)-path of T_1 as well as a (y,z)-path of T_1 contains at least $\alpha+10$ vertices less than an (x,y)-path of T_1.
4. A bridge over an (x,y)-path of T_1 containing the half-edge z has the two touch points B and C (see Fig. 3).
5. T_1 contains exactly 27 black vertices (see Lemma 1) and an (x,y)-path of T_1 omits the black vertex of the structure F (see Fig. 3).

As shown in Fig. 6, we construct the structure T_{i+1} $(i \geq 1)$ using 10 copies T_i^j, $j = 1,2,\ldots,10$ of T_i,

one copy S_1^1 of $S_1 \in M$,

one copy S_2^1 of $S_2 \in M$,

four copies S_3^j, $j = 1,2,3,4$ of $S_3 \in M$,

two copies S_4^j, $j = 1,2$ of $S_4 \in M$,

one copy S_5^1 of $S_5 \in M$, and

10 further white vertices; we have chosen $d = \frac{q-61}{2}$, $e = \frac{q+59}{2}$, $f = \frac{q-65}{2}$ for the sake of clarity.

T_i is of type $(61,36,\frac{q-61}{2})$, $i = 1,2,\ldots$ as can be seen easily.

<u>Lemma 2</u>

A longest path through T_i connecting two half-edges is an (x,y)-path of T_i, and any (x,y)-path of T_i contains at least $\alpha + 10$ vertices more than a (u,z)-path of T_i with $u \in \{x,y\}$.

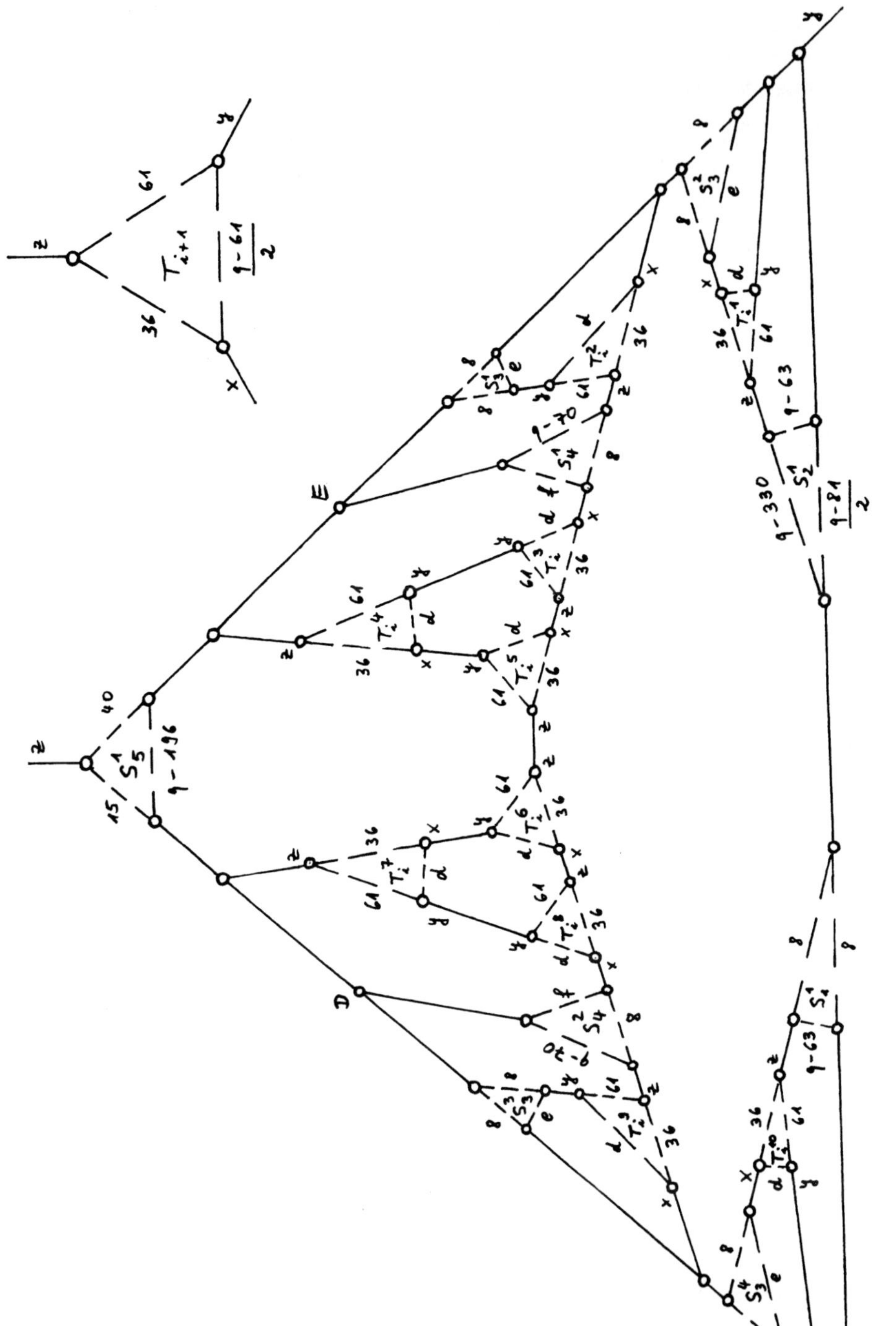

Fig. 6

Proof of Lemma 2 (by induction):

Because $|S| > \alpha + 10$ for any $S \in M'$ a (u,v)-path of T_1 ($u,v \in \{x,y,z\}$) contains more than $25(\alpha + 10)$ vertices (see Fig. 3,4,5). The same property holds for T_i, $i \geq 2$, trivially (see Fig. 6).

For i = 1 Lemma 2 is true because of property 3.
Let us consider an arbitrary longest path L through T_{i+1} connecting any two half-edges (see Fig. 6).
Except the ten copies of T_i the structure T_{i+1} possesses not more than $\alpha + 10$ vertices. For any two half-edges $u,v \in \{x,y,z\}$ there exists a path through T_{i+1} connecting u and v passing all ten copies of T_i, that means that a (u,v)-path of T_{i+1} passes all ten copies T_i^j, $j = 1,2,\ldots,10$ of T_i. Because the statement holds for $i \geq 1$ the longest path L uses as many as possible (x,y)-paths of the ten T_i^j. Let us denote by $|(u,v)^i|$ the length of a (u,v)-path of T_i.
How can L pass the substructure U' of T_{i+1} (see Fig. 7)?

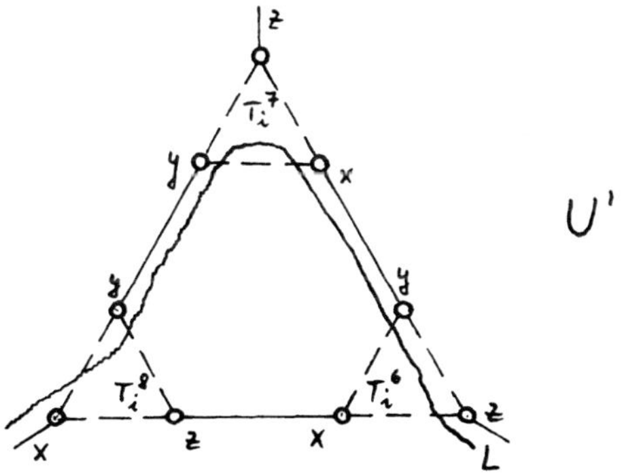

Fig. 7

It is impossible to get an (x,y)-path of all three T_i^j, j = 6,7,8, and the only possibility of getting two (x,y)-paths is shown in Fig. 7. For the substructure U" composed by T_i^3, T_i^4, T_i^5 we have the same situation.
Thus, it is easy to see that L is an (x,y)-path of T_{i+1} using eight (x,y)-paths and two (y,z)-paths of the ten T_i^j. On the other hand, an (u,z)-path of T_{i+1} ($u \in \{x,y\}$) uses only six (x,y)-paths of the ten T_i^j.

Hence, $|(x,y)^{i+1}| \geq 8|(x,y)^i| + 2|(x,z)^i|$ and for $u \in \{x,y\}$

$$|(u,z)^{i+1}| \leq 6|(x,y)^i| + 4 \max_{v\in\{x,y\}} |(v,z)^i| + \alpha + 10$$

and we have

$$|(x,y)^{i+1}| - |(u,z)^{i+1}| \geq 2(|(x,y)^i| - |(w,z)^i|) - \alpha - 10$$

$$\geq 2(\alpha+10) - \alpha - 10 = \alpha + 10 \text{ whereas } |(w,z)^i| := \max_{v\in\{x,y\}} |(v,z)^i| ,$$

and the statement is true for i+1.

Now, let us estimate the number b_i of touch points (belonging to T_i) of a bridge over an (x,v)-path of T_i containing the half-edge z.

Because of property 4 for T_1 we have $b_1 = 2$.
Following the course of an (x,y)-path L of T_{i+1} we can see that L contains neither the z-half-edge of T_i^7 nor of T_i^4.

Furthermore, L does not cross S_5^1, that means that any bridge over L containing z of T_{i+1} contains D and E as touch points and also the 2 b_i touch points of the bridges over the two (x,y)-paths of T_i^7 and T_i^4 containing the half-edges z of T_i^7 and T_i^4, resp. So we have $b_{i+1} = 2\,b_i + 2$, that means that $b_i = 2^{i+1}-2$.

In the following, we construct the sequences $\{H_i\}$ and $\{G_j\}$ of polyhedral graphs, which completes the proof of our theorem:
At first, we construct the polyhedral graph H_i as shown in Fig. 8 with the help of three copies T_i^j (j = 1,2,3) of T_i and two structures S' and S" of type (q-98, q-98, q-98) and $(\frac{q+57}{2}, \frac{q+57}{2}, \frac{q+57}{2})$, resp. S' and S" are fixed for all i.

Let w be a given integer. We choose i sufficiently large such that $3(2^{i+1}-2) \geq w$ and a longest circuit C through H_i contains none of the three z-half-edges.

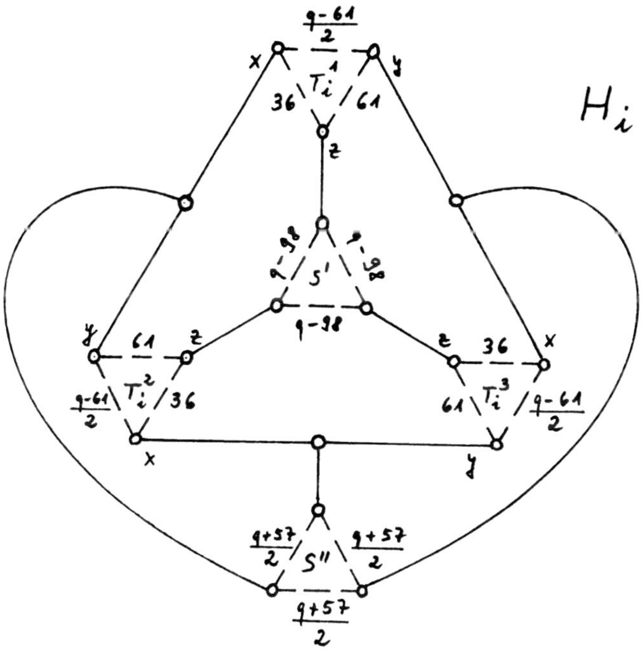

Fig. 8

3. Construction of the sequence $\{G_j\}$

Now, we put $G_1 := H_i$.
We consider the structure U of type (q-3,q-3,q-3) built with the help of three copies W^1,W^2,W^3 of the structure W and three copies R^1,R^2,R^3 of a structure R of type (8,40,q-51), all vertices of R^j (j = 1,2,3) being white, see Fig. 9. U contains exactly 27 black vertices and any longest path through U connecting two of the three half-edges omits exactly one black vertex.
By replacing each black vertex of G_j $(j \geq 1)$ by a copy of the structure U we obtain G_{j+1} and we have $G_j \in \Gamma^{4,q} \cap \Gamma(w)$ for all $j \geq 1$. A short estimation (see [2], [3], [5]) yields

$$\lim_{j\to\infty} \frac{\log c(G_j)}{\log n(G_j)} \leq \log 26/\log 27$$

which proves our Theorem.

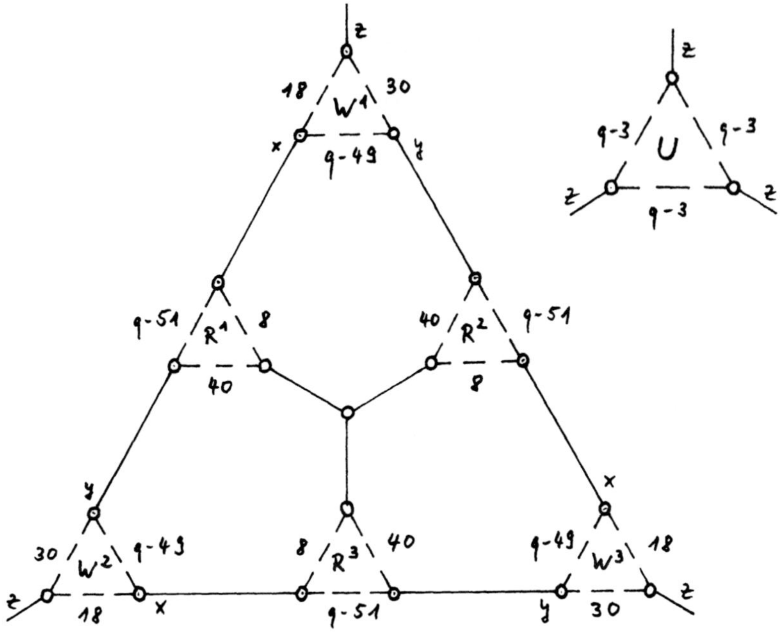

Fig. 9

References

[1] B. Grünbaum and H. Walther, Shortness exponents of families of graphs, J. Combin. Theory 14 (1973), 364-385.

[2] J. Harant, Über den shortness exponent regulärer Polyedergraphen, Dissertation A, TH Ilmenau, 1982.

[3] J. Harant and H. Walther, On a Problem Concerning Longest Circuits in Polyhedral Graphs, Annals of Discrete Math. 41 (1989), 211-220, Els. Sci. Publ. B.V. (North-Holland).

[4] W.T. Tutte, On Hamiltonian circuits, J. London Math. Soc. 21 (1946), 98-101.

[5] H. Walther and H.-J. Voß, Über Kreise in Graphen, VEB Deutscher Verlag der Wissenschaften (1974), 58-62.

Longest Cycles in Circulant Graphs

W. Harnau, Dresden, GDR
D. Jordan, Dresden, GDR

0. Introduction

In 1969 L. LOVASZ /2/ conjectured that every connected vertex transitive graph has a Hamilton path. L. BABAI /1/ proved the validity of this conjecture for graphs with a prime number of vertices. In the same paper he stated, that no useful results on the length of longest cycles in vertex transitive graphs are known. In this paper we generalize the result of /1/ in two directions.
At first BABAI's graphs are proper contained in our set of graphs, for which we secondly prove not only the existence of a Hamilton path, but also the existence of a Hamilton cycle. Furthermore we present the length of the longest cycle of circulant graphs.

1. Basic notations and the main result

Let G=(V,E) be a graph, where V (the set of vertices) is a nonempty finite set with at least 3 elements and E (the set of edges) is a subset of $\binom{V}{2}$ (the set of 2-elementary subsets of V). Consequently, we consider in this paper only undirected, finite graphs without multiple edges and loops.
A graph with n vertices is shortly called n-graph.
A graph G=(V,E) is vertex transitive, iff its automorphism group is transitive on V. A n-graph G=(V,E) is called cir-

R. Bodendiek, R. Henn (Eds.)
Topics in Combinatorics and Graph Theory

culant, iff $V=\{0,1,\dots,n-1\}$ and $(x,y)\in E$ implies $(x\oplus 1,y\oplus 1)\in E$ ($\oplus$ addition by modulo n).

A graph G' is called isocirculant, iff there is a circulant graph G isomorphic to G'. To define a circulant n-graph G, we have not to give all edges of it, because we know by the definition of the circulance of a n-graph, that with an edge (x,y) of G each $(x\oplus a,\ y\oplus a)$ with $a\in\{0,1,\dots,n-1\}$ is an edge of G too.

Let M be a subset of $\binom{\{0,1,\dots,n-1\}}{2}$. We define for $a\in\{0,1,\dots,n-1\}$ by $M\oplus a$ the set $\left\{\{x\oplus a,y\oplus a\}\ :\ \{x,y\}\in M\right\}$.

Let G=(V,E) be a circulant n-graph. A subset D of $\binom{V}{2}$ is called generating set (of the edges) of G, iff $E=\bigcup_{a\in V}(D\oplus a)$.

Such a generating set D is called base (of the edges) of G, iff $D=\emptyset$ or no proper subset of D is a generating set of G. It is obvious that each generating set D of a circulant n-graph G=(V,E) is a subset of E and reducable to a base of G.

Let G=(V,E) be an arbitrary graph. We denote for $i\in V$ by E_i the set of all with i adjacent edges of G:

$$E_i:=\{(i,a)\ :\ a\in V\setminus\{i\}\}\cap E$$

It is obvious too, that in a circulant n-graph G=(V,E) for each $i\in V$ the set E_i is a generating set of G, but in general no base of G.

Let G=(V,E) be a circulant n-graph and $i\in V$. A generating set resp. base D of G is called i-generating set resp. i-base of G, iff $D\subseteq E_i$.

If D is an i-generating set resp. i-base of G=(V,E), then we denote $D':=\{x:\ (i,x)\in D\}$ as abbrevitation for the i-generating set resp. i-base of G . In the following we use only 0-generating sets resp. 0-bases. We denote them shortly by generating set resp. base.

A connected circulant graph G=(V,E) is called critical cc-graph, iff each subgraph G'=(V,E') of G with $E'\subset E$ is not connected or not circulant.

We will prove the following theorem, our main result.

<u>Theorem:</u> Every connected circulant graph has a Hamilton cycle.

Obviously, we receive as

<u>Corollary 1:</u> Every connected isocirculant graph has a Hamilton cycle.

It is obvious, that all isocirculant graphs are vertex transitive. But the converse is not true. The cube is an important counterexample.

TURNER /3/ has proved the following:

<u>Proposition:</u> Every vertex transitive graph with a prime number of vertices is an isocirculant graph.

So we receive as

<u>Corollary 2:</u> Every connected vertex transitive graph with a prime number of vertices has a Hamilton cycle.

We denote the greatest common divisor of the integers $a_1,\dots,a_k$ by $gcd(a_1,\dots,a_k)$.
Since a disconnected circulant n-graph G with the base $D=\{b_1,b_2,\dots,b_t\}$ has $\tau = gcd(n,b_1,b_2,\dots b_t)$ mutual isomorphic components, it follows:

<u>Corollary 3:</u> A isocirculant n-graph $G=(V,E)$, with τ components and more than one edge in a component, has a longest cycle of length $\frac{n}{\tau}$.

2. The proof of the Theorem

If $D=\{b_1,b_2,\dots,b_t\}$ is a base of a circulant n-graph then the proofs of the following two statements are evident:

Lemma 1: If an isocirculant graph G is disconnected then the components of G are mutual isomorphic.

□

Lemma 2: Every circulant n-graph has at least $\gcd(n,b_i)$ edge disjoint cycles of length $\frac{n}{\gcd(n,b_i)}$, $i\in\{1,\dots,t\}$. (For $b_i=\frac{n}{2}$ edges are considered as cycles of length 2.)

□

Furthermore let $\tau = \gcd(n,b_1,b_2,\dots,b_t)$ and $\{\underline{0},\underline{1},\dots,\underline{n-1}\}$ be the partition of $\mathbb{Z}_n$, the ring of integers modulo n, into congruence classes modulo n.

Lemma 3: Every circulant n-graph G with the base D has τ components.

Proof: (1) By Lemma 2, there are $\gcd(n,b_i)$ cycles of length $\frac{n}{\gcd(n,b_i)}$ for each b_i, $i\in\{1,2,\dots,t\}$. Since G is circulant the vertices of these cycles (for one b_i) are labeled with $\underline{0},\underline{1},\dots,\underline{\gcd(n,b_i)-1}$ modulo $\gcd(n,b_i)$ respectively.

(2) Let $g_i=\gcd(n,b_i)$, $i\in\{1,2,\dots,t\}$. Consider the congruence class $\underline{0}$ for each modul g_i. All elements from $\underline{0}$ modulo g_i, $i=1,2,\dots,t$, are included in $\underline{0}$ modulo $\gcd(g_1,g_2,\dots,g_t)$. It is $\gcd(g_1,g_2,\dots,g_t)=\gcd(\gcd(n,b_1),\dots,\gcd(n,b_t))=$ $=\gcd(n,b_1,\dots,b_t)=\tau$.

With (1),(2) and Lemma 1 it is obvious that G has τ components. □

For critical cc-graphs we can conclude:

Corollary 0: (1) Let $D=\{b_1,b_2,\dots,b_t\}$ be a base for a circulant graph G. Then there is a critical cc-graph of G, iff $gcd(n,b_1,b_2,\dots,b_t)=1$.
(2) $gcd(n,b_1,b_2,\dots,b_s)=1$ for each base $D=\{b_1,b_2,\dots,b_s\}$ of critical cc-graphs.
(3) Every connected circulant graph has a critical cc-graph as subgraph.

□

Now using critical cc-graphs we answer the questions: How long are the longest cycles in a circulant graph and is there a Hamilton cycle in a connected circulant graph ?

Theorem: Every connected circulant graph has a Hamilton cycle.

Proof:

(0) By Corollary 0 a connected circulant graph has a critical cc-graph as subgraph. Denote such a n-graph with G and a base of G with $D=\{b_1,b_2,\dots,b_t,b_{t+1}\}$.
It is sufficient to prove that every critical cc-graph is hamiltonian. This is proved by induction over the cardinality of a base D of a critical cc-graph G.

(1) card(D)=1
By Lemma 2 and Corollary 0 is $gcd(n,b_1)=1$ for $D=\{b_1\}$ and therefore there is a cycle of length $\frac{n}{gcd(n,b_1)}=n$.
Furthermore the vertices $0,b_1$ and $0,n-b_1$ are adjacent in these cycle.

(2) card(D')=t
Let $D'=\{b_1,b_2,\dots,b_t\}\subset D$. Then $\tau = gcd(n,b_1,b_2,\dots,b_t)>1$, and τ is the number of components of the circulant subgraph G' of G which is generated by the base D'. The components are mutual isomorphic. Since G' is circulant the vertices of one component are labeled with the elements of one congruence class from $[\underline{0},\underline{1},\dots,\underline{\tau-1}]$ modulo τ respectively.

Let L_h be the component with the vertex h.

We suppose: There is a Hamilton cycle in the component L_0 of the graph G' which is generated by D', in which the vertices $0, b_1$ and $0, n-b_1$ are adjacent.

(i) The element b_{t+1} of D generated the connection of the components from G' in G. Hence the vertices 0, b_{t+1}, $2b_{t+1}, \ldots, (\tau-1)b_{t+1}$ are in different components.

(ii) Since the components are mutual isomorphic there is a Hamilton cycle in each component.

(iii) G' is circulant hence it can be found a Hamilton cycle in the component L_h, $(h=0,1,\ldots,\tau-1)$, in which the following vertices are adjacent:

$$
\begin{array}{lcccc}
L_0 & : & b_1 & -- \quad 0 \quad -- & n-b_1 \\
L_{b_{t+1}} & : & b_1+b_{t+1} & -- \quad b_{t+1} \quad -- & n-b_1+b_{t+1} \\
\cdot & & \cdot & \cdot & \cdot \\
\cdot & & \cdot & \cdot & \cdot \\
\cdot & & \cdot & \cdot & \cdot \\
L_{(\tau-1)b_{t+1}} & : & b_1+(\tau-1)b_{t+1} & -- (\tau-1)b_{t+1} -- & n-b_1+(\tau-1)b_{t+1}
\end{array}
$$

(iv) Now a Hamilton cycle of G can be described in dependence on the number of components of G'.

A) If τ is even then a Hamilton cycle is given by the following sequence of vertices :

$0,\ b_1, \ldots,\ n-b_1,\ n-b_1+b_{t+1}, \ldots,\ b_1+b_{t+1}, \ldots,$
$n-b_1+(\tau-1)b_{t+1}, \ldots,\ b_1+(\tau-1)b_{t+1},\ (\tau-1)b_{t+1},$
$\ldots,\ b_{t+1}, 0$

B) If τ is odd then a Hamilton cycle is given by the following sequence of vertices :

$0,\ b_1, \ldots,\ n-b_1, \ldots,\ n-b_1+(\tau-1)b_{t+1}, \ldots, 0$

□

Obviously, we receive as

<u>Corollary 1:</u> Every connected isocirculant graph has a Hamilton cycle.

□

It is obvious, that all isocirculant graphs are vertex transitiv. But the converse is not true. The cube is an important counterexample.

TURNER /3/ has proved the following:

<u>Proposition:</u> Every vertex transitive graph with a prime number of vertices is an isocirculant graph.

□

So we receive as

<u>Corollary 2:</u> Every connected vertex transitive graph with a prime number of vertices has a Hamilton cycle.

□

Note that a graph G=(V,E), E=∅, is a disconnected isocirculant graph. Moreover a graph G=(V,E) , which has only components that are isomorphic to K_2, is also a disconnected isocirculant graph. Every other disconnected isocirculant graph has more than one edge in each component.
By Lemma 1, Lemma 3 and Corollary 2 follows now:

<u>Corollary 3:</u> An isocirculant n-graph G=(V,E), with τ components and more than one edge in a component, has a longest cycle of length $\frac{n}{\tau}$.

□

3. A generalized conjecture

Let G=(V,E) be a n-graph with V={0,1,...,n-1}. We say, that G is r-pseudocirculant, where r is an arbitrary integer, iff $(x,y)\in E$ implies $(x\oplus r, y\oplus r)\in E$ ($\oplus$ addition by mod n).
A n-graph G' is called r-isopseudocirculant, iff there is a r-pseudocirculant graph G isomorphic to G'.
It is easy to see, that G is a r-pseudocirculant graph, iff G is a gcd(n,r)-pseudocirculant graph.

<u>Conjecture:</u> Every connected r-isopseudocirculant n-graph has a cycle of length $\geq \frac{n}{\gcd(n,r)}$.

In this paper we proved this conjecture for all r with $\gcd(n,r)=1$!

<u>References:</u>

/1/ L.BABAI : Long cycles in vertex transitive graphs, J. Graph Theory, 3, 1979

/2/ L.LOVASZ : Problem 11, Combinatorial structures and their applications, Gordon and Breach, New York, 1970

/3/ J.TURNER : Point-symmetric graphs with a prime number of points, J. Comb. Theorie, 3, 1967

Spanning Trees of the Complete Bipartite Graph

N. Hartsfield, Santa Cruz, USA
J. S. Werth, Washington, USA

Abstract

A new proof that the number of spanning trees of $K_{m,n}$ is $m^{n-1}n^{m-1}$ is presented. The proof is similar to Prüfer's proof of Cayley's formula for the number of spanning trees of K_n.

1 Introduction

The *complete bipartite graph* $K_{m,n}$ is the graph on $m+n$ vertices partitioned into two sets M and N, with m vertices in M and n vertices in N, such that every vertex in M is adjacent to every vertex in N, and no two vertices in the same set are adjacent. In other words, all edges in $K_{m,n}$ have one endpoint in M and the other in N, and every such edge possible actually exists.

A *tree* is a connected acyclic graph. A tree with p vertices has $p-1$ edges. A *spanning tree* of a graph G is a subgraph of G that is a tree and that contains every vertex of G.

Onodera[1] proved that the number of spanning trees of $K_{n_1,n_2,\ldots,n_k}$ is

$$p^{k-2}\prod_{i=1}^{k}(p-n_i)^{n_i-1},$$

where $p=\sum_{i=1}^{k} n_i$, using Kirchoff's matrix tree theorem. This is a beautiful result and implies that the number of spanning trees of $K_{m,n}$ is $m^{n-1}n^{m-1}$.

R. Bodendiek, R. Henn (Eds.)
Topics in Combinatorics and Graph Theory

Scoins[3] proved this more specialized result using sophisticated combinatorial arguments. We sought a shorter proof for this smaller result that did not require the use of either matrices or difficult counting arguments.

2 The Theorem

Theorem 1 *The number of spanning trees of $K_{m,n}$ is $m^{n-1}n^{m-1}$.*

Before we prove the theorem, we prove the following lemma.

Lemma 1 *Given any spanning tree T of $K_{m,n}$, if $m \leq n$, and the vertices are partitioned into the two sets*

$$M = \{a_1, a_2, \ldots, a_m\}$$

$$N = \{b_1, b_2, \ldots, b_k\},$$

where M and N form the natural bipartition of the vertices of $K_{m,n}$, then for some $k, 1 \leq k \leq n$, $deg(b_k) = 1$.

Proof. Let

$$A = \sum_{i=1}^{m} deg(a_i)$$

$$B = \sum_{j=1}^{n} deg(b_j).$$

Notice that $A = B$, since each edge has one endpoint in each of M and N, and that $(A + B)/2 = m + n - 1$, since T is a tree. Thus we have

$$B = \frac{A + B}{2} = m + n - 1.$$

If $\deg(b_j) \geq 2$ for all j, then we have

$$B \geq m + n > m + n - 1.$$

Thus for some k, $\deg(b_k) = 1$.□

3 Proof of the Theorem

As in the proof of the lemma, we assume that $m \leq n$, and that the bipartite sets of the vertices of $K_{m,n}$ are labeled

$$M = \{a_1, a_2, \ldots, a_m\}$$

$$N = \{b_1, b_2, \ldots, b_n\}.$$

We shall establish a one-to one correspondence between the spanning trees of $K_{m,n}$ and the sequences of length $m+n-2$ of the form

$$a_{i_1}, a_{i_2}, \ldots, a_{i_{n-m+1}}, b_{j_1}, a_{i_{n-m+2}}, b_{j_2}, \ldots, a_{i_{n-1}}, b_{j_{m-1}},$$

where the a's are not necessarily distinct, and the b's are not necessarily distinct. There are $m^{n-1}n^{m-1}$ such sequences.

Let T_0 be any spanning tree of $K_{m,n}$. We associate a sequence S_0 with T_0 in the following way.

Step t_1. Since $m \leq n$, by the lemma there is a vertex b_k in T_0 with $\deg(b_k) = 1$. We select the smallest value of a subscript r such that $\deg(b_r) = 1$. The first element of our sequence S_0 will be the neighbor of b_r, call it a_{i_1}.

We now remove b_r and consider $T_0 - b_r$. This is a spanning tree of $K_{m,n-1}$. Now either we have $m \leq n-1$ or $m > n-1$. If $m \leq n-1$, by the lemma there is a b_k in $T_0 - b_r$ such that $\deg(b_k) = 1$, in which case we repeat Step t_1.

Step t_2. After we have repeated the first step $n-m+1$ times, the reduced tree is a spanning tree of $K_{m,m-1}$. Now, by the lemma, there is a vertex a_l such that $\deg(a_l) = 1$. Again, we select the smallest subscript s such that $\deg(a_s) = 1$. The neighbor of a_s is b_{j_1}. We record b_{j_1} as the next element of S_0 and remove a_s from the tree. When we remove a_s, we again have equality in the remaining partite sets, hence we can find a b_k such that $\deg(b_k) = 1$. If we continue to iterate the process, the rest of the sequence S_0 will alternate between a's and b's.

Step t_3. When one edge remains in the tree, we stop.

Thus S_0 consists of first $n-m+1$ a_i's, then alternating a_i's and b_j's to a total length of $m+n-2$. Hence we have associated a sequence S_0 of the appropriate form with our tree T_0.

Now suppose that we are given a sequence S of the form

$$a_{i_1}, a_{i_2}, \ldots, a_{i_{n-m+1}}, b_{j_1}, a_{i_{n-m+2}}, b_{j_2}, \ldots, a_{i_{n-1}}, b_{j_{m-1}},$$

where the a's and b's are elements of the sets

$$M = \{a_1, a_2, \ldots, a_m\}$$

$$N = \{b_1, b_2, \ldots, b_n\}.$$

We associate a tree T with S as follows.

Label $m+n$ vertices by the elements of M and N.

Step s_1. We select the smallest subscript r of an element of N that does not appear as an element of S. The first element of S is a_{i_1}; we connect vertex b_r to vertex a_{i_1}. We remove a_{i_1} from S to obtain S_1, and we remove b_r from N to obtain N_1. If the first element of S_1 is a_{i_2}, we repeat this step with S replaced by S_1 and N replaced by N_1. After repeating step $s_1 n - m + 1$ times, the first element of the reduced sequence is b_{j_1}. (If $m = 1$, then no b's appear, and we are done).

Step s_2. We select the smallest subscript l of an element of M that does not appear in the reduced sequence. We connect vertex a_l to vertex b_{j_1} in the tree, remove a_l from M to obtain M_1, remove b_{j_1} from S_{n-m+1} to obtain S_{n-m+2}, and go to Step s_1.

Step s_3. When no elements of S remain, one element remains in M and one element in N. We connect the two vertices labeled by these elements in the tree. Thus we have associated a tree T with a sequence S. What we must show now is that when a sequence S_0 is associated with a tree T_0, our process then associates T_0 with S_0.

In Step t_1, we find the smallest subscript of a vertex b_k of degree 1 in the tree, remove this vertex, and record its neighbor a_i in the sequence. In Step s_1, we find the smallest subscript of a vertex b_k in N that does not appear in the sequence and connect it to the vertex labeled by the first element a_{i_1} of the sequence, then remove a_{i_1} from the sequence and remove b_k from N. Hence s_1 and t_1 are "inverses" of one another. Similarly, s_2 is the reverse operation to t_2. Finally, in t_3 we stop the process when one edge remains in the tree, and in s_3 we connect the remaining element in M to that in N, so these steps are also "inverses" of each other. Hence, we have a one-to-one correspondence between the spanning trees of $K_{m,n}$ and the sequences described above, proving the theorem. □

4 An Example

We illustrate the process with the tree of Figure 1.

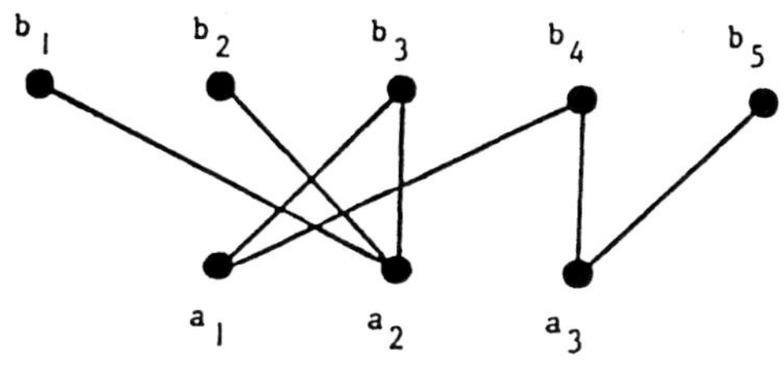

Figure 1:

The smallest subscript of a b vertex with degree one is 1. The neighbor of b_1 is a_2. So the first element of the sequence is a_2. The successive reduced trees and partial sequences are shown in Figure 2.

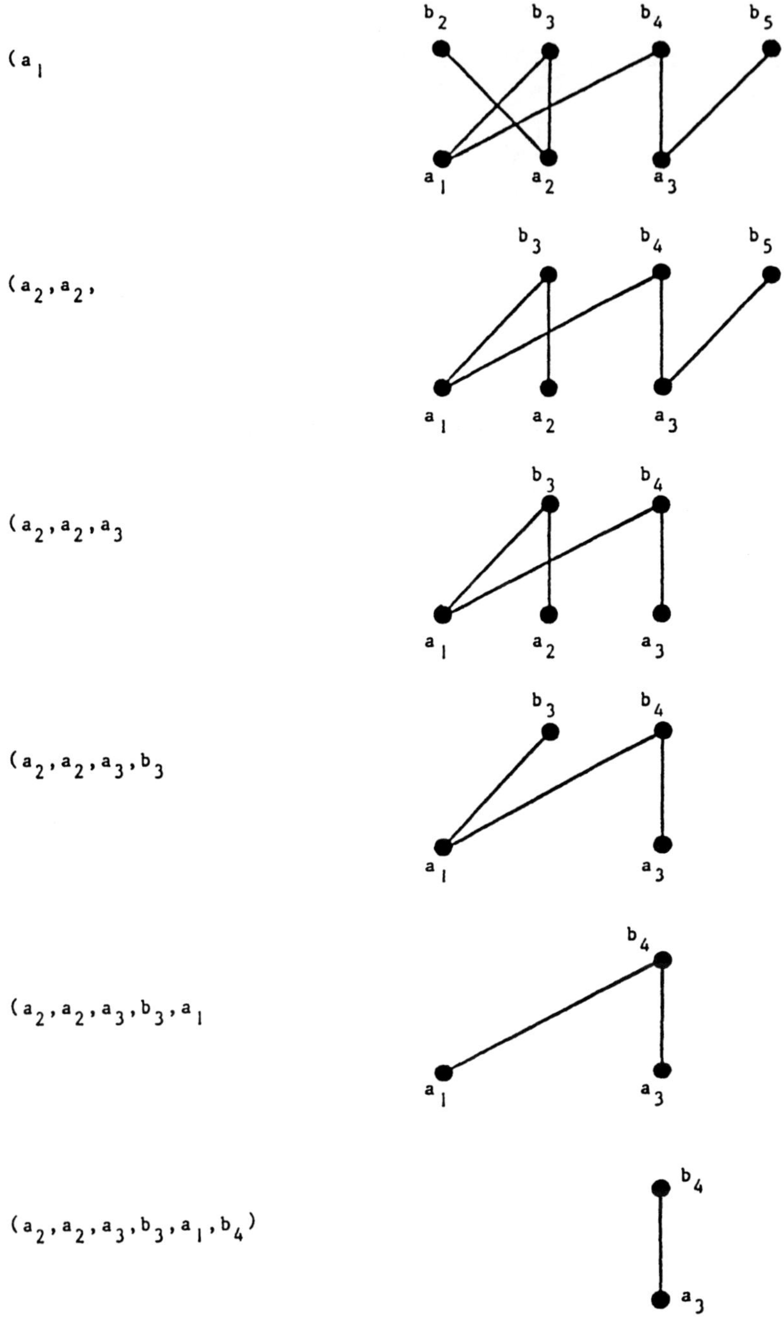

Figure 2:

In Figure 3 we begin with the sequence and derive the tree.

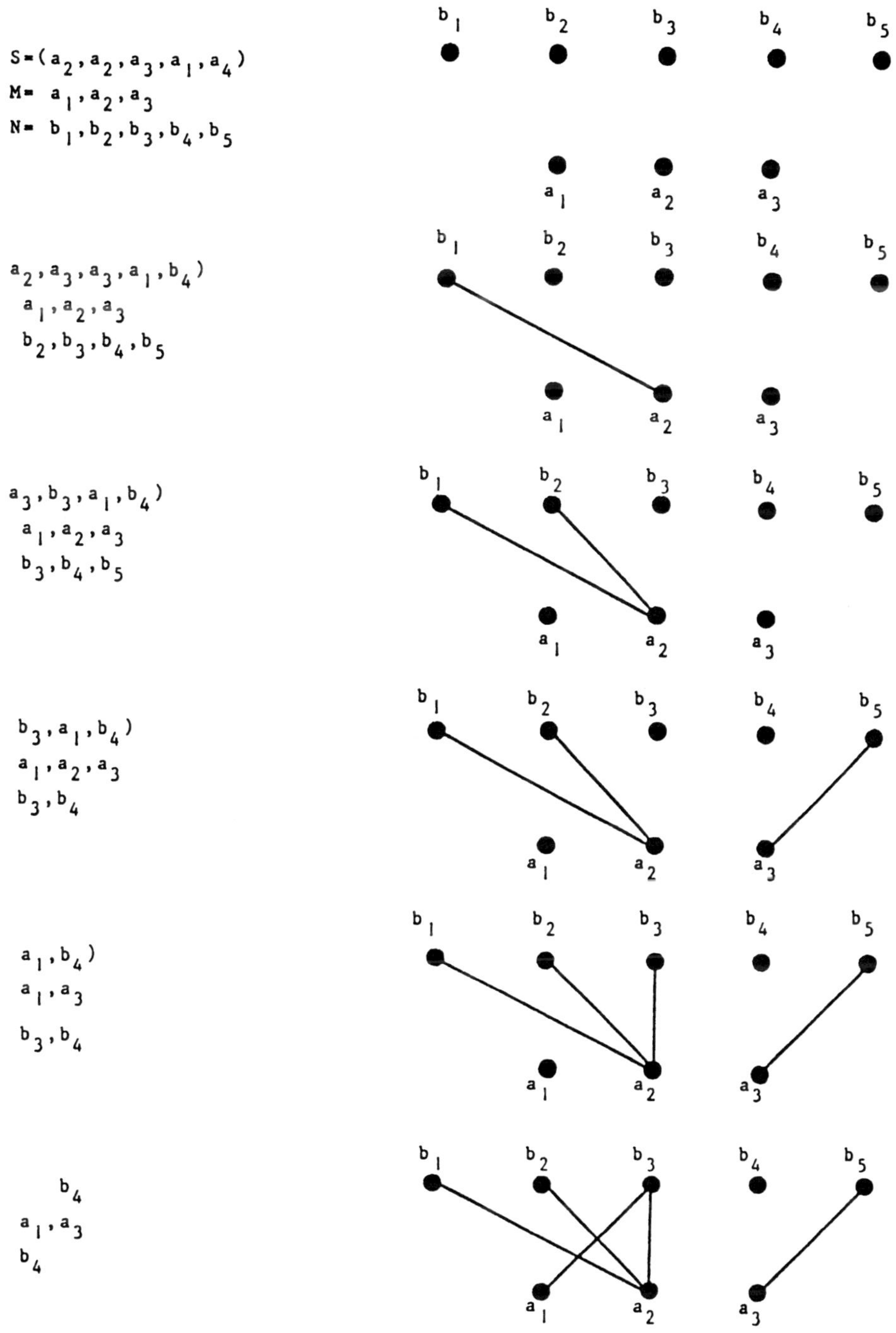

S=(a2,a2,a3,a1,a4)
M= a1,a2,a3
N= b1,b2,b3,b4,b5
b1 b2 b3 b4 b5
a1 a2 a3
a2,a3,a3,a1,b4)
a1,a2,a3
b2,b3,b4,b5
b1 b2 b3 b4 b5
a1 a2 a3
a3,b3,a1,b4)
a1,a2,a3
b3,b4,b5
b1 b2 b3 b4 b5
a1 a2 a3
b3,a1,b4)
a1,a2,a3
b3,b4
b1 b2 b3 b4 b5
a1 a2 a3
a1,b4)
a1,a3
b3,b4
b1 b2 b3 b4 b5
a1 a2 a3
b4
a1,a3
b4
b1 b2 b3 b4 b5
a1 a2 a3

Figure 3: a

a_3
b_4

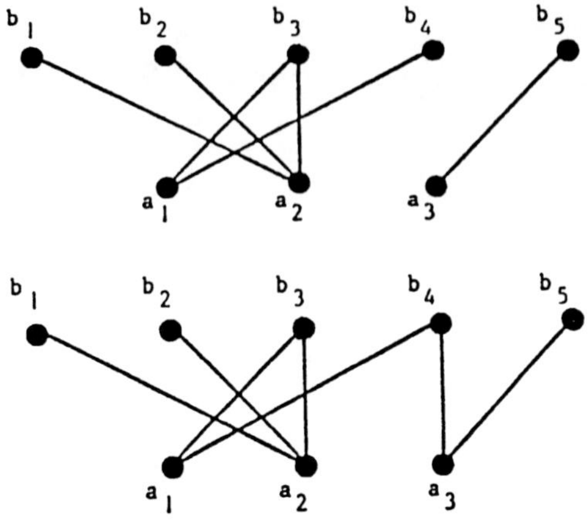

Figure 3: b

References

[1] R. Onadera, On the number of trees in a complete n-partite graph.*Matrix Tensor Quart.*23 (1972/73), 142-146.

[2] H. Prüfer, Neuer Beweiss einer Satzes über Permutationen. *Math. Phys.* **27** (1918), 742-744.

[3] H. I. Scoins, The number of trees with nodes of alternate parity. *Proc. Cambridge Philos. Soc.* **58** (1963), 12-16.

A Combinatorial Theorem Which is Related to the Invariance of the Separating Set for the Plane

E. Harzheim, Düsseldorf, FRG

In this paper we present a theorem, which is related to theorems which were proved in [2]. The starting point of the latter paper was the qualitative part of the separation theorem of Jordan/Brouwer/Alexander, that means the invariance of the (non-) separating set:

(I) If K is a compact subset of $\mathbb{R}^n$ which does not separate $\mathbb{R}^n$, and if f: $K \to \mathbb{R}^n$ is an injective and continuous mapping, then f(K) also does not separate $\mathbb{R}^n$.

Trying to find an elementary proof for (I) the author came across a combinatorial theorem ([2], Theorem 1), which deals with sets which are linked in some prescribed manner, and with the regions into which their union separates the space $\mathbb{R}^n$. Here we establish a theorem for the case n = 2, which is partially sharper than Theorem 1 of [2].

DEFINITION. $\mathbb{N}_0$, $\mathbb{Z}$, $\mathbb{R}$ denotes the set of nonnegative integers, all integers, real numbers respectively; n and k always are positive integers. For sets $M \subset \mathbb{R}^n$ we denote the closure, open kernel, boundary of M by $\overline{M}$, $\mathring{M}$, ∂M respectively.

For $s \in \mathbb{N}_0$ an s-lattice cube of $\mathbb{R}^n$ is a cartesian product of n intervals I_ν, $\nu = 1,..,n$ of $\mathbb{R}$, where each I_ν is of the form $[g_\nu \cdot 2^{-s}, (g_\nu+1)\cdot 2^{-s}]$ with $g_\nu \in \mathbb{Z}$. The set of all s-lattice cubes of $\mathbb{R}^n$ shall be denoted by $\mathcal{L}_s(\mathbb{R}^n)$. A figure is a non-empty subset of $\mathbb{R}^n$ which is a union of s-lattice cubes for some $s \in \mathbb{N}_0$.

A closed interval of $\mathbb{R}^n$ is a set $I_1 \times \ldots \times I_n$, where $I_1,..,I_n$ are closed intervals of $\mathbb{R}$ (which also can degenerate to one point). If M is a non-empty bounded subset of $\mathbb{R}^n$, we define the interval-closure I(M) of M to be the smallest closed interval of $\mathbb{R}^n$ which encompasses M.

R. Bodendiek, R. Henn (Eds.)
Topics in Combinatorics and Graph Theory

If x,y,z are points of $\mathbb{R}^n$ where x is different from y and z, we define arc(y,x,z) to be that number $\alpha \in [0,\pi]$ which is the arc-measure of the angle at x in the triangle which has x,y,z as vertices. (See figure 1.)

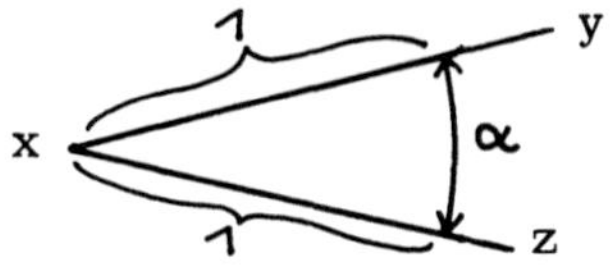

figure 1

For $y,z \in \mathbb{R}^n$ and $\alpha \in (0,\pi]$ we define $\alpha(y,z)$ to be the set $\{y,z\} \cup \{x \in \mathbb{R}^n \setminus \{y,z\} \mid \text{arc}(y,x,z) \geq \alpha\}$. (In figure 2 below $\alpha(y,z)$ is hatched.)

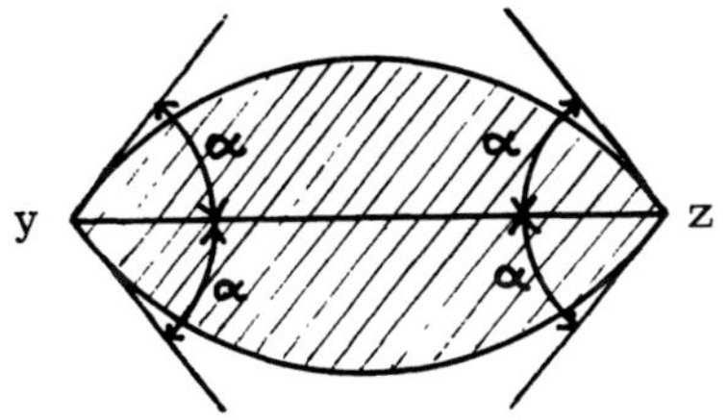

figure 2

More generally we define for $\emptyset \neq M \subset \mathbb{R}^n$ the <u>α-angle-zone</u> of M to be the set $\alpha(M) := \bigcup\{\alpha(x,y) \mid x,y \in M\}$.

<u>EXAMPLE.</u> If M is a segment of $\mathbb{R}^n$, $\frac{\pi}{2}(M)$ is the ball of which M is a diameter segment. (Thales' theorem)

<u>DEFINITION.</u> Let $\mathfrak{A} = (A_i)_{i \in I}$ and $\mathfrak{B} = (B_i)_{i \in I}$ be two families of non-empty sets. If for all $i,j \in I$ we have

a) $A_i \cap A_j = \emptyset \implies B_i \cap B_j = \emptyset$, we call $\mathfrak{B}$ <u>disjointness-homomorphic</u> to $\mathfrak{A}$. If

b) $A_i \cap A_j = \emptyset \iff B_i \cap B_j = \emptyset$, $\mathfrak{A}$ and $\mathfrak{B}$ are said to be <u>equally linked</u>.

These definitions play a role in topological considerations. (See e.g. [2],Theorem 1 and the Remark of [2],p.243.) A typical situation, where this definition can be applied, is described as follows: Let M be a compact subset of $\mathbb{R}^n$. Then for every $s \in \mathbb{N}_0$ we have $M = \bigcup\{M \cap C \mid C \in \mathcal{L}_s\}$. Let $\mathcal{L}_M := \{C \in \mathcal{L}_s \mid M \cap C \neq \emptyset\}$. If now $f: M \to \mathbb{R}^n$ is an injective and continuous mapping, and if two cubes $C,C' \in \mathcal{L}_M$ are disjoint, the sets $f(M \cap C)$, $f(M \cap C')$ are also disjoint. Thus there exists a polyhedron $P(C) \supset f(M \cap C)$ for every $C \in \mathcal{L}_M$ such that for $C,C' \in \mathcal{L}_M$ and $C \cap C' = \emptyset$ also $P(C) \cap P(C') = \emptyset$ holds. Then the family $\mathcal{P} = \{P(C) \mid C \in \mathcal{L}_M\}$ is disjointness-homomorphic to the family $\mathcal{L}_M$, and therefore $\mathcal{P}$ allows conclusions on $\mathcal{L}_M$ and thus on M itself. The family $\mathcal{P}$ has a discrete mathematical structure, - and this can be much simpler than that of M.

Under these conditions one might expect that the bounded regions of $\mathbb{R}^n \setminus \bigcup \mathcal{P}$ become "small" if the bounded regions of $\mathbb{R}^n \setminus M$ are small. Indeed, Theorem 1 of [2] and the following Theorem 1 make statements of that kind.

We define the four quadrants Q_i, $i = 1,..,4$ of $\mathbb{R}^2$ as usual: $(x,y) \in \mathbb{R}^2$ belongs to Q_1 if $x \geq 0 \leq y$, to Q_2 if $x \leq 0 \leq y$, to Q_3 if $x \leq 0 \geq y$, to Q_4 if $x \geq 0 \geq y$. In this context there holds the following theorem (II), which is an immediate consequence of [1], Theorem 4.1:

(II) Let K be a compact subset of $\mathbb{R}^2$ such that the point $\sigma := (0,0)$ is in a bounded region G of $\mathbb{R}^2 \setminus K$. Let $f: K \rightarrow \mathbb{R}^2$ be a continuous mapping such that $f(K)$ does not separate $\mathbb{R}^2$. Then there exist points p,q in K with $p \in Q_1$, $q \in Q_3$, or $p \in Q_2$, $q \in Q_4$ which satisfy $f(p) = f(q)$.

Now we can prove:

THEOREM 1. Let $\{C_1,...,C_k\}$ be a finite set of s-lattice cubes of $\mathbb{R}^2$ for some $s \in \mathbb{N}_0$, and let $\{F_1,...,F_k\}$ be a set of non-empty compact subsets of $\mathbb{R}^2$ such that the following holds: $i,j \in \{1,...,k\}$ and $C_i \cap C_j = \emptyset \Longrightarrow F_i \cap F_j = \emptyset$.

We put $C := C_1 \cup ... \cup C_k$ and $F := F_1 \cup ... \cup F_k$. For $i,j \in \{1,..,k\}$ we call the cubes C_i, C_j neighboring if $C_i \cap C_j \neq \emptyset$ or if there exists a bounded region G of $\mathbb{R}^n \setminus C$ such that the boundary ∂G of G intersects C_i and C_j.

Then every bounded region of $\mathbb{R}^2 \setminus F$ is a subset of $\bigcup \{I(F_i \cup F_j) \mid i,j \in \{1,..,k\}$ and C_i, C_j are neighboring$\}$.

Before starting with the proof we premise a sketch. Bounded regions are hatched.

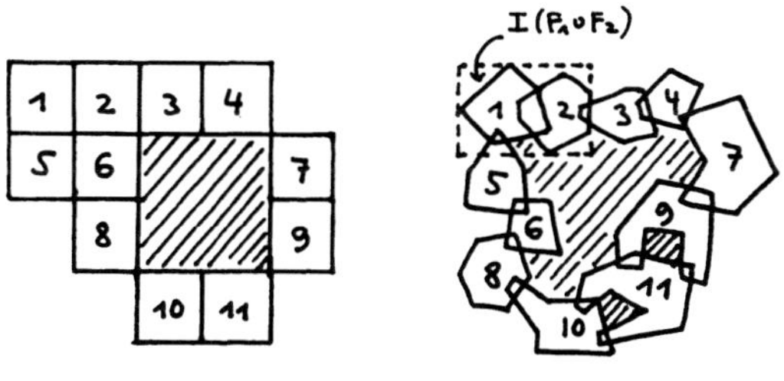

figure 3

PROOF. The proof can be obtained by modifying the proof of [2], Theorem 1. First we consider the special case, where the sets $F_1, \dots, F_k$ are figures which satisfy the condition

(0) For every $i \in \{1,..,k\}$ the set F_i is not a subset of $U_i := \bigcup\{F_j \mid j \in \{1,..,k\} \setminus \{i\}\}$.

Then we choose a number $s \in \mathbb{N}_0$ such that all figures $F_1,..,F_k$ are unions of s-lattice cubes. In addition to this we choose s so large that there exist s-lattice cubes W_i, $i = 1,..,k$, which satisfy $W_i \subset \mathring{F}_i \setminus U_i$. Then $W_1,..,W_k$ are pairwise disjoint.

Now we construct a continuous mapping $h: F \to C$ exactly as in the proof of Theorem 1 of [2], which satisfies

(1) $h(F_i) = C_i$ for $i = 1,\dots,k$.

Let $\mathcal{G}$ be the set of all bounded regions of $\mathbb{R}^2 \setminus C$ and $C^* := C \cup \bigcup\mathcal{G}$. Then, of course, C^* does not separate $\mathbb{R}^2$. Again exactly as in the proof in [2] we construct a surjective continuous mapping $g: C \to C^*$, which satisfies

(3) For $i,j \in \{1,\dots,k\}$ there holds: If $g(C_i) \cap g(C_j) \neq \emptyset$ then we have $C_i \cap C_j \neq \emptyset$ or: There exists a region $G \in \mathcal{G}$ with $\bar{G} \cap C_i \neq \emptyset \neq \bar{G} \cap C_j$, hence: C_i and C_j are neighboring cubes.

Now we put $f := g \circ h$. Then $f: F \to C^*$ is surjective and continuous. Let σ be a point of a bounded region G of $\mathbb{R}^2 \setminus F$. Without loss of generality we can assume, that σ is the origin of $\mathbb{R}^2$. Then, according to (II), there exist two points a,b $\in F$ with $a \in Q_1$ and $b \in Q_3$ or $a \in Q_2$ and $b \in Q_4$, which satisfy $f(a) = f(b)$. We have $a \in F_i$, $b \in F_j$ for some $i,j \in \{1,..k\}$. Thus $f(F_i) = g(C_i)$ and $f(F_j) = g(C_j)$ (- these equalities follow from (1) -) have a non-empty intersection. According to (3) this yields that C_i and C_j are neighboring cubes.

As a consequence of the position of the points a,b in the quadrants we have $\sigma \in I(\{a,b\}) \subset I(F_i \cup F_j)$, and so we have obtained: Every bounded region of $\mathbb{R}^2 \setminus F$ is a subset of $\bigcup\{I(F_i \cup F_j) \mid i,j \in \{1,..,k\}$ and C_i, C_j are neighboring$\}$.

Now we consider the general case where the F_i need not be figures. Then for every $i = 1,\dots,k$ we construct a decreasing sequence $F_{i1} \supset \ldots\ldots \supset F_{im} \supset \ldots$ of figures, which all encompass the compact set F_i, such that there holds:

(4) $F_{im} \subset U_{\frac{1}{m}}(F_i)$ (= the $\frac{1}{m}$- environment of F_i) for all $m \in \mathbb{N}$,

(5) $C_i \cap C_j = \emptyset \Longrightarrow F_{i1} \cap F_{j1} = \emptyset$ (and hence also $F_{im} \cap F_{jm} = \emptyset$ for all $m \in \mathbb{N}$).

(6) For every $m \in \mathbb{N}$ and every $i \in \{1,..,k\}$ the set F_{im} is not a subset of $\bigcup\{F_{jm} \mid j \in \{1,..,k\} \setminus \{i\}\}$.

Then the proof can be finished as in [2], - there we only have to replace $\alpha_{n-1}(\ )$ by $I(\ )$ -).

In the case where C does not separate $\mathbb{R}^2$ we obtain a simpler formulation as corollary:

<u>THEOREM 1'.</u> If in addition to the assumptions of Theorem 1 C does not separate $\mathbb{R}^2$, we obtain that every bounded region of $\mathbb{R}^2 \setminus F$ is a subset of $\bigcup\{I(F_i \cup F_j) \mid i,j \in \{1,..,k\}$ and $C_i \cap C_j \neq \emptyset\}$.

<u>REMARK 1.</u> Theorem 1 of [2], taken for the case n = 2, makes a statement, which differs from that of the above Theorem 1 insofar as it replaces $I(F_i \cup F_j)$ by $\alpha(F_i \cup F_j)$ with $\alpha = \frac{2\pi}{3}$. For a non-empty bounded set $M \subset \mathbb{R}^2$ usually none of the sets $I(M)$, $\alpha(M)$ encompasses the other, and thus from the two theorems none implies the other.

An advantage of the version which uses $I(\ )$ instead of $\alpha(\ )$ is, that in the first case one can generalize the statement to non-metrical spaces. E.g. if we replace in Theorem 1 $\mathbb{R}^2$ by spaces which are the product of two linearly ordered continua (2-fach geordnete Mengen), we obtain a more general statement, whose proof follows from Theorem 1 and a proof technique which was before applied in [3], Theorem 4.2.

<u>REMARK 2.</u> Finally one can raise the question, whether in Theorem 1' the union $\bigcup I(F_i \cup F_j)\ldots$ can be replaced by $\bigcup_{i=1}^{k} I(F_i)$. Indeed, this is not possible: Let $C_1,\ldots,C_9$ be the nine squares of figure 4 (0-lattice cubes of $\mathbb{R}^2$). We put $F_i := C_i$ for $i \in \{1,\ldots,9\} \setminus \{5\}$, and F_5 shall be any 0-lattice cube in $\mathbb{R}^2$ which is different from C_5. Then $F = F_1 \cup \ldots \cup F_9$ separates the plane, and $\mathring{C}_5$ is a bounded region of $\mathbb{R}^2 \setminus F$, but it is not

a subset of $\bigcup\{I(F_i)\mid i = 1,\ldots,9\} = \bigcup\{F_i\mid i = 1,\ldots,9\}$.

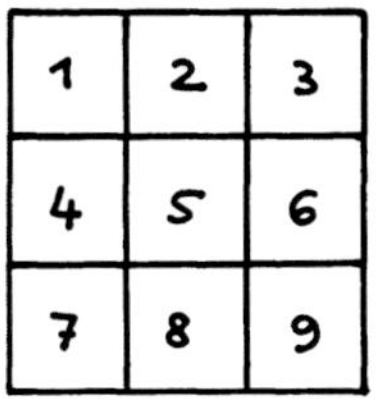

figure 4

Finally we wish to indicate a minor correction to a proof in [2]:

ERRATUM. In [2], page 241, line 2 an additional assumption should be inserted, namely that the sets $F_1,\ldots,F_k$ also satisfy the condition (0) of the present paper. And then the cubes W_i should be constructed so that $W_i \subset \mathring{F}_i \setminus U_i$ holds for $i = 1,\ldots,k$. (Otherwise (1) of [2] wouldn't have to be true.)

In [2], page 242, line 18 the sets $F_{i1},\ldots,F_{im}$ should be chosen in such a way that they also satisfy condition (6) of the present paper. It can easily be seen that this is possible, and thus all theorems of [2] remain valid without any changes.

REFERENCES

[1] Harzheim,E.: Combinatorial generalizations of the Jordan-Brouwer theorem. Ars Combinatoria 25 A (1988),259-268.

[2] Harzheim,E.: A theorem on polyhedra which is related to Alexander's separation theorem. Results in Math.15 (1989), 238-244.

[3] Harzheim,E.: On topological properties of cartesian products of linearly ordered continua. Ann.Discrete Math.23 (1984),171-192.

On Certain Trees in Hypercubes

I. Havel, Praha, CSSR

The aim of this contribution is to give a survey of some basic notions, known facts, conjectures and open problems about embedding trees into hypercubes, and the related problem of spanning trees of hypercubes. A new result in this direction is also presented.

Our graphs are finite, undirected, without loops and multiple edges. The graph Q_n of the n-dimensional hypercube is understood in the usual way, e. g. as follows: Q_n is the graph on 2^n vertices labelled $\{0,1,\ldots,2^n-1\}$ with an edge joining two vertices iff their binary representations differ in a single coordinate. An embedding of the graph G into the graph H is a 1-1 mapping f of $V(G)$ into $V(H)$. An important and commonly studied characteristics of an embedding f – its *dilation* – is defined by

$$\mathrm{dil}(f) = \max_{(u,v)\in E(G)} \varrho_H\left(f(u), f(v)\right).$$

The fact that G can be embedded into H with dilation d, $d \geq 1$ will be written $G \overset{d}{\subseteq} H$; instead of $G \overset{1}{\subseteq} H$ we shall write simply $G \subseteq H$. $G \subseteq H$ clearly means that G is isomorphic to a subgraph of H. In what follows the host graph H of an embedding will always be the hypercube Q_n. If $G \overset{d}{\subseteq} H$ for some n then $|V(G)| \leq 2^n$; therefore it is natural to call $n = \lceil \log_2(|V(G)|)\rceil$ the *optimal dimension* of G (or correspondingly Q_n the *optimal hypercube* for G); the number $\lceil \log_2(|V(G)|)\rceil + 1$ is sometimes called the next-to-optimal dimension of G.

The concepts introduced so far are well-known; they have been studied in various contexts by many authors. A comprehensive survey of the theory of hypercube graphs is e. g. [**4**]. One of the first papers introducing dilation and other characteristics of embeddings is [**14**]. The interest in the study of embeddings of graphs into hypercubes has been increased in recent years especially by the fast development of multi-processor computers based on hypercube architecture.

A graph G is called *cubical* if $G \subseteq Q_n$ for some n. Obviously, every tree is cubical and every cubical graph is bipartite (but e. g. $K_{2,3}$ is an example of a non-cubical bipartite graph). One defines the *cubical dimension* of the graph G by

$$\mathrm{cd}(G) = \min\{\, n \mid G \subseteq Q_n\}.$$

A trivial lower bound is $\mathrm{cd}(G) \geq \lceil \log_2(|V(G)|)\rceil$. Sometimes we can get a better lower bound for $\mathrm{cd}(G)$ as follows. Notice that for any connected bipartite graph G the partition of its vertex set $V(G) = V_1(G) \cup V_2(G)$ into two disjoint parts (the colour classes) is uniquely determined by the requirement that the edges of G are only between $V_1(G)$ and $V_2(G)$. G is called *balanced* if $|V_1(G)| = |V_2(G)|$. Let $\mathrm{lcd}(G)$ denote the smallest n such that $|V_1(G)| \leq 2^{n-1}$ and $|V_2(G)| \leq 2^{n-1}$. Notice that $\mathrm{lcd}(G) \geq \lceil \log_2(|V(G)|)\rceil$. Since Q_n is balanced, we have

$$G \subseteq Q_n \implies |V_1(G)| \leq 2^{n-1} \text{ and } |V_2(G)| \leq 2^{n-1},$$

R. Bodendiek, R. Henn (Eds.)
Topics in Combinatorics and Graph Theory

hence for any connected cubical graph

$$\mathrm{cd}(G) \geq \mathrm{lcd}(G).$$

(Let us note that the assumption of connectivity of G is not very essential here.)

Cubical graphs have been studied (by means of so called cube–critical graphs) in [**2**] and [**3**] among others. [**15**] contains a detailed study of embedding trees into hypercubes with the main result stating the NP–completeness of the problem $\mathrm{cd}(T) \leq k$ for a tree T and an integer k.

One of our main themes is that of spanning trees of hypercubes. It may be easily seen that if a tree T spans Q_n then T is balanced, for the maximum degree $\Delta(T)$ of T $\Delta(T) \leq n$ and $|V(T)| = 2^n$. The question is what should be added to these necessary conditions in order to obtain a sufficient condition for T to span Q_n. [**5**] contains two conjectures in this direction:

CONJ. 1. *Let T be a balanced l–quasistar with 2^n vertices, $l \leq n$. Then T spans Q_n.*

(Here by an l–quasistar we mean any tree homeomorphic to the l–star $K_{1,l}$).

CONJ. 2. *Let T be a balanced tree with maximum degree $\Delta(T) \leq 3$ and 2^n vertices $(n \geq 1)$. Then T spans Q_n.*

Conj. 1 has already been proved and is a theorem. The proof for the case $l = 3$ is easy and is contained in [**5**]. The cases $l = 4$ and $l = 5$ were settled by Nebeský [**12**], the case $l = 6$ was proved by Limaye [**10**] and finally the general case was solved again by Nebeský [**13**].

Conj. 2 has been verified for several classes of trees. It holds for trees with at most 2 vertices of degree 3, for certain modifications of the complete binary tree ([**5**],[**11**]), for one–legged caterpillars ([**7**]) and for generalized one–legged caterpillars with all the legs having length of the same parity ([**8**]). [**9**] is a report on computer verification of Conj. 2 for all trees with at most 16 vertices. Despite many efforts no progress has been made, however, in verifying Conj. 2 for all generalized one–legged caterpillars (i. e. trees of maximum degree 3 with all vertices of degree 3 lying on one path). In general Conj. 2 remains open and there is probably a long way to its full solution. It seems that some new methods will have to be used unless, of course, a counterexample is found. Let us note that the condition $\Delta(T) \leq 3$ seems to be essential; in [**7**] for any $n \geq 4$ a balanced tree T_n of 2^n vertices is constructed with $\Delta(T_n) = 4$ (and even only one vertex of degree 4) such that T_n does not span Q_n.

Let us formulate Conj. 2 in an equivalent way:

CONJ. 3. *Let T be a tree with $\Delta(T) \leq 3$ and $n \geq 1$, let both the parts $V_1(T)$ and $V_2(T)$ of its bipartition contain at most 2^{n-1} vertices. Then $T \subseteq Q_n$.*

To prove the equivalence of Conj. 2 and Conj. 3 is easy. Conj. 3 $\Longrightarrow$ Conj. 2 being trivial, let T be a tree with $\Delta(T) \leq 3$ and such that $|V_1(T)| \leq |V_2(T)| \leq 2^{n-1}$ for some $n \geq 1$. Then it is possible to construct a balanced tree T' fulfilling $T \subseteq T'$, $\Delta(T') \leq 3$ and $|V(T')| = 2^n$, e. g. by using the fact that if $|V_1(T)| < |V_2(T)|$ then there is a $v \in V_2(T)$ with $\deg(v) < 3$. Having constructed T' and using Conj. 2 we get $T' \subseteq Q_n$ and therefore also $T \subseteq Q_n$.

Now we are going to show that Conj. 3 holds for a certain one–dimensional class of generalized one-legged caterpillars.

Definition. Let $m \geq 2$; let the tree A_m be defined as follows: $V(A_m) = \{[i,j]; 1 \leq j \leq i \leq m\}$; $[i,j]$ and $[i',j']$ are adjacent if either $j = j' = 1$ and $|i - i'| = 1$ or $i = i'$ and $|j - j'| = 1$ (cf. Fig. 1).

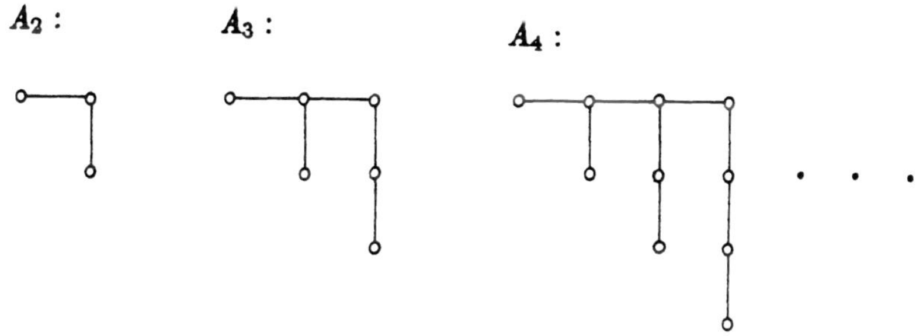

Fig. 1

Clearly A_m has $\binom{m+1}{2}$ vertices and is not balanced (the parts of A_{2k} have k^2 and k^2+k vertices, those of A_{2k+1} have k^2+k and $(k+1)^2$ vertices, respectively). To verify Conj. 3 for the trees A_m means to prove

$$(*) \qquad \mathrm{cd}(A_m) = \mathrm{lcd}(A_m).$$

Let us call the integer m *critical* for q if $q = \mathrm{lcd}(A_m) < \mathrm{lcd}(A_{m+1})$. Since $A_i \subseteq A_{i+1}$ it will suffice to prove $(*)$ just for all m critical for $q = 1, 2, \ldots$. Easy calculations show that the first critical values are 2, 3, 4, 7, 10, 15, 21, 31, ... Further, for any $p \geq 2$, $m = 2^p - 1$ is critical for $2p - 1$. We shall show that $(*)$ is valid for all these values of m:

PROPOSITION 1. *Let* $m = 2^p - 1$, $p \geq 2$. *Then* $\mathrm{cd}(A_m) = \mathrm{lcd}(A_m)$.

PROOF: Let $p \geq 2$ and $m = 2^p - 1$. We shall prove the statement by constructing an embedding of A_m into Q_{2p-1}. The construction follows from two observations stated below which may be easily verified. We use here the notion of the two-dimensional grid $G(m,n)$ – the (cartesian) product of two paths on m and n vertices, $G(m,n) = P_m \Box P_n$.

1. The grid $G(2^{p-1}, 2^{p-1})$ spans Q_{2p-2} and therefore the product $G(2^{p-1}, 2^{p-1}) \Box K_2$ spans Q_{2p-1}.

2. $A_{2^p-1} \subseteq G(2^{p-1}, 2^{p-1}) \Box K_2$.

The first observation is obvious; we illustrate the second in Fig. 2 for the case $p = 3$.

Hence Proposition 1 solves "one half" of the problem of verifying Conj. 3 for generalized caterpillars of the special type A_m. The second half of the problem, i. e. the proof of $(*)$ for those m which are critical for an even q remains open (except for several partial results):

PROBLEM 1. *Is it true that for every* $m \geq 2$ *such that* $\mathrm{lcd}(A_m)$ *is even and* $\mathrm{lcd}(A_m) < \mathrm{lcd}(A_{m+1})$,

$$\mathrm{cd}(A_m) = \mathrm{lcd}(A_m)?$$

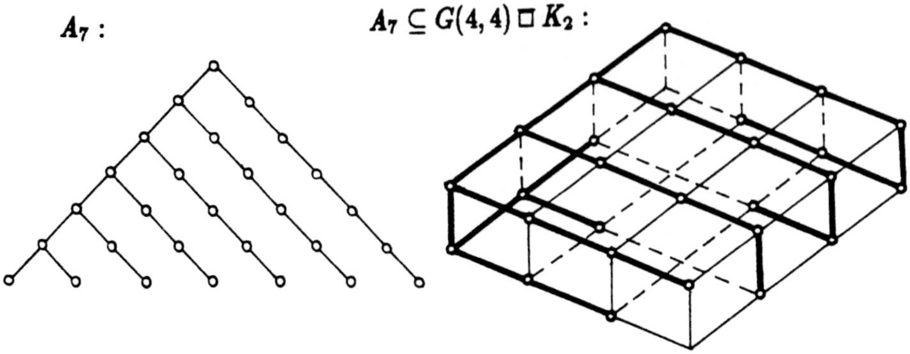

Fig. 2

(The answer is known to be positive for m =2, 4, 10, 21 and 44.)

Remark. The graph A_{2p-1} ($p \geq 2$) from Proposition 1 "almost" spans Q_{2p-1}. It can be modified by adding 2^{p-1} new vertices and the same number of edges to become a tree on 2^{2p-1} vertices. One can see from the proof of Proposition 1 how to add these new vertices and edges so that the construction of embedding described there remains valid. (Cf. Fig. 3, where two possibilities of constructing spanning trees of Q_5 from A_7 are shown as an example).

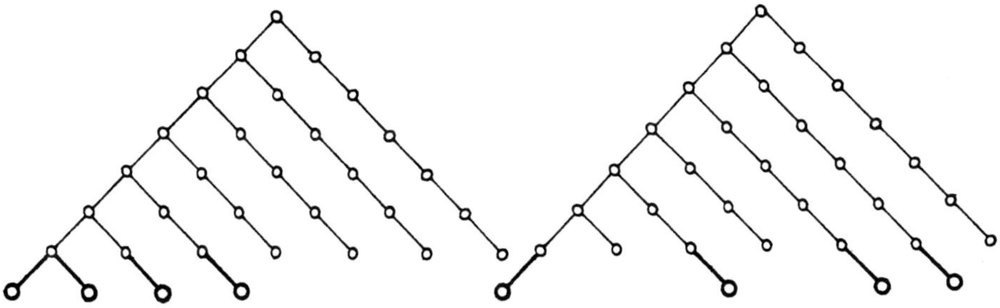

Fig. 3

Two more conjectures, stated in [**15**] are closely related to our Conj. 3:

CONJ. 4. *Any tree with $\Delta(T) \leq 3$ can be embedded into its next–to–optimal hypercube with dilation 1.*

CONJ. 5. *Any tree T with $\Delta(T) \leq 3$ can be embedded into its optimal hypercube with dilation 2.*

Both these conjectures are weaker than Conj. 3. To prove Conj. 3 $\Longrightarrow$ Conj. 4 is easy (cf.[1]), whereas the proof of Conj. 3 $\Longrightarrow$ Conj. 5 is done in [1] by means of a lemma which brings together embeddability with dilation 1 and that with dilation $d > 1$. This lemma makes it possible to prove even more, namely to insert between Conj. 3 and Conj. 5 the following Conj. 6 which is stronger than Conj. 5 but still weaker than Conj. 3:

CONJ. 6. *Any tree T with $\Delta(T) \leq 5$ can be embedded into its optimal hypercube with dilation 2.*

Let us note that the condition $\Delta(T) \leq 5$ on the maximal degree of T is again essential in the following sense: It is possible to construct for any $n \geq 7$ a tree T_n with $\Delta(T_n) = 6$ and $|V(T_n)| = 2^n$ such that $T_n \overset{2}{\not\subseteq} Q_n$ (cf.[1]).

The relationship among the conjectures 2—6 is shown in Fig. 4.

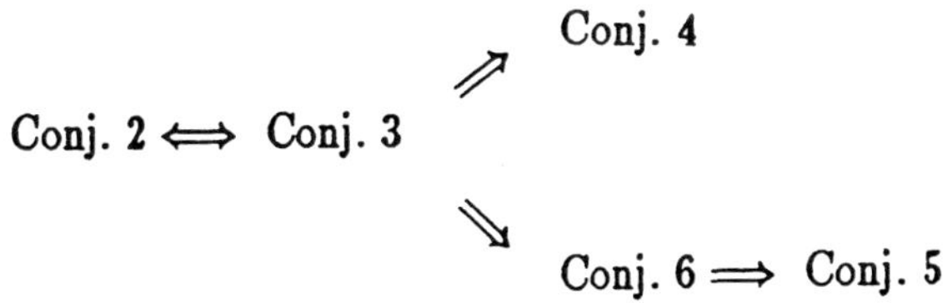

Fig. 4

Using the above mentioned lemma from [1] it is possible to prove that Conj. 5 holds for generalized one–legged caterpillars:

PROPOSITION 2 ([1]). *Any generalized one–legged caterpillar can be embedded into its optimal hypercube with dilation 2.*

Further, it is well known and easy to show that the complete binary tree can be embedded into its optimal hypercube with dilation 2. Using again the lemma and an old result on the cubical dimension of the complete binary tree ([6]) one gets

PROPOSITION 3([1]). *The complete 4–ary tree can be embedded into its optimal hypercube with dilation 2.*

Of course, the maximum degree of the complete 4–ary tree is 5, hence Proposition 3 can be considered to support Conj. 6. An interesting problem arises in this connection; the answer being positive for complete 2–ary and 4–ary trees, what is the situation for complete 3–ary trees:

PROBLEM 2. *Is it true that the complete 3–ary tree can be embedded into its optimal hypercube with dilation 2?*

(If Conj. 6 holds, the answer is positive).

References

1. T. Dvořák, I. Havel, J.–M. Laborde and P. Liebl, *Generalized hypercubes and graph embedding with dilation*, Rost. Math. Kolloq. (to appear).
2. M. R. Garey and R. L. Graham, *On cubical graphs*, J. Combin. Theory B **18** (1975), 84–95.
3. V. A. Gorbatov and A. A. Kazanskij, *Characterization of graphs embedded in n–cube*, Engng. Cybernetics **20** (1983), 96–102.
4. F. Harary, J. P. Hayes and Horng–Jyh Wu, *A survey of the theory of hypercube graphs*, Comput. Math. Applic. **15** (1988), 277–289.
5. I. Havel, *On hamiltonian circuits and spanning trees of hypercubes*, Čas. Pěst. Mat. **109** (1984), 135–152.
6. I. Havel and P. Liebl, *Embedding the dichotomic tree into the cube*, (in Czech with English summary), Čas. Pěst. Mat. **97** (1972), 201–205.
7. I. Havel and P. Liebl, *One–legged caterpillars span hypercubes*, J. Graph Theory **10** (1986), 69–78.
8. I. Havel, J.–M. Laborde and P. Liebl, *Caterpillar–like trees spanning hypercubes*, Čas. Pěst. Mat. (to appear).
9. J.–M. Laborde, *Le plongements des arbres à au plus 16 sommets dans l'hypercube*, Rapport de Recherche 7551, LSD IMAG Grenoble.
10. N. B. Limaye, *6–quasistars and n–cubes*, Rapport de Recherche 563, LSD IMAG Grenoble.
11. L. Nebeský, *On cubes and dichotomic trees*, Čas. Pěst. Mat. **99** (1974), 164–167.
12. L. Nebeský, *On quasistars in n–cubes*, Čas. Pěst. Mat. **109** (1984), 153–156.
13. L. Nebeský, *Embedding m–quasistars into n–cubes*, Czech. Math. Journ. **38** (1988), 705–712.
14. A. L. Rosenberg, *Issues in the study of graph embedding*, in "Graph–theoretic concepts in computer science," H. Noltmeier, ed., LNCS 100, Springer–Verlag, 1981.
15. A. S. Wagner, *Embedding trees in the hypercube*, Technical report 204/87, University of Toronto (1987).

Extending Hall's Theorem

A. J. W. Hilton, Reading, Great Britain
P. D. Johnson Jr., Alabama, USA

1. Introduction

Let G be a finite simple graph. Suppose that to each vertex $v \in V(G)$ there is assigned a finite set (or "list") $C(v)$ of colours (or "symbols"). The general problem is: what conditions on G and the colour-set assignment C guarantee that the vertices of G can be coloured so that each $v \in V(G)$ is coloured with a colour from $C(v)$, and adjacent vertices are coloured differently? Such a colouring of $V(G)$ will be called a C-colouring of G.

Ryser's theorem [7] on extending Latin rectangles to Latin squares can be regarded as a successful imposition of such conditions in the special case when $G = L(K_{n,n})$, the line graph of $K_{n,n}$ (see [6]). Hall's theorem ([5]; see also [8], chapter 5) solves the problem as completely as one could wish when $G = K_n$. This view of Hall's theorem is due to Vizing [10].

There are two well known, currently unsolved problems that conform to our topic. The <u>edge list colouring problem</u> (see [2], [3], [6] and [9]) asks if there is necessarily a C-colouring of G when $G = L(H)$ is a line graph and $|C(v)| \geq \chi(G) = \chi'(H)$ for all $v \in V(G) = E(H)$. <u>Dinitz's problem</u> (see [2], [4], and [6]) is the special case when $H = K_{n,n}$.

The vertex list-colouring problem (just as stated above, except that G is not a line graph) has been settled in the negative (see [2], [3], [4], and [10]). A counter-example is given in Figure 1. It is not the smallest possible - see [4] - but it suits the purposes of this paper.

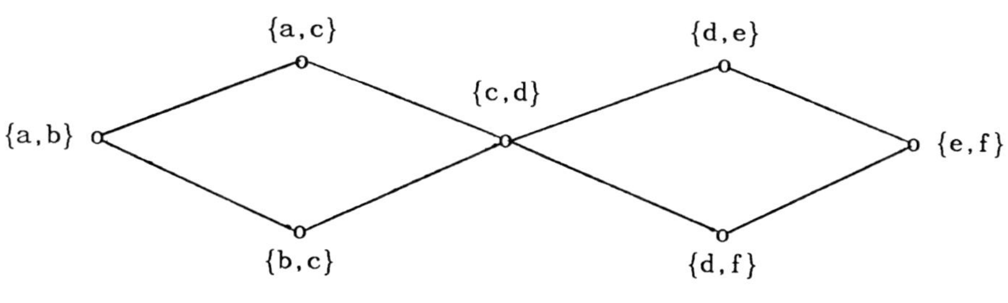

Figure 1

R. Bodendiek, R. Henn (Eds.)
Topics in Combinatorics and Graph Theory

Our aim here is to organize an attack on the general problem by singling out an obviously necessary condition on G and C for C-colourability, and then asking which G are C-colourable whenever that necessary condition is fulfilled; for those that are not, we begin the search for additional hypotheses that will force C-colourability. Our main results are in section 3. The necessary condition referred to, to be defined below, is a straightforward generalization of the condition given in Hall's theorem [5].

2. Definitions and Preliminary Lemmas

Let S be an infinite set and $\mathcal{F}$ the collection of finite subjsets of S. S will be the set of colours, or symbols, at our disposal. Only finitely many members of S will be involved in the discussion at any one time. A vertex colour set assignment to a simple graph G is a function $C: V(G) \to \mathcal{F}$. In the definitions to follow, G will be a finite simple graph and C a vertex colour set assigment to G. When H is a subgraph of G, the restriction of C to V(H) will also be denoted C.

I. A C-colouring of G is a function $\varphi: V(G) \to S$ such that $\varphi(v) \in C(v)$ for all $v \in V(G)$, and, if uv is an edge of G, then $\varphi(u) \neq \varphi(v)$. The restriction of such a φ to any subset of V(G) will continue to be denoted by φ. If φ is a C-colouring of G, $\sigma \in S$, and H is a subgraph of G, let $n(H,\varphi,\sigma) = |\varphi^{-1}(\{\sigma\}) \cap V(H)|$, the number of vertices of H coloured σ by φ. Observe that $|V(H)| = \sum_{\sigma} n(H,\varphi,\sigma)$.

II. Suppose that $\sigma \in S$. A σ-transversal in G under C is an independent set of vertices in $\{v \in V(G); \sigma \in C(v)\}$. If G and C are fixed in the discussion, we may omit the phrase "in G under C".

The maximum size of a σ-transversal will be denoted $t(G,C,\sigma)$, and any σ-transversal of this size will be called a maximum σ-transversal (in G, under C).

III. We will say that the pair (G,C) satisfies Hall's Condition if and only if, for each subgraph H of G,

$$(*)\qquad |V(H)| \leq \sum_{\sigma \in S} t(H,C,\sigma) .$$

Lemma 1. Hall's Condition is necessary for the existence of a C-colouring

<u>of</u> G.

<u>Proof</u>. If φ is a C-colouring of G, and H is a subgraph of G, then, for each $\sigma \in S$, $\varphi^{-1}(\{\sigma\}) \cap V(H)$ is an independent set of vertices in H, so $n(H,\varphi,\sigma) \leq t(H,C,\sigma)$. Therefore,

$$|V(H)| = \sum_{\sigma \in S} n(H,\phi,\sigma) \leq \sum_{\sigma \in S} t(H,C,\sigma) . \qquad \square$$

<u>Lemma</u> 2. (G,C) <u>satisfies Hall's Condition if and only if</u> (*) <u>holds for each connected</u>, <u>induced subgraph</u> H <u>of</u> G.

<u>Proof</u>. The "only if" assertion is trivial.

Suppose that H is a subgraph of G. If H has components $H_1,\ldots,H_k$, then $t(H,C,\sigma) = \sum_{i=1}^{k} t(H_i,C,\sigma)$, for each σ, so (*) holds if it holds with H replaced by each H_i. If $\hat{H}$ is the subgraph of G induced by V(H), then $t(\hat{H},C,\sigma) \leq t(H,C,\sigma)$, for each $\sigma \in S$, and $V(H) = V(\hat{H})$, so (*) holds if it holds with H replaced by $\hat{H}$. $\square$

When G is a complete graph and H is an induced subgraph of G, then H is complete, so $t(H,C,\sigma)$ is zero or one depending on whether or not $\sigma \in \bigcup_{v \in V(H)} C(v)$. Consequently, in this case,

$$\sum_{\sigma \in S} t(H,C,\sigma) = \Big| \bigcup_{v \in V(H)} C(v) \Big| .$$

Thus, in view of Lemma 2, it is straightforward to see that Hall's theorem [5] states that Hall's Condition is sufficient for the existence of a C-colouring of G, when G is complete.

IV. Suppose that m is a non-negative integer. G will be called Hall(m) if and only if Hall's Condition and the requirement that $|C(v)| \geq m$ for all $v \in V(G)$ suffice to ensure the existence of a C-colouring of G.

Clearly all Hall(m) graphs are Hall(m+1).

Note that if (G,C) satisfies Hall's Condition, then C(v) is non-empty for each $v \in V(G)$. (Take H to be the single vertex v, in (*).) Thus Hall(0) and Hall(1) are the same thing, and amount to Hall's Condition being sufficient for the existence of a C-colouring of G. In this case, G will be, simply, Hall.

The <u>Hall</u> <u>number</u> of G, denoted h(G), is the smallest positive integer m such that G is Hall(m). (Thus, G is Hall if and only if $h(G) = 1$.)

V. We will say that G is list-colourable if and only if there is a C-colouring of G for every vertex colour set assignment C satisfying $|C(v)| \geq \chi(G)$. That is, the list-colourable graphs are those for which the (now exploded) vertex list-colouring conjecture holds.

VI. Following [4], G is m-*choosable* if and only if there is a C-colouring of G for every vertex colour set assignment C satisfying $|C(v)| \geq m$. The *choice number* of G, denoted $c(G)$, is the smallest integer m such that G is m-choosable. (Vizing [10] calls the choice number the "upper chromatic number ".)

If G is m-choosable, then, clearly, $m \geq \chi(G)$ and G is Hall(m). The converse holds, by Theorem 1, below. It is straightforward to see that $c(G) \geq \chi(G)$, with equality if and only if G is list-colourable, and that $c(G) \geq h(G)$. Note Corollary 2, below, for further relations among these numbers, and see section 5 for further remarks on m-choosability.

3. Main Results

Theorem 1. *If G is a simple graph, C is a vertex colour set assignment to G, and $|C(v)| \geq \chi(G)$ for all $v \in V(G)$, then (G,C) satisfies Hall's Condition.*

Corollary 1. *G is list-colourable if and only if G is Hall ($\chi(G)$).*

Corollary 2. *If either $c(G) > \chi(G)$, or $h(G) \geq \chi(G)$, then $c(G) = h(G)$.*

Remark 1. It is, perhaps, surprising that there exist graphs G for which $\chi(G)$ is not the smallest integer m with the property that (G,C) satisfies Hall's Condition for every $C\colon V(G) \to \mathcal{F}$ satisfying $|C(v)| \geq m$ for all $v \in V(G)$. See section 5 for examples.

Remark 2. In [6] we proved Theorem 1 for line graphs. The proof given here, below, is not much different from that in [6].

Theorem 2. *Every induced subgraph of a finite Hall(m) graph is Hall(m).*

Remark. Every complete graph is Hall, and thus Hall(m) for all $m \geq 1$. We shall see later that there are non-Hall(m) graphs, for each m. It follows that the word "induced" cannot be deleted from Theorem 2.

Theorem 3. *A simple graph G is Hall if and only if every block of G is a clique.*

(As in [1], a "block of G" is a maximal connected subgraph of G with no cut vertex.)

Corollary 3. If every block of G is a clique, then G is list-colourable.

Remark. Corollary 3 is also a corollary of the following theorem on operations under which the class of list-colourable graphs is closed.

Theorem 4. Suppose that G_1 is a list-colourable graph. The following are list-colourable:

(a) any subgraph of G_1 with the same chromatic number;

(b) any graph obtained by attaching another graph G_2 to G_1 at a set $A = V(G_1) \cap V(G_2)$ of vertices, provided

(i) A is an independent set of vertices in G_2 and

(ii) $c(G_2 - A) \leq \chi(G_1) - |A|$;

(c) any graph obtained by attaching a complete graph at a single vertex of G_1;

(d) any graph obtained by adjoining a new vertex adjacent to every vertex in a clique of G_1.

4. Proofs, and Intermediate and Minor Results

Proof of Theorem 1. Suppose that H is a subgraph of G, and let U be a set of independent vertices of H, of maximum size. Note that $|U|\chi(H) \geq |V(H)|$. Let H_o be the subgraph of H induced by U. Since H_o is an induced subgraph of H, $t(H,C,\sigma) \geq t(H_o,C,\sigma)$ for each $\sigma \in S$. Since H_o is, in fact, a collection of $|U|$ isolated vertices, $t(H_o,C,\sigma) = |\{v \in U;\ \sigma \in C(v)\}|$, for each $\sigma \in S$, and since $|C(v)| \geq \chi(G)$ for each $v \in U$, we have that $\sum_{\sigma \in S} t(H_o,C,\sigma) = \sum_{v \in U} |C(v)| \geq |U|\chi(G)$.

Consequently, $\sum_{\sigma \in S} t(H,C,\sigma) \geq \sum_{\sigma \in S} t(H_o,C,\sigma) \geq |U|\chi(G) \geq |U|\chi(H) \geq |V(H)|$ □

Corollaries 1 and 2 are immediate from Theorem 1 and the definitions.

Proof of Theorem 2. Suppose that G is Hall (m), and that F is an induced subgraph of G. Suppose that C is a vertex colour set assignment to F, such that (F,C) satisfies Hall's Condition, and $|C(v)| \geq m$ for each $v \in V(F)$. Extend C to $V(G)$ by letting $C(v)$ be some finite subset of S with at least $\max(m, |V(G)|)$ elements, for $v \in V(G) \setminus V(F)$. Thus extended, C clearly satisfies $|C(v)| \geq m$ for all $v \in V(G)$. It is also true that (G,C) satisfies Hall's Condition. If H is a subgraph of G with $V(H) \subseteq V(F)$, then H is a subgraph of F (since F is an induced subgraph of G), so the inequality (*) holds because (F,C) satisfies Hall's Condition. If H has a vertex outside V(F), then that one vertex has enough colours assigned to it to imply (*).

Since G is Hall(m), G has a C-colouring. Therefore, so does F; C was arbitrary, so F is Hall(m). □

Lemma 3. Suppose that F is a graph with a Hamiltonian cycle such that there are two vertices a distance two apart on the cycle, which are not adjacent in F. Then F is not Hall.

Proof. The hypothesis implies that the order of F is ≥ 4. In the case $|V(F)| = 4$, Figure 2 (in which the broken line indicates the missing edge) depicts a vertex colour set assignment that shows that F is not Hall.

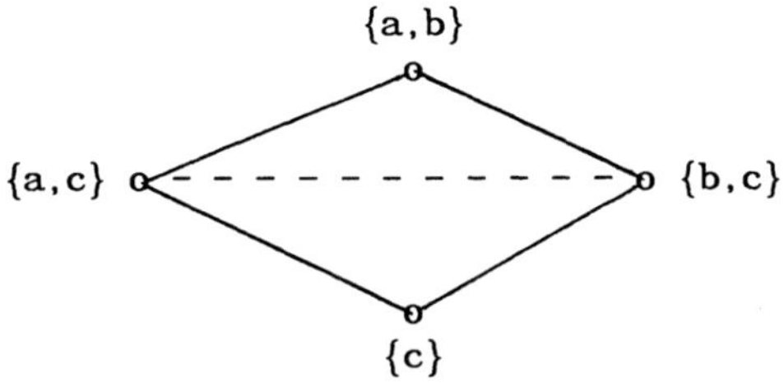

Figure 2

Something very similar was introduced in [6]. The case $|V(F)| \geq 5$ is not much different. Let $v_o, v_1, \ldots, v_k, v_o$, $k \geq 4$, be the Hamiltonian cycle in F, and suppose that $v_o v_2$ is not an edge of F. Assign {c} to v_1, $\{a_1,c\}$ to v_2, $\{a_{i-2}, a_{i-1}\}$ to v_i, $3 \leq i \leq k$, and $\{a_{k-1},c\}$ to v_o, where $a_1,\ldots,a_{k-1}$ and c are distinct members of S.

Let C be the colour set assignment defined above, on V(F). Clearly F has no C-colouring, since setting $\varphi(v_1) = c$ forces $\varphi(v_2) = a_1$, $\varphi(v_3) = a_2$, etc., and finally $\varphi(v_k) = a_{k-1}$, which leaves us in difficulties when it comes to colouring v_o. To see that F and C satisfy Hall's Condition, first observe that any subgraph obtained by deleting an edge on the cycle is C-colourable. Therefore, (*) holds for H, a proper induced subgraph of F, by Lemma 1. It remains to verify (*) when H = F. We have $t(F,C,c) = 2$ and $t(F,C,a_i) = 1$, $i = 1,\ldots,k-1$, whence $\sum_\sigma t(F,C,\sigma) = k+1 = |V(F)|$. □

Lemma 4. *If B is a block, and B is not complete, then B is not Hall.*

Proof. By Theorem 2 and Lemma 3, it suffices to show that B has an induced subgraph F as described in Lemma 3. The existence of F follows from Theorem R of [4], but we present a self-contained proof here.

Since B is a block, but is not complete, B does contain a cycle such that not every unordered pair of vertices on the cycle is an edge of B (see [1], pp. 36-9). If, on any such cycle, there are vertices two apart on the cycle which are not connected by an edge in B, then B is not Hall, by Lemma 3 and Theorem 2. So, suppose that $u_o, u_1, \dots, u_{n-1}, u_o$ is a cycle in B such that not every edge u_iu_j, $0 \le i < j \le n-1$, is in B, and let $k \ge 3$ be the smallest integer ≥ 2 such that $u_iu_{i+k} \notin E(B)$ for some $i \in \{0,\dots,n-1\}$ (where $i+k$ is reduced mod n).

Renumbering the vertices, we may as well assume that $u_ou_k \notin E(B)$ (and $k \le \frac{n}{2}$, so $n \ge 6$); we have that $u_iu_{i+r} \in E(B)$ for $0 \le i \le n-1$, $1 \le r \le k-1$, with all indices reduced mod n. Therefore, $u_o, u_1, u_k, u_2, \dots, u_{k-1}, u_{k+1}, u_{k+2}, \dots, u_{n-1}, u_o$ is a cycle in B, and $v_o = u_o$ and $v_2 = u_k$ are not adjacent in B. □

Lemma 5. Suppose that G_1 is Hall, and that G is obtained by attaching a complete graph to G_1 at a vertex of G_1. Then G is Hall.

Proof. Let k be the order of the complete graph K attached at the vertex v of G_1. We proceed by induction on k. The proof is modelled on the proof of Hall's Theorem in [8], Chap. 5.

If $k = 1$ then $G = G_1$. Suppose that $k > 1$.

Suppose that C is a vertex colour set assignment to G, such that (G,C) satisfies Hall's Condition. Applying (*) when H is a subgraph of K, we have that Cases 1 and 2 below exhaust the possibilities.

Case 1: $\left| \bigcup_{u \in U} C(u) \right| \ge |U| + 1$ for all non-empty subsets U of $V(K) \setminus \{v\}$. In this case, note that since (G_1,C) satisfies Hall's Condition, and G_1 is Hall, there is a C-colouring φ of G_1. For $u \in V(K) \setminus \{v\}$, let $C'(u) = C(u) \setminus \{\varphi(v)\}$. Then, for $U \subseteq V(K) \setminus \{v\}$, $\left| \bigcup_{u \in U} C'(u) \right| \ge |U|$. Thus, K-v and C' satisfy the condition in Hall's theorem, and so K-v has a C'-colouring. Clearly, extending φ to $V(K) \setminus \{v\}$ by means of this C'-colouring gives a C-colouring of G.

Case 2: $\left| \bigcup_{u \in U} C(u) \right| = |U|$ for some non-empty $U \subseteq V(K) \setminus \{v\}$. Let $r = |U| \le k-1$ and let H be the subgraph of $K \subseteq G$ induced by U; H is a copy of K_r. Since (H,C) satisfies Hall's Condition, H has a

C-colouring. Suppose that $a_1,\dots,a_r$ are the colours assigned by this colouring to the elements of $V(H) = U$. Keep in mind that $\{a_1,\dots,a_r\} = \bigcup_{u \in U} C(u)$, since $|\bigcup_{u \in U} C(u)| = |U| = r$.

For $w \in V(K) \setminus U$, set $C'(w) = C(w) \setminus \{a_1,\dots,a_r\}$. For $w \in V(G) \setminus V(K)$, set $C'(w) = C(w)$. We claim that G_2, the subgraph of G induced by $V(G) \setminus U$ (which is also obtainable from G_1 by attaching a copy of K_{k-r} at v), and C' satisfy Hall's Condition. If this holds, then, by induction on K, since $r \geq 1$, G_2 has a C'-colouring, which we can put together with the C-colouring of H to obtain a C-colouring of G (since the only vertices of $V(G) \setminus U$ to which the vertices in U are adjacent are in $V(K) \setminus U$).

Suppose, contrary to the claim, that there is a subgraph H_2 of G_2, which by Lemma 2 we may as well take to be induced and connected, such that

$$\sum_{\sigma \in S} t(H_2,C',\sigma) < |V(H_2)| .$$

Let H_1 be the subgraph of G induced by $V(H_2) \cup U$. Since $\bigcup_{u \in U} C(u) = \{a_1,\dots,a_r\}$, and H is complete, it follows that $t(H_1,C,\sigma) = t(H_2,C',\sigma)$ for $\sigma \in S\setminus\{a_1,\dots,a_r\}$, and $t(H_1,C,a_i) = t(H_2,C',a_i)+1$, $i = 1,\dots,r$. Thus

$$\sum_{\sigma \in S} t(H_1,C,\sigma) = r + \sum_{\sigma \in S} t(H_2,C',\sigma) < r + |V(H_2)| = |V(H_1)|,$$

contradicting the assumption that (G,C) satisfies Hall's Condition. □

<u>Proof of Theorem</u> 3. The necessity is immediate from Lemma 4, Theorem 2, and the observation that every block of G is an induced subgraph of G. For connected graphs, the sufficiency follows by induction on the number of blocks, from Hall's theorem, Lemma 5, and Theorem 2.19, p. 37, of [1], which says that for connected graphs G with more than one block, there is a "last" block, attached to G at a cut vertex. For disconnected graphs, the sufficiency now follows from a trivial consequence of the definitions: if every component of G is Hall(m) then so is G. □

<u>Lemma</u> 6. <u>Suppose that</u> G_1 <u>is Hall</u>(m), $A = V(G_1) \cap V(G_2)$ <u>is an independent set of vertices in</u> G_2 <u>and</u> $c(G_2 - A) \leq m - |A|$. <u>Then</u> $G = G_1 \cup G_2$ <u>is Hall</u>(m).

<u>Proof</u>. Suppose that C is a vertex colour set assignment to G, that $|C(v)| \geq m$ for all $v \in V(G)$, and that G and C satisfy Hall's

Condition. Then G_1 is C-colourable. Let $A = V(G_1) \cap V(G_2) = \{v_1,\dots,v_t)$ where $t = |A|$, and let $a_1,\dots,a_r$, $r \leq t$, be the distinct colours assigned by a C-colouring of G_1 to $v_1,\dots,v_t$. (The case $t = 0$ is trivial.) For each $v \in V(G_2) \setminus A$, let $C'(v) = C(v) \setminus \{a_1,\dots,a_r\}$. Then $|C'(v)| \geq |C(v)|-r \geq m-r \geq m-t \geq c(G_2-A)$, so $G_2 - A$ is C'-colourable. Putting this colouring together with the C-colouring of G_1 (and noting that the only edges among the vertices of $V(G_1) \cap V(G_2)$ are G_1 edges), we obtain a C-colouring of G. □

Proof of Theorem 4. The assertion regarding (a) follows from Corollary 1 and Theorem 2 for induced subgraphs, so it suffices to consider the case when H is obtained from G by removing edges. Since any C-colouring of G is also a C-colouring of H, in this case, the result follows. The assertion regarding (b) follows from Corollary 1 and Lemma 6, and the observation that, if G, G_1, and G_2 are as in Lemma 6, then $\chi(G_1) \leq \chi(G)$. (In fact, it is evident from the proof of Lemma 6 that $\chi(G_1) = \chi(G)$.)

To see (c), suppose that G is obtained from G_1, a list-colourable graph, by attaching a complete graph $G_2 = K_r$ at a vertex v. If $r \leq \chi(G_1)$, then it follows from the validity of (b) that G is list-colourable. If $r > \chi(G_1)$, then $\chi(G) \geq r > \chi(G_1)$. Suppose that C is a vertex colour set assignment to G satisfying $|C(u)| \geq \chi(G)$ for all $u \in V(G)$. Since G_1 is list-colourable, there is a C-colouring of G_1. Remove the colour assigned to v from each set C(u), $u \in V(G_2) \setminus \{v\}$, to obtain a vertex colour set assignment C' to $G_2-v \simeq K_{r-1}$, with $|C'(u)| \geq \chi(G)-1 \geq r-1$ for each $u \in V(G_2) \setminus \{v\}$. Then G_2-v has a C'-colouring, so G has a C-colouring, and is therefore list-colourable. (The proof also shows what can be easily verified directly, that $\chi(G) = \max[r, \chi(G_1)]$.)

The proof of the assertion regarding (d) is similar, with $K_{1,r}$ playing the role of K_r above. If $r \leq \chi(G_1)-1$, the desired conclusion follows from the validity of (b). We cannot have $r > \chi(G_1)$, because G_1 cannot contain a clique of order greater than $\chi(G_1)$. That leaves the case $r = \chi(G_1)$. In this case, G contains a clique of order $\chi(G_1) + 1$, so $\chi(G) \geq \chi(G_1)+1$. The list-colourability of G now follows easily from that of G_1; we omit the details.

5. Remarks and Problems

Hall(2) graphs

First, we will see that every cycle is Hall(2). Suppose that G is a cycle, and that C is a vertex colour set assignment to G such that $|C(v)| \geq 2$ for all $v \in V(G)$, and (G,C) satisfies Hall's Condition. If $|C(u) \cup C(v)| \geq 3$ for adjacent vertices $u, v \in V(G)$, then G is C-colourable. For suppose $a \in C(v)$ and $b, c \in C(u)$, with a,b,c distinct. Colour v with a and go around the cycle moving away from u, colouring each vertex with a colour different from the colour of the preceding vertex. When the crunch comes at u, we find that we can successfully colour u with either b or c.

So we may as well suppose that $|C(v)| = 2$ for all $v \in V(G)$, and that the sets $C(v)$ are all the same, say $C(v) = \{a,b\}$ for all $v \in V(G)$. If $|V(G)|$ is odd, the assumption that (G,C) satisfies Hall's Condition is violated, since then $t(G,C,a) = t(G,C,b) = \frac{|V(G)|-1}{2}$, so we are done with that possibility. If $|V(G)|$ is even, then G is clearly C-colourable. Thus every cycle is Hall(2).

It follows from Theorem 3 that the class of Hall(2) graphs properly contains the class of Hall graphs.

Figure 3 shows a graph which is not Hall(2). Figure 1 provides another example.

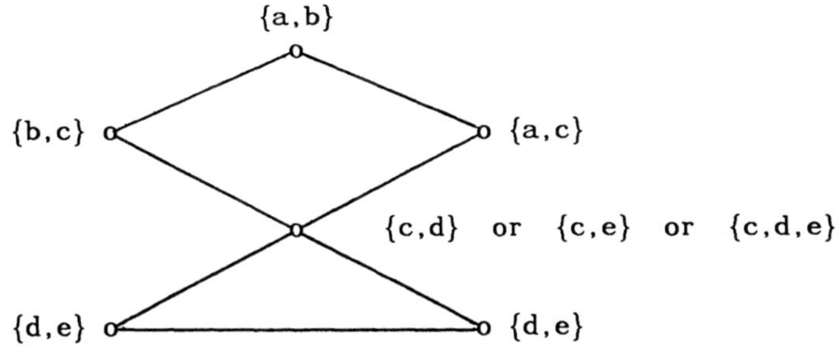

Figure 3

As Figure 2 is the foundation of an extensive class of examples, so are Figures 1 and 3. Graphs prepared in the following way, which we will refer to as AC ("attached cycles") graphs, are not Hall(2). (The proof is omitted; it is similar to that of Lemma 3.) Take two Hamiltonian graphs, G_1 and G_2, and attach them at a vertex v such that, for some Hamiltonian cycle in G_1, the two vertices on either side of v on the cycle are not adjacent in G_1. Throw in any $V(G_1) - V(G_2)$ edges you wish.

The line graph of ($K_{3,3}$-minus-two-independent-edges) is an AC graph, so not every line graph, even of a bipartite graph, is Hall (2). (This was noted in [6], in different terms.)

By Theorem 2, every graph with an AC induced subgraph is not Hall (2). This is a start, but we are still far from a succinct characterization of the Hall (2) graphs. Notice that, by Theorem 1, a graph is 2-choosable if and only if it is bipartite and Hall(2). The 2-choosable graphs are well-described in [4]; consequently, we know which bipartite graphs are Hall(2), and which are not. Among those that are not are several basic types that do not have an AC induced subgraph. We list them here, and refer to [4] for proofs and explanation.

Some graphs that are not Hall (2), with no AC induced subgraph

1. Two vertex-disjoint even cycles connected by a path.
2. Three internally disjoint paths between two distinguished vertices, all of even length with at most one path of length 2, or all of odd length.
3. Four internally disjoint paths of even length between two distinguished vertices.

There are similar classes of non-bipartite, non-Hall(2) graphs, most having no AC induced subgraph.

As a last remark on Hall(2) graphs, note that the example in Figure 3 shows that the statement analogous to Lemma 5, obtained by replacing "Hall" in the lemma by "Hall(2)", is false.

Hall(3) graphs

We repeat here a conjecture given in [6], in different terms.

Conjecture. Every line graph is Hall(3).

The truth of this conjecture would settle the edge list-colouring conjecture, and thus Dinitz's problem.

Hall(m) graphs, for large m, and list-colourability

It is shown in [4] and [10] that there are bipartite graphs with arbitrarily large choice number. It follows from this, as mentioned in [2], that, for any integers k,m satisfying $2 \leq k \leq m$, there exists a graph G with $\chi(G) = k$ and $c(G) = m$. To see this, form a bipartite graph H with $c(H) \geq m$. Form H_1 by adding on a disjoint clique K_k, so that $\chi(H_1) = k$. Then delete edges from H until the choice number falls to exactly m. (Deleting an edge either leaves the choice number the same, or diminishes it by 1.) Taking a little more trouble, we could arrange for G to be connected, but this is not important in the present discussion.

Taking $m = k+1$, and noting Corollaries 1 and 2, we see that G is Hall(k+1) but not Hall(k), so the former class properly contains the latter, when $k \geq 2$. We have previously noted that this is also true for $k = 1$.

A fortiori, there is no m such that every graph is Hall(m). In fact, the Hall(m) graphs are sparse among graphs of large order. From Theorem 2 and the fact that, for each m, there exist non-Hall(m) graphs, it follows that the probability of a random graph G on n vertices being Hall(m) tends to 0 as $n \to \infty$. (This assumes that each edge has fixed probability $p \in (0,1)$ of appearing in G, with these appearances being independent.)

Nevertheless, we conjecture that almost all graphs are list-colourable.

Graphs for which Theorem 1 is not sharp.

For a graph G, let $s(G)$ denote the smallest integer m with the property that (G,C) satisfies Hall's Condition for every $C:V(G) \to \mathcal{F}$ satisfying $|C(v)| \geq m$ for all $v \in V(G)$, and let $s_o(G)$ denote the number defined similarly, except that C is restricted to constant maps, i.e. assignments of the same colour list to the different vertices.

Clearly $s_o(G) \leq s(G)$. Theorem 1 asserts that $s(G) \leq \chi(G)$. When is this inequality strict?

It will expedite the discussion if we first show that $s_o(G) = s(G) = \max \left\{ \left\lceil \frac{|V(H)|}{i(H)} \right\rceil ; \ H \text{ is a subgraph of } G \right\}$, where $i(H)$ denotes the independence number of H, the largest size of a set of vertices independent in H. For short, let $\max_H \left\{ \left\lceil \frac{|V(H)|}{i(H)} \right\rceil \right\} = M(G)$. Suppose that C assigns the same set of m colours, $\{\sigma_1,\ldots,\sigma_m\}$, to each $v \in V(G)$. Then for any subgraph H of G, $t(H,C,\sigma_j) = i(H)$ for each j, so $\sum_{\sigma} t(H,C,\sigma) = mi(H) \geq |V(H)|$ if and only if $m \geq \left\lceil \frac{|V(H)|}{i(H)} \right\rceil$; thus the inequality (*) holds for each H if $m = M(G)$, and fails for some H if $m < M(G)$, so $s_o(G) = M(G)$.

Now suppose that $C: V(G) \to \mathcal{F}$ is any assignment satisfying $|C(v)| \geq m$ for all $v \in V(G)$. By the proof of Theorem 1, the inequality (*) will hold for each subgraph H of G such that

$$(**) \qquad i(H)m \geq |V(H)| .$$

If $m = M(G)$, (**) holds for all subgraphs H of G. Thus $s(G) \leq M(G)$. Thus $s_o(G) = s(G) = M(G)$.

It follows that $s(G) \geq \omega(G)$, the clique number of G, and thus that if $\chi(G) = \omega(G)$, then $s(G) = \chi(G)$.

If either $s(G) = 1$ or $\chi(G) = 1$, then there are no edges in G, so $s(G) = \chi(G)$.

If $s(G) = 2$ then there are no odd cycles in G (for, if there were, then $M(G) \geq 3$), so $\chi(G) = s(G)$. Also, $\chi(G) = 2$ implies that $\chi(G) = s(G)$ (since $s(G) = 1$ implies $\chi(G) = 1$).

For any integer $x \geq 4$, there is a graph G with $\chi(G) = x$ and $s(G) = x-1$. Form G by surrounding K_{x-3} with an odd cycle C_{2r+1}, $r \geq 2$. (Any odd cycle except K_3 will do.) Let each vertex of the cycle be adjacent to each vertex of K_{x-3}. Clearly $\chi(G) = x$. It is straightforward to see that $M(G) = x-1$.

Problem. Does there exist a G with $s(G) \leq \chi(G)-2$? How much less than $\chi(G)$ can $s(G)$ be ?

REFERENCES

1. M. Behzad, G. Chartrand, and L. Lesniak-Foster, Graphs and Digraphs, Prindle, Weber & Schmidt, Boston, 1979.
2. A. Chetwynd and R. Häggkvist, *A note on list-colourings*, Report, Department of Mathematics, University of Stockholm, 1986, no. 17.
3. B. Bollobás and A.J. Harris, *List-colourings of graphs*, Graphs and Combinatorics 1 (1985), 115-127.
4. P. Erdös, A. Rubin, and H. Taylor, *Choosability in graphs*, (Proc. West Coast Conf. on Combinatorics, Arcata, Humboldt State University, 1979) Congressus Numerantium 26(1979), 125-157.
5. P. Hall, *On representatives of subsets*, Jour. London Math. Soc. 10 (1935), 26-30.
6. A.J.W. Hilton and P.D. Johnson Jr., *Ryser's theorem and the list-colouring conjecture*, Research Notes in Mathematics (Pitman), to appear.
7. H. Ryser, *A combinatorial theorem with an application to Latin rectangles*, Proc. Amer. Math. Soc. 2(1951), 550-552.
8. H. Ryser, *Combinatorial Mathematics*, Carus Monograph, The Mathematical Association of America, John Wiley, 1963.
9. B. Toft, *75 Graph Colouring Problems*, presented at the Open University one-day conference, 9 December 1988; Mathematics and Computer Science, Odense University, DK-5230 Odense M, Denmark.
10. V.G. Vizing, *Colouring the vertices of a graph in prescribed colours* (Russian), Diskret. Analiz. No. 29 Metody Diskret. Anal. V Teorii Kodov i Shem (1976) 3-10, 101 (MR58 16371).

On the Coupling Condition and Hamiltonicity

C. Hoede, Enschede, The Netherlands
H. J. Smit, Enschede, The Netherlands

Abstract
The coupling condition is a necessary and sufficient condition for a graph to be hamiltonian. It is shown that the condition also applies to directed graphs and hypergraphs.

1. INTRODUCTION

We use the terminology of [3]. In [4] the following results were derived.

Theorem 1. Let G be a 2-connected graph. The following statements are equivalents.

(i) G is hamiltonian.
(ii) For any pair of cycles there is a cycle in G that contains all points of the two cycles.
(iii) For any pair of cycles C_1 and C_2 and any pair of points P_1 and P_2, contained in C_1 but not in C_2 and in C_2 but not in C_1, respectively, there is a cycle in G that contains the points that C_1 and C_2 have in common as well as P_1 and P_2.

Statement (iii) in this characterization of hamiltonian graphs is called the *Coupling Condition*. As checking hamiltonicity is an NP-complete problem, see [1], this condition is not easily applied and must be seen as of theoretical interest mainly. However, Hoede and Veldman used the coupling condition to prove

Theorem 2. Every non-hamiltonian 2-connected graph can be contracted to θ or to θ^*.

R. Bodendiek, R. Henn (Eds.)
Topics in Combinatorics and Graph Theory

The two graphs θ and θ^* are depicted in Figure 1. $\theta \cong K_{2,3}$ and $\theta^* \cong K_{2,3}$ with a line joining the points of degree 3 added.

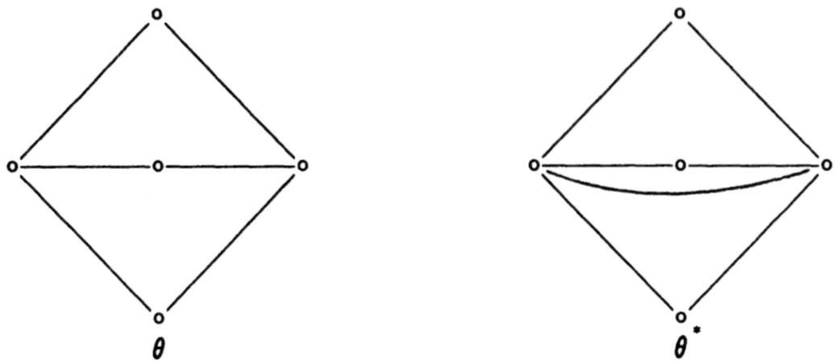

Figure 1

Theorem 2 was considerably extended in [5]. In this paper we will show that the coupling condition also can be used to characterize hamiltonian directed graphs and hypergraphs. For ease of notation we denote by C_1 and C_2 the point sets of these cycles (in stead of $V(C_1)$ and $V(C_2)$). The coupling condition can then be described as

$$P_1 \in C_1 - C_2 \wedge P_2 \in C_2 - C_1 \Rightarrow \exists C_3: (C_1 \cap C_2) \cup \{P_1, P_2\} \subset C_3,$$

this condition holding for any pair of points P_1 and P_2 and any pair of cycles C_1 and C_2 in the graph G.

2. DIRECTED GRAPHS

For a directed graph $\vec{G}$ the (multi-)graph obtained by replacing arcs by lines in called the *underlying undirected graph*. A cycle in the underlying undirected graph determines a *pseudocircuit* in $\vec{G}$.

In a pseudocircuit $\vec{C}$ of $\vec{G}$ two consecutive arcs that are incident with point P are said to be *in phase* if one arc is incident in P and the other is incident out P. If P is either the head or the tail of both arcs the arcs are *out of phase*. So only if all pairs of consecutive arcs are in phase a pseudocircuit is a circuit. As the coupling condition is about point sets we need

Definition 1. A point P of a directed graph $\vec{G}$ is called *in phase on* a subgraph $\vec{H}$ of $\vec{G}$ if and only if the indegree and the outdegree of P in $\vec{H}$ are greater than or equal to 1.

This definition enables us to state that a pseudocircuit $\vec{C}$ of $\vec{G}$ is a circuit if and only if all points of $\vec{C}$ are in phase on $\vec{C}$.

Lemma 1. If a point P of a weakly 2-connected directed graph $\vec{G}$ has indegree and outdegree greater than 1, there exists a pseudocircuit in $\vec{G}$ on which P is in phase.

Proof. Let G be the undirected underlying graph of $\vec{G}$ and let e and f be two lines of G that correspond to an incoming and an outgoing arc of P in $\vec{G}$. As G is 2-connected there exists a cycle C of G that contains e and f. This cycle is underlying a pseudocircuit $\vec{C}$ on which P is in phase. □

We denote by $V_{ph}(\vec{H})$ the set of points of $\vec{H}$ that are in phase.

Theorem 3. A weakly 2-connected directed graph $\vec{G}$ is hamiltonian if and only if each point is in phase on $\vec{G}$ and for any pair of pseudocircuits $\vec{C}_1$ and $\vec{C}_2$ we have that for any pair of points

$$P_1 \in V_{ph}(\vec{C}_1) - V_{ph}(\vec{C}_2) \wedge P_2 \in V_{ph}(\vec{C}_2) - V_{ph}(\vec{C}_1) \Rightarrow$$

$$\exists \vec{C}_3: \{V_{ph}(\vec{C}_1) \cap V_{ph}(\vec{C}_2)\} \cup \{P_1, P_2\} \subset V_{ph}(\vec{C}_3).$$

Proof. The "only if"-part is rather evident. A Hamilton circuit $\vec{C}_3$ implies that all points are in phase on $\vec{G}$ and on $\vec{C}_3$ so that the coupling condition holds. For the "if"-part we suppose that $\vec{G}$ is non-hamiltonian. Let $\vec{C}$ be a pseudocircuit with maximum number of points in phase. Let P′ be a point that does not belong to $V_{ph}(\vec{C})$. P′ is in phase on $\vec{G}$ by assumption.

By Lemma 1 there exists a pseudocircuit $\overrightarrow{C'}$ containing P′ on which P′ is in phase. As $\vec{C}$ has maximum number of points in phase $V_{ph}(\overrightarrow{C'})$ does not contain all points of $V_{ph}(\vec{C})$. Let $P_1 \in V_{ph}(\vec{C})$ be a point for which $P_1 \notin V_{ph}(\overrightarrow{C'})$. Then there exists by assumption a pseudocircuit $\overrightarrow{C''}$ for which

$$V_{ph}(\overrightarrow{C''}) \supset \{V_{ph}(\vec{C}) \cap V_{ph}(\overrightarrow{C'})\} \cup \{P_1, P'\}.$$

Thus $V_{ph}(\overrightarrow{C''})$ has one point more in common with $V_{ph}(\vec{C})$ than $V_{ph}(\vec{C'})$ has.

Repeating this procedure a pseudocircuit is constructed that contains all points of $V_{ph}(\vec{C})$ and point P′ as points in phase. This contradicts the maximality of $\vec{C}$ with respect to the number of points in phase. $\vec{C}$ therefore contains all points of $\vec{C}$ in such a way that all points are in phase, but this means that $\vec{C}$ is a Hamilton circuit. □

3. HYPERGRAPHS

We consider undirected *hypergraphs* H(V,E), where E is a non-empty set of subsets of V, called *hyperlines*. We recall from [2] that a *path* of length k is a hypergraph in a sequence

$$x_1\ E_1\ x_2\ E_2\ x_3 \dots x_k\ E_k\ x_{k+1}$$

of alternating points and hyperlines, that are all distinct, with the property that $x_i \in E_i$ and $x_{i+1} \in E_i$. A *cycle* in a hypergraph arises if $x_{k+1} = x_1$.
The extension of the coupling condition to hypergraphs seems immediate, but there is a problem concerning the concept of hamiltonicity. In a graph on n points a Hamilton cycle can be seen as a set of n lines, n pairs of points. The union of all lines gives the set of all n points and each point acts as a "point of contact" for two lines. Let us consider the hypergraph of Figure 2, that has ten points and nine hyperlines, two of which contain three points. The hyperlines are given by a Venn diagram.

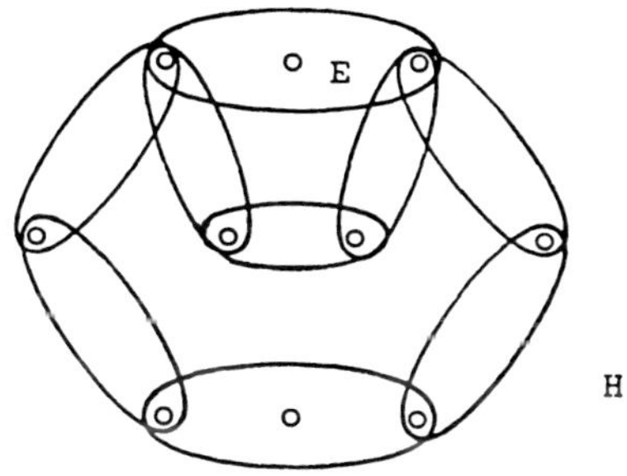

Figure 2

Definition 2. A *contact point* of a hypergraph is a point that is contained in two or more hyperlines of cardinality greater than one.

The hypergraph H in Figure 2 contains two points that are not contact points. Suppose the uppermost hyperline E of three points is deleted, then we obtain a cycle C. This cycle has eight hyperlines and contains all eight contact points. The union of the hyperlines gives all nine points occurring in that hypergraph. The same cycle C contains all eight contact points in H as well. However, the union of the hyperlines of C misses the tenth point i.e. the non-contact point of the hyperline E. This discussion motivates the following definitions.

Definition 3. A hypergraph H(V,E) is called *hamiltonian* if H contains a cycle C with the property that the union of the hyperlines of C is equal to V.

Definition 4. A hypergraph H(V,E) is called *contact-hamiltonian* is H contains a cycle C with the property that C contains all contact points of H.

The hypergraph of Figure 2 is contact-hamiltonian but not hamiltonian.
The coupling condition can easily be extended to the case of hamiltonicity as well as to the case of contact-hamiltonicity. The only finesse is the 2-connectedness of hypergraphs, needed in the proof to guarantee the existence of a cycle containing an arbitrarily given point.

Definition 5. A hypergraph H(V,E) is called *k-connected* if and only if for each pair of points there exist k paths between them that are disjoint on their contact points and on their hyperlines but, possibly, for the first and last lines, in case non-contact points are considered as endpoints.

Note that according to this definition paths may have hyperlines in common when non-contact points are considered. The hypergraph H in Figure 2 is 2-connected. Two paths between the two non-contact points both contain the two hyperlines that are triples.

Lemma 2. In a 2-connected hypergraph H(V,E) each point is contained in a cycle.

Proof. Let x be an arbitrary contact point. Choose one of the hyperlines E incident with x and containing two or more points. This line contains at least one more contact point for if not E would contain a non-contact point y that cannot be connected to a point z, not in E, by two paths disjoint on the contact points. The 2-connectedness of H guarantees the existence of two paths between both contact points that are line disjoint and therefore form a cycle containing x. If x is a non-contact point it is contained in a hyperline E, that for similar reasons as above contains at least two contact points y_1 and y_2. For these two points there exist two line-disjoint paths by the 2-connectedness. One of them may be given by y_1 E y_2, in which case we find a cycle containing x consisting of both paths. If not, one of these paths may be combined with this short path to give a cycle containing x. □

In the following theorem C_1, C_2 and C_3 are point sets that are unions of the hyperlines they contain.

Theorem 5. A 2-connected hypergraph H(V,E) is hamiltonian if and only if for any pair of cycles C_1 and C_2 and any pair of points P_1 and P_2

$$P_1 \in C_1 - C_2 \wedge P_2 \in C_2 - C_1 \Rightarrow \exists C_3 : \{C_1 \cap C_2\} \cup \{P_1, P_2\} \subset C_3.$$

Proof. The proof is strictly analogous to that in [3] for hamiltonian graphs and is left to the reader. See also the proof of Theorem 3. □

In the following theorem C_1, C_2 and C_3 are points sets that are unions of the contact points of the hyperlines they contain.

Theorem 6. A 2-connected hypergraph H(V,E) is contact-hamiltonian if and only if for any pair of cycles C_1 and C_2 and any pair of points P_1 and P_2

$$P_1 \in C_1 - C_2 \wedge P_2 \in C_2 - C_1 \Rightarrow \exists C_3: \{C_1 \cap C_2\} \cup \{P_1, P_2\} \subset C_3.$$

Proof. See the proof of Theorem 5. □

4. DISCUSSION

The coupling condition has been shown to characterize hamiltonicity of graphs, directed graphs and hypergraphs. The theorems 3, 4, 5 and 6 seem to be of importance mainly because they have been found after development of proper concepts liked the distinction between points in and out of phase for directed graphs and that between contact points and non-contact points for hypergraphs. We shall derive a necessary condition for hamiltonian directed graphs that is a consequence of the fact that on a hamiltonian circuit all points have to be in phase.

We replace a point P of a directed graph $\vec{G}$ by three points $P_{i(n)}$, $P_{m(iddle)}$, $P_{0(nt)}$ and join P_m with P_i and P_0 for all points P and join each point P_0 with other points P_i according to the incidence structure of $\vec{G}$. An undirected graph G is obtained. This translation is illustrated in Figure 3.

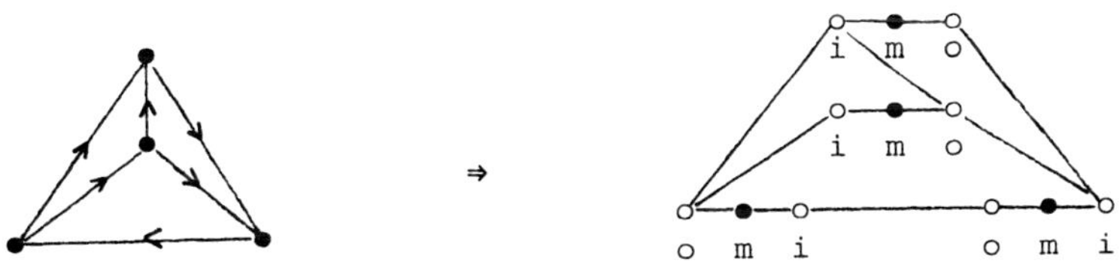

$\vec{G}$ G

Figure 3

It is clear that $\vec{G}$ is hamiltonian if and only if G is hamiltonian. A Hamilton circuit $\vec{C}$ in $\vec{G}$ determines a Hamilton cycle C in G. Conversely a Hamilton cycle C in G must contain the middle vertices P_m, therewith the lines P_iP_m and P_mP_0, that correspond to an incoming and an outgoing arc in $\vec{G}$. The corresponding (pseudo-)circuit $\vec{C}$ in $\vec{G}$ has therefore only points that are in phase. We now focus out attention on the subgraph $A(\vec{G})$ of G induced by the points P_0 and P_i. The components of $A(\vec{G})$ are bipartite graphs. In Figure 3 $A(\vec{G})$ has two components on two and six points respectively. The Hamilton cycle C has lines P_0 P_i in common with these components and contains all points P_0 and P_i. Therefore the components of $A(\vec{G})$ must admit a perfect matching. So we have

Lemma 3. If a directed graph $\vec{G}$ is hamiltonian then the components of $A(\vec{G})$ admit a perfect matching.

Let o(G) denote the number of odd order components of G. The characterization of graphs with perfect matchings given by Tutte [6] states that a graph G(V,E) has a perfect matching if and only if for every $S \subset V$ $o(G-S) \leq |S|$. We were led to this relation between hamiltonicity and perfect matchings by the concept of points in phase. Lemma 3 has as first consequence that a directed graph $\vec{G}$ with a source or a sink is non-hamiltonian as $A(\vec{G})$ contains a component consisting of one point in either case. More generally, if $A(\vec{G})$ has an odd order component then $\vec{G}$ is non-hamiltonian.

REFERENCES

[1] A.V. Aho, J.E. Hopcroft, J.D. Ullman, "The design and analysis of computer algorithms", Addison Wesley, Reading, Massachusetts (1974).
[2] C. Berge, "Graphs and Hypergraphs", North-Holland, Amsterdam (1973).
[3] F. Harary, "Graph Theory", Addison-Wesley, Reading, Massachusetts (1969).
[4] C. Hoede, H.J. Veldman, "On characterization of hamiltonian graphs", Journal of Combinatorial Theory (B) 25 (1978), 47-53.
[5] C. Hoede, H.J. Veldman, "Contraction theorems in hamiltonian graph theory", Discrete Mathematics 34 (1981), 61-77.
[6] W.T. Tutte, "The factorization of linear graphs", Journ. London Math. Soc. 22 (1947), 107-111.
[7] D.J.A. Welsh, "Matroid Theory", Academic Press, London (1976).

Transversals and Matroids

P. Horák, Bratislava, CSSR

1. Introduction

The first application of matroids in transversal theory goes back to the early forties and since then they have played the essential role in this area. As a matter of fact, there are two fundamental results concerning both transversals and matroids. In [9] Rado established a necessary and sufficient condition for a finite family of sets to possess a transversal which is independent in a given matroid. The second result, stated by Edmonds and Fulkerson [3], says that the partial transversals of a finite family of sets form a matroid. The two theorems lie at the very heart of transversal theory and therefore there are many variations and generalizations of them. A comprehensive survey of this field is in [7], for later ones see e.g. [10]. Yet, the generalizations of these classical results seem to go in different directions. In this paper it is shown that by means of k–transversals, introduced by Asratian [1] and originally called compatible transversals, it is possible to obtain "parallel" generalization of them.

To be able to state the results we first slightly extend the concept of a matroid by introducing bag matroids. (Unlike matroids, where the rank function is defined on subsets of an underlying set S the rank function of a bag matroid is defined on bags (= multisets) over S.) Then we prove Rado's type result for k–transversal with given defect which is independent in a bag matroid and show that all partial k–transversals of a finite family of sets form a bag matroid. Actually, the latter is a consequence of a stronger result. Namely, it is shown that there is a way of associating a bag matroid with an arbitrary finite graph through k–matchings, the subgraphs of the maximum degree at most k, such that obtained bag matroids are direct generalization of the well–known matching matroids.

2. Preliminaries

In this part of our paper we recall some definitions and explain the terminology used in subsequent sections.

R. Bodendiek, R. Henn (Eds.)
Topics in Combinatorics and Graph Theory

For the sake of compact formulations, besides the concept of a set, we will also make use of the concept of a bag. Sometimes bags are also called multisets. A bag, like a set, is a collection of elements over some domain, but unlike sets, bags allow multiple occurrences of elements. However, also for bags, the order of the elements in a bag is not important. For a bag B the $|(x,B)|$ function defines the number of occurrences of an element x in a bag B. The cardinality $|B|$ of a bag B is defined by $|B| = \sum_x |(x,B)|$. A bag A is a subbag of a bag B, denoted by $A \subseteq B$, if $|(x,A)| \leq |(x,B)|$. A bag A is a subbag of a bag B, denoted by $A \subseteq B$, if $|(x,A)| \leq |(x,B)|$ for each x. Lastly, for the bag union $A \cup B$, the bag intersection $A \cap B$, the bag difference $A - B$ we have $|(x,A \cup B)| = \max(|(x,A)| , |(x,B)|)$, $|(x,A \cap B)| = \min(|(x,A)| , |(x,B)|)$, $|(x,A - B)| = \max(0,|(x,A)| - |(x,B)|)$. Let E be a set and d be a natural number. Then the bag space $\overline{E}_d$ is the collection of all bags B satisfying, for any $x \in E$, $0 \leq |(x,B)| \leq d$.

Let $r: \overline{E}_d \to N = \{0,1,2,\cdots\}$ be a function with the following properties:

R1) $r(A) \leq |A|$,

R2) if $A \subseteq B$, then $r(A) \leq r(B)$,

R3) r is submodular, i.e. $r(A) + r(B) \geq r(A \cup B) + r(A \cap B)$.

Then the pair $(\overline{E}_d,r)$ is called a bag matroid over E, the bags B of $\overline{E}_d$ with the property $r(B) = |B|$ are called independent. When not specified, the subscript d will often be dropped. It is obvious that the bag matroids are a special case of super–matroids introduced in [2].

To avoid misunderstanding any time we treat bags we will refer to it explicitly.

Next the notion of a partial k–transversal will be introduced. Throughout the paper all considered families of sets will be finite.

Let $\mathcal{A} = (A_i, i \in I)$ be a family of subsets of a set E, let k be a natural number. Suppose that there exists a mapping $f: M \to E$, $M \subseteq I\times\{1,2,\cdots,k\}$, such that $f(i,j) \in A_i$ for each $(i,j) \in M$ and f is injective in both coordinates, i.e. $f(i,j_1) \neq f(i,j_2)$ for $j_1 \neq j_2$ and $f(i_1,j) \neq f(i_2,j)$ for $i_1 \neq i_2$. Then the bag $\{f(i,j), (i,j) \in M\}$ will be called a partial k–transversal of size $|M|$ or of the defect $k.|I| - |M|$, or shortly partial k–transversal. In the case $M = I\times\{1,\cdots,k\}$ the bag $\{f(i,j), (i,j) \in M\}$ will be referred to as (full) k–transversal. Clearly, for k = 1, the concept of a k–transversal coincides with the concept of a transversal defined in the usual way.

Let J be a subset of the set I. Then by $A(J)_k$ we will denote the bag B over E such that $|(x,B)| = \min(k, |\{i, i \in J, x \in A_i\}|)$. For $k = 1$, $A(J)_1 = A(J) = \bigcup_{i \in J} A_i$ which is the standard notation in transversal theory.

Now we state a result of Asratian [1] which in our terminology reads as follows:

Theorem 2.1. Let $\mathcal{A} = (A_i, i \in I)$ be a finite family of sets. Then $\mathcal{A}$ possesses a full k–transversal if and only if $|A(J)_k| \geq k|J|$ for any $J \subseteq I$.

We note that this result can also be read out of Corollary 2.3 from [4]. In [6] it has been proved that in the case of an arbitrary set of indices it suffices to replace the condition "for any $J \subseteq I$" by "for any finite subset J of I".

At the end of this section we recall some graph theoretical notions. Let G be a graph. Then the vertex set, edge set, of G is denoted by V(G), E(G), respectively. For a vertex v of G $\deg_G v$ denotes its degree in G. If G is understood we will abbreviate it by deg v. A subgraph H of G whose vertices have degree at most k will be called a k–matching. Thus, a 1–matching is simply a matching in the usual sense.

3. Independent k–transversals

In this section we state the defect form of Rado's theorem for k–transversals. We start with some auxiliary results.

Lemma 3.1. Let $\mathcal{A} = (A_i, i \in I)$ be a family of subsets of a set E, $\mathcal{M} = (\overline{E}, r)$ be a bag matroid over E, and let k be a natural number. Then, for every $J_1, J_2 \subseteq I$, $r(A(J_1)_k) + r(A(J_2)_k) \geq r(A(J_1 \cup J_2)_k) + r(A(J_1 \cap J_2)_k)$.

The assertion of Lemma 3.1 is well known for $k = 1$. The proof for $k > 1$ is omitted as it is only a technical matter.

Let $\mathcal{A}$, $\mathcal{M}$, k be as above. Then we define the k–defect $d_k(\mathcal{A}, \mathcal{M})$ of $\mathcal{A}$ with respect to $\mathcal{M}$ by $d_k(\mathcal{A}, \mathcal{M}) = \max_{J \subseteq I}(k|J| - r(A(J)_k))$. If $\mathcal{A}$ and $\mathcal{M}$

are understood we will simply denote this number by d_k. A subfamily $\mathscr{C} = (A_i, i \in C)$ of $\mathscr{A}$ will be called k–critical if $r(A(C)_k) = k|C| - d_k$ and for each proper subset C' of C, $r(A(C')_k) > k|C'| - d_k$.

Theorem 3.2. Let $\mathscr{A} = (A_i, i \in I)$ be a finite family of subsets of a set E, $(\overline{E},r)$ be a bag matroid, and let k be a natural number such that $d_k > 0$. Then there exists a unique k–critial subfamily of $\mathscr{A}$.

Proof. From the definition of d_k there exists $J \subseteq I$ such that $r(A(J)_k) = k|J| - d_k$. In addition, for $J = \emptyset$ we have $r(A(\emptyset)_k) = 0|\emptyset| > -d_k$, hence $\mathscr{A}$ contains a k–critical subfamily. Suppose that $\mathscr{A}' = (A_i, i \in J_1)$, $\mathscr{A}'' = (A_i, i \in J_2)$, $J_1 \neq J_2$, are two k–critical subfamilies of $\mathscr{A}$. Then $r(A(J_1 \cap J_2)_k) = k|J_1 \cap J_2| - d_k + n$, where $n > 0$ and from Lemma 3.1 we have $r(A(J_1 \cup J_2)_k) \leq r(A(J_1)_k) + r(A(J_2)_k) - r(A(J_1 \cap J_2)_k)$ $= k|J_1| - d_k + k|J_2| - d_k - k|J_1 \cap J_2| + d_k - n < k|J_1 \cup J_2| - d_k$ which is a contradiction.

The following theorem is the main result of this section.

Theorem 3.3. Let $\mathscr{A} = (A_i, i \in I)$ be a finite family of subsets of a set E, $\mathscr{M} = (\overline{E},r)$ be a bag matroid, and let k be a natural number. Then $\mathscr{A}$ possesses an independent partial k–transversal of defect m, $0 \leq m \leq k|I|$, if and only if, for each $J \subseteq I$, $r(A(J)_k) \geq k|J|$. In particular, $\mathscr{A}$ possesses an independent full k–transversal if and only if, for each $J \subseteq I$, $r(A(J)_k) \geq k|J|$.

Proof. To prove the left to right implication suppose that a bag X is an independent partial k–transversal of $\mathscr{A}$ of defect m given by mapping $f\colon M \to E$, $M \subseteq I\times\{1,2,\cdots,k\}$. Put $\mathscr{B} = (B_i, i \in I)$, where $B_i = \{f(i,j), (i,j) \in M), i \in I$. Clearly, $A_i \supseteq B_i$, $i \in I$, and, for each $J \subseteq I$, $r(A(J)_k) \geq r(B(J)_k) = |B(J)_k| = \sum_j |B_j| \geq k|J| - m$ because $B(J)_k \subseteq X$ is independent and any element of E occurs in at most k of B's.

The converse implication will be proved by induction with respect to the defect m. Thus, suppose first that m = 0, i.e. we need to show

$$\text{if} \quad r(A(J)_k) \geq k|J| \quad \text{whenever} \quad J \subseteq I, \tag{1}$$

then $\mathscr{A}$ possesses an independent full k–transversal. Denote by $\mathscr{B} = (B_i,\ i \in I)$, where $B_i \subseteq A_i$ for $i \in I$, the "minimal" family satisfying (1), i.e. removing an arbitrary element from some of B_i results in a family violating (1). Suppose that there is $j \in I$ with $|B_j| = n > k$, $B_j = \{x_1,\cdots,x_n\}$. From the minimality of $\mathscr{B}$ it follows, for each $1 \leq s \leq n$, that there exists $J_s \subseteq I$ such that $r(B(J_s)_k - x_s) < k|J_s|$. It is obvious that $j \in J_s$ for $s = 1,\cdots,n$, therefore $j \in J = \bigcap_{1 \leq s \leq n} J_s$. This means $|(x_s,B(J)_k)| > 0$, $s = 1,\cdots,n$. At the same time $B(J)_k \subseteq \bigcap_{1 \leq s \leq n} B(J_s)_k \subseteq B(J_s)_k$. Applying n times the fact that if $r(D - x) < r(D)$, then $r(C - x) < r(C)$ for any bags C, D, where $|(x,C)| > 0$ and $C \subseteq D$, we get $r(B(J)_k - B_j) = r(B(J{-}j)_k) = r(B(J)_k) - n$. In particular, it yields $r(B_j) = n$. On the other hand, as $\mathscr{B}$ satisfies (1), $r(B(J_s)_k - x_s) < k|J_s|$ implies $r(B(J_s)_k) = k|J_s|$, $s = 1,\cdots,n$, and by means of Lemma 3.1 it is also true that $r(B(J)_k) = k|J|$. Because $r(B_j) \doteq r(B\{j\})_k) = n > k.1$ we have that $\{j\}$ is a proper subset of J. Hence $J - j \neq \emptyset$ and $r(B(J{-}j)_k) = r(B(J)_k) - n = k|J| - n < k|J{-}j|$ which contradicts (1). Therefore any set B_i of $\mathscr{B}$ is of cardinality $|B_i| = k$. As $r(B(I)_k) \geq k|I| \geq |B(I)_k|$, the fact that for every bag B over E, $|B| \geq r(B)$ implies $r(B(I)_k) = |B(I)_k|$. Hence $T = B(I)_k$ is an independent bag of $(\overline{E},r)$. Consequently, each subbag of I is independent, i.e. $r(B(J)_k) = |B(J)_k| \geq k|J|$ whenever $J \subseteq I$. Hence the assumptions of Theorem 2.1 are fulfilled, therefore $\mathscr{B}$ possesses k–transversal, say S. However, $|B_i| = k$, $i \in I$, which implies T = S, i.e. T is also k–transversal of $\mathscr{B}$ and the proof of the first step of induction is complete.

Next suppose m > 0. It is plain that $d_k \leq m$. If $d_k < m$ then by induction hypothesis $\mathscr{A}$ possesses an independent partial k–transversal of defect d_k and clearly also of defect m. Assume therefore $d_k = m$. Consider a family $\mathscr{A}' = (A_i',\ i \in I)$ of subsets of a set $E \cup x$, where $x \notin E$, which has arisen from $\mathcal{A}$ by adding a new element e to exactly one of sets of the k–critical subfamily $\mathcal{C} = (A_i, i \in C)$. Further, denote by $\mathcal{M}' = (\overline{E \cup x}, r')$ the bag matroid whose independent bags are exactly those of the form $M \cup X$, where M is independent in $\mathcal{M}$ and either $X = x$ or X is the empty set. Obviously, $r'(N) = r(N)$ for $|(x,N)| = 0$, otherwise $r'(N) = r(N) + 1$. Let J be a subset of I of the property $r(A(J)_k) = k|J| - d_k(\mathcal{A},\mathcal{M})$. From Theorem 3.2 $\mathcal{A}$ contains a unique k–critical subfamily. Therefore $J \supseteq C$ and subsequently $r'(A'(J)_k) = r'(A(J)_k \cup \{x\}) = r(A(J)_k) + 1 = k|J| - d_k(\mathcal{A}, \mathcal{M}) + 1$. This means $d_k(\mathcal{A}',\mathcal{M}') = d_k(\mathcal{A},\mathcal{M}) - 1$ and from the induction hypothesis $\mathcal{A}'$ possesses an independent partial k–transversal T of defect m–1. However, the element x occurs in exactly one of the sets of $\mathcal{A}'$, therefore $|(x,T)| \leq 1$ and T–x is an independent partial k–transversal of $\mathcal{A}$ of defect m. This completes the proof.

Putting $k = 1$ Theorem 3.3 reduces to the defect form of Rado's result given in [8]. If the bag matroid $\mathcal{M}$ is free we obtain the defect form of Theorem 2.1. As an immediate consequence we get

Corollary 3.4. The maximal size s_k of an independent partial k–transversal of $\mathcal{A}$ is given by

$$s_k = k|I| - d_k(\mathcal{A}, \mathcal{M}) = k|I| + \min_{J \subseteq I} (r(A(J)_k) - k|J|).$$

It is routine to show, by means of Lemma 3.1, that the function f defined on the subsets J of the index set I by $f(J) = r(A(J)_k) - k|J|$ is submodular. With respect to [5] this means that there exist a polynomial–time algorithm for determining s_k. We also point out that in the case of free bag matroid s_k is the maximum number of edges among k–matchings in a bipartite graph naturally associated with $\mathcal{A}$. This can be obtained also from Theorem 2.2 of [4].

4. Bag matroids

It is well–known that the collection of all subsets of the vertex set of a graph which can be covered by a matching form a matroid. In this part of our paper it will be shown that by means of k–matchings it is possible to associate with every graph a bag matroid which is a generalization of the matching matroid. In particular, we obtain that all partial k–transversals of a family of sets comprise a bag matroid.

Theorem 4.1. Let G be a finite graph, d, k be natural numbers, and r: $\overline{V(G)}_d \to N$ be a function defined by

$r(A) = \max_H \sum_{v \in V(G)} \min(|(v,A)|, \deg_H v)$, where H ranges over all k–matchings of G. Then the pair $(\overline{V(G)}_d, r)$ is a bag matroid.

Proof. It is straightforward that r satisfies R1 and R2. To show the submodularity of r we first introduce one more notation. Let A be a bag of $\overline{V(G)}_d$ and H be a k–matching of G. Then by g(A,H) we will denote the number $g(A,H) = \sum_{v \in V(G)} \min(|(v,A)|, \deg_H v)$. First of all it will be shown that if $A \subseteq B$ are bags from $\overline{V(G)}_d$, then there exists a k–matching H of G such that $g(A,H) = r(A)$ and at the same time $g(B,H) = r(B)$. Let S,T be two k–matchings of G with $g(A,S) = r(A)$, $g(B,T) = r(B)$. If it were $g(A,T) = r(A)$, then T would be the desired subgraph. Suppose that $g(A,T) < r(A)$. Then there exists a vertex v of G with $\deg_T v < \deg_S v$ and $\deg_T v < |(v,A)|$. Now we construct a new k–matching T' of G by adding to T an edge $e = vz$, $vz \in E(S)$, $vz \notin E(T)$ and by removing an edge e' from $T \cup e$, where e' is defined as follows. We will distinguish between two cases. If there is in T an edge zw such that $g(A,T) < g(A,(T \cup e) - zw)$, set $e' = zw$. If all edges f of T incident with z have the property $g(A,T) = g(A,(T \cup e) - f)$, then $g(B,T) = r(B)$ implies $\deg_T z = k$. This yields that there is in T an edge zu such that $zu \notin E(S)$ and we set $e' = zu$. It is plain that $g(B,T') = r(B)$. In the former case, $g(A,T) < g(A,T')$, in the latter $g(A,T) = g(A,T')$ however, $|E(S) \cap E(T)| < |E(S) \cap E(T')|$. Owing to the finiteness of E(T) by repeatedly applying the described procedure of adding and removing edges we arrive at a desired k–matching H of G. With this in hand, suppose X, Y are two bags over V(G). Let H be a k–matching of G satisfying $g(X \cup Y,H) = r(X \cup Y)$ and $g(X \cap Y,H) = r(X \cap Y)$. Then

$r(X \cup Y) + r(X \cap Y) = \sum_{v \in V(G)} \min(\max(|(v,X)|, |(v,Y)|), \deg_H v) +$

$$\sum_{v\in V(G)} \min(\min(|(v,X)|,\ |(v,Y)|),\ \deg_H v) = \sum_{v\in V(G)} \min(|(v,X)|,\ \deg_H v) + \sum_{v\in V(G)} \min(|(v,Y)|,\ \deg_H v) = g(X,H) + g(Y,H) \leq r(X) + r(Y),$$ i.e. r is a submodular function. This establishes the desired result.

It can be easily seen that for k = 1 and arbitrary d the independent sets of $(\overline{V(G)}_d, r)$ are exactly the independent sets of the matching matroid of G and that two bag matroids $(\overline{V(G)}_d, r)$, $(\overline{V(G)}_{d'}, r)$, where $d, d' \geq k$, have the same set of independent bags.

We finish the paper by stating a generalization of the classical result of Edmonds and Fulkerson.

Theorem 4.2. Let $\mathcal{A} = (A_i,\ i \in I)$ be a finite family of subsets of a set E, and let k be a natural number. Then there exists a bag matroid over E such that the collection of its independent bags coincides with the collection of all partial k–transversals of $\mathcal{A}$.

Proof. Consider a bipartite graph G with bipartition (I,E), where ix, $x \in X$, $i \in I$, is an edge of G if $x \in A_i$. Let $\mathcal{M} = (\overline{V(G)}_k, r)$ be the bag matroid from Theorem 4.1. It is easy to see that also the restriction $\mathcal{M}\,|\overline{E}_k$ of $\mathcal{M}$ to the bag space $\overline{E}_k$ is a bag matroid. Let B be a bag over E which is a partial k–transversal of $\mathcal{A}$ given by mapping f: $M \to E$, $M \subseteq I \times \{1,2,\cdots,k\}$. Then for a k–matching H of G with $E(H) = \{ix,\ i \in I,\ x \in X,\ f(i,n) = x$ for some $n \in \{1,\cdots,k\}\}$ we have $r(B) \geq \sum_{v\in V(G)} \min(|(v,B)|, \deg_v H) = \sum_{v\in V(G)} |(v,B)| = |B|$. This means, B is independent in $\mathcal{M}$ and therefore also in $\mathcal{M}\,|\overline{E}_k$. By similar reasoning we get that a bag independent in $\mathcal{M}\,|\overline{E}_k$ is also a partial k–transversal of $\mathcal{A}$.

Acknowledgement:

The final version of the paper was prepared while the author was visiting the Department of Mathematics and Statistics of the University of Nebraska–Lincoln. He would like to express his thanks to the department for its hospitality.

References

1. A.S. Asratian, Compatible systems of distinct representatives (Russian), Diskret. Analiz Vyp. 27(1975), 3–12.

2. F.D.J. Dunstan, A.V. Ingleton, D.J.A. Welsh, Supermatroids, in: Combinatorics (D.J.A. Welsh, D.R. Woodal, eds.), Inst. Math. Appl., (1972), 72–122.

3. J. Edmonds, D. Fulkerson, Transversals and matroid partitions, J. Res. Nat. Bur. Stand. 69B(1965), 147–153.

4. J.E. Graver, W.B. Jurkat, f–factors and related decompositions of graphs, J. Comb. Theory(B)28 (1980), 66–84.

5. M. Grötschel, L. Lovász, A. Schrijver, The ellipsoid method and its consequences in combinatorial optimization, Combinatorica 1(1981), 169–197.

6. P. Horák, A generalization of Hall's theorem, J. Comb. Math. Comb. Comput. 2(1987), 223–233.

7. L. Mirsky, Transversal Theory, Academic Press, 1971.

8. H. Perfect, A generalization of Rado's theorem on independent transversals, Proc. Cambridge Philos. Soc. 66(1969), 513–515.

9. R. Rado, A theorem on independence relations, Quart. J. Math. (Oxford) 13(1942), 83–89.

10. D.J.A. Welsh, Matroid Theory, Academic Press, 1976.

Classification and Construction of Geodetic Block With Diameter Two

M. Jingzhong, Hubei, China

ABSTRACT

In this paper we have been pointed out the classification and construction of geodetic block with diameter two and given the necessary and sufficient condition of the existance of geodetic block with diameter two. So we completely answered the questions of J.G.Stemple offered in the 1974.

Keyword. Geodetic block, Clique, Affine plane.

1. Introduction

We only discuss 2-connected undirected simple graph.

The graph G is geodetic iff G is connected and for any two vertices of G there exactly exist a shortest path connected them.

According to [1] the clique of G is a maximal complete subgraph of G which contains at least three vertices. For any $x\in V(G)$, if there is a clique contains x, then x is said to be clique vertex. Otherwise, x is non-clique vertex. The meaning of terminology and notations without explanation is as it in [2].

J.G.Stemple have been investigated the basic properties of geodetic block with diameter two in detail. The main results can described in the following:

Proposition 1. Suppose G is a geodetic graph, diam(G)=2, and $cl(G)=k\geq 3$. Then 1. The order of the cliques of G are the same, i.e. cl(G); 2. If $xy\in E(G)$ and both x and y are clique vertex, then xy contained in a clique; 3. For any $x\in V(G)$, $d(x)=\Delta(G)$ iff either x is a clique vertex, or the distance between x and every clique vertex is 2; 4. If x is a nonclique vertex and adjacent to a clique vertex, then $d(x)=\Delta(G)-k+2=\delta(G)$; 5. If $\delta(G)\geq 3$, then $\Delta(G)\geq 2(k-1)$; 6. $|V(G)|=n=\Delta(G)\cdot\delta(G)+1$; 7. If every vertices of G are clique vertex, then $\Delta(G)=m(k-1)$ and $m\geq k$ and G is regular graph.

Before the end of [1] J.G.Stemple offered three questions:

R. Bodendiek, R. Henn (Eds.)
Topics in Combinatorics and Graph Theory

(i) Are there 2-connected geodetic graphs of diameter two in which all vertices are clique vertices? (ii) We define the clique incidence graph of G, denoted G^I, by: V(G) is the set of cliques of G; if $H_1, H_2 \in V(G^I)$, then edge $H_1H_2 \in E(G^I)$ if and only if H_1 and H_2 have a common vertex in G. If G is a geodetic block of diameter two, is G^I geodetic? What can be said about the diameter of the connected components of G^I? (iii) For each set of values (n,k,Δ) satisfying $n=\Delta^2-\Delta(k-2)+1$ with $\Delta > 2(k-1)$, does there exist a corresponding geodetic block of diameter two? If yes, is it unique?

K.R.Parthasarathy and N.Srinivasan ([3]) have discussed the constructions of some kind of geodetic block and in [4] they have investigated the properties of some geodetic block with diameter two. Ho-Jin Lee ([5]) have discussed the relations between the geodetic block and orthogonal Latin squares. But the Stemple's three questions haven't get any answer directly up to now.

These are the motivation of this paper.

2. Main Results

The following theorems have described the construction of geodetic blocks with diameter two completely, and at the same time answered the three questions of Stemple completely. For the sake of simple we use G to denote the geodetic block with diameter two, and suppose $cl(G)=k\geq 3$, clique is the maximal subgraph of G.

1. The theorem of nonexistance.

Theorem 1. The graph G that, for every x of V(G) , x is a clique vertex, is not existant.

2. The classification and construction of geodetic block.

J.G.Stemple ([1]) have been pointed out that, the most simple geodetic block with diameter two is:: if G contains the vertices of degree 2, then G is a regular pyramid; if G contains no clique, then G is a Moore graph with diameter two and we at most have four graphs of this kind, i.e. the 5-cycle C_5, Peterson graph, 7 regular Moore graph and may be there is a 57 regular Moore graph.

The remain cases are $\delta(G)\geq 3$ and G contains clique.

Case 1. There is a vertex a of V(G) that, a contained at least two cliques. The construction of G are explained in the following theorems.

Theorem 2. If $\Delta(G)=\lambda(k-1)$, $\lambda\geq 2$ and $\exists\, a\in V(G)$ s.t. a belongs to exactly λ cliques $H_1,H_2,\ldots,H_\lambda$. Then we can not have that, every vertex of $\{x\mid x\in V(H_i),\ i=1,2,\ldots,\lambda\}\setminus\{a\}$ belongs to only one clique.

Theorem 3. If there is a vertex a of V(G) s.t., a belongs to exactly λ cliques and $\lambda\geq 2$, $d(a)=\Delta(G)=\lambda(k-1)+\mu$, then we can not have any nonclique vertex belongs to $\Gamma(a)$,i.e. $\mu=0$ and $\Delta(G)=\lambda(k-1)$.

Theorem 4. Suppose $\Delta(G)=\lambda(k-1)$ and there is a vertex a of V(G) s.t., a belongs to λ cliques exactly. Then when $\lambda\geq 3$, we take any $b\in\Gamma(a)$, we can not have b belongs to μ cliques exactly and $2\leq\mu\leq\lambda-1$.

Theorem 5. Suppose $\Delta(G)=\lambda(k-1)$ and there is a vertex a of V(G) s.t., a belongs to λ cliques $H_1,H_2,\ldots,H_\lambda$ exactly and there is a vertex b of $V(H_\lambda)\setminus\{a\}$ s.t., b belongs to λ cliques $K_1,K_2,\ldots,K_{\lambda-1},H_\lambda$ exactly and $\lambda\geq 2$, then $\lambda=2$ and every vertex of $\{x\mid x\in V(K_i),\ i=1,2,\ldots,\lambda-1\}\setminus\{b\}$ can only belongs to one clique.

Theorem 6. Suppose $\Delta(G)=2(k-1)$ and there is a clique H and two vertices a and b of V(H), $a\neq b$ s.t., both a and b belongs to two cliques. Denote $V(H)=\{b_1,b_2,\ldots,b_{k-2},a,b\}$. Then for $1\leq i\leq k-2$ we have b_i belongs to exactly two clique.

The construction of this kind will described after theorem 8 in detail and then gives the condition of its existance.

Case 2. For any vertex a of V(G) at most a belongs to one clique.

Theorem 7. Suppose $\Delta(G)=k-1+\mu$ and there is a clique H in G s.t.,for any vertex a of V(G) the clique which contains a is only clique H. Then either $\mu=1$ and G contains only this clique (G is a regular pyramid); or $\mu=k$ and G contained k+1 cliques which are disjoint each others.

Theorem 8. Suppose $\Delta(G)=k-1+\mu$, $H_0,H_1,H_2,\ldots,H_{\lambda-1}$ are the cliques contained in G and are disjoint each others, $\lambda\geq 2$, and every vertex of $V(G)\setminus(V(H_0)\cup V(H_1)\cup\ldots\cup V(H_{\lambda-1}))$ is nonclique vertex, then $\mu=k$, $\lambda=k+1$ and there is an affine plane of order k.

Under the condition and the results of theorem 8, if we denote $V(H_0)=\{a_1,a_2,\ldots,a_k\}$, for $1\leq i\leq k$ we denote $\Gamma(a_i)\setminus V(H_0)=X_i=\{x_{i1},x_{i2},\ldots,x_{ik}\}$, and for $1\leq i,j\leq k$ we denote $\Gamma(x_{ij})\cap(V(H_1)\cup V(H_2)\cup\ldots\cup V(H_k))=Y_{ij}$. Then we have that: consider the k^2 vertices of $V=V(H_1)\cup V(H_2)\cup\ldots\cup V(H_k))$ as to compose vertices set, the vertices subset $V(H_1)$, $V(H_2),\ldots,V(H_k)$, $Y_{11},Y_{12},\ldots,Y_{1k},\ldots,Y_{k1}$, $Y_{k2},\ldots,Y_{kk}$ as to compose line set, these amountto k^2+k lines. Each line contains exactly k vertices and each vertex belongs to exactly (k+1) lines. Divide the lines described as above into (k+1) families of lines: $L_0=\{V(H_1),V(H_2),\ldots,V(H_k)\}$, $L_i=\{Y_{i1},Y_{i2},\ldots,Y_{ik}\}$, $1\leq i\leq k$. Then we know any two lines which belongs to the same line family are disjoint each other and every line meet any line of other line family in exactly one vertex. So we really have an affine plane of order k on the vertices set V.

Conversely, from any affine plane of order k we can construct a geodetic block with diameter two which contains exactly (k+1) cliques which are disjoint each other by the process described as above without any difficulty.

Therefore in this case the existance of G is equivalent to the existance of affine plane of order k.

Theorem 9. There is a clique H in G with $V(H)=\{a_1,a_2,\ldots,a_k\}$ s.t., for $1\leq i\leq k$ there is always a clique H_i s.t., $V(H)\cap V(H_i)=\{a_i\}$ and $\Gamma(a_i)\subseteq V(H)\cup V(H_i)$,if and only if there is an affine plane of order k-1.

This is because that, if we take out the clique H from G, then the remain part of G is the graph described by theorem 8. Therefore we know the existance of graph G described in case 1 is equivalent to the existance of affine plane of order k-1.

By now, we have described the classification and the construction of every geodetic block G. Therefore the three questions of J.G.Stemple can answered easily.

Refferences

[1] Stemple,J.G.,Geodetic Graphs of Diameter Two, J. Comb. Theory,B 17(1974), 266-280

[2] Bollobas,B., Extremal Graph Theory, Academic Press, INC. (London) LTD, 1978

[3] Parthasarathy, K.R. and Srinivasan, N., Some General Constructions of Geodetic Block, J. Comb. Theory, B 33(1982), 121-136

[4] Parthasarathy, K.R. and Srinivasan, N., Geodetic Blocks of Diameter Two, Some Structural Properties, ARS Combinatoria, vol. 20 Dec. 49-60

[5] Lee Ho-Jin, A note on Geodetic Graphs of Diameter Two and Their Relation to Orthogonal Latin Squares, J. Comb. Theory, B 22(1977), 165-169

[6] Hoffman, A.J. and Singleton, R.R., On Moor Graphs with Diameters 2 and 3, IBM. J. Res. Develop. 4(1960) 497-504

Graph Distances and Similarity

F. Kaden, Berlin, GDR

1. Similarity between graphs

How similar are the two graphs given in Figure 1? Of course this question has no unique answer. At first we must specify in what sense similarity between graphs is meant.

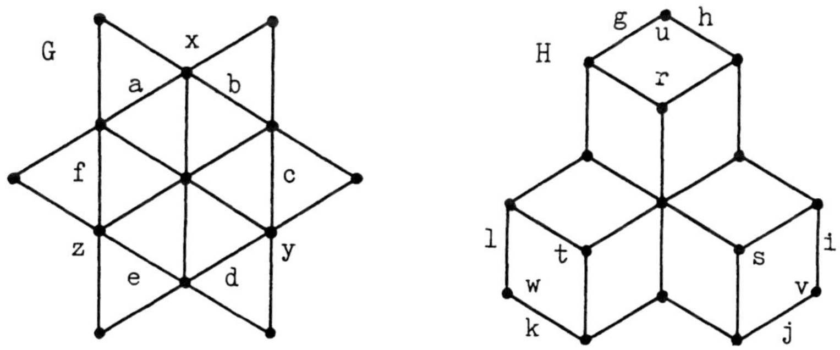

Figure 1. How similar are these two graphs?

In the case of the given graphs it seems adequate to express their difference by the minimal number of edges to be removed in both graphs obtaining isomorphic subgraphs. Because graph G has twelf triangles, no more than two of them having an edge in common, at least six edges must be removed to get a triangle-free graph such as any subgraph of H. Indeed, this is possible by deleting the edges a,b,...,f. These edges are uniquely determined because the outermost triangles contain only one edge in each case that belong to other triangles. Remove the six edges g,h,...,l of the graph H and isolated vertices occuring thereby to leave a subgraph isomorphic to the subgraph of G constructed above. Hence the graphs G and H have the edge-distance six. Now let us consider the graph distance defined by the minimal number of vertices to be removed in both graphs in order to leave isomorphic induced subgraphs. Because graph H has not any triangle, we must delete all the triangles of G by removing as few vertices as possible. After omitting the vertices x,y,z graph G is a tree, that can be obtained as induced subgraph of H when removing the vertices r,s,...,w there. Hence the vertex-difference of the graphs G and H is also equal to six.

R. Bodendiek, R. Henn (Eds.)
Topics in Combinatorics and Graph Theory

Instead of an introduction, we discussed the problem of graph similarity by an example. In the next chapter, we define a family of graph distances based on the approach of ZELINKA 1975 using maximal common induced subgraphs. Dealing with special graph classes we modify that definition by forcing the common subgraphs and supergraphs to fulfil certain properties. We discuss, in which cases the graph distance is a metric, and give counterexamples of the triangle inequality if it is no metric. In the last chapter, we introduce the concept of distance graphs which is appropriate to discuss the problem of isometry concerning the relation between structural defined graph distances and the length of shortest sequences transforming one graph to another one by steps of minimal positive distance. The main results of the paper concern this problem with respect to the distance graphs defined on several finite graph sets and corresponding graph metrics. Other graph distances known from literature accentuate various aspects of graph similarity, for instance GERNERT 1979, SHAPIRO 1982, SOBIK 1983, SANFELIU/FU 1983, and ESHERA/FU 1984.

2. Graph distances

We consider finite undirected graphs without loops and multiple edges. The notion graph is used in the sense of isomorphy classes. Let $n(G)$ be the number of vertices of the graph $G = (V,E)$, whose vertex and edge sets are V and E, respectively. A graph $G' = (V',E')$ is called induced subgraph of the graph $G = (V,E)$ (in terms $G' \leq G$), if there are a subset $U \subseteq V$ of the vertex set V and a bijection $f: U \to V'$ so that $uv \in E$ if and only if $f(u)f(v) \in E'$ for all $u,v \in U$. If $G' \leqq G$ and both graphs are not isomorphic to each other, we write $G' < G$ and call G' a proper induced subgraph of G. Given three graphs H, G, and G' the graph H is called a common subgraph (or supergraph) of G and G' if $H \leqq G$ and $H \leqq G'$ (or $G \leqq H$ and $G' \leqq H$) hold.

Let us consider an arbitrary graph set M containing common subgraphs and common supergraphs of any two graphs $G,G' \in M$. The functions d_M and d'_M called graph distances are defined as follows.

$$d_M(G,G') = \max\{n(G),n(G')\} - \max\{n(H) \mid H \in M,\ H \leqq G,G'\},\quad G,G' \in M$$
$$d'_M(G,G') = \min\{n(H) \mid H \in M,\ G,G' \leqq H\} - \min\{n(G),n(G')\},\quad G,G' \in M$$

The following conditions are sufficient for the graph distances d_M and d'_M to be metrics on the set M.

Condition 1. If $H' \in M$ is a common supergraph of $G,G' \in M$, then there is a common subgraph $H \in M$ of G,G' with $n(H) \geqq n(G) + n(G') - n(H')$.

<u>Condition 2.</u> If $H \in M$ is a common subgraph of $G, G' \in M$, then there is a common supergraph $H' \in M$ of G, G' with $n(H') \leq n(G) + n(G') - n(H)$.

<u>Theorem 1.</u> The functions d_M and d'_M are metrics on the graph set M if Condition 1 or 2 holds, respectively.

Because of the definition we have only to show that the triangle inequalities hold. It is easy to see that the conditions stated above directly imply these inequalities of the functions d_M and d'_M, respectively. On the other hand, they are not necessary. But in all cases considered here, there is a counterexample for the triangle inequality if the corresponding condition fails.

Now we are dealing with the graph sets given in Figure 2 which shows their inclusion scheme. With the help of Theorem 1 we decide which graph distances are metrics. If the sufficient condition fails then Figure 3 gives counterexamples of the corresponding triangle inequality.

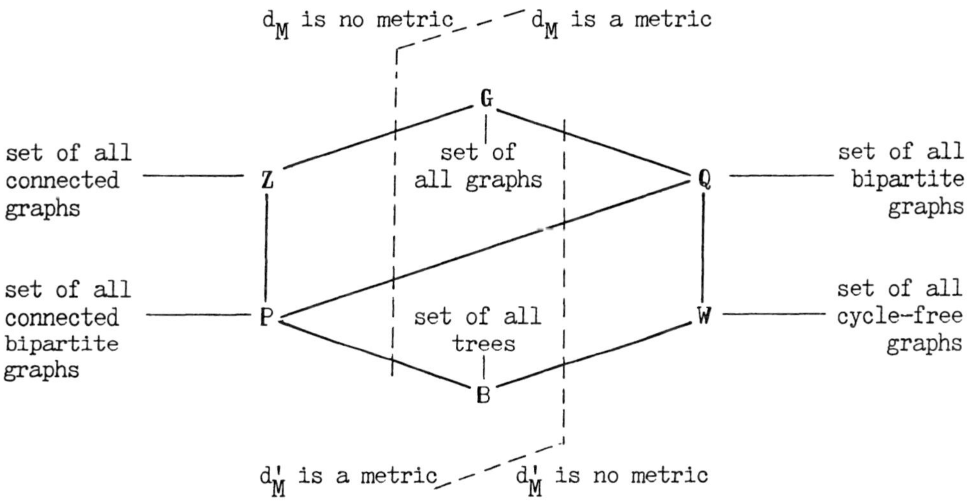

Figure 2. Six graph sets and corresponding graph distances

The Zelinka-distance d_o coincides with the functions d_G and d'_G. Furthermore, the graph distances d_B and d'_B are identical, but different from the Zelinka-distance restricted to the set B of all trees. The latter can be shown by the graphs $K_{1,6}$ and P_7.

Let Sub M be the set of all induced subgraphs of graphs contained in the set M. Given two graph sets M and M' with $M \subseteq M'$ we define $Sub_{M'}M = M' \cap Sub\ M$. Evidently, $M \subseteq M'$ implies $d_{M'} \leq d_M$ and $d'_{M'} \leq d'_M$ on the set M. Furthermore, the distances d_M and $d_{M'}$ are identical on M if $Sub_{M'}M \subseteq M$ holds.

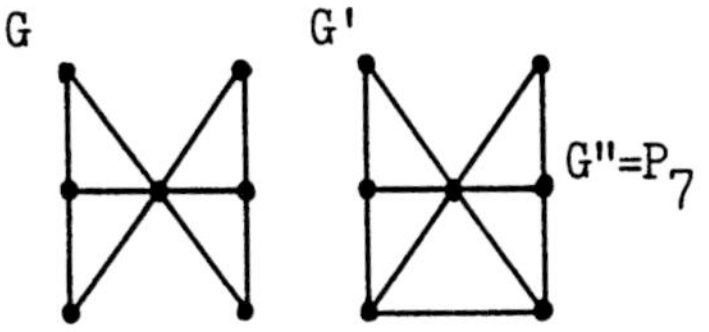

a) d_Z is no metric because of $d_Z(G,G')+d_Z(G',G'')<d_Z(G,G'')$

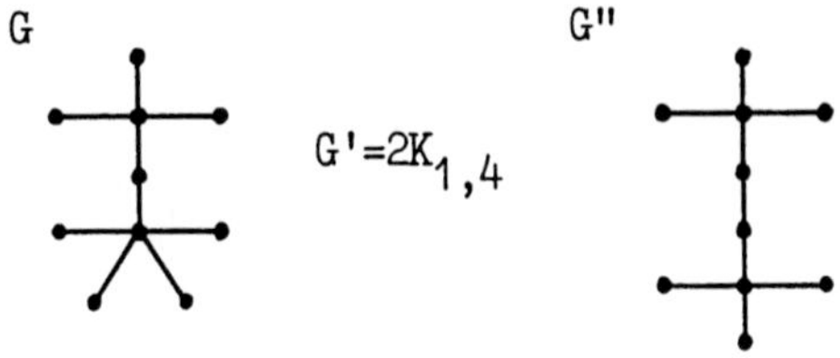

b) d'_Q is no metric because of $d'_Q(G,G')+d'_Q(G',G'')<d'_Q(G,G'')$

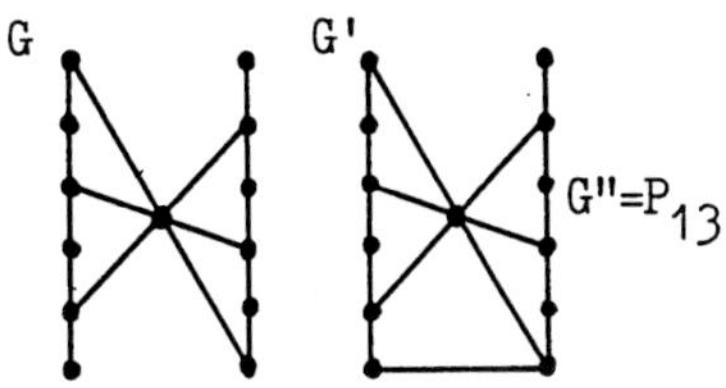

c) d_P is no metric because of $d_P(G,G')+d_P(G',G'')<d_P(G,G'')$

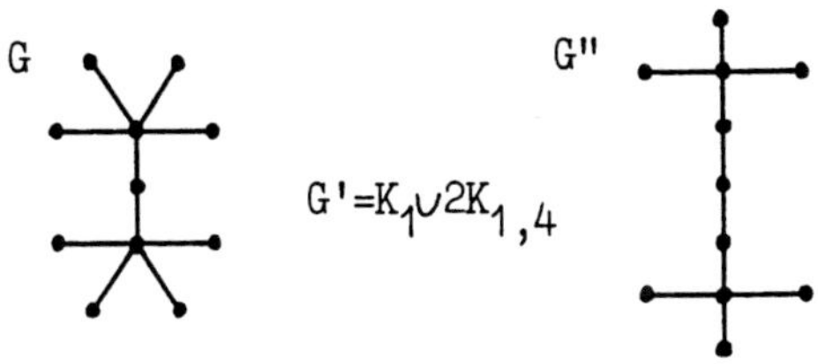

d) d'_W is no metric because of $d'_W(G,G')+d'_W(G',G'')<d'_W(G,G'')$

Figure 3. Counterexamples of the triangle inequality

With respect to the graph distances d'_M let Super M be the set of all graphs H' for which there are graphs $G,G'\in M$ with $G,G'\lneqq H'$ but no graph H satisfies $G,G'\lneqq H<H'$. For two graph sets $M\subseteq M'$ we define $\text{Super}_{M'}M = M'\cap \text{Super } M$. It can easily be proved that the functions d'_M and $d'_{M'}$ are identical on M if $M\subseteq M'$ and $\text{Super}_{M'}M\subseteq M$ hold. The next figure illustrates identities between graph distances as far as they are proved by the above arguments. In all cases indicated by arrows ($d\longrightarrow d'$) the functions d and d' are different but $d\leqq d'$.

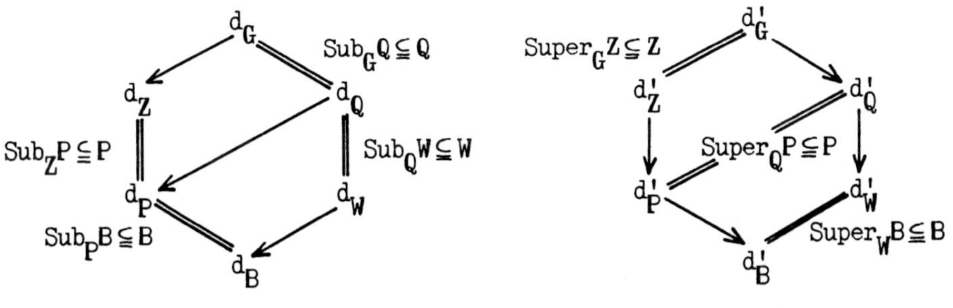

Figure 4. Relations between the graph distances d_M and d'_M, respectively

The graphs G and G" in Figure 3a and 3c show that the graph distances d_G, d_Z and d_Q, d_P are different, respectively. As above, the graphs $K_{1,6}$ and P_7 demonstrate the difference of d_W and d_B. The graphs G and G" in Figure 3b

and 3d can be used to prove that the graph distances d'_G, d'_Q and d'_Q, d'_W are different, respectively. Because the graphs G, G" in Figure 3b are bipartite and connected, they show also that d'_Z and d'_P are different. Let G be the graph obtained from $K_{1,4}$ by subdividing all the edges. Graph H is derived from two even cycles C_6 by identifying one vertex of them. Finally, the graph distances d'_P and d'_B are different because the graphs G and P_9 have the bipartite common supergraph H of eleven vertices, but there is no common supertree of G and P_9 with less than 13 vertices.
The following figure summarizes the relations between the graph metrics d_M and d'_M presented in Figure 4 regarding the facts $d_o = d_G = d'_G$ and $d_B = d'_B$.

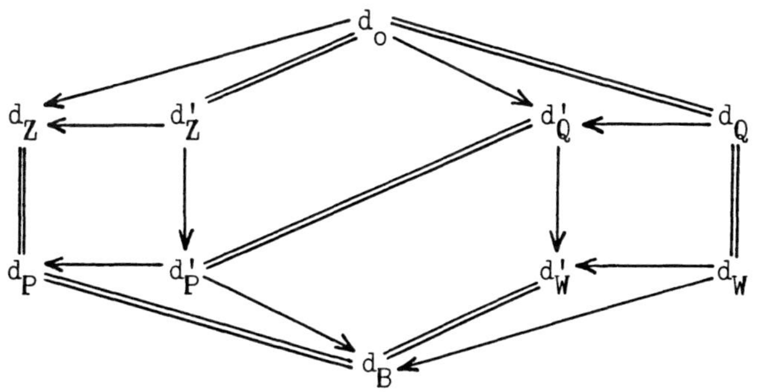

Figure 5. Relations between the graph distances d_M and d'_M

3. Distance graphs

Let (M,d) be a finite metric space and d an integer-valued metric. The corresponding distance graph D(M,d) = (M,E,d) is defined on the vertex set M, so that any two vertices $u,v \in M$ are adjacent in D(M,d) ($uv \in E$) if and only if d(u,v) = 1 holds. Graph-theoretic terms applied to the distance graph D(M,d) are used in the sense of the ordinary graph (M,E). All distance graphs D = D(M,d) to be considered here are connected. Let $\delta_D(u,v)$ be the distance between any two vertices u,v of D which is defined by the length of a shortest path connecting them.
Now we are dealing with the question whether the metrics d and δ_D are identical on the set M. With the help of the triangle inequality for the metric d the relation $d(u,v) \leq \delta_D(u,v)$, $u,v \in M$ is easily proved. The distance graph D = D(M,d) is called isometric if d and δ_D are identical on M. The so-called degree of non-isometry (deg D) of the distance graph D is defined as follows.

$$\deg D = \max\left\{\delta_D(u,v) - d(u,v) \mid u,v \in M\right\}$$

Due to KAY/CHARTRAND 1965 the distance graph $D = D(M,d)$ is isometric if and only if for any two elements $u,v \in M$ with $d(u,v) \geqq 2$ there is an element $w \in M$ different from u and v with $d(u,v) = d(u,w) + d(w,v)$.
Our aim is to decide whether the distance graphs corresponding to the graph sets and metrics introduced in Chapter 2 are isometric, and if not to calculate or estimate their degree of non-isometry.
Let $\mathbf{G}_n$ be the set of all graphs of vertex number n. The corresponding distance graph G_n^o with respect to the Zelinka-distance d_o is isometric for all vertex numbers n (ZELINKA 1975). Now let $\mathbf{Z}_n$ be the set of all connected graphs of vertex number n. We show that the distance graph $Z_n = D(\mathbf{Z}_n, d_o)$ is isometric if and only if the vertex number n is less than ten. Furthermore, it will be proved that the degree of non-isometry of Z_n is equal to one in all other cases.

Theorem 2. The distance graph $Z_n = D(\mathbf{Z}_n, d_o)$ fulfils $\deg Z_n = \begin{cases} 0 & \text{if } n<10 \\ 1 & \text{else.} \end{cases}$

Proof. At first we show that the distance graph Z_n is isometric if $n<10$. Let $G,G' \in \mathbf{Z}_n$ be two connected graphs with $d_o(G,G') = p \geqq 2$. We have to prove that there is a connected graph $G'' \in \mathbf{Z}_n$ with $d_o(G,G'') + d_o(G'',G') = p$. Let H be a common subgraph of G,G' with exactly $n - p$ vertices and a minimal number of components $H_1,\dots,H_k$. Furthermore, let $H = G \cap G'$, $V(G) \setminus V(H) = U = \{u_1,\dots,u_p\}$ and $V(G') \setminus V(H) = V = \{v_1,\dots,v_p\}$. Without loss of generality each vertex of U is adjacent to each vertex of V in the common supergraph $H' = G \cup G' \cup K(U,V)$. The merging of G and G' in H is called trivial if there is a set $W \subsetneqq U \cup V$ different from U and V of the cardinal number p so that every component of H contains a vertex which is adjacent to a vertex of W. Then the subgraph G" of H' induced by $W \cup V(H)$ fulfils $d_o(G,G'') + d_o(G'',G') = p$ as demanded. If H has non-trivial automorphisms, the merging of G and G' may be changed in order to search for a trivial merging. In the case $2p>k$ it is always possible to arrange a trivial merging. For $k \geqq 6$ at least five components of H are isolated vertices and there is a suitable automorphism of H leading to a trivial merging. The only non-trivial cases for $k = 4$ and $k = 5$ are shown in Figure 6. Because of $n<10$ we obtain a trivial merging after using a suitable automorphism of H, or there is a common subgraph of G,G' with $n - p$ vertices and less than k components in contradiction to the assumption. This completes the proof of $\deg Z_n = 0$ for $n<10$.
The graphs G and G' given in Figure 6 present a counterexample for the isometry of the distance graph Z_{10} of the merging type $k = 4$, that can be generalized to $n>10$. These graphs $G,G' \in \mathbf{Z}_{10}$ have Zelinka-distance two, but

there is no connected graph of ten vertices having Zelinka-distance one to both, G and G'. In order to show that any two graphs $G,G' \in Z_n$ of Zelinka-distance p are connected by a path of length p or p + 1 in the distance graph Z_n, we add a vertex adjacent to all vertices of a minimal common supergraph H' of G,G' and choose a suitable sequence of induced subgraphs of that graph. This completes the proof.

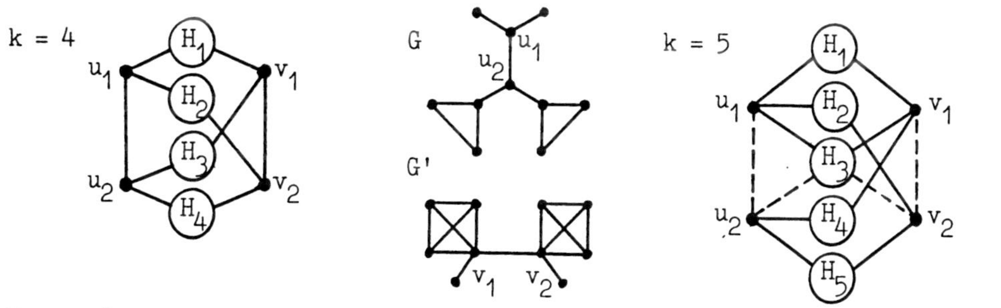

Figure 6. Non-trivial merging and a counterexample for the isometry of Z_{10}

Let P_n be the set of all connected bipartite graphs of vertex number n. The graph metric d'_P is defined by minimal bipartite common supergraphs. It can be shown that the corresponding distance graph $P'_n = D(P_n, d'_P)$ satisfies an equation similar to that given in Theorem 2.

<u>Theorem 3.</u> The distance graph $P'_n = D(P_n, d'_P)$ fulfils $\deg P'_n = \begin{cases} 0 & \text{if } n<11 \\ 1 & \text{else.} \end{cases}$

Sketch of the proof (for a detailed proof see KADEN 1986). By adding one vertex to each vertex class that is adjacent to all vertices of the other class, it is easy to prove that any two bipartite graphs $G,G' \in P_n$ of the distance $d'_P(G,G') = p$ are connected by a path not longer than p + 2 in P'_n. Choosing only one vertex to be added in a subtle way not described here, it is possible to show that there is a path not longer than p + 1 proving $\deg P'_n < 1$. A lot of cases have to be considered to demonstrate the isometry of the distance graph P'_n with $n<11$. These cases are divided in a way similar to that way used in the proof of Theorem 2. In Figure 7 we show only one case, namely a critical one, that leads to a counterexample $G,G' \in P_{11}$ that can be generalized to vertex numbers $n > 11$.

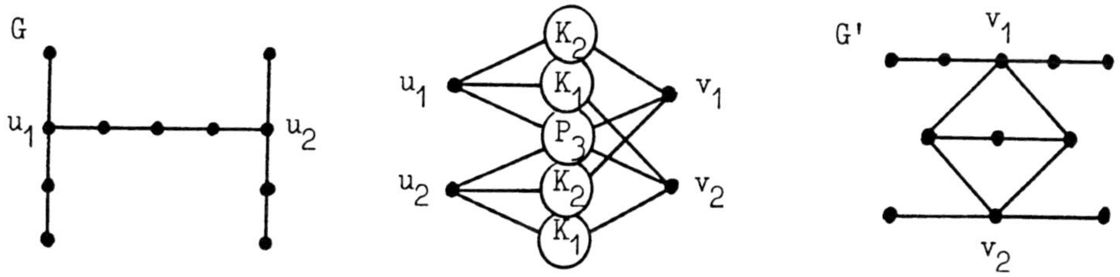

Figure 7. Non-trivial merging and a counterexample for the isometry of P'_{11}

In the case of the set $\mathbf{B}_n$ containing all trees of vertex number n, we have to consider two distance graphs denoted by $B_n = D(\mathbf{B}_n, d_o)$ and $B'_n = D(\mathbf{B}_n, d'_B)$. The latter is isometric for all vertex numbers n (B. Zelinka, unpublished). Finally, we show that the degree of non-isometry of the first one tends to infinity with the vertex number n.

Theorem 4. The distance graph $B_n = D(\mathbf{B}_n, d_o)$ satisfies $\lim \deg B_n/n = 1$.

Proof. Let Δ denote the maximal vertex degree of a graph. Then the following inequality holds for the distance δ defined on the set $\mathbf{B}_n$ by the length of shortest paths in the distance graph B_n.

$$\delta(G,H) \geqq | \Delta(G) - \Delta(H) |, \quad G,H \in \mathbf{B}_n$$

Because of $d_o \leqq d'_B$ the isometric distance graph B'_n is a spanning subgraph of B_n. Let δ' be the distance defined by the length of shortest paths in B'_n. Then we have $d_o(G,H) \leqq \delta(G,H) \leqq \delta'(G,H) = d'_B(G,H) < n$ for all trees $G,H \in \mathbf{B}_n$ and therefore $\deg B_n/n < 1$. For proving the above limit let G_k be the tree having one vertex of distance two to all pendant vertices and the degree of any non-pendant vertex is k, $k > 2$. Furthermore, $G'_k = K_{1,k^2}$ so that both trees G_k and G'_k have the vertex number $n_k = k^2 + 1$. Because of $d_o(G_k, G'_k) = k$ and $\delta(G_k, G'_k) \geqq | \Delta(G_k) - \Delta(G'_k) | = k(k-1)$, a minor sequence of $\deg B_n/n$ tends to one and completes the proof.

For further results about distance graphs on $\mathbf{Q}_n$, $\mathbf{P}_n$, and $\mathbf{W}_n$ see KADEN 1986.

References

M.A. Eshera, K.S. Fu. A graph distance measure for image analysis. IEEE Trans. SMC-14(1984), 398-408.

D. Gernert. Measuring the similarity of complex structures by means of graph grammars. Bull. of the EATCS, Febr. 1979, 3-9.

F. Kaden. Graphmetriken und Isometrieprobleme zugehöriger Distanzgraphen. ZKI-Informationen 1/86, Berlin 1986.

D.C. Kay, G. Chartrand. A characterization of certain ptolomaic graphs. Canad. J. Math. 17(1965), 342-346.

A. Sanfeliu, K.S. Fu. A distance measure between attributed relational graphs for pattern recognition. IEEE Trans. SMC-13(1983), 353-362.

L.G. Shapiro. Organization of relational models. Proc. Intern. Conf. on Pattern Recognition, München 1982, 360-365.

F. Sobik. On some measures of distance between graphs. in: Graphs and Other Combinatorial Topics, ed. M. Fiedler, Teubner-Texte zur Math. 59, Leipzig 1983, 278-285.

B. Zelinka. On a certain distance between isomorphism classes of graphs. Časopis Pěst. Mat. 100(1975), 371-373.

Witt Rings and Semiorderings of Planar Ternary Rings

F. B. Kalhoff, Dortmund, FRG

As we have seen in [3,4,5,6] a good part of the algebraic theory of reduced quadratic forms, orderings, and real places of commutative fields, as presented by, say Prestel [14] or Lam [8,9], goes over to arbitrary planar ternary rings (PTRs). One of the fundamental tools within these theories is the concept of quadratic semiorderings of fields, which allows sophisticated proofs of the representation criterion or the isotropy principle ([1,2]). For these reasons, it seems desirable, to establish and to study the notion of quadratic semiorderings also over arbitrary, formally-real PTRs.

Basic work on this theme is started in this paper. After codifying the notion of quadratic semiorderings in PTRs, we discuss their main properties, describe the relationship between quadratic semiorderings and the Witt rings of PTRs, and finally give a characterization of Pasch-preorderings.

We adopt the notion of a PTR (Ternärkörper) from Pickert [13] and use the terminology of [7]. In particular, for the algebraic theory of orderings and quadratic forms in PTRs, the concepts of halforderings, preorderings, Witt rings, fans, SAP, etc., the reader is referred to [7] or to [3,4,5,6]. Further, for elements a, b, c of a PTR T with $c \neq 0$, let $a+b := T(1,a,b)$, $ab := T(a,b,0)$, and a-b, -b, a/c, $c\backslash a$, be defined by, resp., $(a\text{-}b) + b = a$, $\text{-}b := 0\text{-}b$, $(a/c)c = a$, and $c(c\backslash a) = a$.

In what follows let T be a formally-real PTR, P a fixed preordering of T, Q = Q(T) the intersection of the kernels of all halforderings, and ΣQ the intersection of all preorderings of T.

Definition. A subset S of T is called a P-*semiordering* of T if it fulfills:

(S0) $S \cup \text{-}S \cup \{0\} = T$

(S1) $S \cap \text{-}S = \emptyset$

(S2) $S + S \subset S$

(S3) $P \cdot S \subset S$

(S4) $1 \in S$.

We denote the set of P-semiorderings of T by Y/P, and call the elements of Y := $Y/\Sigma Q$ *quadratic semiorderings* of T.

R. Bodendiek, R. Henn (Eds.)
Topics in Combinatorics and Graph Theory

Note, that any P-semiordering is also a quadratic semiordering, and that in case of a commutative field T, our concept reduces to the notion of Prestel [14].
Throughout this paper, we will make heavily use of the fact, that P is a normal subloop of T^*, and that T^*/P is an abelian group of exponent two. Since P contains the radical of T, (S3) provides us with the following rules of arithmetic from [3,4,5]:

(1) Lemma. *For all nonempty subsets* S,S',S" *of* T *fulfilling* (S3), *and for all elements* a,b,c,d ∈ T, *we have:*

(a)	$S[(a+b)-(a+c)] = S[b-c]$,	$S[(a+b)-(c+b)] = S[a-c]$,
(b)	$S[(-a)+b] = S[b-a]$,	$S[a-(-b)] = S[a+b]$,
	$S[-(a-b)] = S[b-a]$,	$S[(a+b)+c] = S[a-((-c)-b)]$,
(c)	$S[(a-b)c] = S[ac-bc]$,	$S[a(b-c)] = S[ab-ac]$,
(d)	$S[-ab] = S[(-a)b] = S[a(-b)]$,	$S[(-1)ab] = S[((-1)a)b] = S[a(b(-1))]$,
(e)	$S+S' = S'+S$,	$(S+S')+S'' = S+(S'+S'')$,
(f)	$SS' = S'S$,	$(SS')S'' = S(S'S'')$,
(g)	$(S+S')a = Sa+S'a$,	$a(S+S') = aS+aS'$,
(h)	$-(Sab) = S(-ab) = S(-a)b = Sa(-b)$.	

Note, that in view of this lemma, for ΣQ-semiorderings, the axiom (S3) may be replaced by $Q \cdot S \subset S$. Hence we have:

(2) Proposition. *Given a quadratic semiordering* S *of* T, *the rule*

$$a < b \;:\Leftrightarrow\; b-a \in S$$

defines a linear order relation on T *which fulfills:*

(a) $a < b \Rightarrow a+c < b+c$ *and* $c+a < c+b$,

(b) $a < b \Rightarrow qa < qb$ *and* $aq < bq$ *for all* $q \in Q$,

(c) $0 < 1$,

Conversely, any linear order relation on T *fulfilling (a), (b) and (c) is uniquely determined in this way by some quadratic semiordering* S *of* T.

Proof. Making use of (1), it is easily checked that the relation < defined above is a linear order relation on T (for instance, the implication $c-b,\, b-a \in S \Rightarrow c-a \in \Sigma Q[(c-b)-(a-b)] \subset \Sigma Q[S-(-S)] \subset \Sigma Q(S+S) \subset S$ yields transitivity). The claims (a), (b) and (c) follow easily from (S3), (1)(a), (1)(c), and by (S4). Conversely, given a linear order relation < fulfilling (a), (b), and (c), then, clearly, $S := \{t \in T \mid 0 < t\}$ satisfies (S2), (S3), (S4). By (1), we get $t \in S \Leftrightarrow -t \notin S$, and thus (S1) and (S0). □

(3) Proposition. *Let* $<$ *be an order relation induced by a* P*-semiordering* S *of* T. *Then we have for all* $a,b \in T$:

(a)	$a < b$	$\Rightarrow$	$pa < pb$ *and* $ap < bp$ *for all* $p \in P$,		
(b)	$0 < a$	$\Leftrightarrow$	$0 < 1/a$	$\Leftrightarrow$	$0 < a\backslash 1$,
(c)	$0 < a < b$	$\Rightarrow$	$(ba)a < (bb)a$ *and*		$b(aa) < b(ba)$,
(d)	$1 < b$	$\Rightarrow$	$b < bb$,		
(e)	$0 < a < 1$	$\Rightarrow$	$0 < aa < a$,		
(f)	$0 < a < b$	$\Leftrightarrow$	$0 < 1/b < 1/a$	$\Leftrightarrow$	$0 < b\backslash 1 < a\backslash 1$,
(g)	$1 < a$	$\Leftrightarrow$	$0 < 1/a < 1$	$\Leftrightarrow$	$0 < a\backslash 1 < 1$,
(h)	$0 < a < b$	$\Rightarrow$	$0 < aa < bb$, *provided that either* $a \in P$ *or* $b \in P$.		

Proof. (a) follows immediately by (1)(c) and (S3), and (b) by (S3) and $P(1/a) = P(a\backslash 1) = Pa$. To check (c), first note that $(b - a) / (a(b - a)) \in P(1/a) \subset S$ and $a / (a(b - a)) \in P(1/(b - a) \subset S$, by lemma (1). Thus, by (1), (S3), and (S2):

$$\begin{aligned} b(b - a)a &\in P[b / (a(b - a))] = P[((b - a) + a) / (a(b - a))] \\ &\subset P[(P(b - a) + Pa) / (a(b - a))] \\ &= P[\, P((b - a)/(a(b - a)) + P(a/(a(b - a)))] \subset P[S + S] \subset S. \end{aligned}$$

Using (1), we get $(ba)a < (bb)a$ and $b(aa) < b(ba)$. The claims (d) and (e) follow immediately from (c). To check (f) let $x := 1/a$ and $y := b/(xb)$. Then we have $(y\backslash b)/b = x$, and, using (1), we obtain from $0 < a < b$:

$$\begin{aligned} b - y &\in P[b - b/(xb)] = Pxb[b(xb) - b] = Px(xb - 1) = P[x\backslash(xb - 1)] \\ &= P(b - x\backslash 1) = P(b - a) \subset PS \subset S, \end{aligned}$$

and, using (c):

$$\begin{aligned} 1/a - 1/b &\in P[(y\backslash b)/b - 1/b] = Pb[(y\backslash b) - 1] = Pb(b - y)y \\ &= P((bb)y - (by)y) \subset S. \end{aligned}$$

Hence we have $0 < a < b \Rightarrow 1/b < 1/a$. With $b' := b\backslash 1$ and $a' := a\backslash 1$ the implication just proved leads to $0 < a' < b' \Rightarrow 0 < 1/b' < 1/a' \Rightarrow 0 < b < a$, and thus to $0 < a < b \Rightarrow 0 < a\backslash 1 < b\backslash 1$. By (b), also the reverse implications hold. (g) is a specialization of (f). To prove (h) let $a \in P$. Then we clearly have $aa < ba$. Since, using (c), we get $bb - ba \in P(bb - ba) = P(bb - ba)a = P[(bb)a - (ba)a] \subset PS \subset S$, we obtain $ba < bb$, and thus $aa < bb$. Similarly, we find $aa < bb$ in the case $b \in P$. □

As in the classical case, we obtain:

(4) Proposition. *With respect to the topology defined by the subbasic sets* $H(t) := \{S \in Y \mid t \in S\}$, $t \in T^*$, *the set* Y *becomes a compact, totally disconnected Hausdorff space, containing* X, Y/P *and* X/P *as closed subspaces.*

Now we are going to establish the connections between the concept of quadratic semiorderings just developed and the algebraic theory of Witt rings of PTRs. Following [9, chap.14] we start with:

Definition. A nonempty subset $M \subset T$ is called a P-*module*, if M fulfills $0 \in M$, $M + M \subset M$, and $P \cdot M \subset M$. M is called *anisotropic*, if $M \cap -M = \{0\}$.

(5) Proposition. *For a* P-*module* M *of a* PTR T *the following statements are equivalent:*

(a) $M \neq T$

(b) M *is anisotropic*

(c) *for all* $m_1,..., m_n \in M \setminus \{0\}$, *the form* $< m_1,..., m_n >$ *is* P-*anisotropic.*

In particular, a P-*form* $< a_1, ..., a_n >$ *is* P-*anisotropic if and only if the* P-*module* $D_P(< a_1, ..., a_n >) \cup \{0\}$ *is anisotropic.*

Proof (cf. Lam [9, 14.1]). (a)⇒(b): Assume there exists a nonzero $m \in M$ such that $-m \in M$. By [6, 2.1] and (1), we obtain $T = Tm = (P + P(-1))m = Pm + P(-m) \subset M$, contradicting (a).
(b)⇒(c): Assume that $< m_1,..., m_n >$ is P-isotropic. By [6, 3.2], we have $0 = p_1m_1+...+p_rm_r$, for some $r \leq n$, $p_1,...p_r \in P \cup \{0\}$, and with, say $p_1 \neq 0$. Using (1), we get $-p_1m_1 \in Pm_2 + ... + Pm_r \subset M$, contrary to (b).
(c)⇒(a) is immediate, since the form $< 1, -1 >$ is isotropic. □

(6) Corollary. *Let* M *be an anisotropic* P-*module, and* $a \in T^*$. *Then we have:*

(a) $a \notin -M \Leftrightarrow M' := (M + Pa) \cup M$ *is an anisotropic* P-*module*

(b) M *is a maximally anisotropic* P-*module* $\Leftrightarrow$ $M \cup -M = T$.

Proof (cf. Lam [9, 14.2, 14.3]). Clearly, by (1), M' is a P-module. In case M' = T there exist $m \in M$ and $p \in P \cup \{0\}$ such that $-a = m + pa$. We obtain $(-a) - pa = m$, and thus, by (1), $a \in Pa = P[(-1)((-a)/a - p)]a = -P((-a)/a - p)a = -Pm \subset -M$. Thus M' is anisotropic, provided that $a \notin -M$. For the reverse implication of (a), note that $a,-a \in M'$ if $a \in -M$. The claim (b) follows immediately from (a). □

In particular, a set M containing zero is a maximally anisotropic P-module if and only if $M \setminus \{0\}$ or $-M \setminus \{0\}$ is a P-semiordering of T. Thus, within the theory of quadratic semiorderings, the anisotropic P-modules play a role similar to that of the preorderings, and using Zorn's Lemma, we obtain an analogue of Artin's Theorem (cf. [9, 14.10]):

(7) Proposition. *Any anisotropic* P-*module* M *of the* PTR T *is the intersection of the maximally anisotropic* P-*modules of* T *containing* M.

Obviously, we can identify any semiordering $S \in Y/P$ with its signature $sg_S: T^* \to \{1,-1\}$ given by:

$$sg_S(t) = 1 :\Leftrightarrow t \in S \quad \text{and} \quad sg_S(t) = -1 :\Leftrightarrow t \in -S,$$

and for any P-form $\phi = < a_1, ..., a_n >$, we may define the *signature of* ϕ *at* S by:

$$sg_S(\phi) := sg_S(a_1) + ... + sg_S(a_n).$$

Now we can state the desired characterization of the Witt ring $W_P(T)$ in terms of P-semiorderings (cf. [2] or [9, 14.9]):

(8) Theorem. *Let* T *be a formally-real* PTR *and* P *a preordering of* T. *Two* P-*forms* ϕ, ψ *of the same dimension over* T *are* P-*isometric if and only if* $sg_S(\phi) = sg_S(\psi)$ *for all* $S \in Y/P$.

Proof. By definition, it suffices to check the "only if" part. Since two isometric forms of the same dimension are chain equivalent (see [6, 3.3]), we only have to consider the case:

$$\phi = < a, b >, \quad \psi = < a+b, ab(a+b) >,$$

with $a + b \neq 0$. Defining $a' := a + b$ and $b' := ab(a + b)$, and making heavy use of (1), we observe:

$$\begin{aligned} a' &\in Pa + Pb, \\ b' &\in ab(Pa + Pb) = abPa + abPb = Pb + Pa = Pa + Pb, \\ a &\in Pa(a+b)^2 \subset Pa(Pa + Pb)(a+b) = P(P + Pab)(a+b) = (P + Pab)(a+b) \\ &= Pa' + Pb', \text{ and} \\ b &\in Pb(a+b)^2 \subset Pb(Pa + Pb)(a+b) = Pa' + Pb'. \end{aligned}$$

Thus, if $S \in Y/P$ contains a and b, then it also contains a' and b', and vice versa. From this, we easily get $sg_S(\phi) = sg_S(\psi)$. □

(9) Proposition. *A* P-*form* ϕ *is* P-*isotropic if and only if it is indefinite with respect to all* $S \in Y/P$.

Proof. If $|sg_S(\phi)| = \dim(\phi) = n$ for a semiordering $S \in Y/P$, then we have $D_P(\phi) \subset S$ or $D_P(\phi) \subset -S$. In both cases, ϕ is P-anisotropic. Conversely, if ϕ is P-anisotropic, then $T \neq M := D(\phi)_P \cup \{0\}$. Thus, by (7), there exists a maximally anisotropic P-module S', containing M. In view of (6)(b), $S := S' \setminus \{0\}$ or -S is a P-semiordering of T, depending on wether $1 \in S$ or $-1 \in S$. Hence $sg_S(\phi) = n$ or $sg_{-S}(\phi) = -n$. □

Definition. A preordering P of a formally-real planar ternary ring T is called a *Pasch-preordering,* if X/P = Y/P.

(10) Corollary. *A preordering* P *of a formally-real* PTR T *is a Pasch-preordering, if and only if it is a* SAP- *preordering.*

Proof (cf. Lam [9, 16.2]). Assume P is not Pasch. Then there exists a proper P-semiordering S and elements a,b ∈ S with ab ∉ S. Clearly, the form < 1, a, b, -ab > cannot be P-isotropic, and, in view of [7, 2.32], P cannot be SAP. On the other hand, if P is Pasch, then any form < 1, a, b, -ab >, a,b ∈ T^*, is indefinite at every ordering of T, and thus P-isotropic. By [7, 2.32], P is SAP. □

Thus, given a nontrivial fan F, corollary (10) assures the existence of proper F-semiorderings. The following proposition gives a concrete construction of such semiorderings:

(11) Proposition. *Let* F *be a nontrivial fan of the* PTR T *and* P ∈ X/F. *Then the set*

$$S := F \cup -(P\setminus F)$$

is a proper F*-semiordering of* T.

Proof (cf. Lam [9, 14.12]). Since $F \cdot (P\setminus F) = P\setminus F$ and thus $-(-(P\setminus F)) = -(-F(P\setminus F)) = F \cdot P\setminus F = P\setminus F$, the set $S := F \cup -(P\setminus F)$ fulfills (S3), (S0), (S1), and, clearly, (S4). Hence it remains to check that $S + S \subset S$. Therefore, let a,b ∈ S. Then we have $ab \notin -F$, since otherwise $a \in -Fb \subset -F(F \cup -(P\setminus F)) = -F \cup P\setminus F = -S$. By [5, 4.2], we get: $a + b \in Fa + Fb = Fa \cup Fb \subset S$. This shows S ∈ Y/F. Since F is nontrivial, there exists x,y ∈ P such that 1, x, y and xy are distinct modulo F. Then -x, -y, -xy ∈ S, but xy ∉ S, which means $(-x)(-y) \notin S$, by $F(-x)(-y) = Fxy$. Hence S ∉ X/F. □

(12) Remarks. (a) As in the classical case, theorem (8) provides us with a group homomorphism c from $W_P(T)$ into the ring C(Y/P, **Z**) of continuous functions from Y/P into **Z** (with the discrete topology); here c(ϕ) is defined to be that function which maps S ∈ Y/P onto $sg_S(\phi)$.

(b) The calculations given in the proof of (8) show that it is possible to transfer Becker's and Köpping's treatment of reduced quadratic forms given in [2] to our setting. In particular, this would yield a proof of the representation criterion for Witt rings of PTRs which does not rely on Marshall's theory (cf. [6] and [10,11,12]).

(c) For generalizations of the concept of quadratic semiorderings of PTRs, in particular, for the notion of not necessarily quadratic semiorderings (fulfilling only (S0), (S1), (S2), and (S4)) and for its geometric significance, the reader is referred to [15].

References

[1] E. Becker, L. Bröcker. *On the description of the reduced Witt ring.* J. Algebra **52** (1978), 328-346

[2] E. Becker, E. Köpping. *Reduzierte quadratische Formen und Semiordnungen reeller Körper.* Abh. Math. Sem. Univ. Hamburg 46 (1977), 143-177

[3] F. Kalhoff. *Eine Kennzeichnung anordnungsfähiger Ternärkörper.* J. Geometry 31 (1988), 100-113

[4] F. Kalhoff. *Über Präordnungen von Ternärkörpern.* Abh. Math. Sem. Univ. Hamburg, to appear

[5] F. Kalhoff. *Über Stellen und Präordnungen von Ternärkörpern.* Geometriae dedicata **27** (1988), 137-151.

[6] F. Kalhoff. *Spaces of orderings and Witt rings of Planar Ternary Rings.* J. Pure App. Algebra **58** (1989), 169-180

[7] F. Kalhoff. *Anordnungsräume und Wittringe projektiver Ebenen.* Habilitationsschrift Dortmund, 1988/89

[8] T. Y. Lam. *The theory of ordered fields.* Ring Theory and Algebra III (ed. B. MacDonald) Lecture notes in Pure and Appl. Math. **55** (1980), Dekker, New York, 1-152

[9] T. Y. Lam. *Orderings, valuations and quadratic forms.* CBMS Regional Conference Series in Mathematics **52** (1983), American Mathematical Society, Providence

[10] M. A. Marshall. *Classification of finite spaces of orderings.* Canad. J. Math. **31** (1979), 320-330

[11] M. A. Marshall. *The Witt ring of a space of orderings.* Trans. Amer. Math. Soc. **258** (1980), 505-521

[12] M. A. Marshall. *A simple system of axioms for spaces of signatures.* J. Pure Appl. Algebra **57** (1989), 159-164

[13] G. Pickert. *Projektive Ebenen.* Springer, Berlin Heidelberg New York, 2. ed. 1975
(1. ed. 1955)

[14] A. Prestel. *Lectures on formally real fields.* IMPA Lecture notes **22** (1975), Rio de Janeiro (2. ed., Springer, 1984)

[15] H. Tecklenburg. *Stufen der Anordnung in Geometrie und Algebra.* Habilitationsschrift, Hannover 1988/89

Linear Inequalities Describing the Class of Intersecting Sperner Families of Subsets, I

G. O. H. Katona, Budapest, Hungary
G. Schild, Ilmenau, GDR

1. Introduction

Let X be a finite set of n elements. We say that the family $\mathcal{A} = \{A_1, \dots, A_m\}$ of its distinct subsets is *intersecting* if $A_i \cap A_j \neq \emptyset$ holds for any $1 \leq i \leq j \leq m$. A classic theorem of *Erdős Ko* and *Rado* [2] states that an intersecting family $\mathcal{A}$ consisting of k-element subsets, where $k \leq n/2$, has at most $\binom{n-1}{k-1}$ members.

Actually, they proved a more general theorem. The family $\mathcal{A}$ is called *Sperner* (see [9]) if $A_i \subseteq A_j$ never holds for $i \neq j$, that is, $\mathcal{A}$ is inclusion-free. It is proved in [2] that an intersecting Sperner family consisting of subsets of size at most $k(\leq n/2)$, has at most $\binom{n-1}{k-1}$ members. This paper of Erdős, Ko and Rado (together with [9]) started a new area of combinatorics. Here we present only the results significant from our point of view.

The next major step is due to *Bollobás* [1]. Let $p_i = p_i(\mathcal{A})$ denote the number of i-element members of $\mathcal{A}$.

Bollobás's inequality. *If $\mathcal{A}$ is an intersecting Sperner family then the following holds:*

$$\sum_{i \leq n/2} \frac{p_i}{\binom{n-1}{i-1}} \leq 1. \tag{1}$$

Two subsets of size $k > n/2$, always intersect, so it is not interesting to investigate the maximally sized intersecting families consisting of such subsets. However, if all sizes are allowed then the question becomes interesting. *Milner* [7] proved that an intersecting Sperner family has maximally $\binom{n}{\lceil \frac{n+1}{2} \rceil}$ members. It is easy to see that the inequality

R. Bodendiek, R. Henn (Eds.)
Topics in Combinatorics and Graph Theory

of Bollobás implies both forms of the Erdős-Ko-Rado theorem, but it does nt imply Milner's one. This fact led *Greene, Katona* and *Kleitman* [6] to the following inequality:

$$\sum_{i\leq n/2} \frac{p_i}{\binom{n}{i-1}} + \sum_{n/2<j\leq n} \frac{p_j}{\binom{n}{j}} \leq 1. \tag{2}$$

This inequality implies Milner's theorem, but it is too weak to imply Erdős-Ko-Rado. The reason of this annoying situation was shown in [3]. The vector $(p_0, p_1, \ldots, p_n) = (p_0(\mathcal{A}), p_1(\mathcal{A}), \ldots, p_n(\mathcal{A}))$ is called the *profile* of $\mathcal{A}$. The extreme points of the convex hull of the set of profiles of all intersecting Sperner families on n elements were determined in [3]. As a by-product, a large set of inequalities was determined for the intersecting Sperner families. Both (1) and (2) were particular cases in this class of inequalities. However, the class was too wide. Many of the inequalities were consequences of others. The authors of [3] did not notice that a minimal set of inequalities could have been easily deduced. This work is done in Section 2 of the present paper. These inequalities determine the hyperplanes bordering the convex hull of profiles of the set of all intersecting Sperner families.

2. Minimal set of inequalities

If a family $\mathcal{A}$ is intersecting then it does not contain the empty set as a member. Therefore p_0 is zero. We modify the definition of the *profile* omitting this superflous component: $(p_1, \ldots, p_n)$. Consider the set of profiles of all intersecting Sperner families in the n-dimensional Euclidean space, where n is the size of the ground set of the families. Take the convex hull of this set of profiles. The vertices of this convex hull are called briefly the *extreme points* of the class of intersecting Sperner families. The starting point of our investigations is the following theorem:

THEOREM. [3] *The extreme points of the class of intersecting Sperner families are*

$$z = (0, \ldots, 0),$$

$$v_j = \left(0, \ldots, 0, \binom{n}{j}, 0, \ldots, 0\right) \quad (n/2 < j \leq n),$$

$$w_i = \left(0, \ldots, 0, \binom{n-1}{i-1}, 0, \ldots, 0\right) \quad (1 \leq i \leq n/2),$$

$$w_{ij} = \left(0, \ldots, 0, \binom{n-1}{i-1}, 0, \ldots, 0, \binom{n-1}{j}, 0, \ldots, 0\right) \quad (1 \leq i \leq n/2,\ n < i+j).$$

For two vectors x and y we write $x \leq y$ if the inequality holds componentwise. xy denotes the inner product.

A polyhedron P in the n-dimensional Euclidean space is called ***anti-blocking type*** if $P \neq \emptyset$, $x \in P$ implies $0 \leq x$ and $0 \leq y \leq x \in P$ implies $y \in P$ (see [8]). P is ***full*** if it contains elements $(0, \ldots, 0, x_i, 0, \ldots, 0)$ $(0 < x_i)$ for all i $(1 \leq i \leq n)$. It is easy to see that the polihedron determined by the points given in the Theorem (that is, the convex hall of the class of intersecting Sperner families) is full and anti-blocking type.

Define (see [8])

$$A(P) = \{z : 0 \leq z,\ zx \leq 1, \text{for all } x \in P\}.$$

By a theorem of ***Fulkerson*** [4], [5] (see Theorem 9.4 in [8]) $A(P)$ is full and anti-blocking type, again.

An extreme point x of a polyhedron P is called ***essential*** if there is no other extreme point $y \in P, x \leq y$. It is easy to see that an anti-blocking type polyhedron is uniquely determined by its essential extreme points.

A ***bordering hyperplane*** of an n-dimensional polyhedron P is an $n-1$-dimensional hyperplane, given by an equation $a_1x_1 + \ldots + a_nx_n = 1$ or $a_1x_1 + \ldots + a_nx_n = 0$, containing at least n extreme points of P and satisfying the inequality of the same direction for all points of P.

If P is full and anti-blocking type then $x_i = 0$ is a bordering hyperplane for each i and no other bordering hyperplane have 0 on the right hand side. The latter ones are the ***essential bordering hyperplanes***.

LEMMA 1. *$\sum a_ix_i = 1$ is an essential hyperplane of the full anti-blocking type polihedron P iff $a = (a_1, \ldots, a_n)$ is an essential extreme point of $A(P)$.*

PROOF: First we verify that if $\sum a_ix_i = 1$ is a bordering hyperplane of P then $a \in A(P)$. As $0 \in P$, the inequality $ax \leq 1$ must hold for the elements x of P. To prove $a \in A(P)$ we need to show $0 \leq a$, only. The hyperplane contains at least n vertices of P. If all these vertices had 0 as a first component then they would be in the $(n-2)$-dimensional intersection of the hyperplanes $ax = 1$ and $x_1 = 0$. This contradiction ensures the existence of a vertex $x = (x_1, \ldots, x_n)$ with a positive first component, on the hyperplane. $(0, x_2, \ldots, x_n) \in P$ follows by the anti-blocking type property. However $ax = 1, a_1 < 0$ and $0 < x_1$ imply $0x_1 + a_2x_2 + \ldots + a_nx_n > 1$. This contradiction proves $a_1 \geq 0$ and, in general, $a_i \geq 0$.

Let us show now that if $ax = 1$ is a bordering hyperplane then a is an extreme point of $A(P)$. Suppose that it is not true. Then there exists a non-zero n-dimensional vector b such that $a + b \in A(P), a - b \in A(P)$. Hence we have $(a + b)x \leq 1$ and $(a - b)x \leq 1$ for each vertex of P contained in the bordering hyperplane $ax = 1$. $bx \leq 0, -bx \leq 0$ and consequently $bx = 0$ follow for these vertices. This is a contradiction, since these vertices span a hyperplane not containing the origin.

Finish this part of the proof showing that if $ax = 1$ is a bordering hyperplane then a is an essential extreme point. Suppose the contrary: there is a $b \in A(P)$ such that $a \leq b$, $a \neq b$. Let $a_k < b_k$. We saw in the first section of this proof that there must exist a vertex $x \in P$ with equality in $ax = 1$ and satisfying $0 < x_k$. For this vertex $bx > ax = 1$ holds, contradicting $ax \leq 1$. Therefore, if $ax = 1$ is a bordering hyperplane of P, a must be an essential extreme point of $A(P)$.

Now we prove the other direction. Suppose that a is an essential extreme point of $A(P)$ and assume, in an indirect way, that $ax = 1$ is not a bordering hyperplane. By the definition of $A(P)$, a satisfies $ax \leq 1$ for all points x of P, thus the vertices v of P satisfying $av = 1$ are contained in a hyperplane H of dimension k less than $n-1$. Then all vertices v of P satisfying $av = 1$ are linear combinations of the vertices $v_1, \ldots, v_k$ where $av_i = 1\,(1 \leq i \leq k)$. For the other vertices w of P $aw < 1$ holds. Two cases are distinguished.

1)Suppose that $a_j = 0$ implies that there is a vertex $u = (u_1, \ldots, u_n) \in P$ satisfying $au = 1$ and $0 < u_j$ for all j. Then $u' = (u_1, \ldots, u_{j-1}, 0, u_{j+1}, \ldots, u_n) \in P$ holds by the properties of P. The equation $au' = 1$ implies $u' \in H$. Let b be a non-zero solution of the system $bv_i = 0\,(1 \leq i \leq k)$. Since both u and u' are linear combinations of $v_1, \ldots, v_k$, we have $bu = bu' = 0$. These two equations imply $b_j u_j = 0$ and, by our assumption, $b_j = 0$ is true. Then $a \pm \varepsilon b$ consists of non-negative components if $\varepsilon > 0$ is small enough. The equation $(a \pm \varepsilon b)v = 1$ is obvious for all elements of H and $(a \pm \varepsilon b)w < 1$ for all other vertices of P if ε is small enough. Thus $a + \varepsilon b$ and $a - \varepsilon b$ are in $A(P)$, a contradiction.

2)Let $a_j = 0$ and let all vertices $u = (u_1, \ldots, u_n)$ of P satisfying $au = 1$ have $u_j = 0$. Then $a' = a + (0, \ldots, 0, \varepsilon, 0, \ldots, 0)$ (ε is in the jth component) satisfies $a'u = 1$ for these vertices and $a'w < 1$ for other vertices, if ε is small enough. $a' \in A(P)$ is a contradiction. ■

Let P be the convex hull of the points given in the theorem. Then $v_j\ (n/2 < j \leq n)$ and $w_{ij}\ (1 \leq i \leq n/2,\ n < i + j)$ are then essential extreme points of P. $A(P)$ consists of the points $x = (x_1, \ldots, x_n)$ satisfying

$$\binom{n}{j} x_j \leq 1 \quad (n/2 < j \leq n), \tag{3}$$

$$x_1 \leq 1, \tag{4}$$

$$\binom{n-1}{i-1} x_i + \binom{n-1}{j} x_j \leq 1 \quad (2 \leq i \leq n/2,\ j < n,\ n < i + j). \tag{5}$$

By Lemma 1 we have to determine only the essential extreme points of $A(P)$, that is, the set of vectors x determined by (3),(4) and (5).

The following easy lemma enables us to make a reduction. Let $\mathrm{diag}(c_1,\ldots,c_n)$ denote an $n \times n$ matrix with these entries in the main diagonal and 0s otherwise. For a polyhedron P, define

$$\mathrm{diag}(c_1,\ldots,c_n)P = \{\mathrm{diag}(c_1,\ldots,c_n)x : x \in P\}.$$

LEMMA 2. *Suppose that* $c_1,\ldots,c_n > 0$. *If* P *is anti-blocking type then* $\mathrm{diag}(c_1,\ldots,c_n)P$ *has the same property.* x *is an essential extreme point of* P *iff* $\mathrm{diag}(c_1,\ldots,c_n)x$ *is an essential extreme point of* $\mathrm{diag}(c_1,\ldots,c_n)P$.

PROOF: It is trivial. ■

Let $c_i = \binom{n-1}{i-1}$ $(1 \le i \le n/2)$, $c_j = \binom{n-1}{j}$ $(n/2 < j < n)$, $c_n = 1$. By (3),(4) and (5) $S = \mathrm{diag}(c_1,\ldots,c_n)A(P)$ consists of the points y satisfying

$$y_j \le 1 - \frac{j}{n} \quad (n/2 < j < n), \tag{6}$$

$$y_1 \le 1, \quad y_n \le 1, \tag{7}$$

$$y_i + y_j \le 1 \quad (2 \le i \le n/2,\ j < n,\ n < i + j). \tag{8}$$

Now we try to find the essential extreme points of S (as it is anti-blocking type by Lemma 2).

LEMMA 3. *If* $(y_1,\ldots,y_n)$ *is an essential extreme point of* S *then*

$$y_1 = y_n = 1, \tag{9}$$

$$\text{for every } i \ (2 \le i \le n/2) \text{ there is a } j \text{ such that } n - i < j < n \text{ and } y_i + y_j = 1 \tag{10}$$

$$\text{for every } j \ (n/2 < j < n) \text{ either } y_j = 1 - j/n \text{ or there is an } i > n - j \text{ such that } y_i + y_j = 1. \tag{11}$$

PROOF: y_1 and y_n occur in (7) but not in (6) and (8). If their values were < 1, y could be increased, consequently it would not be an essential extreme point. (9) is proved. (10) and (11) can be proved by the same argument: if there is a strict inequality at all occurences of a variable y_i in (6) and (8) then its value can be increased letting the other components unchanged. This contradicts the assumption that y is an essential extreme point. ■

LEMMA 4. *If $y = (y_1, \ldots, y_n)$ is an essential extreme point of S then $y_{i-1} \geq y_i$ $(2 \leq i < n)$.*

PROOF: Suppose that $y = (y_1, \ldots, y_n) \in S$ and $y_{j-1} < y_j$ $(n/2+1 < j < n)$. Then we can show that $(y_1, \ldots, y_{j-2}, y_j, y_j, y_{j+1}, \ldots, y_n) \in S$. (6) holds for the new $(j-1)$th component since $y_j \leq 1 - j/n < 1 - (j-1)/n$. (8) holds for it because $i + j - 1 > n$ implies $i + j > n$. The modified vector is larger than y, so y cannot be an essential extreme point of S.

If $y_{j-1} < y_j$ $(2 \leq j \leq n/2)$ then the same argument is used to ensure the validity of (8). The missing cases are $j = \frac{n+1}{2}$ and $j = \frac{n}{2} + 1$. In the first one of these cases $j = \frac{n-1}{2}$ holds and $y_{\frac{n+1}{2}} + y_k \leq 1$ should be proved for $\frac{n-1}{2} + k > n$. This is, however, a consequence of (6). The case $j = \frac{n}{2} + 1$ can be settled in the same way. ■

LEMMA 5. *If $y = (y_1, \ldots, y_n)$ is an essential extreme point of S then*

$$y_{n-i+1} = 1 - y_i \quad (2 \leq i \leq n/2).$$

PROOF: Suppose the contrary: $y_i + y_{n-i+1} < 1$ holds for some $2 \leq i \leq n/2$. By Lemma 4 we have

$$y_{n-1} \leq y_{n-2} \leq \ldots \leq y_{n-i+1} < 1 - y_i,$$

that is, (10) is violated for this i. ■

LEMMA 6. *$(y_1, \ldots, y_n)$ is an essential extreme point of S iff $y_1 = y_n = 1$, and there are some integers $1 = i_1 < i_2 < \ldots < i_{r+1} = \lceil \frac{n+1}{2} \rceil$ $(1 \leq r)$ such that*

$$y_{i_k} = 1 - \frac{i_k - 1}{n} \quad (1 \leq k \leq r),$$

$$y_i = y_{i_k} \quad (i_k \leq i < i_{k+1},\ 1 \leq k \leq r),$$

$$y_{n-i+1} = 1 - y_i, \quad (2 \leq i \leq n/2) \tag{12}$$

and

$$y_{\frac{n+1}{2}} = \frac{1}{2} - \frac{1}{2n}.$$

PROOF: First, let us prove that the essential extreme points $(y_1, \ldots, y_n)$ must have this form. Lemma 5 and (6) imply

$$y_i \geq 1 - \frac{i-1}{n} \quad (2 \leq i \leq \frac{n}{2}). \tag{13}$$

If n is even, then $y_{\frac{n}{2}} \geq \frac{1}{2} + \frac{1}{n}$ follows by (13) while $y_{\frac{n}{2}+1} \leq \frac{1}{2} - \frac{1}{n}$ is a consequence of (6). Similarly, $y_{\frac{n-1}{2}} \geq \frac{1}{2} + \frac{3}{2n}$ and $y_{\frac{n+1}{2}} \leq \frac{1}{2} - \frac{1}{2n}$ hold for the odd ns. Our conclusion is that these neighbouring ys must be different.

So, if m is defined by $y_2 = y_{m-1} > y_m$ then the inequality $m - 1 \leq \frac{n}{2}$ is obvious.

(13) implies $1 - \frac{1}{n} \leq y_2$. Suppose that $1 - \frac{1}{n} < y_2 < 1$. Then there is an $\varepsilon > 0$ such that both

$$(1, y_2 + \varepsilon, \ldots, y_{m-1} + \varepsilon, y_m, \ldots, 1 - y_{n-m+1}, 1 - y_{n-m+2} - \varepsilon, \ldots, 1 - y_2 - \varepsilon, 1)$$

$$(1, y_2 - \varepsilon, \ldots, y_{m-1} - \varepsilon, y_m, \ldots, 1 - y_{n-m+1}, 1 - y_{n-m+2} + \varepsilon, \ldots, 1 - y_2 + \varepsilon, 1)$$

are in S. This contradiction shows that $(y_1, \ldots, y_n)$ cannot be an essential extreme point. Consequently, y_2 is either $1 - \frac{1}{n}$ or 1. In the first case $i_2 = 2$, in the latter one $i_2 = m$. Repeating this procedure we obtain that the essential extreme points satisfy the conditions of the lemma. (If n is odd then $y_{\frac{n+1}{2}} \leq 1/2 - 1/2n$. Choosing equality, the vector is in S and is larger than the vectors with smaller $y_{\frac{n+1}{2}}$.)

It remained to prove that all these vectors are essential extreme points. Denote their set by Y. No two of them are comparable (in the sense of $\leq$) so it is sufficient to see that they are extreme points, that is, they are not convex linear combinations of other extreme points. In other words, if $y \in Y$, $z^1, \ldots, z^t \neq y$ are different extreme points of S and $\alpha_1, \ldots, \alpha_k$ are positive numbers ($\sum \alpha_i = 1$) then

$$y = \sum_{i=1}^{t} \alpha_i z^i \tag{14}$$

is impossible.

First prove this statement for the case if one of the z^is is a non-essential extreme point. Then this z^i satisfies $z^i \leq y', z^i \neq y'$ for some $y' = (y'_1, \ldots, y'_n) \in Y$. The inequality $y'_i + y'_{n-i+1} < 1$ (or $y'_{\frac{n+1}{2}} < 1/2 - 1/2n$) is a consequence for some i. (14) implies that the same holds for y, that is, $y \notin Y$.

Therefore we may suppose that $z^1, \ldots, z^t \in Y$. An element z of Y is uniquely determined by the sequence $i_1(z) < i_2(z) < \ldots < i_r(z)$ $(r = r(y))$. If $z^1 \neq z^2$ then these sequences are different, so there is an $i_j = i_j(z^1)$ which is not equal to any $i_s(z^2)$ (or the other way around). Clearly we have

$$1 - \frac{i_j - 1}{n} = z^1_{i_j} < z^1_{i_j - 1}$$

and

$$z^1_{i_j} < z^2_{i_j} = z^2_{i_j - 1}$$

by the definition of the sequence i_j. (14) results in $1 - \frac{i_j - 1}{n} < y^1_{i_j} < y^1_{i_j - 1}$, in contradiction with $y \in Y$. ■

To obtain $A(P)$ Lemma 2 is used with $\text{diag}^{-1}(c_1, \ldots, c_n)$. (See the lines after Lemma 2.) The application Lemma 1 gives the final result.

THEOREM. *The essential bordering hyperplanes of the convex hull of the class of intersecting Sperner families are the following ones. Let* $1 = i_1 < \ldots < i_{r+1} = \lceil \frac{n+1}{2} \rceil$ $(1 \le r)$ *be some integers.*

$$\sum_{k=1}^{r} \left(\sum_{i=i_k}^{i_{k+1}-1} \frac{p_i}{\binom{n-1}{i-1} \frac{n}{n-i_k+1}} \right) + \frac{p_{\frac{n+1}{2}}}{\binom{n}{\frac{n+1}{2}}} +$$

$$+ \sum_{k=1}^{r} \left(\sum_{n-i_{k+1}+1<j\le n-i_k+1} \frac{p_j}{\binom{n-1}{j} \frac{n}{i_k-1}} \right) \le 1,$$

where the middle term appears only for odd ns, the terms with $i_1 (= 1)$ *should be taken to be 0 and the term with* $j = n$ *is simply* p_n.

Observe that $i_2 = \lceil \frac{n+1}{2} \rceil$ leads to Bollobás's inequality (with a slight modification) while the case $i_k = k\,(1 \le k < \lceil \frac{n+1}{2} \rceil)$ gives the inequality of Greene, Katona and Kleitman. The total number of inequalities in the theorem is exponential.

REFERENCES

[1] Bollobás, B., Sperner systems consisting of pairs of complementary subsets, *J. Comb. Theory, Ser. B* **15**(1973), 363-366.
[2] Erdős, P., Chao Ko and Rado, R., Intersection theorems for systems of finite sets, *Q. J. Math. Oxf. II Ser.* **12**(1961), 313-318.
[3] Erdős, P.L., Frankl, P. and Katona, G.O.H., Intersecting Sperner families and their convex hulls, *Combinatorica* **4**(1984), 21-34.
[4] Fulkerson, D.R., Blocking and anti-blocking pairs of polyhedra, *Math. Program.* **1**(1971), 168-194.
[5] Fulkerson, D.R., Anti-blocking polyhedra, *J. Comb. Theory Ser. B* **12**(1972), 50-71.
[6] Greene, C., Katona, G.O.H. and Kleitman, D.J., Extensions of the Erdős-Ko-Rado theorem, *SIAM Stud. Appl. Math.* **55**(1976), 1-8.
[7] Milner, E.C., A combinatorial theorem on systems of sets, *J. London Math. Soc.* **43**(1968), 204-206.
[8] Schrijver, A., *Theory of linear and integer programming*, Wiley-Interscience Series in Discrete Mathematics, John Wiley & Sons, Ltd, Chichester, 1986.
[9] Sperner, E., Ein Satz über Untermenge einer endlichen Menge, *Math. Z.* **27**(1928), 544-548.

Integral Drawings of the Complete Graph K_6

A. Kemnitz, Braunschweig, FRG

A drawing $\mathcal{D}(G)$ of a graph G is such a realization of G in the plane that the vertices of G are mapped into different points, called vertices of $\mathcal{D}(G)$, and the edges of G are mapped into Jordan curves connecting the maps of the incident vertices, called edges of $\mathcal{D}(G)$, and two edges of $\mathcal{D}(G)$ have at most one point in common, either a vertex or a crossing.

If all vertices of $\mathcal{D}(G)$ are straight line segments then $\mathcal{D}(G)$ is called line graph or straightline drawing of G.

There is a remarkable result of Wagner and Fáry [5,13] which states that every planar graph can be drawn in the plane such that all edges are straight line segments (see also [14, p.109], [15, p.23], and the incomplete proof of [3]).

A drawing $\mathcal{D}(G)$ of a graph G is called an integral drawing if all edges of $\mathcal{D}(G)$ are straight line segments of integral length.

For any n there exist integral drawings of the complete graph K_n (see for example [1,2,7,10,11]). All these drawings contain at least n-2 vertices which are concyclic. Therefore, Erdös has posed the following question since many years during several mathematical congresses (see also [6, Problem $\mathcal{D}$20]):

Do there exist an integral drawing of the K_6 without four concyclic vertices?

Or reformulating the question in terms of Euclidean geometry one may ask: Do there exist six points in the plane such that no three points are in line, no four points are on a circle, and all mutual distances are intgral?

Recently, an affirmative answer to this question was independently given by Leech (see [9]) and Harborth, Kemnitz [8].

R. Bodendiek, R. Henn (Eds.)
Topics in Combinatorics and Graph Theory

Point sets of cardinality six with the mentioned properties are called integral hexagons in general position. In this note we will give two different constructions which lead to two different classes of integral hexagons in general position.

The hexagons of the first class are composed by six copies of a triangle with rational sides a,b,c and rational medians $\frac{x}{2},\frac{y}{2},\frac{z}{2}$ (see Figure 1). The distances which occur in such an arrangement are $a,b,c,2a,2b,2c,x,y,z$. Such hexagons fulfil the desired properties by multiplying with the lowest common denominator of all distances. They don't contain four concyclic points if a,b,c are mutually different:

Two pairs of opposite points form a parallelogram which has no right angles because of the difference of the sides. Four consecutive points are a trapezium which isn't isosceles by the same argument. A quadrangle consisting of three consecutive points and the opposite one has diagonals $2a$ and x, or $2b$ and y, or $2c$ and z, respectively, and a and x, or b and y, or c and z, respectively, bisect themselves. This implies that the four points are concyclic iff $x=\sqrt{3}a$, or $y=\sqrt{3}b$, or $z=\sqrt{3}c$, respectively, which is impossible for rational distances.

Hexagons of this type are uniquely determined by their containing parallelograms. Therefore, by the Theorem of Apollonius, we get

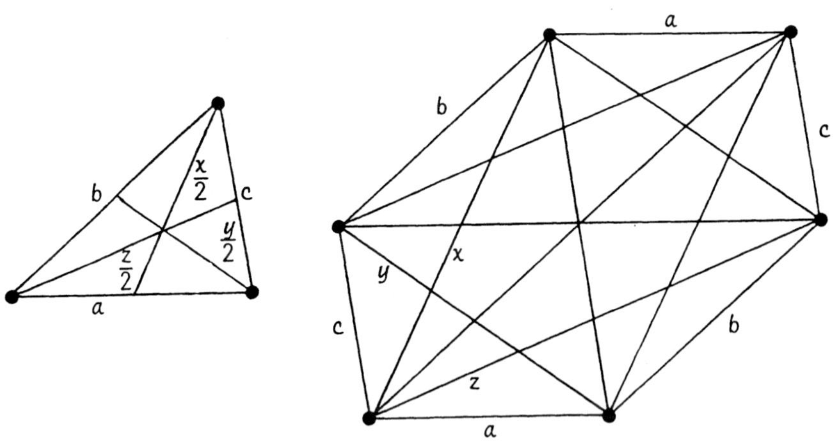

Figure 1.
Hexagon composed by six copies of a triangle.

the following system of quadratic equations

$$x^2=2b^2+2c^2-a^2$$
$$(1) \qquad y^2=2a^2+2c^2-b^2$$
$$z^2=2a^2+2b^2-c^2$$

where a,b,c must fulfil the triangle conditions.

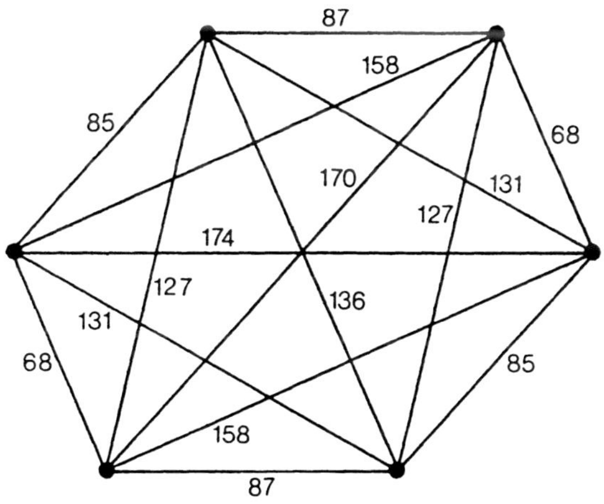

Figure 2.
Minimum integral hexagon in general position.

With the aid of the computer Amdahl 470 of the Technical University Braunschweig we have determined all integer solutions of system (1) with $gcd(a,b,c)=1$, $a>b>c$ and $a<b+c$ up to $a=4000$. There are altogether 39 such solutions, the smallest one is $a=87$, $b=85$, $c=68$, $x=127$, $y=131$, $z=158$ (see Figure 2).

The maximum distance of two points of a point set is called diameter of the set. By a systematical search with the aid of a computer we have proved that this special hexagon with diameter 174 of Figure 2 is the smallest integral hexagon in general position at all (see [8]).

If the 6-tuple (a,b,c,x,y,z) fulfils system (1) in integers then does the 6-tuple $(z,y,x,3c,3b,3a)$ which leads to a non-similar hexagon (see Figure 3).

To an integral hexagon (a,b,c,x,y,z) of (1) a related hexagon

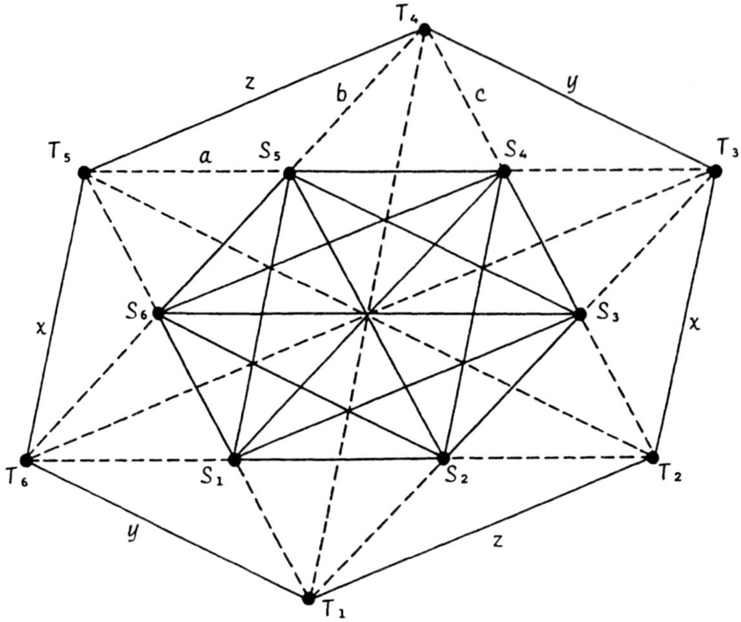

Figure 3.
Hexagon (a,b,c,x,y,z) with vertices S_i and non-similar hexagon $(z,y,x,3c,3b,3a)$ with vertices T_i.

is obtained by inversion with respect to a circle with a radius r with rational r^2 and centred at the triple intersection point of the diagonals (see Figure 4). The inverse hexagon consists of six triangles which are similar but not congruent. After rescaling the occuring distances are $a^2, b^2, c^2, 2ab, 2ac, 2bc, ax, by, cz$.

The question for triangles with rational sides and rational medians, thus the question for rational solutions of system (1), is an old problem. In [4, p.202] one can find several parametric solutions of (1), some of them by Euler, e.g.

$$a=|(m+n)p-(m-n)q|, \quad b=|(m-n)p+(m+n)q|, \quad c=2|mp-nq|$$

with $p=(m^2+n^2)(9m^2-n^2)$, $q=2mn(9m^2+n^2)$ and rational parameters $m,n \neq 0$.

For example, $m=1$, $n=2$ leads to $a=127$, $b=131$, $c=158$ and $m=2$, $n=3$ gives $a=85$, $b=87$, $c=68$, after division by the greatest common divisor.

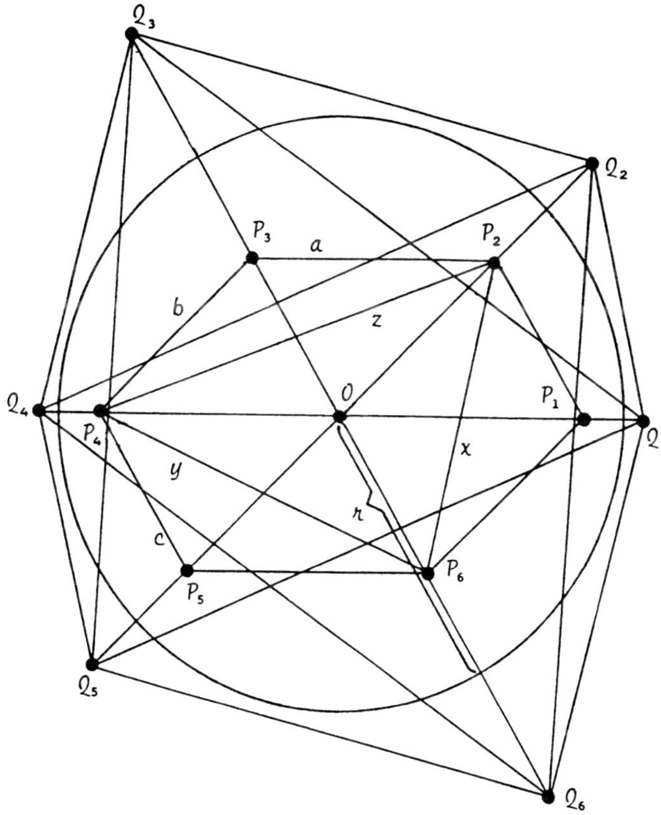

Figure 4.
Hexagon with vertices P_i and inverse hexagon with vertices Q_i.

The following second construction implies another class of integral hexagons in general position.

A quadrangle with two non-adjacent sides of common length is marking off three times in cyclic order. If the sides of common length a or their extensions intersect themselves with an angle $\varphi=60°$ one obtains a hexagon without three collinear points and at most five different distances, say a,b,c,d,e (see Figure 5). If one orients in such a quadrangle the lines containing the two sides of common length a and denotes the origin of these sides by x or $-x$ and y or $-y$, respectively, depending on the position of the origin to the intersection point of the lines one gets the following system of equations by multiple application of the cosine rule

$$b^2 = x^2 + y^2 - xy$$

(2) $$c^2 = a^2 + b^2 + a(2y - x)$$

$$d^2 = a^2 + b^2 + a(2x - y)$$

$$e^2 = a^2 + b^2 + a(x + y)$$

Solutions of (2) in positive integers a,b,c,d,e yield integral hexagons without three collinear points if a,b,c and a,b,d and a,c,e and a,d,e and b,c,e and b,d,e fulfil the triangle conditions.

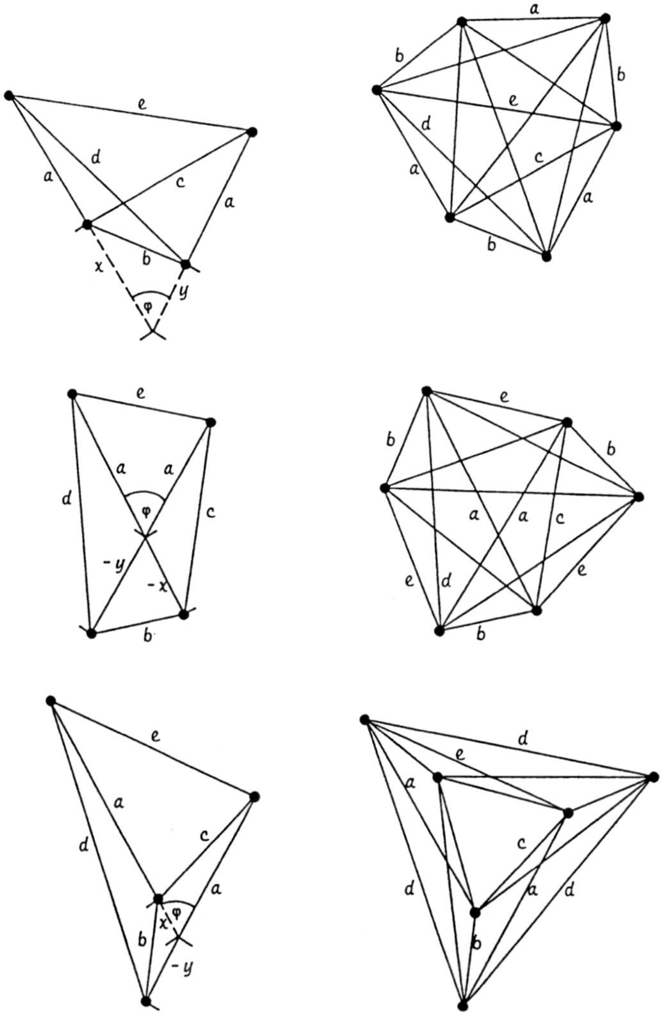

Figure 5.
Quadrangles and hexagons with the same distances a,b,c,d,e.

A necessary condition for integer solutions of (2) is rationality of x and y. Other implications of (2) are, e.g., $a^2+b^2+e^2=c^2+d^2$ and $3(a^2b^2+(a^2+b^2)e^2)=c^4+c^2d^2+d^4$. These equations show that one gets with one solution (a,b,c,d,e) of (2) even 3!2!=12. With some other necessary relations between the distances a systematical investigation of system (2) with the aid of a computer gave as a result that $a=392$, $b=323$, $c=407$, $d=713$, $e=645$ is the smallest integer solution of (2) such that the appropriate hexagon has no four concyclic points (see Figure 6).

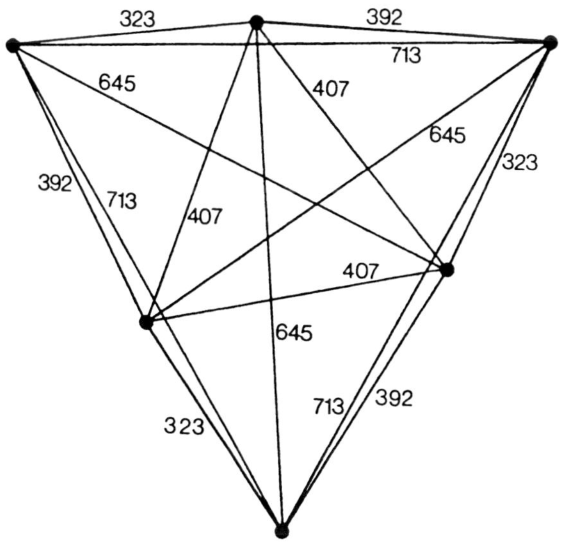

Figure 6.
Integral hexagon in general position with five different distances.

The Theorem of Ptolemy states that in a quadrangle the product of the diagonals equals the sum of the products of two non-adjacent sides if and only if the points of the quadrangle are concyclic. Since the difference of the sum of the products and the product of the diagonals equals 71808, 28490, 66978, 782, and 49910, respectively, none of the five different quadrangles which are contained in the hexagon of Figure 6 is an inscribed one.

Our systematical search with the aid of a computer for integral hexagons in general position gave as a result that there are exactly three such hexagons with a diameter ≤ 320. The two smallest with diameters 174 and 316 are of "Euler type", i.e. consisting of six copies of a triangle with rational sides and medians. The minimum example of such a hexagon of another type has diameter 319 (see Figure 7). This hexagon is a non-convex one.

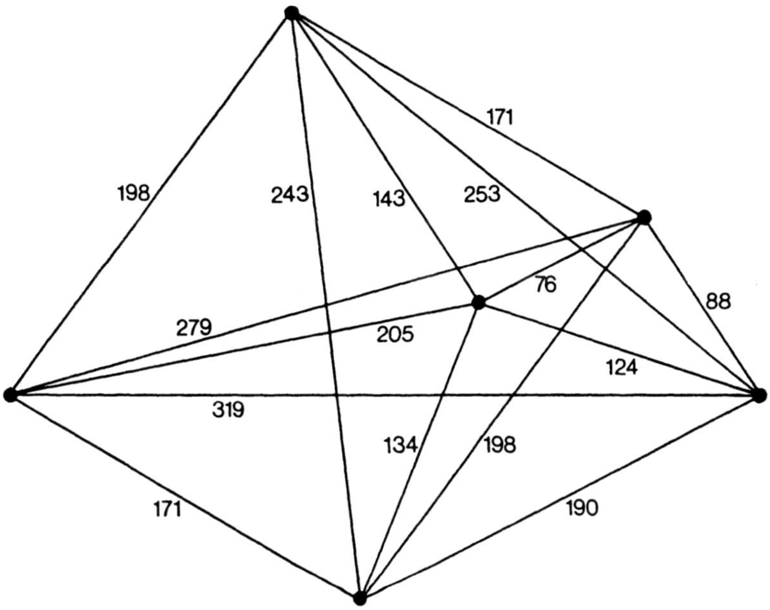

Figure 7.
Minimum non-convex integral hexagon in general position.

In [12] so-called 6-clusters are determined. They are defined as sets of six lattice points in the plane with mutually integral distances such that no three points are collinear and no four points are concyclic, i.e. in addition the coordinates of the vertices are integers. A list of 6-clusters which are found by computers is tabled. The minimum diameter of a 6-cluster is 1886. Is this the smallest integral hexagon in general position with integral area? The analogous problem for n-gons with $n \leq 5$ is solved in [8].

References

[1] M. Altwegg: Ein Satz über Mengen von Punkten mit ganzzahliger Entfernung. Elem. Math. 7 (1952) 56-58.

[2] D.D. Ang, D.E. Daykin and T.K. Sheng: On Schoenberg's rational polygon problem. J. Austral. Math. Soc. 9 (1969) 337-344.

[3] V.W. Bryant: Straight line representation of planar graphs. Elem. Math. 44 (1989) 64-66.

[4] L.E. Dickson: History of the Theory of Numbers, Vol. II: Diophantine Analysis. Reprint. New York, N.Y. 1966.

[5] I. Fáry: On straight line representation of planar graphs. Acta Sci. Math. (Szeged) 11 (1948) 229-233.

[6] R.K. Guy: Unsolved problems in Number Theory. New York, Heidelberg, Berlin 1981.

[7] H. Harborth: On the problem of P. Erdös concerning points with integral distances. Annals New York Acad. Sciences 175 (1970) 206-207.

[8] H. Harborth and A. Kemnitz: Diameters of integral point sets. Coll. Math. Soc. János Bolyai, 48: Intuitive Geometry, Siófok (Hungary) 1985 (eds.: K. Böröczky, G. Fejes Tóth). Amsterdam, Oxford, New York 1987.

[9] J. Lagrange: Points du plan dont les distances mutuelles sont rationnelles. Séminaire de Théorie des Nombres de Bordeaux, 1982-1983, Exposé no. 27, Talence 1983.

[10] D.N. Lehmer: Rational triangles. Annals Math., Ser. 2, 1 (1899/1900) 97-102.

[11] A. Müller: Auf einem Kreis liegende Punktmengen ganzzahliger Entfernungen. Elem. Math. 8 (1953) 37-38.

[12] L.C. Noll and D.I. Bell: n-clusters for $1<n<7$. Math. Comp. 53 (1989) 439-444.

[13] K. Wagner: Bemerkungen zum Vierfarbenproblem. Jahresbericht DMV 46 (1936) 26-32.

[14] K. Wagner: Graphentheorie. Mannheim, Wien, Zürich 1970.

[15] K. Wagner und R. Bodendiek: Graphentheorie I. Mannheim, Wien, Zürich 1989.

On Certain Binomial Sums

W. Klotz, Clausthal-Zellerfeld, FRG

We investigate the following binomial sums:

$$A_p(m,n) = \sum_{k=0}^{n} k^p \binom{m}{k} k!\, m^{n-k},$$

where p, n are nonnegative integers, and m is a real number. These sums arise in complexity estimates of tests for equal entries of n-tuples [2]. We derive recursive formulas (Theorem 1) and asymptotic representations (Theorem 2) of $A_p(m,n) = A_p$.

Theorem 1.

(i) $$A_{p+1}(m,n) = \sum_{r=0}^{p-1} \binom{p}{r}(mA_r - A_{r+1}) - (n+1)^p(m-n)\binom{m}{n}n! + mr_p,$$

where $r_p = m^n$ for $p=0$ and $r_p = 0$ for $p>0$.

(ii) $$A_{p+1}(m,n) = m \sum_{r=1}^{p} (-1)^{r+1}\binom{p}{r}A_{p-r} + (-1)^p m^{n+1} - n^p(m-n)\binom{m}{n}n!.$$

Proof. As the assertion is trivially true for m=0, we may assume $m \neq 0$. By the definition of r_p we have

$$A_p = \sum_{k=1}^{n} k^p \binom{m}{k} k!\, m^{n-k} + r_p.$$

Substitute $\binom{m}{k}k! = \binom{m}{k-1}(k-1)!(m-k+1)$ and take k-1 as a new summation variable.

$$A_p = \sum_{k=0}^{n-1} (k+1)^p(m-k)\binom{m}{k}k!m^{n-k-1} + r_p,$$

$$A_p = \sum_{k=0}^{n} (k+1)^p(m-k)\binom{m}{k}k!m^{n-k-1} - (n+1)^p(m-n)\binom{m}{n}n!m^{-1} + r_p.$$

The last sum is divided into sums S_1 and S_2.

$$S_1 = \sum_{k=0}^{n} (k+1)^p\binom{m}{k}k!m^{n-k} = \sum_{r=0}^{p}\binom{p}{r}\sum_{k=0}^{n} k^r\binom{m}{k}k!m^{n-k} = \sum_{r=0}^{p}\binom{p}{r}A_r,$$

$$S_2 = m^{-1}\sum_{k=0}^{n} k(k+1)^p\binom{m}{k}k!m^{n-k} = m^{-1}\sum_{r=0}^{p}\binom{p}{r}A_{r+1}.$$

R. Bodendiek, R. Henn (Eds.)
Topics in Combinatorics and Graph Theory

This leads to

$$A_p = \sum_{r=0}^{p} \binom{p}{r}(A_r - m^{-1}A_{r+1}) - (n+1)^p (m-n)\binom{m}{n} n!\, m^{-1} + r_p. \qquad (1)$$

Solving this equation for A_{p+1} shows (i).

For p=0 we see from (i):

$$A_1 = m^{n+1} - (m-n)\binom{m}{n} n!. \qquad (2)$$

This confirms (ii) for p=0. So, to prove (ii) we may assume $p \geq 1$. (1) can be written in the following form:

$$A_p = \sum_{r=0}^{p} \binom{p}{r}(A_r - m^{-1}A_{r+1} - n^r (m-n)\binom{m}{n} n!\, m^{-1}) + r_p. \qquad (3)$$

For every nonnegative integer r let

$$b_r = A_r - m^{-1}A_{r+1} - n^r (m-n)\binom{m}{n} n!\, m^{-1},$$

$$A_0^* = A_0 - m^n, \quad A_r^* = A_r \text{ for } r>0.$$

By (2) and (3) we have

$$A_p^* = \sum_{r=0}^{p} \binom{p}{r} b_r \quad \text{for } p=0, 1, 2, \ldots .$$

Applying binomial inversion [1, page 96] shows

$$b_p = A_p - m^{-1}A_{p+1} - n^p (m-n)\binom{m}{n} n!\, m^{-1} = \sum_{r=0}^{p} (-1)^r \binom{p}{r} A_{p-r}^*,$$

which can be solved for A_{p+1} to prove (ii).

<u>Corollary.</u>

1) $A_1(m,n) = m^{n+1} - (m-n)\binom{m}{n} n!,$

2) $A_2(m,n) = mA_0 - A_1 - (n+1)(m-n)\binom{m}{n} n!,$

3) $A_3(m,n) = -mA_0 + (2m+1)A_1 - (n^2-1)(m-n)\binom{m}{n} n!,$

4) $A_2(m,n) + A_3(m,n) = 2m^{n+2} - (n^2+n+2m)(m-n)\binom{m}{n} n!.$

These identities are easily deduced from Theorem 1. Formula 1) for m=n seems to be first proved by Riordan [3]. Identity 4), too, takes an interesting form for m=n:

$$\sum_{k=1}^{n} (k+1)\, k^2 \binom{n}{k} k!\, n^{n-k} = 2n^{n+2}.$$

We need two lemmas to prove an asymptotic estimate for $A_p(m,n)$.

<u>Lemma 1.</u> If $p \geq 2$ then $A_p(m,n)$ has the following form.

$$A_p(m,n) = (-1)^p (mf_p(m)A_0 - g_p(m)A_1) + (m-n)\binom{m}{n} n!\, G_p(m,n),$$

where $f_p(m)$, $g_p(m)$, $G_p(m,n)$ satisfy:

1) $f_p(m)$, $g_p(m)$ are polynomials in m with positive integer coefficients and constant term 1. $G_p(m,n)$ is a polynomial in m and n with integer coefficients and degree p-1.

2) If p is even, then deg f_p=deg g_p=(p-2)/2 and the leading coefficient of f_p is $u_p=1\cdot 3\cdot 5\cdot \ldots \cdot (p-1)$.

3) If p is odd, then deg f_p=(p-3)/2, deg g_p=(p-1)/2 and the leading coefficient of g_p is $v_p=2\cdot 4\cdot 6\cdot \ldots \cdot (p-1)$.

4) $$f_{p+1}(m) = m\sum_{r=1}^{p-2}\binom{p}{r}f_{p-r}(m)+1, \quad g_{p+1}(m) = m\Big(\sum_{r=1}^{p-2}\binom{p}{r}g_{p-r}(m)+p\Big) +1,$$

$$G_{p+1}(m,n) = m\sum_{r=1}^{p-2}(-1)^{r+1}\binom{p}{r}G_{p-r}(m,n) - n^p + (-1)^p.$$

<u>Proof.</u> We abbreviate: $A_p(m,n)=A_p$, $f_p(m)=f_p$, etc. By the corollary, 1), 2), 3) are true for p=2 and p=3. We proceed by induction on p. Assume $p \geq 3$ and suppose A_r has the asserted form for $r = 2,3,\ldots,p$ satisfying 1), 2), 3). We modify recursion (ii) of Theorem 1 by the first identity of the corollary.

$$\begin{aligned} A_{p+1} &= m\sum_{r=1}^{p}(-1)^{r+1}\binom{p}{r}A_{p-r} + (-1)^pA_1 + ((-1)^p-n^p)(m-n)\binom{m}{n}n! \\ &= m\sum_{r=1}^{p-2}(-1)^{r+1}\binom{p}{r}A_{p-r} + (-1)^{p+1}mA_0 + (-1)^p(pm+1)A_1 \\ &\quad + ((-1)^p-n^p)(m-n)\binom{m}{n}n!. \end{aligned}$$

According to the induction hypothesis we have for $1 \leq r \leq p-2$:

$$A_{p-r} = (-1)^{p-r}(mf_{p-r}A_0 - g_{p-r}A_1) + (m-n)\binom{m}{n}n!G_{p-r}.$$

Inserting A_{p-r} in the equation for A_{p+1} yields:

$$\begin{aligned} A_{p+1} &= (-1)^{p+1}\Big(m\Big(m\sum_{r=1}^{p-2}\binom{p}{r}f_{p-r}+1\Big)A_0 - \Big(m\sum_{r=1}^{p-2}\binom{p}{r}g_{p-r}+pm+1\Big)A_1\Big) \\ &\quad + (m-n)\binom{m}{n}n!\Big(m\sum_{r=1}^{p-2}(-1)^{r+1}\binom{p}{r}G_{p-r} + (-1)^p - n^p\Big), \end{aligned}$$

which proves the recursions in 4). The remaining statements 1), 2), 3) follow from 4) by the induction hypothesis. This is immediately clear for 1). To confirm 2) and 3) we notice deg f_{p+1} = deg f_{p-1} + 1 and deg g_{p+1} = deg g_{p-1}+ 1. If p+1 is even, then deg f_{p-1} > deg f_{p-2} and leading coefficient of f_{p+1} = p·leading coefficient of $f_{p-1}=pu_{p-1}=u_{p+1}$. If p+1 is odd, then deg g_{p-1}>deg g_{p-2} and leading coefficient of g_{p+1} = p · leading coefficient of $g_{p-1} = pv_{p-1} = v_{p+1}$.

Let $m>0$ and $a_p(m,n)=A_p(m,n)/m^n$. The following lemma is proved in [2].

Lemma 2. If m is a function of n satisfying $m \geq n$ and $m \log m \leq n^2$ for every sufficiently large n, then

$$a_0(m,n) = (\pi m/2)^{1/2} + O(\log^2 m).$$

Theorem 2. Suppose q is a real constant, $1 \leq q < 2$. Let m be a real function of n satisfying $n \leq m \leq n^q$ for every sufficiently large n.

(i) If p is even and $u_p = 1 \cdot 3 \cdot 5 \cdot \ldots \cdot (p-1)$, then

$$a_p(m,n) = u_p(\pi m^{p+1}/2)^{1/2} + O(m^{p/2} \log^2 m).$$

(ii) If p is odd and $v_p = 2 \cdot 4 \cdot 6 \cdot \ldots (p-1)$, then

$$a_p(m,n) = v_p m^{(p+1)/2} + O(m^{p/2}).$$

Proof. For $p=0$ (i) is true according to Lemma 2 and $u_0 = 1$. Let $p \geq 1$. Applying $1-x \leq \exp(-x)$ for $x \geq 0$ we estimate $f(m,n)=\binom{m}{n} n!/m^n$.

$$f(m,n) = \prod_{k=1}^{n-1} (1-k/m) \leq \exp(-m^{-1} \sum_{k=1}^{n-1} k) = \exp(-n(n-1)/(2m)),$$

$f(m,n) \leq \exp(-n^{\alpha})$ for a real number $\alpha > 0$ and every sufficiently large n. The first identity of the corollary gives

$$a_1(m,n) = m - (m-n)f(m,n) = m + o(1),$$

which is stronger than (ii) for $p=1$. If $p \geq 2$, then Lemma 1 yields

$$a_p(m,n) = (-1)^p(mf_p(m)a_0(m,n)-g_p(m)a_1(m,n)) + (m-n)G_p(m,n)f(m,n).$$

The last product is o(1). Inserting

$$a_0(m,n) = (\pi m/2)^{1/2} + O(\log^2 m) \text{ and } a_1(m,n) = m + o(1)$$

we derive (i) and (ii) by Lemma 1.

References

1. M. Aigner, Combinatorial Theory, Springer, New York, 1979.
2. W. Klotz and W. Lex, Testing n-tuples for equal entries, to appear.
3. J. Riordan, Enumeration of linear graphs for mappings of finite sets, Ann. Math. Statist. 33 (1962), 178-185.

Colouring of Spider Graphs

M. Koebe, Greifswald, GDR

1. Introduction

Intersection graphs are an interesting topic of algorithmic graph theory: On the one side many well - known graph - theoretic problems which are NP - complete for general graphs become polynomially solvable if they are restricted to a class of intersection graphs. Hence studying intersection graphs can give information about the borderline between P and NP - if there is one. On the other side, intersection graphs are very useful to model problems in the natural, social and technical sciences (see e.g. [3]).

In the present paper we investigate a new class of intersection graphs. It is a common generalization of trapezoid and circle graphs. We discuss the colouring and the maximum independent set problem for such graphs.

2. Some classes of intersection graphs

Throughout this paper let $G = (V,E)$ denote a finite undirected graph without loops and multiple edges. $G = G(S)$ is called an <u>S - intersection graph</u> if there is a system of sets S and a bijection Φ between V and S such that two vertices P,Q of V are adjacent iff the sets $\Phi(P)$ and $\Phi(Q)$ in S intersect. The system S is called a <u>representation</u> of G .

With this general notion <u>every</u> graph is an intersection graph [6] . If we assume that S contains only sets of a special type we get restricted graph classes. Their algorithmic properties depend on the kind of sets which can be used to represent them.

R. Bodendiek, R. Henn (Eds.)
Topics in Combinatorics and Graph Theory

We give some examples: If S is a set of (closed) intervals on the real line we get the well - known class of <u>interval graphs</u>. Let K be a circle in the plane, S a system of chords in K with pairwise disjoint end points. Then $G(S)$ is called a <u>circle graph</u>.

Now suppose that there are given a permutation π of $\{1,2,\ldots,n\}$ and a set of n points placed on a line and labelled from left to right by $1,2,\ldots,n$. Suppose further that there is a second line parallel to the first and having n points labelled from left to right by $\pi(1),\pi(2),\ldots,\pi(n)$. If we connect every pair of points $(\pi^{-1}(i),i)$, $i = 1,\ldots,n$; by a straight line we get a set S of segments which defines the <u>permutation graph</u> $G(S)$.

If we have two parallel lines and S is a set of trapezoids such that all of their corner points are positioned on the lines $G(S)$ is called a <u>trapezoid graph</u> (note that a trapezoid can degenerate to a triangle or a segment connecting the two lines). This class was introduced independently in [1] and [2]. If all trapezoids are rectangles we get interval graphs; if all are segments connecting the lines we get permutation graphs. This shows that trapezoid graphs are a common generalization of permutation and interval graphs.

Now we want to define a class of graphs which contains both trapezoid and circle graphs as proper subclasses:

Let K be a circle in the plane, $\mathcal{T} = \{T_1,\ldots,T_n\}$ a system of disjoint point sets on K. T_i and T_j are called <u>alternating</u> if there is no chord of K separating T_i and T_j. In other words, if we choose for each T_i a Steiner tree $s(T_i)$ connecting its elements and embed $s(T_i)$ in the interior of the circle K in an arbitrary way, T_i and T_j alternate iff in every case $s(T_i)$ and $s(T_j)$ intersect. A graph G is called a <u>spider</u> graph if there is a circle K, a system $\mathcal{T}$ of point sets on K and a bijection Φ between the vertex set and $\mathcal{T}$ such that P and Q are adjacent in G iff the corresponding sets $\Phi(P)$ and $\Phi(Q)$ alternate on K.
It is straightforward that we get an equivalent definition by

choosing a system

$S = \{ s(T_i) \mid s(T_i) \text{ minimal Steiner tree for } T_i ,\ T_i \in \mathcal{J} \}$.

Adjacency of vertices now means intersection of Steiner trees.

It is easy to see that the class of spider graphs contains all other classes mentioned above [5] . Fig. 1 shows a spider graph which is neither a trapezoid nor a circle graph, hence this containment is proper.

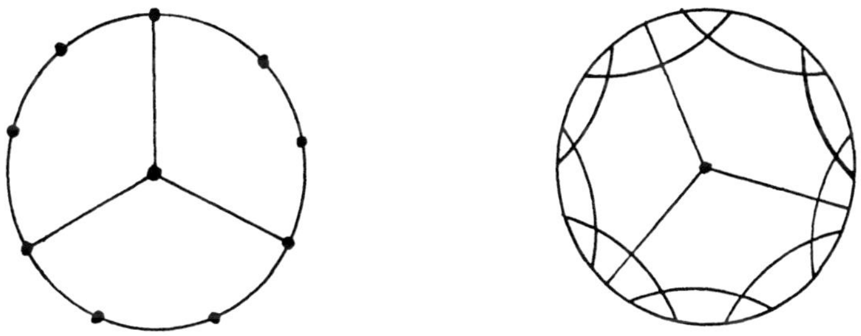

fig. 1

A graph is called <u>chordal</u> if every cycle of length greater than 3 has a chord [3] . In [5] it is shown that all chordal graphs have a spider representation. Let c denote a chordless cycle of length greater than 3 in a spider graph G . Then the edges connecting c and G - c must be "concentrated" in a special way on triplets of consecutive vertices of c . This leads to a characterization and a recognition algorithm for spider graphs (for details see [5]).

Fig. 2 shows two examples of graphs which have no spider representation.

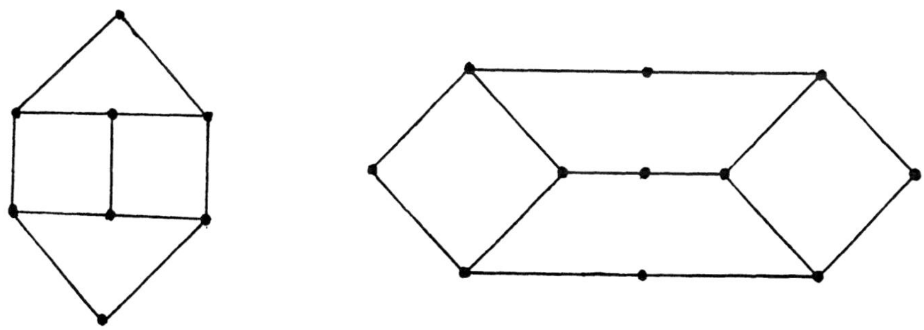

fig. 2

3. Colouring of spider graphs

If we want to investigate colouring properties of spider graphs it should be useful to recapitulate facts about the subclasses mentioned in section 2 concerning this question.

For interval graphs there is an $O(n\log n)$ colouring algorithm [4] . Colouring of permutation graphs has the same complexity (if the defining permutation of the graph is given [3] - otherwise we need $O(n^2)$ steps to determine this permutation [7]). Trapezoid graphs are colourable in $O(nk)$ time [2] , where n denotes the number of vertices and k the number of colours.

An exclusive role play circle graphs: In contrary to the other classes they are not perfect and colouring circle graphs is NP - complete. Even the k - colouring problem, i.e. if we ask for colourability with k colours where k is fixed, is NP - conplete for $k > 3$ [9] (for $k = 3$ this problem is still open [9]). Hence we have the following fact about spider graphs:

<u>Theorem 1.</u> k - colouring of spider graphs is NP - complete for every $k > 3$. □

In [9] further was shown that circle graphs of bounded degree d - i.e. no vertex has more than d adjacent vertices - are colourable in polynomial time. This can be generalized to spider graphs of bounded degree:

<u>Lemma 1.</u> Let $G = (V,E)$ be a spider graph with $n = |V|$ and maximum degree d . Then G has a $2(d+1)$ - separator.

<u>Proof.</u> We follow the proof of [9] in the case of circle graphs and have to show that there is a partition $[A,B,C]$ of V with

(1) $|A| < 2/3\ n$
(2) $|B| < 2/3\ n$
(3) $|C| \leq 2(d+1)$
(4) There is no edge connecting A and B .

If P is a vertex of G we denote by $T(P)$ the set of points on K - called terminals - representing P . P(T) and P(t) mean the vertex represented by T and the vertex which terminal set contains t , respectively. We suppose that all terminals on K and all sets T_i are given in the clockwise ordering. Let $t_1, t_2 \in T(P)$ be two consecutive terminals with respect to this ordering, then $K(t_1, t_2)$ is the segment of K bounded by t_1 and t_2 . A set T(P) of m terminals divides the circle K into m segments which contain no terminal of T(P), except the end points. We say that a vertex P' is contained in a segment $K(t_1, t_2)$ if $T(P') \subset K(t_1, t_2)$. The set

$$U(K(T_1, t_2)) := \{ P' \in V \mid T(P') \subset K(t_1, t_2) \}$$

consists of all vertices which are contained in the segment $K(t_1, t_2)$. U(P) denotes the union over all $U(K_i)$ where K_i is a segment defined by an arbitrary pair of consecutive terminals of T(P) .

There are two possible cases:

case A. There is a vertex P and segments $K_{i_1}, \ldots, K_{i_r}$ with respect to consecutive terminals of P such that

$$1/3\ n < \sum_{j=1}^{r} |U(K_{i_j})| < 2/3\ n .$$

Then Adj(P) - the set of all vertices neighboured to P - is a separator.

case B. There is no vertex which satisfies the condition of case A. Let P be a vertex such that for <u>every</u> set $K_{i_1}, \ldots, K_{i_r}$ of segments with respect to P we have $\sum_{j=1}^{r} |U(K_{i_j})| > 2/3\ n$. Hence $|U(K_i)| > 2/3\ n$ for all K_i with respect to this vertex. Under all vertices with this property we choose P_0 and a segment K_0 such that for all $P' \in U(K_0)$ we have $|U(P_0)| = \bigcup_j |U(K_j)| < 1/3\ n$, where K_j denotes a segment defined by consecutive terminals of P' which is contained in K_0 . Then $\{P_0, P'\}$ is a separator of the required type. □

As a consequence, we get the following result:

<u>Theorem 2.</u> k - colouring of spider graphs with bounded maximum degree is polynomial. □

4. Maximum independent sets

Finally, we want to investigate the maximum independent set problem for spider graphs. Again we use techniques known from circle graphs [8].

Let $G = (V,E)$ be a spider graph with a representation in the circle K. Suppose that the set $\{t^1,\dots,t^s\}$ of all terminals on K is clockwise ordered; m_i denotes the number of terminals in $T(P_i)$. For every pair of indices (i,j) with $1 \leq i < j \leq s$ we define an induced subgraph of G:

$G(i,j) := \langle \{P \in V \mid T(P) \subset K(t^i,t^j)\} \rangle$, where $K(t^i,t^j)$ denotes the segment of the circle between t^i and t^j.

If $i \geq j$ we define $G(i,j) := \emptyset$. $MIS(i,j)$ denotes a maximum independent set of $G(i,j)$. Step by step we calculate all $MIS(i,j)$ starting with small i and j, and ending with $MIS(1,s)$ - a maximum independent set of G.

We inspect K terminal by terminal in the clockwise order starting with t^1. Let t^j denote the last terminal of a set $T(P) = \{t_1,\dots,t_m = t^j\}$ - i.e. all other elements already were inspected. Two cases are possible:

case A. $P \in MIS(i,j)$. Then $T(P)$ divides K into segments $K_1,\dots,K_m$, w.l.o.g. $K_1 = K(t_1,t_2),\dots,K_p = K(t_p,t_{p+1})$ are contained in $K(t^i,t^j)$. All terminals of every $P' \in MIS(i,j)$ must be contained in the same segment. On the other side

$\bigcup_{k=1}^{p} MIS(t_k,t_{k+1}) \cup MIS(t^i,t_1) \cup MIS(t_p,t^j)$ is an independent set in G.

Hence in this case we get $MIS(i,j)$ by adding P to $MIS(i,j-1)$

case B. $P \notin MIS(i,j)$. Here we have $MIS(i,j) = MIS(i,j-1)$.

If the current terminal t^j on K is not the last terminal of a vertex $P(t^j)$ we have $MIS(i,j) = MIS(i,j-1)$, because $G(i,j) = G(i,j-1)$.

<u>Theorem 3.</u> It is possible to determine a maximum independent set of a spider graph in $O(s^2)$ time, s the number of terminals over all vertices of G. □

References

[1] D.G.Corneil, P.A.Kamula: Extensions of permutation and interval graphs, Congr. Numer. 58(1987) 267-275.

[2] I.Dagan, M.C.Golumbic, R.Y.Pinter: Trapezoid graphs and their coloring, Discr. Appl. Math. 21(1988)1, 35-46.

[3] M.C.Golumbic: Algorithmic graph theory and perfect graphs, Academic Press, 1980.

[4] U.I.Gupta, D.T.Lee, J.Y.-T.Leong: An optimal solution for the channel - assignment problem, IEEE Trans. Comput. C-28(1979)11, 807-810.

[5] M.Koebe: Spider graphs - a new class of intersection graphs, to appear.

[6] E.Marczewski: Sur deux proprietes des classes d'ensembles, Fund. Math. 33(1945) 303-307.

[7] J.Spinrad: Transitive orientation in $O(n^2)$ time, in: Proc. 15th Ann. ACM Symp. on Theory of Computing 1983, 457 - 466.

[8] K.J.Supowit: Finding a maximum planar subset of a set of nets in a channel,IEEE Trans. Comp.-Aided Design CAD-6(1987)1, 93-94.

[9] W.Unger: On the k-colouring of circle graphs, in: STACS'88 (R.Cori, M.Wirsing ed.), LNCS 294, Springer 1988, 61-72.

[10] W.Wessel, R.Pöschel: On circle graphs, in: Graphs,Hypergraphs and Applications, Eyba 1984 (H.Sachs ed.), Teubner Leipzig, 1985, 207-210.

A Las-Vergnas Type Theorem for Trees

I. Krasikov, Tel-Aviv, Israel
Y. Roditty, Tel-Aviv, Israel

ABSTRACT

Let G be a connected simple graph. We study the maximal order of a subtree $T \subset G$ with a given maximum degree Δ_T. We obtained conditions on the degree sequence of a graph which are similar to that of Bondy and Las-Vergnas on the Hamilton cycle, and strengthen early results of Win [10] and the authors [4].

1. Introduction

Graphs in this paper are finite, have no loops and multiple edges. By $deg_G\ v$ we denote the degree of a vertex v in a graph G. As usual, the edge and the vertex set of G, are denoted by $E(G)$ and $V(G)$, respectively. We also put $|V(G)| = n$. The set of all vertices adjacent to a vertex x is denoted by $N(x)$.

Let G be a graph with the degree sequence $d_1 \leq d_2 \leq ... \leq d_n$. It is well known that G has a Hamilton cycle if, in particular, one of the following conditions holds:

(i) $d_1 + d_2 \geq n$, Ore [8].

(ii) $d_i \geq i + 1$ for $i \leq \frac{n}{2}$, Posa [9].

(iii) $d_i \leq i,\ d_j \leq j \Rightarrow d_i + d_j \geq n\ (i \neq j)$, Bondy [2].

(iv) $i < j,\ i + j \geq n,\ (v_i, v_j) \notin E(G),\ deg v_i \leq i,\ deg v_j \leq j - 1 \Rightarrow deg v_i + deg v_j \geq n$, Las Vergnas [6].

(v) $\kappa \geq \beta_0$, where κ is the connectivity and β_0 is the independence number of G, Erdos and Chvatal [5].

Similar conditions can be given for the existence of a Hamilton path in a graph.

R. Bodendiek, R. Henn (Eds.)
Topics in Combinatorics and Graph Theory

When a graph is not Hamiltonian it is natural to ask how far it is from being Hamiltonian. One of possible formalizations of this question, suggested by Win [10], is how small can be the maximum degree Δ_T, of a spanning tree T in G?.

It turned out that many known results on Hamilton cycle or path can be extended on spanning trees with a given maximum degree k although they require utterly different proofs which even do not hold for the case $k = 2$ [3], [4], [7], [10], [11].

Thus, Win established a Posa-type conditions on the degree sequence of a graph guaranteeing the existence of a spanning tree with a given value of Δ_T. In [7] Erdos-Chvatal-type conditions on Δ_T, involving the independence number and the connectivity, was given.

More generally, one can ask about the maximum size of a subtree $T \subset G$ with a given maximum degree k. Of course, for $k = 2$ this is just the length of a longest path. Here we present Las-Vergnas and Bondy-type conditions on the degree sequence of a graph implying the existence of a subtree of a given size with a prescribed maximum degree. This strengthen Ore-type conditions obtained in [4].

2. Results

Let m be a positive integer. An independent subset $S \subset V$, $|S| = m$, of vertices of G is called an *m–cluster* if the graph $<G \setminus S'>$ is connected for any $S' \subsetneq S$. Note that any subset of a cluster is also a cluster.

Given a connected graph G and a number $k \geq 3$. A largest tree $T \subset G$ with $\Delta_T = k$ is called *a k–optimal tree* if it has as few endvertices as possible and it does not span G.

We use $X_i = X_i(T)$ for the number of vertices of degree i in T.

Note that, by maximality of T, any vertex of T adjacent to $G \setminus T$ has degree k.

Lemma 1: The set of the endvertices of a k–optimal tree T is a cluster.

Proof: Observe that any subset of endvertices of a k–optimal tree T is not a cut in G. Indeed, any edge e of G adjacent to an endvertex of T is adjacent also to another vertex of T. Since, removing e das not disconnect T it can not disconnect G. So, it is enough to show that the set of the endvertices is independent. Assume the contrary. Let x, y be two adjacent endvertices of T and let z be any vertex of degree more than two in the unique path, $P \subset T$, between x and y. Choose $T' = (T \cup (x, y)) \setminus (t, z)$, where $t \in P$ is a vertex adjacent to z (possibly, $t = x$ or y). But $X_1(T') < X_1(T)$, contradicting the minimality of $X_1(T)$. □

Let v be a vertex, called the root, of degree k in a k–optimal tree T adjacent to $G \setminus T$. Let $B_1, \ldots, B_k$ be the branches of T at v. We agree that $v \notin \bigcup_{i=1}^{k} B_i$.

A set of endvertices $W = W(v) = \{w_1, w_2, ..., w_k\}$, where $w_i \in B_i$, will be called *a representative set* of T. We set $D(W) = \sum_{w \in W} deg_G w$.

Lemma 2: Let T be a k–optimal tree in G. If $(u_i, u_j) \in E(G)$, $u_i \in B_i$, $u_j \in B_j$, $i \neq j$, then $\max \{deg_T u_i , deg_T u_j\} = k$.

Proof: Assume, by the contrary, that $\max (deg_T u_i, deg_T u_j) \leq k - 1$. Then $\Delta_{T'} \leq k$, where $T' = ((T \cup (u_i, u_j)) \setminus e_i) \cup w$, contradicting the maximality of T. □

Lemma 3: Let T be an k–optimal tree in G . Then $|T| \geq D(U) + 1$ for any representative set U of T.

Proof: Denote by e_i, $1 \leq i \leq k$, the edge joining the root v with B_i and let v_i be a vertex in B_i adjacent to v . Let $U = \{u_1, u_2, \cdots u_k\}$ be a representative set of T. Define the following sets:

$$A_i = \{ x \mid deg_T(x) \leq k-1 , x \in B_i \} .$$

$$F_i = \{ x \mid deg_T(x) = k , x \in B_i \} .$$

$$C_i = \{ x \mid deg_T(x) = k , x \notin B_i \} .$$

$N(u_i) \setminus v$ as follows:

Observe that, by lemma 2, $N(u_i) \setminus v = A_i \cup F_i \cup C_i$, and thus, it is a partition of $N(u_i) \setminus v$. Note also that $x \in A_i$ implies $deg_T x \geq 2$ since otherwise inserting (u_i, x) leads to a tree with a smaller number of endvertices.

Let $A = \sum_{i=1}^{k} |A_i|$, $F = \sum_{i=1}^{k} |F_i|$, $C = \sum_{i=1}^{k} |C_i|$.

Denote by $X_k(i)$ the number of vertices of degree k in B_i .

We will establish four inequalities on A, F, C which imply the lemma.

(1) $$1 + F \leq X_k .$$

It follows from the definitions of F, X_k, and the fact that $v \notin \bigcup_{i=1}^{k} B_i$.

(2) $$\frac{k-2}{k-1} C \leq (X_k - 1)(k - 2) .$$

Indeed, $|C_i| \leq X_k - X_k(i) - 1$, hence

$$C = \sum_{i=1}^{k} |C_i| \leq kX_k - \sum_{i=1}^{k} X_k(i) - k = (k - 1)(X_k - 1)$$

by $\sum_{i=1}^{k} X_k(i) = X_k - 1$.

Multiplying both sides of the last inequality by $\frac{k-2}{k-1}$ we obtain (2).

(3) $$D(U) - k \leq A + F + C.$$

To prove (3) observe that

$$|A_i| + |F_i| + |C_i| \geq deg_G\, u_i - 1$$

by Lemma 1. Hence,

$$A + F + C \geq \sum_{i=1}^{k} deg_G\, u_i - k \geq D(U) - k\ .$$

(4) $$A + \frac{C}{k-1} \leq \sum_{i=2}^{k-1} X_i\ .$$

To establish (4) consider for each i, and each $x \in C_i$ the graph $T \cup (u_i, x)$, and define:

$Y_i = \{ y : (x, y)$ is the edge in the unique cycle of $T \cup (u_i, x)$, $y \neq u_i$, $x \in C_i \}$.

Put $Y = \bigcup Y_i$, $Z_i = A_i \cup Y_i$, $Z = \bigcup Z_i$.

Observe that $deg_T\, y = 2$ since otherwise for $T' = (T \cup (u_i, x)) \setminus (x, y)$, $X_1(T') \leq X_1(T)$. Thus, $|Y_i| = |C_i|$ and $|Z| \leq \sum_{i=2}^{k-1} X_i$. Moreover, if $y \in B_j$ then $y \notin A_j$, namely, $(y, u_j) \notin E(G)$ and so, $Y \cap (\bigcup A_i) = \emptyset$. Indeed, $\{T \cup (y, u_j) \cup (u_i, x)\} \setminus (x, y)$ has the unique cycle. By deleting from this cycle any edge, such that one of its endvertices is of degree at least three in T (such an edge does exist, for example the edge incident to x on the unique path in T from x to u_i), we obtain a tree with a smaller value of X_1.

Since each $x \in C_i$, $x \in B_j$ can belong also to any C_l but C_j, it follows $|Y| = |\bigcup C_i| \geq \frac{C}{k-1}$. Now we obtain (4) by

$$\sum_{i=2}^{k-1} X_i \geq |Z| = |Y \cup (\bigcup A_i)| \geq A + |Y| \geq A + \frac{C}{k-1}\ .$$

Adding inequalities (1) - (4) we get

$$D(U) - 1 \leq X_k(k-1) + \sum_{i=2}^{k-1} X_i.$$

Here the RHS does not exceed

$$\sum_{i=2}^{k} (i-1)X_i = 2e(T) - |T| = |T| - 2\ ,$$

and so, $|T| \geq D(U) + 1$. □

An immediate corollary of lemma 3 is an Ore-type bound on the order of a k-optimal tree [4].

Corollary 1: Let $d_1 \le d_2 \le \cdots \le d_n$ be the degree sequence of G. Then $|T| \ge 1 + D(U) \ge 1 + \sum_{i=1}^{k} d_i$. □

More careful arguments show that this bound can be strengthen in the spirit of Las-Vergnas conditions for hamiltonicity.

For, let G be a connected graph whose vertices are indexed by the numbers $1, 2, \ldots, n$. Let $i(v)$ be the index of a vertex v, and let $d_i = deg_G v_i$.

Our main result is:

Main Theorem: Suppose that for any k-cluster $S = \{ v_{i_1}, v_{i_2}, \ldots, v_{i_k} \} \subset G$, $i_1 < i_2 < \cdots < i_k$, such that $d_{i_j} \le i_j - j + 1$, $1 \le j \le k$, it follows $\sum_{j=1}^{k} d_{i_j} \ge m$.

Then G has a tree T, $|T| \ge \min\{ m + 1, n \}$, such that $\Delta_T \le k$.

Proof: Clearly, we may assume that any spanning tree in G has maximum degree more than k. Consider a k-optimal tree T in G.

We fix a root v and a representative set $U = \{u_{i_j}\}$. W.L.O.G we may assume $d_{i_1} \le d_{i_2} \le \cdots \le d_{i_k}$. Furthermore, suppose that $I = \sum_{j=1}^{k} i_j$, is maximal among all choices of T, v and U.

Now it is enough to show that $d_{i_j} \le i_j - j + 1$ for $1 \le j \le k$. Indeed, U is a cluster, and $D(U) \le |T| - 1$, by lemma 3. By the assumption of the theorem, T is not maximal whenever $|T| < \min\{ n, m+1 \}$.

For each $u \in U$ and each edge $(u, x) \in G \setminus T$ consider the graph $G(u, x) = T \cup (u, x)$. Let $u = u_{i_j}$ we define the following sets:

$$Y_j = \{ y : (x, y) \text{ is the edge in the unique cycle of } G(u, x),\ y \ne u \}.$$

Compare the trees T and $T' = G(u, x) \setminus (x, y)$, $y \in Y_j$. Since T' has at most as many endvertices as T it follows that $deg_T y = 2$. Moreover, by maximality of I, we have $i(y) \le j$.

Finally, each vertex in the set $W_j = Y_j \cup (\bigcup_{l<j} d_{i_l})$ has the index less than i_j. By $|Y_j| = d_{i_j} - 1$ and $Y_j \cap (\bigcup_{l<j} d_{i_l}) = \varnothing$ we have $|W_j| = d_{i_j} - 1 + j - 1$, hence $i_j \ge d_{i_j} + j - 1$ and the result follows. □

As a simple corollary of the above theorem we get the following Bondy-type conditions:

Corollary 2: Suppose that for any set of k vertices $\{v_{i_1}, v_{i_2}, \ldots, v_{i_k}\}$, of G, $i_1 < i_2 < \cdots < i_k$, such that $d_{i_j} \le i_j - j + 1$, $1 \le j \le k$ it follows $\sum_{j=1}^{k} d_{i_j} \ge m$. Then G has a tree T, $|T| \ge \min\{m+1, n\}$ such that $\Delta_T \le k$. □

Example: The degree sequence of a graph drawn in Fig.1 is 1, 1, 4, 4, 5, 5, 5, 5. The corresponding values of $(d_i - i - 1)$ are 1, 2, 0, 1, 1, 2, 3, 4. The minimum value of $\sum_{j=1}^{3} d_{i_j}$ subjected the conditions of Corollary 2 is $d_1 + d_2 + d_7 = 7$. Hence, $|T| \ge 8$, i.e. G has a spanning tree with maximum degree 3. Corollary 1 gives here only $|T| \ge 7$.

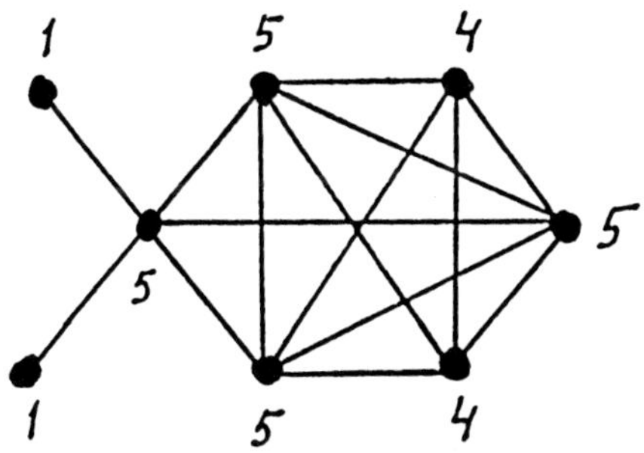

Fig.1

We finish this paper with the following conjecture which, in the case $k = 2$, is just the variant of the Las-Vergnas theorem for Hamilton paths.

Conjecture: Suppose that for any k–cluster, $S = \{v_{i_1}, v_{i_2}, \ldots, v_{i_k}\}$, of G, $i_1 < i_2 < \cdots < i_k$, such that $d_{i_j} \le i_j - j + 1$, $1 \le j \le k$ and $\sum_{j=1}^{k} i_j \ge m$ it follows $\sum_{j=1}^{k} d_{i_j} \ge m$. Then G has a tree T, $|T| \ge \min\{m+1, n\}$ such that $\Delta_T \le k$.

3. References:

[1] J.C. Bermond, Hamiltonian graphs, in Selected Topics in Graph Theory, ed. L. Bienekc and R.J. Wilson, Academic Press, London (1978),127-167.

[2] J.A. Bondy, Large cycles in graphs, Discrete Math. 1 (1971), 121-132.

[3] Y.Caro, I.Krasikov and Y.Roditty, Spanning trees and some edge-reconstructible graphs, Ars Combinatoria 20-A (1985), 109-118.

[4] Y.Caro, I.Krasikov and Y.Roditty, On the largest subtree of a given maximum

degree in a connected graph, submitted.

[5] V. Chvatal and P. Erdos, A note on hamiltonian circuits, Discrete Math. 2 (1972), 111-113.

[6] M. Las Vergnas, Sur une propriete des arbores maximaux dans un graphe, Acad. Sci. Paris. Ser A-B 272 (1971) 1297-1300.

[7] V. Neumann-Lara and E. Rivera-Campo, Spanning trees with bounded degrees, Combinatorica, to appear.

[8] O. Ore, Note on Hamilton circuits, Amer. Math. Monthly 67 (1960) 55.

[9] L. Posa, A theorem concerning hamilton lines, Magyar Tud. Akad. Math. Kutato Int. Kozl. 7 (1962) 225-226.

[10] S. Win, Existenz von Gerusten mit Vorgeschriebenem Maximalgrad in Graphen, Abhandlungen aus dem Mathematichen Seminar der Universitat Hamburg, 43 (1975), 263 - 267.

[11] S. Win, On a connection between the existence of k-trees and the toughness of a graph, Graphs and Combinatorics, vol. 5, 2 (1989) 201-205

Quick Gossiping by Multi-Telegraphs

R. Labahn, Rostock, GDR
I. Warnke, Rostock, GDR

1. Introduction

Entringer and Slater /1/ considered the following communication problem: Suppose each of $n \geq 2$ points (persons) $1, 2, \ldots, n$ knows an item of information which is not known to all the others. They exchange information using telegrams arranged in consecutive rounds whereby in every round:

(1) each point can either send or receive telegrams, i.e. it is impossible for a given point to send some and to receive other telegrams in the same round;

(2) any pair $r, s \in V = \{1, 2, \ldots, n\}$ is allowed to communicate by a telegram, and if s sends a telegram to r, then in that round, r learns all information which s knows at this time;

(3) every point can communicate with at most k different points ($k \geq 1$), i.e. depending on whether it is a sender or a receiver in that round the point can either send at most k telegrams or receive at most k telegrams.

The information exchange stops when every point knows every unit of information, and we shall call it complete then.
This model is a generalization of the telegraph problem which we get for k=1. Because of (2), the underlying network is the complete digraph on the n points in that case. Moreover, this is a modification of the well-known telephone problem where the underlying network is the complete graph, i.e. we have bidirectional information transmission by telephone calls instead of telegrams which proceed information in one direction only. One of the basic combinatorial questions in all the models is how many rounds are needed to complete the information exchange. This has been completely answered for the telephone problem by Knödel /2/: We need at least $\lceil \log_2 n \rceil$ (if n is even) or $\lceil \log_2 n \rceil + 1$ (if n is odd) rounds. Surprisingly the according telegraph problem was unsolved, and Entringer, Sla-

R. Bodendiek, R. Henn (Eds.)
Topics in Combinatorics and Graph Theory

ter /1/ gave the best results we knew. If $T_k(n)$ denotes the minimal number of rounds necessary to complete the information exchange in the model introduced in the beginning, then they proved:

1. $T_k(n) \geqslant \lceil \log_{k+1} n \rceil$. This is rather obvious, since because of (3), the number of points knowing one fixed item can increase at most by the factor k+1 per round.
2. $T_k(n) \leqslant 2 \cdot \lceil \log_{k+1} n \rceil$. During the first $\lceil \log_{k+1} n \rceil$ rounds inform one fixed point about everything, then reverse that process.
3. Let $\{F_t\}$ and $\{P_t\}$ denote the sequences of Fibonacci- and Pell-numbers, i.e. $F_0=0$, $F_1=1$, $F_{t+1}=F_t+F_{t-1}$ and $P_0=0$, $P_1=1$, $P_{t+1}=2P_t+P_{t-1}$ for $t\geqslant 1$, respectively. Then they constructed sequences of telegrams which prove:
   - if $F_{s-1} < \lfloor n/2 \rfloor \leqslant F_s$, then $T_1(n) \leqslant \begin{cases} s+1 & \text{if } n \text{ even} \\ s+3 & \text{if } n \text{ odd} \end{cases}$; (4)
   - if $P_{s-1} < \lfloor n/2 \rfloor \leqslant P_s$, then $T_2(n) \leqslant \begin{cases} s+1 & \text{if } n \text{ even} \\ s+3 & \text{if } n \text{ odd} \end{cases}$.

We should mention that the first construction gives $T_1(n) \leqslant 1.44\ldots \cdot \log_2 n + 2.23\ldots$ for even n.

Furthermore, they conjectured $T_k(n)=\alpha_k \cdot \log_{k+1} n+O(1)$ with $\alpha_k \to 1$ if $k \to \infty$. As a consequence of our results this turns out to be true.

In Section 2 we prove $T_k(n) \geqslant \log_{\tau_k} n$, where $\tau_k=\frac{1}{2}[k+(k^2+4)^{1/2}]$ is the positive root of x^2-kx-1.

By constructing a more general arrangement of telegrams for $k\geqslant 2$ or $k=1$, n even in Section 3, we show that

$$\text{if } F_k(s-1) < \lceil n/(k+1) \rceil \leqslant F_k(s), \text{ then } T_k(n) \leqslant s+1,$$

where $\{F_k(t)\}$ is the sequence defined by $F_k(0)=0$, $F_k(1)=1$ and for $t=1,2,\ldots$, $F_k(t+1)=kF_k(t)+F_k(t-1)$.

This gives $T_k(n) < \log_{\tau_k} n + 2$ for $k\geqslant 2$.

Our construction includes that given by Entringer and Slater for k=1, n even, but it improves their result for k=2 already. The solution to the original telegraph problem, i.e. the proof of the lower bound in the case k=1 has been presented by the authors in /3/.

2. The lower bound

Let $c_m(i)$ denote the number of units of information which the point $i \in V$ knows after the m-th round resp. before round $m+1$. Note that $c_0(i)=1$ for $i=1,\dots,n$. The basic idea is to investigate the growth of the function $f(m) := \Sigma_{i=1}^{n} c_m^2(i)$ for $m=1,2,3\dots$. We start with a technical inequality.

Proposition 1 For any positive integer q, $q \leqslant k$, and arbitrarily chosen real numbers $x, y_1,\dots,y_q$, it holds

$$(x+\sqrt{k}\cdot\Sigma_{i=1}^{q} y_i)^2 + (\Sigma_{i=1}^{q} y_i^2) \leqslant \tau_k^2\cdot(x^2+\Sigma_{i=1}^{q} y_i^2)$$

Proof: We have $(x+\sqrt{k}\cdot\Sigma_{i=1}^{q} y_i)^2 + (\Sigma_{i=1}^{q} y_i^2) =$

$$= (x^2+\Sigma_{i=1}^{q} y_i^2) + 2x\sqrt{k}\cdot(\Sigma_{i=1}^{q} y_i) + k\cdot(\Sigma_{i=1}^{q} y_i)^2$$
$$\leqslant (x^2+\Sigma_{i=1}^{q} y_i^2) + 2x\sqrt{k}\cdot(\Sigma_{i=1}^{q} y_i) + k\cdot q\cdot(\Sigma_{i=1}^{q} y_i^2)$$
$$\leqslant (x^2+\Sigma_{i=1}^{q} y_i^2) + \Sigma_{i=1}^{q}(2x\sqrt{k}y_i + k^2y_i^2). \qquad (4)$$

Now for $i=1,\dots,q$,

$$2x\sqrt{k}y_i + k^2y_i^2 = [2(x\tau_k)(\sqrt{k}y_i) + k^2\tau_k y_i^2] / \tau_k \leqslant$$
$$\leqslant [x^2\tau_k^2 + ky_i^2 + k^2\tau_k y_i^2] / \tau_k = [x^2\tau_k^2 + ky_i^2(1+k\tau_k)] / \tau_k =$$
$$= \tau_k x^2 + k\tau_k y_i^2 \qquad \text{by } \tau_k^2-k\tau_k-1=0.$$

So (4) can be continued:

$$(x+\sqrt{k}\cdot\Sigma_{i=1}^{q} y_i)^2 + (\Sigma_{i=1}^{q} y_i^2) \leqslant$$
$$\leqslant (x^2+\Sigma_{i=1}^{q} y_i^2)+k\tau_k x^2+k\tau_k\cdot(\Sigma_{i=1}^{q} y_i^2) = (1+k\tau_k)(x^2+\Sigma_{i=1}^{q} y_i^2).$$

Using $\tau_k^2=1+k\tau_k$ again the assertion follows immediately. ∎

This will be used to prove the main result of this section.

Lemma 1 $T_k(n) \geqslant \log_{\tau_k} n$.

Proof: Let $R \subseteq V$ and $S \subseteq V$ denote the sets of receivers and senders of round $m+1$, respectively, and for every $i \in S$, let p_i denote the number of points i is communicating with in round $m+1$. From the model it follows $1 \leqslant p_i \leqslant k$.
Moreover, for an arbitrarily fixed $r \in R$ let W be the set of points sending telegrams to r during that round. For $j=m$ and $j=m+1$, we consider $d_j(r) := c_j^2(r) + \Sigma_{s \in W} c_j^2(s)/p_s$.

Then $$\Sigma_{r \in R}\, d_j(r) = \Sigma_{i \in R \cup S}\, c_j^2(i) \qquad (5),$$

because for each $s \in S$, the summand $c_j^2(s)/p_s$ occurs in $d_j(r)$ iff r receives a telegram from s, i.e. exactly p_s times.

Clearly for any $s\in W$, $c_{m+1}(s)=c_m(s)$, because s does not learn anything during round m+1. But r learns everything known to one of the points of W, i.e. $c_{m+1}(r)\leq c_m(r)+\Sigma_{s\in W}\, c_m(s)$. Hence

$$\begin{aligned} d_{m+1}(r) &= c^2_{m+1}(r) + \Sigma_{s\in W}\, c^2_{m+1}(s)/p_s \\ &\leq [c_m(r)+\Sigma_{s\in W}\, c_m(s)]^2 + \Sigma_{s\in W}\, c^2_m(s)/p_s \\ &\leq [c_m(r)+\sqrt{k}\cdot\Sigma_{s\in W}\, c_m(s)/\sqrt{p_s}]^2 + \Sigma_{s\in W}(c_m(s)/\sqrt{p_s})^2. \end{aligned}$$

Since $|W|\leq k$, we may apply Prop. 1 here:

$$d_{m+1}(r) \leq \tau^2_k\cdot[c^2_m(r)+\Sigma_{s\in W}\, c^2_m(s)/p_s] = \tau^2_k\cdot d_m(r).$$

Summing up these inequalities for $r\in R$ and using (5) it follows

$$\Sigma_{i\in R\cup S}\, c^2_{m+1}(i) \leq \tau^2_k\cdot\Sigma_{i\in R\cup S}\, c^2_m(i).$$

But for every $i\in V\setminus(R\cup S)$, obviously $c_{m+1}(i)=c_m(i)<\tau^2_k\cdot c_m(i)$ and therefore we finally get: $f(m+1) \leq \tau^2_k\cdot f(m)$.

By induction this implies $f(t) \leq \tau^{2t}_k\cdot f(0) = n\cdot\tau^{2t}_k$ since $c_0(i)=1$ for $i\in V$. On the other hand, after the last round T of any complete information exchange everybody knows all the n items of information, i.e. $n^3 = f(T) \leq n\cdot\tau^{2T}_k$ and $n \leq \tau^T_k$.

Hence we need at least $\log_{\tau_k} n$ rounds to complete an information exchange.

■

3. The upper bound

At first we will construct a complete information exchange for $k\geq 2$ or for k=1 and even n. Having this the upper bound for $T_k(n)$ can be proved.

Let be $n=q(k+1)-r$ where q, r integer, $0\leq r\leq k$. Note that $q\geq 1$. The basic idea of our scheme is the following.

<u>Round 1</u> Select the q points 1,3,5,...,2q-1 and arrange the kq-r remaining points in q groups V_i (i=1,2,...,q) of at most k points each in an arbitrary way. Then during the first round, for i=1,2,...,q, every point of V_i sends a telegram to 2i-1.

Obviously the basic conditions (1) and (3) are satisfied. After this round each unit of information is known to one of the points 1,3,..., or 2q-1.

Round 2,... In the following rounds only the points of $V_o:=\{1,3,5...,2q-1\}$ and $V_e:=\{2,4,...,2q\}$ send or receive in an alternating way: If t is even, then during the t-th round, V_o contains all the senders and V_e all the receivers, and vice versa if t is odd. This can be organized such that after a certain round s every receiver of that round knows all information. (6)

We will prove that below, suppose for now after round s every point of $V'\in\{V_o,V_e\}$ already knows everything. It will depend on n and k whether $V'=V_o$ or $V'=V_e$.

Round s+1 Select the q points of V' and arrange the kq-r remaining points in q groups V_i' $(i=1,2,...,q)$ of at most k points each in a arbitrary way. Then during this round, for $i=1,2,...,q$, $2i-1$ (if $V'=V_o$) resp. $2i$ (if $V'=V_e$) sends telegrams to every point of V_i'.

Since all the telegrams of round s+1 contain all information, after this round also the points of V-V' know everything, i.e. the information exchange is complete.

Remark 1: We used the point 2q. This point indeed is in V if $k\geq 2$: if $q=1$, clearly $2q=2\leq n$; if $q\geq 2$, then $k\geq 1+1/(q-1)$ and consequently $2q\leq q(k+1)-k\leq q(k+1)-r=n$ since $r\leq k$. But if $k=1$, $2q\leq n=2q-r$ requires $r=0$, i.e. even n. Therefore we cannot apply our construction to the case when k=1 and n is odd. Here Entringer and Slater's result (4) remains the best we know so far. For k=1 and even n, both their and our construction will give the same algorithm.

It remains to define the rounds 2,3,...,s precisely. Let $\{F_k(t)\}$ be the sequence of integers given by $F_k(0)=0$, $F_k(1)=1$ and $F_k(t+1):=kF_k(t)+F_k(t-1)$ for $t=1,2,...$. Obviously k=1 and k=2 give the Fibonacci- and Pell-numbers, respectively. Furthermore, for $t=2,3,...$ and $j=1,2,...,k$, let

$$d_t(j) := 2(j-1)F_k(t-1) + 2F_k(t-2) + (-1)^t. \qquad (7)$$

Note that for given t, all the $d_t(j)$ are pairwise different. Now during round t, $t=2,3,...,s$, every point $z\in V_o$ (if t even) resp. $z\in V_e$ (if t odd) sends telegrams to the k points

$$z+d_t(1)\ ,\ z+d_t(2)\ ,\ ...\ ,\ z+d_t(k), \qquad (8)$$

where of course the sums are taken mod 2q so that they are points of $\{1,2,...,2q\}$. Note that because all $d_t(j)$ are odd numbers, all the receivers belong to V_e if $z\in V_o$ and vice versa. Therefore condition (1) is satisfied. Since equivalently to the above definition (8) we may say that every receiver y gets telegrams exactly from the k points $y-d_t(1)$, $y-d_t(2)$,

..., $y-d_t(k)$, also (3) is fulfilled. This completes the construction and it remains to prove that it satisfies property (6). Next we consider the dissemination of the set J_1 of items which the point 1 knew before the second round.

Proposition 2 For $t=2,3,\ldots$, before round t (or after round t-1) the first $\min\{q;F_k(t-1)\}$ among the points of the sending set and the first $\min\{q;F_k(t-2)\}$ among the points of the receiving set of round t know J_1.

Proof: We proceed by induction over t. In round 2, points of V_o will send to points of V_e. J_1 is known to $1 \in V_o$ only, and indeed $F_k(1)=1$, $F_k(0)=0$. The induction step has to be done for $t=2,3,\ldots$ and it is analogue for even and odd t. We shall present the calculations for odd t here. Moreover we may assume $q>F_k(t-2)$, because otherwise already each point of $V_o \cup V_e$ knows J_1, and nothing remains to prove.
During round t, points of V_e send telegrams to points of V_o. Therefore also before round t+1, the first $\min\{q;F_k(t-1)\}$ points of V_e - this will be the receiving set of round t+1 - know J_1 and the assertion concerning the receiving set is proved.
Now let $x \in V_o$ be arbitrarily chosen such that

$$\min\{2q-1;2F_k(t-2)-1\} = 2F_k(t-2)-1 < x \leq \min\{2q-1;2F_k(t)-1\}$$

and let us take $j := \lceil (x-2F_k(t-2)+1) / 2F_k(t-1) \rceil$ (9).
Since $0<x-2F_k(t-2)+1\leq 2[F_k(t)-F_k(t-2)]=2kF_k(t-1)$ it follows $1\leq j\leq k$, and we may consider $x-d_t(j)$. From (9) we get

$$x-2F_k(t-2)+1 - 2F_k(t-1) \leq 2(j-1)F_k(t-1) < x-2F_k(t-2)+1,$$

and with (7): $x-2F_k(t-1) \leq d_t(j) < x$. Therefore $0<x-d_t(j)\leq 2F_k(t-1)$, and since $x-d_t(j)$ is even and $x-d_t(j)<x<2q$ we finally have $2 \leq x-d_t(j) \leq \min\{2q;2F_k(t-1)\}$.
Consequently, $z:=x-d_t(j)$ is one of the first $\min\{q;F_k(t-1)\}$ points of the sending set V_e. By the induction hypothesis, z knows J_1 already before round t, and because of (8), $x=z+d_t(j)$ receives a telegram from z during round t. So every $x\in V_o$ with $2F_k(t-2)+1 \leq x \leq \min\{2q-1;2F_k(t)-1\}$ knows J_1 before round t+1. By the induction hypothesis, already before round t every $x\in V_o$ with $1\leq x\leq 2F_k(t-2)-1$ knew J_1, i.e. we altogether proved that the first $\min\{q;F_k(t)\}$ points of V_o - this is the sending set of round t+1 - know J_1.

■

Now we are able to determine s. Let V' be the receiving set during round s. On the one hand we demanded that each point of V' after round s knows everything, i.e. particularly J_1. On

the other hand, by setting $t=s+1$ we know from Prop. 2 that after round s, in V' - this would be the sending set of round s+1 - the first $\min\{q;F_k(s)\}$ points know J_1. Therefore we have to define s to be the smallest integer with $F_k(s) \geq |V'|=q$.
It remains to prove that for any $z\in V_0$ the set J_z of items known to the point z before round 2 is known to all the points of V' after that round s. Let $y\in V'$ be chosen arbitrarily and consider the point $y-(z-1) \bmod 2q$. Since $z-1$ and $2q$ are even, $y-(z-1) \bmod 2q$ is even or odd if y is even or odd, respectively. Hence $y-(z-1) \bmod 2q$ belongs to V' and it knows J_1 after round s. But by definition (8) it is obvious that if J_1 is known to a point x, then at the same time J_z is known to the point $x+(z-1) \bmod 2q$. Consequently, $y-(z-1)+(z-1)=y \bmod 2q$ knows all J_z, i.e. every point of V' knows everything after round s and property (6) is proved.
The given construction of a complete information exchange shows

Lemma 2 For $k\geq 2$, or $k=1$ and even n,
if $F_k(s-1) < \lceil n/(k+1)\rceil \leq F_k(s)$, then $T_k(n) \leq s+1$.

To compare this result with the lower bound we use the explicit formula $F_k(t) = [\tau_k^t - (k-\tau_k)^t] / (k^2+4)^{1/2}$, $t=0,1,2,\ldots$.

Proposition 3

If $F_k(s-1) < \lceil n/(k+1)\rceil$, then $\tau_k^{s-1} < n\cdot(k^2+4)^{1/2}/(k+1)$.

Proof: If s=1, then

$$n\cdot(k^2+4)^{1/2}/(k+1) \geq 2\cdot(k^2+4)^{1/2}/(k+1) > 1 = \tau_k^{s-1}.$$

Let be $s\geq 2$ in the following. We have $\lceil n/(k+1)\rceil \geq F_k(s-1)+1$, $n/(k+1) > F_k(s-1)$, and $n \geq (k+1)\cdot F_k(s-1)+1$. Therefore

$$n\cdot(k^2+4)^{1/2}/(k+1) \geq \tau_k^{s-1} - (k-\tau_k)^{s-1} + (k^2+4)^{1/2}/(k+1). \quad (10)$$

Since $\tau_k^2-k\tau_k-1 = \tau_k(\tau_k-k)-1 = 0$ and $\tau_k>1$, it holds $-1<k-\tau_k<0$ and $(k-\tau_k)^{s-1}<\tau_k-k$ because $s\geq 2$. With $\tau_k=[k+(k^2+4)^{1/2}]/2$ one easily gets $\tau_k-k\leq(k^2+4)^{1/2}/(k+1)$. Putting these inequalities into (10) the assertion follows immediately. ∎

With the help of this technical proposition we get

Lemma 3 For $k\geq 2$, or $k=1$ and even $n\geq 2$, $T_k(n) \leq \log_{\tau_k} n + \beta_k$,
where $\beta_k = 2 + \log_{\tau_k}[(k^2+4)^{1/2}/(k+1)]$.

Note that $\beta_1=2.23\ldots$, but for $k\geq 2$, $1.9<\beta_k<2$ and $\beta_k\to 2$ if $k\to\infty$.

In the other multi-telegraph model considered by Entringer and Slater /1/ as well as in the conference-call model considered by Schmitt /4/ for the undirected case (hypergraphs) the minimal number of rounds is basically $\lceil \log_{k+1} n\rceil$. Therefore it might be interesting to give the results in terms of $\log_{k+1} n$. We have $\log_{\tau_k} n = \alpha_k \cdot \log_{k+1} n$, where $\alpha_k=\log_{\tau_k}(k+1)$, $1.44 > \alpha_k > 1$ and $\alpha_k\to 1$ if $k\to\infty$. This means that Entringer and Slater's conjecture on the behaviour of $T_k(n)$ is true, although they conjectured a slightly worse upper bound than we constructed here.

Summarizing the results of Lemmas 1 and 3 we are now able to give an almost complete solution to the original problem:

Theorem For $k\geq 2$, $T_k(n) \in \{\lceil \log_{\tau_k} n\rceil \, ; \lceil \log_{\tau_k} n\rceil+1\}$.

The best known estimates for k=1 are presented in /3/, but for even n, they are also contained in our lemmas.

References

/1/ R.C.Entringer and P.J.Slater,
Gossips and telegraphs,
J. Franklin Inst. 307 (1979) 353-359.

/2/ W.Knödel,
New gossips and telephones,
Discrete Math. 13 (1975) 95.

/3/ R.Labahn and I.Warnke
Quick gossiping by telegraphs,
submitted to: Discrete Math.

/4/ P.Schmitt,
Spreading information by conferences,
Discrete Math. 15 (1976) 305-306.

Numberings on Graphs Having Small Edge Weights

R. Lang, Hamburg, FRG

Numbering problems on graphs have been investigated intensively for the past three decades. One of the best known of these problems, the RINGEL–tree–numbering–problem, remains unsolved to he present. It consists in proving that all trees are graceful ([3],[7]). Another significant numbering problem on graphs is the bandwidth problem ([2]).

In this note, a numbering problem on graphs is considered which is in a certain sense related to the bandwidth problem for graphs. We call our problem Mimimum Weight Problem and prove its NP–Completeness.

1. The Minimum Weight Problem

Let $G=(X,U)$ be a finite, undirected and simple graph with $|X|=n$ vertices. A bijection $f\colon X \longrightarrow \{1,2,3,\ldots\ldots,n\}$ is said to be a **numbering** on G. With $\mathfrak{N}(G)$ we denote the set of all numberings on G. By means of f we define an edge weight function $g_f\colon U \longrightarrow \mathbb{N}$ by

(1) $$g_f(u) = g_f(P,Q) = f(P) + f(Q) \text{ for all } u=(P,Q) \in U.$$

$g_f(u)$ is the **weight of the edge u relative to f.**
The maximum edge weight

(2) $$g_f(G) = \max\{ g_f(u) \mid u \in U \}$$

is called the **weight of G relativ to f.**
The problem is to find a numbering $f \in \mathfrak{N}(G)$ so that

(3) $$g_{f^*}(G) \leq g_f(G) \text{ for all } f \in \mathfrak{N}(G).$$

R. Bodendiek, R. Henn (Eds.)
Topics in Combinatorics and Graph Theory

We call such a numbering f^* an **optimal numbering on G** and

(4) $$g(G) = \min\{ g_f(G) \mid f \in \mathfrak{N}(G) \}$$

the **weight of G**.
Let us call our numbering problem **Minimum Weight Problem** for graphs.

If f is a numbering of G with weight $g_f(G)$ and if the vertices of G are ordered according to this numbering f, then the adjacence matrix of G (Fig.1) has the form

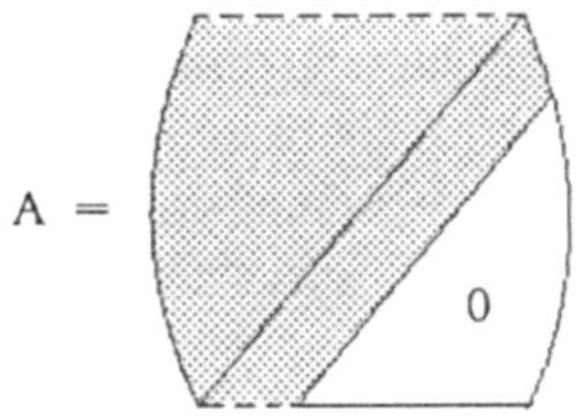

Fig.1

where $a_{ik} = 0$, if $i+k > g_f(G)$.

This problem differs from the well-known bandwidth problem for graphs in that in (1) "+" stands instead of "-" ([2]).

The bandwidth problem ist NP—hard ([6]), and now the question arises:

"*Is the Minimum Weight Problem perhaps easier*?"

We shall see that is not so. The Minimum Weight Problem is also NP—hard.

It is possible to formulate the Minimum Weight Problem also for hypergraphs:

If $H=(X,\mathfrak{E})$ is a hypergraph (see [1]) and $\mathfrak{N}(H)$ the set of all vertex-numberings on H, then

$$g(H) = \min_{f \in \mathfrak{N}(H)} \max \{ g_f(E_i) \mid E_i \in \mathfrak{E} \}$$

with

$$g_f(E_i) = \sum_{P \in E_i} f(P)$$

is to be found.

The Minimum Weight Problem for hypergraphs is also NP—hard, because it contains the Minimum Weight Problem for graphs as a special case.

Example 1: A hypergraph with a optimal numbering f and weight g(H)=15

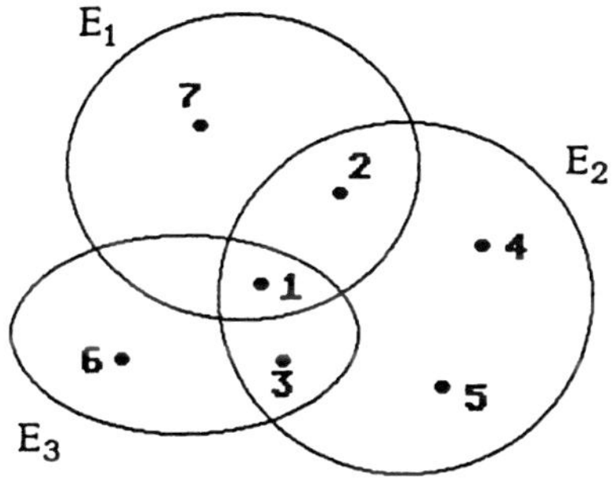

Fig.2

It is easy to see that the following simple statements hold:

1. If K_n is the complete graph with $n>1$ vertices, then
$$g(K_n) = 2n - 1 .$$
2. If $K_{m,n}$ is the complete bipartite graph with $m \geq n$, then
$$g(K_{m,n}) = m + 2n .$$
3. If G is a tree with $n>1$ vertices, then
$$g(G) = n + 1.$$
4. If G is a connected graph with minimum degree δ, then
$$g(G) \geq n + \delta .$$
5. If α is the vertex independence number of G, then
$$2n - 2\alpha + 1 \leq g(G) \leq 2n - \alpha .$$

2. The NP–Completeness of the Minimum Weight Problem

First we give the decision version of the Minimum Weight Problem:

MIMIMUM WEIGHT:

INPUT: Graph $G=(X,U)$, $K \in \mathbb{N}$

QUESTION: Is there a bijection $f: X \longrightarrow \{1,2,....,|X|\}$ so that for all edges $(P,Q) \in U$,
$$f(P) + f(Q) \leq K \ ?$$

We prove the

THEOREM: MIMINUM WEIGHT is NP–complete.

Proof: Obviously, MIMIMUM WEIGHT is contained in $\mathcal{NP}$.
We show that that 3–SAT $\propto$ MINIMUM WEIGHT.

We apply a polynomial transformation used by R. KARP [5] for his proof of the NP–Completeness of VERTEX COVER (see also [4]).

Let I be any instance of 3–SAT given by a set $V=\{v_1,v_2,\ldots,v_n\}$ of boolean variables and a set $\mathfrak{C}=\{c_1,c_2,\ldots,c_m\}$ of clauses. Each clause has the form

$$c_i=\{x_1^i,x_2^i,x_3^i\} \text{ with } x_1^i,x_2^i,x_3^i \in \{v_1,v_2,\ldots,v_n\} \cup \{\overline{v}_1,\overline{v}_2,\ldots,\overline{v}_n\},$$

and in each such clause a variable is contained only once and at no time together with its negation.
To I we assign a graph $G=(X,U)$ in the following manner:

$$X=V \cup \overline{V} \cup A$$

with

$$V=\{v_1,v_2,\ldots,v_n\} \text{ and } \overline{V}=\{\overline{v}_1,\overline{v}_2,\ldots,\overline{v}_n\},$$

and with a set of new vertices

$$A=\{a_{ik} \mid 1\le i\le m,\ 1\le k\le 3\}.$$

Moreover,

$$U = U_1 \cup U_2 \cup U_3$$

with

$$U_1 = \{(v_i,\overline{v}_i) \mid 1\le i\le n\},$$

$$U_2 = \{(a_{ik},a_{il}) \mid 1\le k,l\le 3,\ 1\le i\le m,\ k\neq l\},$$

$$U_3 = \{(a_{ik},x) \mid x\in V \cup \overline{V},\ a_{ik} \in A,\ x=x_k^i\},$$

and we set

$$K = 2n + 4m + 2\,.$$

Example 2: If I is given by $V=\{v_1,v_2,v_3,v_4\}$ and

$$\mathfrak{C} = \{\{v_1,v_2,\overline{v}_3\},\{\overline{v}_1,\overline{v}_2,v_3\},\{v_1,\overline{v}_3,\overline{v}_4\}\}$$

then Fig.3 shows the corresponding graph $G=(X,U)$.

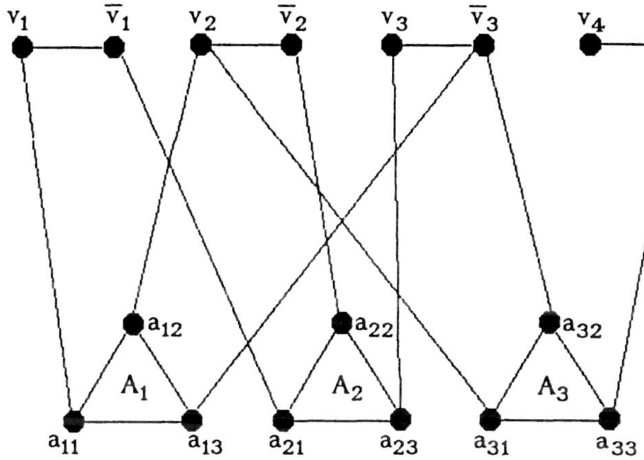

Fig.3

We call a vertex $x \in V \cup \overline{V}$ **V-vertex** and a vertex $a \in A$ **A-vertex**. It is well-known that I is Yes-instance for 3–SAT iff G possesses a vertex cover X_1 with $|X_1| = n + 2m$ (see [4]). If X_1 is a vertex cover of G, then $X_0 = X \setminus X_1$ is an independent set of G. And if X_1 is a minimum vertex cover of G, then each $x \in X_1$ is adjacent to at least one vertex $y \in X_0$.

We prove: I is a Yes-instance for 3–SAT iff for the graph G corresponding to I is $g(G) \leq 2n + 4m + 2$.

Let I now be a Yes-instance for 3–SAT and X_1 a minimum vertex cover of the corresponding graph G with $|X_1| = n + 2m$. We call the vertices of X_1 **white** vertices and those of $X_0 = X \setminus X_1$ **black** vertices.

The example 2 is a Yes-instance of 3–SAT. The white vertices of the graph in Fig.4 are a minimum vertex cover for G.

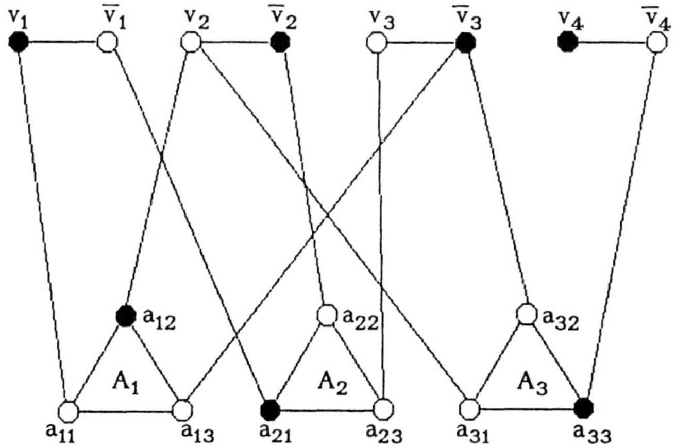

Fig. 4

We shall place the vertices of G onto the line of numbers (this corresponds in natural way a numbering f of G) so that for each edge $(P,Q) \in U$ holds

$$g_f(P,Q) = f(P) + f(Q) \le 2n+4m+2 = K.$$

If all edges of U_3 are deleted from G, then G is split into n+m components, i.e. into n **V-components** containing only V-vertices and m **A-components** containing only A-vertices. Each V-component contains one white and one black vertex, each A-Component contains two white and one black vertices. Each A-component is connected with 3 V-components, more exactly: a black A-vertex is connected with a white V-vertex, a white A-vertex is connected with either a white or a black V-vertex. Each A-component has either no or one or two black V-vertices as neighbours. Thus we distinguish **A^0-, A^1-** and **A^2–components**, according to the quantity of the adjacent white V-vertices.

In the following we distinguish two cases according to $m \ge n$ or $m < n$. In each case the white vertices of G are placed onto the positions 1,2,....,n+2m, the black vertices of G are placed onto the positions n+2m+1, n+2m+2,...,2n+3m (Fig.5).

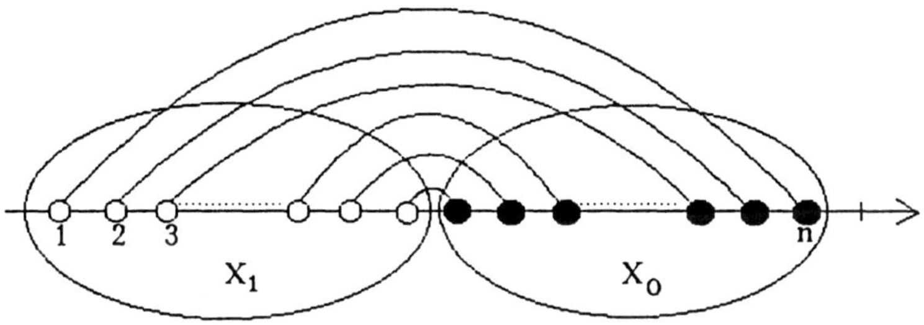

Fig.5

Case 1: $m \ge n$

We assume the A-components are ordered as sequence $A_1,A_2,...,A_m$ in which first all A^0-components are specified, then all A^1-components and at least all A^2-components.
We begin with the successive embedding of the A^0-components into the line of numbers. Let A_i be an A^0-component (Fig.6a). The white A-vertices of A_i are assigned to the greatest numbers $< \frac{K}{2}$, and the black vertex of A_i is assigned to the lowest number $\ge \frac{K}{2}$ which hitherto have not been used (Fig.6b).

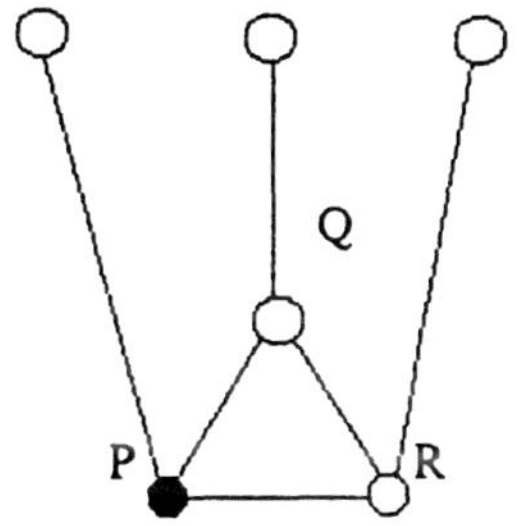

Fig.6a

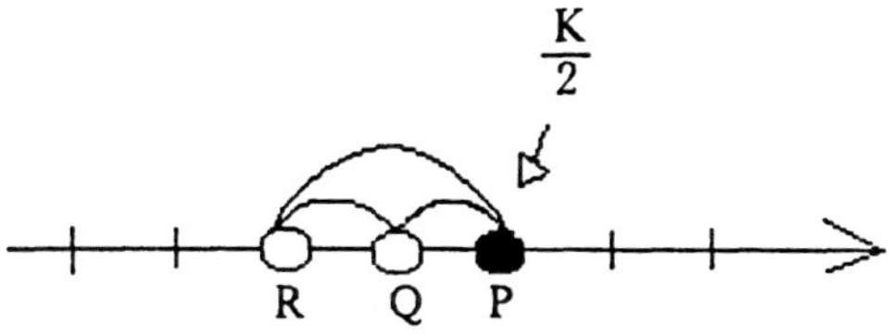

Fig.6b

After embedding of all A^0-components, the A^1-components are embedded together with their black V-neighbours. Let A_i be an A^1-component and P that white A_i-vertex adjacent to a black V-vertex A (Fig.7a). We construct the white-black alternating path

R, Q, P, A

and embed the vertices of that path from left to right according to the rule:

(R) *A white vertex is assigned the greatest free number* $< \frac{K}{2}$, *a black vertex the lowest free number* $\geq \frac{K}{2}$.

(See also Fig.7b).

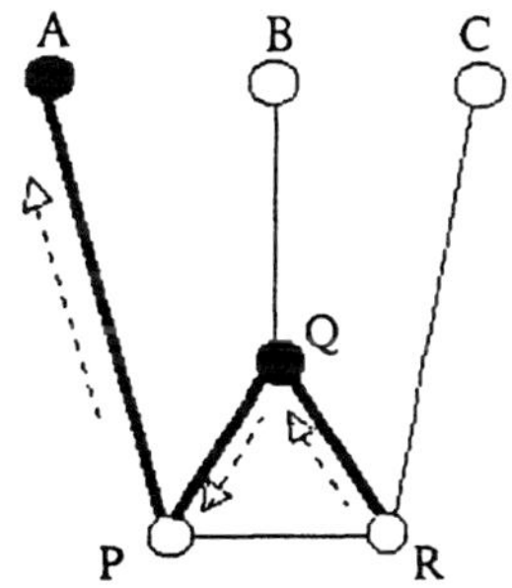

Fig.7a

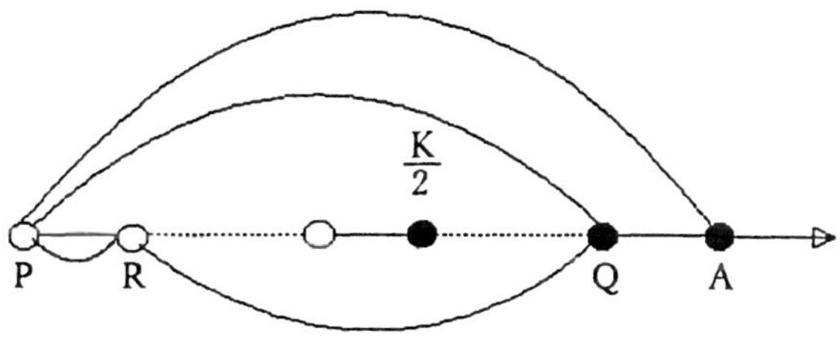

Fig.7b

Obviously the weights of the edges between vertices embedded until now is at all times $\leq$ K (proof by induction).

However, it is possible that in the course of the embedding of A^1-components such components appear, the black V-neighbours of which are already embedded. Such components are considered at this time simply as A^0-components. In the same way an A^2-component, black V-neighbours of which are already placed, is treated as A^1- or A^0—component, resp.

Hencefort, we assume only A^2-components are to be embedded, the black V-neighbours of which are all unplaced (Fig.8a). Then we construct a black-white alternating path of the length 4, e.g.

A,P,Q,R,C

and embed the vertices of this path from left to right according the rule (**R**) (Fig.8b).

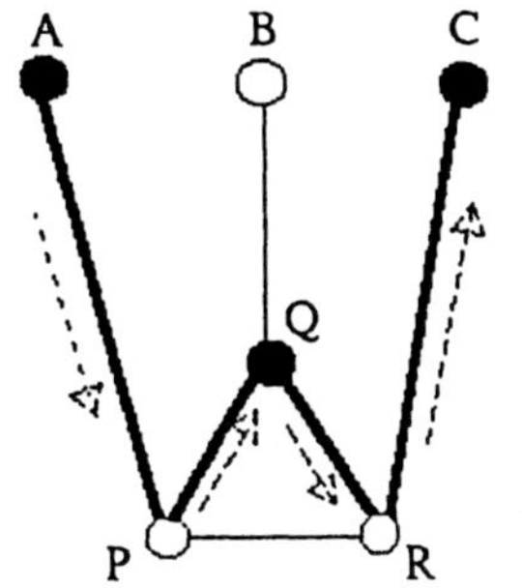

Fig. 8a

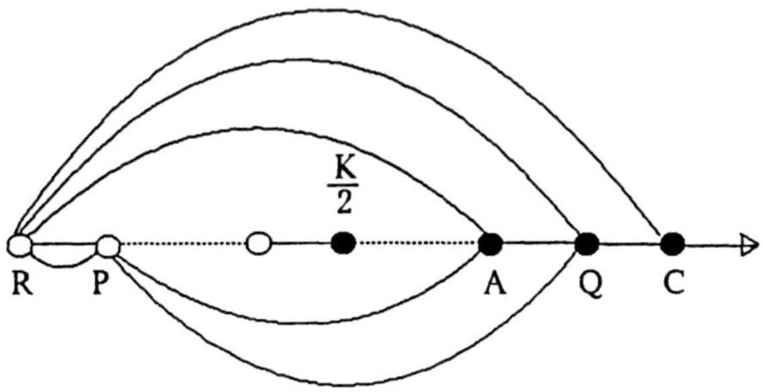

Fig. 8b

In the course of embedding of A^2-components, difficulties can arise in that more black than white vertices are to be embedded which can possibly lead to edge weights > K. These difficulties can be avoided if the embedding of the A^2-components is carried out in a suitable order. We call a sequence of A-components

$$A_{i_1}, A_{i_2}, \ldots\ldots, A_{i_k}$$

suitable, if in the course of the embeddding of the components of the sequence from left to right after each step the number of embedded black vertices is at most 1 greater than the number of the white vertices thus far embedded, and if all edge weights are $\leq$ K.

The following lemma holds:

LEMMA: *If G is a graph corresponding to a Yes-instance of 3–SAT with $m \geq n$, then there is a suitable sequence of all A-components.*

Proof of the lemma:

Assume the lemma is false. Let $A_1, A_2, \ldots\ldots, A_i$ be a suitable sequence of A-components of G with maximal length and $i < m$. Assume these components are already suitablv embedded into the line of numbers. Then the components $A_{i+1}, \ldots\ldots, A_m$ are all A^2-components, the black V-neighbours of which are as yet all unplaced. Therefore, 2i white A-vertices and 2i+1 black vertices are embedded. The components $A_{i+1}, \ldots\ldots, A_m$ are adjacent to n-i-1 unplaced black V-neighbours. Because $m \geq n$, the number of unplaced black V-vertices is smaller than the number of unplaced A-components. Therefore, among the unplaced black V-vertices there are such which are adjacent to at least two unplaced A-components.

Let $l > 0$ be the smallest integer so that l unplaced A-components have at most l-1 black V-neighbours in common. It is $l \leq m-i$.
W.l.o.g. we assume that $A_{i+1},.....,A_{i+l}$ have this property. We show that there is a suitable sequence for the components $A_1, A_2, ...,A_i, A_{i+1},.....,A_{i+l}$.

Let $\tilde{v}$ be a black V-neighbour of the components $A_{i+1},.....,A_{i+l}$ with maximal valence $deg(\tilde{v}) > 1$. Then at first we embed an A-component adjacent to $\tilde{v}$ - e.g. A_{i+1} -, after that we embed as many further A-components from the group $A_{i+2},.....,A_{i+l}$ as possible.

Assume we can embed in this way in suitable manner $j \leq l$ components. It is necessary $j = l$. Since in the case $j < l$, 2j white A-vertices and 2j+1 black vertices are embedded, among them j+1 black V-vertices, and we have left $2(l-j)$ white A-vertices and at most $l-1-(j+1) = l-j-2$ black V-vertices, in contradiction to the minimality of l. Hence $j = l$.

There is also a suitable embedding of the A-components $A_{i+1},.....,A_{i+l}$. W.l.o.g. we assume this sequence suitable. In the course of embedding these components in the suitable sequence 2l white and at most 2l-1 black vertices are therefore embedded. Hence the sequence

$$A_{i+1},.....,A_{i+l}, A_1, A_2, ...,A_i$$

is also suitable, in contadiction to the choise of i. Thus the lemma is proven.

From the lemma follows the suitable embeddability of all A-components including the black V-neighbours of the A-components. At the end of the embedding procedure the embedding of all white V-vertices remains. But these can be placed in arbitrary manner onto the positions 1,2,.....,n.

Case 2: $m < n$

In this case it is not always possible to embed the vertices of G onto the line of numbers so that the white V-vertices get the numbers 1,2,.....,n (Fig.9).

But we can reduce this case to case 1 in the following way:
If G has more V-components than A-components, then there are white V-vertices nonadjacent to any black A-vertex.

If x is such a white V-vertex and y his black V-neighbour, then x is to be placed first onto the greatest free position $< \frac{K}{2}$ and y onto the lowest free position $\geq \frac{K}{2}$. In this manner, all white V-vertices without black A-neighbours are placed, together with their black V-neighbours. If there are no such white V-vertices, then the number n' of the unplaced V-components is smaller or at most equal to the number m' of the A-components which are to be embedded, and it is to proceed according to case 1.

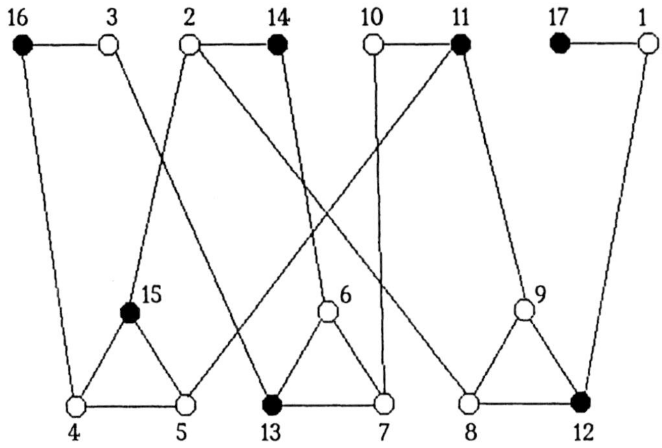

Fig.9

It remains to prove the inverse directtion:

> Each Non-instance of 3-SAT corresponds to a graph G with $g(G) > 2n+4m+2 = K$.

If I is a Non-instance of 3–SAT with n variables and m clauses, each vertex cover of the corresponding graph G contains at least $n+2m+1$ vertices, therefore also a minimum cover of G.

Let X_1 be a minimum vertex cover of G. Assume there is a numbering f of G so that

$$g_f(G) \leq 2n+4m+2 = K.$$

We put

$$S = \{ P \mid P \in X \text{ and } f(P) < \frac{K}{2} \},$$

$$T = \{ P \mid P \in X \text{ and } f(P) \geq \frac{K}{2} \}.$$

T is an independent set of G, then otherwise would be $g_f(G) > K$. Therefore S is a vertex cover of G with $|S| < |X_1|$, in contradiction to the minimality of X_1.

Thus the NP-completeness of MINIMUM WEIGHT is proven.

REFERENCES:

[1] Berge,C., "Graphs and Hypergraphs", Wiley, New York 1973

[2] Chinn,P.Z., Chvàtalova,J., Dewdney,A.K., Gibbs,N.E., "The Bandwidth Problem for Graphs and Matrices – A Survey", J. Graph Theory **6**, (1982), 223-254

[3] Gallian,J.A., "A Survey: Recent Results, Conjectures, and Open Problems in Labeling Graphs", J. Graph Theory **13** (1989), 491–504

[4] Garey,M.R., Johnson,D.S., "Computers and Intractability. A Guide to the Theory of NP-Completeness", Freeman, San Francisco 1979

[5] Karp,R.K., "Reducibility among Combinatorial Problems", in: R.E.Miller, J.W.Thatcher (Eds.): "Complexity of Computer Computations", Plenum Press, New York (1972), 85–103

[6] Papadimitriou,C.H., "The NP-Completeness of the Bandwidth Minimization Problem", Computing **16**, (1976), 263-270

[7] Ringel,G., "Problem 25" in: M.Fiedler(Ed.):"Theory of Graphs and its Applications", Proceedings of the Symposium held in Smolenice in June 1963, Publishing House of the Czech.Acad.of Science, Prague, (1964), p.162

On Vertexminimal Graphs With Radius r and Connectivity 2m

G. Lassmann, Berlin, FRG

Abstract:

What is the minimal number of vertices in a graph with prescribed radius and with prescribed connectivity? In this paper we show following result: The minimal graph with radius r, r>7, and even connectivity 2m has 2m(r-1)+2 vertices and is unique .

The graphs considered are finite, non-directed, without loops and multiple edges. The distance d(a,b) between two vertices a, b is the length of the shortest path between a and b. The diameter D of a vertexset M is the maximal distance between two vertices of M. A vertex from G with minimal maximal distance to all vertices of the vertexset M belong to the center of M in G. That distance is the radius $r_G(M)$ of M in G. N(a) denotes the neighborhood of the vertex a in G.

A 2m-connected graph with radius r has, according to the Menger theorem, at least 2m(r-1)+2 vertices. A graph G with connectivity 2m, radius r and 2m(r-1)+2 vertices is therefore 2m-regular and has diameter r. Hence every vertex lies in the center of G. We shall give a construction of a graph, which fulfills these conditions.

Let be the vertex e_0 an arbitrary vertex of G. e_0 lies, like all vertices, in the center of G. Thus there is one unique vertex e_r with distance r to e_0. These two vertices are, according to the Menger theorem, connected by exactly 2m disjoint mainpaths W_i of the length r. We call the set of the vertices with distance j to e_0 the sphere S_j. The unique vertex, which lies on the mainpath W_i and in S_j, is called $e_{i,j}$. We denote the graph, which is spanned by the edges incident to at least one vertex of S_j, by M_j.

Proposition

If M_j is connected, then S_j has in M_j at most radius 4.

R. Bodendiek, R. Henn (Eds.)
Topics in Combinatorics and Graph Theory

Let the graph M_j be connected. M_j has at most 6m vertices. Let $x,y \in S_j$ be vertices with maximal distance d in M_j.

If $d \leq 2$ the proposition follows immediately. Every vertex $v \in S_j$ has 2m neighbors in M_j, therefore some from the set $M=N(x) \cup N(y) \cup x \cup y$ (M has 4m+2 vertices).

I. $2<d\leq 4$ There is a vertex z, which is adjacent both to $N(x) \cup x$ and $N(y) \cup y$. An arbitrary vertex $v \in S_j$ has the distance $d(z,v) \leq d(z,x)+d(x,v) \leq 4$. (respectively $d(z,y)+d(y,v) \leq 4$)

II. $d>4$ First we show: $d \leq 6$.

Let P be the shortest path between x and y. Assume there is a vertex $z \in P$ with $d(x,z) \geq 3$ and $d(y,z) \geq 3$. Because vertices from S_j have neighbors in M, the vertex z must be from $S_{j-1} \cup S_{j+1}$. In M_j there is no adjacency between vertices from $S_{j-1} \cup S_{j+1}$. Therefore the neighbors z_x, z_y of z on P are from S_j and z_x has at most distance 2 to x (z_y respectively to y).

a) d=6, z already exist. Let v be an arbitrary vertex from S_j. v is adjacent to M, i.e. to $N(x) \cup x$. Hence v is not adjacent to $N(y) \cup y$. The vertices v, $z_x \in S_j$ have further no common neighbor with $y \in S_j$ (P would be shorter). For that reason there remain at most 6m-(2m+1)=4m-1 different possible neighbors, so v and z_x have common neighbors. We have $d(z,v) \leq d(z,z_x)+d(z_x,v) \leq 3$.

b) d=5, z_x and z_y are adjacent. At least one of them, let's take z_x, is in S_j. Let $v \in S_j$ be an arbitrary vertex, we will show that $d(z_y,v) \leq 4$. v is adjacent to M. If v is adjacent to $N(x) \cup x$, then v and z_x have no common neighbor with y (P would be shorter). Hence v and z_x have common neighbors. Thus $d(z_y,v) \leq d(z_y,z_x) + d(z_x,v) \leq 3$. If v is adjacent to $N(y) \cup y$, then we have $d(z_y,v) \leq d(z_y,y)+d(y,v) \leq 4$.

<u>Proposition</u>

If $z \in S_q$ is a vertex of the center of S_j ($1 \leq j \leq r-1$) in subgraph H of G, then the radius hold: $r_H(S_j) \geq \min\{j+q,2r-j-q)\}$.

Assume there is a sphere S_j, $1 \leq j \leq r-1$, which holds $r(S_j) < \min\{j+q,2r-j-q)\}$. All vertices of G lie on mainpaths and they are attainable from z. All vertices except S_{r-q} have distance less than r to z. An arbitrary vertex $e_{i,r-q}$ lies on mainpath W_i and has at most distance r-j-q respectively j+q-r to $e_{i,j}$.

(1.case $j+q \leq r$) $d(e,e_{i,r-q}) \leq d(z,e_{i,j})+d(e_{i,j},e_{i,r-q})$

$$\leq r(S_j)+r-q-j < j+q+r-q-j = r.$$

(2.case $j+q>r$) $d(e,e_{i,r-q}) \leq d(z,e_{i,j})+d(e_{i,j},e_{i,r-q})$

$$\leq r(S_j)+j+q-r < 2r-q-j+j+q-r = r.$$

Hence the graph with center z has a radius less than r.

<u>Corollary</u>

If $r>5$, then M_j, $3\leq j\leq r-3$, is not connected.

If the radius is greater than 5, then M_k, $k=[r/2]_+$, has two components with m vertices of S_k each. The m vertices of the first component lie on m mainpaths. We denote these mainpaths with the numbers 1 to m, the other mainpaths with the numbers m+1 to 2m. We call the inner vertices of the mainpaths 1 to m, the <u>left side</u>. The inner vertices of the other mainpaths belong to the <u>right side</u>.

From now on the radius r is always greater than 7.

From a simple argument using the 2m-connectivity follows, that if M_j $(1<j<r-1)$ is not connected, then M_j has two components. Each component contains m vertices of S_j. Therefore two vertices in S_j from the same component have at most distance 2.

<u>Proposition</u>

No vertex from S_j, $2\leq j\leq r-2$, is adjacent to the other side.

Assume $x\in S_j$ from the left side is adjacent to $y\in S_j$ or $y\in S_{j+1}$ from the other side. Without loss of generality let be $j\leq k=[r/2]_+$. All vertices except those from Sr-j have distance less than r to x. The vertices from the left side of S_{r-j} have at most distance $r-2j+2<r$ to x, because in the left component of M_k there is at most distance 2 between the mainpaths of the left side. An arbitrary vertex $v\in S_{r-j}$ on mainpath W_q from the right side holds (let y lie on W_p) $d(x,v) \leq d(x,y)+d(y,e_{p,k})$ $+d(e_{p,k},e_{q,k})$ $+d(e_{q,k},e_{q,r-q}) \leq 1+k-j+2+r-j-k \leq r-3j+3 < r$ for $j>1$. It turns out that the radius is less than r.

If the distance from an arbitrary vertex $e_{i,j}$ to the vertex $e_{k,l}$ of the same side is $|i-k|$, then we call $e_{k,l}$ <u>detourless</u> attainable from $e_{i,j}$. We have a <u>detour</u> if the distance is greater.

<u>Proposition</u>

The maximal detour between vertices of the same side, which lie in different spheres S_k and S_j, $3\geq j,k\geq r-3$, is 1.

Let $x\in S_j$ and $y\in S_{j+1}$ be from the same side. Assume, we have detour 2 from x to y. Hence $N(x)\cap N(y)$ is empty. Because x has at most m neighbors in S_{j-1} and y has at most m neighbors in S_{j+2}, both have at least m neighbors in $(S_j-x)\cup(S_{j+1}-y)$. Since

$|(S_j-x)\cup(S_{j+1}-y)| = 2m-2$ the vertices x and y have two neighbors in common.

Proposition

In $G(S_1)$ (respectively in $G(S_{r-1})$) are at least m-1 independent edges between the sides.

Assume, in $G(S_1)$ there are less than m-1 independent edges between the sides. If we remove e_0 and all vertices of the left side of S_1 incident to the independent edges, and all vertices of the right side of S_2, then the remaining graph consists of at least two components, because the vertices of the right side of S_1 are not connected with the remaining other vertices.

Theorem

An arbitrary vertex ek,l is detourless attainable from $e_{i,j}$ of the same side, l>j (respectively l<j), iff $e_{i,j}$ is adjacent to $e_{k,j+1}$ (respectively $e_{k,j-1}$).

We consider the case l>j, the proof of the case l<j is similar.

Assume, there is a vertex $e_{k,j-1}$, not adjacent to $a:=e_{i,j}$ but $e_{k,l}$ is detourless attainable from a. The vertex a has m_1 neighbors in S_{j-1} and m_3 neighbors in S_{j+1}. Because a has at most m-1 neighbors in S_j, we have $m_1+m_3 \geq m+1$. Therefore at least m_1+1 vertices from the sphere S_1 of the same side are detourless attainable from a. At least m_3 vertices from the sphere S_{r-1} of the same side are detourless attainable from a. In S_1 (respectively in S_{r-1}) there are at least m-1 independent edges between the different sides. Hence we have at least m_1 vertices in the other side of S_1 with distance j to a and we have at least m_3-1 vertices of the other side of S_{r-1} with distance r-j to a. Let p be the antipodal vertex to a. The vertex p has k_1 neighbors in S_{r-j-1} and k_3 neighbors in S_{r-j+1}. Because p has at most m-1 neighbors in S_{r-j}, it holds $k_1+k_3 \geq m+1$. Neither the m_3-1 vertices from S_{r-1} with distance r-j to a nor the m_1 vertices from S_1 can be detourless attainable from p, otherwise the distance between a and p is less than r. Hence p is not adjacent to m_3-1 vertices from S_{r-j+1} and m_1 vertices from S_{r-j-1}. So we have $2m-m_1-m_3+1 \geq k_1+k_3$. But we know $m_1+m_3 \geq m+1$ und $k_1+k_3 \geq m+1$, so it follows $2m-(m+1)+1 \geq m+1$ and therefore $m \geq m+1$ gives a contradiction.

Corollary

- If the vertex e from S_j has m_1 neighbors in S_{j-1} and m_3 neighbors in Sj+1, then the antipodal vertex p(e) (with distance r to e) has exactly $m-m_1+1$ neighbors in S_{r-j-1} and $m-m_3+1$ neighbors in S_{r-j+1}.

- In $G(S_1)$ (respectively in $G(S_{r-1})$) there are exactly $m-1$ independent edge between the sides.
- The graph spanned by the vertices of each side of S_j is a simplex.

We denote the set of the vertices $N(a) \cap S_j$ with $N_j(a)$.

Theorem

- If a and b are vertices of the same side of S_j $(1<j<r-1)$, they hold either $N_{j+1}(a) \subset N_{j+1}(b)$ and $N_{j-1}(b) \subset N_{j-1}(b)$ or $N_{j+1}(b) \subset N_{j+1}(a)$ and $N_{j-1}(a) \subset N_{j-1}(b)$.
- If a, b are from the same side of S_1: Either $N_2(a) \subset N_2(b)$ and $N_1(b) \subset N_1(a)$ or $N_2(b) \subset N_2(a)$ and $N_1(a) \subset N_1(b)$.
- If a, b are from the same side of S_{r-1}: Either $N_{r-2}(a) \subset N_{r-2}(b)$ and $N_{r-1}(b) \subset N_{r-1}(a)$ or $N_{r-2}(b) \subset N_{r-2}(a)$ and $N_{r-1}(a) \subset N_{r-1}(b)$.

I. Assume, the vertices a, $b \in S_j$ hold neither $N_{j+1}(a) \subset N_{j+1}(b)$ nor $N_{j+1}(b) \subset N_{j+1}(a)$.

1.Case $1 \le j < r-3$

In S_{j+1} there are m_1 vertices adjacent to a, but not adjacent to b. The set of the mainpaths containing these vertices is called H_1. In S_{j+1} there are m_3 vertices adjacent to b, but not adjacent to a. The set of the mainpaths containing these vertices is called H_3. The set of the mainpath containing the m_2 vertices adjacent to a and to b is called H_2. The remaining mainpaths belong to H_4. We know from the assumption that $m_1>0$ and $m_3>0$. Between the following sets of vertices there is no adjacency, otherwise we have detourless attainable mainpaths with no adjacency from S_{j+1} to a or b $(j<k \le r-2)$: No adjacency from $H_1 \cap S_k$ to $H_3 \cap S_{k+1}$ or $H_4 \cap S_{k+1}$, from $H_3 \cap S_k$ to $H_1 \cap S_{k+1}$ or $H_4 \cap S_{k+1}$, from $H_2 \cap S_k$ to $H_1 \cap S_{k+1}$, $H_3 \cap S_{k+1}$ or $H_4 \cap S_{k+1}$. Hence $H_1 \cap S_{j+2}$ can only be connected with $H_1 \cap S_{j+1}$, $H_1 \cap S_{j+3}$, $H_2 \cap S_{j+2}$, $H_2 \cap S_{j+3}$, $H_3 \cap S_{j+2}$, $H_4 \cap S_{j+1}$ and $H_4 \cap S_{j+2}$. The sum of the quantities is: $m_1+m_1+m_2+m_2+m_3+m-m_1-m_2-m_3+m-m_1-m_2-m_3 = 2m-m_3 < 2m$. So we have a cutting set of vertices with less than 2m vertices.

2.Case $r-3 \le j \le r-1$ The argumentation is similar to the first case.

II. Assume it holds $N_{j+1}(a)=N_{j+1}(b)$ and $N_{j-1}(a)=N_{j-1}(b)$ for $a,b \in Sj$ $(2 \le j \le r-2)$. The vertices a,b have $m+1$ neighbors in S_{j+1} and S_{j-1} and they have $m-2$ neighbors in S_j. So we have a cutting set of vertices with less than 2m vertices.

Theorem

The vertexminimal 2m-connected graph $G_{2m,r}$ with radius r, r<7, has 2m(r-1)+2 vertices and is unique.

Using the propositions and the theorem above, we give the construction of the graph. The vertices e_0 and e_r are connected by 2m mainpaths of length r. We renumber the mainpaths of every side. All vertices of each sides of the spheres S_j, $1 \leq j \leq r-1$, are adjacent. Because there are only m-1 independent edges between the sides, there must be a vertex in every side of S_1 which is not adjacent to the other side. These vertices must be adjacent to all m vertices of S_2 of the same side.

Left side: We denote the mainpath of the left side, which contains the vertex with m neighbors in S_2, with W_m. Each vertex of the left side of S_1 has a different number of neighbors in S_2. We denote the mainpath containing the specific vertex with k neighbors in S_2 with W_k. This vertex is, according to the last theorem, adjacent to the vertices in S_2, which lie on the mainpaths W_k to W_m. This holds for the vertices of the spheres S_2, ..., S_{r-2} too: The vertex $e_{k,j}$ from S_j on W_k is adjacent to the k vertices $e_{1,j+1}, ..., e_{k,j+1}$ from S_{j+1}.

Right side: We denote the mainpath containing the vertex with k neighbors in S_2 with W_{m+k}. This vertex is adjacent to the vertices in S_2, which lie on the mainpaths W_{m+1} to W_{m+k}. The vertex $e_{k,j}$ from S_j on W_{m+k} is adjacent to the k vertices $e_{m+1,j+1}$, ..., $e_{m+k,j+1}$ from S_{j+1}.

The vertex $e_{k,1}$ from the left side of S_1 has k neighbors in S_2, m-1 neighbors from the left side of S_1, is adjacent with e_0 has therefore exactly m-k neighbors on the right side of S_1. Hence ek,1 is adjacent to $e_{m+1,1}$, .., $e_{2m-k,1}$. A similar argument shows that $e_{k,r-1}$ is adjacent to $e_{2m-k+2,r-1}$, ..., $e_{2m,r-1}$.

Proof of the properties:

1. Radius

The higher spheres of the mainpaths W_1 to W_i are from an arbitrary vertex $e_{i,j}$ (i.e. $i \leq m$) detourless attainable, also the lower spheres of W_i to W_m. In S_1 exactly m-i vertices of the other side have distance j-1 to $e_{i,j}$, these are the vertices $e_{m+1,1}$ to $e_{2m-i,1}$. The vertices $e_{2m-i+2,r-1}$ to $e_{2m,r-1}$ from the other side of S_{r-1} have distance r-j to $e_{i,j}$. The mainpath W_{2m-i+1} is not detourless attainable from these two vertexsets. Hence the vertex $e_{2m-i+1,r-j}$ has to $e_{i,j}$ the distance (using S_1) j-1+1+r-j-1+1=r or (using S_{r-1}) r-j+(r-1-(r-j))+1=r. So we have radius r.

2. Connectivity

We connect (uniquely) the mainpaths using e_0, e_r and the independent edges between the sides to m disjoint circuits. Every vertex lies on one circuit. Every vertex has two adjacent neighbors on all other circuits. Let a,b be arbitrary not adjacent vertices. On every circuit we have two disjoint paths from the two neighbors of a to the two neighbors of b. If N(a) and N(b) are not disjoint, these paths have length 0.

Hence we have 2m open disjoint paths between a and b, and the graph $G_{2m,r}$ is 2m-connected.

We have chosen an arbitrary vertex e_0. Our construction starting with another vertex gives the same graph, so it follows:

Corollary

The graph $G_{2m,r}$ is vertex-transitiv.

Embedding Schemes and the Jordan Curve Theorem

C. H. C. Little, Palmerston North, New Zealand
A. Vince, Florida, USA

ABSTRACT

We survey several combinatorial embedding schemes that encode and generalize the concept of a map, and attempt to unify them. We also show that two combinatorial generalizations of the Jordan curve theorem arising from such schemes are in fact equivalent.

1. INTRODUCTION

An embedding of a graph in a surface S is called a *2-cell embedding* if all the regions are 2-cells. A *map* is classically defined, for example in Coxeter and Moser [CM], as a decomposition of a surface into simply connected regions by an embedded graph. A map, in turn, is a topological generalization of the boundary of a polyhedron; the boundary complex of a polyhedron can be viewed as a map on the sphere. Over the past few decades, numerous combinatorial schemes have been introduced to encode and generalize the concept of a map. Embedding schemes have been used successfully to investigate coloring problems, the genus of a graph and the classification of highly symmetric graphs and maps. Section 2 gives a brief survey of some of these schemes, and an attempt is made to unify them. The exposition is roughly historical.

R. Bodendiek, R. Henn (Eds.)
Topics in Combinatorics and Graph Theory

Tutte $[T_1]$ $[T_2]$ $[T_3]$ and Stahl $[S_1]$ use algebraic schemes to give a combinatorial generalization of the Jordan curve theorem. The classical Jordan curve theorem states that every simple closed curve γ separates the 2-sphere into two connected components having γ as their common boundary. Little and Vince [LV] give elementary graph theoretic versions of his theorem. The results in Section 3 of this paper state that Stahl's version and a version of Little and Vince are equivalent. Thus the algebraic abstraction is not more general than the graph theoretic version. The idea is to translate back and forth between schemes, in the spirit of Section 2.

2. EMBEDDING SCHEMES

Associated with each of the schemes described in this section is (1) a set of faces of various dimensions, (2) the notion of incidence of two faces and (3) a topological space given in terms of a cell complex. Of course, the faces of a classical map are vertices, edges and regions and the topological space is the surface in which the graph is embedded. Each of the following embedding schemes encodes these attributes, and some generalize them.

Rotation schemes.

Graphs will be connected and finite. Let S be a compact, connected, orientable surface without boundary. Later some of these restrictions on S will be relaxed. The first combinatorial scheme was used by Dyck [D] in 1888 and Heffter [H] in 1891 and formalized by Edmonds [E] in 1960. A further exposition is given by Biggs and White [BW]. Let Γ be a graph with vertex set V, edge set E and no loops or multiple edges. For each $v \in V$ let ρ_v be a cyclic permutation of the vertices adjacent to v, and let $\rho = \{\rho_v\}_{v \in V}$. Then the ordered pair (Γ, ρ) is called a *rotation scheme*. The motivation for this definition is as follows. Given an edge $(u, v) = (u_0, v_0)$ of Γ, construct a cycle consisting of successive edges: $(u, v) = (u_0, v_0), (u_1, v_1), \ldots, (u_m, v_m) = (u, v)$ where

$$u_{i+1} = v_i$$
$$v_{i+1} = \rho_{v_i}(u_i)$$

for all $i \in \{0, 1, \ldots, m-1\}$, where $i + 1$ is taken modulo m. Another cycle containing edge (u, v) is similarly obtained by starting with $u_0 = v$ and $v_0 = u$.

Thus each edge lies on two such cycles. Regarding each cycle as the boundary of a polygonal 2-cell and gluing together identical edges of these 2-cells, we obtain a surface in which Γ is embedded. Conversely, for any 2-cell embedding of a graph Γ in an orientable surface, there are cyclic permutations ρ_v at each vertex naturally induced by the orientation of the surface, clockwise or counterclockwise. With this set ρ of permutations, it can be shown that the pair (Γ, ρ) induces the original embedding of Γ.

The following alternative interpretation of the rotation scheme ties in better with the more abstract schemes described below. Consider the barycentric subdivision of graph Γ on the surface S. For each edge (u, v) of Γ there corresponds a cluster of four simplices in the barycentric subdivision, as in Figure 1. Instead of regarding ρ_v as acting on the neighbors of v, regard ρ_v as acting on such clusters of four triangles. The surface can be obtained by "gluing" together the clusters according to straightforward rules applied to the cyclic permutations of the rotation scheme. This viewpoint is more than a heuristic convenience in the schemes that follow.

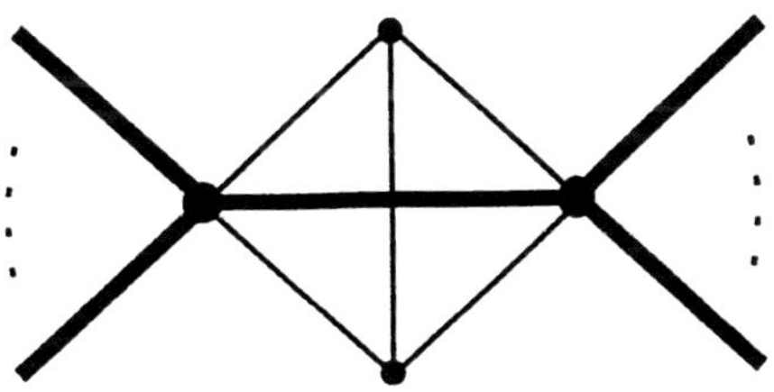

Fig. 1

Notice that in a rotation scheme, vertices and edges and their incidence are described in terms of two different kinds of structures: a graph Γ to describe the vertices and edges and permutations to describe the 2-dimensional faces. In the next scheme this is unified - all faces are defined in terms of permutations.

Orientable maps.

In 1979 Tutte [T_1] defined an *orientable map* M as a pair (θ, Q) of permutations acting on a finite set X of elements, called *darts*, such that each cycle of θ is of length 2 and the group generated by θ and Q acts transitively on X. Similar schemes had been previously introduced in the late 1960's by Jacques [J] and Cori [C_1] [C_2]. The same notion has also been used by Jones and Singerman [JS]. The *vertices*, *edges* and *faces* of M are defined as the cycles of Q, θ and $Q \circ \theta$, respectively. Vertices and edges of M are *incident* if they have a dart in common. For vertices and faces, and for edges and faces, incidence is defined similarly. The motivation for these definitions is as follows. Given a graph Γ (possibly having loops or multiple edges) embedded in an orientable surface, to each edge of Γ associate two opposite directed edges, the "darts". Then θ is the permutation of the dart set that interchanges each pair of opposite darts, and the cycles of Q are the "rotations" acting on the darts at each vertex.

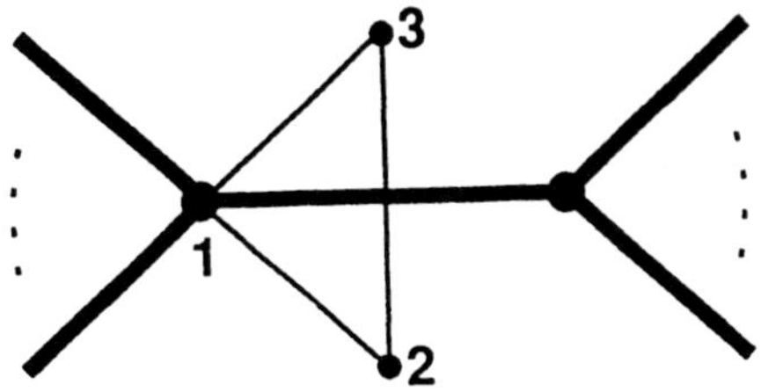

Fig. 2

In terms of the barycentric subdivision of the graph on the surface S, a dart can equally well be regarded as a pair of simplices, as in Figure 2. With this idea as motivation, the surface S of an orientable map M = (θ, Q) can be defined directly by simple rules for identifying edges of darts. More precisely, given permutations θ and Q acting on X, start with the disjoint union of triangular "darts", one for each element of X. Label the three vertices of each dart 1, 2, 3. If θx = y, then dart x is joined to dart y along the edge labeled 23. If Qx = y then the edge of dart x labeled 12 is identified with the edge of y labeled 13 (1 to 1 and 2 to 3). The transitivity requirement in the definition of an orientable map is equivalent to the connectedness of the surface S. That the barycentric subdivision of any orientable map can be decomposed into darts (adjacent pairs of simplices) is a consequence of the assumption that there is a coherent orientation on the set of simplices.

The restriction that the surface be without boundary can be dropped by allowing θ to have fixed points. Now we merely assume that θ is an involution, i.e., has order two. The assumption that the surface be connected can be dropped by removing the restriction that the permutations generate a transitive group. The assumption that the surface be orientable can also be dropped by not grouping the simplices of the barycentric subdivision into pairs. Thus a *map* M is given in [V_1] as a triple (θ_0, θ_1, θ_2) of involutions acting on a set X such that $\theta_0 \circ \theta_2 = \theta_2 \circ \theta_0$. Let G be the group generated by the three involutions. The *vertices*, *edges* and *faces* of M are the orbits of the subgroups $G_0 = \langle\theta_1, \theta_2\rangle$, $G_1 = \langle\theta_0, \theta_2\rangle$, and $G_2 = \langle\theta_0, \theta_1\rangle$, respectively. Two faces are *incident* if the corresponding orbits have non-empty intersection. The requirement that $\theta_0 \circ \theta_2 = \theta_2 \circ \theta_0$ guarantees that each edge is incident to at most two at most two vertices. To obtain the surface S associated with M, start with the disjoint union of 2-simplices, one for each element of X. Label the three vertices of each such simplex 0, 1, 2. Identify edges of the simplices according to the thetas. More precisely, if $\theta_i(x) = y$, then identify simplex x and simplex y along the edge opposite the vertex labeled i. The motivation for these definitions is as follows. Given a graph embedded in a surface, X can be regarded as the set of simplices in the barycentric subdivision, as in Figure 3. Then the three vertices of such a simplex can be label 0, 1, or 2 according to whether the vertex in the barycentric subdivision corresponds to a vertex (0-dimensional), edge (1-dimensional) or face (2-dimensional) of the embedding. The involution θ_i takes a simplex x to the adjacent simplex y that shares a common edge not labeled with i. The simplices in the orbits of the subgroups G_0, G_1 or G_2 correspond to simplices about the barycenter of a vertex, edge or face of the embedding, respectively. Incidence in the abstract map corresponds to incidence in the embedding.

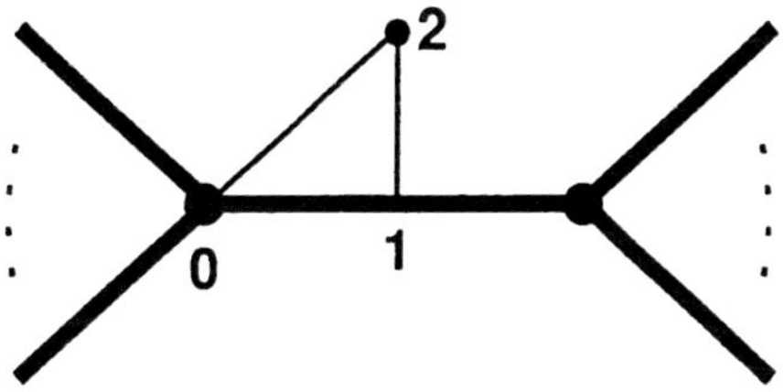

Fig. 3

Permutation pairs.

In 1980 Stahl [S_1] gave a generalization of an orientable map (θ, Q). He defined a *permutation pair* as a pair (P, Q) of permutations P and Q acting on a finite set X, allowing P to be an arbitrary permutation rather than restricting it to be an involution. The same definition had been used earlier by Walsh in 1975 [W] and interpreted as a hypergraph embedding on an orientable surface. The definitions of vertex, edge, face and incidence are the same as for an orientable map. However, an edge may be incident to more than two vertices, as in a hypergraph. The surface is obtained by associating with each element of X a quadrilateral with vertex labels 1, 2, 3, 4. If Px = y then edge 12 in quadrilateral x is identified with edge 14 in quadrilateral y. If Qx = y then edge 34 of x is identified with edge 32 of y.

Combinatorial maps.

A scheme of Vince [V_1] is slightly more general than a permutation pair in that it allows for non-orientable surfaces. A (2-dimensional) *combinatorial map* is a triple (θ_0, θ_1, θ_2) of involutions acting on a finite set X. The only difference between this and the definition above is that the restriction $\theta_0 \circ \theta_2 = \theta_2 \circ \theta_0$ is dropped. In the orientable case, Stahl's P and Q correspond to $\theta_0 \circ \theta_2$ and $\theta_1 \circ \theta_2$, respectively. There is no reason to restrict the dimension. An *n-dimensional combinatorial map* is an (n+1) - tuple (θ_0, θ_1, ..., θ_n) of involutions acting on a finite set X. Let G be the group generated by these involutions. For i = 0, 1, ..., n, the faces of dimension i are defined as the orbits of the subgroup G_i generated by the set of all thetas except θ_i. Incidence is defined as for a map. This scheme now generalizes certain higher dimensional cell complexes studied by Grünbaum [Gr], Coxeter and Shephard [CS] and others, and is almost equivalent to the incidence polytopes of Danzer and Schulte [DS] [S]. The underlying topological space is obtained, as in the case of a surface, by identifying X with a set of vertex labeled n-dimensional simplices. If $\theta_i(x) = y$ then simplex x is identified with simplex y along the codimension 1 face opposite the vertex labeled i.

N-graphs.

The scheme described next is the dual of that in the preceding paragraph. An *n-graph* is a finite graph Γ, regular of degree n, together with an edge coloring with n colors such that incident edges receive distinct colors. Let X denote the set of vertices and {0, 1, 2, ..., n-1} the set of colors. Let Γ_i denote the subgraph of Γ induced by all edges not colored i. Define a *face of dimension i* as a connected component of Γ_i. Two faces are *incident* if the corresponding subgraphs of Γ have non-empty intersection. An (n-1)-dimensional topological space is obtained from Γ by taking the disjoint union of (n-1)-dimensional simplices, one for each vertex of Γ, and labeling the n vertices of each such simplex with {0, 1, ..., n-1}. Identify simplex x to simplex y along the codimension 1 face opposite the vertex labeled i if and only if vertices x and y are joined in Γ by an edge labeled i. An n-graph is dual to a combinatorial map in the sense that the dual 1-skeleton of the simplicial complex Δ associated with a combinatorial map is an n-graph Γ. In turn, the simplicial complex associated with the n-graph Γ is the original complex Δ. The notion of an n-graph was introduced independently by Ferri and Gagliardi [FG] in a topological context, by Vince [V_2] in a combinatorial context and by Lins [L] and Robertson [R] for the 2-dimensional case.

Geometries.

The topological spaces associated with all the above schemes are "thin", in the sense that each codimension 1 face of the complex is incident to at most 2 facets. (Recall that a *facet* is a highest dimensional cell in the complex.) For example, in dimension 2 the spaces are surfaces and in higher dimensions they are pseudomanifolds. We can remove this "thinness" restriction by allowing edge colorings of a graph that are not necessarily proper, i.e., a vertex can be incident to more than one edge of the same color. This allows for Tits buildings, chamber complexes and other structures associated with groups and their geometries. We will not go into this topic here.

3. DISCRETE JORDAN CURVE THEOREMS.

Two discrete versions of the Jordan curve theorem are stated below. Theorem 1 [LV] is formulated in terms of cycles of a graph embedded on an orientable surface. Theorem 2 [S_2] is formulated in terms of 3-graphs as discussed in Section 2. The relevant definitions are given below.

THEOREM 1.

Let G be a graph embedded in an orientable surface S. Every cycle separates if and only if S is a sphere.

THEOREM 2.

Let H be a bipartite 3-graph. Every normal semicycle separates if and only if $\alpha(H) = 0$.

Concerning the combinatorial definition of "separates" in Theorem 1, let $C = \{c_1, c_2, ..., c_r\}$ be the set of boundaries of the faces of the embedding. For any cycle c in G, let $E'(c)$ be the set of edges of G - c incident on a vertex of c. Define $e_1 \sim e_2$ if e_1 and e_2 are edges of $E'(c)$ belonging to the same cycle of C. The transitive closure of the relation $\sim$ is then an equivalence relation. It is not hard to show [LV] that there are at most two equivalence classes, denoted $L'(c)$ and $R'(c)$, for any cycle c. If only one equivalence class exists, we call it $R'(c)$, and we define $L'(c) = \emptyset$. A cycle c in G is said to *separate* if $L(c) \cap R(c) = \emptyset$ where $L(c)$ and $R(c)$ are the sets of vertices reachable from edges of $L'(c)$ and $R'(c)$, respectively, by paths not containing a vertex in $V(c)$. The sets $L(c)$ and $R(c)$ are intuitively the "inside of c" and "outside of c".

The terminology of 3-graphs in Theorem 2 is somewhat more involved. Let the three colors of a 3-graph H be red, yellow and blue and call the two-colored polygons *bigons*. Following Stahl's terminology for permutation pairs, we call a cycle Z in H a *semicycle* if it contains a blue edge and any red-yellow bigon in H has at most two vertices incident with blue edges of Z. It is a *normal* semicycle if there is no blue edge of Z adjacent to edges of Z of distinct color. The *necklace* N(Z) of Z is the set of red-yellow bigons Z meets, and the *beads* of Z are the bigons of its necklace. If Z is a semicycle of H, its *conjugate* Z^* is defined as $Z + \cup N(Z)$ where + denotes symmetric difference. Note that Z^* is also a semicycle; $N(Z^*) = N(Z)$; $Z^{**} = Z$; and Z^* is normal if Z is normal. Define $L(Z) = Z \cap (\cup N(Z))$ and $R(Z) = Z^* \cap (\cup N(Z))$. Hence $L(Z^*) = R(Z)$ and $R(Z^*) = L(Z)$. Call L(Z) and R(Z) the *sides* of Z.

Let D be a non-trivial path in H whose terminal edges are blue, and suppose that each red-yellow bigon in H has at most two vertices incident with blue edges of D. Then D is a *semipath* of H. Necklaces, beads, and conjugates are defined as for semicycles. In addition, a red-yellow bigon is called a *terminal bead* if it contains a terminal vertex of D, and the elements of N(D) are called *internal beads* of D.

A semicycle Z and semipath D are said to *miss* if they are disjoint and no internal bead of D is a bead of Z. If Z and D miss, then D is said to *link* the sides of Z if one terminal vertex is incident with L(Z) and the other with R(Z). We say that Z *separates* if no semipath links its sides.

Given a pair (G, S) consisting of a graph G embedded in a surface S, a 3-graph H(G, S) can be constructed as follows. First construct the barycentric subdivision Δ of the embedding of G in S , and label each vertex of Δ with 0, 1 or 2 depending on whether it corresponds to a vertex, edge or face of the embedding. Then H is the dual graph of Δ. An edge e of H is colored red, blue or yellow depending on whether e crosses in Δ an edge labeled 01, 12, or 20, respectively. Note that all red-blue bigons in H are quadrilaterals. It is known [V_3] that S is orientable if and only if H(G, S) is bipartite.

Conversely, given a bipartite 3-graph H, a graph G(H) together with an embedding in a surface S(H) can be constructed as follows. The vertices of G(H) are the yellow-red bigons of H, and two vertices are joined by an edge in G(H) for each blue edge in H joining the corresponding yellow-red bigons. (Thus each chord of a red-yellow bigon in H yields a loop in G(H).) There is a unique embedding of H (and hence of G(H)) on a surface S(H) such that the bigons of H are the faces of the embedding. This embedding can be described using the rotation scheme given in Section 2. Let $\{V_1, V_2\}$ be a bipartition of H. For each vertex v of V_1 let ρ_v be the rotation induced by the cyclic permutation (red, yellow, blue) of the colors; for each vertex v of V_2 let ρ_v be induced by the inverse permutation (blue, yellow, red).

Theorems 1 and 2 are equivalent in the following sense. Theorems 3 and 4 below state that it is possible to translate back and forth between the language of graphs and cycles and the language of 3-graphs and normal semicycles in such a way that Theorems 1 and 2 are merely restatements of the same result in a different terminology. Here

$$\alpha(H) = 1 - \frac{m - n}{2}$$

where m is the number of bigons in H and n is half the number of vertices of H.

THEOREM 3.

Let G be a graph embedded in an orientable surface S.

(1) Every cycle of G separates if and only if every normal semicycle of H(G, S) separates.

(2) S is a sphere if and only if $\alpha(H(G, S)) = 0$.

THEOREM 4.

Let H be a bipartite 3-graph.

(1) Every normal semicycle of H separates if and only if every cycle of G(H) separates.

(2) $a(H) = 0$ if and only if S(H) is a sphere.

PROOF OF THEOREM 3.

Let Z be any normal semicycle in H that does not separate, and let D be a semipath that links the sides of Z. The semicycle Z induces a cyclic sequence of red-yellow and blue-red bigons, each sharing a red edge with its successor. This chain corresponds to a cycle c of incident vertices and edges in G. Similarly, to the semipath D in H there corresponds a path p in G. That D links the sides of Z implies that p links R(c) and L(c). Therefore c does not separate.

Conversely let p be a path in G linking R(c) and L(c) for some cycle c in G that does not separate. The cycle c can be regarded as an alternating cyclic sequence of vertices and edges, which corresponds to an alternating cyclic sequence of red-yellow and blue-red bigons in H such that consecutive bigons have a common edge. The blue-red bigons are actually quadrilaterals. Thus a normal semicycle Z in H can be formed by choosing (arbitrarily) one of the two blue edges in a blue-red bigon, moving to the adjacent yellow edge of the next (red-yellow) bigon in the sequence, then progressing around that bigon until the first blue edge in the next (blue-red) bigon is reached, and continuing in this fashion. Similarly, to the path p in G there corresponds a semipath D in H. Again the fact that p links R(c) and L(c) implies that D links the sides of Z. Hence Z does not separate.

Concerning statement (2), let v, e, and f denote the numbers of vertices, edges and faces, respectively, of the graph G embedded in S. Then for $H = H(G, S)$ we have $m(H) = v + e + f$, $n(H) = 2e$ and $m - n = v - e + f$. Therefore S is a sphere if and only if $m - n = 2$, hence if and only if $\alpha(H) = 0$.

PROOF OF THEOREM 4.

Let Z be a normal semicycle in H that does not separate, and let D be a semipath that links the sides of Z. In Z each bead corresponds to a vertex of G(H) and each blue edge to an edge of G(H), and so Z corresponds to a cycle c in G(H). Similarly to the semipath D there corresponds a path p in G(H). That D links the sides of Z implies that p links R(c) and L(c), so that c does not separate.

Conversely let p be a path in G linking R(c) and L(c) for some cycle c in G that does not separate. To the cycle c there corresponds in H a sequence of red-yellow bigons joined by blue edges. The red-yellow bigons are the beads of a normal semicycle uniquely defined up to conjugacy. Similarly to the path p there corresponds a semipath D. Again the fact that p links R(c) and L(c) implies that D links the sides of Z. Hence Z does not separate.

Concerning statement (2) let v, e and f denote the numbers of vertices, edges and faces, respectively, of the graph G(H) embedded in S(H). Then $m(H) = v + f$ and, since the number of blue edges of H is half the number of vertices of H, we have $n(H) = e$ and $m - n = v - e + f$. Now the proof is exactly the same as in Theorem 3.

REFERENCES

[BW] N L Biggs and A T White, Permutation Groups and Combinatorial Structures, Cambridge University Press, London, 1979.

[C_1] R Cori, Graphes planaires et systèmes de parenthèses, Centre National de la Recherche Scientifique, Institut Blaise Pascal, 1969.

[C_2] R Cori, Un code pour les graphes planaires et ses applications, Asterisk 27, 1975.

[CM] H S M Coxeter and W O J Moser, Generators and Relations for Discrete Groups, Springer-Verlag, New York, 1980.

[CS] H S M Coxeter and G C Shephard, Regular 3-complexes with toroidal cells, J. Combinatorial Theory B22, 1977, 131-138.

[DS] L Danzer and E Schulte, Regulare Inzidenzkomplexe I, Geometriae Dedicata 13, 1982, 295-308.

[D] W Dyck, Beitrage sur Analysis Situs, Math. Ann. 32, 1888, 457-512.

[E] J R Edmonds, A combinatorial representation for polyhedral surfaces, Notices Amer. Math. Soc. 7, 1960, 646.

[F] M Ferri, Una rappresentazione delle n-varieta topologiche tringolabili mediante grafi (n+1)-colorati, Boll. Un. Mat. Ital. 13B, 1976, 250-260.

[FG] M Ferri and C Gagliardi, Crystallization moves, Pacific J. Math. 100, 1982, 85-103.

[G] C Gagliardi, A combinatorial characterization of 3-manifold crystallizations, Boll. Un. Mat. Ital. 16A, 1979, 441-449.

[Gr] B Grünbaum, Regularity of graphs, complexes and designs, Colloques internationaux C.N.R.S., No 260 - Problème combinatoire et théorie des graphes, Orsay, 1976, 191-197.

[H] L Heffter, Über das Problem der Nachbargebiete, Math. Ann. 38, 1891, 477-508.

[J] A Jacques, Sur le genre d'une paire de substitutions, C R Acad. Sci. Paris 367, 1968, 625-627.

[JS] G A Jones and D Singerman, Theory of maps on orientable surfaces, Proc. Lond. Math. Soc. (3) 37, 1978, 273-301.

[L] S Lins, Graph-encoded maps, J. Combinatorial Theory B32, 1982, 171-181.

[Lit] C H C Little, Cubic combinatorial maps, J. Combinatorial Theory B44, 1988, 44-63.

[LV] C H C Little and A Vince, Discrete Jordan curve theorems, J. Combinatorial Theory B, to appear.

[R] N Robertson, Graphs Minimal under Girth, Valency and Connectivity, PhD thesis, University of Waterloo.

[S] E Schulte, Regular Inzidenzkomplexe II and III, Geometriae Dedicata 14, 1983, 33-79.

[S_1] S Stahl, Permutation-partition pairs: a combinatorial generalization of graph embeddings, Trans. Amer. Math. Soc. 259, 1980, 129-145.

[S_2] S Stahl, A combinatorial analog of the Jordan curve theorem, J. Combinatorial Theory B35, 1983, 28-38.

[T_1] W T Tutte, Combinatorial oriented maps, Can. J. Math. 31, 1979, 986-1004.

[T_2] W T Tutte, From topology to algebra, J. Graph Theory 19, 1986, 331-337.

[T_3] W T Tutte, What is a map?, in New Directions in Graph Theory, F. Harary, ed., Academic Press, New York, 1973.

[V_1] A Vince, Combinatorial Maps, PhD thesis, University of Michigan, 1981.

[V_2] A Vince, Combinatorial maps, J. Combinatorial Theory B34, 1983, 1-21.

[V_3] A Vince, N-graphs, Discrete Math., to appear.

[W] T R S Walsh, Hypermaps versus bipartite maps, J. Combinatorial Theory B18, 1975, 155-163.

Subgraph Packing - A Survey

M. Loebl, Bonn, FRG
S. Poljak, Prague, CSSR

1. Introduction

The research on factors of graphs concentrated mostly on factors satisfying certain local degree conditions, like regular factors or factors with degrees within prescribed intervals. More recently, also other kinds of factors have been investigated. Here we survey results on factors with prescribed components. Let $\mathcal{F}$ be a family of graphs. A graph G is said to have an $\mathcal{F}$ - *factor* if it has a factor each component of which is isomorphic to a member of family $\mathcal{F}$. The $\mathcal{F}$ - *factor problem* is to decide whether a given graph G admits an $\mathcal{F}$ - factor. The $\mathcal{F}$ - *packing problem* is to find a subgraph of maximum order which admits an $\mathcal{F}$ - factor.

In particular, one important case surveyed here concerns the solution of the K_2, H - packing problem, i. e. packing by copies of a single graph H and copies of K_2.

When writing this survey we restricted ourselves in two further aspects: We are motivated by the complexity of the packing problems, and the underlying graph G is general. Many results omitted here, like the existence of factors ensured by specific condition given on G, can be found in the survey by Akiyama and Kano [AK].

We assume that a reader is familiar only with basic notions of matching theory.

2. Some definitions and notation

Let G be a graph. *A packing* is a collection of vertex disjoint subgraphs of G. Let $\mathcal{F}$ be a family of prescribed graphs. Then packing Q is called $\mathcal{F}$ - *packing*

R. Bodendiek, R. Henn (Eds.)
Topics in Combinatorics and Graph Theory

if each element of Q is isomorphic to a graph from $\mathcal{F}$. A subset $X \subset V(G)$ is called *saturated* by a packing Q if each vertex of X belongs to a subgraph of Q. A graph G is called $\mathcal{F}$ - *perfect* if there is an $\mathcal{F}$ - packing Q which saturates all vertices of G. A graph G is called $\mathcal{F}-$ *critical* if G is not $\mathcal{F}$ - perfect but $G-x$ is whenever a vertex x is deleted. Hence a graph is *perfectly matchable* if it is $\mathcal{F}$ - perfect and a graph is *hypomatchable* if it is $\mathcal{F}$ - critical for $\mathcal{F}=\{K_2\}$. A packing is *maximum* if it saturates maximum number of vertices of G.

νG denotes the size of maximum matching of a graph G. Let i be a positive integer. Then S_i denotes a star on $i+1$ vertices.

3. K_2, H - packing

A perfect matching (1 - factor) is a factor all components of which are $K_2's$. Hence a natural extension is to ask about factors all components of which are isomorphic to a larger graph. This question was considered by Hell and Kirkpatrick (1983).

3.1 Theorem *[HK2] If H has a connected component distinct from K_1 and K_2 then H - factor problem is NP - complete.*

The content of the above theorem tells that all nontrivial cases of H - factor problem are difficult. The next natural and already more interesting subject appeared in the K_2, H - factors where each component is isomorphic to a given connected graph H or to K_2. The complexity of K_2, H - factor problem was completely classified by Loebl and Poljak (1987).

We define *Propeller* as a graph obtained from a hypomatchable graph W by adding a disjoint copy (cr) of K_2 and by joining c by at least one edge to W. The smallest example of a propeller is a path of length 2 (S_2 star).

3.2 Theorem *[LP3] Let H be a connected graph which is either perfectly matchable or hypomatchable or a propeller. Then K_2, H - factor problem is polynomial. On the other hand the K_2, H - factor problem is NP - complete for all other connected graphs H.*

The case when H is perfectly matchable is trivially equivalent to a matching. A polynomial time algorithm for the case with H hypomatchable was given by Cornuejols, Hartvigsen and Pulleyblank (1982). They also obtained the following good characterisation theorem.

3.3 Theorem *[CHP] Let H be a hypomatchable graph. A graph G admits a*

K_2, H - factor if and only if the number of K_2, H - critical components of $G - A$ is a most $|A|$ for each $A \subset V(G)$.

We mention another case ($K_2, S_2 - factor$) solved earlier in section 4. An efficient solution for K_2, H - factor problem for H equal to a propeller was given by Loebl and Poljak (1987). The algorithm is based on the following good characterisation. Given a graph G, a subset A of its vertices, and a family χ of K_2, H - critical components of $G - A$, define an auxiliary bipartite graph $B = B(A, \chi)$ on the vertex set $A \cup \chi$, and with the edges (v, K) where $v \in A$ is adjacent in G to a vertex of $K \in \chi$. Further define the subgraph $B_0 = B_0(A, \chi)$ of B as follows. The edge (v, K) belongs to B_0 if $K \cup v$ contains a copy of $H - r$ where $c = v$ and $K - W$ is perfectly matchable (c, r and W were introduced above in the definition of a propeller).

3.4 Theorem *[LP3] Let H be a propeller. A graph G admits a K_2, H - factor if and only if $\nu B + \nu B_0 \geq |\chi|$ for every set $A \subset V(G)$ and every subset χ of the K_2, H - critical components of $G - A$.*

In order to establish that theorems 3.3 and 3.4 provide a good characterisation it is necessary to show that K_2, H - critical graphs can be efficiently recognised for H hypomatchable and a propeller. This follows from results presented in section 6. Let us mention that partial results for the NP - completeness part of 3.2 were obtained by Hell and Kirkpatrick (1984) who proved the statement in case H is 2 - connected and every its maximum matching omits at least 3 vertices. Our NP - completeness proof is rather complicated and it is based on a notion of a packing with respect to a special position of the copies of H in G (see [LP2, LP3]).

4. Packing by Larger Families

In this section we consider packings by larger families of graphs. One nice example is packing by a sequential set of stars where $\mathcal{F} = \{S_1, S_2, \ldots, S_k\}$. This case has been studied by Hell and Kirkpatrick (1982, 1986).

4.1 Theorem *[HK1, HK4] Let $k \geq 2$. A graph G admits an $\{S_1, S_2, \ldots, S_k\}$ - factor if and only if $G - A$ has at most $k \cdot |A|$ isolated vertices for every $A \subset V(G)$. Conversly if $k \geq 2$ and $S_k \in \mathcal{F} \not\subseteq \{S_1, S_2, \ldots, S_k\}$ then the $\mathcal{F}$ - factor problem is NP - complete.*

The first part of the theorem was proved independently by Amahashi and Kano (1982). It has been observed in [HK4]that the result follows also from a theorem of Las Vergnas (1978). Theorem 4.1 is generalised by Loebl and Poljak (1987) for sequential star packings where the number of stars and their sizes are bounded.

4.2 Theorem *[LP3] The following problem can be solved in polynomial time. Let $c_1 \geq c_2 \geq \ldots \geq c_r \geq 0$ be integers. Find maximum $\{S_1, S_2, \ldots, S_{r+1}\}$ packing Q satisfying the system of inequalities*

$$\sum_{i=k}^{r}(i-k+1)f_i(Q) \leq \sum_{i=k}^{r} b_i, k = 1, \ldots, r,$$

where $f_i(Q)$ denotes the number of stars S_{i+1} in Q.

Packings by sequential sets of stars and K_2, H - packings with H hypomatchable or a propeller are good cases of the $\mathcal{F}$ - factor problem. A natural question is to find possibly largest families $\mathcal{F}$ for which $\mathcal{F}$ - factor problem is still efficiently solvable. The above good cases can be mixed together in the working notion of a *good family* (see [LP3]). A k - propeller is any graph obtained from a star S_{k+1} by replacing k its end - vertices by some hypomatchable graphs (called blades of the k - propeller). The center of the star becomes the center of the k - propeller and $(k+1)$-th-end-vertex of the star is called the root of the k - propeller. A *closed family* of propellers (of a graph G) is a collection of subgraphs of G which generalizes the notion of a sequential set of stars. For details see [LP3]. A *good family* in a graph G is a collection of some hypomatchable subgraphs of G, a closed family of propellers of G, and all copies of K_2 in G. Let $\mathcal{F}$ be a good family of subgraphs of a graph $G = (V, E)$. Let χ be a collection of mutually disjoint subsets of V, and x be a vertex not in $U\chi$. We define cap (x, χ), the capacity of x with respect to χ, as the maximum k for which there is a $(k-1)$ - propeller $H \in \mathcal{F}$ satisfying

i. H is centered at x,

ii. the root r and the blades $B_1, \ldots, B_{k-1}$ lie in k distinct sets $X_0, X_1, \ldots, X_{k-1}$

iii. $X_i - H$ has a perfect matching for each $i = 0, \ldots, k-1$.

4.3 Theorem *[LP3] Let G be a graph and $\mathcal{F}$ be a good family of subgraphs of G. Then there is a packing $Q \subset \mathcal{F}$ which saturates all vertices of G if and only if $|\chi| \leq \sum_{x \in A} cap(x, \chi)$ for every subset A of $V(G)$ and every family χ of $\mathcal{F}$ - critical components of $G - A$.*

The above good characterisation enables to solve the $\mathcal{F}$ - packing problem for a graph G and a good family $\mathcal{F}$ of its subgraphs in time polynomial in the sizes of G and $\mathcal{F}$. Again, a recognition of $\mathcal{F}$ - critical graphs is necessary (see [LP3]). A special case when the family consists only of hypomatchable graphs has been formulated in [CHP].

4.4 Theorem *[CHP] Let $\mathcal{F}$ be a family of hypomatchable subgraphs of a graph G. Then there is a packing $Q \subset \mathcal{F}$ which saturates all vertices of G if and only if the number of $\mathcal{F}$ - critical components of $G - A$ is at most $|A|$ for each $A \subset V(G)$.*

We conclude this section with the problem of classification of complexity of $\mathcal{F}$ - factor problems which is still far from being solved.

5. Fractional matching and 2 - factors

In this section we survey some results concerning packing by families consisting of cycles and K_2.

(a) $\{K_2, C_3, C_5, \ldots,\}$ - **packing.**

Let $\mathcal{F} = \{K_2, C_3, C_5, \ldots,\}$, i.e. $\mathcal{F}$ consists of K_2 and all odd cycles. This packing is usually called either *fractional matching* or *2-matching*. These terms comes from the linear program formlation

$$max \sum x_e \text{ subject to } x(\delta(v)) \leq 2, v \in V, \text{ and } x \geq 0,$$

because all basic feasible solutions of the above system are integral, and correspond to packing by copies of K_2 (edges of value 2) and odd cycles (edges of value 1), see [Ba]. The good characterization of graphs with a $\{K_2, C_3, C_5, \ldots\}$ - factor is given by the following theorem.

5.1 Theorem *A graph $G = (V, E)$ has a $\{K_2, C_3, C_5, \ldots\}$ - factor if and only if $|A|$ is a least the number of isolated vertices of $G \setminus A$ for every $A \subseteq V$.*

Let us mention that Theorem 5.1 can be regarded as a special case of Theorem 4.4, since K_1 is the only $\{K_2, C_3, C_5, \ldots\}$ - critical graph. Maximum $\{K_2, C_3, C_5, \ldots\}$ - packing can also be constructed in polynomial time [P].

Several properties of $\{K_2, C_3, C_5, \ldots\}$ - packing has been studied. We mention here two results.

5.2 Theorem *[MST] Let Q and Q' be two maximum $\{K_2, C_3, C_5, \ldots\}$ - packings such that each of them uses maximum possible number of copies of K_2. Then Q and Q' here the same number of odd cycles of each length.*

Theorem 5.2 can easily be derived from Theorem 6.1 below.

Another result which can be derived from "theory of economical packing" presented in section 6 is a theorem due to Balas [B]and Uhry [U].

5.3 Theorem *([B], [U]) Let Q be a maximum $\{K_2, C_3, C_5, \ldots\}$ - packing which uses minimum number of odd cycles. Let M be a matching obtained from Q by replacing each odd cycle by a maximum matching in it. Then M is a maximum matching of G.*

It has been observed in [LP3] that the above property of odd cycles can be extended to a broader class of graphs, and later proved in [KP] that the characterization is complete.

Let us say that a graph H has *Balas-Uhry property* if, for every graph G, and for every maximum K_2, H - packing Q using minimum number of H - copies, the matching M obtained from Q by replacing each H - copy by a maximum matching in it is a maximum matching of G.

Every odd cycle has Balas-Uhry property, and it has been proved in [CH] that also every hypomatchable graph has the property. Complete characterization is given by the following theorem.

5.4 Theorem *([LP3],[KP]) A connected graph H has Balas-Uhry property if and only if one of the following holds.*

(i) H is perfectly matchable;

(ii) for every vertex $x \in V(H)$ there is a maximum matching such that either x or a neighbour of x is the only exposed vertex.

(b) 2-matching without short cycles.

Let $\mathcal{F} = \{K_2, C_{2k+1}, C_{2k+3}, C_{2k+5}, \ldots\}, k \geq 2$. (The case $k = 1$ has been discussed in the previous paragraph.) Since $\mathcal{F}$ is a hypomatchable family, the characterization of Theorem 4.4. remains valid also for this case. However the polynomial time algorithm does not directly follow from the general case, since $\mathcal{F}$ is not a finite family. Fortunately, there is a polynomial time algorithm by [CP1] to recognize $\{K_2, C_{2k+1}, C_{2k+3}, C_{2k+5}, \ldots\}$ - critical graphs for every fixed k.

(c) $\{C_3, C_4, C_5, \ldots\}$ - packing.

Here we consider packing by a family $\mathcal{F}$ consisting of cycles of all lengths (cycles of an even length were not necessary in paragraphs (a) and (b) since they could be replaced by copies of K_2). A $\{C_3, C_4, C_5, \ldots\}$ - factor is usually called a *2-factor*.

5.5 Theorem *([T]) A graph $G = (V, E)$ admits a 2-factor if and only if*

$$|A| \geq \frac{1}{2}(|K| + \chi) \text{ for every } A \subseteq V$$

where K is the set of vertices of degree at most 1 in $G \setminus A$, and χ is the number of components of $G\setminus A$ which contain odd number of vertices from K.

A polynomial time algorithm for 2-factor problem can be derived from the algorithm for a more general problem of *perfect simple b-matching* with $b = 2$.

The algorithm for maximum $\{C_3, C_4, C_5, \ldots\}$ - packing cannot be directly obtained from the 2-factor problem. However, it is possible to reduce it to *maximum weight perfect matching* problem by gadget replacement (cf. Section 8).

A constrained 2-factor problem has been considered by Kratochvil and Poljak [KrP]. A *transition system* $T(v)$ at vertex v is a set of pairs of edges incident to v. It is convenient to view $T(v)$ as a graph. A cycle packing Q is said to be *compatible* with $T(v)$ if the edge pair incident to v in Q belongs to $T(v)$.

Compatible 2-factor problem.
Instance: A graph $G = (V, E)$ a transition systems $T(V)$ for every $v \in V$.
Question: Does G admit a 2-factor compatible with each $T(v)$?

5.6 Theorem *The compatible 2-factor problem is polynomially solvable in the following two cases.*

(i) Each transition system $T(v)$ is a complete multipartite graph;

(ii) Each transition system $T(v)$ is a subgraph of C_4, i.e. one of

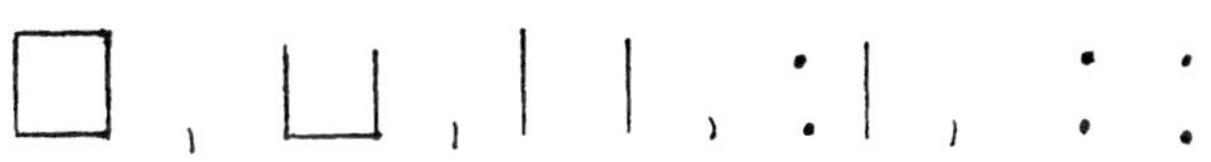

The problem is NP - complete for any other fixed family of transition systems distinct from (i) and (ii). (In particular, (i) and (ii) cannot be mixed.)

The situation slightly differs when we are interested only in maximum compatible $\{C_3, C_4, C_5, \ldots\}$ - packing. Here only the above case (i) is polynomially solvable, while case (ii) moves among the NP - complete ones.

(d) 2-factors without short cycles.

Situation becomes more complex if the cycles of short length are not allowed to be used in a 2-factor. The problem of $\{C_k, C_{k+1}, C_{k+2}, \ldots\}$ - factor is NP - complete for $k \geq 6$ by a result of Papadimitriu reported in [CP2]. It has been recently proved by Hartvigsen [H] that $\{C_4, C_5, C_6, \ldots\}$ - factor problem is polynomial. We do not know whether the result can be extended to $\{C_4, C_5, C_6, \ldots\}$ - packing. Due to a discussion from [HKKK], further possible candidates for polynomial solution are the families $\{C_5, C_6, C_7, \ldots\}$ and $\{C_3, C_5, C_6, C_7, \ldots\}$ (without C_4). The evidence is the following theorem.

Let $L \subset \{3,4,\ldots\}$ be a non-empty set, and $L^- = \{3,4,\ldots\} \setminus L$. An *L-restricted 2-factor* is a 2-factor whose cycle lengths belong to L.

5.7. Theorem *([HKKK]) The following holds if and if only if L^- is a subset of $\{3,4\}$: If $uv \in E(G)$ and if both $G \setminus u$ and $G \setminus v$ have an L-restricted 2-factor, then G has it as well.*

5.8. Theorem *([HKKK]) When $L^- \not\subset \{3,4\}$ the L - restricted 2-factor problem is NP - hard.*

The NP - completeness of the odd cycles restricted 2-factor has been proved earlier by Vornberger [V].

6. Economical Packing

The principal method of the matching theory is the augmenting paths argument. If M_1 and M_2 are two matchings, then connected components of $M_1 \cup M_2$ are paths and cycles. On the contrary, the connected components of $Q_1 \cup Q_2$ of two $\mathcal{F}$ - packing Q_1 and Q_2 can be rather complicated even for well solvable cases of the $\mathcal{F}$ - packing problem. Fortunately, the property of the matching can be extended to *economical packings* by edges and hypomatchable graphs.

Let $\mathcal{F} = \{K_2\} \cup \chi$ where χ is a family of hypomatchable graphs. We will call such a family *hypomatchable.* Every $\mathcal{F}$ - packing Q can be split into two parts $Q_e \cup Q_h = Q$ where Q_e consists of copies of K_2 and Q_h of copies of hypomatchable graphs. A packing is called *economical* if there is no packing Q' such that

(i) $V(Q') \supsetneq V(Q)$ and $Q'_h \subseteq Q_h$ or

(ii) $V(Q') = V(Q)$ and $Q'_h \subsetneq Q_h$.

This means that there is no other packing using the same copies of hypomatchable graphs, or less, and saturating the same set (or more) vertices. In particular, every maximum packing which uses minimum number of hypomatchable graphs is economical.

Given two packings Q and Q', we say that C is a *component of* $Q \cup Q'$ if C is a connected component of the graph formed by the edges of $E(Q) \cup E(Q')$. We say that a component C is of *type* (k,l), or a (k,l) - component, if C contains exactly k and l hypomatchable graphs of Q and Q', respectively. The next theorem describes all possible types of components of a union of two economical packings.

6.1. Theorem *([LP1]) Let Q and Q' be two economical $\mathcal{F}$ - packings where $\mathcal{F} = \{K_2\} \cup \mathcal{X}$ and $\mathcal{X}$ is a family of hypomatchable graphs. The only possible types of components of $Q \cup Q'$ are $(0,0), (0,1), (1,0)$, and $(1,1)$. Moreover (i) the $(0,0)$ - components are even cycles or even paths; (ii) if C is a $(1,0)$ - component, then Q covers all vertices of C, and Q' all but one; (iii) if C is a $(1,1)$ - component then both Q and Q' cover the whole C.*

Theorem 6.1 enable to give surprisingly simple proofs to several known theorems. One of them is a good characterization of $\mathcal{F}$ - critical graphs.

6.2. Theorem *([CHP], [CH]) Let $\mathcal{F}$ be a hypomatchable family. A graph G is $\mathcal{F}$ - critical if and only if G is hypomatchable, and does not admit an $\mathcal{F}$ - factor with exactly one component distinct from K_2.*

Further corollaries of Theorem 6.1 are Theorems 5.2 and 5.3 on fractional matching, and the gadget classification presented in Section 8.

7. Matroids induced by packing subgraphs

Edmons and Fulkerson (1965) proved that the maximal subsets of vertices of a graph saturated by a matching form the bases of a matroid. Some other instances of $\mathcal{F}$ - packing problem also bear this property. If e.g. $\mathcal{F}$ consists of K_2 and some hypomatchable graphs then it was proved by Cornuejols and Hartvigsen (1986). We characterized the graphs H for which K_2, H - packing problem is matroidal.

7.1. Theorem *[LP1] A connected graph H has the property that the maximal subsets of vertices of any graph, saturated by some K_2, H - packing, form the bases of a matroid if and only if H is perfectly matchable or hypomatchable or a1 - propeller. Moreover the maximal subsets of vertices of a graph saturated by an $\mathcal{F}$ - packing form the bases of a matroid in case $\mathcal{F}$ is a good family.*
Comparing this with 4.3 we present a conjecture.

Conjecture [LP1] For a finite family $\mathcal{F}$ of connected graphs the following are equivalent.

i. $\mathcal{F}$ - packing problem is polynomially solvable,

ii. The maximal subsets of vertices of any graph G saturated by an $\mathcal{F}$ - packing forms the bases of a matroid.

We remark that this equivalence does not take place when graphs of $\mathcal{F}$ may be unconnected or $\mathcal{F}$ infinite.

Finally we present a weighted version of second part of 7.1 proved by Lemos (1989).

7.2. Theorem *[Le] Let $\mathcal{F}$ be a good family of subgraphs of a graph G. Let f be a weight function on the vertices of G satisfying*

i. $f(v) + f(w) \neq U$ for each pair of vertices of G,

ii. the center of each propeller of $\mathcal{F}$ has a positive weight.

Then the subsets of vertices of G saturated by a packing Q with maximum total weight form the bases of a matroid.

8. General Factor Problem

Let G be a graph and let, for each vertex i of $V(G), B_i$ be a subset of $\{0, 1, \ldots, d_G(i)\}$. $d_G(i)$ denotes the degree of node i in G. The *general factor problem* asks to find a factor F such that $d_F(i) \in B_i$ for each i of $V(G)$. This problem was formulated by Lovász who solved important special cases (see e.g.[Lo]) and formulated a conjecture solved by Cornuejols (1988):

A set B_i is said to have a *gap of length* $p \geq 1$ if there exists an integer k of B_i such that $(k+1), \ldots, (k+p)$ does not belong to B_i and $(k+p+1)$ belongs to B_i. A set may have more gaps.

8.1 Theorem *[C] The general factor problem with the sets B_i having all gaps of length at most one is polynomially solvable.*

The general factor problem with gaps greater than one is known to be NP - complete.

Special instances of general factor problem may be reduced to K_2, K_3 - factor problem (see [C], [Lo]). A reduction is defined locally at each node. A *gadget* consists of a graph $H = (V \cup \{v_1, \ldots, v_k\}, E \cup \{e_1, \ldots, e_k\})$ with $v_i \in e_i$ and $d_G(v_i) = 1, i = 1, \ldots, k$, and of a family $\mathcal{F}$ of triangles of G.

Given a vertex i of a graph G, the gadget *represents* the general factor condition B_i if

i. $k = d_G(i)$,

ii. for $J \subseteq \{1, \ldots, k\}$, the graph $(V \cup \{v_j; j \in J\}, E \cup \{e_j; j \in J\})$ admits an $\mathcal{F}$ - factor if and only if $|J| \in B_i$.

Given a gadget representing each condition B_i, one can easily transform general factor problem to K_2, K_3 - factor problem. Pulleyblank posed a question whether there exist conditions B_i that cannot be represented by a gadget. This was solved in [L]where the classification of representable conditions is given.

8.2 Theorem *[L] A condition B may be represented by a gadget if and only if*

i. *B is an interval, i.e.* $B = \{p,\ p+1, \dots,\ p+r\}, r \geq 0,$

ii. *B is an intersection of an interval with a parity condition ,i.e.* $B = \{p,\ p+2, \dots,\ p+2r\}, r \geq 1,$

iii. $B = \{p,\ p+2,\ p+3, \dots,\ p+r\}, r \geq 3,$

iv. $B = \{p,\ p+1, \dots,\ p+r-2,\ p+r\}, r \geq 3,$

v. $B = \{p,\ p+2,\ p+3, \dots,\ p+2r-2,\ p+2r\}, r \geq 3.$

It might be possible that some further instances of general factor problem can be reduced to other efficiently solvable $\mathcal{F}$ - factor problems by constructing more general gadgets.

References

[AAE] Akiyama, J., Avis, D., Era, H.: On a {1.2} - factor of a graph, *TRU Math. 16 (1980) 97 - 102.*

[AK] Akiyama, J., Kano, M.: Factors and factorizations of graphs - A survey, *J. Graph Theory 9 (1985) 1 -42.*

[AmK] Amahashi, A., Kano, A.: On factors with given components, *Discrete Math. 42 (1982) 7 - 26.*

[B] Balas, E.: Interger and fractional matchings, Studies on graphs and discrete programming (P. Hansen ed.), *Annals of Discrete Math. 11 (1981) 1 - 13.*

[Ba] Balinski, M.L.: Integer programming: Methods uses and computation, *Management Sci. 12 (1965) 253 - 313.*

[C] Cornuejols, G.: General factors on Graphs, *J. Comb. Th. B (1988).*

[CH] Cornuejols, G., Hartvigsen, D.: An extension of matching theory, *J. Comb. Th. B 40 (1986) 285 - 296.*

[CP1] Cornuejols, G., Pulleybank, W.R.: Critical graphs matchings and tours or a hierarchy of relaxations for the traveling salesman problem, *Combinatorica 3 (1983) 35 - 52.*

[CP2] Cornuejols, G., Pulleybank, W.R.: A matching problem with side conditions, *Discrete Math. 29 (1980) 135 - 159.*

[CHP] Cornuejols, G., Hartvigsen, D., Pulleybank, W.R.: Packing subgraphs in a graph, *Oper. Res. Letters 1 (1982) 139 - 143.*

[E] Edmonds, J.: Path, trees and flowers, *Canadian J. Math. 17 (1965) 449 - 467.*

[EF] Edmonds, J., Fulkerson, D.R.: Transversals and matroid partition, *J. Res. Nat. Bur. Standards 69B (1965) 147 - 153.*

[EJ] Edmonds, J., Johnson, E.: Matching, Euler tours, and the Chinese postman, *Math. Programming 5 (1973) 88 - 124.*

[H] Hartvigsen, D.: Extensions of Matching Theory, *Ph. D. Thesis, Carnegie - Mellon University 1984.*

[HK1] Hell, P., Kirkpatrick, D.: Star factors and star packings, *TR 82-6, Dept. Comp. Sci., Simon Fraser University 1982.*

[HK2] Hell, P., Kirkpatrick, D.: On the complexity of general graph factor problem, *SIAM J. Computing (1983) 601 -609.*

[HK3] Hell, P., Kirkpatrick, D.: Packing by cliques and finite families of graphs, *Discrete Math. 49 (1984) 118 - 133.*

[HK4] Hell, P., Kirkpatrick, D.: Packings by complete bipartite graphs, *SIAM J. Alg. Disc. Math. 7 (1986) 113 - 129.*

[HKKK] Hell, P., Kirkpatrick, D., Kratochvil, J., Kříž, I.: On restricted two - factors, *SIAM J. Discrete Math.*

[KP] Kano, M., Poljak, S.: Graphs with Balas - Uhry property, *KAM Series 89-137, Charles University 1989.*

[KrP] Kratochvil, J., Poljak, S.: Compatible 2 - factors, *KAM Series 88 - 88, Charles University 1988.*

[La] Las Vergnas, M.: An extension of Tutte's 1 - factor Theorem, *Discrete Math. 23 (1978) 241 - 255.*

[Le] Lemos, M.: Matroids induced by packing weighted subgraphs, *Universidade Federal de Pernambuco 1989.*

[LP1] Loebl, M., Poljak, S.: On matroids induced by packing subgraphs, *J. Comb. Th. B 44 (1988) 338 - 354.*

[LP2] Loebl, M., Poljak, S.: Bipartite packing, *"Finite and Infinite Sets" (V. T. Sos ed.), Proc. of the 7th Hung. Coll. on Combinatorics, Eger 1987, pp. 375 - 384.*

[LP3] Loebl, M., Poljak, S.: Efficient subgraph packing, *KAM Series 87 -50, Charles University 1987.*

[L] Loebl, M.: Gadget classification,, *KAM Series 89 - 138, Charles University 1987.*

[Lo] Lovasz, L.: Antifactors of graphs, *Periodica Math. Hung. 4 (1973) 121 - 123.*

[LoP] Lovasz, L., Plummer, M.: Matching Theor, *Academiai Kiado/ North - Holland 1986.*

[MST] Mühlbacher, J.R., Steinparz, F.X., Tinhofer, G.: On certain classes of fractional matchings, *Discrete Applied Math. 9 (1984) 235 - 244.*

[P] Pulleybank, W.R.: Dual integrality in b - matching problems, *Math. Programming Study 12 (1980) 176 - 196.*

[S] Sebõ, A.: Multi - antifactors of graphs, *Technical Report, Eötvös Lorand University, Budapest 1986.*

[T] Tutte, W.T.: The factors of graphs, *Canad. J. Math. 4 (1952) 314 - 329.*

[V] Vornberger, O.: Complexity of path problems in graphs, *Ph.D. Thesis, Universität GH - Paderborn 1979.*

[U] Uhry, J.P.: Sur le Problème du Couplage Maximal, *R.A.I.R.O. 9 (1975) V-3, 13 - 20.*

On the Radius of Random Subgraphs of the n-Cube

K. Mahrhold, Warnemünde, GDR
K. Weber, Warnemünde, GDR

Abstract: Dyer, Frieze and Foulds (cf. [1]) introduced a general model of random subgraphs G_n of the n-cube graph Q_n. We determine the radius $R(G_n)$ of these random subgraphs G_n of Q_n for fixed probabilities p_v, p_e satisfying $1/2 \leq p_v p_e < 1$: With probability tending to one as n tends to infinity we have

$$R(G_n) = n - 1 \quad \text{if } 1/2 < p_v < 1,$$
$$R(G_n) = n \quad \text{if } p_v = 1 \text{ and } p_e > 1/2,$$
$$n - 1 \leq R(G_n) \leq n \quad \text{if } p_v = 1 \text{ and } p_e = 1/2 \text{ and}$$
$$n - 2 \leq R(G_n) \leq n - 1 \quad \text{if } p_v = 1/2 \text{ and } p_e = 1.$$

1. Introduction

The n-cube Q_n is the graph consisting of the 2^n vertices $x = (x_1, x_2, \dots, x_n)$, $x_i \in \{0,1\}$, and the $n\,2^{n-1}$ edges between vertices differing in exactly one coordinate. Let $G_n = (V_n, E_n)$ be the random subgraph of Q_n produced as follows: V_n is randomly sampled from the vertex set of Q_n so that $P(x \in V_n) = p_v$ independently for each vertex $x \in Q_n$. E_n is now randomly sampled from the set of edges induced by V_n in Q_n so that $P(xy \in E_n) = p_e$ independently for each induced edge xy (cf. [1]). Dyer et al. determined the threshold probabilities for s-connectivity of G_n, s = 1,2, ... In [8] we describe the component structure of G_n for fixed $p = p_v p_e$. In [2] bounds are given for other graph theoretic parameters of G_n as minimum and maximum degree, independence number and related parameters, thickness and genus. In the sequel we are concerned with the radius $R(G_n)$ of the random graphs G_n for fixed probabilities p_v, p_e with $1/2 \leq p_v p_e < 1$. The following theorem is the main result of this paper. (If G_n is not connected then $R(G_n)$ means the radius of the largest component of G_n (cf. Theorem 2).)

Theorem 1. Put $p = p_v p_e$, and let p_v and p_e be fixed.

(i) Let $p = 1/2$. Then with probability tending to one as $n \to \infty$

$$n-2 \leq R(G_n) \leq n.$$

R. Bodendiek, R. Henn (Eds.)
Topics in Combinatorics and Graph Theory

Moreover,

$$n-2 \leqq R(G_n) \leqq n-1 \qquad \text{if } p_e = 1, \tag{1}$$

$$n-1 \leqq R(G_n) \leqq n \qquad \text{if } p_v = 1, \tag{2}$$

$$R(G_n) = n-1 \qquad \text{if } 1/2 < p_v, p_e < 1. \tag{3}$$

(ii) Let $p > 1/2$. Then with probability tending to one as $n \to \infty$

$$n-1 \leqq R(G_n) \leqq n.$$

Moreover,

$$R(G_n) = n \qquad \text{if } p_v = 1, \tag{4}$$

$$R(G_n) = n-1 \qquad \text{if } p_v < 1. \tag{5}$$

Inequality (1) was shown by Saposhenko ([3], [4]). The bounds (2) and (4) are contained in [7]. (Actually Toman proved the upper bound (10) (cf. Section 2) for $p_e = 1$ ([5]) and for $p_v = 1$, $p_e = 1/2$ ([6]). But the result (4) requires (10) for $p_v = 1$, $p_e > 1/2$ too.)

2. Notation and preliminaries

All limits asymptotics, etc., are considered as $n \to \infty$. For two sequences $\alpha = \alpha(n)$ and $\beta = \beta(n)$ we write $\alpha \sim \beta$ if $\alpha/\beta \longrightarrow 1$, $\alpha \precsim \beta$ if $\alpha = O(\beta)$ and $\alpha \asymp \beta$ if $\alpha \precsim \beta$ and $\beta \precsim \alpha$. Everywhere $\omega = \omega(n)$ denotes a sequence tending to infinity arbitrarily slowly. By log we denote the logarithm to base 2. We say that a property of G_n holds almost surely (a.s.) if this property holds with probability tending to one. In this paper the random variables are defined on random graphs G_n. They are non-negative and integer-valued. The expectation and the variance of the random variable X (for convenience we omit the index n) are denoted by EX and D^2X, respectively. The inequality of Markov implies

$$P(X < \omega EX) \geqq 1 - 1/\omega \to 1, \tag{6}$$

and in particular

$$P(X = 0) \longrightarrow 1 \quad \text{if} \quad EX \longrightarrow 0 \tag{7}$$

since X is non-negative and integer-valued. From the inequality of Chebyshev we conclude that there is an $\varepsilon = \varepsilon(n) \to 0$ such that

$$P(|X-EX| < \varepsilon EX) \longrightarrow 1 \quad \text{if} \quad D^2X = o((EX)^2). \tag{8}$$

In the latter case we will use the short notation $X \sim EX$ a.s. For a graph G with the vertex set V the order $|G|$ of G is the number $|V|$ of vertices of G. Instead of $x \in V$ we write also $x \in G$.

Note that the order $|G_n|$ of the random graph G_n is binomially distributed with parameters 2^n and p_v such that $|G_n| \sim p_v 2^n = E|G_n|$ a.s. An x,y-path connects the vertices x and y, its length is the number of edges in it. The distance r(x,y) between the vertices $x,y \in Q_n$ is the customary Hamming distance, i.e. the number of coordinates in which x and y differ or, in other words, the minimal number of edges in an x,y-path in Q_n. For $x, y \in G_n$ we denote by $\varrho(x,y)$ the distance between x and y in the graph G_n, i.e. the minimal length of an x,y-path in G_n. Put $\varrho(x,y) = \infty$ if x and y belong to different components of G_n. For a vertex set $U \subseteq V_n$ and a vertex $x \in V_n$ define $\varrho(x,U)$ as the minimum over all vertices $y \in U$ of the distances $\varrho(x,y)$. The radius $R(G_n)$ is the minimum over all vertices $x \in G_n$ of the maximum distance from that vertex x to any of the others in G_n. The diameter $D(G_n)$ is the maximum distance between two vertices in G_n. The Hamming sphere $S_t(x)$ of radius t centered at x is defined as the vertex set $\{ z \in Q_n \mid r(x,z) \leq t \}$. For $x \in Q_n$ denote by $\bar{x}$ the opposite vertex in Q_n, i.e. the vertex with $r(x,\bar{x}) = n$. In all what follows the probabilities p_v and p_e are fixed and $p = p_v p_e$.

Following [4] and [5] we proved the following two lemmata.

<u>Lemma 1.</u> ([8]) Denote by $\Psi(n)$ the conditional probability that the random subgraph G_n does not contain a path of length n joining the vertices $\underline{0} = (0, \ldots, 0)$ and $\underline{1} = (1, \ldots, 1)$ of Q_n provided that $\underline{0}$ and $\underline{1}$ are vertices of G_n. Then for every fixed probability $p = p_v p_e$, $0 < p < 1$, we have $\Psi(n) \asymp (1-p)^n$, i.e., in particular, there exists a positive constant Ψ_p such that

$$\Psi(n) \leq \Psi_p (1-p)^n \tag{9}$$

for all n = 1,2, ...

<u>Lemma 2.</u> ([8]) Denote by $\varphi(n,k)$ the conditional probability that for a fixed pair of vertices $x,y \in G_n$, $r(x,y) = k$, $k = 1,2, \ldots, n$, the inequality $\varrho(x,y) > k + 4$ is fulfilled (in particular, this inequality is fulfilled if x and y belong to different components of G_n). Then for every fixed probability $p = p_v p_e$, $0 < p < 1$, there exists a positive constant φ_p such that

$$\varphi(n,k) \leq \varphi_p (1-p)^n \tag{10}$$

for all n = 1,2, ... and k = 1,2, ..., n.
Let the random variable B count the number of pairs (x,y) of vertices $x,y \in G_n$, $x \neq y$, for which the inequality $\varrho(x,y) > r(x,y) + 4$ is fulfilled. Then for the expectation of B we have

$$EB \leq 2^n \sum_{k=1}^{n} \binom{n}{k} \varphi(n,k)$$

and therefore by (10)

$$EB \leq \varphi_p \left[4(1-p)\right]^n. \qquad (11)$$

Moreover, we use the following connectivity properties of random graphs G_n which are immediate from [1]. (These properties are also immediate consequences of (11), cf. [8].)

Theorem 2. ([1], [8]) For fixed $p = p_v p_e > 1/2$ the random graph G_n is a.s. connected. If p = 1/2 then G_n has a.s. only isolated vertices outside the largest component; the number of isolated vertices is Poisson distributed with mean $\lambda = p_v$, i.e., P(G_n is connected) $\sim$ P(G_n has no isolated vertex) $\sim e^{-p_v}$. (Let us appoint that $R(G_n)$ means the radius of the main component of G_n.)

3. Proof of Theorem 1

We start by showing that $R(G_n) \leq n-1$ a.s. if $p \geq 1/2$ and $p_v < 1$. Recall that $|G_n| \sim p_v \; 2^n$ a.s.
Denote by $A_1 = A_1(G_n)$ the set of vertices $x \in G_n$ for which there is a vertex $y \in G_n$ at distance $k \geq 0.8\,n$ with $\varrho(x,y) \neq r(x,y) = k$ (i.e., there is no x,y-path of length k in G_n). Now the expectation of the cardinality $|A_1|$ is upperestimated using Lemma 1:

$$E|A_1| \leq p_v \; 2^n \sum_{k \geq 0.8n} p_v \binom{n}{k} \Psi(k) \lesssim 2^n \sum_{k \geq 0.8n} \binom{n}{k} (1-p)^k$$

$$\leq 2^n \sum_{k \geq 0.8n} \binom{n}{k} 2^{-k}$$

$$\leq 2^{n-0.8n} \sum_{k \geq 0.8n} \binom{n}{k} \leq 2^{0.2n + H(0.8)} = o(2^n),$$

where $H(u) = -u \log u - (1-u) \log (1-u)$ is, as usually, the entropy function, and H(0.8) = 0.721 ... . Denote by $A_2 = A_2(G_n)$ the set of those vertices $x \in G_n$ for which there exist more than log n vertices $y \in G_n$ such that $\varrho(x,y) > r(x,y) + 4$ is fulfilled.

Then by (11) and (6) we conclude $|A_2| \log n \leq B \lessapprox \omega\, 2^n$ and consequently $|A_2| \leq 2^n/\log\log n$ a.s.

Finally denote by $A_3 = A_3(G_n)$ the set of vertices $x \in G_n$ with $\bar{x} \notin G_n$. Then $E|A_3| = p_v(1-p_v)2^n$, and $|A_3| \sim E|A_3|$ a.s. is easy to show by estimating the variance of $|A_3|$ and using (8).

Putting $A = A(G_n) = (V_n \cap A_3)\setminus(A_1 \cup A_2)$ we obtain $|A| \sim p_v(1-p_v)2^n$ a.s. But for any $x \in A$ we have $\varrho(x,y) \leq n-1$ for all vertices $y \in G_n$ (belonging to the main component of G_n if $p = 1/2$, cf. Theorem 2). Indeed if $r(x,y) \geq 0.8n$ then $\varrho(x,y) = r(x,y) \leq n-1$ since $x \in A_3 \setminus A_1$. If $r(x,y) < 0.8n$ then let $A_x = A_x(G_n)$ be the set of vertices $z \in G_n$ for which $\varrho(x,z) > r(x,z) + 4$ is fulfilled. Since $x \notin A_2$ we know that $|A_x| \leq \log n$. If $y \in V_n \setminus A_x$ then $\varrho(x,y) \leq r(x,y) + 4 < 0.8n + 4 \leq n-1$ for sufficiently large n. If $y \in A_x$ then (provided that y belongs to the main component of G_n) there is a vertex $x' \in V_n \setminus A_x$ with $\varrho(x',y) = \varrho(V_n \setminus A_x, y)$. Furthermore we have $\varrho(x',y) \leq \log n$ since all inner vertices of a shortest y,x'-path in G_n belong to A_x too. Thus $\varrho(x,y) \leq \varrho(x,x') + \varrho(x',y) \leq r(x,x') + 4 + \varrho(x',y) \leq r(x,y) + r(y,x') + 4 + \varrho(x',y) \leq r(x,y) + 2\varrho(x',y) + 4 \leq 0{,}8n + 2\log n + 4 \leq n-1$ for sufficiently large n.

On the other hand we have $R(G_n) \geq n-1$ a.s. if $p_v > 1/2$ since a.s. there is no $S_1(z)$ without vertices of the main component of G_n. Indeed by Theorem 2 outside the main component G_n contains a.s. less than ω isolated vertices, where $\omega = \omega(n) \to \infty$ arbitrarily slowly. Thus the conditional expected numer of spheres $S_1(z)$ without vertices of the main component is not greater than

$$2^n \sum_{i=0}^{\omega} \binom{n+1}{i} p_v^{\,i}(1-p_v)^{n+1-i} \leq 2^n (n+1)^{\omega}\ (1-p_v)^{n-\omega}\ .$$

For $p_v > 1/2$ the last expression tends to zero. (Actually, for $p > 1/2$ we get only one summand $2^n(1-p_v)^{n+1}$ as this conditional expectation, cf. Theorem 2.)

For $p_v = p = 1/2$, $p_e = 1$ we derive $R(G_n) \geq n-2$ a.s. analogously. Letting $a = \binom{n}{2} + n+1 = |S_2(z)|$ the conditional expected number of spheres $S_2(z)$ without vertices of the main component is not greater than $2^n \sum_{i=0} \binom{a}{i} p_v^{\,i}(1-p_v)^{a-i} \leq 2^{n-a}\, a^{\omega} \leq 2^{n-\frac{n^2}{2}} \cdot (n^2)^{\omega} \to 0$.

Thus with probability tending to one for all vertices $x \in G_n$ the spheres $S_2(\bar{x})$ contain a vertex of the main component of G_n.

Moreover, by the standard technique (cf. e.g. [9]) one can show that for $p_v = p = 1/2$, $p_e = 1$ the number of Hamming spheres $S_1(z)$ with $S_1(z) \cap V_n = \emptyset$ or $S_1(z) \cap V_n = \{z\}$ is Poisson distributed with mean $\lambda = 1$. Thus $P(R(G_n) = n-1) \gtrsim e^{-1}$ in this case.

Finally it remains open the case $p_v = 1$, i.e. $|V_n| = 2^n$, and G_n is a spanning subgraph of Q_n. That $R(G_n) \leq n$ a.s. in this case may be derived as the upper bound $R(G_n) \leq n-1$ for $p_v < 1$. The only difference is that we do not need the set $A_3(G_n)$ (now we have $\bar{x} \in G_n$ for all $x \in G_n$). For $p > 1/2$ G_n is a.s. connected (cf. Theorem 2) and thus $R(G_n) \geq n$ and we obtained (4). (A simpler proof of (4) is given in [7]: Fix a vertex $x \in G_n$. Then by (10) the expectation of the number of vertices $y \in G_n$ satisfying $\varrho(x,y) > r(x,y) + 4$ is at most

$$\sum_{k=1}^{n} \binom{n}{k} \varphi(n,k) \leq \varphi_p [2(1-p)]^n$$

which tends to zero for $p > 1/2$. Hence $\varrho(x,y) \leq r(x,y) + 4 \leq n$ if $r(x,y) \leq n-4$. But there are a.s. no vertices $y \in G_n$ with $n-3 \leq r(x,y) \leq n$ such that there is no x,y-path of length $r(x,y)$ in G_n since by (9) the expectation of the number of these vertices is at most

$$\sum_{k=n-3}^{n} \binom{n}{k} \Psi(k) \leq \Psi_p \sum_{k=n-3}^{n} \binom{n}{k} (1-p)^k \lesssim n^3 (1-p)^{n-3} \longrightarrow 0$$

if $p > 1/2$.

Let $p = p_e = 1/2$, $p_v = 1$. Then by Theorem 2 outside the main component G_n contains only isolated vertices. But the probability that there is a sphere $S_1(x)$, where all $z \in S_1(x)$ are isolated vertices of G_n, is not greater than $2^n(1-p)^{n(n-1)+n}$ $= 2^{-n^2+n} \longrightarrow 0$. Thus a.s. every $S_1(x)$ contains a vertex of the main component of G_n, and consequently $R(G_n) \geq n-1$ a.s. and (2) is shown. Since for $p = p_e = 1/2$, $p_v = 1$ the number of isolated vertices of G_n is Poisson distributed with mean $\lambda = 1$ (cf. [9]) we may conclude that $P(R(G_n) = n) \gtrsim e^{-1}$.

Remark: For fixed $p > 3/4$ by (11) we get $EB \to 0$ and so by (7) $B = 0$ a.s., i.e. $\varrho(x,y) \leq r(x,y) + 4$ for all vertices $x,y \in G_n$. If $n-3 \leq r(x,y) \leq n$ then even $\varrho(x,y) = r(x,y)$ since

$$2^n \sum_{k=n-3}^{n} \binom{n}{k} \Psi(k) \lesssim 2^n n^3 (1-p)^{n-3} < 2^n n^3 \left(\frac{1}{4}\right)^{n-3} \longrightarrow 0$$

(cf.(9)). Thus if $p > 3/4$ is fixed for the diameter we have

$$D(G_n) = n \quad \text{a.s.}$$

Finally let us mention that $D(G_n) = n+1$ a.s. if $p = p_v = 1/2$ and $p_e = 1$ was shown by Saposhenko (cf. [3], [4]).

References

[1] M. Dyer, A. Frieze, L. Foulds. On the strength of connectivity of random subgraphs of the n-cube. Annals of Discr. Math. 33 (1987) 17-40

[2] U. Konieczna, K. Weber. Graph theoretic parameters of random subgraphs of the n-cube. Dresdner Reihe zur Forschung 9/88 (Proceedings "Graphentheorie und ihre Anwendungen", Stadt Wehlen, 16.-20.5.88) 25-28

[3] A.A. Saposhenko. Metric properties of almost all Boolean functions. Diskretny Analiz 10, 91-119 (1967) (in Russian)

[4] -. Geometric structure of almost all Boolean functions. Problemy Kibernetiki 30, 227-261 (1975) (in Russian)

[5] E. Toman. Geometric structure of random Boolean functions. Problemy Kibernetiki 35, 111-132 (1979) (in Russian)

[6] -. On the probability of connectedness of random subgraphs of the n-cube. Math. Slovaca 30 (1980) 3, 252-265 (in Russian)

[7] -. Estimation of the radius of a random subgraph of the n-cube. Acta Math. Univ. Comen. 40/41 (1982) 87-92

[8] K. Weber. On components of random subgraphs of the n-cube. (submitted)

[9] -. Poisson convergence in the n-cube. Math. Nachr. 131 (1987) 49-57

A Result in Combinatorial Matroid Theory

D. Marcu, Bucharest, Romania

Defining the orthogonality in a matroid, we show that if an element is orthogonal to a set A, then it is also orthogonal to the span $\bar{A}$ of A .

The matroid theoretic terminology, used in this note, is according to the standard literature (see, e.g., [2,3,4]). Let E be a finite set and M := M(E,r) a <u>matroid</u> on E, with r as the <u>rank function</u> ($r : 2^E \longrightarrow \mathbb{N}$, where 2^E is the power set of E and $\mathbb{N}$ the set of nonnegative integers). A subset $S \subseteq E$ is called <u>independent</u>, if $r(S) = |S|$, where $|S|$ denotes the cardinality of S . Let F(M) be the family of independent sets of M . A <u>basis</u> of M is a maximal set in F(M). A subset of E, which is not independent, is called <u>dependent</u> , and a minimal dependent subset of E is a <u>circuit</u> .

We shall denote by C(M) the family of circuits of M . If B is any basis of M and $f \in E-B$, then there exists an unique $C \in C(M)$ containing f and otherwise only elements of B. This circuit C is called <u>fundamental</u> with respect to f and B and will be denoted by C(f,B).

For every $S \subseteq E$, $\bar{S}$ denotes the <u>span</u> of S, that is,

$$\bar{S} = \{e \in E : r(S \cup \{e\}) = r(S)\} .$$

Let B be a fixed basis of M and $e \in E$, arbitrarily chosen. Obviously, $e \in \bar{B}$. Then, we may consider the family of sets

$$D(e,B) = \{S \subseteq B : e \in \bar{S}\}.$$

We shall denote by B(e) a minimal (with respect to the inclusion of sets) element of D(e,B) . One can easily check that

$$\text{if } r(\{e\}) = 0, \text{ then } B(e) = \emptyset,$$

R. Bodendiek, R. Henn (Eds.)
Topics in Combinatorics and Graph Theory

$$\text{if } e \in B, \text{ then } B(e) = \{e\}.$$

Suppose that $e \notin B$ and let $C(e,B)$ be the fundamental circuit with respect to e and B. Obviously, by definition, $B(e) \cup \{e\}$ is a circuit of M contained in $B \cup \{e\}$ and, therefore, $B(e) \cup \{e\} = C(e,B)$. Thus, $B(e)$ is uniquely defined.

Let B be a fixed basis of M and $A \subseteq E$. We shall consider the set $O(B,A)$, defined by

$$O(B,A) = \{e \in E : B(e) \cap B(a) = \emptyset, \text{ for all } a \in A\}.$$

We say that an element $e \in E$ is <u>orthogonal to A with respect to B</u>, if $e \in O(B,A)$.

<u>LEMMA 1.</u> Let M be a matroid, B a basis of M and $b \in B$. Then, $B' = (B - \{b\}) \cup \{a\}$ is a basis of M if and only if $b \in B(a)$.

<u>Proof.</u> If $a \in B$, then $B' = B$ and $B(a) = \{b\}$. If $a \notin B$, then $B'(b) \subseteq B'$ and $B'(b) \cup \{b\} \notin F(M)$. Thus, $a \in B'(b)$ and $B(a) = (B'(b) \cup \{b\}) - \{a\}$, by uniqueness.

Conversely, if $a \in B$, then $b = a$ and so, $B' = B$. If $a \notin B$ and B' is dependent, then there exists $A \subseteq B'$, such that A is minimal dependent. Since B is independent, then $a \in A \subseteq (B - \{b\}) \cup \{a\}$, so that $B(a) = A - \{a\}$ and $b \notin B(a)$, that is, a contradiction. Thus, B' is independent, $|B| = |B'|$ and, therefore, B' is a basis of M.

<u>LEMMA 2.</u> Let M be a matroid, B a basis of M, $a \in E-B$, $b \in B(a)$ and $B' = (B - \{b\}) \cup \{a\}$. Then, for every $e \in E$, we have

$$B'(e) = B(e), \text{ if } b \notin B(e), \tag{1}$$

$$B'(b) = (B(a) \cup \{a\}) - \{b\}, \tag{2}$$

$$B'(a) = \{a\}, \tag{3}$$

$B(a) + B(e) \subseteq B'(e) \subseteq B(a) \cup B(e) \cup \{a\}$, if $b \in B(e)$, $b \neq e$, where $X + Y$ denotes the symmetric difference of X and Y. (4)

Proof. B' is a basis of M, by Lemma 1 .

(1) If $b \notin B(e)$, then $B(e) \subseteq B'$ and, hence, $B'(e) = B(e)$.

(2) The proof of Lemma 1 shows that $B'(b) = (B(a) \cup \{a\}) - \{b\}$.

(3) By definition.

(4) Right hand side. The minimal dependent sets form a "circuit" matroid (see [1, Proposition 2]). Since $b \in B(e) \cup \{e\}$ and $b \in B(a) \cup \{a\}$, then there exists a minimal dependent set C, with $e \in C \subseteq B(a) \cup B(e) \cup \{a,e\} - \{b\} \subseteq B' \cup \{e\}$. Hence, $B'(e) = C - \{e\} \subseteq B(a) \cup B(e) \cup \{a\}$. Also, $a \in B'(e)$, by Lemma 1, since $B = (B' - \{a\}) \cup \{b\}$.

Left hand side. For $z \in B(e) - B(a)$, $B'' = (B - \{z\}) \cup \{e\}$ is a basis, by Lemma 1 and, since $z \notin B(a)$, $B(a) = B''(a)$, by (1) . On the other hand, $B'(e) \cup \{e\}$ is a circuit containing a, and if $z \notin B'(e)$, then $B'(e) \cup \{e\} \subseteq B'' \cup \{a\}$, giving $b \in B(a) = B''(a) = B'(e) \cup \{e\} - \{a\}$, that is, a contradiction. For $z \in B(a) - B(e)$, $B'' = (B - \{z\}) \cup \{a\}$ is a basis, by Lemma 1 and, since $z \notin B(e)$, $b \in B''(e) = B(e)$, by (1) . If $z \notin B'(e)$, then $B'(e) \subseteq (B(e) \cup B(a) \cup \{a\}) - \{z\} \subseteq B''$. Since $B'(e) \cup \{e\}$ is a circuit in $B'' \cup \{e\}$, we have $b \in B'(e) = B''(e)$, that is, a contradiction . Thus, $B(e) + B(a) \subseteq B'(e)$.

THEOREM. Let M be a matroid, B a basis of M, $A \subseteq E$ and $e \in O(B,A)$. Then, $e \in O(B,\bar{A})$.

Proof. If $B(e) = \emptyset$, then there is nothing to prove. If $f \in \bar{A}$ with $B(e) \cap B(f) \neq \emptyset$, then $f \notin A$ and there exists a circuit C, such that $f \in C = A' \cup \{f\}$, with $A' \subseteq A$ and $A' \in F(M)$ .Then, $e \in O(B,A')$ and $e \notin O(B,\bar{A'})$ and $e \in O(B,A)$. So, it is sufficient to prove the theorem for sets $A \in F(M)$.

If the theorem is not true, then there exists a triple (e,A,B), where $e \in E$, B is a basis of M, $A \in F(M)$, $0 \leq |A - B|$, $e \in O(B,A)$ and $e \notin O(B,\bar{A})$. Suppose that (e,A,B) is such a triple. If $A \subseteq B$, then

$$\bigcup_{a \in A} B(a) = A$$

and, for $f \in \bar{A} - A$, there exists a circuit C with $f \in C = A' \cup \{f\}$ and $A' \subseteq A$. Hence, $B(f) = A' \subseteq A$ and $B(e) \cap B(f) = \emptyset$, that is, a contradiction. Thus, for such triple, we have $0 < |A - B|$ and we may choose one with $|A - B|$ least.

Take $a_0 \in A - B$. Since $\{a_0\} \in F(M)$, then there exists $b \in B(a_0)$ and, by hypothesis, $b \notin B(e)$. By Lemma 1, $B' = (B - \{b\}) \cup \{a_0\}$ is a basis of M and, by (1), $B(e) = B'(e)$. Now, $a_0 \notin B'(e)$, for otherwise, by (4), with B' and B interchanged, $B'(a_0) + B'(e) = B(e) = B'(e)$, giving $B'(a_0) = \{a_0\} = \emptyset$, that is, a contradiction. For the remaining $a \in A$, again, using (4), $B'(a) \subseteq B(a) \cup B(a_0) \cup \{a\}$, so that $B'(e) \cap B'(a) = \emptyset$ for all $a \in A$. Thus, $e \in O(B',A)$ and, by the minimality of $|A - B|$, $e \in O(B',\bar{A})$.

Now, take $f \in \bar{A}$. If $b \notin B(f)$, then, by (1), $B'(e) \cap B'(f) = B(e) \cap B(f) = \emptyset$. If $b \in B(f)$, then, by (4), we have $B(a_0) + B(f) \subseteq B'(f)$ and $B'(e) \cap B'(f) = B(e) \cap (B(a_0) + B(f)) = B(e) \cap B(f) = \emptyset$. In all cases, $B(e) \cap B(f) = \emptyset$ and so, $e \in O(B,\bar{A})$, that is, a contradiction to the existence of a triple (e,A,B).

Thus, the theorem is proved.

REFERENCES

1. J. Edmonds, Minimum partition of a matroid into independent subsets, J. Res. Nat. Bur. Stand., 69B (1965), 67-72 .
2. R. von Randow, Introduction to the Theory of Matroids, Lecture Notes in Economics and Mathematical Systems, Springer-Verlag, 1975.
3. W. T. Tutte, Lectures on matroids, J. Res. Nat. Bur. Stand., 69B (1965), 1-47 .
4. D. J. A. Welsh, Matroid Theory, Academic Press, 1976.

On Graphs Embeddable With Short Faces

R. Nedela, Banská Bystrica, CSSR
M. Skoviera, Bratislava, CSSR

Introduction and Preliminaries

Genus embeddings of graphs whose genus is known are, for the most part, either triangular or quadrilateral, or differ from these in several exceptional faces. In order to determine all orientable surfaces in which such graphs have 2-cell embeddings it is sufficient to compute their maximum genus. So far, it has been known that every graph which triangulates some closed surface is upper embeddable, that is, its maximum genus is equal to $\lfloor \beta/2 \rfloor$, where β is its Betti number (Khomenko and Glukhov [2], Nebeský [3]). The main objective of the present paper is to generalize this result. We shall show that every graph admitting a 2-cell embedding in in some closed surface so that the length of every face does not exceed four is upper embeddable.

Besides the fact that this result determines the maximum genus of most graphs with known genus, it has an interesting corollary that every 2-cell embedding of a loopless non-upper embeddable graph has a face with length at least five.

Before proceeding to our main result we recall some definitions and known facts. By a *surface* we mean a closed surface, possibly non-orienta-

R. Bodendiek, R. Henn (Eds.)
Topics in Combinatorics and Graph Theory

ble. An embedding $i:G \longrightarrow S$ of a graph G on a surface S is always assumed to be cellular. In this paper we shall deal with maximum-genus embeddings, that is, with embeddings in orientable surfaces having the minimum number of faces. The *maximum genus* of a connected graph G, $\gamma_M(G)$, is the maximum among the genera of all orientable surfaces in which G has 2-cell embeddings. It is well known [4] that for every connected graph G it holds that $\gamma_M(G) \le \lfloor \beta(G)/2 \rfloor$, where $\beta(G) = |E(G)| - |V(G)| + 1$ is the Betti number of G. Moreover, if the equality holds, G is called *upper embeddable* and its maximum genus embedding has one or two faces according as $\beta(G)$ is even or odd, respectively. The following characterization of upper embeddable graphs is due, independently, to Jungerman [1] and Xuong [6]. Let T be a spanning tree of a graph G. A connected component of the cotree G-E(T) is said to be *even* (*odd*) if it has even (odd) number of edges. By [1, 6], a connected graph G is upper embeddable if and only if it admits a spanning tree T such that G-E(T) has at most one odd component. Such a spanning tree will be referred to as *optimal*. Note that the number of odd components in the cotree G-E(T) has always the same parity as $\beta(G)$.

Using the above Jungerman - Xuong characterization, the maximum genus of many graphs has been determined. In particular, it has been shown by Payan and Xuong [5] that every cyclically 4-edge connected graph is upper embeddable. This fact will be useful in the proof of our main result.

Main Result and Proof

Let i: $G \longrightarrow S$ be a 2-cell embedding of a connected graph G in a surface S, orientable or not. A face of the embedding i will be called *short* if its boundary has length at most four (repeated edges are counted twice). The boundary ∂F of a short face F need not be a cycle in G, and if it is not, F will be called *degenerated*. If each face of a 2-cell embedding is short (possibly degenerated) then the embedding itself is said to be a *short-face embedding*.

To encode a face boundary we use the cyclic sequence of edges in the order as they appear when traversing the boundary in some direction, e.g. $\partial F = (xyzt)$. The orientations of edges induced by the direction of traversing will be suppressed as they are irrelevant for our purposes.

Assume that $i\colon G \longrightarrow S$ is a short face embedding and e is an edge of G. If e is replaced on S by a bigon with boundary (ee'), where e' is a new edge parallel to e, then a short face embedding of $G+e'$ is obtained. Conversely, if $G+e'$has a short-face embedding, with a bigonal face (ee'), then so has G. This simple observation will be crucial in the proof of the following theorem which is the main result of this paper:

Theorem. *Let G be a connected loopless graph which has a short-face embedding in some closed surface. Then G is upper embeddable.*

Proof. Let $i\colon G \longrightarrow S$ be a short-face embedding of G in a surface S. In order to prove that G is upper embeddable we employ induction on the number of edges of G.

It is easily checked that the statement of the theorem holds for graphs with small number of edges. To accomplish the induction step, consider a graph G satisfying the conditions of the theorem but not covered by the induction basis. If G is cyclically 4-edge connected then by [5] it is upper embeddable. Thus we may suppose that G is not cyclically 4-edge connected. We shall distinguish two cases (and a number of subcases) according as G does or does not contain a cut-vertex.

Case 1. G contains a cut-vertex. Let v be any cut-vertex of G. The embedding i necessarily contains a face F the boundary of which includes edges belonging to different blocks at v. Since G has no loops, F has length and its boundary ∂F has the form $(x_1y_1x_2y_2)$ where both (x_1y_1) and (x_2y_2) are closed walks based at v belonging to different blocks. If F is a degenerated face then, say, (x_1y_1) is a pendant edge e traversed in both directions. The restriction $i|G-e$ of i is, obviously, a short-face embedding of $G-e$ in S, so, by induction hypothesis, $G-e$ is upper embeddable. This, of course, remains valid after giving back the pendant edge to $G-e$.

Hence, G is upper embeddable as well. Now assume that the face F is non-degenerated. Under these conditions, the boundary of F is homeomorphic to two cycles identified at a vertex, namely to (x_1y_1) and (x_2y_2) identified at v. It follows that both (x_1y_1) and (x_2y_2) are bounding cycles in S. This allows us to perform the following surgery. For fixed $k = 1$ or 2 we cut S along (x_ky_k) to obtain two separate bordered surfaces, both bounded by a copy of (x_ky_k). One of these does not contain the other of the two cycles (x_1y_1) and (x_2y_2), and we denote it by S'_k. Capping the boundary of S'_k by a 2-cell we obtain a closed surface S_k together with a subgraph $G_k \subseteq G$ cellularly embedded in it. It is clear from the construction that $G_1 \cup G_2 = G$ and $G_1 \cap G_2 = v$. Observe, however, that G_k $(k = 1,2)$ may well be composed of many blocks, some of which are incident with v. Each face G_k in S_k is either a face of G or the bigon (x_ky_k). Thus the embedding of G_k in S_k obtained in this way is again a short-face embedding. Let T_k be an optimal spanning tree of G_k, guaranteed by the result of Jungerman [1] and Xuong [6]. If G_1 or G_2 has even Betti number then $T = T_1 \cup T_2$ is an optimal spanning tree of G. In this case G is upper embeddable, as required. So assume that both G_1 and G_2 have odd Betti number. Form a new graph G'_k $(k = 1, 2)$ by adding to G_k a new edge z_k parallel to both x_k and $y_k \neq x_k$. Note that G'_k has even Betti number. Since G'_k admits a short-face embedding and has fewer edges than G, we can use the induction hypothesis to deduce that G'_k contains a spanning tree T'_k such that G'_k-E(T'_k) consists of even components. Obviously, at most one of the edges x_k, y_k, and z_k belongs to T'_k, say x_k. Deleting z_k from G'_k results in exactly one odd component C_k in G_k-E(T'_k), namely the one containing the edge y_k and, consequently, the vertex v. In G-E$(T'_1 \cup T'_2)$ the two odd components C_1 and C_2 merge into one even component $C_1 \cup C_2$, so that the cotree G-E$(T'_1 \cup T'_2)$ consists of even components. Thus G is upper embeddable, and Case 1 is complete.

Case 2. G contains no cut-vertex. Since G is not cyclically 4-edge connected, there exists a cycle-separating edge cut-set B with $|B| < 4$. Choose B to be minimum and let G_1 and G_2 be the connected components of G-B.

First of all we shall analyse how B can be placed on the surface S.

Let $\mathcal{F}$ be the set of all faces of the embedding $i\colon G \longrightarrow S$ meeting B. Since B is a cut-set, each face in $\mathcal{F}$ meets the set B twice. Thus there are at most three faces in $\mathcal{F}$. Moreover, they are all non-degenerated for G has no cut-vertex. Let H be the bordered surface obtained from $\mathcal{F}$ by identifying the faces in pairs along the edges of B. From the above observations it readily follows that H is homeomorphic to a cylinder, with some subgraph $K \subseteq G$ embedded in it. Abusing notation slightly, we can write $K = B \cup \partial H$. Removing from S the interior of H, the surface splits into two bordered surfaces S'_1 and S'_2 (containing embedded subgraphs G_1 and G_2 of G, respectively), each with one boundary component. Spanning the boundary of S'_k with a 2-cell yields a closed surface S_k together with a 2-cell embedding $i_k\colon G_k \longrightarrow S_k$, $k = 1, 2$. Let F_k be the face of the embedding i_k obtained by capping the boundary of S'_k. Clearly, $\partial H = \partial F_1 \cup \partial F_2$. As we shall see later, each of these two faces is a short one, implying that both i_1 and i_2 is a short-face embedding.

Note that the cut-set B contains at least two independent edges x and y, and possibly another edge z. Let W be the set of end-vertices of x and y. Now we shall distinguish several cases subcases depending on the structure of K, in particular on the number of common vertices of z and W.

At the very beginning of our case analysis we shall deal with the situation when B consists of two edges. It is readily seen that under these conditions K is uniquely determined. Thus we have just one case with $|B| = 2$ to consider.

Case 2.1. $B = \{x, y\}$. In this case H consists of two quadrilaterals, both containing x and y. Writing their boundaries as (xe_1yf_1) and (xe_2yf_2), where e_k and f_k belong to G_k ($k = 1, 2$), it is clear that $\partial F_1 = (e_1f_1)$ and $\partial F_2 = (e_2f_2)$. This in turn implies that both $i_1\colon G_1 \longrightarrow S_1$ and $i_2\colon G_2 \longrightarrow S_2$ are short-face embeddings. Let G'_k be the one of G_k and $G_k - e_k$ which has even Betti number. By the induction hypothesis, there exists a spanning tree T_k in G'_k such that $G'_k - \mathrm{E}(T_k)$ consists of even components. Set $T = T_1 \cup y \cup T_2$. Then $G - \mathrm{E}(T) = (G'_1 - \mathrm{E}(T_1)) \cup (G'_2 - \mathrm{E}(T_2)) \cup (B - y) \cup D$, where D is a subset of $\{e_1, e_2\}$, possibly empty. Since $(B-y) \cup D$ forms a connected subgraph in G, the cotree $G - \mathrm{E}(T)$ contains at most one odd component. Thus G is upper embeddable.

Next we shall consider four cases in which z is incident with two vertices in W. In the first of them (2.2), z is parallel to one of x or y, say to x. The other possibility is that z is incident with both x and y. This possibility leads, up to symmetry, to three cases depending on the lengths of faces incident with z. It is easily checked that these faces are either two triangles (2.3) or a triangle and a quadrilateral (2.4) or two quadrilaterals (2.5).

Case 2.2. $B = \{x, y, z\}$, where z is parallel to x. It follows that there exist edges e_k and f_k in G_k $(k = 1, 2)$ such that H consists of two quadrilaterals with boundaries (xe_1ye_2) and (zf_1yf_2), respectively, and of a bigon with boundary (xz). Similarly as above, $\partial F_k = (e_kf_k)$, $k = 1, 2$. Arguing analogously as in Case 2.1 we deduce that G is upper embeddable.

Case 2.3. $B = \{x, y, z\}$, where z is incident with both x and y and lies in two triangles on H. In this case there exist edges e_k and f_k in G_k $(k = 1, 2)$ such that H comprises the faces with boundaries (xe_1ye_2), (yf_1z) and (zxf_2). It follows that $\partial F_k = (e_kf_k)$, whence $i_k\colon G_k \longrightarrow S_k$ is a short-face embedding for $k = 1, 2$. The rest of this case can be handled similarly as in the previous cases.

Case 2.4. $B = \{x, y, z\}$, where z is incident with both x and y and lies in a triangle and a quadrilateral. Without loss of generality we may assume that the boundaries of faces on H may be written as (xe_1ye_2), (yzf_1) and (zf_2g_2x), where e_k, f_k and g_k belong to G_k for $k = 1, 2$. Since $\partial F_1 = (e_1f_1)$ and $\partial F_2 = (e_2f_2g_2)$, both $i_1\colon G_1 \longrightarrow S_1$ and $i_2\colon G \longrightarrow S_2$ are short-face embeddings. By our induction hypothesis, G_k admits a spanning tree T_k such that G_k-E(T_k) has at most one odd component, $k = 1, 2$. If either G_1 or G_2 has even Betti number then $T = T_1 \cup y \cup T_2$ is an optimal spanning tree of $G_1 \cup y \cup G_2 = G-\{x, z\}$. Since x and z are adjacent edges, T is an optimal spanning tree of $(G-\{x,z\}) \cup \{x, z\} = G$ as well. Thus G is upper embedda- ble. Now assume that both G_1 and G_2 have odd Betti number. In this case consider the graphs $G'_1 = G_1-e_1$ and $G'_2 = G_2+e'_2$, where e'_2 is is a new edge parallel to e_2. Note that both G'_1 and G'_2 admit short-face embeddings and have even Betti number. Since both of them have fewer edges than G has, from our induction hypothesis it follows that there exists a spanning tree T'_k in G'_k ($k = 1, 2$) such that $G'_k - E(T'_k)$ consists of even components. Without loss of generality we may suppose that e'_2, being parallel to e'_2, does not belong to T'_2. Set $T' = T'_1 \cup z \cup T'_2$. Note that the cotree $G - E(T')$ can be obtained from $L = (G'_1 \cup G'_2 \cup z) - E(T')$ by replacing the edge e'_2 with the path xe_1y. Since all components of L are even and e'_2 is replaced by a path of length three joining the same vertices, $G - E(T')$ consists of even components. Thus G is again upper embeddable.

Case 2.5. $B = \{x, y, z\}$, where z is incident with both x and y and lies in two quadrilaterals. Then there exist edges e_k, f_k and g_k in G_k ($k = 1, 2$) such that H consists of the faces with boundaries (xe_1ye_2), (yf_1g_1z) and (zxf_2g_2). Thus $\partial F_k = (e_kf_kg_k)$ which implies that $i_k : G_k \to S_k$ is a short-face embedding for $k = 1, 2$. By the induction hypothesis, G_k admits an optimal spanning tree T_k. If G_1 or G_2 has even Betti number then the same arguments as in the corresponding situation of Case 2.4 yield that $T_1 \cup y \cup T_2$ is an optimal spanning tree of G, implying that G is upper embeddable. Now assume that both G_1 and G_2 have odd Betti number. Consider the graphs $G'_k = G_k + e_k$ ($k = 1, 2$), where the edge e'_k is parallel to e_k. Since both G'_1 and G'_2 have fewer edges than G and admit short-face embeddings, we can use our induction hypothesis to deduce that G'_k ($k = 1, 2$) contains a spanning tree T'_k such that $G'_k - E(T_k)$ consists of even components. We may clearly suppose that e'_k is not contained in T'_k. It follows that $G_k - E(T'_k)$ contains exactly one odd component C_k and this component is incident with some end-vertex of e_k. Obviously, the edge z has no ,more than two end-vertices in $C_1 \cup C_2$. If z has two end-vertices in $C_1 \cup C_2$ then set $T = T'_1 \cup y \cup T'_2$. If z has just one end-vertex in $C_1 \cup C_2$ then one of x and y , say y, has no end in $C_1 \cup C_2$. In this case set $T = T'_1 \cup y \cup T'_2$. Finally, if no end-vertex of z belongs to $C_1 \cup C_2$, set $T = T'_1 \cup z \cup T'_2$. In all these cases it is easily verified that $G - E(T)$ consists of even components, implying that G is upper embeddable. This concludes Case 2.5.

Further we shall consider the situation when the edge z is incident with exactly one vertex in W. Up to symmetry this leads to two cases according as z lies and a quadrilateral (2.6), or in two quadrilaterals (2.7).

Case 2.6. $B = \{x, y, z\}$, where z is incident with x but not with y and lies in a triangle and a quadrilateral. Then there exist edges e_1 and f_1 in G_1, and e_2, f_2 and g in G_2 such that the faces on H have the boundaries (xe_1ye_2), (yf_1zf_2) and (zxg_2). It follows that $\partial F_1 = (e_1f_1)$ and $\partial F_2 = (e_2f_2g_2)$ which implies that $i: G \to S$ is a short-face embedding for $k = 1, 2$. The rest of the proof in this case can be patterned on that of Case 2.4.

Case 2.7. $B = \{x, y, z\}$, where z is incident with x but not with y and lies in two quadrilaterals. Then there exist edges e_1 and f_1 in G_1, and e_2, f_2, g_2 and h_2 in G_2 such that H consists of the faces with boundaries (xe_1ye_2), (yf_1zf_2) and (zxh_2g_2). Thus $\partial F_1 = (e_1f_1)$ and $\partial F_2 = (e_2f_2g_2h_2)$ and so $i_k: G_k \to S_k$ is a short-face embedding for $k = 1, 2$. The rest is similar to Case 2.4.

The last possibility to be considered concerns the situation when z is independent of both x and y. This obviously leads to a single case.

Case 2.8. $B = \{x, y, z\}$, where x, y and z are independent edges. In this case there exist edges e_k, f_k and g_k in G_k $(k = 1, 2)$ such that H consists of three quadrilaterals with boundaries (xe_1ye_2), (yf_1zf_2) and (zg_1xg_2). Consequently, $\partial F_k = (e_kf_kg_k)$ which implies that $i_k: G_k \to S_k$ is a short-face embedding for $k = 1, 2$. Let T_k be an optimal spanning tree of G_k guaranteed by the induction hypothesis. First assume that at least one of G_1 and G_2 has even Betti number. Clearly, among the edges on $\partial F_1 = (e_1f_1g_1)$ there exists one that does not belong to T_1, say e_1. Setting $T = T_1 \cup z \cup T_2$ it is easy to see that T is an optimal spanning tree of $G-\{x, y\}$. Since both x and y are incident with $e_1 \in G-\{x, y\}-E(T)$, they belong to the same component of $G-E(T)$. Therefore, T is an optimal spanning tree of G, yielding that G is upper embeddable. It remains to consider the case when both G_1 and G_2 have odd Betti number. Consider the graph $G_1' = G_1+e_1'$, where e_1' is a new edge parallel to e_1. Since G_1' admits a short-face embedding and has even Betti number, the induction hypothesis implies that G_1' contains a spanning tree T_1' such that $G_1'-E(T_1')$ consists of even components. Thus G_1-E(T'_1) contains exactly one odd component C_1, and this component is incident with some end-vertex of e_1. Denote this vertex by v_1. (If both ends of e_1 belong to C_1, choose v_1 to be any of them.) Among the edges of the triangle $(e_1f_1g_1)$ there is one not incident with v_1, say f_1. Now consider the graph $G_2 = G_2+f'_2$, where f'_2 is a new edge parallel to f_2. Similarly as for G'_1, there exists a spanning tree T'_2 of G'_2 such that G'_2-E(T'_2) has exactly one odd component C_2, and this component is incident with some end-vertex of f_2. Denote this end-vertex by v_2. (Again, if f_2 has both ends in C_2, choose v_2 to be any of them.) Finally there exists in B an edge t is incident with neither v_1 nor v_2. Set $T' = T'_1 \cup t \cup T'_2$. Note that G-E(T) = (G_1-E(T'_1)) $\cup$ (G_2-E(T'_2)) $\cup$ ($B-t$) and recall that the only odd compo-nent of G_k-E(T'_k) is incident with v_k (k = 1, 2). Since one of the edges in $B-t$ is incident with v_1 and the other with v_2, G-E(T') consists of even components. Hence, G is upper embeddable.

It can easily be verified that the above cases cover all the possibilities. Thus the induction step is complete and our theorem is proved.∎

Corollaries and Remarks

The following corollary immediately follows from our Theorem. However, the original proofs in [2] and [3] are based on completely different ideas.

Corollary 1. [2, 3] *Every connected graph having a triangular embedding in some closed surface is upper embeddable.*∎

The next obvious consequence of our Theorem is new.

Corollary 2. *Every connected graph having a quadrilateral embedding in some closed surface is upper embeddable.*∎

Of course, it is understood that graphs considered in the above corollaries have no loops.

Although the following corollary is just a simple rephrasing of our Theorem, it is interesting for its own sake, as it says something about all 2-cell embeddings of a non-upper embeddable graph.

Corollary 3. *Let* G *be a connected loopless graph that is not upper embeddable. Then every 2-cell embedding of* G *in a closed surface contains some face of length at least five.*∎

It is easy to see that the graph G_n obtained from the cycle of length $n \geq 3$ by adding one loop to every vertex is not upper embeddable. On the other hand G_n embeds cellularly in the torus with all faces of length four. This shows that the assumption about loops in Corollary 3 and our Theorem cannot be dropped.

It is natural to ask whether our Theorem cannot be generalized by allowing a short-face embedding to have faces of length at most k, for some $k \geq 5$ instead of $k = 4$. Observe, however, that the graph H_n obtained from G_n defined above by subdividing each loop by one additional vertex admits an embedding in the torus with all faces of length six. Thus the only generalization, with multiple edges allowed, we can expect has $k = 5$. To this end, no counterexample is known.

Further generalizations would only be possible restricting ourselves to simple graphs. In this case perhaps $k = 7$.

References

[1] M. Jungerman, A characterization of upper embeddable graphs. *Trans. Amer. Math. Soc.* **241** (1978) 401-406.

[2] N. P. Khomenko and A. D. Glukhov, On upper embeddable graphs. In: *Graph Theory* (in Russian), Izd. Akad. Nauk Ukrain. SSR, Kiev, 1977, pp. 85-87.

[3] L. Nebeský, Every connected, locally connected graph is upper embeddable. *J. Graph Theory* 5 (1981) 197-199.

[4] E. A. Nordhaus, B. M. Stewart and A. T. White, On the maximum genus of a graph. *J. Combinatorial Theory* **11** (1971) 258-267.

[5] C. Payan and N. H. Xuong, Upper embeddability and connectivity of graphs. Discrete Math. **27** (1979) 71-80.

[6] N. H. Xuong, How to determine the maximum genus of a graph. *J. Combinatorial Theory Ser.* **B** **26** (1979) 216-227.

On Cyclic Representations of Triples by Pairs

J. Novák, Liberec, CSSR

1. Introduction

Definition 1.1. The cyclic representation $\mathcal{R}(n,3,2)$ of triples by pairs is a system of pairs on a set $E = \{1,2,\dots,n\}$ such that a) any triple $T \subset E$ contains (is represented by) a pair $D \in \mathcal{R}(n,3,2)$, b) the system $\mathcal{R}(n,3,2)$ posseses the automorphism $\pi(i)=i+1, i=1,2,\dots,n$, where the numbers $i+1$ are reduced modulo n.

Denote the minimum of $|\mathcal{R}(n,3,2)|$ by $Q(n)$. We shall investigate the problem of findind $Q(n)$ for odd numbers n. When n is even, we have $Q(n)=(n/2).(n/2-1)$, as we can easily deduce from [7]. Other related papers to our problem are [1], [2], [3], [4], [5], [6].

Suppose $n=2m+1$. Denote $\binom{E}{k}$ the system of all k-subsets of the set E, $1<k<n$. It is well known that $\binom{E}{3}$ can be partitioned into disjoint classes $\mathcal{T}$ of triples. Each class $\mathcal{T}$ arises from any triple $T=\{a,b,c\} \subset E$ by iterating n−1 times the mapping π. If $n \not\equiv 0 \pmod 3$ then each class $\mathcal{T}$ contains n triples, otherwise one class contains n/3 triples, the others contain n triples.

R. Bodendiek, R. Henn (Eds.)
Topics in Combinatorics and Graph Theory

Assign to each class $\mathcal{T}$ a sole triple $\{d_1, d_2, d_3\}$, so called the triple of differences. The differences d_i are defined in the following way. Let $\{a, b, c\} \in \mathcal{T}$. Denote $d_1' = |a-b|$, $d_2' = |b-c|$, $d_3' = |a-c|$. If $d_i' > m$ then put $d_i = n-d_i'$, else $d_i = d_i'$. In this way we obtain three differences not different of necessity for which either one difference equals to the sum of remaining two, or $\sum d_i = n$.

We shall consider the differences to be also differences of pairs $\{a,b\}, \{b,c\}, \{a,c\}$ contained in the triple $\{a,b,c\}$.

2. Triples of differences

Definition 2.1. Let $M=\{1,2,\dots,m\}$. A system $\mathcal{F}(m)$ of triples $\{x,y,z\}$, where $x,y,z \in M$, is called a system of triples of differences on the set M, iff either one element of the triple equals to the sum of remaining two elements, or $x+y+z = 2m+1 = n$.

Theorem 2.1. a) The system $\mathcal{F}(m)$ equals to the system of triples of differences assigned to the classes $\mathcal{T}$ of $\binom{E}{3}$, where $n=2m+1$, as it was introduced in section 1. b) It is further:

$|\mathcal{F}(m)| = (m^2 + m)/3$ for $n \not\equiv 0 \pmod 3$,

$|\mathcal{F}(m)| = (m^2 + m + 1)/3$ for $n \equiv 0 \pmod 3$.

Proof. a) Given a triple $\{x, y, z\} \in \mathcal{F}(m)$ assume $z = x+y$. Form the triple $T \in \binom{E}{3}$, $T = \{1, 1+x, 1+x+y\}$. Determine their differences: $d_1' = x = d_1$, $d_2' = y = d_2$, $d_3' = x+y = z \leq m$. Hence $d_3 = d_3' = z$.

Assume $x + y + z = n$. Form the triple $T = \{1, 1+x, 1+x+y\}$. Determine their differences: $d_1' = x = d_1$, $d_2' = y = d_2$, $d_3' = x+y = n-z > m$. Hence $d_3 = n - d_3' = z$.

Conversely let $\{d_1, d_2, d_3\}$ be a triple of differences assigned to a class $\mathcal{T} \subset \binom{E}{3}$. Then $\{d_1, d_2, d_3\} \in \mathcal{F}(m)$ as it follows from section 1.

b) Let $x \in M$. If $2x \leq m$ or $2x > m$ then the triple $\{x, x, 2x\}$ or the triple $\{x, x, n-2x\}$ belong to the system $\mathcal{F}(m)$, respectively. If $n \equiv 0 \pmod 3$, then the triple $\{n/3, n/3, n/3\}$ also belongs to the system $\mathcal{F}(m)$. Let $\{x, y\}$ be a pair of different elements of M. This pair occurs allways in two triples of the system $\mathcal{F}(m)$. Assume $y > x$ and set $y - x = z_1$. Hence $z_1 + x = y$. Thus the triple $\{x, z_1, y\} \in \mathcal{F}(m)$. Set $x+y = z_2$. For $z_2 \leq m$, the triple $\{x, y, z_2\} \in \mathcal{F}(m)$, for $z_2 > m$, the triple $\{x, y, n-z_2\} \in \mathcal{F}(m)$.

So we obtain two different triples of the system $\mathcal{F}(m)$, because $z_1 \neq z_2$ and $z_1 \neq (n-z_2)$, as we can easily see.

Further determine cardinality of the system $\mathcal{F}(m)$. Suppose $n \not\equiv 0 \pmod 3$. At first we have m triples of the form xxz with $z \neq x$, for $x = 1, 2, \ldots, m$. These triples contain m pairs of different elements x, z.

Number of remaining pairs of different elements is equal to $2 \cdot \binom{m}{2} - m = m^2 - 2m$. Thus we obtain altogether $(m^2 - 2m)/3 + m = (m^2 + m)/3$ triples of $\mathcal{F}(m)$.

Suppose $n \equiv 0 \pmod 3$. Then we have m-1 triples of the form xxz, where $z \neq x$, and one triple of the form xxx. These triples contain m-1 pairs of different elements. The number of remaining pairs of different elements is equal to $\binom{m}{2} - m + 1 = m^2 - 2m+1$. Thus we obtain altogether $(m^2 - 2m + 1)/3 + m = (m^2 + m + 1)/3$ triples of $\mathcal{F}(m)$.

<u>Theorem 2.2.</u> Problem of finding $Q(n)=\text{minimum } |\mathcal{R}(n,3,2)|$ for an odd n (see section 1.) is equivalent to problem of minimum representation of all triples of system $\mathcal{F}(m)$ by elements of the set $M=\{1,2,\ldots,m\}$.

Proof. It is clear that a representation $\mathcal{R}(n,3,2)$ can be decomposed into disjoint classes each having n pairs. Let $\mathcal{R}(n,3,2)$ be minimum with q classes of pairs. Let $\mathcal{T}$ be any class of triples of the system $\binom{E}{3}$. Then there is a class $\mathcal{D} \subset \mathcal{R}(n,3,2)$ of pairs whose pairs are contained in the triples of the class $\mathcal{T}$. All pairs of the class $\mathcal{D}$ have the same difference $d \in M$, which must be contained among three differences assigned to the class $\mathcal{T}$. Thus we obtain a representation of the triples of $\mathcal{F}(m)$ by elements of M. Show this representation to be minimum.

Suppose that there is a representation $\{r_1,\ldots,r_h\}$ of the system $\mathcal{F}(m)$ such that $h < q$. Then all triples of any class $\mathcal{T} \subset \binom{E}{3}$ having differences $\{d_1,d_2,d_3\}$ are represented by the pairs of a class $\mathcal{D}$ having a difference r_j representing the triple $\{d_1,d_2,d_3\} \in \mathcal{F}(m)$ according to Theorem 2.1. Thus h classes of pairs suffice to represent system $\binom{E}{3}$. We obtain a contradiction with the assumption about q.

One can easily see that every minimum representation $\{r_1,\dots,r_h\}$ of system $\mathcal{F}(m)$ by elements of the set M leads necessarily to a minimum cyclic representation of the system $\binom{E}{3}$ by pairs. Each element r_j leads to a class of pairs having difference r_j.

Theorem 2.3. Let $t = [n/3]$. All triples of system $\mathcal{F}(m)$ can be represented by elements $i \in M$, $i=1,2,\dots,t$.

Proof. Arrange triples of $\mathcal{F}(m)$ into a sequence so that we put triples containing the element 1 as first in the sequence, then we put remaining triples containing the element 2, and so on, at the end remaining triples containing the element t. There is no triple containing all elements greater than t, as we can easily deduce from definition of system $\mathcal{F}(m)$.

Example. Let us introduce the arrangement of triples $\mathcal{F}(7)$. We have n=15, t=5. The considered arrangement is the following:

1 1 2   1 2 3   1 3 4   1 4 5   1 5 6   1 6 7   1 7 7

2 2 4   2 3 5   2 4 6   2 5 7   2 6 7

3 3 6   3 4 7   3 5 7   3 6 6

4 4 7   4 5 6

5 5 5 .

3. Estimates for $|\mathcal{R}(\mathcal{F}(m))|$

Denote a representation of the system $\mathcal{F}(m)$ by elements of the set M by $\mathcal{R}(\mathcal{F}(m))$.Denote the minimum number of elements of such a representation by q(m).For q(m) the following theorem is valid.

Theorem 3.1. Let $m \neq 2$. Then the following estimate holds:

$m/2 < q(m) \leq [(2m+1)/3]$.

Proof.Theorem 2.3. implies $q(m) \leq t = [(2m+1)/3]$.Now consider the lower bound. Let m be an even integer, then put $p = m/2$, otherwise put $p=(m-1)/2$.

Suppose that p elements of $M = \{1,2,\dots,m\}$ represent the system $\mathcal{F}(m)$ and denote such elements a_i, $i=1,2,\dots,p$. Suppose $a_1 < a_2 < \dots < a_p$. It follows from the construction of the system $\mathcal{F}(m$ that every element a_i represents just m triples of $\mathcal{F}(m)$, every pair $a_i a_j$ $(i \neq j)$ is contained in two triples of $\mathcal{F}(m)$.Form p disjoint classes of triples of $\mathcal{F}(m)$ and denote them by $\mathcal{A}_i$, i=1,2, ...,p.

Put all triples containing the element a_1 into the class $\mathcal{A}_1$, then put all triples containing the element a_2 in the system $\mathcal{F}(m) - \mathcal{A}_1$ into the class $\mathcal{A}_2$, and so on. The last class $\mathcal{A}_p$ is formed by all triples which contain the element a_p in the system $\mathcal{F}(m) - \bigcup_1^{p-1} \mathcal{A}_i$. This element a_p occurs at most twice in each class $\mathcal{A}_i$, $i=1,2,\dots,p-1$, therefore it occurs at most in $2p-2=m-2$ triples of $\mathcal{F}(m)$. Thus the class $\mathcal{A}_p$ contains two triples at least. Analogously the other classes $\mathcal{A}_i$ are not empty as well.

Let $1 \leq j \leq p$. Then every element a_j must be contained in at most two triples of the class $\mathcal{A}_k$, where $1 \leq k \leq j-1$. Therefore the number of such triples with the element a equals to

2(j-1) at most. Thus number of triples in the class $\mathcal{A}_j$ equals at least to m-2(j-1). Determine the minimum number of triples in the classes $\mathcal{A}_1, \mathcal{A}_2, \dots, \mathcal{A}_p$, where $|\mathcal{A}_1| = m$.

$$\sum_{j=1}^{p} m - 2(j-1) = m.p - (2+4+\dots+2p-2) = mp - p^2 + p = m^2/4 + m/2$$

(if m is even) or

$m^2/4 + m/2 - 3/4$ (if m is odd).

Considered triples with the element a_j can also be identical in the case when two precedent elements a_i, a_k, $i, k < j$, form a triple $a_i a_k a_j$ of the system $\mathcal{F}(m)$. Then the cardinality of $\mathcal{A}_j$ increases by one. The number of such triples with the element a_j equals $[(j-1)/2]$ at most. This consideration is valid for $j = 2, \dots, p$.

Let p be an odd number. Determine the sum $\sum_{j=2}^{p} [(j-1)/2] = 0 + 1 + 1 + 2 + 2 + 3 + 3 + \dots + (p-1)/2 = (p^2 - 2p + 1)/4$.

If p is an even number then we obtain $\sum_{j=2}^{p} [(j-1)/2] = 0 + 1 + 1 + 2 + 2 + \dots + (p-2)/2 + (p-2)/2 = (p^2 - 2p)/4$. Therefore $|\mathcal{A}| = |\mathcal{A}_1 \cup \mathcal{A}_2 \cup \dots \cup \mathcal{A}_p| \le m^2/4 + m/2 + (p^2 - 2p + 1)/4$ for an odd p, or $|\mathcal{A}| \le m^2/4 + m/2 + (p^2 - 2p)/4$ for an even p, if m is even.

Considering 4 cases: m: even, p: odd; m: even, p: even; m: odd, p: odd; m: odd, p: even; we obtain 4 upper estimates for $|\mathcal{A}|$:

1) $(5m^2 + 4m + 4)/16$, 2) $(5m^2 + 4m)/16$, 3) $(5m^2 + 2m - 3)/16$,

4) $(5m^2 + 2m - 7)/16$.

But the cardinality of the system $\mathcal{F}(m)$ is either $(m^2 + m)/3$ or $(m^2 + m + 1)/3$. We can easily see that $|\mathcal{F}(m)| > |\mathcal{A}|$ for $m \neq 2$. Hence the result that the elements $a_1, \dots, a_p$ do not represent all triples of the system $\mathcal{F}(m)$. Thus it is necessary that $q > m/2$ for an even number m and $q > (m-1)/2$ for an odd number m. In the second case it must be $q \geq (m+1)/2 > m/2$.

<u>Theorem 3.2.</u> Let $m \equiv 2 \pmod 5$. Then the estimate $q(m) \leq (3m-1)/5$ holds.

Proof. Choose $x \in M = \{1, 2, \ldots, m\}$ such that $x \equiv 1$ or 4 or $5 \pmod 5$.

Denote by R the subset of M consisting of all selected x ,

$$R = \{1, 4, 5, 6, 9, 10, 11, \ldots, m-3, m-2, m-1\}.$$

Obviously $|R| = (3m-1)/5$. We shall show that the set $N = M - R = \{y;\ \ y \in M,\ y \equiv 2 \text{ or } 3 \pmod 5\}$ is an independent set of elements for the system $\mathcal{F}(m)$. This means that no triple of $\mathcal{F}(m)$ has all its elements in the set N.

Let $a, b \in N$, $a \equiv 2 \pmod 5$, $b \equiv 3 \pmod 5$.

Triples of the system $\mathcal{F}(m)$ have either a type aac, or bbc, or abc, where c denotes the third element of the triple counted by one from following formulas:

1) $c = 2a$, 2) $c = 2m+1 - 2a$, 3) $c = 2b$, 4) $c = 2m+1 - 2b$, 5) $c = a+b$, 6) $c = 2m+1 -(a+b)$.

Since $2m+1 \equiv 0 \pmod 5$, we obtain following values for the third element in determined cases:

1) $c \equiv 4 \pmod 5$, 2) $c \equiv 1 \pmod 5$, 3) $c \equiv 1 \pmod 5$,

4) $c \equiv 4 \pmod 5$, 5) $c \equiv 0 \pmod 5$, 6) $c \equiv 0 \pmod 5$.

Therefore the third element belongs to the set R.

When the triple of $\mathcal{F}(m)$ has a form zzz, then $z = (2m+1)/3$. Hence $2m+1 \equiv 0 \pmod{15}$, $z \equiv 0 \pmod 5$. Thus $z \in R$.

We obtain the result that the set R represents all triples of $\mathcal{F}(m)$.

Theorem 3.3. Let $m \equiv 5 \pmod{11}$. Then the estimate $q(m) \leq (7m-2)/11$ holds.

Proof. Form a subset $R \subset M$, $R = \{x;\ x \equiv i \pmod{11},\ i \in H = \{1,4,5,6,7,10,11\}$. We shall show that the set N=M-R is an independent set for the system $\mathcal{F}(m)$. $N = \{y;\ y \equiv j \pmod{11}, j \in \{2,3,8,9\}\}$. Analogously as in Theorem 3.2. we verify that we can form no triple of $\mathcal{F}(m)$ from elements of the set N, thus it is independent. Hence the set R represents all triples of $\mathcal{F}(m)$. The cardinality of R equals to $(7m-2)/11$.

Conjecture 3.1. If $m \equiv 2 \pmod 5$, then $q(m) = (3m-1)/5$. This conjecture was verified for m=7,12,17,22 and 32.

Known values of q(m) are given by following results:

m =	1	2	3	4	5	6	7	8	9	10	11	12	13	14	15	16	17	18	19	20	21	22	32;
q(m)=	1	1	2	3	3	4	4	5	6	7	7	7	9	9	10	10	10	?	?	?	?	13	19.

REFERENCES

[1] A.E.BROUWER,M.VOORHOEVE: Turán Theory and the Lotto Problem, Mathematical centre tracts 106(1979) 99-105.

[2] D.de CAEN, On Turan's hypergraph problem, University of Toronto, Canada, 1982.

[3] A.V.KOSTOCHKA, A class of constructions for Turan's (3-4)-pro-

blem, Combinatorica 2(2) (1982) 187-192.

[4] J.NOVÁK, On a conjecture of Turán, Čas.pěst.mat. 106(1981) Praha 127-137.

[5] G.RINGEL, On extremal problems in graph theory, in Proceedings of the Symp. held in Smolenice 1963, ČSAV Praha.

[6] P.TURÁN, Research problems, Magyar Tud.Akad.Mat.Kutato Int. Kozl. 6(1961) 417-423.

[7] P.TURÁN, On the theory of graphs, Colloq. Math. 3(1954) 19-30.

On the Steiner Periphery and Steiner Eccentricity of a Graph

O. R. Oellermann, Durban, South Africa
H. C. Swart, Durban, South Africa

Abstract

Let G by a connected graph and S a nonempty set of vertices of G. Then the Steiner distance $d(S)$ of S is the minimum number of edges in a connected subgraph of G that contains S. If $n \geq 2$ is an integer, and G is a graph with at least n vertices, then the n-eccentricity $e(n;v)$ of a vertex v is defined as $\max\{d(S) | S \subseteq V(G),\ |S| = n \text{ and } v \in S\}$. The n-radius $\text{rad}_n G$ is the minimum n-eccentricity over all vertices of G while the n-diameter is the maximum n-eccentricity over all vertices of G. The subgraph induced by those vertices with n- eccentricity $\text{rad}_n G$ is called the n-centre $C(n;G)$ of G and the subgraph induced by those vertices with n- eccentricity $\text{diam}_n G$ is called the n-periphery $P(n;G)$ of G. A vertex v is called n-eccentric if there exists a vertex u in $C(n;G)$ and a set S of n vertices that contains both u and v such that $d(S) = e(n;u) = \text{rad}_n G$. The subgraph of G induced by all n-eccentric vertices of G is called the n-eccentricity of G and is denoted by $EC(n;G)$. It is shown that for a tree T of order at least n, $P(n;T) = EC(n;T)$. Further, it is shown that all possible set-inclusion relations between $P(n;G)$ and $EC(n;G)$ may occur if G is not a tree.

(Graph Theory terminology not presented here can be found in [2].) Let G be a graph. The *distance* $d_G(u,v)$ (or $d(u,v)$) between a pair u,v of vertices in G is the length of a shortest $u-v$ path in G if such a path exists; otherwise $d_G(u,v) = \infty$. The *eccentricity* of a vertex v, denoted by $e_G(v)$ (or $e(v)$), is defined as $\max\{d(v,u) | u \in V(G)\}$. The *radius* of G, rad $G = \min\{e(v) | v \in V(G)\}$ and the *diameter*, diam $G = \max\{e(v) | v \in V(G)\}$. The *centre* $C(G)$ of G is the subgraph induced by the vertices of G with eccentricity rad G, while the *periphery* $P(G)$ of G is the subgraph induced by the vertices with eccentricity diam G. A vertex v of G is called an *eccentric* vertex if there is a vertex u in $C(G)$ such that $d(u,v) =$ rad $G = e(u)$. The subgraph induced by all eccentric vertices of G, $EC(G)$ is called the *eccentricity of* G.

Observe that the distance between a pair of vertices in a connected graph G is the minimum size among all subgraphs of G that contain u and v. This observation suggests a generalization of distance. Suppose that S is a nonempty set of vertices of a graph G. Then the *Steiner distance* $d_G(S)$ (or $d(S)$) of S is the minimum number of edges in a connected subgraph of G that contains S, if such a subgraph exists, otherwise $d(S) = \infty$. The literature abounds with results relating to the problem of efficiently

R. Bodendiek, R. Henn (Eds.)
Topics in Combinatorics and Graph Theory

trying to determine $d(S)$ in a given graph; see for example [7]. Note that if G is a connected graph and H is a subgraph of minimum size that contains a set S, then H is necessarily a tree, called a *Steiner tree* for S. This motivates the following definitions. In particular, for an integer $n \geq 2$, the *n-eccentricity* $e_G(n;v)$ or $e(n;v)$ of a vertex v is defined as $\max\{d(S)|S \subseteq V(G), |S| = n \text{ and } v \in S\}$, the *n-radius* $\text{rad}_n G$ of G as $\min\{e(n;v)|v \in V(G)\}$ and the *n-diameter*, $\text{diam}_n G$ as $\max\{e(n;v)|v \in V(G)\}$. The *n-centre* $C(n;G)$ is the subgraph induced by the vertices of a minimum n-eccentricity and the *n-periphery* $P(n;G)$ is the subgraph induced by the vertices of maximum n-eccentricity in G. A vertex v is *n- eccentric* if there exists a vertex u in $C(n;G)$ and a set S of n vertices (of G) that contains both u and v such that $d(S) = e(n;u)$. The subgraph induced by the *n-eccentric* vertices is called the *n-eccentricity of* G and is denoted by $EC(n;G)$.

In this paper we study relationships between the n-periphery and n-eccentricity of connected graphs. Since trees are the simplest connected graphs we begin by comparing their n- peripheries and n-eccentricity. In [1] Buckley and Lewinter established the following result.

Theorem A If T is a tree, then

$$EC(T) = P(T).$$

Therefore $EC(2;T) = P(2;T)$ for all trees T. It is well-known that the centre of a tree is isomorphic to K_1 or K_2. In [5] it was shown that the n-centre of a tree is connected (i.e., n-centres of trees are trees) and the following characterization of n-centres of trees was established.

Theorem B A tree is the n-centre of a tree if and only if it has at most $n-1$ end-vertices.

If S is a nonempty set of vertices in a tree T, then T contains a unique Steiner tree for S which we denote by T_S. If S is a set of n vertices of T with $d(S) = \text{diam}_n T$, then S is called an *n-diameter set* for T. In [4] it is shown that for $n \geq 3$, $P(n-1;T) \subset P(n;T)$ for all trees T. Actually the proof established the following stronger result which we will find useful.

Theorem C Suppose $n \geq 3$ is an integer and T a tree of order at least n. If S_{n-1} is an $(n-1)$- diameter set, then there exists an n-diameter set S_n such that $S_{n-1} \subseteq S_n$.

Further, the first of the next two results is established in [5] and the other one in [3].

Theorem D Let $n \geq 2$ be an integer and T a tree of order at least n. Then $C(n;T) = C(n;T_{S_n})$ for every n-diameter set S_n of T.

Theorem E Let $n \geq 3$ be an integer and T a tree of order at least n. Then $\mathrm{diam}_{n-1}T = \mathrm{rad}_n T$.

If T is a tree with at most $n-1$ end-vertices, then it follows from Theorem B that $C(n;T) = T$. To determine $C(n;T)$ if T has at least n end-vertices, the following concept is defined in [5]. Let $n \geq 3$ be a fixed integer. If T has at least n end-vertices, then the (1st) *derivative* of T with respect to n, T_n^1 is the tree obtained by deleting the end-vertices from T. Suppose now that the kth derivative of T with respect to n, $T_n^{(k)}$ has been defined where $k \geq 1$. If $T_n^{(k)}$ has at least n end-vertices, then the $(k+1)$st *derivative* of G with respect to n is defined as the derivative of $T_n^{(k)}$. So $T_n^{(k)} = (T_n^{(k)})_n^1$. Now let $T_n^0 = T$. Suppose m is the first integer such that $T_n^{(m)}$ has at most $n-1$ end-vertices. Then it is shown in [5] that $C(n;T) = T_n^{(m)}$.

We are now prepared to present our main result which is a generalization of Theorem A.

Theorem 1 Let $n \geq 3$ be an integer and T a tree of order at least n. Then $P(n;T) = EC(n;T)$.

Proof If T has at most $n-1$ end-vertices, then, by Theorem B, $T = C(n;T) = P(n;T)$. Let v be any vertex of T. If S is any set of n vertices that contains v and all the end-vertices of T, then S must contain a vertex $u \neq v$. So u belongs to $C(n;T)$ and $d(S) = e(n;u) = \mathrm{rad}_n T$. Hence $v \in EC(n;T)$ which implies that $T = EC(n;T)$, i.e. $P(n;T) = EC(n;T)$.

Suppose now that T has at least n end-vertices. We show first that $P(n;T) \subseteq$

$EC(n;T)$. Let v be a vertex of $P(n;T)$ and S_n an n-diameter set containing v. Then S_n is a subset of the end-vertices of T. So every vertex of S_n is an end-vertex of T_{S_n}. For each vertex x in S_n let P_x be a shortest path from x to a vertex of degree at least 3 in T_{S_n} and ℓ_x its length. Suppose u is a vertex of S_n so that $\ell_u = \min\{\ell_x | x \in S_n\}$.

By the discussion preceding Theorem 1, $C(n;T_{S_n})$ is $(T_{S_n})_n^{(\ell_u)}$. So by Theorem D, $(T_{S_n})_n^{(\ell_u)}$ is $C(n;T)$. Therefore, $\text{diam}_n T = q(C(n;T)) + n\ell_u$.

We show next that $\text{rad}_n T = q(C(n;T)) + (n-1)\ell_u$. Let S'_{n-1} be any $(n-1)$-diameter set. Then, by Theorem C, there exists an n-diameter set S'_n that contains S'_{n-1}. Let a be the vertex of $S'_n - S'_{n-1}$. For each vertex z in S'_n let ℓ'_z be the length of a shortest path from z to a vertex of degree at least 3 in $T_{S'_n}$. Then necessarily $\ell'_a = \min\{\ell'_z | z \in S'_n\}$; otherwise there exists a set of $n-1$ vertices of T whose distance exceeds that of $S'_{n-1} = S'_n - \{a\}$, which is not possible. As before $\text{diam}_n T = q(T_{S'_n}) = q(C(n;T)) + n\ell'_a$. Thus $\ell'_a = \ell_u$, and $\text{diam}_{n-1} T = q(T_{S'_n}) - \ell'_a = \text{rad}_n T$. So $\text{rad}_n T = \text{diam}_{n-1} T = \text{diam}_n T - \ell_u = q(C(n;T)) + (n-1)\ell_u$.

Let $y \neq v$ be a vertex of S_n, and suppose P_y is a $y - w$ path. Observe that w is a vertex of $(T_{S_n})_n^{(\ell_u)}$. So w is in $C(n;T)$. Further, $(S_n - \{y\} \cup \{w\})$ is a set of n vertices that contains both v and w such that $d(S_n - \{y\} \cup \{w\}) = q(T_{S_n}) - \ell_u = \text{diam}_n T - \ell_u = \text{rad}_n T$. Therefore $v \in V(EC(n;T))$. So $P(n;T) \subset EC(n;T)$.

We show now that $EC(n;T) \subset P(n;T)$. Assume, to the contrary, that there is a vertex v that belongs to $EC(n;T)$ but not to $P(n;T)$. Observe that v is an end-vertex of T, since T has at least n end-vertices. Let u be a vertex in $C(n;T)$ for which there exists a set S of n vertices that contains both u and v with $d(S) = e(n;u) = \text{rad}_n T$. Note that u is an end-vertex of T_S; otherwise u belongs to $T_{S-\{u\}} (= T_S)$ and, by Theorem E, v belongs to an $(n-1)$-diameter set, namely $S - \{u\}$ which, by Theorem C, is contained in some n-diameter set. However, this contradicts the fact that v does not belong $P(n;T)$.

Let S' be an n-diameter set such that $|S' \cap S|$ is as large as possible. Then S' consists of end-vertices of T. Since v belongs to S and is *not* contained in an n- diameter set, it follows from Theorem C, that $S' - S \neq \emptyset$. In fact, since $u \notin S'$, $|S' - S| \geq 2$. Let ℓ_v be the length of a shortest path P_v from v to a vertex w of degree at least 3 in T_S. Let $P_v : (v =) v_0, v_1, ..., v_m (= w)$ be the $v - w$ path. Observe that the shortest path from every vertex y in $S' - S$ to a vertex of T_S must have length at most ℓ_v; otherwise $e(n;u) > q(T_S) = \text{rad}_n T$, which is not possible. Let $P : (u =) u_0, u_1, ..., u_k$ be a shortest path from u to a vertex of degree at least 3 in T_S. Let T_1 and T_2 be the two components of $T - u_{k-1}u_k$, and assume that u belongs to T_2. By our choice of P, u is the only vertex of S in T_2. Since u belongs to $T_{S'}$, by Theorem D, it follows that T_2 contains a vertex, z say, of S' such that the $z - v$ path in T contains P.

We show now that no vertex of P_v except possibly w belongs to $T_{S'}$. Clearly $v_0 = v$ does not belong to $T_{S'}$. Then there exists a vertex $a \in S' - S$ such that v_i belongs to the $a - u$ path in $T_{S'}$. We may assume that i is the smallest integer such that v_i belongs to some $a - u$ path of $T_{S'}$ where $a \in S' - S$, i.e. if $j < i$ then v_j does not belong to $T_{S'}$. The $v_0 - v_i$ path must have length less than the $a - v_i$ path; otherwise if this is not the case, then $d(S' - \{a\} \cup \{v_0\}) \geq d(S') = \text{diam}_n T$ and therefore $S' - \{a\} \cup \{v_0\}$ must be an n-diameter set that has more vertices in common with S than S'. Observe that the $a - v_i$ path has no vertex of T_S as internal vertex; otherwise if b is such a vertex then there exist two $v_i - b$ paths in T, which is not possible. However, then $d(S \cup \{a\} - \{v\}) > d(S) = e(n; u)$ which is not possible. Therefore no internal vertex of P_v belongs to $T_{S'}$.

Suppose now that T_1 contains a vertex a of $S' - S$. We show that a shortest path Q from a to a vertex of degree at least 3 in $T_{S'}$ does not contain a vertex of T_S as internal vertex. Suppose $Q : (a =)a_0, a_1, ..., a_r$ and that some a_j belongs to T_S. Choose j to be as small as possible. Then there exists an end-vertex $x \in S - S'$ of T such that a_j belongs to the $x - u$ path in T_S. As in the case of v we can show that a_j is *not* an internal vertex of a shortest path from x to a vertex of degree at least 3 in T_S, and no internal vertex of the $x - a_j$ path belongs to $T_{S'}$. However, then $d(x, a_j) \geq d(a, a_j)$ which implies that $d(S' \cup \{x\} - \{a\}) \geq d(S') = \text{diam}_n T$. So $S' \cup \{x\} - \{a\}$ must be an n-diameter set that has more vertices in common with S than S'. This is not possible. Therefore T_1 contains no vertex of $S' - S$. Hence T_2 must contain at least two vertices of $S' - S$. Let z be a vertex of $S' - S$. Since the shortest path from z to a vertex of T_S has length at most ℓ_v and since u belongs to $T_{S'}$, the length from z to a vertex of degree at least 3 in $T_{S'}$ can have length at most ℓ_v. However, this fact together with an earlier observation shows that $d(S' - \{z\} \cup \{v\}) \geq d(S') = \text{diam}_n G$. So $S' - \{z\} \cup \{v\}$ must be an n-diameter set that has more vertices in common with S than S'. This is not possible. Hence $EC(n; T) \subset P(n; T)$. □

In [6] it was shown that there exist graphs for which the periphery is properly contained in its eccentricity and vice versa. So Theorem A cannot be extended to graphs in general. We show now that Theorem 1 does not hold for graphs in general.

Theorem 2 For every positive integer $n \geq 3$ there exists a graph G_n such that $P(n; G_n) \not\subset EC(n; G_n)$.

Proof Consider the complete bipartite graph $K(n,n)$ with partite sets $U = \{u_1, u_2, ..., u_n$ and $V = \{v_1, v_2, ..., v_n\}$. Join a new vertex v to every vertex in U and delete the edges of the type $u_i v_i$ for $1 \leq i \leq n$. Finally, subdivide, exactly once, each edge of the type $u_i v_j$ for $1 \leq i, j \leq n$ and $i \neq j$. Let w_{ij} be the vertex of degree 2 that is adjacent with u_i and v_j. Let G_n be the resulting graph. Then it can be shown that $\text{rad}_n G_n = 2n - 1$, $\text{diam}_n G_n = 2n + 2, C(n; G_n) = \langle\{v\}\rangle, P(n; G_n) = \langle V \rangle$ and $EC(n; G_n) = \langle V \cup \{w_{ij} | 1 \leq i, j \leq n \text{ and } i \neq j\}\rangle$. □

Theorem 3 For every positive integer $n \geq 3$ there exists a graph H_n such that $EC(n; H_n) \not\subset P(n; H_n)$.

Proof Let $Q_1, Q_2, ..., Q_{n-1}$ be $n - 1$ cycles of length 5 where $Q_i : v_{i,0}, v_{i,1}, ..., v_{i,4}, v_{i,0}$. Let H_n be obtained by identifying the $n-1$ vertices $v_{1,0}, v_{2,0}, ..., v_{n-1,0}$ in a single vertex v_0. Then it can be shown that $\text{rad}_n H_n = 2n - 2$, $\text{diam}_n H_n = 2n - 1, C(n; H_n) = \langle\{v_0\}\rangle$. Further $EC(n; H_n) = \langle\{v_{i,2}, v_{i,3} | 1 \leq i \leq n-1\}\rangle$ and $P(n; H_n) = H - v_0$. So $EC(n; H_n) \not\subset P(n; H_n)$. □

Theorem 4 For every positive integer $n \geq 3$ there exists a graph F_n such that $P(n; F_n) \not\subset EC(n; F_n)$ and $EC(n; F_n) \not\subset P(n; F_n)$.

Proof Let $R_1, R_2, ..., R_{n-1}$ be $n - 1$ cycles of length 7 where $R_i : v_{i,0}, v_{i,1}, ..., v_{i,6}, v_{i,0}$ for $1 \leq i \leq n - 1$. Let F_n be obtained by identifying the $n - 1$ vertices $v_{1,0}, v_{2,0}, ..., v_{n-1,0}$ in a single vertex v_0 and then joining a new vertex u to $v_{1,1}$. Then again it can be shown that $C(n; F_n) = \langle\{v_0\}\rangle, EC(n; F_n) = \langle\{v_{i,3}, v_{i,4} | 1 \leq i \leq n - 1\}\rangle$ and $P(n; F_n) = \langle\{v_{i,3}, v_{i,4} | 2 \leq i \leq n-1\} \cup \{u, v_{1,4}\}\rangle$. Since u belongs to $P(n; F_n)$ but not $EC(n; F_n)$ and $v_{1,3}$ belongs to $EC(n; F_n)$ but not $P(n; F_n)$ the theorem now follows. □

References

[1] F. Buckley and M. Lewinter, Minimal graph embeddings, eccentric vertices and the peripherian. To appear.

[2] G. Chartrand and L. Lesniak, *Graphs and Digraphs* 2nd *Edition.* Wadsworth & Brooks/Cole, Monterey, CA (1986).

[3] G. Chartrand, O.R. Oellermann, S. Tian and H.B. Zou, *Casopis Pěst. Mat.* To appear.

[4] M.A. Henning, O.R. Oellermann and H.C. Swart, On vertices with maximum Steiner eccentricity in graphs. Submitted for publication.

[5] O.R. Oellermann and S. Tian, Steiner centers in graphs. *J. Graph Theory.* To appear.

[6] K.B. Reid and Weizhen Gu, Peripheral and eccentric vertices in graphs. Submitted for publication.

[7] P. Winter, Steiner problem in networks: A Survey. *Networks* 17 (1987) 129-167.

Cycles Containing Three Consecutive Edges in 2k-Edge-Connected Graphs

H. Okamura, Osaka, Japan

We consider finite undirected graphs possibly with multiple edges but without loops. Let G be a graph and let V=V(G) and E=E(G) be the set of vertices and edges of G respectively. We allow repetition of vertices (but not edges) in a path and cycle. $\lambda(G)$ denotes the edge-connectivity of G. For $X,Y \subset V(G)$, with $X \cap Y=\emptyset$, $\partial(X,Y)$ denotes the set of edges between X and Y. $\partial(X)$ denotes $\partial(X,V(G)-X)$, and is called a cut. A cut $\partial(X)$ is called an n-cut if $|\partial(X)|=n$, and will be called nontrivial if $|X| \geq 2$ and $|V(G)-X| \geq 2$. We set $e(X,Y):=|\partial(X,Y)|$ and $e(X)=|\partial(X)|$. For an $f \in E(G)$, V(f) denotes the set of end vertices of f. Cycles containing two (adjacent) edges in 2k-(k-)edge-connected graphs reducing the edge-connectivity at most two are investigated in [3] ([2] and Mader [1]). We here consider cycles containing three consecutive edges.

Theorem 1. Suppose that $k \geq 2$ is an even integer, $\lambda(G) \geq k$, f_1, f_2, f_3 are edges, $u_1 \in V(f_1) \cap V(f_2)$, $u_2 \in V(f_2) \cap V(f_3)$, and (P_1) below holds.

(P_1) There is no k- or (k+1)-cut containing f_1, f_2, f_3.

Then there exists a cycle C containing f_1, f_2, f_3 such that $\lambda(G-E(C)) \geq k-2$.

It is easy to see that (P_1) is a necessary condition for the result. For odd k, Theorem 1 is not true in the following graph.

$V(G)=\{u_1, u_2, v_1, v_2\}$,
$e(v_1,v_2)=0$, $e(u_1,u_2)=1$, $e(u_i,v_i)=(k+1)/2$ $(i=1,2)$,
$e(u_1,v_2)=e(u_2,v_1)=(k-1)/2$, $f_1 \in \partial(u_1,v_1)$, $f_2 \in \partial(u_1,u_2)$,
and $f_3 \in \partial(u_2,v_2)$.

Theorem 2. Suppose that $k \geq 2$ is an even integer, $\lambda(G) \geq k$, $s \neq t$ are vertices, $f_1 \neq f_2$ are edges, either

R. Bodendiek, R. Henn (Eds.)
Topics in Combinatorics and Graph Theory

(1) $V(f_1)=\{s,u_1\}$, and $f_2 \in \partial(u_1)$,

or

(2) $s \in V(f_1)$ and $t \in V(f_2)$,

and (P_2) below holds.

(P_2) There is no k-cut $\partial(X)$ such that $f_1, f_2 \in \partial(X)$ and $|X \cap \{s,t\}|=1$.

Then there exists a path P between s and t containing f_1 and f_2 such that $\lambda(G-E(P)) \geq k-2$.

We show that Theorem 1 implies Theorem 2. Let $H:=(V(G),E(G)\cup g)$, where g is a new edge between s and t. Then by Theorem 1, H has a cycle C containing f_1, f_2, and g, and C-g is a required path for G.

The following result is proved for each integer $k \geq 2$ in [4]. In the same way as above, the even case follows from Theorem 1.

Cororally 1. If $k \geq 2$ is even, $\lambda(G) \geq k$, f is an edge, and s,t are vertices, then there exists a path P between s and t such that $f \notin E(P)$ and $\lambda(G-E(P)-f) \geq k-2$.

Let $x,y \in V(G)$. $\lambda(x,y)$ denotes the maximal number of edge-disjoint paths between x and y. We set $N(x):=\{\, a \in V(G)-x \mid e(a,x) > 0\}$, and $\Gamma(G,k):=\{Z \subset V(G) \mid$ for each $a,b \in Z$, $\lambda(a,b) \geq k\}$. G/X denotes the graph obtained from G by contracting X, and for $a \in X$, we denotes the corresponding vertex in G/X by $\tilde{a}$. For $a,b \in N(X)$ with $a \neq b$, $f \in \partial(x,a)$, and $g \in \partial(x,b)$, $G_x^{a,b}$ and $G^{f,g}$ denote the graph $(V(G),(E(G)\cup h)-\{f,g\})$, where h is a new edge between a and b.

2. Proof of Theorem 1

Lemma 1 ([2] and Mader [1]). If $\lambda(G) \geq k \geq 2$, $s \in T \in \Gamma(G,K)$, and $f_1, f_2 \in \partial(s)$, then there exists a cycle C such that $f_1, f_2 \in E(C)$ and $T \in \Gamma(G-E(C),k-2)$.

Lemma 2 ([3, Lemma 6(2)]). Suppose that $k \geq 2$ is even, $V(G)=W_1 \cup W_2$ (disjoint union), and each $v \in W_2$ has even degree. If $V(G)=X_1 \cup X_2$ (disjoint union), $x_i \in X_i \cap W_1$ $(i=1,2)$, $e(X_1) \leq k+1$, and $(X_i \cap W_1) \cup \tilde{x}_j \in \Gamma(G/X_j,k)$ for $(i,j)=(1,2),(2,1)$, then $W_1 \in \Gamma(G,k)$.

By simple counting we have

Lemma 3. If $X,Y \subset V(G)$, then

$$e(X-Y) + e(Y-X)=e(X)+e(Y)-2e(X\cap Y,\overline{X\cup Y}),$$
$$e(X\cap Y)+e(X\cup Y)=e(X)+e(Y) - 2e(X-Y,Y-X).$$

In Theorem 1 if $f_1,f_3 \in \partial(u_1,u_2)$, then by (P_1) $\lambda(G-\{f_1,f_2\})\geq k$, and by Lemma 1 $G-\{f_1,f_2\}$ has a cycle C such that $f_3 \in E(C)$ and $\lambda(G-\{f_1,f_2\}-E(C))\geq k-2$. If $f_1\in \partial(u_1,u_2)$, $f_3 \in \partial(u_2,v)$, and $v\neq u_1$, then by (P_1) $e(u_2)\geq k+2$ and $\lambda(G^{f_2,f_3})\geq k$, and so by Lemma 1 G^{f_2,f_3} has a cycle C such that $\{f_1,f_2\}\subset E(C)$ and $(G^{f_2,f_3}-E(C))\geq k-2$. Therefore to prove Theorem 1, it suffices to prove

Lemma 4. Suppose that $k\geq 2$ is even, $V(G)=T\cup W_1\cup W_2$ (disjoint union), $T=\{u_1,u_2,v_1,v_2\}$, $|T|=4$, $V(G)-W_2\in \Gamma(G,k)$, $|W_2|\leq 1$, $e(W_2)$ is even, $f_i \in \partial(u_i,v_i)$ $(i=1,2)$, $g \in \partial(u_1,u_2)$, and property (P) below holds.

(P) For each $X\subset V(G)$ with $\partial(X)\supset \{f_1,g,f_2\}$ (i.e. $X\cap T=\{u_1,v_2\}$ or $\{u_2,v_1\}$), $e(X)\geq k+2$.

Then there is a cycle C such that $\{f_1,f_2,g\}\subset E(C)$ and $V(G)-W_2\in \Gamma(G-E(C),k-2)$.

Proof. Let G be a counterexample with $|E(G)|$ as small as possible.

(1) If $\partial(X)$ is a nontrivial k-or (k+1)-cut, then $X\cap T=\{u_1,v_1\}$ or $\{u_2,v_2\}$.

Proof. We may let $|X\cap T|\leq 2$. If $|X\cap T|=2$, then by (P) we may let $X\cap T=\{u_1,u_2\}$. By Lemma 1 G/X has a cycle C such that $\{f_1,f_2\}\subset E(C)$ and $V(G/X)-W_2\in \Gamma(G/X-E(C),k-2)$. By Lemma 2 $V(G)-W_2\in \Gamma(G-E(C)-g,k-2)$. Thus $|X\cap T|\leq 1$, and G/X has a required cycle C. If $X\cap T=\{v_1\},\{v_2\}$, or $\emptyset$, and $E(C)\cap \partial(X)\neq \emptyset$, then we can extend C to a required cycle for G by using Lemma 1 in G/X.

(2) $e(u_i)\leq k+1$ $(i=1,2)$.

Proof. Assume $e(u_1)\geq k+2$. By (P) and (1) $V(G)-W_2\in \Gamma(G^{f_1,g},k)$, and using Lemma 1 in $G^{f_1,g}$ the result follows.

(3) $\{u_1,u_2\}$ does not separates v_1 from v_2.

Proof. If otherwise, by (2) $e(\{u_1,u_2\}) \leq 2(k+1)-2=2k$, and so by (1) $V(G)=T$, $e(u_i)=k+1$, and $e(v_i)=k$ $(i=1,2)$. Thus $e(v_1,u_i)=k/2$ $(i=1,2)$ and $e(\{v_1,u_2\})=k+1$, contrary to (P).

(4) There is no nontrivial k- or (k+1)-cut.

Proof. Assume that $\partial(X)$ is a nontrivial k- or (k+1)-cut. By (1) we may let $X \cap T=\{u_1,v_1\}$. By (3) there is an edge $h \in \partial(X)-\partial(u_1)-\partial(u_2)$. By (1), in G/X, for edges h, g, f_2, (P) holds. Thus G/X has a cycle C_1 such that $\{h,g,f_2\} \subset E(C_1)$ and $V(G/X)-W_2 \in \Gamma(G/X-E(C_1),k-2)$. Similarly $G/\overline{X}$ has a cycle C_2 containing f_1, g, h, and $C_1 \cup C_2$ is a required cycle for G.

(5) $W_2=\phi$, and for each $x \in W_1$, $e(x)$ is odd.

Proof. Let $x \in W_2$ if $W_2 \neq \phi$, and let $x \in W_1$ and $e(x)$ be even if $W_2=\phi$. Set $N(x)=\{y_1,\dots,y_n\}$. By (4) for each $1 \leq i<j \leq n$, $V(G)-W_2 \in \Gamma(G_x^{y_i,y_j},k)$. If for some i,j (P) holds in $G_x^{y_i,y_j}$, then the result holds in $G_x^{y_i,y_j}$, and it also holds in G. Thus for each $1 \leq i<j \leq n$,

$$S_{i,j}:=\left\{ X \subset V(G)-x \;\middle|\; \begin{array}{l}\{y_i,y_j\} \subset X,\ e(X) \leq k+3, \text{ and} \\ X \cap T=\{u_1,v_2\} \text{ or } \{u_2,v_1\}\end{array} \right\} \neq \phi .$$

Let $X_1 \in S_{1,2}$ and $|X_1|=\max_{Y \in S_{1,2}} |Y|$. We may let $X_1 \cap T=\{u_1,v_2\}$ and $y_3 \notin X_1$. Let $X_2 \in S_{1,3}$. Then $X_1-X_2 \neq \phi$ by the chice of X_1.

Case 1. $X_2 \cap T=\{u_1,v_2\}$.

$$e(X_1 \cap X_2,\overline{X_1 \cup X_2}) \geq e(u_1,\{v_1,u_2\})+e(v_2,u_2)+e(y_1,x) \geq 4.$$

Thus by Lemma 3

$$e(X_1-X_2)+e(X_2-X_1) \leq 2k+6-8=2k-2,$$

a contradiction.

Case 2. $X_2 \cap T=\{u_2,v_1\}$.

By Lemma 3

$$e(X_1-X_2) + e(X_2-X_1) \leq 2k+6-2=2k+4,$$
$$e(X_1 \cap X_2)+e(X_1 \cup X_2) \leq 2k+6-6=2k.$$

Thus by (4) $e(X_1-X_2)=e(X_2-X_1)=k+2$ and $e(X_1)=e(X_2)=k+3$. Then $e(X_1 \cap X_2) \equiv e(X_1) + e(X_1-X_2) \equiv 1 \pmod 2$, and so by Lemma 3

$e(X_1 \cup X_2)$ is odd. Hence $e(X_1 \cap X_2) \geq k+1$ and $e(X_1 \cup X_2) \geq k+1$, a contradiction.

(6) If $x, y \in W_1$, then $e(x,y)=0$.

Proof. Let $h \in \partial(x,y)$. By (4) and (5) $\lambda(G-h) \geq k$. Thus (P) does not hold in G-h, and so in G for some $X \subset V(G)$, $h \in \partial(X)$, $X \cap T = \{u_1, v_2\}$ and $e(X)=k+2$. We may let $x \in X$ and $y \notin X$. By (4) $V(G)-u_1 \in \Gamma(G^{f_1,g},k)$. Let $H := G^{f_1,g}/X$, then $\lambda(H) \geq k$, and in H (P) holds for f_1 (=g), f_2, and h. Thus H has a cycle C_1 such that $\{f_1, f_2, h\} \subset E(C_1)$ and $\lambda(H-E(C_1)) \geq k-2$. Similarly $G^{f_2,g}/\overline{X} =: K$ has a cycle C_2 such that $\{f_2, f_1, h\} \subset E(C_2)$ and $\lambda(K-E(C_2)) \geq k-2$. $C_1 \cup C_2$ is a required cycle for G.

Now $e(v_1,v_2)=0$. If for some $x \in W_1$, $e(x,v_i) > 0$ $(i=1,2)$, then the result holds. Thus by (6) for each $x \in W_1$, $N(x) \subset \{u_1, u_2, v_i\}$ (i=1 or 2), contrary to (3).

References

1. W. Mader, Paths in graphs, reducing the edge-connectivity only by two, Graphs Combin. 1 (1985), 81-89.
2. H. Okamura, Paths and edge-connectivity in graphs, J. Combin. Theory Ser. B 37 (1984), 151-172.
3. H. Okamura, Paths in k-edge-connected graphs, J. Combin. Theory Ser. B 45 (1988), 345-355.
4. H. Okamura, Every 4k-edge-connected graph is weakly 3k-linked, Graphs Combin, to appear.

Graph Distance and Euclidean Distance on the Grid

J. Pach*, New York, USA
R. Pollack**, New York, USA
J. Spencer, New York, USA

Abstract

Given a connected graph $G = (V, E)$, $V = \mathbf{Z}^2$, on the lattice points of the plane, let $d_G(p,q)$ and $d(p,q)$ denote the graph distance and the Euclidean distance between p and q respectively. In this note we prove that for every $\epsilon > 0$ there is a graph $G = G_\epsilon$ and a constant $d = d_\epsilon$ such that

$$|d_G(p,q) - d(p,q)| < \epsilon d(p,q)$$

for every pair $p, q \in V$ with $d(p,q) \geq d$. It remains open whether or not there is a graph G and a suitable constant K which satisfies

$$|d_G(p,q) - d(p,q)| < K$$

for all $p, q \in \mathbf{Z}^2$.

1 Defining the Graph

Let G_0 denote the graph on $\mathbf{Z}^2$ which can be obtained by joining two integer points with an edge if and only if their Euclidean distance is 1. The idea for the construction stems from the observation that, although $\sup \frac{d_G(p,q)}{d(p,q)} = \sqrt{2}$, if the straight line pq is almost parallel to the x−axis or to the y−axis, then the graph distance in G_0 and the Euclidean distance are very close to each other. Therefore we are going to define a large number of square lattices $\Lambda_i \subseteq \mathbf{Z}^2$ with various side lengths so that the directions of their axes are fairly uniformly distributed in $[0, 2\pi)$. For every i, we shall construct a graph G_i similar to G_0, whose main nodes are the elements of Λ_i, and we shall

*Is also at the Mathematical Institute of the Hungarian Academy of Sciences and is supported in part by NSF grant CCR-8901484 and Hungarian SF grant OTKA-1814
**Supported in part by NSF grant CCR-8901484 and NSA grant MDA904-89-H–2030

R. Bodendiek, R. Henn (Eds.)
Topics in Combinatorics and Graph Theory

glue these graphs together. The most important property of G_i will be that $d_{G_i}(p,q)$ and $d(p,q)$ do not differ too much, whenever $p, q \in \Lambda_i$ and the line determined by p and q is almost parallel to one of the axes of Λ_i.

To be more precise, we need some notation. Let $\epsilon > 0$ be fixed, and choose a set of vectors (a_i, b_i), $1 \le i \le N_\epsilon$, with the properties

(i) a_i, b_i are positive integers;

(ii) $\{\frac{b_i}{a_i} : a \le i \le N_\epsilon\}$ is well distributed in $[-1, +1]$ in the sense that for every; $x \in [-1, +1]$ there exists an i with $|x - \frac{b_i}{a_i}| < \frac{\epsilon}{100}$;

(iii) the length S_1 of (a_1, b_1) is at least K_ϵ, and $S_i = \sqrt{a_i^2 + b_i^2} > 5S_{i-1}$ for $2 \le i \le N_\epsilon$ (where $K_\epsilon > 50$ is a constant which will be fixed later).

Let $\Lambda_i \subseteq \mathbf{Z}^2$ denote the square lattice generated by (a_i, b_i), i.e.,

$$\Lambda_i = \{m(a_i, b_i) + n(-b_i, a_i) : m, n \in \mathbf{Z}^2\}, 1 \le i \le N_\epsilon$$

For simplicity, let

$$(m,n)_i = m(a_i, b_i) + n(-b_i, a_i) = (ma_i - nb_i, mb_i + na_i),$$

and let $C_i(m,n)$, the cell of $(m,n)_i$ be defined as the convex hull of $\{(m,n)_i, (m+1,n)_i, (m,n+1)_i, (m+1,n+1)_i\}$. Further, let $(m,n)_i^*$ denote a point $(m',n')_{i-1}$ which is closest to $(m,n)_i$ and whose cell $C_{i-1}(m',n') \subset C_i(m,n)$, $(i \ge 2)$.

Assume recursively that for every $j < i$ and for every pair of integers (m,n) we have already defined three paths $Pr_j(m,n)$, $Pu_j(m,n)$ and $P_j(m,n)$ which satisfy

1. $Pr_j(m,n)$ connects $(m,n)_j$ to $(m+1,n)_j$,
   $Pu_j(m,n)$ connects $(m,n)_j$ to $(m,n+1)_j$,
   $P_j(m,n)$ connects $(m,n)_j$ to $(m,n)_j^*$;
2. $Pr_j(m,n)$ and $Pu_j(m,n)$ have length $\lceil S_j \rceil$, while the length of $P_j(m,n)$

   is $\lceil d((m,n)_j, (m,n)_j^*) \rceil$;
3. All of these paths are internally disjoint from one another and from all previously defined paths, i.e., they can only meet at their endpoints;
4. All internal vertices of $Pr_j(m,n)$, $Pu_j(m,n)$ and $P_j(m,n)$ are in the interior of the cell $C_j(m,n)$.

Observe that at this point of the construction the number of nonisolated vertices in $C_i(m,n)$ is at most

$$\sum_{j<i}(\frac{S_i+2\sqrt{2}S_j}{S_j})^2 4\lceil S_j\rceil < \sum_{j<i}(\frac{2S_i}{S_j})^2 5S_j = 20S_i^2\sum_{j<i}\frac{1}{S_j} < \frac{S_i^2}{2},$$

provided that $S_1 \geq 50$. Hence, there are plenty of isolated vertices left in $C_i(m,n)$ that can be used to form the paths $Pr_i(m,n)$, $Pu_i(m,n)$ and $P_i(m,n)$ preserving the above properties.

Suppose that we have carried out the construction for all $i \leq N_\epsilon = N$, and consider the graph consisting of all paths already defined. For each point $p \in \mathbf{Z}^2$ which remained isolated, we add an edge to the nearest point of Λ_i.

2 Estimating the Distances

Since the length of every path which begins and ends at a vertex of some Λ_i is at least as large as the Euclidean distance between its endpoints, and every internal vertex belongs to some Λ_i,

$$d_G(p,q) > d(p,q) - 2\sqrt{2}S_N.$$

Bounding $d_G(p,q)$ from above is just a bit more complicated.

Lemma 1 *For every point $(m,n)_N \in \Lambda_N$, the graph G contains a path $P(m,n)$ of length at most S_N, all of whose vertices are in the cell $C_N(m,n)$ and which visits at least one vertex p_j of each Λ_j, $1 \leq j \leq N$.*

Proof: Consider the path, $P(m,n)$, obtained by concatenating $P_N(m,n)$, $P_{N-1}(m',n'), P_{N-2}(m'',n''),\ldots,P_2(m^{(N-2)},n^{(N-2)})$, as defined in (1)–(4) above. Its length is clearly at most $\sum_{i<N}\lceil 2S_i\rceil < S_N$, and it visits exactly one vertex of each Λ_i, $1 \leq i \leq N$. □

Lemma 2 *For any two points $x = (m_x,n_x)_i$ and $y = (m_y,n_y)_i$ of Λ_i*

$$d_G(x,y) < (1+\delta)(1+\frac{1}{S_i})d(x,y)$$

where $\delta = \min(\frac{n_x-n_y}{m_x-m_y},\frac{m_x-m_y}{n_x-n_y}) \leq 1.$

Proof: Using only paths Pr_i and Pu_i, we obtain that

$$\begin{aligned} d_G(x,y) &\leq (|m_x - m_y| + |n_x - n_y|)\lceil S_i \rceil \\ &\leq (1+\delta)\sqrt{(m_x - m_y)^2 + (n_x - n_y)^2} S_i (1 + \frac{1}{S_i}) \\ &= (1+\delta)(1+\frac{1}{S_i}) d(x,y), \end{aligned}$$

as desired.□

Lemma 3 *Given any integer point $p \in \mathbf{Z}^2$ and $1 \leq i \leq N$, there is a point $p_i \in \Lambda_i$ such that $d_G(p,p_i) < 10S_N$ and $d(p,p_i) \leq \sqrt{2}S_N$.*

Proof: Let $p \in C_N(m,n)$. Clearly, there is a $j \leq N$ such that p has graph distance at most $2\lceil S_j \rceil$ from some $q_j \in \Lambda_j \cap C_N(m,n)$. Let p_j (p_i) be the vertex of Λ_j (Λ_i) which is on $P(m,n)$, and whose existence is guaranteed by Lemma 1. By Lemma 2 there is a path in G from q_j to p_j whose length is at most $3d(p_j,q_j) \leq 3\sqrt{2}S_N < 5S_N$. Thus

$$\begin{aligned} d_G(p,p_i) &\leq d_G(p,q_j) + d_G(q_j,p_j) + d_G(p_j,p_i) \\ &\leq 2\lceil S_j \rceil + 5S_N + S_N \\ &< 10S_N. \end{aligned}$$

Since $p, p_i \in C_N(m,n)$, $d(p,p_i) \leq \sqrt{2}S_N$ readily follows. □

To establish the result stated in the abstract it is sufficient to prove the following

Theorem 1 *There exists a constant d_ϵ such that*

$$d_G(p,q) < (1+\epsilon)d(p,q)$$

for every pair $p,q \in \mathbf{Z}^2$ whose Euclidean distance exceeds d_ϵ.

Proof: Given $p,q \in \mathbf{Z}^2$, choose i so that $\tan\theta$, the slope of the line determined by p and q, differs from $\tan\theta_i = \frac{b_i}{a_i}$ by less than $\frac{\epsilon}{100}$. Choose $p_i = (m_1,n_1)_i$ and $q_i = (m_2,n_2)_i$ as in Lemma 3. Let α denote the angle between the lines pq and p_iq_i. Then $\sin\alpha \leq \frac{\sqrt{2}S_N}{d(p,q)/2}$, whence

$$\tan\alpha \leq \frac{\sqrt{2}S_N}{\sqrt{\frac{d^2(p,q)}{4} - 2S_N^2}} = \sqrt{\frac{8S_N^2}{d^2(p,q) - 8S_N^2}}.$$

Obviously,

$$\begin{aligned} \left|\frac{n_2 - n_1}{m_2 - m_1}\right| &< \tan(|\theta - \theta_i| + \alpha) \\ &< \tan(\frac{\epsilon}{100} + \alpha) \\ &< \frac{\epsilon}{10}, \end{aligned}$$

provided that $d(p,q)$ is so large that $\tan\alpha < \frac{\epsilon}{100}$. Thus, by Lemma 2,

$$d_G(p_i, q_i) < (1 + \frac{\epsilon}{10})(1 + \frac{1}{S_i})d(p_i, q_i).$$

On the other hand, Lemma 3 implies that

$$\begin{aligned} d_G(p,q) &\leq d_G(p,p_i) + d_G(p_i,q_i) + d_G(q_i,q) \\ &\leq 20S_N + (1 + \frac{\epsilon}{10})(1 + \frac{1}{S_i})(d(p,q) + 2\sqrt{2}S_N) \\ &\leq (1+\epsilon)d(p,q), \end{aligned}$$

if S_1 and $d(p,q)$ are sufficiently large. □

Our thanks are due to Paul Erdös, who communicated the problem discussed in this note to us. The question originated in computer science. If the Euclidean metrics could be approximated by graph metrics, then we could run efficient graph algorithms for solving geometric problems on-screen.

About the Complexity of Some Homomorphism Problems on Graphs

A. Quilliot, Aubière, France

Sometimes a graph may be considered as the discretization of some continuous structure, and it may be interesting to settle some problems, which appear to be mainly combinatorial in nature, in a discrete context. We deal here with the extension of homomorphism problems, the fixed point and the pursuit games problem, subwriting the connections which may exist between these questions, and discussing their respective complexity.

I COMPLEXITY OF SOME HOMOMORPHISM PROBLEMS ON GRAPHS

It happens sometimes that we may consider a graph as a discretization of some continuous structure. Then it can be of interest to study some problems which seem to be mainly combinatorial (for instance, the fixed point problem), inside a discrete context.

II THE EXTENSION OF HOMOMORPHISM PROBLEM : A GENERAL RESULT

A homomorphism between two oriented graphs $G = (X,\vec{E})$ and $H = (Y,\vec{F})$ is a function f from X to Y such that if $\overrightarrow{x,y} \in E$ then $\overrightarrow{f(x),f(y)} \in F$. If $G = (X,\vec{E})$ and $H = (Y,\vec{F})$ are two oriented graphs, if f is a function from $A \subset X$ to Y, then an instance of the extension of homomorphism problem (EHO) is defined by the question: Can f be extended into a homomorphism from G to H?

<u>Remark</u> : The above concept may be restricted to symmetric graphs (simple graphs). It follows that EHO is NP-complete since colouring the vertices of a simple graph G with n colours is also finding a homomorphism from G to the complete graph K_n.

<u>The Helly Property</u> : A family V of subsets of a set E satisfies the Helly property if for any $V' \subset V$ such that : $\forall v,w \in V'$, $v \cap w \neq \phi$ then also $\bigcap_{v \in V'} v \neq \phi$.

If we set **W** = {finite sequences in {0,1}} then with any path T from

R. Bodendiek, R. Henn (Eds.)
Topics in Combinatorics and Graph Theory

$x \in X$ to $y \in X$ in a graph $G = (X,\vec{E})$ we may associate $\sigma(T) \in W$ defined by :

an arc e in T induces 1 in $\sigma(T)$ if it is oriented like T (from x to y) and -1 otherwise.

If $x,y \in X, \sigma \in W$, then we set : $d_G(x,y) = \{\sigma(T), T \text{ paths from } x \text{ to } y\}$;

$B_G(x,\sigma) = \{y \in X / \sigma \in d_G(x,y)\}$;

and say that $G = (X,\vec{E})$ is a Helly graph if it is connected and if the $B_G(x,\sigma), x \in X, \sigma \in W$, satisfy the Helly property.

Extension theorem : If $G = (X,\vec{E})$ and $H = (Y,\vec{F})$ are two oriented graphs (eventually infinite), if f is a function from $A \subset X$ to Y and if H is a Helly graph, then f can be extended into a homomorphism from G to H if and only if :

$$\forall x,y \in A,\ d_G(x,y) \subset d_H(f(x),f(y)). \quad (*)$$

Proof : The side $\Rightarrow$) of the above equivalence is obvious. Conversely we only need to prove that if $x_0 \in X-A$, then $y_0 \in Y$ exists such that $(*)$ still holds with $A \cup \{x_0\}$ and f extended by $f(x_0) = y_0$. The rest will be mere induction technique.

Thus y_0 must be such that : $\forall x \in A,\ d_G(x_0,x) \subset d_H(y_0,f(x))$ which means :

$$\forall x \in A,\ \sigma \in d_G(x,x_0), y_0 \in B_H(f(x),\sigma).$$

But if $x,x' \in A$, $\sigma \in d_G(x,x_0)$, $\sigma' \in d_G(x',x_0)$, then the concatenation σ'' of σ and the reverse of σ' is in $d_G(x,x') \subset d_H(f(x),f(x'))$, which means $B_H(f(x),\sigma) \cap B_H(f(x'),\sigma') \neq \phi$. Since H is a Helly graph, we may choose y_0 in $\bigcap_{x \in A, \sigma \in W} B_H(f(x),\sigma) \neq \phi$. □

III THE EXTENSION OF HOMOMORPHISM PROBLEM FOR SIMPLE GRAPHS (SEHO)

N.B. : From now on, we mean by simple graph any non oriented graph which is loop complete (provided with all the possible loops).

The above result yields two possible interpretations, when the considered graphs are representative of order relations, and when they are simple graphs (symmetric and loop complete). In this last case, $d_G(x,y)$ becomes the shortest path distance between x and y in $G, B_G(x,n)$ the ball with center x and radius $n \in N$. We get :

Extension theorem for simple graphs : If $G = (X,E)$ and $H = (Y,F)$ are two simple graphs, (eventually infinite), if f is a function from $A \subset X$ to Y and if H is a Helly graph, then f can be extended into a homomorphism from

G to H if and only if

$\forall x,y \in A,\ d_H(f(x),f(y)) \leq d_G(x,y)$.

Remark : Any tree is a Helly graph; also any product of Helly graphs is a Helly graph. (The product $G \otimes H = (X.Y,T)$ of $G = (X,E)$ and $H = (Y,F)$ is defined by :

$[(x,y),(x',y')] \in T \Leftrightarrow [x,x'] \in E$ and $[y,y'] \in F$).

Recognizing whether a simple graph $G = (X,E)$ is a Helly graph can be done in $O(|X|^4)$ elementary operations (see E. Pesch and H. Bandelt [2]).

Theorem 2 : The restriction SEHO of EHO to simple graphs is NP-complete.

Proof : Let us consider $G = (X,\vec{E})$, $H = (Y,\vec{F})$, $A \subset X$, f from X to Y defining an instance of EHO, and do the following construction : any $e = \overrightarrow{x,y} \in \vec{E} \cup \vec{F}$ is replaced by an elementary path x,u_e,v_e,w_e,y and we add to G and H, three vertices U,V,W, all the edges $[U,u_e]$, $[V,v_e]$, $[W,w_e]$, $e \in \vec{E} \cup \vec{F}$ and all the possible loops. Then we get two simple graphs $\bar{G}$ and $\bar{H}$ and set : $\bar{A} = A \cup \{U,V,W\}$; $\forall x \in A$, $\bar{f}(x) = f(x)$;

$\bar{f}(U) = U$; $\bar{f}(V) = V$; $\bar{f}(W) = W$.

One easily checks that an extension of f from G to H exists if and only if an extension of $\bar{f}$ from $\bar{G}$ to $\bar{H}$ exists. Thus EHO is a special case of SEHO.
□

Theorem 3 : If $G = (X,E)$, $H = (Y,F)$, $A \subset X$ and f from A to Y define an instance of SEHO, and if H is a tree, then the following algorithm solves this instance in $O(|E| + |Y|)$ elementary operations.

Algorithm :

- Choose $y_0 \in Y$, orient H as an oriented tree with root y_0, and for every $y \in Y-y_0$, compute its predecessor pred(y) and its distance r(y) to y_0 (it can be done in $O(|Y|)$).
- Mark:= ϕ ; Not contr; B:=A ;

  While Mark $\neq$ X and Not Contr do

  begin

  Choose $x \in B \cap$ (X-Mark) such that r(x) is maximal;

  If $y \in B$ adjacent to x and such that $[f(y),f(x)] \notin F$ exists then Contr else for any $y \notin B$ adjacent to x do

    begin

    B:=B $\cup$ {y}; f(y) :=pred(x);

    end;

```
        Mark:= Mark ∪{x};
        end;
```

Easy verification. □

IV THE FIXED POINT AND PURSUIT GAME PROBLEMS

A simple graph $G = (X,E)$ satisfies the k-fixed point property ($k \in N$) if for any homomorphism f from G^k to G there exists a vertex $x = (x_1,..,x_k)$ in G^k and $i \in 1...k$ such that $[x_i,f(x)] \in E$.
Solving the k-fixed point problem (k-FP) means to recognize if a given graph G satisfies the k-fixed point property.
The k-pursuit game J(G,k) is played by a coalition of k pursuers $A_1,...,A_k$ and an evader B who moves in turn along the edges of G; the A_is choose first their initial position and try to catch the evader.
Solving the k-pursuit game problem (k-PG) means to determine who wins J(G,k). These two problems are closely related : if there exists a homomorphism f from G^k to G such that : $\forall x \in G^k$, $i \in 1,...,k, d_G(x_i,f(x)) \geq 2$, then the evader B wins J(G,k) by moving at every time to $f(x_i,...,x_k)$ when the $x_i, i \in 1...k$ are the current positions of the $A_i, i \in 1...k$.
Still, while checking the polynomiality of k-PG is easy (through dynamic programming), one may observe that :

Theorem 4 : 1-FP is NP-complete.

Proof : We polynomially reduce 3SA (3-satisfiability problems) to 1-FP as follows :
Given an instance of 3SA defined by a variable set V and a 3-factor set S; if $C = s_1 \vee s_2 \vee s_3$ (where $s_i = x$ or $\bar{x}$, $x \in V$), then we set $A(C) = \{\sigma \in \{0,1\}^3$ which can be considered as 3-uples of values for s the $s_i, i...3\}$, and we denote by σ_0 the element (0,0,0). If $C,C' \in S, \sigma \in A(C)$, $\sigma' \in A(C')$, we say that σ and σ' are compatible if they induce the same values for the elements of X.

Then we set : $X = \{(\sigma,C), C \in S, \sigma \in A(C)\}$;

$E_1 = \{[(\sigma_0,C),(\sigma,C')], C \neq C'\}$;

$E_2 = \{[(\sigma,C),(\sigma',C')], C,C' \in S, \sigma \in A(C), \sigma' \in A(C'),$ σ and σ' compatible$\}$;

$E = E_1 \cup E_2$;

and we define a simple graph G = (X,E). One can then easily check that the equation $\bigcap_{C \in S} C = 1$ has a solution if and only if there exists a homomorphism f from G to itself such that : $\forall x \in X, d_G(x,f(x)) \geq 2$.

If $n \in N$, let us call I_n the elementary chain $(0,1,2,\ldots,n)$; a simple graph $G = (X,E)$ is said to be contractible if there exists $n \in N$, $x_0 \in X$ and a homomorphism f from $G \otimes I_n$ to G such that $\forall x \in X, f(x,0) = x_0$ and $f(x,n) = x$.

Let us say that $y \in X$ dominates $x \in X$ if any vertex of G adjacent to x is also adjacent to y; one can see that G is contractible if and only if X can be written $X = \{x_0,\ldots,x_p\}$ in such a way that : $\forall i \in 0,\ldots,p-1$, there exists $y_i \in \{x_0,\ldots,x_i\}$ which dominates x_{i+1} in the subgraph of G induced by $\{x_0,\ldots,x_{i+1}\}$ and denoted by $G_{\{x_0,\ldots,x_{i+1}\}}$.

From this notion of dismantability by irreductibles first introduced by I. Rival [13] for the partially ordered sets, we deduce :

Theorem 5 : (See A. Quilliot (1978) [11], A. Nowakowski, P. Winkler (1982) [7]).

The pursuer wins J(G,1) if and only if G is contractible.

Corollary : Any contractible graph G satisfies the 1-fixed point property.
From the Extension theorem we also get :

Theorem 6 : Any Helly graph $G = (X,E)$ is contractible

Proof : Let us set n = diameter of G and $A = \{(x,n),(x,e), x \in X\} \subset X.\{0\ldots n\}$, and choose $x_0 \in X$. Since G is a Helly graph, the extension theorem can be applied to the function f defined by : $\forall x \in X, f(x,n) = x$ and $f(x,0) = x_0$.

The notion of contractibility (or 1-contractibility) can be extended as follows :

A connected graph $G = (X,E)$ is k-contractible ($k \geq 2$), if X can be written $X = X_0 \vee \ldots \vee X_p$ in such a way that :

$\forall i \in 0,\ldots p$, the subgraph G_{X_i} of G induced by X_i is (k-1)-contractible;

$\forall i \in 0,\ldots p-1$, there exists a retraction from $G_{X_0 \cup \ldots \cup X_{i+1}}$ to $G_{X_1 \cup \ldots \cup X_i}$;

(A retraction r from $G = (X,E)$ to a subgraph $G_A (A \subset X)$ is a homomorphism r from G to G_A such that : $\forall x \in A$, $r(x) = x$; this concept, first introduced by Borsuk [3] for topological spaces, has been widely studied in discrete context by I. Rival [13,6], P. Hell [5], E. Pesch [8] and others).

Theorem 7 : If $G = (X,E)$ is k-contractible ($k \geq 2$), then k pursuers win the game $J(G,k)$; as a consequence G satisfies the k-fixed point property.

Proof : We proceed by induction on k and the length p of the decomposition $X = X_0 \cup \ldots \cup X_p$. Let r be a retraction from G to G_{X-X_p}; once the A_is (pursuers) caught $r(B)$ (B is the evader), one of them (say A_k) follows $r(B)$, while the others move in G_{X_p} and catch B (if B moves out of X_p, he is caught by A_k).

Remark : This control technique permits to get the fact that if G is planar then the pursuers win $J(G,3)$ (M. Aigner and M. Fromm (1984) [1]), and that if the genus of G is k, then the pursuers win $J(G,2k+3)$ (A. Quilliot [12]).

Still, while recognizing the k-contractibility of a graph can be done in polynomial time if $k = 1$, (obvious), it remains an open problem for $k \geq 2$.

Conjecture : Any planar graph is 3-contractible.

BIBLIOGRAPHY

1 M. Aigner and M. Fromm : "A game of cops and robbers"; Disc Applied Math.8 (1984) pp.1-12.

2 H. Bandelt and E. Pesch : "Dismantling absolute retracts of reflexive graphs "European Jour. Comb.

3 K. Borsuk : "Sur les rétractes"; Fund. Math.17 (1931) pp.152-170.

4 M. Garey and D. Johnsson : "Computers and intractability; A guide to the theory of NP-completeness", Freeman and Co., San Francisco (1979).

5 P. Hell : "Retractes de graphes", Ph.D. Thesis, Univ. Montréal (1972).

6 R. Nowakowski and I. Rival : "Fixed edge theorem for graphs with loops", Jour. Graph Th 3 (1979) pp.339-350.

7 R. Nowakowski and P. Winkler : "Vertex to vertex pursuit on a graph" Disc Math 43 (1983) pp.235-239.

8 E. Pesch : "Retracts of graphs"; Math Syst.in Economics 110, Athenam (1988).

9 A. Quilliot : "An application of the Helly property to posets", J.C.T.A. 35 (1983) pp. 185-198.

10 A. Quilliot : "On the Helly property working as ...", J.C.T.A. 40 (1985) pp.186-193.

11 A. Quilliot : "Jeux positionnels", Thèse 3ème Cycle Paris VI (1978).

12 A. Quilliot : "A short note about pursuit ...", J.C.T.B. 38 (1985) pp.89-93.

13 I. Rival : "A fixed point theorem for partially ordered sets", J.C.T.A. 21 (1976).

On an Inequality of Sperner

A. Rausche, Greifswald, GDR

1. Introduction

Let n be an integer and let N be an n-element set. We will assume $N=\{1,2,\ldots,n\}$. We consider subsets of the set $\binom{N}{k}$ of all k-element subsets of N. Furthermore, we define the lower shadow $\Delta\mathcal{F}$ of a family $\mathcal{F}\subseteq\binom{N}{k}$ by $\Delta\mathcal{F}=\{F:\exists\, G\in\mathcal{F},\ F\subset G,\ |G\setminus F|=1\}$.

In the following lemma of Sperner [1] we can see an easy relationship between $|\mathcal{F}|$ and $|\Delta\mathcal{F}|$:

Lemma 1: If $\mathcal{F}\subseteq\binom{N}{k}$ then $|\Delta\mathcal{F}|\geqslant\frac{k}{n-k+1}\,|\mathcal{F}|$. Equality holds if and only if $\mathcal{F}=\emptyset$ or $\mathcal{F}=\binom{N}{k}$.

This gives a first lower bound for $\frac{|\Delta\mathcal{F}|}{|\mathcal{F}|}$. But the inequality is not good for families $\mathcal{F}$, which sizes are small relative to $\binom{n}{k}$. An important theorem in this field is the Kruskal-Katona-Theorem [2], [3]:

Theorem 1: Let $\mathcal{F}\subseteq\binom{N}{k}$ and $|\mathcal{F}|=\binom{a_k}{k}+\binom{a_{k-1}}{k-1}+\ldots+\binom{a_t}{t}$ with $a_k>a_{k-1}>\ldots>a_t\geqslant t\geqslant 1$.

Then $|\Delta\mathcal{F}|\geqslant\binom{a_k}{k-1}+\binom{a_{k-1}}{k-2}+\ldots+\binom{a_t}{t-1}$.

Now an useful corollary [4] can be proved.

R. Bodendiek, R. Henn (Eds.)
Topics in Combinatorics and Graph Theory

Corollary 1: Let $\mathcal{F}\subseteq\binom{N}{k}$ and $|\mathcal{F}|\leqslant\binom{m}{k}$ with $m\in\mathbb{N}$. Then

$$|\Delta\mathcal{F}| \geqslant \frac{k}{m-k+1}\,|\mathcal{F}|, \text{ i.e. } \frac{|\Delta\mathcal{F}|}{|\mathcal{F}|} \geqslant \frac{\binom{m}{k-1}}{\binom{m}{k}}.$$

Before proving corollary 1 let us repeat some notations. Consider the ordering $<_1$ on $\binom{N}{k}$ which is defined by $X<_1 Y$ iff $\max\{v:v\in X-Y\}<\max\{v:vY-X\}$. Let $C(m,k)$ denote the first m sets of $\binom{N}{k}$ relative to the ordering $<_1$ and $C\mathcal{F}=C(|\mathcal{F}|,k)$.

Then it is not difficult to verify that the Kruskal-Katona-Theorem is equivalent to

Theorem 2: Let $\mathcal{F}\subseteq\binom{N}{k}$. Then $|\Delta\mathcal{F}| \geqslant |\Delta(C\mathcal{F})|$.

Next, let us give a proof of corollary 1.

Proof of corollary 1:

$|C(\Delta\mathcal{F})| = |\Delta\mathcal{F}| \geqslant |\Delta(C\mathcal{F})|$ because of theorem 2.

With $|\mathcal{F}| \leqslant \binom{m}{k}$ we obtain $C\mathcal{F} \subseteq \binom{M}{k}$, where $M=\{1,2,\dots,m\}$. By lemma 1 we have $|\Delta(C\mathcal{F})| \geqslant \frac{k}{m-k+1}\,|C\mathcal{F}| = \frac{k}{m-k+1}\,|\mathcal{F}|$ □

2. The results

Is it possible to improve the inequality of corollary 1 for various conditions of $|\mathcal{F}|$?

Before we can state and prove our theorem we need the following Lemma and notations:

Lemma 2: i.) For all $x,y,z,w\in\mathbb{R}^+$

$$1.)\quad \frac{x}{y} \leqslant \frac{z}{w} \Leftrightarrow \frac{x}{y} \leqslant \frac{x+z}{y+w} \leqslant \frac{z}{w}$$

$$2.)\quad \frac{x}{y} \leqslant \frac{x+z}{y+w} \Leftrightarrow \frac{x}{y} \leqslant \frac{x+z}{y+w} \leqslant \frac{z}{w}$$

$$3.)\quad \frac{x+z}{y+w} \leqslant \frac{z}{w} \Leftrightarrow \frac{x}{y} \leqslant \frac{x+z}{y+w} \leqslant \frac{z}{w}$$

ii.) For all $x,y,z,w\in\mathbb{R}^+$ with $y>w>0$

$$\frac{x}{y} \leqslant \frac{x-z}{y-w} \Leftrightarrow \frac{x}{y} \geqslant \frac{z}{w}$$

Now we use $k,m\in\mathbb{N}^+$ with k fix and $m:=|\mathcal{F}|=\sum_{i=t}^{k}\binom{a_i}{i}$ with $a_k>a_{k-1}>\ldots>a_t\geqslant t\geqslant 1$. By theorem 1 we obtain $|\Delta\mathcal{F}|\geqslant\Delta m:=\sum_{i=t}^{k}\binom{a_i}{i-1}$.

Now we are able to state our main result.

<u>Theorem 3</u>: Let $\mathcal{F}\subseteq\binom{\mathbb{N}}{k}$ with $m=|\mathcal{F}| \in [1,\binom{a}{k}+\binom{b}{k-1}]$ and $a>b\geqslant b-1\geqslant 1$. Then

i.) for $b\leqslant\frac{k-1}{k}a-\frac{1}{k}$ is $\dfrac{|\Delta\mathcal{F}|}{|\mathcal{F}|}\geqslant\dfrac{\binom{a}{k-1}}{\binom{a}{k}}$

ii.) for $b\geqslant\frac{k-1}{k}a-\frac{1}{k}$ is $\dfrac{|\Delta\mathcal{F}|}{|\mathcal{F}|}\geqslant\dfrac{\binom{a}{k-1}+\binom{b}{k-2}}{\binom{a}{k}+\binom{b}{k-1}}$

3. Proof of theorem 3

i.) If $m\leqslant\binom{a}{k}$ and corollary 1, we have $\dfrac{\Delta m}{m}\geqslant\dfrac{\binom{a}{k-1}}{\binom{a}{k}}$. Now let $\binom{a}{k}+1\leqslant m\leqslant\binom{a}{k}+\binom{b}{k-1}$. Then $\dfrac{\Delta m}{m}=\dfrac{\binom{a}{k-1}+\sum_{i=t}^{k-1}\binom{b_i}{i-1}}{\binom{a}{k}+\sum_{i=t}^{k-1}\binom{b_i}{i}}$ with $b_{k-1}\leqslant b-1$.

$b\leqslant\frac{k-1}{k}a-\frac{1}{k}$ iff $\frac{b+1}{k-1}\leqslant\frac{a+1}{k}$, iff $\frac{k-1}{b-k+2}\geqslant\frac{k}{a-k+1}$ and iff

$$\frac{\binom{b}{k-2}}{\binom{b}{k-1}}\geqslant\frac{\binom{a}{k-1}}{\binom{a}{k}}. \tag{1}$$

By theorem 1 and corollary 1 for (k-1) we know

$$\frac{\sum_{i=t}^{k-1}\binom{b_i}{i-1}}{\sum_{i=t}^{k-1}\binom{b_i}{i}} \geqslant \frac{\binom{b}{k-2}}{\binom{b}{k-1}}$$ and with (1) holds $$\frac{\sum_{i=t}^{k-1}\binom{b_i}{i-1}}{\sum_{i=t}^{k-1}\binom{b_i}{i}} \geqslant \frac{\binom{a}{k-1}}{\binom{a}{k}}.$$ Then from

lemma 2i.)1.) we infer $\frac{\Delta m}{m} \geqslant \frac{\binom{a}{k-1}}{\binom{a}{k}}$.

ii.) $b \geqslant \frac{k-1}{k}a - \frac{1}{k}$ iff $\frac{\binom{a}{k-1}}{\binom{a}{k}} \geqslant \frac{\binom{b}{k-2}}{\binom{b}{k-1}}$. By lemma 2i.) 1.) we have

$$\frac{\binom{a}{k-1}}{\binom{a}{k}} \geqslant \frac{\binom{a}{k-1}+\binom{b}{k-2}}{\binom{a}{k}+\binom{b}{k-1}}. \tag{2}$$

Since $\frac{\Delta m}{m} \geqslant \frac{\binom{a}{k-1}}{\binom{a}{k}}$ for $m \leqslant \binom{a}{k}$ we consider the interval

$\binom{a}{k}+1 \leqslant m \leqslant \binom{a}{k}+\binom{b}{k-1}$. It is necessary to decompose this interval according to the further proof.

Therefore let $m \in \left[\binom{a}{k}+\binom{b-1}{k-1}, \binom{a}{k}+\binom{b}{k-1}\right]$ and

$$\frac{\Delta m}{m} = \frac{\binom{a}{k-1}+\binom{b-1}{k-2}+\sum_{i=t}^{k-2}\binom{b_i}{i-1}}{\binom{a}{k}+\binom{b-1}{k-1}+\sum_{i=t}^{k-2}\binom{b_i}{i}}$$ with $b_{k-2}<b-1$. Using the assumption it

is $\frac{\binom{b}{k-2}}{\binom{b}{k-1}} \leqslant \frac{\binom{a}{k-1}}{\binom{a}{k}}$ and with lemma 2i.)1.) we have

$$\frac{\binom{b}{k-2}}{\binom{b}{k-1}} \leqslant \frac{\binom{a}{k-1}+\binom{b}{k-2}}{\binom{a}{k}+\binom{b}{k-1}}. \qquad (3)$$

By corollary 1 we obtain $\frac{\Delta m'}{m'} := \frac{\binom{b-1}{k-2}+\sum_{i=t}^{k-2}\binom{b_i}{i-1}}{\binom{b-1}{k-1}+\sum_{i=t}^{k-2}\binom{b_i}{i}} \geqslant \frac{\binom{b}{k-2}}{\binom{b}{k-1}}$. Hence

$-\Delta m'\binom{b}{k-1} \leqslant -m'\binom{b}{k-2}$ and we get $\binom{b}{k-1}\left[\binom{b}{k-2}-\Delta m'\right] \leqslant \binom{b}{k-2}\left[\binom{b}{k-1}-m'\right]$

with $\binom{b}{k-1}-m'>0$ because of the definition of m'. It follows

$\frac{\binom{b}{k-2}-\Delta m'}{\binom{b}{k-1}-m'} \leqslant \frac{\binom{b}{k-2}}{\binom{b}{k-1}}$ and by the help of (3) $\frac{\binom{b}{k-2}-\Delta m'}{\binom{b}{k-1}-m'} \leqslant \frac{\binom{a}{k-1}+\binom{b}{k-2}}{\binom{a}{k}+\binom{b}{k-1}}$.

This holds [by lemma 2ii.)] iff

$$\frac{\binom{a}{k-1}+\binom{b}{k-2}}{\binom{a}{k}+\binom{b}{k-1}} \leqslant \frac{\binom{a}{k-1}+\binom{b}{k-2}-\binom{b}{k-2}+\Delta m'}{\binom{a}{k}+\binom{b}{k-1}-\binom{b}{k-1}+m'} = \frac{\binom{a}{k-1}+\binom{b-1}{k-2}+\sum_{i=t}^{k-2}\binom{b_i}{i-1}}{\binom{a}{k}+\binom{b-1}{k-1}+\sum_{i=t}^{k-2}\binom{b_i}{i}} = \frac{\Delta m}{m} \qquad (4)$$

Let $x' \in \mathbb{N}$ satisfying $x' \leqslant \frac{k-1}{k}a-\frac{1}{k}$ and $x'+1 \geqslant \frac{k-1}{k}a-\frac{1}{k}$. By (4) we have that for $m \in \left[\binom{a}{k}+\binom{x'+i}{k-1}, \binom{a}{k}+\binom{x'+i+1}{k-1}\right]$ and for all i;

$i=0,\dots,b-x'-2,b-x'-1$; holds $\frac{\Delta m}{m} \geqslant \frac{\binom{a}{k-1}+\binom{x'+i+1}{k-2}}{\binom{a}{k}+\binom{x'+i+1}{k-1}}$ since

statement (4) is true for arbitrary b with $b \geqslant \frac{k-1}{k}a-\frac{1}{k}$. Our considered interval $\binom{a}{k}+1 \leqslant m \leqslant \binom{a}{k}+\binom{b}{k-1}$ can be decomposed

into $m \in \left[\binom{a}{k}, \binom{a}{k}+\binom{x'}{k}\right] \bigcup_{i=0}^{b-x'-1} \left[\binom{a}{k}+\binom{x'+i}{k-1}, \binom{a}{k}+\binom{x'+i+1}{k-1}\right]$. Therefore we have previously investigated the intervals of the second part and by theorem 3i.) we know that $\frac{\Delta m}{m} \geqslant \frac{\binom{a}{k-1}}{\binom{a}{k}}$ for the first part of the interval.

Now we have to show only: How is the relation between the minimums of the several intervals of the second part?

In order to show that the desired minimum lies at the right border we need the following function $\varphi(y)=\frac{\binom{a}{k-1}+\binom{y}{k-2}}{\binom{a}{k}+\binom{y}{k-1}}$ with $y\in[x'+1,x'+2,\dots,b]$ and $y\geqslant\frac{k-1}{k}a-\frac{1}{k}$. Now the open problem is: When does $\varphi'(y)\leqslant 0$ hold? $\varphi'(y)$ denotes the first derivative of $\varphi(y)$ and

$$\varphi'(y)=\frac{\binom{y}{k-2}'\left[\binom{a}{k}+\binom{y}{k-1}\right]-\binom{y}{k-1}'\left[\binom{a}{k-1}+\binom{y}{k-2}\right]}{\left[\binom{a}{k}+\binom{y}{k-1}\right]^2}.$$

It is necessary to investigate the inequality $\binom{y}{k-2}'\left[\binom{a}{k}+\binom{y}{k-1}\right]-\binom{y}{k-1}'\left[\binom{a}{k-1}+\binom{y}{k-2}\right] \leqslant 0$. We get

$$\frac{1}{(k-2)!}\Big((y-k+3)\dots(y-1)y\Big)'\left[\binom{a}{k}+\binom{y}{k-1}\right]$$

$$-\frac{1}{(k-1)!}\Big((y-k+2)s\Big)'\left[\binom{a}{k-1}+\binom{y}{k-2}\right] \leqslant 0 \quad \text{with } s:=(y-k+3)\dots(y-1)y$$

and we obtain

$$(k-1)s'\left[\binom{a}{k}+\binom{y}{k-1}\right]-s\left[\binom{a}{k-1}+\binom{y}{k-2}\right]-(y-k+2)s'\left[\binom{a}{k-1}+\binom{y}{k-2}\right]\leqslant 0.$$

Since $s>0$ we must show only

$(k-1)s'\left[\binom{a}{k}+\binom{y}{k-1}\right]-(y-k+2)s'\left[\binom{a}{k-1}+\binom{y}{k-2}\right]\leqslant 0$ and since $s'>0$

$(k-1)\left[\binom{a}{k}+\binom{y}{k-1}\right]-(y-k+2)\left[\binom{a}{k-1}+\binom{y}{k-2}\right]\leqslant 0$. Our assumption was

$$\frac{\binom{a}{k-1}}{\binom{a}{k}} \geqslant \frac{\binom{a}{k-1}+\binom{y}{k-2}}{\binom{a}{k}+\binom{y}{k-1}} \geqslant \frac{\binom{y}{k-2}}{\binom{y}{k-2}}$$ and we can estimate

$$(k-1)\left[\binom{a}{k}+\binom{y}{k-1}\right]-(y-k+2)\left[\binom{a}{k-1}+\binom{y}{k-1}\right]$$

$$\leqslant (k-1)\frac{\binom{y}{k-1}}{\binom{y}{k-2}}\left[\binom{a}{k-1}+\binom{y}{k-2}\right]-(y-k+2)\left[\binom{a}{k-1}+\binom{y}{k-2}\right]$$

$$= \left[\binom{a}{k-1}+\binom{y}{k-2}\right]\left[(k-1)\frac{y-k+2}{k-1}-(y-k+2)\right]=0.$$

We have got $\varphi'(y)\leqslant 0$, i.e. $\varphi(y)$ is monotonically decreasing for $y\in[x'+1, \quad ,b]$ and $y\geqslant\frac{k-1}{k}a-\frac{1}{k}$. Under consideration of (2) we conclude our proof. □

<u>Remark 1</u>: If $b=\frac{k-1}{k}a-\frac{1}{k}$ then $\frac{\binom{a}{k-1}}{\binom{a}{k}}=\frac{\binom{b}{k-2}}{\binom{b}{k-1}}$ and with corollary 2i.)1.) is $\frac{\binom{a}{k-1}}{\binom{a}{k}}=\frac{\binom{a}{k-1}+\binom{b}{k-2}}{\binom{a}{k}+\binom{b}{k-1}}$. So we obtain the same result in theorem 3i.) and in theorem 3ii.).

References

[1] Sperner,E. Ein Satz über Untermengen einer endlichen Menge. Math.Z. 27, 544-548 (1928).

[2] Kruskal,J.B. The number of simplices in a complex. Mathematical optimization techniques, pp. 251-278, Univ. of California Press, Berkeley, Calif., (1963). MR 27-4771.

[3] Katona,G.O.H. A theorem of finite sets. Theory of Graphs

(Proc. Colloq., Tihany, 1966), pp. 187-207. Academic Press, New York, (1968). MR 45-76.

[4] Frankl,P. On Sperner families satisfying an additional condition. J. Combinatorial Theory Ser. A(20), 1-11 (1976).

Counting Perfect Matchings in Lattice Graphs

H. Sachs, Ilmenau, GDR

This is partly joint work with Khaled Al-Khnaifes from Damascus who investigated relations between graph theory and linear algebra in his doctoral dissertation [1].

1. Introduction

Let G be an undirected graph. A matching M of G is a subset of the edge set of G whose elements are pairwise disjoint; M is called perfect if it covers all vertices of G. Matching theory, in particular the theory of perfect matchings, is a well-developed part of graph theory: see, e.g., [11].

Let m(G) denote the number of perfect matchings contained in a finite graph G. m(G) plays an important role in some areas of chemistry (benzenoid hydrocarbons, Pauling's bond order; see, e.g., [2, 13]) and physics (crystal surfaces, the dimer problem; see, e.g., [10, 12, 14, 15]). Therefore, it is not only of theoretical interest to find efficient algorithms for calculating m(G). Such algorithms have been developed for some types of graphs of practical relevance, namely, (subgraphs of) hexagonal and square systems (Figure 1), see [2 - 8] .

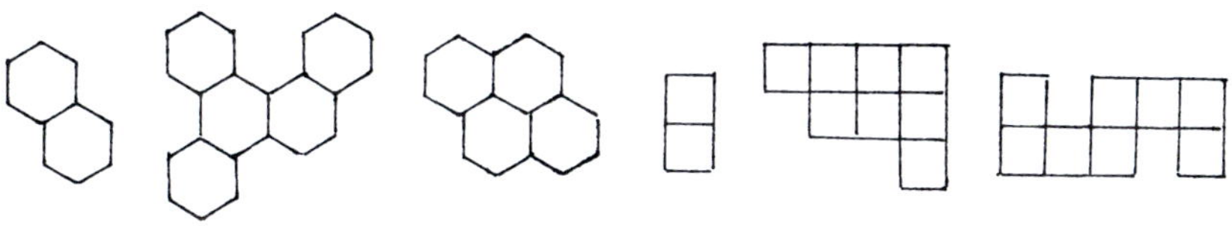

Figure 1. Some hexagonal and square systems

The methods used can be applied to much larger classes of graphs. The author has selected a result on plane "balanced lattice graphs" which seems to him particularly beautiful and may also have applications in physics and chemistry. The proofs (which are contained in [1]) will be published elsewhere; the

R. Bodendiek, R. Henn (Eds.)
Topics in Combinatorics and Graph Theory

general scheme is as follows. Using a well-known theorem of P. W. Kasteleyn [9, 10], for the graphs G considered m(G) can be expressed as the determinant of some matrix whose order equals half the number of vertices of G. By simple methods similar to those used for hexagonal and square systems (see [7, 8]), the order of the determinant is considerably reduced.

2. Definitions; Notation

A subset S of the Euclidean plane is called doubly periodic if its characteristic function is doubly periodic, i.e., if there are non-collinear vectors a, b such that $S+a = S+b = S$.

A point lattice is a doubly-periodic set which has no accumulation points. A lattice network is a doubly-periodic two-connected locally finite undirected plane graph whose vertex set is a point lattice. (Figures 2, 3a – 9a.)

Let G be a bipartite graph with bipartition $B = B(G)$, $W = W(G)$; the elements of B and W will be referred to as the black and white vertices of G, respectively.

Let N be a lattice network and P one of its fundamental parallelograms. N is called balanced if it is bipartite and $|B(N) \cap P| = |W(N) \cap P|$. (Figures 4 – 9.)

Let C be a circuit contained in a lattice network N. The plane subgraph G of N spanned by the vertices on and in the interior of C is called a lattice graph (with respect to N); C is its boundary. G is balanced if N is balanced and $|B(G)| = |W(G)|$. (Figures 1, 4b – 9b.)
Put $n(G) := |B(G)| = |W(G)|$.

A perfect matching D of a balanced network N is called a distinguished matching (briefly, a d-matching) if it satisfies the following conditions:

(i) D is subject to the same double periodicity as N,

(ii) for any face F of N, the number of edges of D which lie on the boundary C_F of F is one less than half the length of C_F. (Figures 4a – 9a.)

The edges of D are called the d-edges of N.

Remark. It is an open problem to determine under what conditions a given balanced lattice network possesses a distinguished perfect matching.

3. A Theorem on Balanced Lattice Graphs

Let N be a balanced lattice network, F a face of N with boundary C_F, D a d-matching of N, and G a balanced lattice graph with respect to N. Direct all d-edges from their black to their white end vertex and all other edges from their white to their black end vertex, thus turning each edge into an arc and N, C_F, D, G into $\vec{N}$, $\vec{C}_F$, $\vec{D}$, $\vec{G}$, respectively. Clearly,

(a) $\vec{N}$ has no sources or sinks,

(b) the number of arcs on $\vec{C}_F$ whose direction is in accordance with the positive orientation of C_F is odd.

Further, it is not too difficult to prove that

(c) $\vec{N}$ contains no directed circuit.

(See Figures 4 – 9.)

Let $s = s(\vec{G})$ be the number of sources of $\vec{G}$: because of (c), s is positive. Clearly, all sources of G are white, all sinks are black, and the number of sinks equals the number of sources; by (a), all sources and sinks lie on the boundary C of G.

Because of (b), the orientation of the arcs is of the "Kasteleyn type": thus, using this orientation, Kasteleyn's theorem can immediately be applied to G. Condition (c) enables the reduction procedure developed in [7, 8] (see [1] for a general setting) to be applied to Kasteleyn's skew determinant, thus reducing the order of the determinant eventually to be calculated from n(G) (this is half the number of vertices of G) to $s(\vec{G})$ (which is no bigger - and, frequently, much smaller - than half the number of vertices on the boundary C of G).

The algorithm.

If u is an arc of $\vec{G}$, denote by $\sigma(u)$ and $\tau(u)$ the vertex from which it springs and in which it terminates, respectively.

Let $\{a_1, a_2, \ldots, a_s\}$ and $\{b_1, b_2, \ldots, b_s\}$ be the sets of sources and sinks of G. Put

$$\underline{p}(a_k) := (\delta_{k1}, \delta_{k2}, \ldots, \delta_{ks}) \quad , \quad k = 1, 2, \ldots, s$$

where $\delta_{ii} = 1$, $\delta_{ij} = 0$ if $i \neq j$.

For any vertex x of G which is not a source, define the vector

$\underline{p}(x) = (p_1(x), p_2(x), \ldots, p_s(x))$ recursively by the formula

$$\underline{p}(x) := \sum_{\tau(u)=x} \underline{p}(\sigma(u)) .$$

(Note that multiple arcs may occur.)

Clearly, $p_k(x)$ is the number of paths issuing from source a_k and terminating in x.

Note that the vectors $\underline{p}(x)$ can easily be determined following step by step the paths issuing from the sources. (See Figures **4b – 9b**.)

Put

$$p_{ik} := p_k(b_i) , \quad P := (p_{ik}) \quad i,k = 1,2,\ldots,s .$$

Theorem. $m(G) = |\det P|$.

For some examples see the next section.

It should be mentioned that a similar algorithm and a similar theorem can also be formulated for certain classes of nonbipartite lattice graphs, e.g. for triangular and pentagonal lattice graphs (Figures 2,3) (see [1]) - however, these cannot be reported here.

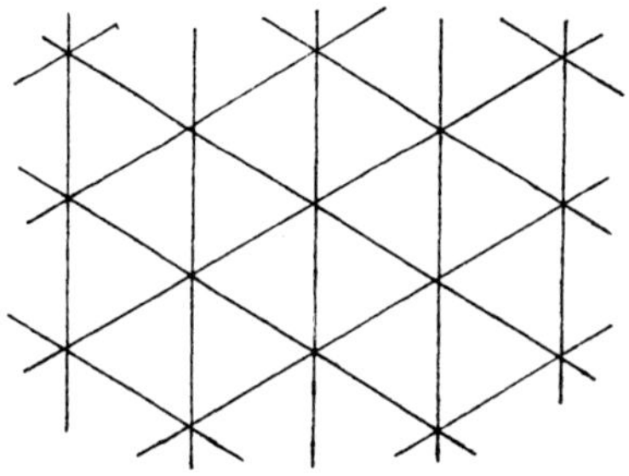

Fig. 2

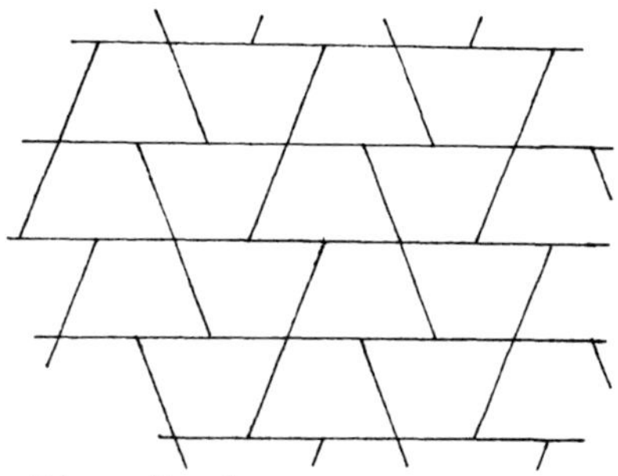

Fig. 3a.1

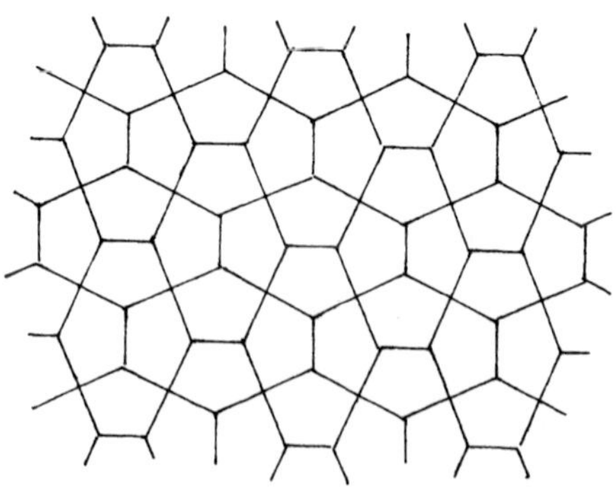

Fig. 3a.2

Figure 2.
Triangular lattice network

Figure 3a.
Two pentagonal lattice networks with isomorphic graphs

4. Some Examples

Sources: ◎ Sinks: ◉

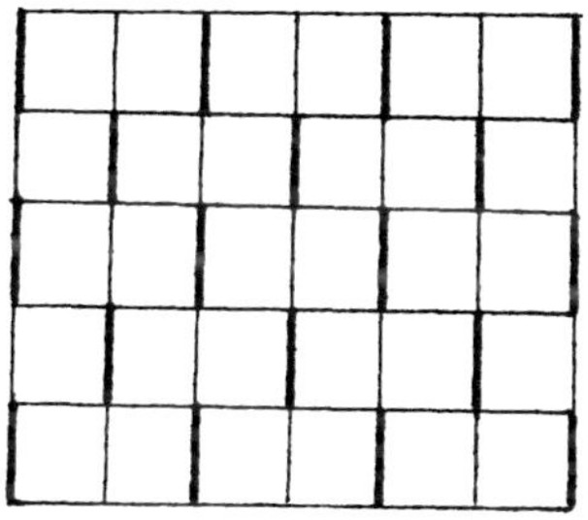

Fig. 4a.1

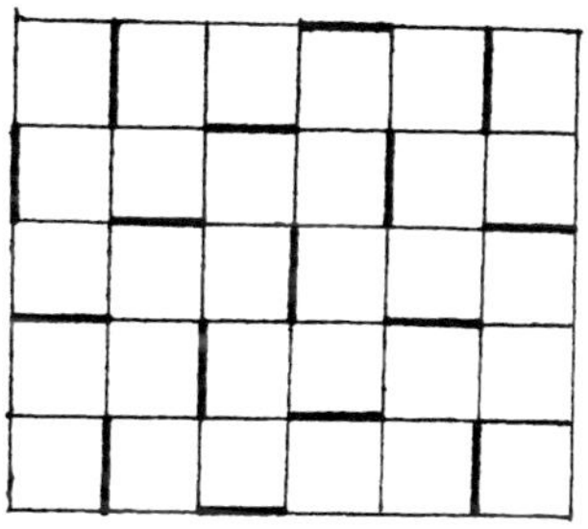

Fig. 4a.2

Figure 4a.
The square lattice network with two essentially different d-matchings

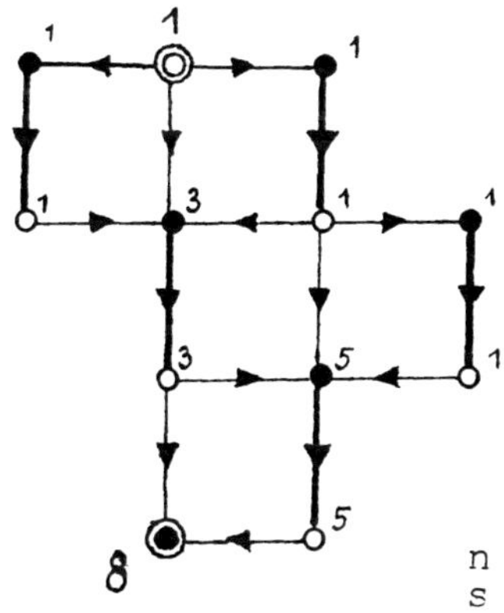

n = 6
s = 1

m = 8

Fig. 4b

Figure 5a.
The hexagonal lattice network

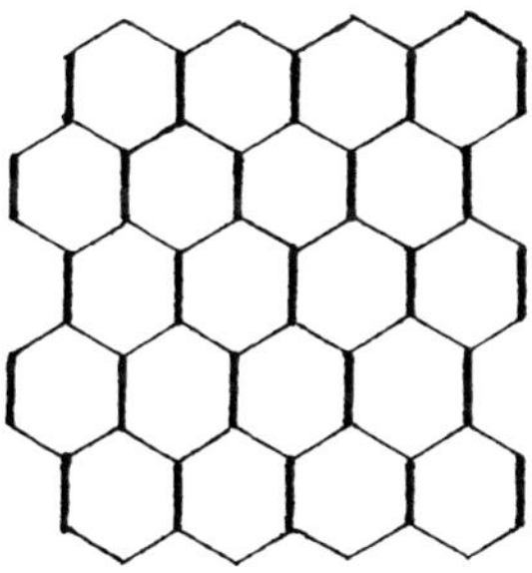

Fig. 5a

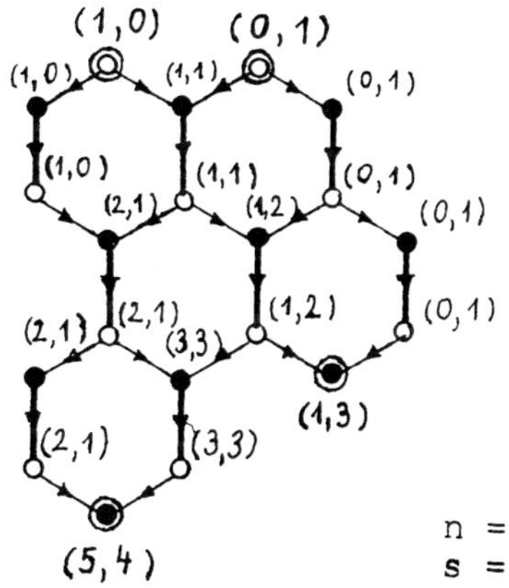

n = 10
s = 2

$$m = \left|\det\begin{pmatrix}5 & 4\\ 1 & 3\end{pmatrix}\right| = 11$$

Fig. 5b

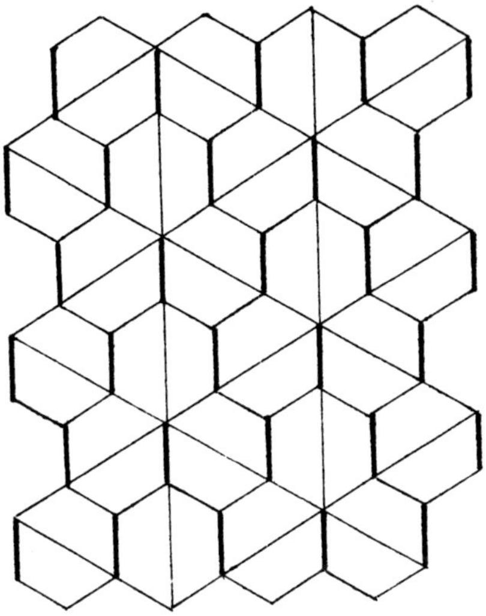

Fig. 6a

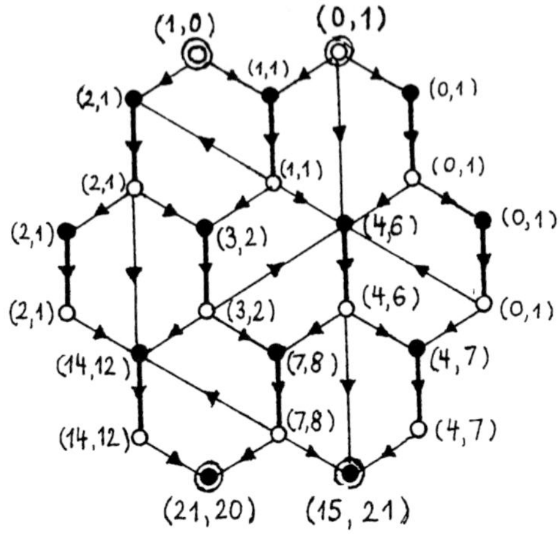

n = 12
s = 2

$$m = \left| \det\begin{pmatrix} 21 & 2C \\ 15 & 21 \end{pmatrix} \right| = 141$$

Fig. 6b

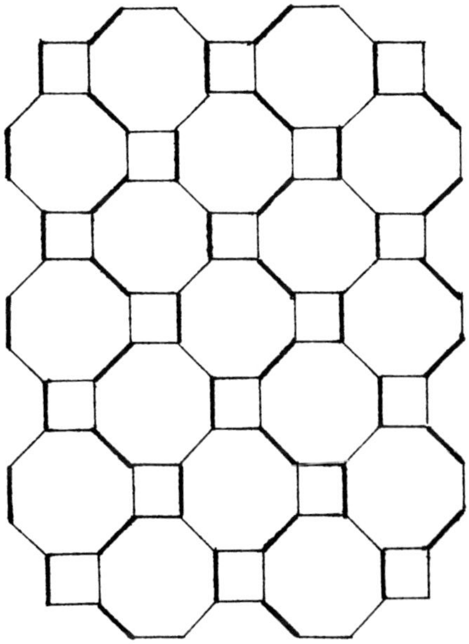

Fig. 7a

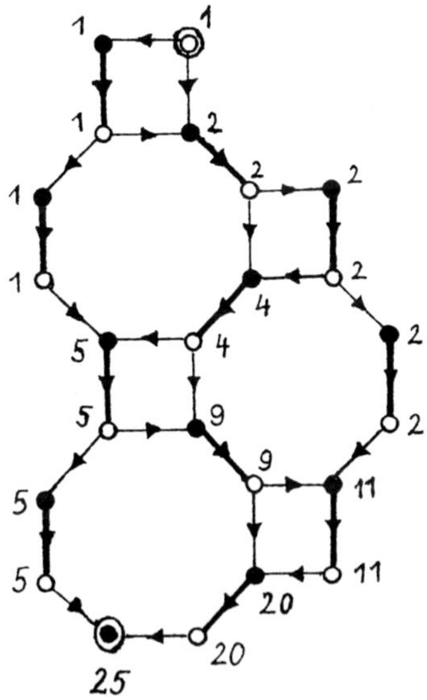

n = 12
s = 1
m = 25

Fig. 7b

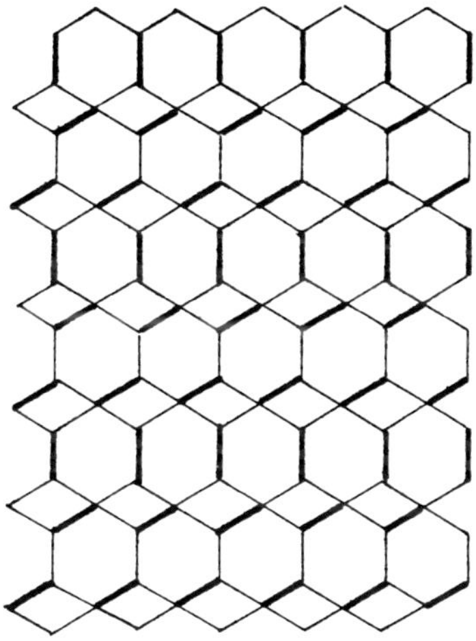

Fig. 8a

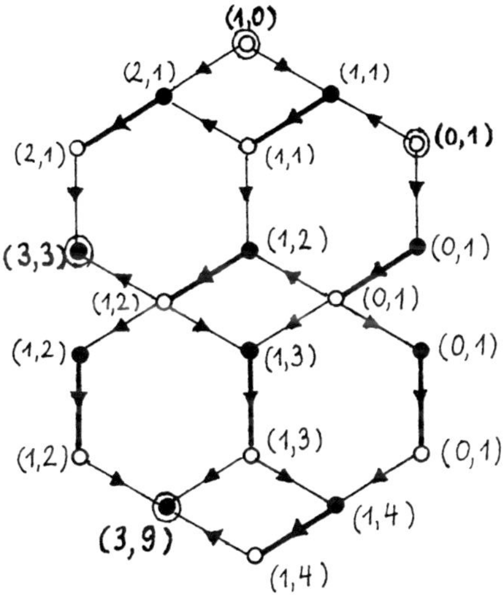

n = 10
s = 2

$$m = \left|\det\begin{pmatrix} 3 & 3 \\ 3 & 9 \end{pmatrix}\right| = 18$$

Fig. 8b

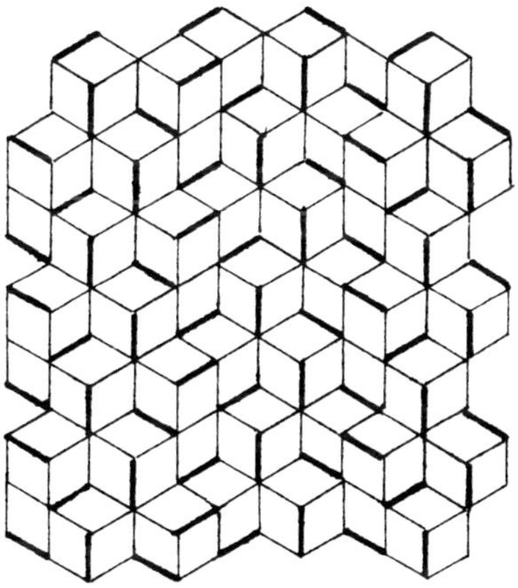

Fig. 9a

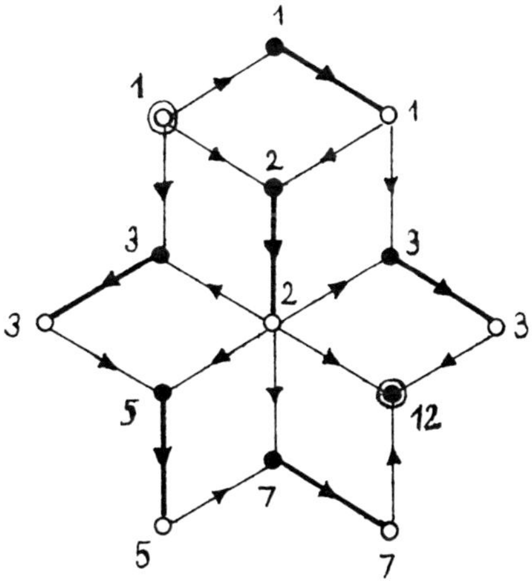

n = 8
s = 1

m = 12

Fig. 9b

REFERENCES

[1] Al-Khnaifes, Kh.: Graphentheoretische Methoden zur Lösung linearer Gleichungen und ihre Anwendung auf Probleme inner- und außerhalb der Graphentheorie. Doctoral Dissertation, Technische Hochschule Ilmenau 1988.

[2] Cyvin, S.J. and I. Gutman: Kekulé Structures in Benzenoid Hydrocarbons (Lecture Notes in Chemistry 46). Springer-Verlag Berlin 1988.

[3] Gordon, M. and W.H.T. Davison: Theory of resonance topology of fully aromatic hydrocarbons. J. Chem. Phys. 20 (1952), 428 - 435.

[4,5] John, P. and J. Rempel: Counting perfect matchings in hexagonal systems.
John, P. and H. Sachs: Calculating the number of perfect matchings and Pauling's bond orders in hexagonal systems whose inner dual is a tree.
In: Graphs, Hypergraphs and Applications (Teubner-Texte zur Mathematik 73). B.G. Teubner Leipzig 1985; pp. 72 –91.

[6] John, P. and H. Sachs: Calculating the numbers of perfect matchings and of spanning trees, Pauling's bond orders, the characteristic polynomial, and the eigenvectors of a benzenoid system. To appear in: Advances in the Theory of Benzenoid Hydrocarbons (ed. I. Gutman and S.J. Cyvin) (Topics in Current Chemistry 153). Springer-Verlag Berlin.

[7] John, P. und H. Sachs: Wegesysteme und Linearfaktoren in hexagonalen und quadratischen Systemen. In: Graphen in Forschung und Unterricht (ed. R. Bodendiek et al.). Verlag B. Franzbecker, Salzdetfurth 1985; pp. 85 - 101.

[8] John, P., H. Sachs and H. Zernitz: Counting perfect matchings in polyominoes with an application to the dimer problem. Zastosowania Matematyki (Applicationes Mathematicae) 19 (1987), 465 - 477.

[9] Kasteleyn, P.W.: The statistics of dimers on a lattice. Physica 27 (1961), 1209 - 1225.

[10] Kasteleyn, P.W.: Graph Theory and Crystal Physics. In: Graph Theory and Theoretical Physics (ed. F. Harary), Academic Press London 1967; pp. 43 - 110.

[11] Lovász, L. and M.D. Plummer: Matching Theory. Akadémiai Kiadó, Budapest 1986. North-Holland, Amsterdam 1986.

[12] Montroll, E.W.: Lattice Statistics. In: Applied Combinatorial Mathematics (ed. E.F. Beckenbach). John Wiley and Sons, New York 1964; pp. 96 - 143.

[13] Pauling, L.: The Nature of the Chemical Bond. Cornell University Press, Ithaca 1960.

[14] Percus, J.K.: Combinatorial Methods (Appl. Math. Sci. 4). Springer-Verlag New York 1971; pp. 89 - 123.

[15] Temperley, H.N.V. and M.E. Fisher: Dimer problems in statistical mechanics. An exact result. Phil. Mag. 6 (1961), 1061 - 1063.

Genus - Minimal Edges and Kuratowski Subgraphs of a Graph

J. Širaň, Bratislava, CSSR

1. Introduction

An edge e of a graph G will be said to be *genus-minimal* if $\gamma(G-e) < \gamma(G)$ or $\tilde{\gamma}(G-e) < \tilde{\gamma}(G)$ where γ and $\tilde{\gamma}$ denote the genus and the crosscap number (= nonorientable genus), respectively. A graph G will be called *minimal* with respect to a closed surface S if G does not embed on S but every proper subgraph of G embeds on S. In particular, every edge of a minimal graph (with respect to some closed surface) is genus-minimal.

By a *Kuratowski subgraph* of a graph G we understand any subdivision SK_5 of the complete graph K_5 or $SK_{3,3}$ of the complete bipartite graph $K_{3,3}$ in G. The *Kuratowski cover number of* G, *Kcn(G)*, is the smallest number of Kuratowski subgraphs of G whose union is G; if no such covering exists then we define Kcn(G) = ∞ . Thus, Kcn(G) < ∞ if and only if every edge of G is contained in a Kuratowski subgraph of G.

As a consequence of a very general result, Robertson and Seymour [7] proved that, for any closed surface S, the set of pairwise non-homeomorphic graphs minimal with respect to S is finite; an independent proof for orientable surfaces was found by Bodendiek and Wagner [3]. Nevertheless, very little is

R. Bodendiek, R. Henn (Eds.)
Topics in Combinatorics and Graph Theory

known about the structure of minimal graphs. In this context, a natural measure of complexity of such graphs seems to be their Kuratowski cover number. Vollmerhaus [11] conjectured that if G is a minimal graph with respect to S then Kcn(G) does not exceed $2\gamma(S)+1$ or $\tilde{\gamma}(S)+1$ according as S is orientable or not. As regards the orientable case, the same has been conjectured by Glover (cf. [4, p. 53]). On the other hand, by the result of Archdeacon [1], the conjecture of Vollmerhaus is true for the projective plane. But apart from this, no general result on the Kuratowski cover number of minimal graphs is known. In fact, it is even not known whether Kcn(G) is finite for every minimal graph G with respect to some surface [2].

Towards this end, we prove the following result.

Theorem 1. *Let e be a genus-minimal edge of a 3-connected graph G. Then e is contained in a Kuratowski subgraph of G.*

It is remarkable that if we restrict ourselves to 4-connected non-planar graphs then the above assertion is true, irrespective of whether e is minimal or not [8]. Therefore, at the first glance it is not obvious which topological reasons should force a genus-minimal edge of a graph of lower connectivity to be contained in a Kuratowski subgraph. Our method of proving Theorem 1 will be based on a detailed study of the combinatorial and topological properties of a possible counterexample; this will be carried out in Sections 2 and 3. The final Section 4 contains a summarization and a few remarks.

2. Contracting an uncovered edge

Our aim in this section will be to investigate the structure of a 3-connected graph G which contains an edge e not covered by any Kuratowski subgraph of G. Without any extra topological assumption we show that the graph G/e, obtained from G by contracting the edge e (and suppressing parallel edges that might thus have arisen), must be planar. This fact will enable us to gain more information about such a graph G.

We start with two useful technical results which have already been mentioned in the literature.

Lemma 2. *Let e be an edge of a graph G such that G/e is a block. If G/e contains an* SK_5 *then e belongs to a Kuratowski subgraph of G.*

Proof. See [9].□

Lemma 3. *Let G be a 3-connected non-planar graph containing an edge e = uv such that G/e is planar and G-u-v is a block. Then e belongs to a Kuratowski subgraph of G.*

Proof. This is a direct consequence of the method used in [10, proof of Theorem 3.2].□

Now we state and prove the already announced result on the planarity of G/e.

Lemma 4. *Let G be a 3-connected non-planar graph and e an edge of G which is not contained in any Kuratowski subgraph of G. Then the contracted graph G/e is planar.*

Proof. It is clearly sufficient to consider the case when G has at least 6 vertices. By [5], such a 3-connected non-planar graph G contains an $SK_{3,3}$. Since e belongs to no $SK_{3,3}$ in G, it follows from [8, Lemma 2] that G/e cannot contain any $SK_{3,3}$. But Lemma 2 implies that G/e cannot contain any SK_5, either. Consequently [6], G/e must be planar.□

As indicated in the beginning, much more can now be said about the structure of our graph G. Let us first sum up some facts concerning the connectivity $\varkappa$ of G and its subgraphs.

Corollary 5. *Let G be a 3-connected non-planar graph and e = uv an edge of G which does not belong to any Kuratowski subgraph of G. Then:*

(a) G/e is planar,

(b) $\varkappa(G) = 3$,

(c) $\varkappa(G/e) = 2$, and

(d) $\varkappa(G-u-v) = 1$.

Proof. All these assertions are immediate consequences of Lemmas 3 and 4.□

Further, we shall be interested in the block-cutvertex structure of the (obviously planar) graph H = G-u-v. For the sake of convenience, a block in H will be called *pendant* if it contains exactly one cutvertex of H. A block isomorphic to K_2 will be said to be *trivial*; any block other than K_2 will be called *non-trivial*.

Let G satisfy all the assumptions of Corollary 5. Let Γ be a plane embedding of G/e such that the vertex w corresponding to the contracted edge e lies in the outer face of the embedding. We shall now examine pendant blocks in the plane embedding Γ-w of the graph H = G-u-v.

Lemma 6. *Let B be a non-trivial pendant block of Γ-w. Let C be the boundary cycle of B and x the unique cutvertex of Γ-w contained in B. Then, no vertex in the subgraph B-C is adjacent to u or v in G. Moreover, only the following two cases can occur:*

(a) The path C-x is a union of two paths, say P_1 and P_2, intersecting in a single vertex y, and such that P_1 (resp. P_2) contains a vertex adjacent to u (resp. v) but no vertex of P_1-y (or P_2-y) is adjacent to the vertex v (or u), respectively.

(b) C-x contains exactly two vertices y and z both adjacent to u as well as to v. Then, no other vertex of C can be adjacent to one of the vertices u or v. In addition, x then belongs to precisely two blocks of H, one of them being B and the other being trivial.

Proof. Clearly, G contains an x-u or x-v path P such that P and C have only the vertex x in common. Let F be the subgraph of G induced by C and the vertices u and v. All the matter concerning (a) and the second part of (b) is based on the following observation [10, proof of Theorem 3.2]: If the cycle C contained four vertices t_1, t_2, t_3 and t_4 (in that cyclic order) such that t_1 and t_3 were joined (= adjacent or connected by P) to u and t_2, t_4 to v then the edge e would belong to an $SK_{3,3}$ in F ∪ P. As for the first and third assertion of (b), if C-x contained three vertices adjacent to both u and v, or if there were two such vertices in C-x and two paths Q_1: x-u and Q_2: x-v in G with $Q_1 \cap Q_2 = Q_1 \cap C = Q_2 \cap C = x$ then the edge e = uv would be contained in an SK_5 in F or $F \cup Q_1 \cup Q_2$, respectively. The latter implies that x cannot be incident with a further non-trivial block of H (recall that G is 3-connected!) or to three blocks of H. Finally, the fact that no vertex of B-C can be adjacend to u or v is a direct consequence of the planarity of Γ.□

As regards trivial pendant blocks of H = G-u-v ≅ Γ-w, it is worth noting that every pendant vertex of H is necessarily adjacent to both u and v in G (because of the 3-connectivity).

3. Embeddings of graphs with an uncovered edge

Throughout this section, G will denote a 3-connected non-planar graph containing an edge e = uv which lies in no Kuratowski subgraph of G. The structure of such graphs has already been investigated in the previous section. Now we shall focus on the existence of particular genus embeddings of these graphs. Our point of departure will be a simple observation.

Lemma 7. *Let the graph $H = G-u-v$ contain a trivial pendant block. Then every embedding of the graph $G-e$ can be extended to an embedding of G in the same surface (orientable or not).*

Proof. By our assumption, if y is a pendant vertex in H then y is a common neighbour of both u and v in G and $\deg_G y = 3$. But then we may insert the edge e in any given embedding of G-e along an appropriate "side" of the uyv path on the surface in question.□

Let us proceed with considering a slightly more complicated case. Let B be a non-trivial pendant block of the graph H such that B is isomorphic to K_3. Let x be the unique cutvertex of H contained in B, and let $f = yz$ be the edge of B which is not incident with x.

Lemma 8. *Let each of the vertices y and z be adjacent to both u and v in G. Then, every embedding $i: G-e-f \longrightarrow S$ can be extended to an embedding $j: G \longrightarrow S$, regardless of the orientability of the surface S.*

Proof. It follows from Lemma 6(b) that $\deg_G x = 3$. Applying the procedure described in the preceding proof, we may extend any embedding $i: G-e-f \rightarrow S$ to $i_f : G-e \rightarrow S$ by drawing the edge $f = yz$ along the yxz path in S. After having done this, the edges uy and vy have obviously remained consecutive in a local rotation at the vertex y with $\deg_{G-f} y = 3$. Thus, it is possible to draw in i_f the edge e along the path uyv to obtain the required embedding of the whole graph G in S. □

We are now ready to pass to the main result on genus embeddings of our graph G; the blocks of the graph $H = G-u-v$ will be assumed arbitrary.

Lemma 9. $\gamma(G-e) = \gamma(G)$ *and* $\tilde{\gamma}(G-e) = \tilde{\gamma}(G)$.

Proof. Let us employ induction on the number of edges

of G. We take Lemmas 7 and 8 for the base of the induction. Consider first the case when H contains a non-trivial pendant block B of the type referred to in Lemma 6(a). Replace B by a single edge (i.e., by a trivial block) xy where x is the unique cutvertex in B and y is adjacent to both u and v, obtaining thereby a new graph G_o. Note that G_o is homeomorphic to a subgraph of G; the same holds for G_o-e. Clearly, we may apply the induction hypothesis to G_o (or simply use Lemma 7) which yields $\gamma(G_o-e) = \gamma(G_o)$ and $\tilde{\gamma}(G_o-e) = \tilde{\gamma}(G_o)$. However, the planarity of B together with the properties listed in Lemma 6(a) guarentee that the edge xy can be "blown up" to obtain an embedding of the whole graph G in the same surface where G_o or G_o-e have been embedded. Consequently, $\gamma(G-e) = \gamma(G)$ and $\tilde{\gamma}(G-e) = \tilde{\gamma}(G)$.

It remains to investigate the case when B is a pendant block of the type (b) from Lemma 6 and such that B has at least four vertices. Let x, y and z be as in the statement of Lemma 6(b). If there is a vertex t in B which is connected in B to x, y and z by three pairwise disjoint (up to t) non-trivial paths, then let us replace B by the graph $K_{1,3}$ with vertices x, y, z and t where t will be the vertex of degree 3. We have thus constructed a new graph G_1. Applying now induction hypothesis (or, again Lemma 7) we see that $\gamma(G_1-e) = \gamma(G_1)$ and similarly for the non-orientable genus. As above, it is a matter of routine to check that embeddings of both G_1-e and G_1 can be expanded to embeddings of G-e and G into the same surface, as required. Finally, if there is no such vertex t in B then, necessarily, B is a union of three connected planar graphs B_{xy}, B_{yz} and B_{xz} such that $B_{xy} \cap B_{xz}$ = x, $B_{yz} \cap B_{xz}$ = z, and $B_{xy} \cap B_{yz}$ = y. Now, replace B by the triangle xyz, obtaining thereby a graph G_2. By virtue of Lemma 8 used in place of the induction hypothesis, both G_2 and G_2-e have genus embeddings on the same surface. The conclusion that the latter embeddings are extendable to embeddings of G and G-e (without changing the genus of the surface, orientable or not) is a consequence of Lemma 6(b) and the above facts.□

4. Proof of Theorem 1 and concluding remarks

Summing up the results established in the preceding two sections, we have proved:

Corollary 10. *Let e be an edge of a 3-connected graph G such that e does not belong to any Kuratowski subgraph of G. Then e cannot be a genus-minimal edge.*

This last assertion is, however, equivalent to Theorem 1. We have thereby completed the proof of our main result.□

As regards the already discussed conjectures of Vollmerhaus [11] and Glover (cf. [4, p. 53]), we have the following cosequence.

Corollary 11. *Let G be a 3-connected graph which is minimal with respect to a closed surface. Then Kcn(G) < ∞ .*□

The above two corollaries are easily seen to be equivalent mathematical statements. It would be worth trying to extend them also to graphs of connectivity smaller than 3. A particularly nice way how to do it could be based on accordingly refined versions of the known results on the genus of 2-amalgamations of graphs.

References

[1] D. Archdeacon, A Kuratowski theorem for the projective plane, J. Graph Theory 5(1981)243-246.

[2] D. Archdeacon, personal communication.

[3] R. Bodendiek, K. Wagner, Solution to König's graph embedding problem, Math. Nachrichten 140(1989)251-272.

[4] J.L. Gross, T.W. Tucker, "Topological Graph Theory", Wiley, New York, 1987.

[5] D.W. Hall, A note on primitive skew curves, Bull. Amer. Math. Soc. 49(1943) 939-937.

[6] K. Kuratowski, Sur le probléme des courbes gauches en topologie, Fund. Math. 15(1930)271-283.

[7] N. Robertson, P. Seymour, Graph minors VIII. A Kuratowski theorem for general surfaces, submitted.

[8] J. Širáň, Edges and Kuratowski subgraphs of non-planar graphs, Math. Nachrichten 113(1983)187-190.

[9] J. Širáň, R. Bodendiek, Planarity - an introductory survey (in German), in: Graphentheorie II (K. Wagner and R. Bodendiek, Eds.), Wissenschaftsverlag Mannheim-Wien-Zürich, 1989, pp.

[10] C. Thomassen, Planarity and duality of finite and infi- infinite graphs, J. Combinatorial Theory (B) 29(1980)244-271.

[11] W. Vollmerhaus, A characterization of minimal graphs which are embeddable (non-embeddable) in a given 2-dimensional manifold, in: Combinatorical Structures and their Applications (Proc. Conf.Calgary, June 1989) Gordon and Breach, New York - London - Paris, 457-458.

From Tree Path-Factors and Doubly Exponential Sequences to a Binomial Inequality

Z. Skupien, Krakow, Polen

1. Introduction

The aim of this note is to report on a research that led to the formulation of the following constrained inequality between products of some binomial coefficients.

<u>Theorem</u> (Skupień [10]). If p and d_i's are variable integers, $p \geq 1$ and each $d_i \geq 0$, which satisfy the constraint $\Sigma_{i=1}^{p} d_i = s = \text{const} \geq 2$ and, moreover,

$q = \lfloor (s+2)/5 \rfloor$ and $r = s-5q$ (whence $|r| \leq 2$) then

$$\Pi_{i=1}^{p} \binom{d_i}{2} \leq \begin{cases} \binom{s}{2} & \text{for } 2 \leq s \leq 7, \\ \binom{5}{2}^{q-|r|} \binom{5+\operatorname{sgn} r}{2}^{|r|} & \text{for } s \geq 8, \end{cases} \tag{1.1}$$

the equality holds iff both sides are identical up to the ordering of factors.

Proof is very simple and follows from the very definition of a binomial coefficient and the following nine inequalities where a and b are integers: $\binom{a}{2} < \binom{a-4}{2}\binom{4}{2}$ for $a \geq 8$,

$\binom{a}{2}\binom{b}{2} < \binom{a+1}{2}\binom{b-1}{2}$ if $0 < a < b-1$, $\quad \binom{3}{2}^2 < \binom{6}{2}$,

$\binom{3}{2}\binom{4}{2} < \binom{7}{2}$, $\quad \binom{4}{2}^3 < \binom{6}{2}^2$, $\quad \binom{6}{2}\binom{7}{2} < \binom{4}{2}^2\binom{5}{2}$,

$\binom{7}{2}^2 < \binom{4}{2}\binom{5}{2}^2$, $\quad \binom{6}{2}^3 < \binom{4}{2}^2\binom{5}{2}^2$, $\quad \binom{a}{2} < \binom{a+1}{2}$ if $a > 0$.

R. Bodendiek, R. Henn (Eds.)
Topics in Combinatorics and Graph Theory

In fact, the following properties of the right-hand side of (1.1) are implied. (i) Upper symbols in factors are from {2, ..., 7}; (ii) Upper symbols in any two factors differ by at most 1; (iii) If there are two factors then all upper symbols are from {4, 5, 6}, each of 4 and 6, however, can appear at most twice. Hence the number q is the number of factors if $s \geq 3$ and, for $s \geq 8$, r represents factors with the upper symbol 5 + sgn r as stated in Theorem. □

Remarks. 1. Sharpness of the result has implications in some extremal counting problems or extremal characterization problems. For instance, if $s = n-1 \geq 3$, it plays a key role in characterizing [10] n-vertex forests with the maximum number of maximum path-factors (i.e., spanning subgraphs of maximum size and with each component being a path (inclusive of K_1)). Those forests are some trees, namely, special trees of stars.

2. Independently (though earlier), related results involving general binomial coefficients $\binom{d}{h}$ where $d \in \mathbb{N}$ or $d \in \mathbb{R}_+$ are obtained by Tomescu [14] in the course of his maximizing the number of maximal cliques in some (uniform) (h+1)-hypergraphs on n vertices, n = s. In particular, the inequality (1.1) for $s \geq 8$ can be seen as a concise account of a part of [14, Corollaire 1]. The Tomescu's paper stems from the paper [6] which presents the complete solution for graphs (h = 1). That solution is based on a result which is an analog of our Theorem for h = 1. Namely, for $s \geq 2$, the right-hand side of the corresponding inequality is in our notation exactly $3^{q-|r|}(3+\operatorname{sgn} r)^{|r|}$ or, additionally for $s \equiv 1 \pmod 3$, $3^{q-1}\,2^2$ where $q = \lfloor (s+1)/3 \rfloor$ and $r = s-3q$ (whence $|r| \leq 1$). Consequently, for $n = s \geq 2$, the complete multipartite graph $K\left(\{3\}^{q-|r|} \times \{3+\operatorname{sgn} r\}^{|r|}\right)$ and, additionally, $K\left(\{3\}^{q-1} \times \{2\}^{2}\right)$ if $n \equiv 1 \pmod 3$ is/are the only n-vertex graph(s) with the maximum number of maximal cliques. In the Tomescu's paper the uniqueness of the right-hand side is left unsettled, though asymptotics as $n \to \infty$ or $h \to \infty$ are presented there instead.

2. Generalizations

The following results can be proved. First a continuous version of the generalized problem under consideration. Let s, x $\in \mathbb{R}$, h $\in \mathbb{N}$, h = const $\geq$ 1, s = const > 0, x $\neq$ 0, and x $\geq$ h-1 and let

$$f(x) = \binom{x}{h}^{s/x} \quad \text{where} \quad \binom{x}{h} = x\,(x-1)\,\ldots\,(x-h+1)/h!.$$

Then there exists $\max_x f(x)$ which is $f(x_h)$ for exactly one $x = x_h$ which is independent of s and such that f is strictly increasing ($f \nearrow$) and strictly decreasing to 1 ($f \searrow 1$) in intervals $(h-1,\ x_h]$ and $[x_h,\ \infty)$, respectively. Moreover, the sequence $x_h \nearrow$ and, for $h \to \infty$,

$$x_h \cong 2h + \frac{1}{2} \ln h \qquad \text{(Tomescu [14])},$$

$$= 2h + \frac{1}{2} \ln (h\pi/e) + O(\frac{1}{h} \ln^2 h) \qquad \text{(Schinzel [7])},$$

$$\approx 2h + \Sigma_{i\geq 0} \frac{1}{h^i} \Sigma_{j=0}^{i+1} c_{ij} \ln^j h\pi$$

where c_{ij} are constants which can be recursively computed, e.g., $(c_{01}, c_{00}) = (1, -1)/2$, $(c_{12}, c_{11}, c_{10}) = (1, 1, 0)/8$, $(c_{23}, c_{22}, c_{21}, c_{20}) = (4, 9, 20, 3)/192$ etc.

As for a general discrete problem, recall that $\binom{x}{h}$ for x > h-1 is (convex and) strictly log-concave whence, for integers a and b,

$$\binom{a}{h}\binom{b}{h} < \binom{a+1}{h}\binom{b-1}{h} \quad \text{if } a \leq b-2 \text{ and, for } h > 1,\ h-1 \leq a.$$

Let

$$\nu = \nu_h = \max\,\{n \in \mathbb{N}: \binom{n-1}{h}^n < \binom{n}{h}^{n-1}\},$$

$$\rho = \rho_h = \max\,\{j \in \mathbb{N}: \binom{\nu-1}{h}^j \binom{\nu}{h}^{\nu-2j+1} > \binom{\nu+1}{h}^{\nu-j}\},$$

$R = \nu-1-\rho$, $q = q(s, h) = \lfloor (s+\rho)/\nu \rfloor$, and let $r = s-q\nu$. Then ν is the unique integer out of $\lfloor x_h \rfloor$ and $\lceil x_h \rceil$ such that $\max_{x\in\mathbb{N}} f(x) = f(\nu)$. Next, $0 \leq \rho$, $R \leq \nu-1$, and $-\rho \leq r \leq R$.
The main result is that, for each h, there is an integer s_h, $h < s_h \leq \nu^2 + \nu$, which is the minimum of the integer $\tilde{s}$ such that, for each integer $s \geq \tilde{s}$ and all sequences (d_i) of nonnegative integers d_i with $\Sigma_i\, d_i = s$,

$$\Pi_i \binom{d_i}{h} \le \binom{\nu}{h}^{q-|r|} \binom{\nu+\mathrm{sgn}\ r}{h}^{|r|}$$

where equality, for $h > 1$, holds iff both sides are identical up to the ordering of factors. The proof depends on the fact proved by Schinzel [7] that the sides of the inequality in the definition of ρ are mutually distinct for each integer $j \in [0, \nu]$ if $h > 1$.

Hence $\rho = \lfloor (\log \binom{\nu}{h} - \nu \log \frac{\nu+1}{\nu+1-h}) / \log \frac{\nu\ (\nu+1-h)}{(\nu+1)\ (\nu-h)} \rfloor$. Next, for $1 \le h \le 11$, $\nu_h = 2h+1$ and $s_h = R\nu-\rho$ where $\rho_h = 1$ for h= 1, 9, 10; $\rho_h = 2$ for h = 2, 3, 6, 7, 8; $\rho_4 = 3 = \rho_5$ and $\rho_{11} = 0$. Therefore $s_1 = 2$ and $s_2 = 8$ which agrees with what is in Introduction. Moreover, if h = 12, 13 then $\nu_h = 2h+2$ and $s_h = (\rho-1)(\nu-1)$ where $\rho_{12} = 25$ and $\rho_{13} = 26$.

3. Motivation and Some Terminology

It is a research into maximum path-factors that was a stimulus and gave a hint to our Theorem. We are going to present a more detailed account of the final part of that research but a motivation and background follow first. Because a path-factor, F, is a forest with the maximum degree $\Delta \le 2$ therefore F is maximum iff the number of components, k(F), is as small as possible, i.e., $k(F) = \pi(G)$, the vertex-path partition number of the graph G. Notice that, for a nonhamiltonian G on three or more vertices, $\pi(G)$ coincides with the Hamiltonian completion number, hc(G), and the Hamiltonian shortage, $s_H(G)$, in G. Determining $\pi(G)$ is a hard combinatorial problem because it includes both the decision problem TRACEABILITY (whether $\pi(G) = 1$) which is NP-complete and co-TRACEABILITY. Another (harder) problem formulated by the present author in 1974 [8] consists in finding the family, $\mathcal{P}(G)$, of all maximum path-factors of G. Let

$$\varphi(G) = |\mathcal{P}(G)|.$$

Both the problems restricted to trees (or forests) remain non-trivial. Moreover, trees are important herein. For instance, an approximate solution to determining hc(G), presented in [1] (which in fact estimates $\pi(G)$), for a connected graph G consists of a heuristics for a spanning tree, T, and of finding $\pi(T)$ as an approximation to $\pi(G)$. Finding $\pi(T)$ as well as finding a maximum path-factor in T are known [5, 3] to have linear complexities.

The family $\mathcal{P}(T)$ for a tree T can be found either by using a simple linear program [2, cf. 9] with a totally unimodular matrix or by applying a recursive method [13]. That method proves fruitful and inspiring. It includes a linear time procedure for classifying [12] vertices and edges of T with respect to their appearance in $\mathcal{P}(T)$. In particular, a vertex is either a terminal (i.e., an endvertex (or hanging, that is, of degree 0 or 1) in some or all elements of $\mathcal{P}(T)$) or always an inner vertex. An edge is compulsory, forbidden and free iff it appears in all, in no and in some but not all path-factors from $\mathcal{P}(T)$, respectively. A vertex is called a branching vertex if its degree is at least three. Given a tree (or forest) T, let $\tilde{T}$ be the union of all paths whose each endvertex is a branching vertex of T. Then each endvertex, x, of $\tilde{T}$ is called a hanging branching vertex of T. Then, for $d(x, \tilde{T}) = 1$, the edge, e_x, of $\tilde{T}$ incident to x is called the edge of attachment of x in T. If $e_x = xv$ and the vertex v is a terminal in T then the edge xv is free, otherwise xv is forbidden.

Once a method for generating all of $\mathcal{P}(T)$ was described, it remained to estimate its worst case asymptotical performance. To this end, a sequence (T_i) of trees T_i on n_i vertices with $n_i \to \infty$ was constructed such that $\varphi(T_i)$ was bounded below by an increasing exponential function of n_i. This showed an intrinsic intractability of the problem. Nevertheless, the problem of determining the structure of forests F on n vertices with the largest possible number $\varphi(F)$ was posed and next solved in [10]. It is a construction of two sequences of trees and related estimates of φ which are our main subject.

4. Estimates for Parametric Nonlinear Recurrences

Let d be an integer, $d \geq 3$. Let θ_d consist in multiplying the number of endvertices by d-1; namely, θ_d establishes a bijection between endvertices and disjoint sets of d-1 new vertices each and makes each endvertex adjacent to d-1 new associated vertices. Put $T_0 = v$, a single vertex, and use iterates to get

$$T_i = \theta_d^{2i}(T_0), \quad T_i' = \theta_d^{2i-1}(T_0).$$

Then v is called the central vertex of each T_i and T_i'. Moreover, none of T_i and T_i' contains a forbidden edge and the order of T_i is

$$n_i = \Sigma_{j=0}^{2i} (d-1)^j. \tag{4.1}$$

Let $s_i = \varphi(T_i)$ and let w_i be the number of path-factors in T_i such that their size is maximum but the central vertex v is an endvertex. Hence $s_i > w_i$ for $i > 0$. Let s_i' and w_i' correspond to T_i'. Now T_i' (resp. T_i) can be viewed as the union of v and d-1 copies of T_{i-1} (resp. T_i') together with edges joining v to all central vertices of those copies. Hence and because v is a terminal in each T_i and in no T_i',

$$\begin{cases} s_i' = \binom{d-1}{2} w_{i-1}^2 s_{i-1}^{d-3} \\ w_i' = (d-1)\, w_{i-1}\, s_{i-1}^{d-2} \end{cases} \quad \begin{cases} s_i = w_i + \binom{d-1}{2} w_i'^2 s_i'^{d-3} \\ w_i = s_i'^{d-1} + (d-1)\, w_i'\, s_i'^{d-2} \end{cases} \tag{4.2}$$

with $w_0 = 1 = s_0$. Letting

$$Q_i = s_i/w_i \quad \text{for } i \geq 0 \tag{4.3}$$

one gets

$$Q_i = Q_{i-1} + \frac{d}{2d-2} + \frac{(d-2)^2}{(2d-2)\,((2d-2)Q_{i-1}+d-2)} \quad \text{for } i \geq 1 \tag{4.4}$$

where $Q_0 = 1$. Hence $Q_i \nearrow$ as $i \nearrow$. From (4.2) one gets

$$w_i = w_{i-1}^{(d-1)^2}\, G_{i-1} \tag{4.5}$$

where

$$G_j = Q_j^{(d-3)(d-1)} \binom{d-1}{2}^{d-2} \left(\binom{d-1}{2} + (d-1)^2 Q_j\right) \tag{4.6}$$

whence $G_j \nearrow$ as $j \nearrow$. Therefore $w_i \geq w_{i-1}^{(d-1)^2} G_k$ for each $i > k$ with equality for i= k+1. However, the following is known (cf. [4, Ch. 5]) on doubly exponential sequences. If $y_{i+1} = p\ y_i^r$

where $r>1$ and $p>0$ then $\exists c$: $y_i = c^{r^i}/p^{1/(r-1)}$. Therefore, for $i \geq k$, $w_i \geq \tilde{c}_k^{(d-1)^{2i}}/G_k^{1/d(d-2)}$ with equality for $i = k$:

$$\tilde{c}_k = w_k^{1/(d-1)^{2k}} G_k^{1/d(d-2)(d-1)^{2k}} \tag{4.7}$$

Therefore, using (4.1) and (4.3), and defining $s(n_i) = s_i$, $w(n_i) = w_i$ and $c_k = \tilde{c}_k^{(d-2)/(d-1)}$, one gets

$$s(n) \geq c_k^{\,n}\, Q_k\, c_k^{1/(d-2)}/G_k^{1/d(d-2)} \tag{4.8}$$

where $n = n_i$, $i \geq k$ and equality holds if $n = n_k$. Moreover, $c_k = w_k^{(d-2)/(d-1)^{2k+1}} G_k^{1/d(d-1)^{2k+1}}$ by (4.7); whence

$$= (G_0^{1/(d-1)^3} \cdots G_{k-1}^{1/(d-1)^{2k+1}})^{d-2}\, G_k^{1/d(d-1)^{2k+1}} \tag{4.9}$$

for $k > 0$ (by iterating (4.5)); and

$c_0 = G_0^{1/d(d-1)} = [\left(\frac{d-1}{2}\right)^{d-1} (d-2)^{d-2}\, (3d-4)]^{1/d(d-1)} > 1$, whence

$c_{k+1}/c_k = (G_{k+1}/G_k)^{1/d(d-1)^{2k+3}} > 1$ and therefore $c_k \nearrow$ as $k \nearrow$. Thus $s(n)$ possibly grows at least exponentially as $n \nearrow$.

To prove boundedness of c_k we use the following easy lemma.

L1. $\quad ax+b \leq (a+b)\exp((x-1)/2) \quad$ for $x \geq 1$

(with equality iff $x = 1$) provided that $b \geq a > 0$. □

$$Q_i \leq Q_m \exp\frac{i-m}{2} \quad \text{for } i \geq m \geq 1. \tag{4.10}$$

Proof. By (4.4), $(Q_i - Q_{i-1}) \searrow$ and

$Q_i \leq Q_{i-1} + (Q_m - Q_{m-1})$ for $i \geq m \geq 1$ whence

$\leq (Q_m - Q_{m-1})(i-m+1) + Q_{m-1}$ for $i \geq m-1 \geq 0$, whence

$\leq Q_m \exp\frac{i-m}{2}$ for $i \geq m \geq 1$; by L1 provided that $Q_{m-1} \geq Q_m - Q_{m-1}$. The last condition is easily checked for $m = 1$; for $m \geq 2$ it follows from $Q_i \leq Q_{i-1} + (Q_{m-1} - Q_{m-2})$ where $i \geq m-1 \geq 1$ by putting $i = m$. □

Hence and from (4.6) one has $G_j \leq G_m \exp[(j-m)(d-2)^2/2]$ for $j \geq m \geq 1$. Hence, because the sum of exponents at G_j's in (4.9) over $j \geq m$ is $1/d(d-1)^{2m+1}$, (4.9) gives

$c_k \leq c_m \exp(D_m^{k-1} + E_m^k)$ for $k \geq m \geq 1$ where

$$D_m^{k-1} := \frac{(d-2)^3}{2} \sum_{j=m}^{k-1} \frac{j-m}{(d-1)^{2j+3}}, \qquad E_m^k := \frac{(d-2)^2\,(k-m)}{2d(d-1)^{2k+1}}.$$

To find D_m^∞, put $D_m^\infty = D_m^{k-1} + R_{mk}$ where, for $k \ge m$,

$R_{mk} := (d-2)^3 \sum_{j=k}^{\infty} (j-m)/2(d-1)^{2j+3} = f'_{mk}(1)$ where $f_{mk}(x)$ $= \dfrac{(d-2)^3}{2(d-1)^{2k+1}} \dfrac{x^{k-m}}{(d-1)^2-x}$ for $|x| < (d-1)^2$, as the sum of a geometric series. Hence, $R_{mk} = E_m^k + \dfrac{d-2}{2d^2\,(d-1)^{2k+1}}$. Therefore

$D_m^\infty = R_{mm} = \dfrac{d-2}{2d^2(d-1)^{2m+1}}$ and $c_k \le c_m \exp D_m^\infty =: B_m$, $k \ge m \ge 1$.

It appears that it is enough to calculate c_1 and the upper bound B_1 to show that $c_\infty(d)$, the limit of c_k as $k \to \infty$, attains a local maximum at d=5, see the following table. This phenomenon may depend on our construction of trees; still, it may hint that n-vertex trees with bounded maximum degree Δ have the largest possible number of maximum path-factors.

d	c_0	c_1	B_1
3	1.31-	1.33+	1.34+
4	1.48-	1.507-	1.510+
5	1.53-	1.550+	1.552-
6	1.53+	1.548-	1.549-
7	1.52-	1.529+	1.530-

5. Conclusion

Given a p-sequence $\underline{d}$, $\underline{d} = (d_i)_{i=1}^p$, of integers d_i with each $d_i \ge 3$, let $\underline{d}(\hat{j})$ denote a (p-1)-sequence obtained from $\underline{d}$ by deleting the jth term d_j, $1 \le j \le p$. Given two disjoint graphs G and H with an endvertex each, let $G \cdot H$ denote the union $G \cup H$ in which an endvertex of G is identified with one of H. A <u>tree of stars</u>, denoted $TS(\underline{d})$, is defined by recursion. If p=1 and $\underline{d} = d$ then $TS(d) = K(1, d)$, a star with d rays. If $p \ge 2$ then, for each j such that $1 \le j \le p$, each tree $TS(\underline{d}(\hat{j})) \cdot TS(d_j)$ is considered as a $TS(\underline{d})$.

The following result from [11] is of basic importance.

$$\varphi(T \cdot \tau) = \varphi(T)\,\varphi(\tau) \quad (= \varphi(T \cup \tau)) \tag{5.1}$$

for any disjoint trees T and τ. Hence $\varphi(TS(\underline{d}))$ is the left=hand side of (1.1). A result in [10] related to (5.1) and concerning a hanging branching vertex shows that among n-vertex forests φ attains its maximum at some $TS(\underline{d})$ whose order is $n = 1 + \sum_i d_i$. By our Theorem, for $n = 1+s$, $q = \lfloor\frac{n+1}{5}\rfloor$ and $r = n-1-5q$,

$$\mathrm{TS}(\underline{d}) = \begin{cases} \mathrm{TS}(n-1) & \text{if } 4 \le n \le 8, \\ \mathrm{TS}\left(\{5\}^{q-|r|} \times \{5+\operatorname{sgn} r\}^{|r|}\right) & \text{if } n \ge 9. \end{cases}$$

References

1. S.E. Goodman, S. Hedetniemi and P.J. Slater: Advances on the Hamiltonian completion problem. J. Assoc. Comput. Mach. 22, 352-360 (1975)
2. A. Kania: Problem of determining all maximal path-factors of a tree by means of linear programming. In: Graphs, Hypergraphs and Block Systems, Proc. Zielona Góra 1976 Symp., ed. by M. Borowiecki and Z. Skupień, pp. 113-118. Poland: Zielona Góra 1976.
3. S. Kundu: A linear algorithm for the hamiltonian completion problem. Inform. Process. Lett. 5, 55-57 (1976)
4. H. Levy and F. Lessman: Finite Difference Equations. London: Pitman 1958 [Polish translation: PWN 1966]
5. S. Mitchell, S. Hedetniemi and S. Goodman: Some linear algorithms on trees. In: Proc. 6th S-E Conf. Comb. Graph Theory and Computing, pp. 467-483 (1975)
6. J.W. Moon and L. Moser: On cliques in graphs. Israel J. Math. 3, 23-28 (1965)
7. A. Schinzel: Personal communications (1987)
8. Z. Skupień: Path partitions of vertices and hamiltonity of graphs. In: Recent Advances in Graph Theory, Proc. Prague 1974 Conf., ed. by M. Fiedler, pp. 481-491. Praha: Academia 1975
9. Z. Skupień: Finding all maximal path factors of a tree and related problems. In: Combinatorics I-II, Proc. Keszthely 1976 Colloq., ed by A. Hajnal and Vera T. Sós, pp. 1031-1043. Colloq. Math. Soc. J. Bolyai 18, North-Holland 1978
10. Z. Skupień: On counting maximum path-factors of a tree. In: Algebra und Graphentheorie, Proc. Siebenlehn 1985 Conf., pp. 91-94. GDR: Bergakademie Freiberg, Sekt. Math. 1986
11. Z. Skupień and W. Zygmunt: Generating path factors of a tree. Zeszyty Nauk. Uniw. Jagielloń. Prace Mat. 19, 169-173 (1977)
12. Z. Skupień and W. Zygmunt: On vertices and edges in maximum path-factors of a tree. Fund. Math. 109, 89-101 (1980)
13. Z. Skupień and W. Zygmunt: Generating all maximum path= factors of a tree. In: Algebra, Combinatorics and Logic in Computer Science, Proc. Györ 1983 Colloq., ed. by J. Demetrovics et al., pp. 711-717. Colloq. Math. Soc. J. Bolyai 42, North-Holland 1985
14. I. Tomescu: Le nombre maximum de cliques et de recouvrements par cliques de hypergraphes chromatiques complets. Discrete Math. 37, 263-277 (1981)

A Characterization of Point-Colour-Symmetric Hypergraphs

S. Sze-Chin, Singapore

1. INTRODUCTION

A hypergraph X is a finite non-empty set V(X) together with a collection E(X) of distinct non-empty subsets of V(X). Elements of V(X) are called vertices and those of E(X) edges of the hypergraph X. An edge consisting of only one vertex is called a loop.

All hypergraphs considered in this paper are without loops.

Let X and Y be two hypergraphs. A bijection $\theta : V(X) \longrightarrow V(Y)$ with the property that $B \in E(X)$ iff $B\theta \in E(Y)$ is called an isomorphism from X onto Y, and in this case X and Y are isomorphic. An isomorphism from a hypergraph X onto itself is called an automorphism. A hypergraph X is said to be point-symmetric if for every $x,y \in V(X)$, there exists an automorphism θ of X such that $x\theta = y$. Some characterizations of point-symmetric hypergraphs can be found in [3].

Point-colour-symmetric graphs are graphs with high symmetrical properties. A characterization of point-colour-symmetric graphs can be found in [2]. In this paper we shall extend the above characterization to hypergraphs.

2. POINT-COLOUR-SYMMETRIC HYPERGRAPH

A hypergraph X is called an edge-coloured hypergraph if its edges are painted with colours. We shall denote by C(X) the set of colours used to paint the edges and by c(B) the colour of the edge B.

Let X and Y be two edge-coloured hypergraphs. A bijection $\theta : V(X) \longrightarrow V(Y)$ is called an isomorphism from X onto Y if the following two conditions are satisfied :

(i) $B \in E(X) \Longleftrightarrow B\theta \in E(Y)$,

(ii) for every $A, B \in E(X)$, $c(A) = c(B) \Longleftrightarrow c(A\theta) = c(B\theta)$.

R. Bodendiek, R. Henn (Eds.)
Topics in Combinatorics and Graph Theory

Moreover, if $c(B) = i$ and $c(B\theta) = j$, then we say that the isomorphism θ carries colour i to colour j. An isomorphism from an edge-coloured hypergraph X onto itself is called an automorphism of X. An edge-coloured hypergraph X is said to be point-colour-symmetric if the following two conditions are satisfied :

(PCS1) For every $x,y \in V(X)$, there exists an automorphism θ of X such that $x\theta = y$ and θ carries each colour to itself.

(PCS2) For any two colours $i,j \in C(X)$, there exists an automorphism of X carrying colour i to colour j.

Obviously point-colour-symmetric hypergraphs are point-symmetric. If the condition (PCS2) is replaced by the following stronger condition (PCS3), then the edge-coloured hypergraph X is said to be strongly point-colour-symmetric.

(PCS3) For any permutation π of C(X), there exists an automorphism α of X carrying colour i to colour $i\pi$ for each $i \in C(X)$.

In order to characterize a point-colour-symmetric hypergraph, we need the following hypergraph. Let G be a finite group with identity e and H a subgroup of G. Let Γ be a family of non-empty subsets of G each containing e and of cardinality at least 2 such that for every $B \in \Gamma$ and every distinct $b_1, b_2 \in B$, $b_1b_2^{-1} \notin H$. Let γ be a mapping from Γ onto a set I of colours. Let $T = \{\theta_{ij} | i,j \in I\}$ be a family of automorphisms of the group G. Assume the following conditions hold :

(1) $H\theta_{ij} = H$ for every $i,j \in I$, and

(2) for every $i,j \in I$, there exists a one-to-one mapping δ_{ij} from I onto itself such that for each $k \in I$, $k\delta_{ij} = k'$ with $i\delta_{ij} = j$, and for each $B \in \Gamma$ with $B\gamma = k$, $B\theta_{ij} = B' \in \Gamma$ and $B'\gamma = k'$.

Now consider the collection G/H of all left cosets of H in G. Take the following family of subsets of G/H :

$$\Lambda = \{\{H, b_1H, \ldots, b_mH\} | \{e, b_1^{-1}, b_2^{-1}, \ldots, b_m^{-1}\} \in \Gamma\},$$

and define

$$R_\Lambda = \{gA \mid g \in G,\ A \in \Lambda\}$$

to obtain a hypergraph X^* with $V(X^*) = G/H$ and $E(X^*) = R_\Lambda$. We shall use colour i to paint the edge $gA \in R_\Lambda$ if $A = \{H, b_1H, \ldots, b_mH\}$ with $B = \{e, b_1^{-1}, b_2^{-1}, \ldots, b_m^{-1}\} \in \Gamma$ and $B\gamma = i$. We thus obtain an edge-coloured hypergraph X^*. To show that this edge-coloured hypergraph X^* is indeed well-defined, we first note that Λ is obviously well-defined since $b_iH \neq b_jH$ for $i \neq j$. Next suppose $\{x_0H, x_1H, \ldots, x_nH\} = \{y_0H, y_1H, \ldots, y_nH\} \in R_\Lambda$. Then

$$\{x_0H, x_1H, \ldots, x_nH\} = gA = g\{H, b_1H, \ldots, b_nH\}$$

for some $g \in G$ and $\{e, b_1^{-1}, \ldots, b_n^{-1}\} = B \in \Gamma$ with $B\gamma = i$, say. Then gA has colour i. Without loss of generality, we assume that

$$x_0H = gH \quad \text{and} \quad x_iH = gb_iH, \quad i = 1,2,\ldots,n.$$

Then $\qquad x_0 = gh_0$ and $x_i = gb_ih_i$, for some $h_i \in H$, $i = 1,2,\ldots,n$.

We can also assume that

$$x_iH = y_iH \quad \text{for every } i = 0,1,\ldots,n.$$

Then

$$y_i = x_id_i \quad \text{for some } d_i \in H, \quad i = 0,1,\ldots,n.$$

and

$$\{y_0H, y_1H, \ldots, y_nH\} = \{(gh_0d_0)H, (gb_1h_1d_1)H, \ldots, (gb_nh_nd_n)H\}$$
$$= g\{H, b_1H, \ldots, b_nH\}$$

which shows that the edge $\{y_0H, y_1H, \ldots, y_nH\}$ is also painted with colour i. It follows that X^* is a well-defined edge-coloured hypergraph.

THEOREM 1. *An edge-coloured hypegraph X is point-colour-symmetric iff it is isomorphic to an edge-coloured hypergraph X^* described above.*

PROOF. We shall first show that the edge-coloured hypergraph X^* with $V(X^*) = G/H$ and $E(X^*) = R_\Lambda$ satisfies the conditions (PCS1) and (PCS2).

To prove (PCS1), let $xH, yH \in G/H$. Let $z = yx^{-1}$. Define $\phi_z : G/H \to G/H$ by putting $(uH)\phi_z = (zu)H$, for each $uH \in G/H$. It is easily seen that ϕ_z is an automorphism of the edge-coloured hypergraph X^* such that $(xH)\phi_z = yH$ and that it carries each colour to itself.

To prove (PCS2), we define, for each $i,j \in I$, the mapping $\phi_{ij} : G/H \longrightarrow G/H$ by putting $(uH)\phi_{ij} = (u\theta_{ij})H$ for each $uH \in G/H$. Obviously the mapping ϕ_{ij} is a bijection. Furthermore, let

$$gA = g\{H, b_1H, \ldots, b_mH\} \in R_\Lambda$$

where $B = \{e, b_1^{-1}, b_2^{-1}, \ldots, b_m^{-1}\} \in \Gamma$ with $B\gamma = i$.
Then

$$\begin{aligned}(gA)\phi_{ij} &= \{gH, gb_1H, \ldots, gb_mH\}\phi_{ij} \\ &= \{(g\theta_{ij})H, [(g\theta_{ij})(b_1\theta_{ij})]H, \ldots, [(g\theta_{ij})(b_m\theta_{ij})]H\} \\ &= (g\theta_{ij})\{H, d_1H, \ldots, d_mH\}\end{aligned}$$

where $d_\ell = b_\ell\theta_{ij}$, $\ell = 1,2,\ldots,m$. Since $B \in \Gamma$ with $B\gamma = i$, we have $\{e, d_1^{-1}, \ldots, d_m^{-1}\} = B\theta_{ij} \in \Gamma$ with colour j. It follows that $(gA)\phi_{ij} \in R_\Lambda$ and has colour j. Hence the edge-coloured hypergraph X^* described above is point-colour-symmetric.

Conversely, let X be a point-colour-symmetric hypergraph with $C(X) = I$. Let G be the group of all automorphisms of X, carrying each colour to itself. Fix any $p \in V(X)$ and let $H = \{\theta \mid \theta \in G, p\theta = p\}$. Then H is a subgroup of G. For each edge containing p, say $\{p, v_1, \ldots, v_m\} \in E(X)$, there exist $b_i \in G$ $(i=1,2,\ldots m)$ such that $v_i = pb_i^{-1}$, that is, $\{p, v_1, \ldots, v_m\} = \{p, pb_1^{-1}, \ldots, pb_m^{-1}\} \in E(X)$. Then we obtain a set $\{e, b_1^{-1}, \ldots, b_m^{-1}\} = B$, say, where e is the identity of the group G. Note that for every distinct $b_i^{-1}, b_j^{-1} \in B$, $b_i^{-1}b_j \notin H$. Otherwise $pb_i^{-1}b_j = p$ or $pb_i^{-1} = pb_j^{-1}$ which implies $v_i = v_j$. Now let

$$\Gamma = \left\{\{e,b_1^{-1},\ldots,b_m^{-1}\} \mid b_i \in G,\ i = 1,2,\ldots,m, \text{ and } \{p,pb_1^{-1},\ldots,pb_m^{-1}\} \in E(X)\right\}.$$

Define a mapping $\gamma : \Gamma \longrightarrow I$ as follows : for each $B = \{e,b_1^{-1},\ldots,b_m^{-1}\} \in \Gamma$,

$$B\gamma = i \quad \text{iff } \{p,pb_1^{-1},\ldots,pb_m^{-1}\} \text{ has colour } i.$$

Note that γ is a surjection.

Next we shall show there is a family $T = \{\theta_{ij} \mid i,j \in I\}$ of automorphisms of the group G which satisfies conditions (1) and (2). For any two colours $i,j \in I$, there exists an automorphism α_{ij} of the point–colour–symmetric hypergraph X which carries colour i to colour j. Let $p\alpha_{ij} = q$. There exists an automorphism θ of X such that $q\theta = p$ and that each colour is carried to itself. Let $g_{ij} = \alpha_{ij}\theta$. Then g_{ij} is an automorphism of X carrying p to p and colour i to colour j. Define $\theta_{ij} : G \to G$ by putting $x\theta_{ij} = g_{ij}^{-1} xg_{ij}$, for each $x \in G$. Evidently, θ_{ij} is an automorphism of the group G. In this way we obtain a family $T = \{\theta_{ij} | i,j \in I\}$ of automorphisms of G. To show that T satisfies condition (1), let $i,j \in I$. Then for any $h \in H$, we find $p(h\theta_{ij}) = p(g_{ij}^{-1} hg_{ij}) = p$, and it follows that $H\theta_{ij} \subseteq H$. On the other hand, let $h \in H$. Since $p(g_{ij} hg_{ij}^{-1}) = p$, we have $g_{ij}hg_{ij}^{-1} = h' \in H$ or $h = g_{ij}^{-1} h'g_{ij} = h'\theta_{ij} \in H\theta_{ij}$ which implies $H \subseteq H\theta_{ij}$. Hence $H\theta_{ij} = H$.

To prove that T satisfies condition (2), let $i,j \in I$. We have seen that the automorphism g_{ij} of X carries colour i to colour j, colour k to colour k′ say, and $pg_{ij} = p$. Thus g_{ij} induces a one–to–one mapping δ_{ij} from I onto itself such that $i\delta_{ij} = j$ and $k\delta_{ij} = k'$. Now let $B \in \Gamma$ with $B\gamma = k$. Then we have $B\theta_{ij} = B' \in \Gamma$ and $B'\gamma = k'$. Indeed suppose $B = \{e, b_1^{-1}, \ldots b_m^{-1}\}$. Then $A = \{p, pb_1^{-1}, \ldots, pb_m^{-1}\} \in E(X)$ with $c(A) = k$. We find

$$B\theta_{ij} = \left\{e, g_{ij}^{-1} b_1^{-1} g_{ij}, \ldots, g_{ij}^{-1} b_m^{-1} g_{ij}\right\}$$

and

$$\left\{p, p(g_{ij}^{-1} b_1^{-1} g_{ij}), \ldots, p(g_{ij}^{-1} b_m^{-1} g_{ij})\right\}$$

$$= \left\{p, pb_1^{-1}, \ldots, pb_m^{-1}\right\}g_{ij} \in E(X)$$

which has colour k′. Hence $B\theta_{ij} = B' \in \Gamma$ and $B'\gamma = k'$.

Now take

$$\Lambda = \left\{\{H, b_1H, \ldots, b_mH\} \middle| \{e, b_1^{-1}, \ldots, b_m^{-1}\} \in \Gamma\right\}$$

and define

$$R_\Lambda = \{gA | g \in G, A \in \Gamma\}.$$

The edge $gA \in R_\Lambda$ will have colour i if $A = \{H, b_1H, \ldots, b_mH\}$ with

$\{e, b_1^{-1}, \ldots, b_m^{-1}\}\gamma = i$. Thus we obtain an edge-coloured hypergraph X^* whose $V(X^*) = G/H$ and $E(X^*) = R_\Lambda$.

To show that X^* is isomorphic to the given edge-coloured hypergraph X, we define a mapping $\phi : G/H \longrightarrow V(X)$ as follows :

$$(uH)\,\phi = pu^{-1} \quad \text{for each} \quad uH \in G/H.$$

It is easily verified that the mapping ϕ is a bijection. Moreover we note that

$$A^* = g\{H, b_1H, \ldots, b_mH\} \in R_\Lambda \quad \text{with} \quad \{e, b_1^{-1}, \ldots, b_m^{-1}\} \in \Gamma$$

has colour i iff

$$A = \{p, pb_1^{-1}, \ldots, pb_m^{-1}\} \in E(X) \text{ has colour } i,$$

which is true iff

$$\begin{aligned} Ag^{-1} &= \{p, pb_1^{-1}, \ldots, pb_m^{-1}\}g^{-1} \\ &= \{gH, gb_1H, \ldots, gb_mH\}\phi \in E(X) \text{ has colour } i. \end{aligned}$$

This shows that ϕ is an isomorphism from X^* onto X. This completes the proof.

It seems not easy to use Theorem 1 to construct some point-colour-symmetric hypergraphs. We shall therefore introduce a much simpler method of construction which actually gives a strongly point-colour-symmetric hypergraph.

Let $\{X_i \mid i \in \Delta\}$ be a family of hypergraphs indexed by a finite set Δ. Let G^* be the cartesian product of the vertex sets $V(X_i)$, $i \in \Delta$. Elements g^* of G^* are to be represented by $g^* = (g_i)_{i\in\Delta}$, where g_i denotes the ith component of g^*. Let

$$\begin{aligned} R^* = \Big\{ &\{g_1^*, g_2^*, \ldots, g_n^*\} \mid g_j^* = (g_i^{(j)})_{i\in\Delta} \in G^* \text{ for each} \\ &j = 1,2,\ldots,n, \text{ and there exists a } k \in \Delta \text{ such that} \\ &\{g_k^{(j)} \mid j = 1,2,\ldots,n\} \in E(X_k) \text{ and } g_i^{(\ell)} = g_i^{(m)} \\ &\text{for all } \ell, m = 1,2,\ldots,n \text{ and all } i \neq k \Big\}, \end{aligned}$$

where n may vary. Then the hypergraph Y with vertex set $V(Y) = G^*$ and edge set $E(Y) = R^*$ is called the cartesian product of the family $\{X_i \mid i \in \Delta\}$ of hypergraphs. If each X_i is an edge-coloured hypergraph, then Y also becomes an edge-coloured hypergraph, whose edges are painted as follows : Let $A = \{g_1^*, g_2^*, \ldots, g_n^*\} \in R^*$. Then there exists a $k \in \Delta$ such that $A_k = \{g_k^{(j)} \mid j = 1,2,\ldots,n\} \in E(X_k)$ and

$$g_i^{(\ell)} = g_i^{(m)} \quad \text{for all } \ell, m = 1,2,\ldots,n \text{ and all } i \neq k.$$

If A_k has colour r in X_k, then A will be painted with colour r in Y.

THEOREM 2. *Let $\{X_i \mid i \in \Delta\}$ be a family of hypergraphs each being isomorphic to a fixed point-symmetric hypergraph X. For each $i \in \Delta$, paint the edges of X_i with colour i. Then the cartesian product Y of the family $\{X_i \mid i \in \Delta\}$ of edge-coloured hypegraphs is a strongly point-colour-symmetric hypergraph.*

PROOF. Let $g^* = (g_i)_{i \in \Delta}$ and $h^* = (h_i)_{i \in \Delta}$ be any two elements of G^*. For each $i \in \Delta$, there exists an automorphism θ_i of X_i such that $g_i\theta_i = h_i$. Define $\theta : G^* \to G^*$ as follows. For each $x^* = (x_i)_{i \in \Delta} \in G^*$, $x^*\theta = (x_i\theta_i)_{i \in \Delta}$. It is not difficult to see that the mapping θ is an automorphism of Y carrying g^* to h^* and each colour to itself. Hence condition (PCS1) is satisfied.

To show that Y satisfies (PCS3), let π be any permutation of Δ. By hypothesis each X_i (without colour) is isomorphic to X, the isomorphism being α_i say. Then we can define a mapping $\phi : G^* \to G^*$ as follows. For each $g^* = (g_i)_{i \in \Delta} \in G^*$, we put $g^*\phi = (x_i)_{i \in \Delta}$, where $x_i = g_{(i)\pi^{-1}} \alpha_{(i)\pi^{-1}} \alpha_i^{-1}$. Then ϕ is a bijection. Moreover we observe that $A^* = \{g_1^*, g_2^*, \ldots, g_n^*\} \in R^*$ with colour k iff there exists a $k \in \Delta$ such that $A_k = \{g_k^{(1)}, g_k^{(2)}, \ldots, g_k^{(n)}\} \in E(X_k)$ and $g_i^{(\ell)} = g_i^{(m)}$ for all $\ell, m = 1,2,\ldots,n$ and all $i \neq k$; this is equivalent to $A_k\alpha_k\alpha_{(k)\pi}^{-1} \in E(X_{(k)\pi})$ and $g_i^{(\ell)}\alpha_i\alpha_{(i)\pi}^{-1} = g_i^{(m)}\alpha_i\alpha_{(i)\pi}^{-1}$ for all $\ell, m = 1,2,\ldots,n$ and all $i \neq k$, which in turn is equivalent to $A^*\phi \in R^*$ with colour $k\pi$. This shows that condition (PCS3) is also satisfied. Hence the cartesian product Y of the family $\{X_i \mid i \in \Delta\}$ of edge-coloured hypergraphs is strongly point-colour-symmetric.

REFERENCES

1. Berge C., *Graphs and Hypergraphs*, North–Holland, Amsterdam, 1973.

2. Chen C. C. and Teh H. H., Construction of point–colour–symmetric graphs, *J. Comb. Theory Ser. B.* 27, 160–167, (1979).

3. Teh H. H. and Shee S. C., *Algebraic Theory of Graphs*, Lee Kong Chian Institute of Mathematics and Computer Science, Nanyang University, Singapore 1976.

A Linear Algorithm for the Pathwidth of Trees

P. Scheffler, Berlin, GDR

1. Introduction

The pathwidth is a graph parameter only recently studied but closely related to other characteristics of graphs like tree, band- or cutwidth, interval thickness or search number ([S]). The graphs considered here are finite, undirected and simple. First the preliminaries are given. Section 2 contains our main results on the pathwidth of trees, the basis for the algorithm described in section 3.

Def. 1: ([RS])

A path-decomposition of a graph G is a sequence of subsets of the node set $\mathfrak{X} = (X_1, \ldots, X_r)$ such, that holds

(1) $\bigcup_{i=1}^{r} X_i = V(G)$.

(2) For any edge $(u,v) \in E(G)$ of the graph there is a set X_i $(1 \le i \le r)$ containing both ends of the edge $(\{u,v\} \subseteq X_i)$.

(3) For all triple $1 \le i < j < k \le r$ holds $X_i \cap X_k \subseteq X_j$.

The width of a path-decomposition is $p(G,\mathfrak{X}) := \max_{1 \le i \le r} |X_i| - 1$.

The pathwidth of a graph is $p(G) := \min_{\mathfrak{X}} p(G,\mathfrak{X})$.

As examples notice the following facts: All simple paths P_n have pathwidth 1 and so have the stars $K_{1,n}$. A simple cycle C_n has pathwidth 2, the complete graph K_n has pathwidth n-1 and the k∘n-grid has pathwidth k, if $k \le n$.

The interval thickness $\Theta(G)$ of a graph G is the minimum over the clique numbers of all interval graphs having a subgraph G.

Theorem 1: ([S])

For all graphs the pathwidth $p(G)$ is equal to the interval thickness $\Theta(G)$ minus one.

Theorem 2: ([KF])

The problem to determine the interval thickness of a graph is NP-complete.

R. Bodendiek, R. Henn (Eds.)
Topics in Combinatorics and Graph Theory

Hence, it seems to be hopeless to look for an efficient algorithm for the pathwidth problem in the general case. Nevertheless, for a fixed value p somewhat is possible to do, namely we proved:

Theorem 3: ([S])

For any fixed integers p and d there is an algorithm deciding in time $O(n^{p+1})$ for any given graph with maximum degree at most d, whether its pathwidth is not greater than p.

2. The pathwidth of trees

Let T be a tree and t any of its nodes. Then every connected component of $T \setminus \{t\}$ is called a branch of T at t, it is denoted by T_{ts} if it contains the node s (usually a neighbour of t). If the tree has root w, for every edge (t,s) the node t is called the father of s (and s the son of t) if t lies on the path from s to w. The maximal subtree of T rooted at t is denoted by T^t.

Now we are going to prove the crucial theorem on the pathwidth of trees, the basis for the proposed algorithm. It will be applied also to get a bound for the pathwidth of trees and to calculate the pathwidth of complete binary trees.

Theorem 4:

For any integer $p \geq 1$ and any tree T holds:
$p(T) \geq p + 1$ iff there exists a node $t \in V(T)$ with at least three branches T_{tv}, T_{tu} and T_{tw} such that $p(T_{tv}) \geq p$, $p(T_{tu}) \geq p$ and $p(T_{tw}) \geq p$.

Proof:

($\Longleftarrow$) If the pathwidth p of T is the same as the pathwidth of its subtree T_{tv} then in every optimal (i.e. width-p-) path-decomposition of T there must exist a set X_i with cardinality p + 1 contained in the node set of T_{tv}. Assume there is a node $t \in V(T)$ with three branches of width p, but the pathwidth of T is also p. Consider an optimal path-decomposition of T and the three sets X_i, X_j and X_k contained in the three branches at t (T_{tv}, T_{tu} and T_{tw} respectively). Let be $i < j < k$. Then the nodes of X_i and X_k belong to the same connected component of $T \setminus X_j$ (since $t \notin X_j$), but this contradicts the definition of path-decompositions.

($\Longrightarrow$) Let T have pathwidth p+1. Every node t has at least one branch of pathwidth p or more, since otherwise a path-decomposition of width p for T could be constructed from optimal decompositions of the branches at t by writing them one after another and adding to every set X_i the node t. Consider the edge set $F := \{(s,t): p(T_{st}) \geq p$ and $p(T_{ts}) \geq p\}$ in T. It is not empty since a tree has an edge more than nodes. If there is a node with three incident edges in F we are done. So assume F to consist of some paths. Actually, F must be connected: Otherwise, there would exist an edge (s,t) in $T \setminus F$ on a path between two components of F, but then both branches T_{st} and T_{ts} would contain a subtree of pathwidth at least p, what implies that the edge (s,t) belongs to F in contradiction to its choice. Let the path F consist of the nodes $t_1, \ldots, t_k$ in this order. Then every node t_i (for $2 \leq i \leq k-1$) has exactly two branches of pathwidth p or more, namely $T_{t_i t_{i-1}}$ and $T_{t_i t_{i+1}}$. The nodes t_1 and t_k have exactly one branch of pathwidth p or more each, namely $T_{t_1 t_2}$ rsp. $T_{t_k t_{k-1}}$. All other branches at a node t_i (for $i = 1, \ldots, k$) having smaller pathwidth than p combine with t_i to a subtree T^{t_i}, it has (as shown above) a path-decomposition $(X_1^i, \ldots, X_{r_i}^i)$ of width p or less with node $t_i \in X_j^i$ for all j. By writing them one after another with intermediate sets $\{t_i, t_{i+1}\}$ we get a path-decomposition for the whole tree T of width p: $(X_1^1, \ldots, X_{r_1}^1, \{t_1, t_2\}, X_1^2, \ldots, X_{r_k}^k)$. This contradicts the assumption. □

<u>Theorem 5:</u>

The pathwidth of a tree with n nodes is less than $\log_3(2n+1)$.

<u>Proof:</u>

By induction, relying on theorem 4, we will prove that in T for all $i = 0, \ldots, p$ there exists a node set S_i of cardinality $|S_i| = (3^i - 1) / 2$ such that the forest $T \setminus S_i$ has at least 3^i connected components with pathwidth $(p - i)$ or greater. This is true for $i = 0$. Let for some i the node set S_i be given. Apply theorem 4 to every of the 3^i connected components of $T \setminus S_i$ with pathwidth at least p-i. We get a set S_{i+1} with cardinality $3^i + (3^i - 1) / 2 = (3^{i+1} - 1) / 2$ and 3^{i+1} connected components as claimed. Apply this for $i = p-1$. Every component contains at least one edge. So we get for the number of nodes in T: $n \geq (3^{p-1} - 1) / 2 + 2 \circ 3^{p-1} = 5/6 \circ 3^p - 1/2$. Consequently, we have $p \leq \log_3(2n+1) + 1 - \log_3 5$. □

Theorem 6:

The complete binary tree of height k has pathwidth $\lceil k/2 \rceil$.

Sketch of the proof: (All details are given in [S])

By induction on the height k. Consider two cases: (i) $k = 2l + 1$ and (ii) $k = 2l$. Combine the path-decompositions of the two height-(k-1)-subtrees (case (i)) or of the four height-(k-2)-subtrees (case (ii)) rsp. to a path-decomposition of the whole tree. In case (i) apply theorem 4 to a son of the root to get the lower bound. □

3. The algorithm

We prove the following theorem presenting the algorithm. Owing to the limited space only the ideas can be given here, details are contained in [S].

Theorem 7:

There is a linear-time algorithm to determine the pathwidth p(T) of a given tree.

The given tree is undirected, but it is oriented during the work of the algorithm in such a way that over all branches at a node that containing its father is one with maximum pathwidth. This is done by finding all nodes t with a subtree rooted at t of pathwidth p in the order of increasing p. All these nodes are stored in a FIFO-list QUEUE(p) for every $p = 1, \dots, \log_3(2n+1)$ while the subtree T^t has not been merged with its father. The kernel of the algorithm is the recursive procedure MERGE calculating the pathwidth of a tree obtained by joining two subtrees by an edge given the pathwidth of the two subtrees. Unfortunately, it is not possible to determine correctly the pathwidth of the join given only the pathwidths of the two subtrees. Denote the pathwidth of a subtree T^t rooted at node t by $p_1(t)$. Define in addition the parameter $p_2(t)$ to be equal to zero if there is such an optimal path-decomposition of T^t that the root t is an element of the first set ($t \in X_1$) and let be $p_2(T) := 1$ otherwise. Moreover, we need the number of branches at t having maximal path-width. Store this in a parameter $p_3(t) := |\{s: t = FATHER(s) \text{ and } p_1(s) = p_1(t)\}|$. Obviously, it is $0 \le p_3(t) \le 2$. The triple $\bar{p}(t) = (p_1(t), p_2(t), p_3(t))$ is called the path-decomposition-vector of T^t. We will use the following facts proved by applying theorem 4 (see [S]).

Lemma 1:

Let the tree T be obtained from the subtrees T^t and T^s joined by the edge (s,t). Then the pathwidth p(T) satisfies: $\max\{p_1(t), p_1(s)\} \le p(T) \le \max\{p_1(t), p_1(s)\} + 1$.

Lemma 2:

Let T^t be an oriented tree with root t and with the path-decomposition-vector $\bar{p} = (p_1, p_2, p_3)$. Then holds:

(i) If $p_3 = 0$, then also $p_2 = 0$ and T^t has a path-decomposition of width p_1 with $t \in X_i$ for all i.

(ii) If $p_2 = 0$, then there is no node v in the tree T^t possessing two sons of pathwidth $p_1(s) = p_1$.

(iii) If $p_3 = 2$, then $p_2 = 1$.

(iv) If $p_2 = 1$, then there is exactly one node v in the tree T^t possessing two sons of pathwidth p_1, all other nodes have at most one such son.

The node v in a subtree possessing two sons of maximal pathwidth plays an important role: It may get yet a third branch of maximal pathwidth later, when a new subtree T^s is added. Then the pathwidth of T is no longer equal to p_1 but increases by one. The new branch is $T_{vFATHER(v)} = T \setminus T^v$. To take account of this we store also the path-decomposition-vector of this subtree and link it by a pointer REST(t) to the vector of T^t. A chain of pointers may occur there, but it is no longer than $p_1(t)$ since the rest subtrees must have decreasing pathwidths. According to the following rules the procedure MERGE calculates the path-decomposition-vector of the join of two given subtrees in time O(p), where p is the pathwidth of the largest subtree:

Lemma 3:

Let the tree T be obtained from the subtrees T^t and T^s joined by the edge (s,t). Let $\bar{p}(t)$ and $\bar{p}(s)$ be the path-decomposition-vectors of the two subtrees. Then the new path-decomposition-vector $\bar{n}(t)$ of the whole tree T is as follows:

1. case: $p_1(s) > p_1(t)$

1a) $p_2(s) = 0 \implies \bar{n}(t) := (p_1(s), 0, 1)$

1b) $p_3(s) = 2 \implies \bar{n}(t) := (p_1(s), 1, 1)$

1c) $p_2(s) = p_3(s) = 1 \implies$

$$\bar{n}(t) := \begin{cases} (p_1(s) + 1, 0, 0) & , \text{ if } p_1(T \setminus T^v) = p_1(s) \\ \bar{p}(s) & , \text{ otherwise} \end{cases}$$

<u>2. case:</u> $p_1(s) = p_1(t)$

2a) $p_2(t) = 1$ or $p_2(t) = 1 \Longrightarrow \bar{n}(t) := (p_1(s) + 1, 0, 0)$

2b) $p_2(s) = p_2(t) = 0$ and $p_3(t) = 0 \Longrightarrow \bar{n} := (p_1(s), 0, 1)$

2c) $p_2(s) = p_2(t) = 0$ and $p_3(t) = 1 \Longrightarrow \bar{n} := (p_1(s), 1, 2)$

<u>3. case:</u> $p_1(s) < p_1(t)$

3a) $p_2(t) = 0 \Longrightarrow \bar{n}(t) := \bar{p}(t)$

3b) $p_3(t) = 2 \Longrightarrow \bar{n}(t) := \bar{p}(t)$

3c) $p_2(t) = p_3(t) = 1 \Longrightarrow$

$$\bar{n}(t) := \begin{cases} (p_1(t) + 1, 0, 0) & \text{, if } p_1(T \setminus T^v) = p_1(t) \\ \bar{p}(t) & \text{, otherwise.} \end{cases}$$

```
procedure MERGE((t,p̄), (s,q̄), (t,n̄))
// Input: p̄ = p̄(t) - path-decomposition-vector of a subtree T^t;
          q̄ = q̄(s) - path-decomposition-vector of a subtree T^s //
// Output: n̄ - path-decomposition-vector of the merged tree
                       T^t ∪ T^s ∪ ({t,s},{(t,s)}) rooted at t //
    // 1. case: the branch T_ts determines pathwidth //
 (1) if q1 > p1 then
       if q2 = 0 then n̄:= (q1, 0, 1) endif
       if q3 = 2 then n̄:= (q1, 1, 1); REST(t):= p̄ endif
       if q2 = 1 and q3 = 1 then
         call MERGE((t,p̄), (s,REST(s)), (t,r̄))
         if r1 = q1 then n̄:= (q1 + 1, 0, 0)
                    else n̄:= (q1, 1, 1); REST(t):= r̄
         endif
     endif
    // 2. case: equal sized branches T_ts and T_st //
 (2) if q1 = p1 then
       if q2 = 1 or p2 = 1 then n̄:= (q1 + 1, 0, 0) endif
       if q2 = 0 and p2 = 0 and p3 = 0 then n̄:= (q1, 0, 1)
       endif
       if q2 = 0 and p2 = 0 and p3 = 1 then n̄:= (q1, 1, 2)
       endif
     endif
    // 3. case: the branch T_st determines pathwidth //
 (3) if q1 < p1 then
       if p2 = 0 or p3 = 2 then n̄:= p̄ endif
       if p2 = 1 and p3 = 1 then
         call MERGE((t,REST(t)), (s,q̄), (t,r̄))
         if r1 = p1 then n̄:= (p1 + 1, 0, 0)
                    else n̄:= p̄; REST(t):= r̄
         endif
       endif
 (4) return((t,n̄))
```

P_5:

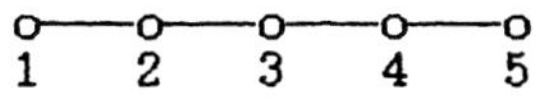

a) $\bar{p}(2) = (1, 0, 1)$

b) $\bar{p}(3) = (1, 1, 2)$

Figure 1: The path-decomposition-vector of P_5 depending on the orientation: a) The root is node 2. b) The root is node 3.

T:

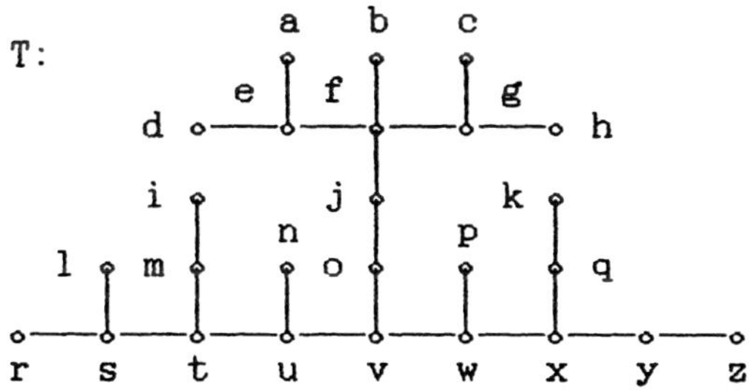

$\bar{p} = (0, 1, 2)$ for the leaves $\{a,b,c,d,h,i,k,l,n,p,r,z\}$.

$\bar{p} = (1, 0, 0)$ for the nodes $\{e,g,m,q,s,y\}$.

$\bar{p} = (1, 1, 2)$ for f, t and x.

$\bar{p}(j) = (1, 1, 1)$ and $REST(j) = (0, 1, 2)$.

$\bar{p} = (2, 0, 0)$ for o, u and w.

$\bar{p}(v) = (3, 0, 0)$; $p(T) = 3$.

$\mathcal{X} = (\{a,d,e,f\}, \{f,g,h,c\}, \{f,b,j,o\}, \{o,v,u,n\}, \{u,v,t,m\}, \{v,t,m,i\}, \{v,t,s,r\}, \{v,s,l\}, \{v,w,p,x\}, \{x,k,q,y\}, \{y,z\})$

Figure 2: A tree T with the path-decomposition-vectors found by the algorithm and its optimal path-decomposition (root is v)

In addition the main programm MINPATHWIDTH is used, organizing the calls of the procedure MERGE for all edges $(s,t) \in E(T)$ in the needed order. The path-decomposition-vector of any node is initialized to $\bar{p} := (0, 1, 2)$. For the leaves this is the final vector. Their fathers can also get an easy treatment. Obviously, for all nodes t all sons of which are leaves we have $\bar{p}(t) = (1, 0, 0)$. This will be used in step (1). Notice that the procedure MERGE is called at most once for any node s, so we get time complexity $\sum_s O(p(T_s))$ in worst case, applying theorem 5 this gives $\sum_s O(\log|V(T_s)|) = O(n)$. For the correctness proof mention yet, that at least two nodes are contained in the QUEUEs until the last step. During the last step (when $i = n - 1$) the path-decomposition-vector of the root is updated by merging the last of its branches to T^w.

After finishing this manuscript we were informed that Ellis, Sudborough and Turner obtained a similar result considering the vertex separation of graphs. Our linear algorithm can be modified so that an optimal path-decomposition is also constructed.

```
procedure MINPATHWIDTH(T)
// Input: tree T with  n = |V(T)|  ;  output: pathwidth p(T) //

     // Initialization //
 (0) for  all  t ∈ V(T)  do  p̄(t):= (0, 1, 2); i:= 0  endfor

     // Preparation: leaves are trivial branches //
 (1) for  all leaves  s   do
       i:= i+1; if  i<n  then  label  s  endif
       the only node with  (s,t) ∈ E(T)  denote by t
       FATHER(s):= t; p̄(t):= (1, 0, 0)
       if  t has only one unlabeled neighbour  v   then
         FATHER(t):= v
         t ⟶ QUEUE(1)
       endif
     endfor

     // Treatment of the inner nodes of the tree //
 (2) while  i < n  do
       p:= the smallest number of a nonempty QUEUE
       take  s  out of QUEUE(p)
       i:= i+1; label s; t:= FATHER(s)

       // Updating //
 (3)   call MERGE((t,p̄), (s,q̄), (t,p̄))

       // p_1(t) is the pathwidth of the whole subtree  T^t //
 (4)   if  t has only one unlabeled neighbour  v  then
         FATHER(t):= v; t ⟶ QUEUE(p_1(t))
       endif
     endwhile

     // Reading the pathwidth of the root //
 (5) w:= the only unlabeled node; FATHER(w):= NIL; p:= p_1(w)
     return("The pathwidth of T is",p)
     end
```

4. References

[KF] Kashiwabara T., Fujisawa T.: NP-completeness of the problem of finding a minimum-clique-number interval graph containing a given graph as a subgraph: Proc. ISCAS 1979, 657-660

[RS] Robertson N., Seymour P.D.: Graph minors. I. Excluding a forest: J. Combin. Theory B 35 (1983), 39-61

[S] Scheffler P.: Die Baumweite von Graphen als ein Maß für die Kompliziertheit algorithmischer Probleme: AdW der DDR, K.-Weierstraß-Institut für Mathematik, Berlin (R-MATH-04/89) 1989

[Y] Yannakakis M.: A polynomial algorithm for the min-cut linear arrangement of trees: J. Assoc. Comput. Mach. 32 (1985) 4, 950-988

The Time Travelling Problem

G. Schild, Ilmenau, GDR

The Problem

For some practical applications (see [1]) a problem is interesting wich can be illustrated in a not nessesarily physically correct way like follows: Suppose an organization for time travels wants to carry out some trips into the past. There shall be visited the points of time $x_1,x_2,\dots,x_f$ where $x_1<x_2<\dots<x_f$. In order to execute these trips there are time travellers in r stations $s_1,s_2,\dots,s_r$ (their number is not limited in any station). But only the station can bring a traveller into the past. A traveller in the point x_i can jump only in the forward direction, that means jumping to a point x_j is possible only if $x_i<x_j$ (wich means $i<j$) or he can return to his station into the present time. Taking a traveller from station s_k into point x_i there are c^k_{0i} energy units nessesary, jumping from x_i to x_j a traveller needs c_{ij} energy units and jumping from x_i to s_k he needs c^k_{i0} units. We want to find a set of time travels such that every point x_i is visited exactly one times and such that the amount of energy is as small as possible.

In [1] there was shown that this problem can be solved by a polynomial algorithm if r=1. Here it will be shown that it is NP-hard already if r=2. In order to do this let $T_{f,r}$ be the directed graph with vertex set

$$V(T_{f,r})=\{s_1,s_2,\dots,s_r,x_1,x_2,\dots,x_f\}$$

and arc set

$$A(T_{f,r})=\{(s_k,x_i),\ (x_i,s_k),\ (x_i,x_j)\mid k=1,\dots,r;\ i,j=1,\dots,f;\ i<j\}.$$

A directed cycle of $T_{f,r}$ containing exactly one of the vertices $s_1,s_2,\dots,s_r$ will be called a (time-) travel. If we want to emphasize that it contains vertex s_k we will call it a k-travel. A set of travels ψ will be called a travelling plan if every node x_i, $i=1,\dots,f$, is contained in exactly one of the travels of ψ.

Example:

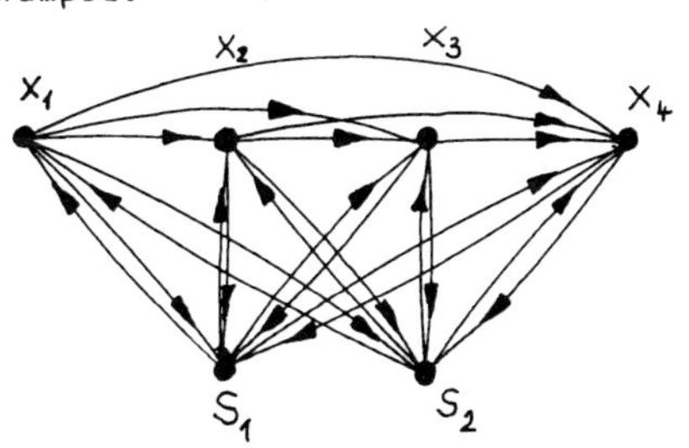

$T_{4,2}$

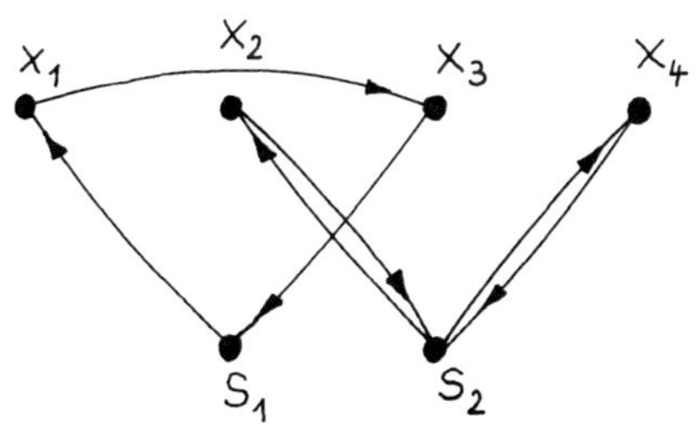

a possible ψ

R. Bodendiek, R. Henn (Eds.)
Topics in Combinatorics and Graph Theory

Now we define a cost function on the arcs of $T_{f,r}$: (s_k,x_i) gets the costs c^k_{0i}, (x_i,s_k) gets the costs c^k_{i0} and (x_i,x_j) gets the costs c_{ij}. Then the problem is to find a travelling plan ψ of $T_{f,r}$ with $c(\psi)$ minimal, where $c(\psi)$ denotes the sum of the costs of all arcs of the travels of ψ.

Complexity results

At first we consider the case r=1. The subgraph of $T_{f,r}$ spanned by the nodes $x_1,x_2,\dots,x_f$ has no directed cycle and that's why every directed cycle of $T_{f,1}$ is a travel. Hence we look after a cost minimal set of dicycles such that every vertex x_i is contained in exactly one of these cycles. This problem can be formulated as a circulation problem (see [1]) and can be solved by known algorithms (see e.g. [3]).

Now let's have r=2. In order to make the proofs more clear we do not only consider travelling plans on $T_{f,r}$ but also on subgraphs of $T_{f,r}$. This is no generalization because some arcs beeing not in the subgraph can be "forbidden" by assigning sufficiant high costs to them. We therefore consider the following decision problem:

(P_1): Given a natural number f, a graph G such that $V(G)=V(T_{f,2})$, $A(G)\subset A(T_{f,2})$, a cost function $c:A(G)\longmapsto R$ and a number $K\in R$. Is there a travelling plan ψ on G such that $c(\psi)\leq K$?

In order to make the proofs more clear we also consider a more general problem. It may happen that time travellers of different stations need a different quantity of energy in order to jump from x_i to x_j. We can describe this situation by constructing a graph $S_{f,r}$ consisting of r different copies of $T_{f,1}$:

$$V(S_{f,r})=\{x^k_i,\ s^k \mid k=1,\dots,r;\ i=1,\dots,f\}$$

$$A(S_{f,r})=\{(s^k,x^k_i),\ (x^k_i,s^k),\ (x^k_i,x^k_j) \mid k=1,\dots,r;\ i,j=1,\dots,f;\ i<j\}.$$

On this graph a time travelling plan $\varkappa$ is a set of directed cycles (time travels) such that for every i=1,...,f the following property holds: Exactly one of the nodes $x^1_i,x^2_i,\dots,x^r_i$ is contained in a cycle of $\varkappa$.

Example:

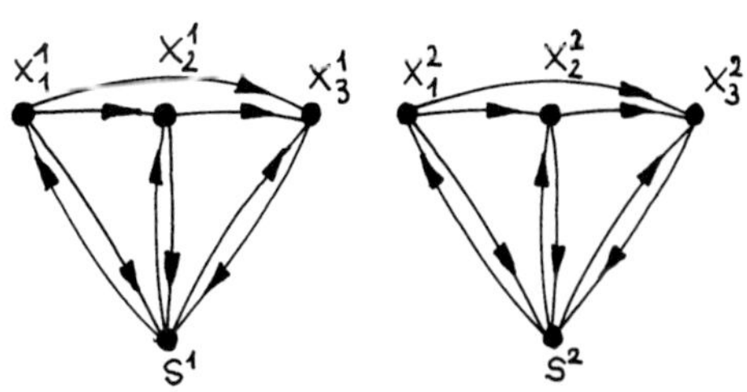

$S_{3,2}$

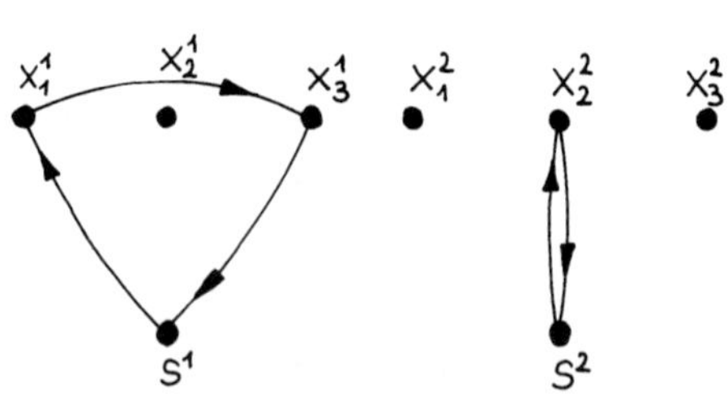

a possible $\varkappa$

Similar to (P_1) we can formulate a second problem:

(P_2): Given a natural number f, a graph H such that $V(H)=V(S_{f,2})$, $A(H)\subset A(S_{f,2})$, a cost function $d:A(H)\longmapsto R$ and a number $K\in R$. Is there a travelling plan $\varkappa$ on H such that $d(\varkappa)\leq K$?

Obviously every instance of (P_1) can be formulated as an instance of (P_2). But the reverse is also true:

<u>Theorem 1:</u> If (P_1) is polynomially solvable then so is (P_2).

<u>Proof:</u> Suppose that we have an instance of (P_2). We construct a graph G satisfying $V(G)=V(T_{4f,2})$, $A(G)\subset A(T_{4f,2})$ such that there is a one to one correspondence between the travelling plans $\varkappa$ of H and ψ of G. Then we define a cost function c on the arcs of G such that $c(\psi)=d(\varkappa)$ for corresponding ψ and $\varkappa$. From this the assertion follows.

Let G be the Graph with vertex set $V(G)=V(T_{4f,2})$ and arc set constructed as followes.

We define a function $u:V(H)\longmapsto V(G)$ by

$$u(s^1)=u(s^2)=s_1,\ u(x_i^1)=x_{f+i},\ u(x_i^2)=x_{2f+i},\ i=1,\dots,f.$$

Then let

$$A(G)=\{\ (u(y),u(z))\ |\ (y,z)\in A(H)\ \}\cup A_0$$

where

$$A_0=\{\ (s_2,x_i),\ (x_i,x_{f+i}),\ (x_i,x_{2f+i}),\ (x_{f+i},x_{3f+i}),\ (x_{2f+i},x_{3f+i}),\ (x_{3f+i},s_2)\ |\ i=1,\dots,f\ \}.$$

Obviously if $Z=(s^k,x_{i_1},x_{i_2},\dots,x_{i_l},s^k)$ is a k-travel of H (k=1,2), then $Z'=(u(s^k),u(x_{i_1}),u(x_{i_2}),\dots,u(x_{i_l}),u(s^k))$ is a 1-travel of G. Furthermore every 1-travel of G is of the form Z' with Z being a k-travel of H. In this way we get a one-to-one correspondence between the travels of H and the 1-travels of G.

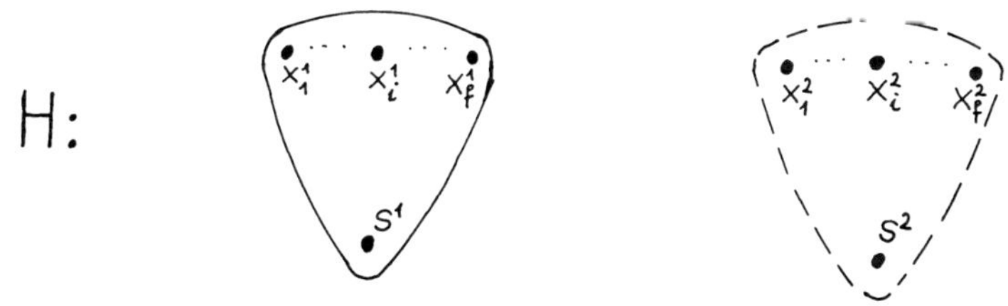

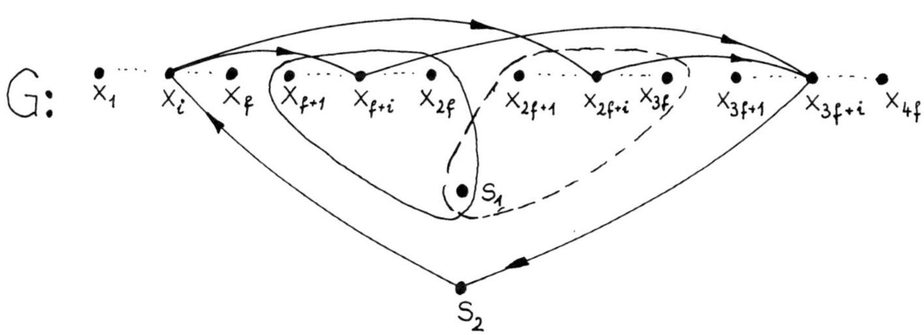

Now consider a travelling plan χ of H. We construct a travelling plan ψ of G from it.

At first using the correspondence explained above we get the set ψ' of all 1-travels of G corresponding to the travels of χ. Because χ is a travelling plan of H, ψ' has the property that for every $i=1,\dots,f$ exactly one of the two vertices x_{f+i}, x_{2f+i} is contained in a cycle of ψ'. We complete ψ' to form a travelling plan ψ of G in the following way:

For $i=1,\dots,f$ if x_{f+i} is in a cycle of ψ' then add the 2-travel $(s_2,x_i,x_{2f+i},x_{3f+i},s_2)$ to ψ' and if x_{2f+i} is in a cycle of ψ' then add the 2-travel $(s_2,x_i,x_{f+i},x_{3f+i},s_2)$. One can easily check that the result is a travelling plan of G.

On the other hand let ψ be a travelling plan of G, ψ' the set of 1-travels of ψ and χ the set of travels of H corresponding to the travels of ψ. Then χ is a travelling plan of H. To see this, consider at first the vertex x_f. x_f is the head of only one arc, namely (s_2,x_f). From this it follows that x_f is contained in a 2-travel Z_f of ψ. The next vertex in Z_f can only be x_{2f} or x_{3f} and because x_{2f} and x_{3f} have only x_{4f} and s_1 as successors, there is $Z_f=(s_2,x_f,x_{2f},x_{4f},s_2)$ or $Z_f=(s_2,x_f,x_{3f},x_{4f},s_2)$. The conclusion is that exactly one of the nodes x_{2f}, x_{3f} is contained in a 2-travel and the other is in a 1-travel. Additionally, none of the arcs (x_{f+i},x_{2f}) and (x_{2f+i},x_{3f}), $i=1,\dots,f-1$, is contained in a 2-travel of ψ.

Next we consider x_{f-1}. It is by similar reasons contained in a 2-travel Z_{f-1} and taking into consideration that the arcs (x_{2f-1},x_{2f}) and (x_{3f-1},x_{3f}) are not contained in a 2-travel, it follows that $Z_{f-1}=(s_2,x_{f-1},x_{2f-1},x_{4f-1},s_2)$ or $Z_{f-1}=(s_2,x_{f-1},x_{3f-1},x_{4f-1},s_2)$.

Continuing these considerations we see that for every $i=1,\dots,f$ exactly one of the two nodes x_{f+i}, x_{2f+i} is contained in a 1-travel of ψ. From this it follows that χ is a travelling plan of H.

Now constructing the cost function c for G is easy: Set $c(u(y),u(z))=d(y,z)$ for $(y,z)\in A(H)$ and the costs of arcs being in A_0 to zero. Then the indicated properties follow.

Now we prove

<u>Theorem 2:</u> (P_2) is NP-complete.

<u>Proof:</u> It is easy to see that (P_2) is in the class NP. We will reduce the independend set problem to it:

(IS): Given an undirected Graph U and a number $K\in R$, does there exist an independent set M in U such that $|M|\geq K$?

Here an independent set is a set M of vertices of U such that none of the edges of U has both endnodes in M. **(IS)** is known to be NP-complete (see [2]).

Let U have n vertices and m edges. We denote the vertices and the edges by numbers i and l ($i=1,\dots,n$; $l=1,\dots,m$). We may assume that U has no isolated vertex (an isolated vertex is contained in every maximal independent set and that's why if U would contain such a node i, then we could consider the independent set problem for $U\setminus\{i\}$ and the number K-1).

We set f=3m and construct a graph H satisfying $V(H)=V(S_{f,2})$. $A(H)\subset A(S_{f,2})$ such that there is a one to one correspondence between the independent sets of U and the travelling plans of H. Then we define a cost function d on the arcs of H such that

$$|M|=-d(\psi) \qquad (*)$$

for corresponding M and ψ. Hence there exists an independent set with cardinality not less than K if and only if there exists a travelling plan on H with costs not greater than -K and the assertion follows.

First we assign to every inzident pair of a node i and an edge l a natural number q_{il} such that $m<q_{il}\leq 3m$ and $q_{il}\neq q_{jp}$ for $i\neq j$ or $l\neq p$ (this is possible because there are exactly 2m such inzident pairs).

Now for an i, $1\leq i\leq n$ let $l_1,l_2,\ldots,l_h$ be the edges of U inzident to i, $q_k=q_{il_k}$ and without restriction of generality $q_1<q_2<\ldots<q_h$. We can define a 1-travel Z_i as

$$Z_i=(s^1,x^1_{q_1},x^1_{q_2},\ldots,x^1_{q_h},s^1).$$

Because we have no isolated vertex in U, every Z_i is well defined. For every i=1,...,n the arcs of Z_i shall be contained in H.

For every l=1,...,m let i and j be the two endnodes of the edge l and $q_{il}<q_{jl}$. Then the arcs (s^2,x^2_l), $(x^2_l,x^2_{q_{il}})$, $(x^2_l,x^2_{q_{jl}})$, $(x^2_{q_{il}},x^2_{q_{jl}})$, $(x^2_{q_{il}},s^2)$ and $(x^2_{q_{jl}},s^2)$ shall be in H. Other arcs are not in H.

U:

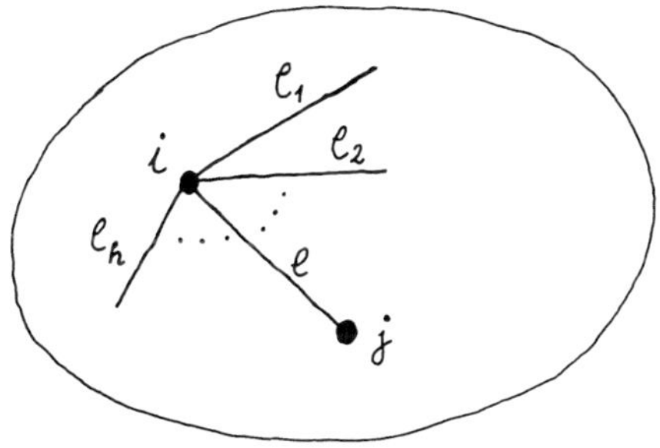

H:

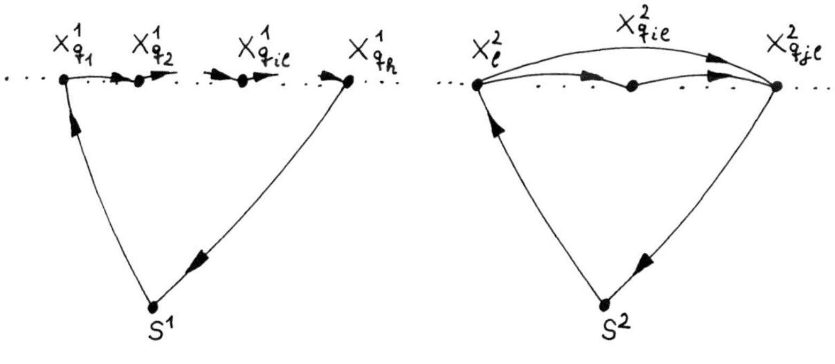

For every independent set M there can be defined a set of 1-travels $\varkappa'$ of H by

$$i \in M \text{ if and only if } Z_i \in \varkappa'. \qquad (**)$$

This $\varkappa'$ can be completed to form a travelling plan as follows. Let i and j be the endnodes of the edge l and $q_{il} < q_{jl}$. There are three cases possible:

1. $i \in M$, $j \notin M$. Then add the 2-travel $(s^2, x_l^2, x_{q_{jl}}^2, s^2)$ to $\varkappa'$.
2. $i \notin M$, $j \in M$. Then add the 2-travel $(s^2, x_l^2, x_{q_{il}}^2, s^2)$ to $\varkappa'$.
3. $i \notin M$, $j \notin M$. Then add the 2-travel $(s^2, x_l^2, x_{q_{il}}^2, x_{q_{jl}}^2, s^2)$ to $\varkappa'$.

Doing this for every edge we get a travelling plan $\varkappa$.

On the other hand let $\varkappa$ be a travelling plan of H and $\varkappa'$ be the set of 1-travels of $\varkappa$. We can define a set M of vertices of U observing instruction (**). We show that M is independent.

Let l be an arbitrary edge of U and i,j be its endnodes. Because x_l^1 is an isolated point in H, x_l^2 must be contained in some (2-) travel Z of $\varkappa$. The succeeding vertex of x_l^2 in Z can only be $x_{q_{il}}^2$ or $x_{q_{jl}}^2$. But this means that one of the two nodes $x_{q_{il}}^1$, $x_{q_{jl}}^1$ is not contained in any travel of $\varkappa$ and that's why one of the two travels Z_i, Z_j is not contained in $\varkappa$. From this it follows that one of the two nodes i, j is not contained in M, wich shows the independence of M.

Setting the costs of exactly one arc of Z_i to -1 for every $i=1,\dots,n$ and setting the costs of all other arcs of H to zero we obtain a cost function with the demanded property (*).

References

[1] G.Ehnert, J.Harant, H.Walther: Ein in Polynomialzeit arbeitender Algorithmus zur Minimierung einer skalarisierten Zielfunktion in einem Kraftverkehrsbetrieb mit einem Depot, einem Fahrzeugtyp und Aufträgen mit unterschiedlichen Anforderungen an die zu transportierenden Mengen. Proceedings of the Conference on Graph Theory held in Eyba, October 1.- 5., 1984. Teubner 1985.

[2] M.R.Garey, D.S.Johnsson: Computers and Intractability: a guide to the Theory of NP-Completeness. Freeman, San Francisco, 1979.

[3] E.L.Lawler: Combinatorial Optimization: Networks and Matroids. Holt, Rinehart and Winston, 1976.

An Aperiodic Triple of Prototiles

P. Schmitt, Wien, Austria

Summary. Based on R. M. Robinson's well-known aperiodic set of six prototiles an aperiodic triple of prototiles is constructed. Further applications of the method are indicated.

0. Introduction

A **tile** is a subset of the plane which is homeomorphic to the closed unit disk. A **tiling** is a family of tiles which cover the plane and whose interiors are pairwise disjoint. A set of **prototiles** is a finite set of tiles. It **admits** a tiling if there is a tiling such that each of its tiles is congruent to one of the prototiles. Two tilings are **equal** if they are congruent. A tiling is periodic if there are two non-parallel translations which are symmetries of the tiling. A set of prototiles is **aperiodic** if it admits a tiling, but does not admit a periodic tiling.

Several examples of aperiodic sets are known (see [3] for a detailed treatment, or [1], [2]). An aperiodic pair has been constructed by Penrose (see [4]), but it is not known whether there exists a **single** prototile which forms an aperiodic set. The aperiodic triple to be described below is derived from an aperiodic set of six prototiles found by Robinson [5] (see Figure 1). It is shown that the behaviour of this set can be simulated by a set of only three tiles.

1. The construction

Let $\mathcal{S} = \{S_1, \ldots, S_n\}$ be a (finite) set of prototiles such that each tile S_i is obtained from the same square by adding small projections to the edges and/or deleting small indentations from the edges. Assume further that the distribution of these bumps and dents is such that, in any tiling, the tiles are necessarily fitted against each other edge to edge, i.e., that any tiling admitted is topologically equivalent to the regular tiling by squares. The tiles in Figure 1 may be considered as an example, if the bumps and dents at the corners are neglected.

Under these conditions, and as far as the tilings admitted, and their properties, are concerned, the individual form and the distribution of the projections and indentations do not matter. Only the fitting conditions realized by them, and the symmetry properties of the edges (and the tiles), are important: It is sufficient to know of each side, which sides of which tiles are complementary to it, i.e., can be fitted against it or its

R. Bodendiek, R. Henn (Eds.)
Topics in Combinatorics and Graph Theory

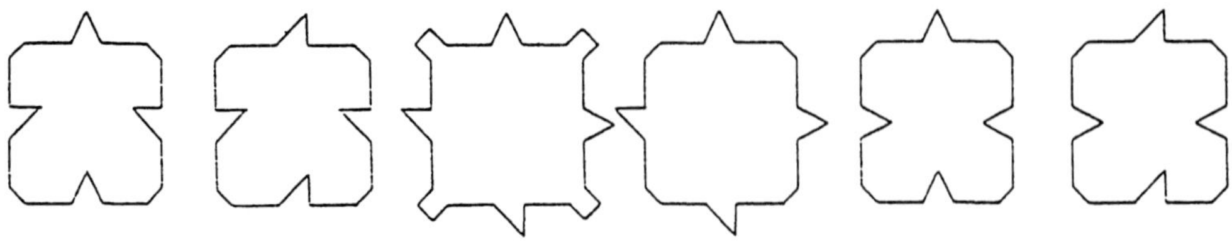

Fig. 1. The aperiodic set of prototiles found by Raphael M. Robinson.

mirror-image without gaps or overlaps. Robinson's tiles, for instance, can be completely described as follows: There is a symmetric edge a, and its complement $\bar{a}$, and an asymmetric edge b^+, and its complement $\bar{b}^+$, and their mirror-images b^- and $\bar{b}^-$; the four sides of the squares are (counterclockwise) $a\bar{b}^-\bar{a}\bar{b}^+$, $b^+\bar{b}^-\bar{b}^+\bar{b}^+$, ab^+b^-a (twice, since corner bumps and dents are still neglected for the moment), $a\bar{a}\bar{a}\bar{a}$, $b^+\bar{a}\bar{b}^+\bar{a}$. Therefore, without loss of generality, we may assume that the projections and indentations are all placed in an arbitrarily small interval at the middle of each side.

A regular $4n$-gon can be thought as the intersection of n squares (see Figure 2(a)). If these squares are replaced by the tiles $S_1, \ldots, S_n$, then — according to the last remark — we may assume that all the projections and indentations are placed on the sides of the resulting $4n$-gon T.

Now let us consider any tiling by $\mathcal{S}$. Each tile S is congruent to some S_i and therefore can be replaced by an appropriately rotated copy of T such that the four sides generated by S_i coincide with the corresponding parts of the sides of S. Thus a packing of copies of T, as shown in Figure 2(c), is obtained where touching sides are complementary. Conversely, from each such packing a tiling by $\mathcal{S}$ is obtained if each copy of T is replaced by one of the S_i according to its orientation.

This correspondence between tilings by $\mathcal{S}$ and "interlocking" packings of copies of T is one-to-one if all the prototiles are asymmetric. It is still almost one-to-one if some, or all, the prototiles have symmetries: The packing still uniquely determines a tiling, but when, in a tiling, a symmetric tile S is replaced by T, then ambiguities arise if T does not possess the same symmetries: It is still uniquely determined which four of the $4n$ sides of T are used, but for the position of the tile several possibilities exist, depending on the number of symmetries.

The holes around the former corners — which are left uncovered by the packings of T — (approximately) are non-convex $4(n-1)$-gons. Due to the different positions of T, and the resulting various different distributions of projections and indentations on the sides of the $4(n-1)$-gons, they are not necessarily congruent to each other. But the intersection, say C, of all possible corner shapes is a prototile such that a copy of it can be placed into each of these holes. Obviously, some small holes are still left uncovered where an indentation is not filled by an appropriate projection. In general these holes may have many different shapes, but — as we shall see — if the realization of the fitting conditions is suitably chosen, then a single additional prototile, say L (for lid), suffices.

To achieve this, a construction similar to the keys and keyholes in [6] is used. A modification has become necessary because both a side and its complement must fit against the sides of the corner-tile C. Furthermore, the symmetry properties of the sides have to be retained.

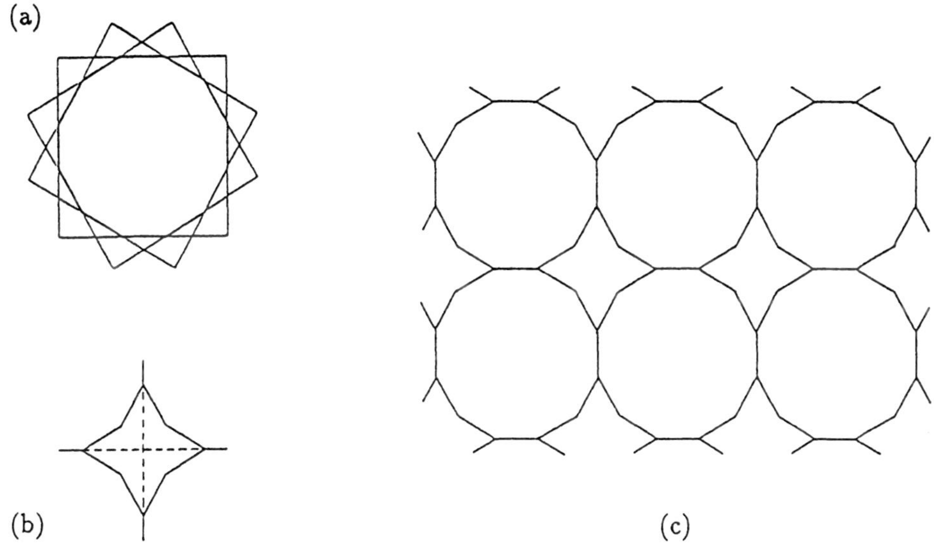

Fig. 2. The construction of the triple corresponding to a set of square tiles.

(a) The 12-gon T corresponding to three square tiles. (b) The corner tile C corresponding to three square tiles. (The shape of the edges is not shown in the Figure.) (c) A packing of copies of the 12-gon T — or a tiling by T and C, if the shape of the sides is neglected.

As the lid-tile L an octagon, as that shown in Figure 3(c), can be chosen. Note that it has the symmetries of a rectangle and does not admit a tiling. In fact, the actual shape of the tile does not matter as long as it has these properties. The size of the lid has to be sufficiently small to allow the necessary deletions from T and C to be carried out. Only one shape p of projections and the corresponding shape $\bar{p}$ of indentations is used, both obtained by adding respectively deleting half of the lid L (see Figure 3).

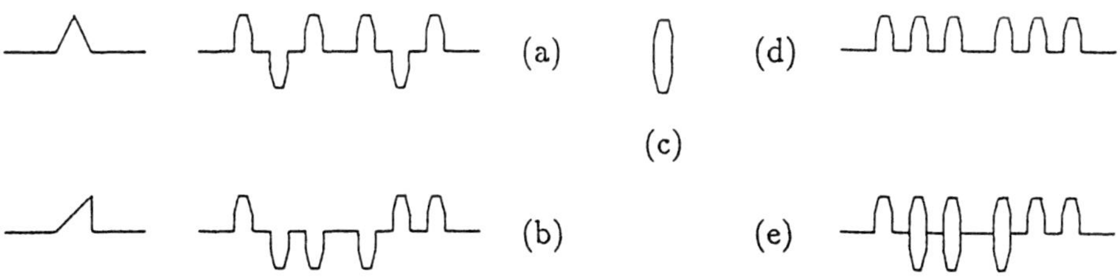

Fig. 3. The realization of the fitting conditions of Robinson's tiles:

(a) the symmetric side a, (b) the asymmetric side b^+, (c) the lid L, (d) the pattern used for the sides of C, (e) the pattern of (b) fitted against the pattern of (d) — the holes left can be filled using copies of L.

Patterns of p's and $\bar{p}$'s are used to realize the fitting conditions. Figure 3 shows how this can be done in the case of Robinson's tiles. Here patterns of six units (p or $\bar{p}$) are used. The pattern on the sides of the corner-tile C consists of six copies of $\bar{p}$ which can either be filled by a copy of p, or — if fitted against another copy of $\bar{p}$ — by a copy of the lid L.

In general, patterns of $4n$ units can be used, each consisting of $2n$ copies of p and $2n$ copies of $\bar{p}$. This ensures that a pattern can only be fitted against its complement — in any other case a copy of p would have to fit against another copy of p, and that obviously is not possible. From these $\binom{4n}{2n}$ patterns $\binom{2n}{n}$ are symmetric, forming $\binom{2n}{n}/2$ pairs of complementary patterns; likewise there are $\binom{2n}{n}$ centrally symmetric patterns (pairs of patterns which are symmetric to each other); the rest of the patterns is asymmetric and forms groups of four patterns (a pattern, its mirror-image, and the two complementary patterns). Therefore any number of patterns with prescribed symmetry is possible if n is sufficiently large. One type of side, however, has no equivalent pattern: the straight line which fits against itself and the mirror image of itself. The only pattern which has both these properties is that consisting only of copies of $\bar{p}$ — but this pattern fits against any other pattern (as already discussed). If such a side occurs, the following procedure can be used: Choose a mirror-symmetric projection m, and its complement $\overline{m}$. For every occurrence of the straight line the corresponding tile S_i is replaced by two tiles which only differ in the form of one side: One tile has the projection m, the other the indentation $\overline{m}$, instead of the straight line.

Therefore, with the tiles C and L, each of the packings can be extended to a tiling (in a unique way). Furthermore, it is easy to see that any tiling admitted by T, C, and L is an extension of a packing. Consequently, there is a correspondence between the tilings admitted by $\mathcal{S}$, and the tilings admitted by $\mathcal{T} = \{T, C, L\}$. Again, it is one-to-one if all the initial prototiles S_i are asymmetric. Furthermore, to each symmetry of a tiling by $\mathcal{T}$ there belongs an induced symmetry of the (unique) corresponding tiling by $\mathcal{S}$. If all the initial prototiles are asymmetric then this correspondence is also one-to-one. Moreover, for each periodic tiling by $\mathcal{S}$ there is a corresponding periodic tiling by $\mathcal{T}$: T, C, and L are aperiodic if and only if $S_1, \ldots, S_n$ are aperiodic.

2. An aperiodic triple.

To complete the construction of an aperiodic triple we need a means to simulate the effect of the corner bumps and dents of the tiles shown in Figure 1: Because of these additional modifications, from the four tiles meeting in a corner precisely one must have a bump. To obtain an equivalent effect using T, C, and L, from the $4n$-gon T the corners are cut, as shown in Figure 4. The resulting $8n$-gon need not be regular, but it must have the symmetries of the original $4n$-gon. The corner-tile C must be altered accordingly, the result being an $(8n-4)$-gon, as shown in Figure 4(b). These new sides are used to carry the information concerning the bumps: A pattern of four units serves this purpose.

If, on an original tile S_i, two sides meet in a corner with a bump then on two of the new sides of T (those which are adjacent to and between the counterparts of the two sides of S_i) the nearest unit is chosen to be a p, and the next a $\bar{p}$. Conversely, if the two sides of the original tile meet in a corner with a dent, then the nearest unit is chosen to be a $\bar{p}$, and the next a p (see Figure 4(c)). In this way all the patterns on the new sides are defined.

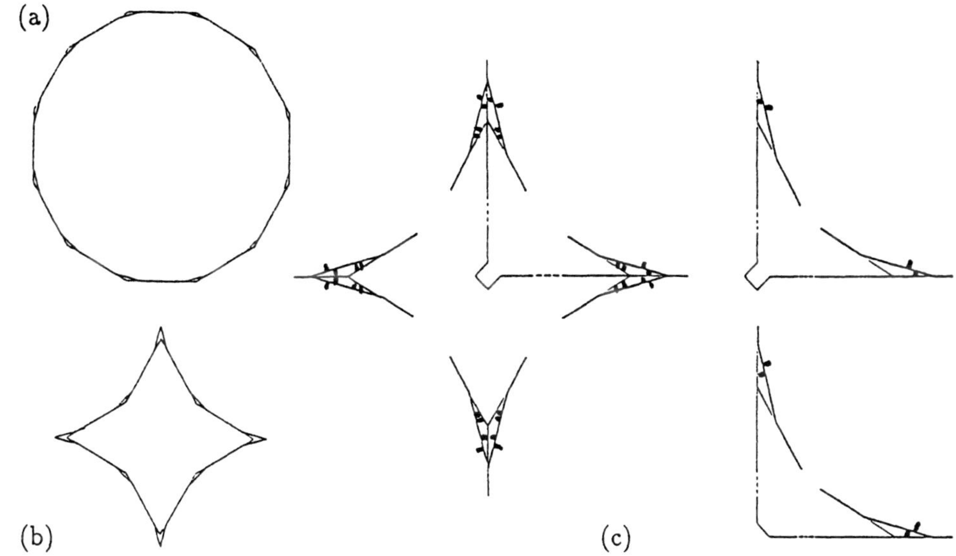

Fig. 4. The modifications necessary for simulating Robinson's tiles:

(a) cutting the corners of T (shown on the tile from Fig. 2(a)), (b) the corresponding modification of the corner-tile C, (c) the simulation of corner bumps and dents.

On the corner tile C, on three of the four arcs between the four "spikes", only the six units nearest to the endpoints are projections p, all the others are $\bar{p}$'s, on the fourth arc the second from each endpoint is a p, all the others are $\bar{p}$'s (see Figure 4(c)).

Assuming these choices, it is easy to verify that from the four copies of T which touch a corner-tile C precisely one must be in a position corresponding to a bump.

Therefore, if the constructions described in the previous section and in this section are applied to Robinson's tiles then three prototiles (a modified 48-gon, a modified 44-gon, and an octagon), and a correspondence between the tilings admitted, are obtained. Since four of the six prototiles are mirror-symmetric this correspondence is not one-to-one. But since it preserves non-periodicity the triple constructed is aperiodic, too.

3. Remarks.

(a) Polymorphic triples.

It is easy to find n (asymmetric) squares $S_1, \ldots, S_n$ such that each of it admits a unique tiling, and no further tiling is admitted by the set (e.g., if the sides of S_i are chosen to be $a_i b_i \bar{a}_i \bar{b}_i$ for some arbitrarily, but differently shaped asymmetric sides). The construction of Section 1 yields three prototiles which admit precisely n (periodic) tilings. In addition, if all the S_i are centrally symmetric, or have 4-fold rotational symmetry, then the same is true for T, as it is for C and L, and for the tilings, as well. Furthermore, in this case, the one-to-one correspondence is maintained. (See [6] and [7] on the subject of polymorphic tilings.) Examples of σ-morphic triples can be derived from some of the examples described in [8] (cf. Remark (c) below).

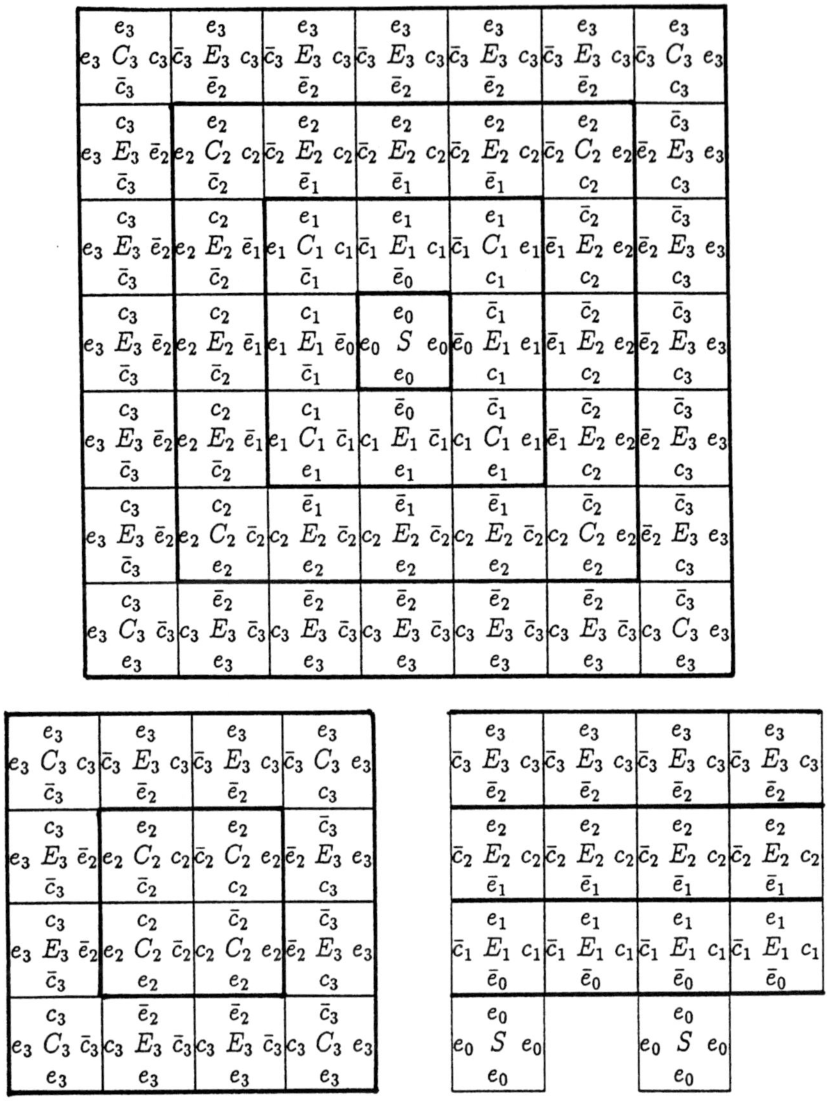

Fig. 5. The three types of patches admitted by the tiles described in Remark (b).

(b) A triple related to Heesch's problem.

Consider the following set of squares: S with sides $e_0 e_0 e_0 e_0$, E_i with sides $\bar{e}_{i-1} c_i e_i \bar{c}_i$, and C_i with sides $c_i e_i e_i \bar{c}_i$, $i = 1, \ldots, n$, where e_i and c_i denote differently shaped asymmetric sides. This set admits a patch of $2n+1$ by $2n+1$ squares. The patch is formed by the central stone S and n square rings: The i-th ring has copies of C_i at the four corners, and copies of E_i form the edges. (If S and E_1 are removed from the set then it only admits a patch of $2n$ by $2n$ squares.) Consequently, the triple constructed from these tiles admits a patch where the central copies of T, C, and L are completely surrounded by n rings of tiles, but it does not admit a larger patch containing an $(n+1)$-th ring. Note, however, that these prototiles tile over an infinite strip which is n tiles wide, and various other tilings made up from (subsets of) such strips (cf. Figure 5). Using more prototiles — e.g., $(2n+1)^2$ different tiles — this strip can be avoided, too. (For details on Heesch's Problem see [3], p. 155f: The example described above proves $f(n,1)=\infty$ for $n \geq 3$. *Remark:* The definition of $f(n,m)$ given there is not completely clear. What is meant by assuming that *every* patch containing m tiles can be surrounded at least $f(n,m)$-times? Even using three squares there are patches which cannot be completely surrounded though, of course, the squares admit a tiling!)

(c) Variations.

The construction of an aperiodic triple can be based on any aperiodic set of Wang tiles (see [3]). Furthermore, similar constructions based on regular tilings by triangles or by hexagons, instead of by squares, are possible. They lead to variations of the examples described above.

(d) Packings.

The correspondence between tilings and packings can also be used to construct disks which have special packing properties (see [9]).

4. References.

[1] M. Gardner, *Scientific American*, Jan. 1977.

[2] B. Grünbaum and G. C. Shephard, Some Problems on Plane Tilings, in: *The Mathematical Gardner* (D. A. Klarner, ed.). Boston: Prindle, Weber & Schmidt/ Belmont: Wadsworth Intern. 1981. pp. 167–196.

[3] B. Grünbaum and G. C. Shephard, Tilings and Patterns. W.H. Freeman and Co., New York 1986.

[4] R. Penrose, Pentaplexity. *Eureka* 39, 16–22 (1978). Reprinted in: *The Math. Intelligencer* 2, 32–37 (1979).

[5] R. M. Robinson, Undecidability and Nonperiodicity for Tilings of the Plane, *Inventiones math.* 12, 177–209 (1971).

[6] P. Schmitt, Sets of Tiles which Admit a Prescribed Number of Tilings. *Geometriae Dedicata* 21, 123–144 (1986).

[7] P. Schmitt, Polymorphic Pairs of Prototiles. *Sitzungsberichte der Österr. Akad. d. Wiss., Mathem.-Naturw. Kl., Abt. II.* 197, 305–313 (1988).

[8] P. Schmitt, σ-Morphic Sets of Prototiles. *Discrete Comput. Geometry* 2, 271–295 (1987).

[9] P. Schmitt, Disks With Special Properties of Densest Packings. *Discrete Comput. Geometry* (to appear).

Representation of Graphs by Integers

R. Schnabel, Kiel, FRG

In this paper by graph we always mean a graph without loops and without multiple lines. Such a graph will be conceived as a pair (V,ρ) where V is a set and ρ is an irreflexive symmetric relation on V. The elements of V are the vertices and ρ represents the adjacency. Concerning other basic concepts we will use the common terminology.

Trying to describe graphs by integers we may ask:

> Is there a set T of integers with the following property: The vertices of each finite graph can be labeled by different numbers such that two arbitrary vertices are adjacent if and only if the difference of their labels belongs to T ?

In the third part of this paper it will be shown that such a set T does exist. In part 1 and 2 the tools of the theory of group graphs will be given.

1. Group graphs

Since the theory of group graphs is well-known here we just give the most basic elements (cf. chap. 6.1.2 of [4]).

1.1 Let H be a group and let T be a subset of H such that $1 \notin T$, $T^{-1} \subseteq T$. The relation ρ_T defined by

$$x \, \rho_T \, y \text{ , if } xy^{-1} \in T \text{ (for all } x,y \in H)$$

obviously is irreflexive and symmetric. (H,ρ_T) is called a group graph on H.

A group graph is not just regular but even homogeneous in the sense that its automorphism group acts transitively on the vertices, since the right-translations of the group are automorphisms of the group graph, of course (also cf. [1]). Conversely, not every homogeneous graph is a group graph. E.g. it is easy to prove that the five regular polyeder graphs are homogene-

R. Bodendiek, R. Henn (Eds.)
Topics in Combinatorics and Graph Theory

ous but the dodekaeder graph is not a group graph. Now the following characterization holds which serves as a definition for group graphs in [1]:

1.2 A graph is a group graph if and only if it is non-empty and its automorphism group contains a subgroup acting strictly transitively on the vertices.

Even as simple a case as the cyclic group of order 6 produces several interesting examples for group graphs. Defining the addition mod 6 on $\mathbb{Z}_6 = \{0,...,5\}$ we get the following group graphs $(\mathbb{Z}_6,\rho_T)$:

$T = \emptyset$ 6 points	$T = \{3\}$ 3 edges	$T = \{2,4\}$ 2 triangles	$T = \{1,5\}$ hexagon
$T = \{2,3,4\}$ prisma	$T = \{1,3,5\}$ complete bigraph $V_{3,3}$	$T = \{1,2,4,5\}$ oktaeder	$T = \{1,2,3,4,5\}$ complete graph V_6

2. Addresses of graphs

Since every graph can be extended to a complete graph (and the latter is a group graph) every graph is embeddable in a group graph. Of a more restrictive character is the embedding as an induced subgraph of a group graph. Here an *induced subgraph* of a graph G is an edge-complete subgraph of G (cf. [2], chap. 2,sec. 1). Hence the following construction suggests itself:

2.1 Let (H,ρ_T) be a group graph and let V be an arbitrary subset of H. Then $G(H,V,T) := (V,\rho_T|_V)$ is an induced subgraph of (H,ρ_T), a so-called *group subgraph on H*.
Hence the group subgraphs are precisely the induced subgraphs of group graphs.

It is easy to show:

2.2 For every group subgraph G(H,V,T) the following holds:
a) For each monomorphism α of H and each $h \in H$ we have $G(H,V\alpha\cdot h,T\alpha) \simeq G(H,V,T)$.
b) With $D(V) := V\cdot V^{-1}\setminus\{1\}$ we have $G(H,V,T) = G(H,V,T\cap D(V))$.

In [1] it is proved that every finite graph is a group subgraph on an appropriate group. Here the following stronger proposition holds:

2.3 Every finite graph is isomorphic to a group subgraph on the group $\mathbb{Z}$ of the integers.

Proof. Let $G := (V,\rho)$ a finite graph. Choose a labeling of the elements of V, for instance $V = \{v_1, v_2, \dots, v_n\}$ with $n = |V|$. Set $P := \{2^i \mid i = 1,2,\dots,n\}$ and $T := \{2^i - 2^j \mid v_i \rho v_j\}$. We claim that $G \simeq G(\mathbb{Z},P,T)$, i.e. $(V,\rho) \simeq (P,\rho_T)$. To prove this it will be shown that $\psi: V \to P, v_i \mapsto 2^i$ is an isomorphism of (V,ρ) onto (P,ρ_T).

Evidently ψ is bijective. ψ preserves the adjacency: By definition of ψ, T, ρ_T we have that $x \rho y$ implies $x\psi \rho_T y\psi$ for all $x,y \in V$. Conversely, let $v_i, v_j \in V$ such that $v_i\psi \rho_T v_j\psi$. Then $2^i \rho_T 2^j$, i.e. $2^i - 2^j \in T$. By definition of T there exist k,m such that $v_k \rho v_m$ and $2^k - 2^m = 2^i - 2^j$. Since $0 \notin T$ we have $k \neq m$, $i \neq j$, and this evidently implies $k = i$, $m = j$, hence $v_i \rho v_j$ which completes the proof.

For the systematic and economic description of types of finite graphs the following method seems to be useful:

2.4 Let A be a finite non-empty subset of the set $\mathbb{N}$ of all positive integers. Let $V(A)$ be defined as the set of all odd integers $\leqq \max A$ of $\mathbb{N} \setminus A$, and set $T(A) := A \cup -A$. Then set $G(A) := G(\mathbb{Z}, V(A), T(A))$.
For a finite graph G a finite non-empty subset A of $\mathbb{N}$ such that $G \simeq G(A)$ will be called an *address* of G.

From 2.3 we are going to:

2.5 Every finite graph has an address.

Proof. The empty graph has the address $\{1\}$. Now let G be a finite non-empty graph. By 2.3 G is isomorphic to a group subgraph $G(\mathbb{Z},V,T)$. By the finiteness of V there exists an odd integer $u \in \mathbb{N}$ such that $|2x| < u$ for all $x \in V$. Set $V' := 2V+u$ and $T' := 2T$. According to 2.2a we have $G \simeq G(\mathbb{Z},V',T')$. Obviousely, since V' consists of odd positive integers there exists a finite subset A of $\mathbb{N}$ such that $V(A) = V'$ and $T(A) \cap D(V') = T' \cap D(V')$. Hence using 2.4 and 2.2b we get $G \simeq G(\mathbb{Z},V',T') = G(\mathbb{Z},V',T' \cap D(V')) = G(\mathbb{Z},V',T(A) \cap D(V'))$ $= G(\mathbb{Z},V',T(A)) = G(\mathbb{Z},V(A),T(A)) = G(A)$, i.e. A is an address of G.

Evidently a finite graph has many addresses. Of course, it is desirable to find shortest addresses consisting of smallest numbers. For example {8,9,10,11,12,16,18,20} is an address of the hexaeder graph.
To clearify the notion "address of a graph" the following table contains as small as possible addresses for all graphs having at most 4 vertices:

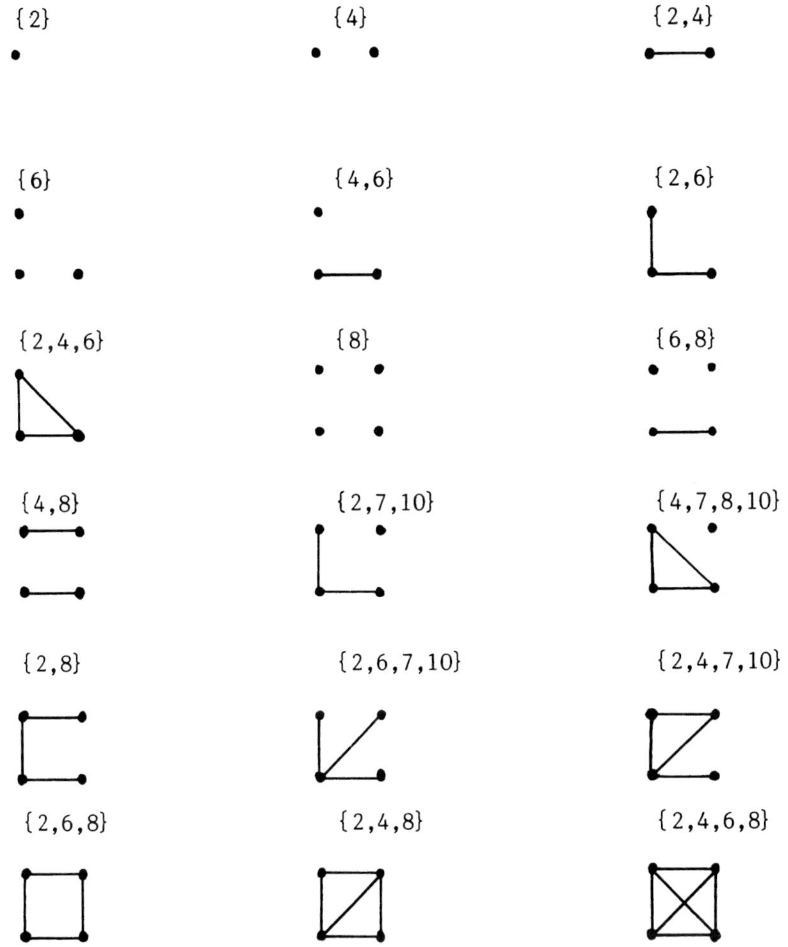

3. An universal group graph on $\mathbb{Z}$

Now we can formulate the question from the introduction in a more concise way and just give the answer as a general characterization theorem:

3.1 There is a group graph on $\mathbb{Z}$ which contains all finite graphs as induced subgraphs up to isomorphy.

Proof. According to 2.5 it is sufficient to construct a group graph $G(\mathbb{Z},T)$ which contains all graphs $G(A)$ as induced subgraphs up to isomorphy where A is an arbitrary finite non-empty subset of $\mathbb{N}$.
The set of all finite non-empty subsets of $\mathbb{N}$ is countable. Let $n \mapsto A_n$ be such a counting function. We shall construct subsets V_n, T_n of $\mathbb{Z}$ for each $n \in \mathbb{N}$ and an appropriate subset T of $\mathbb{Z}$ with the following properties:

(1) $G(A_n) \simeq G(\mathbb{Z},V_n,T_n)$ for all $n \in \mathbb{N}$

(2) $T \cap D(V_n) = T_n \cap D(V_n)$ for all $n \in \mathbb{N}$.

Then using 2.2b, (2), (1) we have for each $n \in \mathbb{N}$:
$G(\mathbb{Z},V_n,T) = G(\mathbb{Z},V_n,T\cap D(V_n)) = G(\mathbb{Z},V_n,T_n\cap D(V_n)) = G(\mathbb{Z},V_n,T_n) \simeq G(A_n)$, i.e. for every graph $G(A_n)$ the group graph $G(\mathbb{Z},T)$ contains an induced subgraph which is isomorphic to $G(A_n)$. This was to prove.
Now the construction of V_n, T_n, T:
Using the notation of 2.4 for each $n \in \mathbb{N}$ set

$d_n := \max(D(V(A_n))\cup\{0\}) + 1$, $p_n := d_1 \cdot d_2 \cdot \ldots \cdot d_n$ (and $p_0 := 1$)

$V_n := p_{n-1} \cdot V(A_n)$, $T_n := p_{n-1} \cdot T(A_n)$.

From these definitions it can be derived immediately that all d_n, p_n are positive integers. Hence using 2.2a and 2.4 we get condition (1).
Further we claim:

(d) $m \neq n$ implies $D(V_m) \cap D(V_n) = \emptyset$ (for all $m,n \in \mathbb{N}$)

Proof: Let $m,n \in \mathbb{N}$ such that $m < n$.
Assume that there exists an element $z \in D(V_m) \cap D(V_n)$.
According to the definition of V_i obviously $D(V_i) = p_{i-1} \cdot D(V(A_i))$ holds.
Hence there are $x \in D(V(A_m))$, $y \in D(V(A_n))$ such that $p_{m-1}x = z = p_{n-1}y$.
This implies $p_{m-1}|x| = p_{n-1}|y|$. On the other hand using the inequalities $m \leq n-1$, $1 \leq |y|$, $|x| < d_m$ and the definition of p_i we obtain
$p_{m-1}|x| < p_{m-1}d_m = p_m \leq p_{n-1} \leq p_{n-1}|y|$, a contradiction.

Now set $T := \bigcup_{n\in N}(T_n \cap D(V_n))$. With the help of (d) condition (2) directly follows. This completes the proof.

Final remark:

The proof of theorem 3.1 is constructive in a strict sense, but the set T is very complicated and its gradual construction is laborious. A graph $G(\mathbb{Z},T)$ which can be defined in a more simple way is unfortunately not known.
The problem solved by 3.1 is similar to the following one which arose in [3]:

Let $\mathbb{R}^2$ be the additive group of the 2-dimensional real vectorspace and let T be the set of all points $x \neq 0$ such that $|x|$ is rational. Which graphs can be represented as induced subgraphs of $G(\mathbb{R}^2, T)$?

Although this problem seems to be more difficult than that of 3.1, it is to hope that the methods evolved in this paper could help solving it.

References

1 Chen,C.C. and Teh,H.H., Embedding theorems of homogeneous graphs. Bull.Math.Soc.Nanyang Univ. 1965, 21-30.

2 Harary,F., Graph Theory. Reading (Massachusetts) 1969.

3 Ringel, G., 250 Jahre Graphentheorie. Graphen in Forschung und Unterricht (Festschrift K.Wagner), Hrsg. R.Bodendiek, Kiel 1985, 136-152.

4 Schnabel,R., Elemente der Gruppentheorie. Stuttgart 1984.

Special Systems of Linear Equations and Graphs of Convex Polytopes

W. Schöne, Karl-Marx-Stadt, GDR

Let be $\bar{\mathcal{A}} = (\bar{a}_1, \bar{a}_2, \ldots, \bar{a}_n)$ a Gale transform in the euclidean space $\bar{E}^{n-d-1}$ with $\dim \operatorname{aff} \bar{\mathcal{A}} = n-d-1$, $2 \leq d \leq n-1$ and the following property. For every open halfspace H of $\bar{E}^{n-d-1}$ with $0 \in \operatorname{bd} H$ the condition

$$\operatorname{card}\left\{ k : \bar{a}_k \in H \right\} \geq 2 \tag{1}$$

is valid. We consider the equations

$$\sum_{k=1}^{n} \bar{a}_k x_k = \mathcal{O} \qquad \sum_{k=1}^{n} x_k = 1 , \tag{2}$$

where x_k are real variables, which we interpret as coordinates of a rectangular Cartesian coordinate system with the origin 0 in the euclidean space E^n. The equations (2) are equivalent to the system

$$A\,x = b \tag{3}$$

with the $(n-d) \times n$ - matrix A which has the form

R. Bodendiek, R. Henn (Eds.)
Topics in Combinatorics and Graph Theory

$$A = \begin{pmatrix} \bar{a}_1 & \bar{a}_2 & \dots & \bar{a}_n \\ 1 & 1 & \dots & 1 \end{pmatrix}, \; \bar{A} = (\bar{a}_1\ \bar{a}_1 \dots \bar{a}_n), \; \sum_{k=1}^{n} \bar{a}_k = \theta ,$$

where $\bar{A}$ is a submatrix of A. The vector $b \in E^{n-d}$ is given by $b^T = (0\ 0 \dots 0\ 1)$. The rank of A is equal n-d. We denote by $X^+(A)$ the set of all non negative solutions of the system (3). We understand under a solution $x^I \in X^+(A)$ with $I \subsetneq N = \{1,2,\dots,n\}$ a n-dimensional vector with the coordinates $x^I_k = 0$ for $k \in I$ and $x^I_k > 0$ for $k \in N \setminus I$. Specially for $I = \{i\}$ or $I = \{j,k\}$ we write x^i or $x^{j,k}$. Two solutions x^I, x^J of $X^+(A)$ are said to be equivalent ($x^I \sim x^J$) if I = J. So we get classes of equivalent solutions in $X^+(A)$. We denote such a class by (I,A) respectively (i,A) or (jk,A). Let be $X^o = \{X^o_i : i \in N\}$, where X^o_i is the unitpoint of the coordinate axis x_i then $S_{n-1} = \operatorname{conv} X^o$ is a regular euclidean simplex of dimension n-1. The affine hull of S_{n-1} is the subspace E^{n-1} of E^n given by $\sum_{k=1}^{n} x_k = 1$. The system (3) determines a subspace E^d of E^{n-1} which contains the centre $O_n(\frac{1}{n},\frac{1}{n},\dots,\frac{1}{n})$ of S_{n-1}. Furthermore (3) has the property that a solution $x^i \in X^+(A)$ exists for every $i \in N$. This property is equivalent to condition (1). The subspace E^{n-d-1} of E^{n-1} may be the orthogonal complement of E^d in E^{n-1}. It holds $E^d \cap E^{n-d-1} = \{O_n\}$. Now we consider two orthogonal projections $\varphi_d : E^{n-1} \longrightarrow E^d$ and $\varphi_{n-d-1} : E^{n-1} \longrightarrow E^{n-d-1}$. We know by a theorem of Mac Mullen (|1| , p. 1168) that the n-tuple $\overline{X} = (\overline{X}_1, \overline{X}_2, \dots, \overline{X}_n)$ with $\overline{X}_k = \varphi_{n-d-1}(X_k^o)$ is a Gale transform of the n-tuple $X = (X_1, X_2, \dots, X_n)$ with $X_k = \varphi_d(X_k^o)$ for $k = 1, 2, \dots, n$. It is easy to prove that $\overline{X}$ and $\overline{\mathcal{A}}$ are iso- isomorphic (2 , p. 89). Moreover $\overline{X}$ and $\overline{\mathcal{A}}$ are both Gale transforms of vert $\overline{p}_d$, where the d-polytope $\overline{p}_d$ arises by

$$\overline{p}_d = \varphi_d(S_{n-1}) \quad . \tag{4}$$

There exists also another d-polytope $\overline{p}_d^*$ which we get by intersection.

$$\overline{p}_d^* = E^d \cap S_{n-1} \tag{5}$$

We can obtain $\overline{p}_d^*$ from $\overline{p}_d$ by a polarity with respect to the imaginary hypersphere of dimension d-1 with the centre O_n and the radius $r_n = i\,\frac{1}{n}$ ($i^2 = -1$). If we write $p_d \approx p_d'$ or $p_d \underset{\alpha}{\approx} p_d'$, we mean that p_d and p_d' are combinatorially or affinely equivalent. The following lemmas are valid.

Lemma 1

To every d-polytope p_d with n vertices exists a d-polytope $\overline{p}_d \underset{\alpha}{\approx} p_d$ which arises from S_{n-1} by $\overline{p}_d = \varphi_d(S_{n-1})$, |4| p. 78 .

Lemma 2

The set $X(I) = \{X_k : k \in I\} \subset E^d$ is the vertex set of a face of the d-polytope $\overline{p}_d = \varphi_d(S_{n-1})$ iff there exists a solution $x^I \in X^+(A)$, |6| p. 282 .

We denote (2) or (3) as a (normed) $\overline{p}$-system (polytope system) of the type (n,d) and $\overline{p}_d$ as the projected, $\overline{p}_d^*$ as the cut polytope of this p-system. The d-polytope $\overline{p}_d$ has n vertices but $\overline{p}_d^*$ has n facets. Now we denote the vertices of the graph

G of $\bar{p}_d$ with 1,2,...,n. There exists a unique solution x^I of $X^+(A)$, where I is the vertex set of a facet of $\bar{p}_d$ in G. To every vertex set I of a facet of $\bar{p}_d$ corresponds such a unique solution $x^I \in X^+(A)$. We denote the vertices X_k^* for $k = 1, 2, \ldots, f_{d-1}$ of $\bar{p}_d^*$ in its graph G^* with the corresponding vertex sets I_k of the facets of $\bar{p}_d$. Two vertices I_j, I_k of G^* are joined by an edge u_l^* ($l = 1,2,\ldots,f_{d-2}$) iff $U_l = I_j \cap I_k$ is the vertex set of a subfacet of $\bar{p}_d$ in G. Therefore we mark every edge u_l^* of G^* with its corresponding vertex set U_l. To every point X of an edge u_l^* belongs a radiusvector $x^{U_l} \in X^+(A)$. The set of all these solutions form an equivalence class (U_l, A) in $X^+(A)$..Now we transform the p-system (3) in a new p-system

$$A' x' = b' \qquad (6)$$

of the type (n+1,d) by adding a new row and a new column in in the matrix A. Replacing x, b, A, $\bar{a}_1$, $\bar{a}_k$ by x', b', A', $\bar{a}'_1$, $\bar{a}'_k$ and writing $(\bar{a}'_{n+1}, a'_{n+1-d,n+1})^T$ for the new column we put $x' = (x'_1, x'_2, \ldots, x'_{n+1})$, $b' = (0,0,\ldots,0,1)^T \in E^{n+1-d}$, $\bar{a}'_1 = (\alpha \bar{a}_1^T, c_1)^T$, $\bar{a}'_k = (\bar{a}_k^T, c_k)^T$ for $k = 2,3,\ldots,n$, $\bar{a}'_{n+1} = ((1-\alpha)\bar{a}_1^T, -c)^T$ with $\alpha = 1$ or $\alpha > 1$, $c = \sum_{k=1}^{n} c_k > 0$ $a'_{n+1-d,n+1} = 1$ and get the system (6). In this paper we consider only the case $\alpha = 1$. Further we regard the hyperplane $h(x) = \sum_{k=1}^{n} c_k x_k = 0$ in the space E^n. We choose h(x) so, that some vertices x^{I_ν} of $\bar{p}_d^*$ for $\nu \in \Gamma = \{1,2,\ldots,r\}$ are contained in the open halfspace $H^- = \{x \in E^n : h(x) < 0\} \subset E^n$. The other vertices x^{I_μ} for $\mu \in \bar{\Gamma} = \{r+1, r+2, \ldots, f_{d-1}\}$ and the point O_n lie in the open halfspace $H^+ = \{x \in E^n : h(x) > 0\} \subset E^n$. The subgraph G_q^r of G^* induced by the vertex set $\{I_\nu : \nu \in \Gamma\}$ of $V(G^*)$ shall be connected and be a proper subgraph of the graph of the facet F_n^* of $\bar{p}_d^*$ given by $x_n = 0$ in (3). Let be $U = \{U_k : k = 1,2,\ldots,q\} = E(G_q^r)$ the set of edges of G_q^r. Then our choice of $h(x)$ secures that condition (1) is also valid for (6), because the equation

$$x' = \lambda \binom{x}{0} + (1-\lambda)\binom{y}{0} \in X^+(A'), \quad \lambda = \frac{q}{q+1}, \quad q = -\frac{h(y)}{h(x)}$$

holds for two points $X, Y \in \bar{p}_d$ with $X \in H^+$ and $Y \in H^-$. If we put $x = x^{I_\nu}$, $\nu \in \Gamma$ and $y = x^{I_\mu}, \mu \in \bar{\Gamma}$, where (I_ν, I_μ) is an edge of G^*, we get $x' = {}'x^I \in X^+(A')$ with $I = (I_\nu \cap I_\mu) \cup \{n+1\}$. For $X, Y \in \operatorname{int} \bar{p}_d^*$ we find $x' = {}'x^{n+1} \in X^+(A')$. As $O_n \in H^+$, we we get $c = \sum_{i=1}^{n} c_i > 0$ and therefore it is rank $A' = n+1-d$. So the system (6) is a p-system of the type $(n+1, d)$. Now we can define some operations which we call p-operations (polytope operations). First we define the p-operations $\omega_{1,U}$ and $\omega^*_{1,U}$. We write

$$\omega_{1,U}\, \bar{p}_d = \bar{p}_d', \quad \omega^*_{1,U}\, \bar{p}_d^* = \bar{p}_d^{*\prime}, \qquad (7)$$

where $U = E(G_q^r)$, $\bar{p}_d'$ and $\bar{p}_d^{*\prime}$ are the projected and the cut polytope of (6) and the index 1 is written because we have $\alpha = 1$. The polytope $\bar{p}_d'$ has n+1 vertices and $\bar{p}_d^{*\prime}$ has n+1 facets. The vertex set of the graph $G^{*\prime}$ of $\bar{p}^{*\prime}$ consists of the vertices I_μ with $\mu \in \bar{\Gamma}$ and the vertices $I = (I_\nu \cap I_\mu) \cup \{n+1\}$ where $(I_\nu \cap I_\mu)$ is an edge of G^* with $\nu \in \Gamma, \mu \in \bar{\Gamma}$. The index n is not contained in any I_μ or I. The edge set of $G^{*\prime}$ is given by all edges $(I_\alpha \cap I_\beta)$ of G^* with $\alpha, \beta \in \bar{\Gamma}$ and all edges $(I_\nu \cap I_\mu)$ of G^* with $\nu \in \Gamma, \mu \in \bar{\Gamma}$ and the new edges $((I_\nu \cap I_\mu) \cap (I_{\nu'} \cap I_{\mu'})) \cup \{n+1\}$ where $\nu, \nu' \in \Gamma$ and $\mu, \mu' \in \bar{\Gamma}$ and $(I_\nu \cap I_\mu) \cap (I_{\nu'} \cap I_{\mu'})$ is the vertex set of a (d-3)-face F of $\bar{P}_d$ in G. As $\bar{P}_d'$ is dual to $\bar{P}_d^{*\prime}$, the d-polytope $\bar{P}_d'$ contains subfacets $F^s_{\nu\mu}$ of $\bar{P}_d$, which have the vertex sets $I_\nu \cap I_\mu$ in G with $\nu \in \Gamma, \mu \in \bar{\Gamma}$. We remove all facets which contain subfacets with the vertex sets $U_k \in U$ in G from the boundary complex $B(\bar{P}_d)$. The vertex sets of these facets are given by the sets I_ν with $\nu \in \Gamma$ in G. We denote the remaining facets of $\bar{P}_d$ with F_μ. In G the facets F_μ have the vertex sets I_μ with $\mu \in \bar{\Gamma}$ and belong to $\bar{P}_d'$. Under these facets F_μ there are some which contain at least one subfacet $F^s_{\nu\mu}$. On the other hand every subfacet $F^s_{\nu\mu}$ is contained in exactly one remaining facet F_μ of $\bar{P}_d$. Therefore in $B(\bar{P}_d')$ we find new facets F given in G by vertex sets of the type $I = (I_\nu \cap I_\mu) \cup \{n+1\}$ with $\nu \in \Gamma, \mu \in \bar{\Gamma}$. If G^r_q is the graph of a k-face of $\bar{P}_d^*$ with $k = 0,1,2,\ldots,d-2$, then $I = \bigcup_{\nu=1}^{r} I_\nu$ is the vertex set of a (d-k-1)-face of $\bar{P}_d$ in G. Then we write $\vartheta_{1,I}$ and $\vartheta^*_{1,I}$ instead of $\vartheta_{1,U}$ and $\vartheta^*_{1,U}$. In this case $\bar{P}_d'$ contains all facets of $\bar{P}_d$ whose vertex sets in G does not contain I as a

subset. Moreover $\bar{p}_d'$ gets new facets given in G by vertex sets of the type $J \quad n+1$. In this case J is the vertex set of a subfacet which belongs to a facet of $\bar{p}_d$ containing I. In order to construct the corresponding p-system for $\omega_{1,I}\bar{p}_d$ and $\omega^*_{1,I}\,\bar{p}^*_d$ we put $c_k = a$ for $k \in I$, $c_k = -b$ for $k \in N \setminus I$ with $a,b \in R$, $a \gg b > 0$ and $c = |I|a - (n-|I|)\,b > 0$,[7].
Further we can define other p-operations, so for example $\omega_{x_n}, \omega^*_{x_n}, \omega_p, \omega^*_p$ and $\omega_b, \omega^*_b = \omega_{pr}$. For the p-operation ω_{x_n} we remove the vertex X_n and all faces incident with X_n from $\bar{p}_d$ and get the d-polytope $\bar{p}_d'$ if $\bar{p}_d$ is not a d-pyramid with the apex X_n by

$$\bar{p}_d' = \omega_{x_n}\bar{p}_d = \operatorname{conv}\left(\pi\{X_1,X_2,\ldots,X_{n-1}\}\right), \qquad (8)$$

where π is the central projection from E^{n-1} in $E^{n-2} = \{X \in E^{n-1} : x_n = 0\}$. Analogously we find

$$\bar{p}^{*\prime}_d = \omega^*_{x_n}\,\bar{p}^*_d = \operatorname{conv}\left(\pi \bigcap_{k=1}^{n-1} \bar{H}^*_k\right), \qquad (9)$$

where $\bar{H}^*_k$ is the closed halfspace of E^d which is bounded by the supporting hyperplane h^*_k of $\bar{p}^*_d$ containing the facet F^*_k $k = 1,2,\ldots,n$ of $\bar{p}^*_d$. The p-operations ω_p or ω^*_p transform the d-polytope $\bar{p}_d$ or $\bar{p}^*_d$ in the (d+1)-pyramid with the apex X_{n+1} and the basis $\bar{p}_d$ or in the (d+1)-pyramid with the apex X^o_{n+1} and the basis $\bar{p}^*_d$. The p-operation ω_b transforms the d-polytope $\bar{p}_d$ in the (d+1)-bipyramid $\bar{p}'_{d+1}$ with the apices X_{n+1}, X_{n+2} and the basis $\bar{p}_d$. The p-operation $\omega^*_b = \omega_{pr}$ transforms the d-polytope $\bar{p}^*_d$ with n facets in the (d+1)-polytope $\bar{p}^{*\prime}_{d+1}$ with n+2 facets where $\bar{p}^{*\prime}_{d+1} \approx z_{d+1}$ and z_{d+1} is a (d+1)-prism. As propositions we give the following definitions:

1) A p-operation ω shall be called realizable for a d-polytope p_d if there exists a d-polytope $p_{d'}$, so that $\omega\bar{p}_d = \bar{p}_{d'}$ where $\bar{p}_d$ and $\bar{p}_{d'}$ are the projected polytopes of the corresponding p-systems.

2) If ω is realizable for all d-polytopes, then ω is called d-realizable.

   For example ω_b, ω^*_b, ω_p, ω^*_p are d-realizable.

3) We call a p-operation ω combinatorially d-realizable, if ω is d-realizable and if $p_d \approx p'_d$ then also $\omega\,\bar{p}_d \approx \omega\,\bar{p}'_d$ is valid.

   The p-operations $\omega_p, \omega^*_p, \omega_b, \omega^*_b, \omega_{1,I}, \omega^*_{1,I}, \omega_{1,U}, \omega^*_{1,U}$ are combinatorially d-realizable, if U is a set of consecutive edges contained as a subset in the edge set of a 2-face of $\bar{p}_d$.

4) We call a not combinatorially d-realizable p-operation ω partly d-realizable for a class of combinatorially equivalent d-polytopes, if there exists in this class a subclass of affinely equivalent d-polytopes so that ω is realizable for all polytopes of this subclass.

5) We call a p-operation ω not realizable for a class of combinatorially equivalent d-polytopes, if there exists

no d-polytope in this class for which ω is realizable. In |3| B. Grünbaum and V.P. Sreedheran enumerate all combinatorial types of simplicial 4-polytopes with 7 and 8 vertices. They obtain 5 different types $p_1^7, p_2^7, \ldots, p_5^7$ and 37 types $p_1^8, p_2^8, \ldots, p_{37}^8$. For the combinatorial equivalence class of p_5^7 the p-operation $\omega_{1,U}$ with $U = \{(1,2,5),(1,4,5),(2,5,6)\}$ is not realizable, otherwise we would obtain that the combinatorial 3-sphere M in |3| is polytopal. It is an interesting result of this paper, that we can get all combinatorial types of simplicial 4-polytopes with 5,6,7 and 8 vertices from the 4-simplex only by p-operations $\omega_{1,I}$ and $\omega_{1,U}$ which are combinatorially d-realizable. We use the same notations as in |3| . Then we find the following formulas for $n = 5$: $p_1^6 = \omega_{1,I} S_4$, $p_2^6 = \omega_{1,J} S_4$ with $I = \{1,2\}$, $J = \{3,4,5\}$ and $S_4 = \{1,2,3,4,5\}$. The 4-polytop p_1^6 is a simplicial bipyramid and p_2^6 combinatorially equivalent to the cyclic 4-polytope with 6 vertices. For $n = 6$ we find $p_1^7 = \omega_{1,I} p_1^6$, $I = \{2,3,5,6\}$; $p_2^7 = \omega_{1,I}\, p_1^6$, $I = \{2,3,5\}$; $p_3^7 = \omega_{1,I}\, p_1^6$, $I = \{1,2\}$; $p_4^7 = \omega_{1,I}\, p_2^6$, $I = \{5,6\}$; $p_5^7 = \omega_{1,U}\, p_2^6$, $U = \{(3,5,7),(2,5,6)\}$. Now we consider the case $n = 7$. From p_1^7 we get: $p_1^8, p_2^8, p_3^8, p_{13}^8, p_{17}^8$ by ω_{1,I_k} with $I_1 = \{1,2,5,6\}$, $I_2 = \{1,2,4,5\}$, $I_3 = \{1,2,3,4\}$, $I_{13} = \{1,2,3\}$, $I_{17} = \{1,3\}$. From p_2^7 we get $p_4^8, p_5^8, p_6^8, p_9^8, p_{10}^8$, p_{14}^8 , p_{15}^8 , p_{22}^8 by ω_{1,I_k} with $I_4 = \{1,2,4,6\}$, $I_5 = \{2,4,6,8\}$,

$I_6 = \{1,2,5,6\}$, $I_9 = \{1,4\}$, $I_{10} = \{2,3,5\}$, $I_{14} = \{3,4,7\}$, $I_{15} = \{4,6,7\}$, $I_{22} = \{2,4\}$. From p_3^7 we get: $p_7^8, p_{16}^8, p_{34}^8, p_{26}^8$ by ω_{1,I_k} and $\omega_{1,U}$ with $I_7 = \{1,3,5,7\}$, $I_{16} = \{3,5,6\}$, $I_{34} = \{2,3\}$, $U = \{(2,3,7),(2,3,5)\}$. From p_4^7 we get: p_8^8, $p_{18}^8, p_{19}^8, p_{20}^8, p_{21}^8, p_{28}^8, p_{29}^8$ by ω_{1,I_k} and $\omega_{1,U}$ with $I_8 = \{2,4,6,7\}$, $I_{18} = \{3,6,7\}$, $I_{19} = \{1,6,7\}$, $I_{20} = \{1,3,7\}$, $I_{21} = \{3,4,5\}$, $I_{28} = \{1,3\}$, $U = \{(1,4,6),(4,6,7)\}$. From p_5^7 we get: $p_{11}^8, p_{12}^8, p_{23}^8, p_{24}^8, p_{25}^8, p_{27}^8, p_{30}^8, p_{31}^8, p_{32}^8, p_{33}^8, p_{35}^8$, p_{36}^8, p_{37}^8 by ω_{1,I_k} and $\omega_{1,U}$; with $I_{11} = \{3,4,6\}$, $I_{12} = \{1,2,3,4\}$, $I_{23} = \{1,5,6\}$, $I_{24} = \{2,3,4\}$, $I_{25} = \{1,6,7\}$, $I_{27} = \{1,4\}$, $I_{30} = \{1,2\}$, $U^{31} = \{(1,2,3),(2,3,7)\}$, $U^{32} = \{(1,2,7),(1,6,7)\}$, $U^{33} = \{(1,5,6),(1,5,7)\}$, $U^{35} = \{(1,3,7),(1,4,7),(1,5,7)\}$, $U^{36} = \{(1,2,7),(1,3,7),(1,4,7)\}$, $U^{37} = \{(1,2,7),(1,3,7),(1,6,7)\}$.

Literature References

|1| Grünbaum, B. Polytopes, graphs, and complexes. Bull. of the Am. Math. Soc. Vol.76, No.6, Nov.1970, p.1131-1201.

|2| Grünbaum, B. Convex polytopes. John Wiley&Sons, London-New York-Sydney 1967, 465 p.

|3| Grünbaum, B. and Sreedheran, V.P. An Enumeration of Simplicial 4-polytopes with 8 Vertices. Journ. of Comb.Theor. Vol.2,No. 4, June 1967, p.437-467.

|4| Schöne, W. Über die Existenz simplizialer d-polytope mit vorgeschriebener Seitenstruktur. Beitr. z. Algebra u. Geometrie 8 (1979) p.67-69.

|5| Schöne, W. Basiskreissysteme, Unterraumschnitte regulärer Simplexe u. konvexe Polytope. 2. Koll. f. Geometrie und Komb. 1983 Karl-Marx-St. Vortragsheft, Teil 2,p.171-178, erhältlich TU Karl-Marx-Stadt, Sektion Mathematik.

|6| Schöne, W. Konvexe Polytope, Unteraumschnitee reg. Simplexe u.l.G.S. Wiss.Z. TH Karl-Marx-St. 26(1984)2,p.

|7| Schöne, W. Fortschritte bei d.Konstrukt.konvexer Polyt. u. ihrer Graphen.3.Koll. Geom.u.Komb.1987,Vortr.p.51-52

On 2-Embeddable Graphs

H. Schumacher, Kiel, FRG

0. Introduction

Let there be a graph without loops G and a closed surface F of the characteristic N.[1] G is called embeddable in F if G can be represented (drawn) on F in such a way that none of G's edges is crossing over another edge of G. In more general terms, G is called n-embeddable in F ($n \in \mathbb{N}$), if G can be represented on F in such a way that each edge of G crosses a maximum of n other edges of G. On each edge of G, all intersections (points where edges are crossing) are counted; if one edge is crossed i times by another, this will be counted i times, if an edge is crossed by j multiple edges $e_1, e_2, \ldots, e_j$ [2], this will be counted j times.

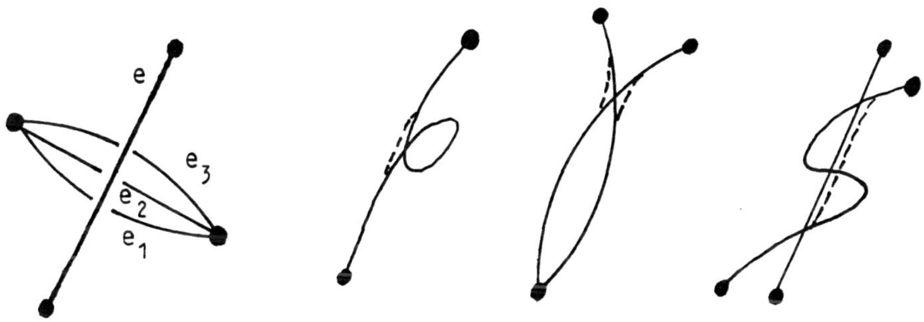

Figure 1

If G is n-embeddable in F, there is an n-embedding of G in F with a minimal number of intersections. In that case, none of G's edges is crossing itself; two different edges with one vertex in common do not cross either, and two different edges without a vertex in common cross once at the most.

1) Concerning the basic terms in graph theory and topology, please see [11] and [10]. This paper is only dealing with finite graphs.

2) See [11], page 10 .

R. Bodendiek, R. Henn (Eds.)
Topics in Combinatorics and Graph Theory

Otherwise, there would be an n-embedding of G in F with less intersections (Figure 1).

G is called n-embeddable in F without 2-gon, if there is an n-embedding of G in F, where no 2-cell [3] of F will have two multiple edges e_1 and e_2 as its boundaries when dissected by G.

It is known that for graphs 0-embeddable in F without 2-gon $\kappa \leq 3(\alpha - N)$[4] holds; α being the number of vertices and κ the number of edges of G. For graphs 1-embeddable in F without 2-gon, $\kappa \leq 4(\alpha - N)$ has been proved in 8 .

This paper will show that for graphs 2-embeddable in F without 2-gon, $\kappa \leq 5(\alpha - N)$ holds, and that there are graphs 2-embeddable in F without 2-gon with $\kappa = 5(\alpha - N)$. Further, it will be proved that there are no graphs 3-embeddable without 2-gon in the surface of the sphere[5] F_0 with $\kappa = 6(\alpha - N)$; this means that the analogies drawn between the graphs 0-, 1-, and 2-embeddable in F ends in the case of the graphs 3-embeddable in F.

The n-chromatic number $\chi_n(F)$ of a closed surface F is the maximum of the chromatic numbers $\chi(G)$, applied to all graphs G n-embeddable in F. For the 2-chromatic number $\chi_2(F)$, the inequality $\kappa \leq 5(\alpha - N)$ is then developed into "Heawood's Inequality":

$$\chi_2(F) \leq \left[\frac{11 + \sqrt{121 - 40N}}{2} \right] \quad ^{6)}$$

Finally, the 2-chromatic number for the projective plane F_1^*, Klein's Bottle F_2^*, the torus F_1 and the surface F_5 (closed orientable surface with $N = -8$) will be determined.

1. An Upper Limit for the Number of Edges

Theorem 1: If G is a graph without loops with α vertices and κ edges, which is 2-embeddable without 2-gon in the closed surface F with the characteristic N, then $\kappa \leq 5(\alpha - N)$ holds.

3) See [10], page 50.

4) See [4], page 61 or [6] page 55.

5) See [10], page 7.

6) The notation [x] means the largest integer not greater than x.

Proof: Let G be a graph without loops which is 2-embedded without 2-gon in F with a minimal number of intersections. Let G be edge-maximal concerning this 2-embedding; i.e. in the 2-embedding of G in F, two vertices $a \neq b$ of G cannot be connected by a new edge without a 2-gon as a result. If a is a vertex of G, the 2-embedding will determine a cyclic order of the edges of G incident in a. In this cyclic order, there will not be three successive edges of G with intersections. The reason is as follows: assume there are three successive edges with intersections (a,b_1), (a,b_2), (a,b_3), there will be a first intersection x on the middle edge (a,b_2) from a.

Let (c,d) be the edge which has this intersection x in common with (a,b_2) (Figure 2).

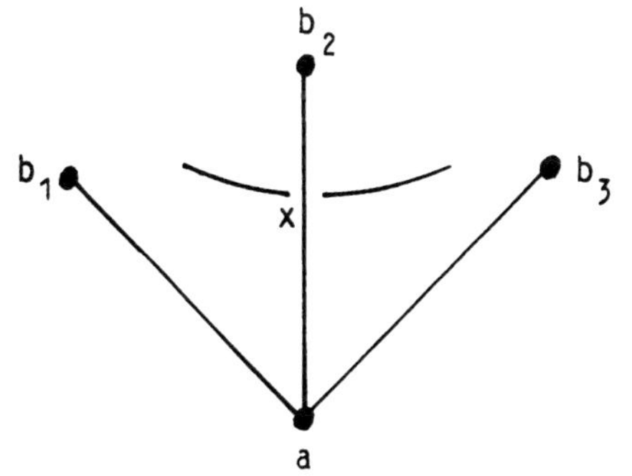

Figure 2

Since there is a minimum number of intersections, c and d are not identical with a, and since at least one of the two demiedges (c,x) and (d,x) has no further intersection point, either a and c or a and d can be connected in F without any intersections. In this, no 2-gon can result; otherwise (a,b_1), (a,b_2) and (a,b_3) would not be three successive edges with intersections in the cyclic order of the edges of G incident in G. This is a contradiction to the edge-maximality.

If all edges with intersections are deleted from G, the result will be a graph G' which is 0-embeddable in F. If α is the number of vertices and κ is the number of edges of G, and if κ^* is the number of edges with intersections of G, κ' the number of edges of G' and λ' the number of areas the surface F is divided into by G', then $\kappa = \kappa' + \kappa^*$ and $\alpha - \kappa' + \lambda' \geq N$ [7] holds with N being the characteristic of F. If F' is an area of F after F

7) See [4],page 61 and 62.

has been divided by G', the boundary of F' will consist of (one or several) closed edge sequences of G'. If i is the number of these closed edge sequences (edges occurring twice will be counted twice), i is called the boundary length of F'.
In Figure 3, the torus is divided by a graph into the tree areas F',F", F''' with the boundary lengths 3, 8, 7.

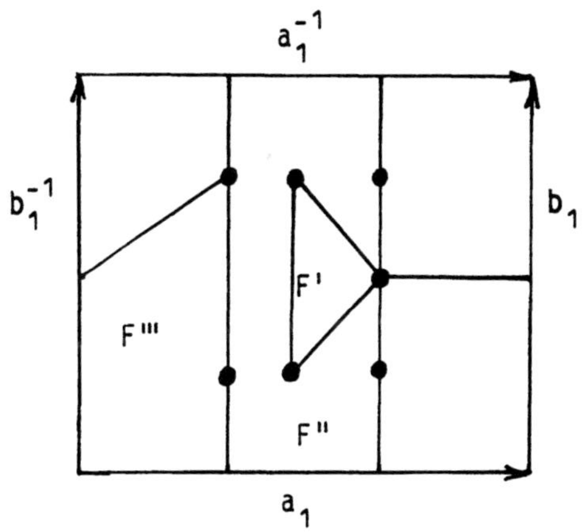

Figure 3

If the number of areas with the boundary length i is called f_i, then $\lambda' = \Sigma f_i$ and $2\kappa' = \Sigma i f_i$ in which all i are summed up for which there are areas with that boundary length. Since in the edge sequence/s of the boundary of an area with the boundary length i, there is a maximum of i vertices,(a vertex can well occur several times), and since in the cyclic order of each vertex a of G there is a maximum of two successive edges with intersections, the number of edges with intersections is in each area of the boundary length i either less than i or equalling i. Since G is without loops and 2-gons and edge-maximal, it also follows that i is more than or equalling 3, and that in an area with the boundary length 3, there are no edges with intersections, in an area with the boundary length 4, there is a maximum of 2 edges with intersections. If one estimates the number of edges of G with the help of the areas, then

$$\begin{aligned} 2\kappa &\le 3f_3 + (2 \cdot 2 + 4)\, f_4 + \sum_{i>5} (2i + i)\, f_i \\ &\le 5 \sum_{i\ge 3} i f_i - 10 \sum_{i \ge 3} f_i \\ &= 10\kappa' - 10\lambda' \\ &\le 10(\alpha - N). \end{aligned}$$

Thus, theorem 1 is proved.

2. Optimum Graphs [8)]

A graph without loops and 2-gons which is 2-embeddable in F and for which $\kappa = 5(\alpha - N)$ holds, is called an optimum graph 2-embeddable. From the proof of theorem 1, it follows that every optimum graph 2-embeddable in F divides the surface F into several pentagons, in which case all diagonals (edges with intersections) belong to each pentagon. A pentagon, or more generally speaking, an n-gon ($n \in \mathbb{N}$) is a 2-cell with a circle of the length n as its boundary. The division of F by G into pentagons further results in the fact that the degree of each vertex of G can be divided by three. Since a graph optimum 0-embeddable in F (with $\kappa = 3(\alpha - N)$) divides the surface into triangles [9)], and since a graph optimum 1-embeddable in F (with $\kappa = 4(\alpha - N)$) divides the surface into quadrilaterals [10)], this regularity is continued by the graphs optimum 2-embeddable in F. The analogy, however, ends with the 3-embeddable graphs. As there is a graph G, 3-embeddable in F, which divides the surface F into hexagons, in each of these hexagons, at least one diagonal will be missing. Thus, it immediately follows for the number of vertices and edges of G that $\kappa < 6(\alpha - N)$.

Optimum graphs 0-, 1-, or 2-embeddable in the sphere F_0 with a minimum number of vertices are the tetrahedron (if one only permits vertices with degree > 2), the cube with all the diagonals in each of its quadrilaterals, or the dodekahedron with all diagonals in each of its pentagons.[11)] The complete graph K_{10} is the graph with the minimum number of vertices which is optimum 2-embeddable in the projective plane F_1^*. As is shown in Figure 4, K_{10} is optimum 2-embeddable in F_1^*. One will only have to identify identically marked vertices and those edges the end vertices of which are identically marked, and then draw all diagonals into each of the six pentagons.

---

8) See [2].

9) See [6], page 57.

10) See [1], [5], [9].

11) See [3].

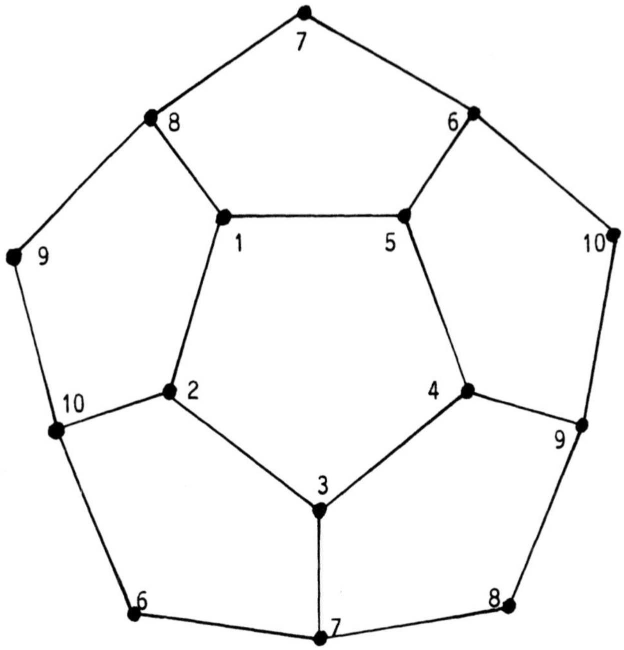

Figure 4

Complete graphs optimum 2-embeddable in the surface F are of maximum importance for the determination of the chromatic number $\chi_2(F)$. There can only be complete, optimum 2-embeddable graphs for certain surfaces F.

Theorem 2: Let F be a closed surface of the characteristic N. If there is then a complete graph G which is optimum 2-embeddable in F, then

$$N = 3 - \frac{n(5n-1)}{2} \quad \text{mit } n \in \mathbb{N} \text{ und } n \equiv 1 \bmod 3$$

or

$$N = 3 - \frac{n(5n+1)}{2} \quad \text{mit } n \in \mathbb{N} \text{ und } n \equiv 2 \bmod 3.$$

Proof: If G is a complete graph with α vertices, which is optimum 2-embeddable in the surface F with the characteristic N, it follows that

$$\alpha = \frac{11}{2} + \frac{1}{2}\sqrt{121-40N}.$$

Since α is a positive integer and N is an integer, it follows that

$$\sqrt{121-40N} \equiv 1(10) \quad \text{or} \quad \sqrt{121-40N} \equiv -1(10).$$

Then there is a positive integer n with $N = 3 - \frac{n}{2}(5n-1)$ or $N = 3 - \frac{n}{2}(5n+1)$. Since in addition to this, the degree of each vertex of G can be divided by 3, it is a fact that $\alpha \equiv 1(3)$. From this, it follows that $N \equiv 1(3)$ and further, $n \equiv 1(3)$ if $N = 3 - \frac{n}{2}(5n-1)$ and $n \equiv 2(3)$, if $N = 3 - \frac{n}{2}(5n+1)$.

After the projecitve plane, the next surface F for which there is a complete graph optimum 2-embeddable in F, is a surface of the characteristic -8, and the appropriate complete graph is K_{16}. Figure 5 is showing a polygonal dissection [12] of the closed orientable surface F_5 (sphere with 5 handles) into the 16 pentagons with 40 vertices and 40 edges, i.e. with the characteristic $N = 16 - 40 + 16 = -8$.

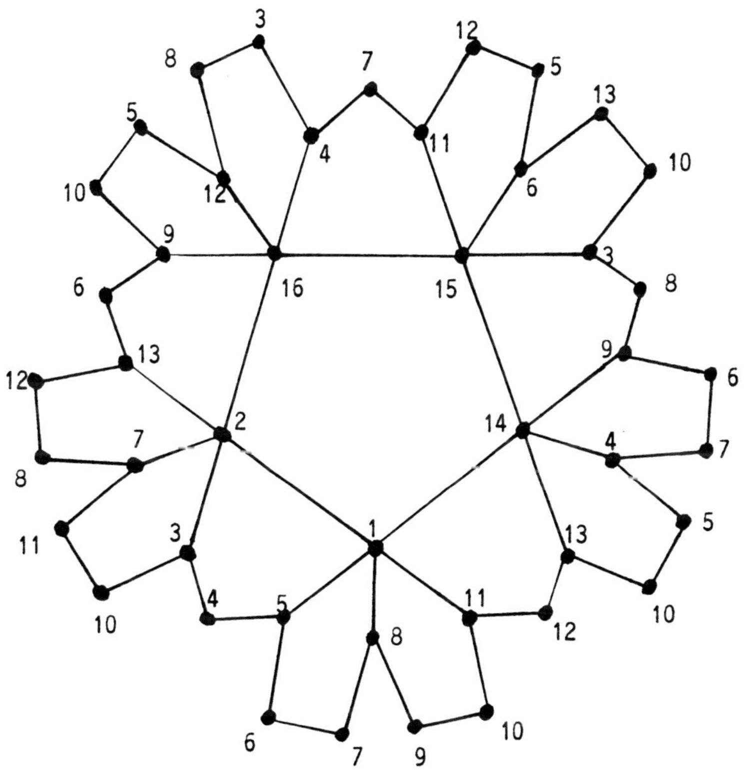

Figure 5

12) See [10], page 130-142.

In Figure 5, all identically marked vertices and edges with identically marked endvertices can of course be identified. If one then also draws all diagonals in each of the 16 pentagons, it becomes obvious that the vertices are all adjacent to each other. Thus, one has the optimum 2-embedding of K_{16} in F_5.

For each surface F, incomplete graphs can be named which are optimum 2-embeddable in F. Figure 6 is showing a graph optimum 2-embeddable in the torus F_1. If one now cuts a pentagon out of the torus which is divided into

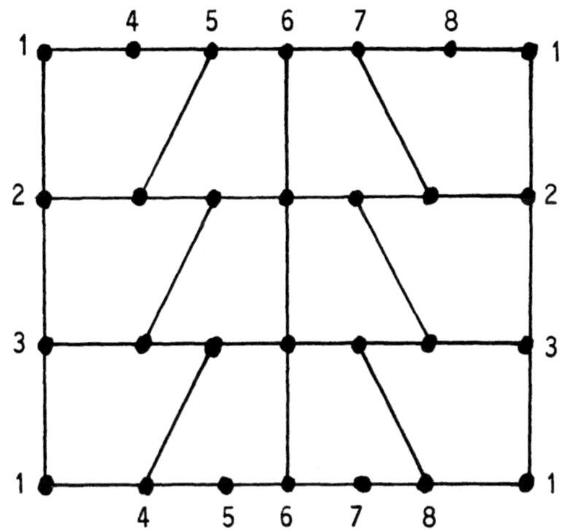

Figure 6

pentagons, one has the handle surface with boundary. This can be successively attached to the torus, the sphere with two handles etc., from each of which a pentagon has been cut out, too, by identifying the two boundaries accordingly. This way, one will get a graph optimum 2-embeddalbe in F for each closed orientable surface F.

The same prodecure can be applied to nonorientable closed surfaces. Here, one starts from the projective plane F_1^* in which the K_{10} is embedded (Figure 4). If one cuts a pentagon from the projective plane (which has been divided into pentagons by K_{10}), one will end up with the Möbius strip (with boundaries).

This can be attached to the projective plane, Klein's Bottle etc. - from all of which a pentagon has been cut out too - so that for each nonorientable surface F, one receives a graph optimum 2-embeddable in F.

3 "Heawood's Inequality"

From theorem 1, an upper limit for the 2-chromatic number $\chi_2(F)$ (F being unequal to the sphere) can be developes, analogously to the respective theorems about graphs 0- or 1-embeddable in a surface F [13].

The limit of the chromatic number then only depends on the characteristic N of the surface F. For this, one has to look at a critical graph G (a graph G is called critical if each proper subgraph has a chromatic number smaller than that of G). Then $(\chi-1)\,\alpha < 2\kappa$ holds. If additionally, G fulfills the prerequisite of theorem 1, the result is $(\chi-1)\,\alpha \leq 10(\alpha-N)$. Since it always holds that $\chi(G) \leq \alpha$, it follows in case that $N \leq 0$ holds, $\chi^2 - 11\chi + 10N \leq 0$, thus

$$(\chi - \frac{11+\sqrt{121-40N}}{2})\;(\chi - \frac{11-\sqrt{121-40N}}{2}) \leq 0.$$

In this case, the right factor of the inequality is more than 0, so that

$$\chi \leq \frac{11+\sqrt{121-40N}}{2} \quad \text{holds.}$$

For N = 1 and N = 2, the result of $(\chi-1)\alpha \leq 10(\alpha-N)$ is only $\chi \leq 0$.

Since every (finite) graph G has a critical subgraph with the same χ as G, $\chi_2(F)$ is the maximum of $\chi(G)$, (only) applied to all critial graphs 2-embeddable in F. This leads to

Theorem 3: a) If F_p $(p \neq 0)$ is a closed orientable surface with the genus $p(N = 2 - 2p)$, it follows that

$$\chi_2(F_p) \leq \left[\frac{11+\sqrt{41+80p}}{2}\right]$$

b) if F^*_q is a nonorientable closed surface with the genus q $(N = 2 - q)$, it follows that

$$\chi_2(F^*_q) \leq \left[\frac{11+\sqrt{41+40q}}{2}\right].$$

13) See [6], [7].

4 Determination of the Chromatic Number χ_2 for the Surfaces F_1^*, F_2^*, F_1 and F_5

For the projective plane F_1^*, it follows that $\chi_2(F_1^*) \leq 10$. And since K_{10} is 2-embeddable in F_1^* (Figure 4), the immediate result will be $\chi_2(F_1^*) = 10$.

In the case of the torus F_1 and of Klein's Bottle F_2^*, theorem 3 results in $\chi_2(F_1) \leq 11$ and $\chi_2(F_2^*) \leq 11$.

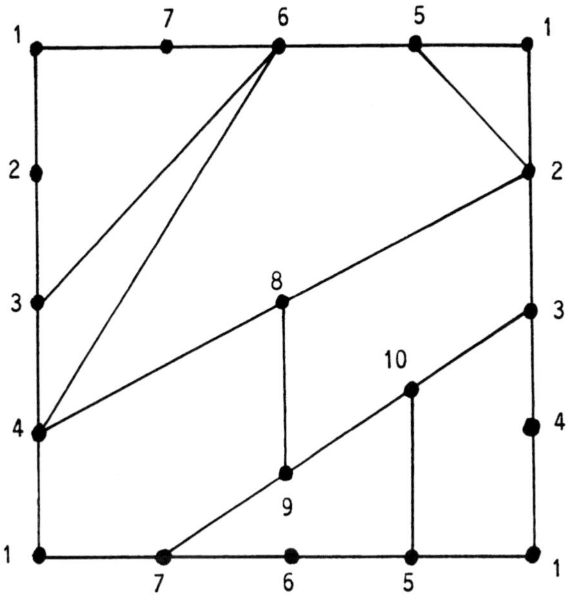

Figure 7

Since K_{10} is also 2-embeddable in F_1 (Figure 7) and since each graph 2-embeddable in F_1^* is also 2-embeddable in F_2^*, the result is $10 \leq \chi_2(F_1) \leq 11$ and $10 \leq \chi_2(F_1^*) \leq 11$. If we assume now that there is a critical graph G with $\chi(G) = 11$, which is 2-embeddable in F_1 or F_2^* respectively, then each vertex v of G has a degree $\gamma(v,G) \geq 10$. From this, it can be followed with the help of theorem 1, that G is optimum 2-embeddable in F_1 or F_2^* respectively, also that $\gamma(v,G) = 10$ for each vertex v of G. This is a contradiction because 3 does not divide $10 = \gamma(v,G)$. Thus, $\chi_2(F_1) = \chi_2(F_2^*) = 10$ holds.

For the surface F_5, according to theorem 3, $\chi_2(F_5) \leq 16$ holds. Since **Figure** 5 is showing a 2-embedding of K_{16} in F_5, the immediate result is $\chi_2(F_5) = 16$.

5 References

[1] R.Bodendiek/H.Schumacher/K.Wagner: Bemerkungen zu einem Sechsfarbenproblem von G.Ringel. Abh.Math.Sem. Universität Hamburg 53(1983), S. 41-52.

[2] R.Bodendiek/H.Schumacher/K.Wagner: Über 1-optimale Graphen. Math.Nachr. 117(1984), S. 1-17.

[3] R.Bodendiek/H.Schumacher: Geometrie ohne Längen und Winkel. MU 4(1986), S. 22-47.

[4] G.Ringel: Färbungsprobleme auf Flächen und Graphen. VEB Deutscher Verlag der Wissenschaften, Berlin 1959.

[5] G.Ringel: Ein Sechsfarbenproblem auf der Kugel. Abh.Math.Sem. Universität Hamburg 29(1965), S. 107-117.

[6] G.Ringel: Map Color Theorem. Grundlagen der mathematischen Wissenschaften, Band 29, Springer Verlag, Heidelberg (1974).

[7] G.Ringel: A Nine Color Theorem for the Torus and the Klein Bottle. Offprints from the Theory of Allications of Graphs, Edited by Dr. Gary Chartrand, Copyringht 1981 by John Wiley & Sons, INC.

[8] H.Schumacher: Ein 7-Farbensatz 1-einbettbarer Graphen auf der projektiven Ebene. Abh.Math.Sem. Universität Hamburg 54(1984), S. 5-14.

[9] H.Schumacher: Zur Struktur 1-planarer Graphen. Math.Nachr. 125 (1986), S. 291-300.

[10] H.Seifert/W.Threlfall: Lehrbuch der Topologie. Teubner Verlag, Leipzig (1934).

[11] K.Wagner: Graphentheorie. Mannheim 1970, B.I. Hochschultaschenbücher 248/248*.

On an Application of the Boolean Differential Calculus to Digital System Theory

B. Stiefel, Mittweida, GDR

1. Introduction

Modern microelectronics providing low cost complex digital devices has at least equalized the previously most important task to minimize the expense of devices with other criteria such as dynamic properties, the ability of decomposition, test comfortability and others. With the existence of several criteria enjoying equal rigths the design of switching systems will become a optimization process. The Boolean differential calculus has turned out to be a suitable means for optimization in an analytic way /1/, /2/, /3/, /4/.
This paper involves two central points of interest: firstly, reduction of any kind of derivations of Boolean functions to the derivation with exclusive disjunction operation and

R. Bodendiek, R. Henn (Eds.)
Topics in Combinatorics and Graph Theory

secondly, application of the differential equation based on this derivation to the analytic description of the stability of binary systems

2. Partial Derivation of Boolean Functions

By extension every term the kind $f\ (x_1,\ldots,x_i,\ldots,x_n)$ of $(x_1,\ldots,\dot{\bar{x}}_i,\ \ldots,\ x_n)$ or shortly $f\ (x_i)\ \circ\ f(\bar{x}_i)$ is designated as partial derivation of the Boolean function $f\ (x_1,\ \ldots,\ x_i,\ \ldots,\ x_n)$ with respect to the variable x_i, where "o" denotes any binary operation. Because of the fact that there are ten different of such binary operations ten different kinds of partial derivations will be resulted, too. The solution of the Boolean equation

$$a_o \oplus a_1\ f\ (x_i) \oplus a_2\ f\ (\bar{x}_i)\ x \oplus a_3\ f\ (x_i)\ f\ (\bar{x}_i) =$$
$$b_o \oplus b_1\ f\ (\underline{x}\ ) \oplus b_2\ \ c_o \oplus c_1\ f\ (x_i) \oplus c_2\ f\ (\bar{x}_i) \oplus$$
$$c_3\ f\ (x_i)\ f\ (\bar{x}_i)\ \ \oplus$$
$$b_3\ f\ (\underline{x})\ \ c_o \oplus c_1\ f\ (x_i) \oplus c_2\ f\ (\bar{x}_i) \oplus c_3\ f\ (x_i)\ f\ (\bar{x}_i)$$

regarding the constants a_i, b_j, c_k shows that only the linear derivation can form all others derivations by combination with the original function.

Because of that in the following only the partial derivation based on exclusive disjunction operation defined as

$$\frac{\partial f(\underline{x})}{\partial x_i} = f(x_i) \oplus f(\overline{x}_i)$$

is used.

Vectorial derivations and such referring to complete subspaces are given by several authors /1/, /2/. The connection of these derivations with the partial one is given by the following equations:

Vectorial derivation:

$$\frac{\partial f}{\partial \underline{x}} = \frac{\partial f}{\partial x_1} \oplus \ldots \oplus \frac{\partial f}{\partial x_n} \oplus \frac{\partial^2 f}{\partial x_1 \partial x_2} \oplus \frac{\partial^n f}{\partial x_1 \ldots \partial x_n}$$

Vectorial minimum:

$$\min_{\underline{x}} f(\underline{x}) = f(\underline{x}) \left[\frac{\partial f}{\partial x_1} \oplus \ldots \oplus \frac{\partial f}{\partial x_n} \oplus \frac{\partial^2 f}{\partial x_1 \partial x_2} \oplus \ldots \oplus \frac{\partial^n f}{\partial x_1 \ldots \partial x_n}\right]$$

Vectorial maximum:

$$\max_{\underline{x}} f(\underline{x}) = f(\underline{x}) \vee \left[\frac{\partial f}{\partial x_1} \oplus \ldots \frac{\partial f}{\partial x_n} \oplus \frac{\partial^2 f}{\partial x_1 \partial x_2} \oplus \ldots \oplus \frac{\partial^n f}{\partial x_1 \ldots \partial x_n}\right]$$

$\triangle$ -Operator:

$$\triangle_{\underline{x}} f(\underline{x}) = \frac{\partial f}{\partial x_1} \vee \ldots \vee \frac{\partial f}{\partial x_n} \vee \frac{\partial^2 f}{\partial x_1 \partial x_2} \vee \ldots \vee \frac{\partial^n f}{\partial x_1 \ldots \partial x_n}$$

n-ary minimum

$$\min_{\underline{x}}^{n} f(\underline{x}) = f(\underline{x})\, \overline{\frac{\partial f}{\partial x_1}} \cdots \overline{\frac{\partial f}{\partial x_n}}\; \overline{\frac{\partial^2 f}{\partial x_1 \partial x_2}} \cdots \overline{\frac{\partial^n f}{\partial x_1 \cdots \partial x_n}}$$

n-ary maximum

$$\max_{\underline{x}}^{n} f(\underline{x}) = f(\underline{x}) \vee \frac{\partial f}{\partial x_1} \vee \dots \vee \frac{\partial f}{\partial x_n} \vee \frac{\partial^2 f}{\partial x_1 \partial x_2} \vee \dots \vee \frac{\partial^n f}{\partial x_1 \cdots \partial x_n}$$

3. Boolean Partial Differential Equation (BPDE)

The most common form of a BPDE is given by

$$F\left(f(\underline{x}), \frac{f}{x_1}, \dots, \frac{{}^n f}{x_1 \cdots x_n}\right) = k \qquad (1)$$

where F and f are Boolean functions and k is a binary konstant.

The solution of a BPDE is the set of those functions transmitting together with their derivations (1) into identity. The solving of BPDE is connected with the inversion operation of the derivation formation - with the integration:

$$f(\underline{x}) = \int g(\underline{x})\, dx_i = x_i g(\underline{x}) \oplus \alpha(\underline{x}^*) \qquad (2)$$

Where $g(\underline{x})$ is the integrand that must satisfy the integration condition $\frac{\partial g}{\partial x_i}=0$ and $\alpha(\underline{x}^*)$ is the integration constant - an arbitrary function also independent of x_i (x^* is a vector of variables $x_1, \ldots, x_{i-1}, x_{i+1}, \ldots x_n$).
Generalizing (2) a expansion in a series producing the connection between the 2^{n-1} derivations and the function itself accurate to one additive constant is obtained:

$$f(\underline{x}) = a \oplus x_1 \frac{\partial f}{\partial x_1} \oplus \ldots \oplus x_n \frac{\partial f}{\partial x_n} \oplus x_1 x_2 \frac{\partial^2 f}{\partial x_1 \partial x_2} \oplus \ldots \oplus$$

$$\oplus x_1 \ldots x_n \frac{\partial^n f}{\partial x_1 \ldots \partial x_n}$$

This expansion in series is the main instrument for solving BPDE /1/. A group of problems at which the advantages of the Boolean differential calculus emerge in comparison with other analytic methods in the selection of functions as one way of function optimization. Concerning this some BPDE are to be given as examples for an analytic closed form of representation of classes of Boolean functions with the help of which differential equation systems can be built for optimization of functions

Linear functions /5/: $\bigvee\limits_{\binom{n}{2}} \frac{\partial^2 f}{\partial x_i \partial x_j} = 0$

Monoton functions: $\bigvee\limits_{\binom{n}{2}} (x_i \oplus x_j) \frac{\partial f}{\partial x_i} \frac{\partial f}{\partial x_j} = 0$

symmetric functions: $\bigvee_{\binom{n}{2}} (x_i \oplus x_j)\left(\frac{\partial f}{\partial x_i} \oplus \frac{\partial f}{\partial x_j} \oplus \frac{\partial^2 f}{\partial x_i \, \partial x_j}\right) = 0$

cyclic functions:

$$\bigvee_{i=1}^{n} \bigwedge_{j=1}^{n} (x_i \oplus x_j) \frac{\partial f}{\partial x_i} \oplus \frac{\partial f}{\partial x_j} \oplus \frac{\partial^2 f}{\partial x_i \, \partial x_j} = 0$$

zero-reproducing functions:

$$f(\underline{x}) \oplus x_1 \frac{\partial f}{\partial x_1} \oplus \dots \oplus x_n \frac{\partial f}{\partial x_n} \oplus x_1 x_2 \frac{\partial^2 f}{\partial x_1 \partial x_2} \oplus \dots \oplus$$
$$\oplus\, x_1 \dots x_n \frac{\partial^n f}{\partial x_1 \dots \partial x_n} = 0$$

one-reproducing functions:

$$f(\underline{x}) \oplus \bar{x}_1 \frac{\partial f}{\partial x_1} \oplus \dots \oplus \bar{x}_n \frac{\partial f}{\partial x_n} \oplus \bar{x}_1 \bar{x}_2 \frac{\partial^2 f}{\partial x_1 \partial x_2} \oplus \dots$$
$$\oplus\, \bar{x}_1 \dots \bar{x}_n \frac{\partial^n f}{\partial x_1 \dots \partial x_n} = 0$$

self-dual functions /5/:

$$\frac{\partial f}{\partial x_1} \oplus \dots \oplus \frac{\partial f}{\partial x_n} \oplus \frac{\partial^2 f}{\partial x_1 \partial x_n} \oplus \dots \oplus \frac{\partial^n f}{\partial x_1 \dots \partial x_n} = 0$$

4. Oscillation equation for combinatorial networks

Besides the observality and controllability /6/ the stability is a central category of the binary system theory. For combinatorial we define the stability as follows:

Definition: A combinatorial network realizing the function $f(\underline{x})$ is stable regarding the variable x_i for all vectors $\underline{x}$ being a solution of the equation

$$\left[f(\underline{x}) \oplus x_i\right] \frac{\partial f}{\partial x_i} = 0 \qquad (3)$$

The so defined criterion of stability may be used in two ways. In the first place we consider (3) as Boolean equation according to a given network with a known switching function $f(\underline{x})$. Solving the equation for the case of unstability, that means the rigth-hand side of (3) equals 1, we obtain the condition on which an oscillation occurs by the connection of the network output and entry (when the signal running time is long enough). This phenomenon may be used for system self-testing /7/.

In the second place we can concieve (3) as BPDE whose set of solutions contains all stable or oscillating functions according as the forcing function of (3) equals "1" and "0" respectirely. This approach can be utilized for the analysis of the stability of sequential network analyzing the combinatorial network originated by breaking the feedback by means of equation (3). The methodology of stability investigations and self-testing developed here for the case of one variable can be easy extended to vectorial approach /7/.

References

/1/ B. Stiefel: Untersuchung Boolescher partieller Differentialgleichungen mit Anwendungen bei der Optimierung von Schaltfunktionen. Dissertation A. Ingenieurhochschule Mittweida, 1980

/2/ B. Stiefel: Über Iterierte einer Booleschen Funktion. Wissenschaftliche Zeitschrift der W.-P.-Universität Rostock., 6/1981

/3/ B. Stiefel: Dekomposition Boolescher Funktionen durch Anwendung von Booleschen Differentialgleichungen. Nachrichtentechnik 29 (1979) H. 8

/4/ s. B. Akers: On a theory of Boolean Functions. J. SIAM, 7 (1959) H. 4.

/5/ D. Bochmann, C. Posthoff: Binäre dynamische Systeme. Akademie Verlag. Berlin 1981.

/6/ B. Stiefel, B. Müller: Zur Beobachtbarkeit in kombinatorischen Systemen. Wiss. Zeitschrift d. IH Mittweida 1/1982

/7/ S. Albert, B. Stiefel: Wertung von off-line-Prüfverfahren hinsichtlich Anwendung zur Selbstprüfung von digitalen Systemen. Nachrichtentechnik 35 (1985) H. 8

Equiareal Sets in $\mathbb{R}^d$

G. W. Teumer, Greifswald, GDR

Seidel and Blokhuis investigated in [1,2] the problem of finding an upper bound for the cardinality of the so-called equidistant sets. These are subsets of the unit sphere in $\mathbb{R}^d$ such that the distance between any two of its vectors is always the same.
Since the area of a parallelogram in $\mathbb{R}^d$, spanned by two vectors, is a function of two variables, namely of the length of these vectors and of the angle between them, it turns out that the problem of equidistance sets can be embedded into a more general one.
Let us consider a finite subset $\mathfrak{M}$ of vectors in the real Euclidean d-dimensional space $\mathbb{R}^d$.
Definition 1: $\mathfrak{M}$ is called equiareal iff the area of each parallelogram, spanned by any pair of vectors of $\mathfrak{M}$, is always the same.□
Clearly, if all vectors are of equal length $\mathfrak{M}$ is a equidistant set.
The problem now is to find upper bounds for the cardinality of equiareal sets. First of all some examples of such sets should be given. S will denote the area of the parallelograms.
Example 1: $d = 2$

$\underline{m}_1 = (1,0)$, $\underline{m}_2 = (x,S)$, $\underline{m}_3 = (1+x,S)$, where $x \in \mathbb{R}$
It can be proved that $\mathfrak{M}$ is maximal.
Example 2: $d = 3$

$S^2 = 4/5$, $a = 1/\sqrt{5}$, $b = 1/2(1 + a)$, $c = 1/2(1 - a)$

$\underline{m}_1 = (-a,\ b,\ \sqrt{c})$, $\underline{m}_2 = (-a,\ b,\ -\sqrt{c})$ $\underline{m}_3 = (a,\ c,\ \sqrt{b})$

$\underline{m}_4 = (a,\ c,\ -\sqrt{b})$ $\underline{m}_5 = (a,\ 2a,\ 0)$ $\underline{m}_6 = (1,\ 0,\ 0)$

All parallelograms are rhombuses. Under these condition $\mathfrak{M}$ is maximal and moreover, it is unique up to stressing or shrinking of all its vectors, i.e. up to a homothetic map.
Example 3: $d = 3$

R. Bodendiek, R. Henn (Eds.)
Topics in Combinatorics and Graph Theory

$S^2 = 3/4$

$\underline{m}_1 = (1, 0, 0)$, $\underline{m}_2 = (0, S, 0)$, $\underline{m}_3 = (1/2, 0, S)$

$\underline{m}_4 = (-1/2, 0, S)$

Among all parallelograms spanned by any two of these vectors there is a rectangle having one side of length 1, i.e. the length is maximal. Under these conditions $\mathfrak{M}$ is maximal.

Thus, the following cases are to be considered:

i) There is no restriction at the shape of the parallelograms

ii) All parallelograms are squares

iii) All parallelograms are rhombuses

iv) Among the parallelograms there is a rectangle of maximal length.

Finally, from the examples above the question arises whether S can be choosen arbitraryly or not if the set $\mathfrak{M}$ is maximal.

Lemma 1: Denote by $G(\underline{m}_i, \underline{m}_j)$ the Gram-matrix of the vectors $\underline{m}_i$ and $\underline{m}_j$. $\mathfrak{M}$ is an equiareal set iff $\det G(\underline{m}_i, \underline{m}_j) = S^2$ for all vectors $\underline{m}_i, \underline{m}_j \in \mathfrak{M}$.

The proof follows immidiately since the area of the parallelogram spanned by $\underline{m}_i$ and $\underline{m}_j$ is equal to $|\underline{m}_i|^2 |\underline{m}_j|^2 |\sin \alpha|$ □

Consequently, if $G(\mathfrak{M})$ denotes the Gram-matrix of all vectors of $\mathfrak{M}$ then $\mathfrak{M}$ is equiareal exactly if all principal minors of second order of $G(\mathfrak{M})$ equal S^2.

The property to be an equiareal set is invariant under any homothetic map. After having carried out a homothetic map the value of the area only will be altered. Hence, one can normalize the equiareal sets:

Definition 2: An equiareal set is called normalized iff all its vectors are embedded in the unit ball and at least on of them is a unit vector.□

For the rest of the paper all equiareal sets are assumed to be normalized. These sets allow a very striking geometrical description: Let $\mathfrak{M}$ be a normalized equiareal set. Then at least one of its vectors, say $\underline{m}_1$, is a unit vector. The coordinate system can be choosen in such a way that $\underline{m}_1$ lies on one of the axes. Supposed, the coordinate system is orthogonal. Thus, for any vector $\underline{m}$ of the set it holds that

$\underline{m} = \alpha\, \underline{m}_1 + \underline{n}$, where $\underline{n} \perp \underline{m}_1$ and

$\det G(\underline{m}, \underline{m}_1) = |\underline{n}|^2 = S^2.$

Thus, the orthogonal projection of $\mathfrak{M}$ onto the hyperplane $\underline{m}_1 = \underline{0}$ yields a subset of the (d-2)-dimensional sphere of radius S together with its centre.

Theorem 1: The cardinality of an equiareal set in $\mathbb{R}^d$ cannot exceed $\binom{d+1}{2}$.

Proof: Define a homogenous polynomial of degree 2 by $P_{\underline{m}_i}(\underline{x}) \equiv G(\underline{x},\underline{m}_i)$ and consider the system F containing all these polynomials where $\underline{m}_i$ runs through the set $\mathfrak{M}$. The theorem is proved when the system F is linearly independent because each of these polynomials is homogenous of degree 2 in d variables. It is well-known that the dimension of the space of homogenous polynomials in d variables and of degree k equals $\binom{d+k-1}{k}$.

Thus, consider the equation $\sum a_{\underline{m}_i} P_{\underline{m}_i}(\underline{x}) \equiv 0$.

For $\underline{x} = \underline{m}_i$ one obtains $S^2 \sum a_{\underline{m}_j} = 0$ for $j \neq i$

This means that all the $a_{\underline{m}_i}$'s are the same and in accordance with the equation above they are equal to 0.□

The bound is met for d = 2 and for d = 3 (c.f. example 1 and example 2)

Assumed now that all parallelograms in question are squares. Since at least one of the vectors is a unit vector so all vectors are unit vectors. All these vectors are mutually orthogonal. Clearly, their maximal number cannot exceed d.

Theorem 2: If $\mathfrak{M}$ is an equiareal set in $\mathbb{R}^d$ such that all parallelograms spanned by any two of its vectors are squares then card($\mathfrak{M}$) cannot exceed d. The bound is met by any orthonormal basis in $\mathbb{R}^d$. □

Theorem 3: Assumed that all parallelograms associated with an equiareal set $\mathfrak{M}$ in $\mathbb{R}^d$ are rhombuses. If v denotes the cardinality of $\mathfrak{M}$, then the following statements hold:

i) If $v = \binom{d+1}{2}$ then $S^2 = 1 - 1/(d+2)$

ii) Assumed $d \geq 4$. If $v \geq 2d+1$, then $(1-S^2)^{-1/2}$ is an odd integer

iii) If $d \geq 4$ and $v = \binom{d+1}{2}$ then d+2 is an odd square integer.

Proof: Consider the following system of homogenous polynomials of degree 2 in d variables:

$F_{\underline{m}_i}(\underline{x}) = (\underline{m}_i, \underline{x})^2, \quad \underline{m}_i \in \mathfrak{M}\ ; \quad F_0(\underline{x}) = (1-S^2)(\underline{x}, \underline{x})$ and

$F = \{F_{\underline{m}_i}\} \cup \{F_0\}$ and let us consider the following equation:

$$\sum a_{\underline{m}_i} F_{\underline{m}_i}(\underline{x}) + a_0 F_0(\underline{x}) \equiv 0 \quad \text{for all } \underline{x} \in \mathbb{R}^d$$

$\underline{x} = \underline{m}_j$ yields: $\quad (1-S^2) \sum a_{\underline{m}_i} + a_{\underline{m}_j} + (1-S^2)a_0 = 0 \quad$ for $i \neq j$

This leads to $a_{\underline{m}_i} = A$ for all i.

Consequently, $A(1-S^2)(v-1) + A + (1-S^2)a_0 = 0$

If S = 1, then the rhombuses are squares. This case was settled by theorem 2. Therefore, a_0 can be eleminated by the last equation. Thus,

$$A \{ \sum F_{\underline{m}_i}(\underline{x}) - F_0(\underline{x})[(1-S^2)v + S^2]/(1-S^2) \} \equiv 0$$

Let $\underline{x}_k$ be a vector of an orthogonal basis in $\mathbb{R}^d$. Then $F_{\underline{m}_i}(\underline{x}_k)$ is equal to the square of the k-th coordinate of $\underline{m}_i$, i.e.

$F_{\underline{m}_i}(\underline{x}_k) = m_{i,k}^2$ and $F_0(\underline{x}_k) = 1-S^2$. Hence,

$$A \,[\sum_i m_{i,k}^2 - (1-S^2)v - S^2] = 0 \quad \text{for } k = 1,2,\dots,d$$

Since all vectors of $\mathfrak{M}$ are of the same length they are unit vectors. Thus, summing up over k yields:

$$A\,[v - d(1-S^2)v - S^2 d\,] = 0$$

If the term in brackets does not vanish, the system F will be linearly independent. Then $v+1 \le \binom{d+1}{2}$

Consequently, if $v = d(d+1)/2$ the system F has to be dependent. The expression in the brackets is equal to zero exactly when $S^2 = 1 - 1/(d+2)$. Thus, i) is proved.

Now let us prove ii). Set $a = \sqrt{1-S^2}$. Consider the matrix $C = a^{-1}(G(\mathfrak{M}) - E)$. Since $g_{ii} = (\underline{m}_i, \underline{m}_i) = 1$ and $|g_{ij}| = |(\underline{m}_i, \underline{m}_j)| = a$, the entries of C are zero on the diagonal and ± 1 off the diagonal. Because v > d, one of the eigenvalues of C equals $-a^{-1}$. This is a root of the characteristic polynomial of

C whose coefficients are all integers. Hence $-a^{-1}$ is an algebraic number. Denote its multiplicity by m. Assumed, b is an algebraic conjugate number to $-a^{-1}$. Then b is an eigenvalue of C with the multiplicity m, too. On the other hand,
$\det(C + a^{-1}E) = a^{-1}\det G(\mathfrak{M})$. Thus, the multiplicity of $-a^{-1}$ is equal to $v - \operatorname{rank} G(\mathfrak{M})$. Consequently, $m \geq v - d$ and $2m \geq 2v - 2d \geq v + 1$ which contradicts the fact that 2m cannot exceed v. Therefore, all algebraic conjugate numbers to $-a^{-1}$ are equal to $-a^{-1}$. That means that $-a^{-1}$ is a rational number. The leading coefficient of the characteristic polynomial of C is 1. Hence, any rational root should be an integer.
Finally, consider the matrix $B = 1/2\,(C + E - J)$, where J is the all-one matrix. The dimension of the eigenspace spanned by all eigenvectors of C with respect to $-a^{-1}$ equals $m = v - \operatorname{rank} G(\mathfrak{M})$ which is larger than 1. The dimension of the eigenspace spanned by the eigenvectors of J with respect to 0 equals $v - 1$. Therefore one can find an vector $\underline{x}$ belonging to the intersection of these two eigenspaces. For this vector it holds
$B\,\underline{x} = 1/2(1 - a^{-1})\,\underline{x}$. Thus, $1/2(1 - a^{-1})$ is an rational eigenvalue of B. By the same reasons like above this has to be an integer, i.e. $-a^{-1}$ is an odd integer. ii) is proved.
iii) follows immediately: According to i) if $v = d(d+1)/2$ then $S^2 = 1 - 1/(d+2)$. On the other hand, ii) says that $1/(1-S^2) = (2l + 1)^2$ for a certain integer l. Putting these together, one obtains that $d+2 = (2l + 1)^2$. □
Clearly, S cannot exceed 1. But if S is naer 1, v cannot be maximal:

Theorem 4: Assumed all parallelograms, associated with a normalized set $\mathfrak{M}$ are rhombuses. If $S^2 \geq 1 - 1/d$, then
$v \leq S^2 d/(1+dS^2-d)$.

Proof: Consider the matrix C as in the proof before and set

$a = \sqrt{1-S^2}$. The rank r of $G(\mathfrak{M})$ does not exceed d and all eigenvalues of this matrix are non-negative. Denote them by λ_i and those of C by σ_i. It holds $\sigma_i = a^{-1}(\lambda_i - 1)$. Especially, for i>r it holds $\sigma_i = -1/a$. For the trace of C one obtains:

$$0 = \operatorname{Tr}(C) = \sum_1^v \sigma_i = \sum_1^r \sigma_i - (v-r)/a \quad \text{or}$$

$$\sum^r \sigma_i = (v-r)/a \geq (v-d)/a$$

1

Applying Cauchy-Schwartz' inequality yields

$(v-d)/a \le \sum_i \sigma_i \le \sum_i |\sigma_i| \le (r \sum_i \sigma_i^2)^{1/2} \le \sqrt{d}\,(\sum_i \sigma_i^2)^{1/2}$

Thus, $\dfrac{(v-d)^2}{a^2} \le d \sum_1^d \sigma_i^2$

On the other hand, $Tr(C^2) = v^2 - v = \sum_1^v \sigma_i^2 = (v-d)/a^2 + \sum_1^d \sigma_i^2$.

Combining the last two equations delivers the desired result.□

Corollary : If all parallelograms associated with $\mathfrak{M}$ are rhombuses and $v \ge 2d+1$ then $S^2 = 1 - 1/m^2$, with $9 \le m^2 \le 2d-1$,and m is an odd integer.

Proof: According to theorem 3 there is an odd integer m such that $S^2 = 1 - 1/m^2$. Consider $G(\mathfrak{M})$. Clearly, its trace equals v. If λ_i are its eigenvalues one gets: $\binom{v}{2} S^2 = \sum \lambda_i \lambda_j$ with $1 < i < j \le d$

Thus, $v(v-1)S^2 = 2 \sum \lambda_i \lambda_j = (\sum \lambda_i)^2 - \sum \lambda_i^2 \le (d-1) \sum \lambda_i^2 =$
$(d-1)[v + (v^2 - v)(1-S^2)]$

Therefore, $1-1/m^2 = S^2 \le v(d-1)/[d(v-1)]$.

Hence, $m^2 \le d\,[1+(d-1)/(d+1)] < 2d$ □

Finally, if one of the parallelograms, associated with $\mathfrak{M}$, is a rectangular having a side of length 1, then there is a unit vector, say $\underline{m}_1$, orthogonal to a certain vector, say $\underline{m}_2$. Thus, $S^2 = G(\underline{m}_1, \underline{m}_2) = |\underline{m}_2|^2$. Take any vector $\underline{m}_i$, $i \ge 3$, of $\mathfrak{M}$:

$S^2 = G(\underline{m}_i, \underline{m}_2) = S^2 |\underline{m}_i|^2 - (\underline{m}_i, \underline{m}_2)^2$. Hence,

$S^2 (1 - |\underline{m}_i|^2) = -(\underline{m}_i, \underline{m}_2)^2$. Since $\mathfrak{M}$ is normalized, the left-hand side is non-negative. Consequently, all vectors of $\mathfrak{M}$ except $\underline{m}_2$ are unit vectors. These vectors are orthogonal to $\underline{m}_2$. Therefore, $\mathfrak{M} - \{\underline{m}_2\}$ is an equiareal set in $\mathbb{R}^{d-1}$ such that all parallelograms are rhombuses. Applying the proof of theorem 4 one gets

Theorem 5: Let $\mathfrak{M}$ be a normalized equiareal set such that among the parallelograms associated with $\mathfrak{M}$ there is a rectangular having a side of length 1. Then the following statements hold:

i) $v \le \binom{d}{2} + 1$

ii) If $S^2 \ge 1 - 1/(d+1)$ then $v \le 2 + (d-2)/[S^2(d-1) - (d-2)]$

iii) Assumed, $d \ge 5$. If $v \ge 2d$, then $S^2 = 1 - 1/m^2$, where m is an odd integer and $9 \le m^2 \le 2d-3$

iv) If $v = 1 + \binom{d}{2}$, then there is an odd integer m with $9 \le m^2 \le d+1$ such that $S^2 = 1 - 1/m^2$. □

The problem of maximal sets having the property that the volume of all k-dimensional cubs, spanned by any k vectors of these sets, is always the same is considered in [3].

I would like to express my gratitude to Professor Gronau for his helpful discussions on this subject.

References :

1. A.Blokhuis and J.J.Seidel, Few-distance sets in $\mathbb{R}^{p,q}$, Symposia Mathematica, vol XXVIII, 145-158

2. A.Blokhuis, "Few-distance sets", Thesis, CWI Tract 7, P.O.B. 4079, Amsterdam, The Netherlands, 1984

3. A.Mohammed and G.Teumer,Equiareal k-sets, to appear in Matematica Africa

On the Piaget Graph

C. Thies, Kiel, FRG

1 Introduction

Jean Piaget (1896-1980) is the founder of genetic epistemology and psychology in Geneva. His scientific interest was devoted to the question of the rise and development of human knowledge. His statements on the genesis of knowledge in children are world-renowned. 'Piaget puts the diversity of the correlations of living thinking into classes and relations' (Aebli, 1978,608). To explain the cognitive structures of children, Piaget uses the grouping concept, which he describes in both colloquial terms and with mathematical terminology and graph-theoretical methods (Fig. 1).

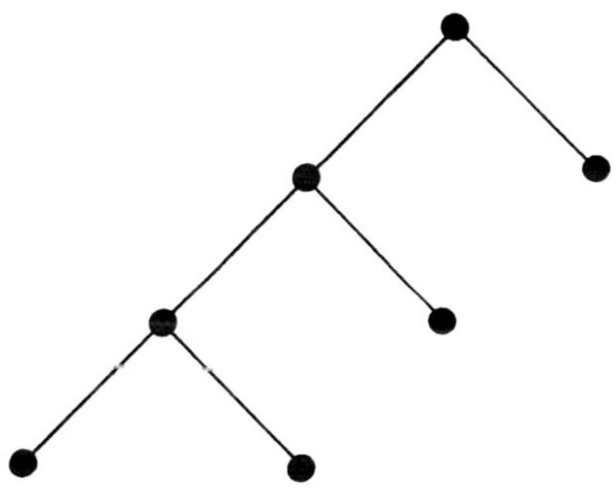

Figure 1

According to this, groupings are characteristically structured systems of operations with the essential attributes of "composition", "reversibility", "associativity", "identity" and "tautology".

Piaget calls the elements of grouping "operations"; these can be interpreted as internalised and systemised actions and have, according to Piaget, developed from pre-operatory schemes, which in turn have their roots in sensori-motor schemes.

R. Bodendiek, R. Henn (Eds.)
Topics in Combinatorics and Graph Theory

Piaget's grouping formalisations do not meet formal demands. They are furthermore difficult to interpret. Many authors have therefore repeatedly endeavoured to improve on the original version.

Since the Seventies, suggestions for a more precise specification and formalisation have been submitted by psychologists, mathematicians and mathematics methodologists. In general, the following can be ascertained in the analysis of these grouping formalisation:

- Piaget perceives the qualitative diversity of correlations of living thinking in classes and relations. This dual aspect is not formalised.
- The term "tautology", which is significant for the grouping, is either sometimes explained completely differently from Piaget, or not taken into consideration at all.
- The given formalisations do not permit varying operation combinations.

The following is a mathematically correct formalisation which adheres closely to Piaget's specifications. Piaget indicates several groupings which can basically be reduced to two types of grouping, class-grouping and relation-grouping. Informally, both can be characterised as semi-lattices with a universal lower (or upper) bound element, in each of which the reversibility is declared apt and compatible with the semi-lattice structure.

2 The Class-grouping

Piaget employs in [4] the term "contingent element" to characterise the class-grouping. For two elements x, y of a set M, y is called the contingent element to x, if $x \neq y$ is valid, and there is a $z \in M$ with $x \lhd z$ and $y \lhd z$. Thereby the relation $\lhd$ "is direct precursor of" is, as always, explained as follows: $x \lhd y$ exactly, when $x < y$ and there is no $v \in M$ with $x < v < y$.

Definition 1: Let M be a non-empty set, the elements of which are called "states". Moreover, let $G \subseteq \{S|S \subseteq M\}$ be a non-empty set. The elements of **G** are called "operations".

With these notations, the poset $(\mathbf{G},\leq)$ is called a class-grouping if the following holds:

(a) $(\mathbf{G},\leq)$ is an upper semi-lattice with a universal lower bound 0.

(b) A contingent element $S' \in G$ exists for every $S \in G$ which is neither minimal or maximal.

3 Observations on the Class-grouping

An upper semi-lattice with a universal lower bound 0 is generally known as a commutative semi-group in which 0 is a neutral group element. Thus the four Piaget grouping attributes "composition", "associativity", "identity" and "tautology" result from (a). Thereby the operation combination (composition) for $A,B \in \mathbf{G}$ is explained by establishing

$$A + B := \sup(A,B).$$

The associativity can be perceived directly; for the identity $0 \in \mathbf{G}$, the following is valid for all $A \in \mathbf{G}$

$$A + 0 = 0 + A = \sup(A,0) = \sup(0,A) = A.$$

The tautology can be described for all $A \in \mathbf{G}$ by

$$A + A = \sup(A,A) = A.$$

The reversibility results from (b); it is explained as follows for all $X,X',Y \in \mathbf{G}$ with $X \lhd Y$ and $X' \lhd Y$:

$$Y - X = X' :\iff Y = X' + X = \sup(X,X')$$
$$Y - X' = X :\iff Y = X + X' = \sup(X,X')$$

Furthermore, let $X - X = 0$ be analogous for all $X \in \mathbf{G}$.

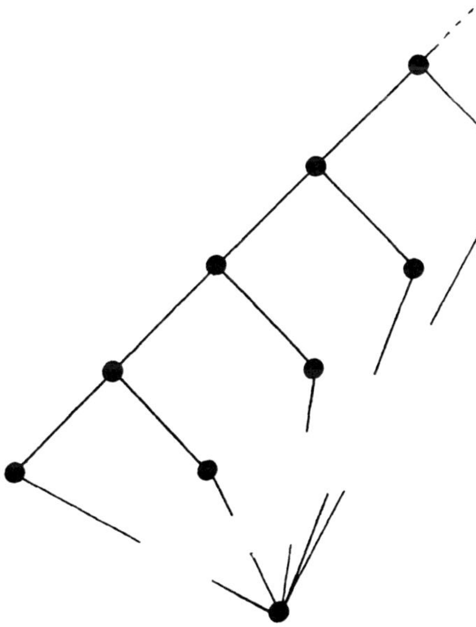

Figure 2

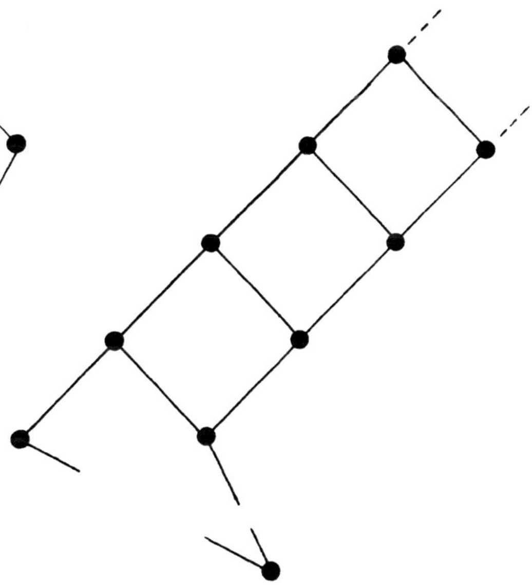

Figure 3

From a mathematical point of view, a class-grouping can be similarly defined as a lower semi-lattice with a universal upper bound element.

An analysis of the definition of the class-grouping **G** yields the following possibilities of representing it as a graph (Hasse-diagram) (Fig. 2-4).

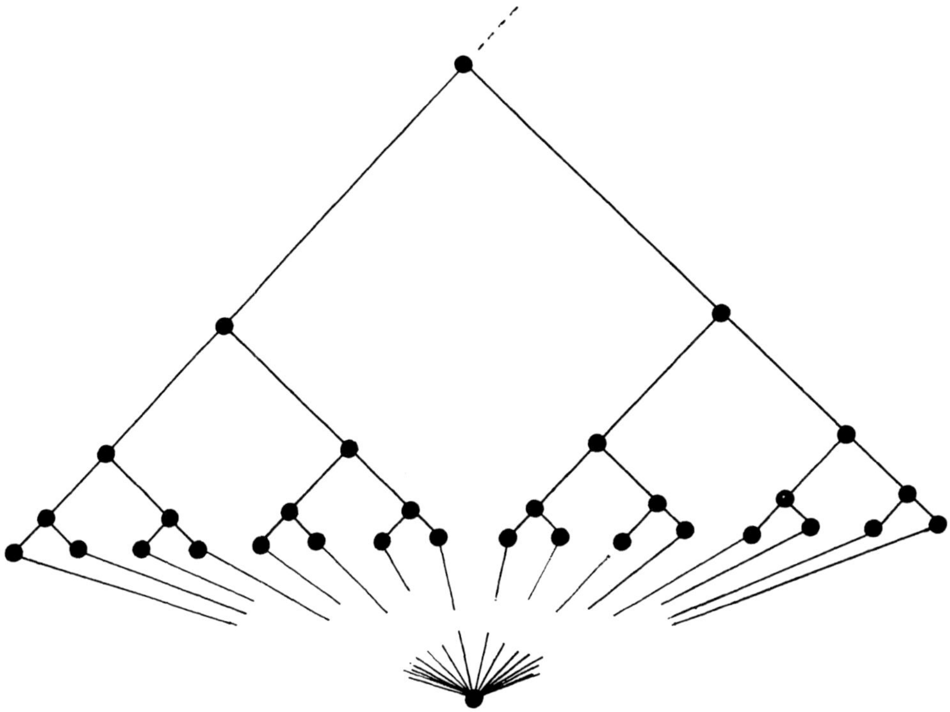

Figure 4

4 The Relation-grouping

According to Piaget, with the exception of the reversibility, the relation-grouping evinces the same structure as the class-grouping.

Definition 2: Let M be a non-empty set, the elements of which are called "states". Moreover, let $\mathbf{G'} \subseteq \{RT(S) \mid S \subseteq M \times M\}$ be a non-empty set of reflexive and transitive relations on M. (RT(S) is the reflexive and transitive closure of the relation S on M).
The elements of **G'** are called "operations", the relations $S \subseteq M \times M$ as in [9], elementary operations. With these notations, the poset $(\mathbf{G'},\leq)$ is called

a relation-grouping if the following holds:

(a) $(G',\leq)$ is an upper semi-lattice with a universal lower bound element 0.

(b) An element $S^{-1} \in G'$ exists for every $S \in G'$.

5 Observations on the Relation-grouping

In the same way as for the class-grouping, the four Piaget group attributes "composition", "associativity", "identity" and "tautology" can be deduced from (a) with the establishing of $S+T := \sup(S,T)$ for $S, T \in G'$; the reversibility results directly from (b).

6 References

[1] Aebli,H.: Entwicklungspsychologie zur Theorie der kognitiven Sozialisation. In: Die Psychologie des 20. Jahrhunderts, Zürich, 1978.

[2] Bodendiek,R/Schumacher,H.: Zur Formalisierung des Piagetschen Gruppierungsbegriffs. In: Graphen in Forschung und Unterricht. Bodendiek,H., Schumacher,H., Walther,G. (Hrsg.), Bad Salzdetfurth 1985.

[3] Piaget,J.: Psychologie der Intelligenz, Olten 1971.

[4] Piaget,J.: Die Entwicklung des Erkennens I. Das mathematische Denken, Stuttgart 1975.

[5] Piaget,J.: Jean Piaget über Jean Piaget, München 1981.

[6] Thies,C.: Eine neue Formalisierung des Piagetschen Gruppierungsbegriffs. In: Beiträge zum Mathematikunterricht 1984.

[7] Thies,C.: Zur Kritik an Piagets Gruppierungsformalisierung. In: Graphen in Forschung und Unterricht. Bodendiek,R., Schumacher,H., Walther,G. (Hrsg.), Bad Salzdetfurth 1985.

[8] Thies,C.: Eine neue relationentheoretische Gruppierungsformalisierung. In: Graphen in Forschung und Untericht. Bodendiek,R., Schumacher,H., Walther,G. (Hrsg.), Bad Salzdetfurth 1985.

[9] Wittmann,E.: Zum Begriff "Gruppierung" in der Piagetschen Psychologie. In: Beiträge zum Mathematikunterricht, 1973, 203-222.

[10] Wittmann,E.: Natural Numbers and Groupings. In: Educational Studies of Mathematics 6, 1975, 53-75.

On a Characterization of Closure Operators by Identities on Semigroups

R. Thron, Halle, GDR
J. Koppitz, Halle, GDR

1. INTRODUCTION AND RESULTS

For a semigroup S it is well known, that the operator C on the power set P(S) of S, which assigns to every U $\in$ P(S) the subsemigroup C(U) of S generated by U, is a closure operator (cf. [2], [4]). Let T $\in$ P(S). Then U is called to be a generating set of T with respect to C if and only if $U \subseteq T$ and $T \subseteq C(U)$. Let GEN(T) be the system of all generating sets of T. To give a more detailed description of the generating sets of T the so-called isolated sets of T are considered. I is called to be isolated in T if and only if $\emptyset \neq I \subseteq T$ and $I \cap C(T \setminus I) = \emptyset$. Let ISO(T) be the system of all I such that I is isolated in T. Then there holds

PROPOSITION 1. Let S be a semigroup and $T \subseteq S$. Then the following statements are equivalent:

(0) U $\in$ GEN(T).

(1) $U \subseteq T$ and from I $\in$ ISO(T) it follows $U \cap I \neq \emptyset$. □

For cardinal numbers n and m let $ISO_n(T)$ be the system of all I $\in$ ISO(T) with $|I| < n + 1$ and K(n,m) be the class of all semigroups S such that for each $T \subseteq S$ with $|T| < m$ the

R. Bodendiek, R. Henn (Eds.)
Topics in Combinatorics and Graph Theory

following holds: GEN(T) is the set of all $U \subseteq T$ such that $U \cap I \neq \emptyset$ for each $I \in ISO_n(T)$.

Example 1. Let S_n be the free rectangular band generated by $n \geq 1$ elements. Then there hold $x^2 = x$ and $xyz = xz$ for all $x,y,z \in S_n$ (cf.[3]). Therefore, $S_n \in K(n,\aleph_0)$ and $S_n \notin K(n-1,\aleph_0)$.

Example 2. Let p be a prime number and S_p be a group of order p. Then $S_p \in K(p-1,\aleph_0)$ and $S_p \notin K(p-2,\aleph_0)$.

PROPOSITION 2. Let S be a semigroup and C be the related closure operator. Then for each natural number $n \geq 1$ the following statements are equivalent:

(0) $S \in K(n,\aleph_0)$.

(1) If $I,T \subseteq S$, $|I| \leq n+1$, $|T| < \aleph_0$, $I \cap C(T) = \emptyset$ and $i \in C(T \cup \{j\})$ for $i,j \in I$, then $|I| < n+1$. □

The aim of the present paper is to characterize classes K of semigroups for any given natural number $n \geq 1$ such that $K \subseteq K(n,\aleph_0)$. In [5] there are characterized all varieties V of semigroups such that $V \subseteq K(1,\aleph_0)$. However, it makes no sense to describe the classes $K(n,\aleph_0)$ for $n \geq 2$ by varieties $V \subseteq K(n,\aleph_0)$ because

PROPOSITION 3. For each variety V of semigroups and for each natural number $n \geq 1$ the following holds: If $V \subseteq K(n,\aleph_0)$, then $V \subseteq K(1,\aleph_0)$. □

Therefore, the classes $K(n,\aleph_0)$ with $n \geq 2$ will be described by classes $K \subseteq K(n,\aleph_0)$ which are not necessarily varieties.

For that purpose let N be the set of the natural numbers and

$$X := \{x,y,z\} \cup \{x_i : i \in N\} \cup \{y_i : i \in N\}$$

be a set of variables. For each $Y \subseteq X$ let Y^+ be the free semigroup on Y. If f is a function from X into S, then let $\overline{f}$ be the homomorphism from X^+ into S such that $\overline{f}(w) = f(w)$ for each $w \in X$. Furthermore, let $\overline{f}\langle A\rangle := \{(\overline{f}(p), \overline{f}(q)) : (p,q) \in A\}$ for each $A \in P(X^+ \times X^+)$. Let SEM be the class of all semigroups and $d(S) := \{ (s,s): s \in S \}$ for $S \in SEM$. Let $\varphi \subseteq P(X^+ \times X^+) \times SEM$ be the following relation such that $A \varphi S$ if and only if $\bar{f}\langle A \rangle \cap d(S) \neq \emptyset$ for each function f from X into S. For $\mathfrak{A} \subseteq P(X^+ \times X^+)$ let

$$MOD(\mathfrak{A}) := \{ S: \text{ If } A \in \mathfrak{A}, \text{ then } A \varphi S \} .$$

Obviously, any class V of semigroups is a variety if and only if there exists some $\mathfrak{A} \subseteq P(X^+ \times X^+)$ such that $V = MOD(\mathfrak{A})$ and $|A| = 1$ for each $A \in \mathfrak{A}$.

Let $(p,q) \in X^+ \times X^+$. Then (p,q) is called to be regular if and only if exactly the same variables appear on both sides p and q. Any $A \in P(X^+ \times X^+)$ is called to be regular if and only if each $(p,q) \in A$ is regular. Let K be a class of semigroups. Then K is called to be regular if and only if there exists some $\mathfrak{A} \subseteq P(X^+ \times X^+)$ such that $K = MOD(\mathfrak{A})$ and for each $A \in \mathfrak{A}$ the following holds: A is regular and $|A| < \aleph_0$.

For natural numbers $n,k \geq 1$ let

$$F_1(n,k) := \{ (x_i \dots x_0)(x_n \dots x_0)^k z : 0 \leq i \leq n \} ,$$

$$F_2(n,k) := \{ z(x_0 \dots x_n)^k (x_0 \dots x_i) : 0 \leq i \leq n \} ,$$

$$F_3(n,k) := \{ (x_i \dots x_0)(x_n \dots x_0)^k z (y_0 \dots y_n)^k (y_0 \dots y_i) : 0 \leq i \leq n \} ,$$

$$A_i(n,k) := \{ (p,q) : (p,q) \in F_i(n,k) \times F_i(n,k), \ p \neq q \}$$

for $i = 1,2,3$ and

$$\mathfrak{A}(n,k) := \{ A_1(n,k), \ A_2(n,k), \ A_3(n,k) \} .$$

<u>PROPOSITION 4.</u> For each natural number $n \geq 1$ and for each regular class K of semigroups the following statements are equivalent:

(0) $K \subseteq K(n, 3n+4)$.

(1) $K \subseteq K(n, \aleph_0)$.

(2) There exists a natural number $k \geq 1$ such that $K \subseteq MOD(\mathfrak{A}(n,k))$. ▯

2. PROOFS OF THE PROPOSITIONS

Proof of PROPOSITION 1. Assume that (0) holds. If there exists some $I \in ISO(T)$ such that $U \cap I = \emptyset$, then $I \cap C(T \setminus I) = \emptyset$ and $U \subseteq T \setminus I$. Therefore $I = I \cap T \subseteq I \cap C(U) = \emptyset$, contradicting $I \neq \emptyset$. Assume that (1) holds. If $T \nsubseteq C(U)$, then $(T \setminus C(U)) \cap C(U) = \emptyset$ with $(T \setminus C(U)) \neq \emptyset$. Therefore, $(T \setminus C(U)) \in ISO(T)$ and $(T \setminus C(U)) \cap U = \emptyset$, a contradiction. ∎

Proof of PROPOSITION 2. Assume that (0) holds. Let $I, T \subseteq S$, $|I| \leq n+1$, $|T| < \aleph_0$, $I \cap C(T) = \emptyset$ and $i \in C(T \cup \{j\})$ for $i,j \in I$. If $|I| = n+1$, then $J \notin ISO_n(T \cup I)$ for each nonempty $J \subseteq I$ and $\bigcup\{J \setminus I: J \in ISO_n(T \cup I)\} \in GEN(T \cup I)$, $I \cap (\bigcup\{J \setminus I: J \in ISO_n(T \cup I)\}) \neq \emptyset$, a contradiction. Assume that (1) holds. Let $J, T \subseteq S$, $|T| < \aleph_0$, $n+1 \leq |J| < \aleph_0$ and $J \cap C(T) = \emptyset$. Then there exists some $U \subseteq J$ with $0 \neq |U| < |J|$ and $U \cap C(T \cup (J \setminus U)) = \emptyset$. Otherwise it follows that $i \in C(T \cup \{j\})$ for $i,j \in J$, contradicting (1). Consequently, if $T \subseteq S$ and $|T| < \aleph_0$, then for each $J \in ISO(T)$ there exists some $I \in ISO_n(T)$ such that $I \subseteq J$. Therefore, $S \in K(n, \aleph_0)$. ∎

If $\mathfrak{A} \subseteq P(X^+ \times X^+)$, then let $F(\mathfrak{A})$ be the set of all $\bar{f}\langle A \rangle$ such that $A \in \mathfrak{A}$ and f is a function from X into X^+ and $D(\mathfrak{A})$ be the set of all $B \in P(X^+ \times X^+)$ such that $B \cap A \neq \emptyset$ for each $A \in \mathfrak{A}$. If $A \in P(X^+ \times X^+)$, then let $\sigma(A)$ be the least congruence relation σ on X^+ with $A \subseteq \sigma$. If $\mathfrak{A} \subseteq P(X^+ \times X^+)$, then let $\Sigma(\mathfrak{A})$ be the set of all $\sigma(A)$ with $A \in D(F(\mathfrak{A}))$. If $K \subseteq SEM$, then let

$$EQ(K) := \{A: \text{ If } S \in K, \text{ then } A \,\varphi\, S\}.$$

LEMMA 1. If $\mathfrak{A} \subseteq P(X^+ \times X^+)$, then the following statements hold:

(0) For each function g from X into X^+ the following implication holds:
If $A \in EQ(MOD(\mathfrak{A}))$, then $\bar{g}\langle A \rangle \in EQ(MOD(\mathfrak{A}))$.

(1) If $\sigma \in \Sigma(\mathfrak{A})$, then $X^+/\sigma \in MOD(\mathfrak{A})$.
(2) $D(\Sigma(\mathfrak{A})) = EQ(MOD(\mathfrak{A}))$.

Proof. (0) Let $A \in EQ(MOD(\mathfrak{A}))$, f be a function from X into S and g be a function from X into X^+. Then $\overline{f}\overline{g}$ is a function from X into S with $\overline{\overline{f}g} = (\overline{f}\,\overline{g})$ and $(\overline{f}\,\overline{g})\langle A\rangle \cap d(S) \neq \emptyset$. Therefore $\overline{g}\langle A\rangle \in EQ(MOD(\mathfrak{A}))$.

(1) Let $\sigma \in \Sigma(\mathfrak{A})$, $A \in \mathfrak{A}$ and f be a function from X into X^+/σ. Then there exists a function g from X into X^+ such that $g(w) \in f(w)$ for $w \in X$ and therefore $\overline{f}(p) = [\overline{g}(p)]_\sigma$ for $p \in X^+$. From $\overline{f}\langle A\rangle = \{([\overline{g}(p)]_\sigma, [\overline{g}(q)]_\sigma) : (p,q) \in A\}$ and $\overline{g}\langle A\rangle \cap \sigma \neq \emptyset$ it follows that $\overline{f}\langle A\rangle \cap d(X^+/\sigma) \neq \emptyset$ and $X^+/\sigma \in MOD(\mathfrak{A})$.

(2) Obviously, $D(\Sigma(\mathfrak{A})) \subseteq EQ(MOD(\mathfrak{A}))$. If $A \in EQ(MOD(\mathfrak{A}))$, $\sigma \in \Sigma(\mathfrak{A})$ and f is a function from X into X^+/σ such that $f(w) = [w]_\sigma$ for $w \in X$, then $\overline{f}\langle A\rangle \cap d(X^+/\sigma) \neq \emptyset$ and $A \cap \sigma \neq \emptyset$. Therefore, $EQ(MOD(\mathfrak{A})) = D(\Sigma(\mathfrak{A}))$. $\Box$

LEMMA 2. Let $\mathfrak{A} \subseteq P(X^+ \times X^+)$ such that $|A| < \aleph_0$ for each $A \in \mathfrak{A}$. Then the following implication holds:
If $MOD(\{\{(xy,yx)\}\}) \nsubseteq MOD(\mathfrak{A})$, then there exist natural numbers k,l such that $1 \leq k < l$ and $\{(x^k,x^l)\} \in EQ(MOD(\mathfrak{A}))$.

Proof. Let $MOD(\{\{(xy,yx)\}\}) \nsubseteq MOD(\mathfrak{A})$. Then there exists some $A \in \mathfrak{A}$ such that $A \notin EQ(MOD(\{\{(xy,yx)\}\}))$. Without loss of generality let $A \subseteq \{x_1,\dots,x_n\}^+ \times \{x_1,\dots,x_n\}^+$ for some natural number $n \geq 2$. Since $|A| < \aleph_0$ there exist a natural number $m \geq 1$ and a function f from X into X^+ such that $f(x_n) = x_{n-1}^m$, $f(w) = w$ for each $w \in (X \setminus \{x_n\})$ and $\overline{f}\langle A\rangle \notin EQ(MOD(\{\{(xy,yx)\}\}))$, $\overline{f}\langle A\rangle \in EQ(MOD(\mathfrak{A}))$, $\overline{f}\langle A\rangle \subseteq \{x_1,\dots,x_{n-1}\}^+ \times \{x_1,\dots,x_{n-1}\}^+$. By induction one gets some $B \in EQ(MOD(\mathfrak{A}))$ such that $B \subseteq \{x_1\}^+ \times \{x_1\}^+$, $B \notin EQ(MOD(\{\{(xy,yx)\}\}))$ and $|B| < \aleph_0$. Since $|B| < \aleph_0$ it follows that there exist natural numbers k,l such that

$1 \leqq k < l$ and $\{(x^k, x^l)\} \in EQ(MOD(\mathfrak{A}))$. □

For natural numbers $n,k \geqq 1$ let

$$\bar{F}_i(n,k) := \{x_0,\dots,x_n\}^+ \cup F_i(n,k)$$

for $i = 1,2$,

$$\bar{F}_3(n,k) := (\{x_0,\dots,x_n\} \cup \{y_0,\dots,y_n\})^+ \cup F_3(n,k),$$

$$\bar{A}_i(n,k) := \{(p,q) : (p,q) \in \bar{F}_i(n,k) \times F_i(n,k),\ p \neq q\}$$

for $i = 1,2,3$ and

$$\bar{\mathfrak{A}}(n,k) := \{\bar{A}_1(n,k), \bar{A}_2(n,k), \bar{A}_3(n,k)\}.$$

<u>LEMMA 3.</u> Let $\mathfrak{A} \subseteq P(X^+ \times X^+)$ such that $|A| < \aleph_0$ for each $A \in \mathfrak{A}$. Then for each natural number $n \geqq 1$ the following implication holds: If $MOD(\mathfrak{A}) \subseteq K(n,3n+4)$, then there exists a natural number k such that $\bar{\mathfrak{A}}(n,k) \subseteq EQ(MOD(\mathfrak{A}))$.

<u>Proof.</u> Let $MOD(\mathfrak{A}) \subseteq K(n,3n+4)$ and p be a prime number such that $p \geqq n+2$ and S_p be a group of order p. Then $S_p \in MOD(\{\{(xy,yx)\}\})$ and $S_p \notin K(n,3n+4)$. Therefore $MOD(\{\{(xy,yx)\}\}) \nsubseteq MOD(\mathfrak{A})$. By LEMMA 2 there exist natural numbers k,l such that $1 \leqq k < l$ and $\{(x^k,x^l)\} \in EQ(MOD(\mathfrak{A}))$. Let i be a natural number such that $1 \leqq i \leqq 3$. Then $\bar{A}_i(n,k) \cap \sigma \neq \emptyset$ for each $\sigma \in \Sigma(\mathfrak{A})$: Otherwise there exists some $\sigma \in \Sigma(\mathfrak{A})$ such that $\bar{A}_i(n,k) \cap \sigma = \emptyset$. Let f be a function from X into X^+/σ such that $f(w) = [w]_\sigma$ for $w \in X$. Then let

$$S(n,k,j) := \{\bar{f}(p) : p \in \{x_0,\dots,x_n\} \cup F_j(n,k)\}$$

for $j = 1,2$ and

$$S(n,k,3) := \{\bar{f}(p) : p \in \{x_0,\dots,x_n\} \cup \{y_0,\dots,y_n\} \cup \cup F_3(n,k)\}.$$

Consequently, $S(n,k,i) \subseteq X^+/\sigma \in MOD(\mathfrak{A}) \subseteq K(n,3n+4)$ and $|S(n,k,i)| < 3n+4$. Since $\bar{A}_i(n,k) \cap \sigma = \emptyset$ it follows from $I \in ISO_n(S(n,k,i))$ that $I \cap \{\bar{f}(x_0),\dots,\bar{f}(x_n)\} \neq \emptyset$.

Therefore, $\{\bar{f}(x_0),\ldots,\bar{f}(x_n)\} \in GEN(S(n,k,i))$, contradicting $\bar{A}_i(n,k) \cap \mathfrak{S} = \emptyset$.

Finally, from $\bar{A}_i(n,k) \cap \mathfrak{S} \neq \emptyset$ for each $\mathfrak{S} \in \Sigma(\mathfrak{A})$ and for each natural number i with $1 \leqq i \leqq 3$ it follows

$$\bar{\mathfrak{A}}(n,k) \subseteq EQ(MOD(\mathfrak{A})). \; \square$$

Proof of PROPOSITION 3. Let V be a variety of semigroups such that $V \subseteq K(n,\aleph_0)$. Then there exists some $\mathfrak{A} \subseteq P(X^+ \times X^+)$ such that $|A| = 1$ for each $A \in \mathfrak{A}$ and $V = MOD(\mathfrak{A})$. Obviously, for each $A \in EQ(MOD(\mathfrak{A}))$ there exists some $B \in EQ(MOD(\mathfrak{A}))$ such that $B \subseteq A$ and $|B| = 1$. Then by LEMMA 3 there exist (p_1,q_1) and (p_2,q_2) such that

$$\{(p_1,q_1)\}, \{(p_2,q_2)\} \in EQ(MOD(\mathfrak{A}))$$

and $(p_1,q_1) \in \bar{A}_1(n,k)$, $(p_2,q_2) \in \bar{A}_2(n,k)$ for some natural number $k \leqq 1$. Finally, by substitution it follows that there exists some natural number $m \geqq 1$ such that

$$\{(zx^m,x^m)\} \in EQ(MOD(\mathfrak{A})) \text{ or } \{(x^mz,x^m)\} \in EQ(MOD(\mathfrak{A}))$$

or $\{\{(x^m,x^{m+1})\},\{(y(xy)^m,(xy)^mx)\}\} \subseteq EQ(MOD(\mathfrak{A}))$. By [5] one gets $V \subseteq K(1,\aleph_0)$. $\square$

Obviously, from LEMMA 1 it follows

LEMMA 4. Let $\mathfrak{A} \subseteq P(X^+ \times X^+)$ such that each $A \in \mathfrak{A}$ is regular. Then for each $A \in EQ(MOD(\mathfrak{A}))$ there exists some regular $B \in EQ(MOD(\mathfrak{A}))$ such that $B \subseteq A$. $\square$

Proof of PROPOSITION 4. Obviously, if (1), then (0). Assume that (0) holds. Then by LEMMA 3 and by LEMMA 4 it follows (2). Finally, assume that (2) holds. Let $S \in MOD(\mathfrak{A}(n,k))$, $I,T \subseteq S$, $|I| \leqq n+1$, $|T| < \aleph_0$, $I \cap C(T) = \emptyset$ and $i \in C(T \cup \{j\})$ for $i,j \in I$ with $I = \{i_0,\ldots,i_n\}$. Then the following holds: $i_0 \in C(T \cup \{i_n\})$, $i_m \in C(T \cup \{i_{m-1}\})$ for $1 \leqq m \leqq n$. Without loss of generality the cases (I) or (II) or (III) are possible:

(I) There hold the following decomposition: $i_0 = s_n i_n$ with

$s_n \in C(T \cup \{i_n\})$, $i_m = s_{m-1}i_{m-1}$ with $s_{m-1} \in C(T \cup \{i_{m-1}\})$ for $1 \leqq m \leqq n$. By substitution one gets

$$I = \{(s_m \dots s_0)(s_n \dots s_0)^k i_0 \ :\ 0 \leqq m \leqq n\}$$

and $|I| < n+1$.

(II) There hold the following decompositions: $i_0 = i_n r_n$ with $r_n \in C(T \cup \{i_n\})$, $i_m = i_{m-1}r_{m-1}$ with $r_{m-1} \in C(T \cup \{i_{m-1}\})$ for $1 \leqq m \leqq n$. By substitution one gets

$$I = \{i_0(r_0 \dots r_n)^k(r_0 \dots r_m) \ :\ 0 \leqq m \leqq n\}$$ and $|I| < n+1$.

(III) There hold the following decompositions: $i_0 = s_n i_n r_n$ with $s_n, r_n \in C(T \cup \{i_n\})$, $i_m = s_{m-1}i_{m-1}r_{m-1}$ with $s_{m-1}, r_{m-1} \in C(T \cup \{i_{m-1}\})$ for $1 \leqq m \leqq n$. By substitution one gets

$$I = \{(s_m \dots s_0)(s_n \dots s_0)^k i_0 (r_0 \dots r_n)^k (r_0 \dots r_m) : 0 \leqq m \leqq n\}$$

and $|I| < n+1$.

By PROPOSITION 2 it follows $S \in K(n, \aleph_0)$. Therefore, if (2), then (1). □

REFERENCES

[1] Edelman, P.H., Meet-distributive lattices and the anti-exchange closure, Algebra Universalis 10(1980), 290-299.

[2] Grätzer, G., Universal Algebra, New York, Heidelberg, Berlin 1979.

[3] Petrich, M., Lectures in semigroups, Berlin 1977.

[4] Schmidt, J., Einige grundlegende Begriffe und Sätze aus der Theorie der Hüllenoperatoren. In: Ber. Math. Tagung, Berlin 1953.

[5] Thron, R., Kittler, D., On semigroups with least generating sets, Semigroup Forum Vol. 36(1987), 341-350.

Symmetries of Group-Triangulations

H.-J. Voss, Dresden, GDR

Many authors dealt with groups and graphs. Here we shall assign to a finite group triangulations of some oriented (closed) surfaces in the sense of H. Sachs [1] . This assignment was intensively studied by W. Voss [9] – [14] . A survey on old and new results is given in H.-J. Voss [6] .
The present paper is devoted to the automorphisms of group-triangulations. It continues the investigations of H.-J. Voss [7] .

1. Concepts. Let G be a finite group.
A triple (x,y,z) of elements $x, y, z \in G$ is said to be regular if $x \neq y \neq z \neq x$ and $xyz = e$. With $xyz = e$ also $yzx = e$ and $zxy = e$. If (x,y,z) is a regular triple then (y,x,z') with $z' = (yx)^{-1}$ is regular, too.
To each regular triple (x,y,z) an oriented triangle $D(x,y,z)$ with arcs (x,y), (y,z), (z,x) and vertices x, y, z is assigned so that $D(x,y,z) = D(x',y',z')$ iff $(x',y',z') \in \{(x,y,z), (y,z,x), (z,x,y)\}$; otherwise $D(x,y,z)$, and $D(x',y',z')$ are disjoint (Fig. 1.1) .

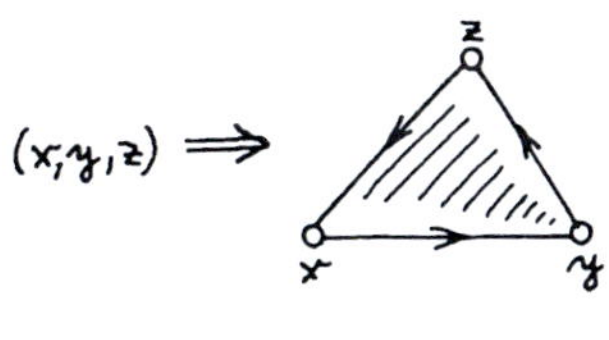

Fig. 1.1

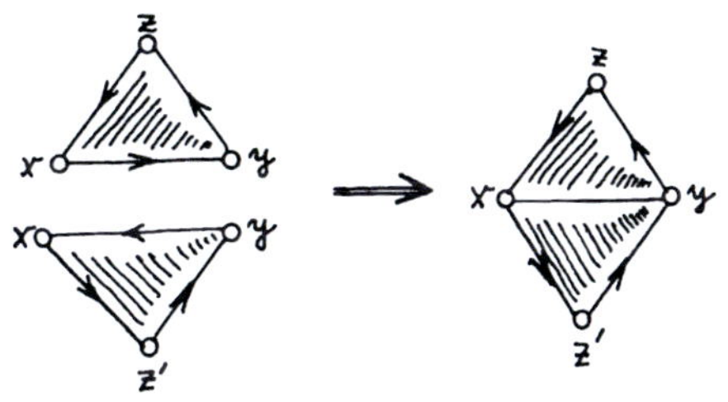

Fig. 1.2

In the set Σ of all so obtained triangles an arc occurs at most once. If (x,y) is in some triangle of Σ then (y,x) is in a triangle of Σ , too.

R. Bodendiek, R. Henn (Eds.)
Topics in Combinatorics and Graph Theory

In the sense of combinatorial topology opposite directed arcs (x,y), (y,x) are identified so that an edge [x,y] is obtained (Fig. 1.2) . Identifying step by step all opposite directed arcs triangulations of closed oriented surfaces are obtained. Thus to each finite group G a set of topological invariants of these triangulations is assigned. We ask:

> Let P be a property of these invariants. Which property of the groups corresponds to the property P of these invariants and vice versa?

If the elements x, y of the regular triple (x,y,z) commute then $xyz = yxz = e$ and the triangulation T is formed by the two triangles D(x,y,z) and D(y,x,z) only. Hence T is the sphere with a 3 - cycle on it which triangulates that sphere.

Consequently, in the following we shall only consider <u>non - commutative</u> elements of G . Really, two non - commutative elements x, y form with $(xy)^{-1}$ a regular triple. Thus to each pair of non - commutative elements $x, y \in G$ a so called group-triangulation T(x,y) of an oriented surface S(x,y) corresponds, where to each vertex a group element of G is assigned. The octahedron T((12)(34), (1234)) with $(12)(34)\cdot(1234) \neq (1234)\cdot(12)(34)$ is presented as an example in Fig. 1.3 .

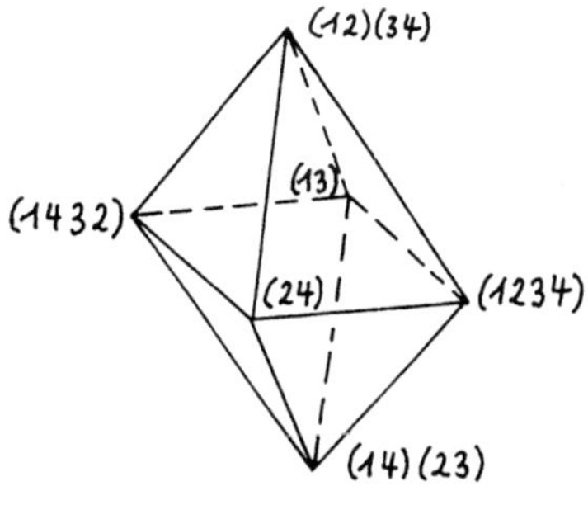

Fig. 1.3

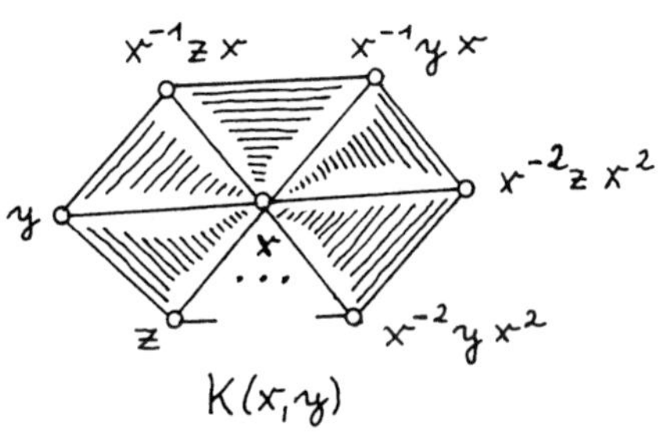

Fig. 1.4

The subgroup of G generated by x and y is denoted by G(x,y) . In T(x,y) any two different triangles are either disjoint or have precisely one vertex in common or have precisely one edge and its end vertices in common. Such triangulations are said to be simple.

The basic disc $K(x,y)$ of $T(x,y)$ with centre x through y consists of the edge $[x,y]$ and all triangles of $T(x,y)$ incident with this vertex x (see Fig. 1.4) .
An automorphism a of a triangulation is a triple of 1 - 1 mappings of the vertex set, the edge set and the triangle set, onto the vertex set, the edge set and the triangle set, respectively, preserving incidence.
The inner automorphism f_x of G with $f_x(g) = x^{-1} g x$ for each $g \in G$ induces a sense preserving automorphism of $T(x,y)$ so that $K(x,y)$ is "rotated clockwise over two sectors", i.e., if $y = y_1, y_2, \dots, y_s, y = y_1$ are the neighbours of x which will be met in this order on the bounding cycle of $K(x,y)$ by a clockwise walk then $f_x(x) = x$ and $f_x(y_i) = y_{i+2}$ for all $1 \leq i \leq s$, where the indices are to be taken modulo s .
These automorphisms have the property: if $D(a,b,c)$ is a face of $T(x,y)$ then $f_a f_b f_c$ is the identity e because
$f_a f_b f_c(g) = f_c(f_b(f_a(g))) = (abc)^{-1} g\, abc = g$ for all $g \in G$.
Let $A^2(T) = A^2(T(x,y))$ denote the subgroup of the automorphism group of $T(x,y)$ generated by f_x, f_y . Let X, Y, Z denote the vertices of $D(x,y,z)$ labelled by x, y, z, resp. Then the triangle $D(x,y,z)$ is denoted by $D(X,Y,Z)$, too.
A consequence of our observations is: if $h \in G(x,y)$ then the inner automorphism f_h of G is defined by $f_h(g) = h^{-1} g h$ for all $g \in G(x,y)$. Hence this automorphism of $T(x,y)$ can be realized by a sequence of automorphisms $f_x, f_x^{-1}, f_y, f_y^{-1}$ inducing clockwise or anticlockwise rotations of the basic discs $K(x,y)$ and $K(y,x)$ over two sectors.

2. Results

First group - triangulations have been studied by W. Voss in [9] - [14] . These results of W. Voss and results of St. Ulbricht and H.-J. Voss ([3],[4]) have been summarized by H.-J. Voss in [6] (Unfortunately this paper contains a few misprints, theorem 5.47 is true for the sphere only, and theorem 5.50 is incorrect). Further investigations have been accomplished by H.-J. Voss in [7] and [8] .
Up to this time results have been obtained in the following

directions.

1°. The genus of a group

The genus of a finite group G is the maximum genus of the triangulations of the set $\{T(x,y)/\ x,\ y \in G \wedge xy \neq yx\}$.
The genus of some groups and some infinite series of groups has been determined by W. Voss in [10], [12] and [14] (see also H.-J. Voss [6]) .
It seems to be clear that the greater the genus the more complex is the group. Really, it turns out that a assertion on the genus of a group is an assertion on the lattice of its subgroups. For this problem a solution for genus 0 will be presented in this paper.

2°. The automorphism group of T(x,y), the center of G(x,y) and properties of the elements x, y

The subgroup $A^2(T)$ of the automorphism group of T(x,y) has been investigated in H.-J. Voss [7] , and will be considered in this paper.
In group theory a so called degree of non - commutativity of two elements of a group or of a whole group has not been defined til now.
Two elements $x, y \in G$ are perhaps of a small degree if $xy = yx\,h$, where $h \neq e$ is an element of the center of G .
In a forthcoming paper I shall deal with group elements having this property.
The non - commutativity properties of a pair of elements x, y are perhaps represented by the properties of the quotient group $G(x,y)/\mathfrak{Z}(x,y)$, where $\mathfrak{Z}(x,y)$ is the center of G(x,y).
It could be proved that $G(x,y)/\mathfrak{Z}(x,y)$ is isomorphic to $A^2(T(x,y))$. This quotient group has been investigated in W. Voss [13] , H.-J. Voss [8] , and will be considered in this paper.

3°. The valencies of the vertices of T(x,y)

It can easily be shown that $A^2(T)$ has at most three vertex orbits O_X, O_Y, O_Z in T(x,y) , containing the vertices X , Y , Z , respectively. Therefore, all vertices have one of the three valencies v_X, v_Y, v_Z of the vertices X, Y, Z, respectively. With Euler's formula a fundamental lemma has been proved (W. Voss in [10]):

$$\left(\frac{1}{v_X} + \frac{1}{v_Y} + \frac{1}{v_Z} - \frac{1}{2} \right) f = 2 - 2P ,$$

where f and P denote the number of triangles and the genus of T(x,y) , respectively.
The dependence of the degrees of T(x,y) on the properties of non - commutative elements $x, y \in G$ has been investigated in [6] .

4°. The conjugacy classes of G(x,y) and the group-triangulation T(x,y). The multiplicity of group elements on T(x,y)

Elements of G being labels of some vertices of the same vertex - orbit are conjugates in G(x,y) . Moreover, each conjugate of x, y, z is a label of at least one vertex of O_X , O_Y , O_Z , respectively (W. Voss in [10]).
Let $m(O_X:g)$ denote the number of vertices of the vertex-orbit O_X labelled by g . Then

$$m(O_X:g) = m(O_X:x) = \frac{q}{v_X} \frac{|N(x)|}{|\mathfrak{z}(x,y)|} ,$$

where N(x) is the normalizer of x in G(x,y) , and q is the number of face - orbits of the group $A_X^2(T)$ of all automorphisms of $A^2(T)$ fixing X (W. Voss in [13] and H.-J. Voss in [7]).

5°. The inversion of the assignment group → triangulation of a closed oriented surface

A class of simple triangulations of closed oriented surfaces has been characterized which contain all group triangulations. This has been used to find new group-triangulations. Thus all group triangulations of genus 0 have been found (St. Ulbricht and H.-J. Voss [3] and [4] , H.-J. Voss [6], and this paper).

6°. The direct product of groups

In order to investigate direct products of groups, it has been introduced a certain "product" of two triangulations with respect to a fixed edge in each of them (St. Ulbricht [5]) .

7°. Quotient groups and quotient triangulations

In [6] and [8] I have introduced a quotient triangulation T' of a triangulation T having the following property: If G is a finite group, N a normal subgroup of G and (x,y,z) a regular triple of the group G such that (xN,yN,zN) is a regular triple of the quotient group $G' = G/N$ then $T'(xN,yN)$ is a quotient triangulation of $T(x,y)$.

Here in this paper I shall contribute to 1° and 2°.

In [13] W. Voss has proved that the subgroup $A^2(T)$ of the automorphism group of $T(x,y)$ is isomorphic to $G(x,y)/\mathfrak{Z}(x,y)$.

In section 3 we shall construct to each triangulation $T(x,y)$ a 2 - cell decomposition $L(x,y)$ of $S(x,y)$ so that up to isomorphism $A^2(T)$ is a subgroup of the automorphism group of $L(x,y)$. In $L(x,y)$ we shall define a fundamental region F of $A^2(T)$, i. e., a face F of $L(x,y)$ such that

(i) each automorphism of $A^2(T)$ maps F onto a face of $L(x,y)$, and

(ii) to each face R of $L(x,y)$ there is precisely one automorphism $r \in A^2(T)$, which maps F onto R.

From $L(x,y)$ an imbedding of a Cayley - colour - graph of $A^2(T)$ in $S(x,y)$ can be derived.

Since $A^2(T) \cong G(x,y)/\mathfrak{Z}(x,y)$ this is an imbedding $\mathcal{L}(x,y)$ of a Cayley - colour - graph of $G(x,y)/\mathfrak{Z}(x,y)$ in $S(x,y)$, too.

A plane group is a group of genus 0.

In section 4 all plane groups will be characterized: a group G is a plane group iff all subgroups $G(x,y)$ generated by two elements $x, y, xy \neq yx$, have the property: $G(x,y)/\mathfrak{Z}(x,y)$ is a dihedral group, or is isomorphic to the symmetry group A_4 of the regular tetrahedron, or to the symmetry group S_4 of the regular octahedron, or to the symmetry group A_5 of the regular icosahedron.

3. Automorphism groups of group - triangulations

Here the investigations of [7] will be continued. In [7] I have shown that in the class of all group - triangulations T there are four different types of automorphism groups $A^2(T)$ depending on the three vertex orbits O_X, O_Y, O_Z of $A^2(T)$. We consider the Cases I - IV.

<u>Case I.</u> $O_X \neq O_Y \neq O_Z \neq O_X$.

This condition holds iff the valencies v_X, v_Y, v_Z of T are even integers and T has a suitable 3 - colouring of its vertices ([7]) . Consequently, T has two distinct face orbits. An example is the octahedron - triangulation T((12)(34),(1234)) of Fig. 1.3 .

In Case I to each O_X, O_Y - edge [U,V] there is precisely one automorphism $\sigma_{U,V}$ mapping [X,Y] onto [U,V] . This automorphism maps $D(X,Y,Z) \cup D(Y,X,Z')$ onto $D(U,V,W) \cup D(V,U,W')$. We delete in T all O_X, O_Y - edges and obtain a quadrangulation Q(X,Y) . Let F(X,Y) denote the quadrangle of Q(X,Y) obtained by deleting [X,Y] from $D(X,Y,Z) \cup D(Y,X,Z')$.

<u>Theorem 3.1.</u> The quadrangle F(X,Y) is the fundamental region of $A^2(T)$, i. e.,

(i) each automorphism of $A^2(T)$ maps F(X,Y) onto an quadrangle of Q(X,Y) , and

(ii) to each quadrangle R of Q(X,Y) there is precisely one automorphism $r \in A^2(T)$, which maps F(X,Y) onto R.

We remark Theorem 3.1 is also true if the elements of the regular triple (x,y,z) are commutative. Then the quadrangulation Q(X,Y) has precisely one quadrangle, namely F(X,Y). We assign to each quadrangle R the automorphism r (unique by (ii)) . Obviously, to F(X,Y) the identity e is assigned. W. Voss has shown in [13] (see also [7]) that

$$A^2(T(x,y)) \cong G(x,y)/\mathfrak{Z}(x,y) .$$

Hence we can replace each automorphism of $A^2(T)$ by an element of $G(x,y)/\mathfrak{Z}(x,y)$. From sections 1 and 2.2° it follows: the quadrangle R is labelled by the coset $p\mathfrak{Z}$ iff the automorphism of $A^2(T)$ mapping F(X,Y) onto R is induced by the inner automorphism $f_{\bar{p}}$, $\bar{p} \in p\mathfrak{Z}$, where by definition $f_{\bar{p}}(g) = (\bar{p})^{-1} g \bar{p}$ for all $g \in G(x,y)$ (note: $f_{p_1} = f_{p_2}$ iff p_1 and p_2 are in the same coset of $\mathfrak{Z}$). We shall denote this automorphism of Q(X,Y) and the corresponding inner automorphism of G(x,y) by the same symbol $f_{\bar{p}}$.

Next we form the dual 2 - cell decomposition $\mathcal{L}(x,y)$ of Q(X,Y): To each face L of Q(X,Y) a vertex V of $\mathcal{L}(x,y)$ is assigned, where two vertices of $\mathcal{L}(x,y)$ are joined by an edge

iff the corresponding faces have a common edge. The vertex V is labelled by the label of L , i. e. , by the coset of $\mathfrak{z}$ assigned to L .
We remark that each edge of $\mathcal{L}(x,y)$ has the valency 4 .
The vertex - sets O_X, O_Y, O_Z of $Q(X,Y)$ are the face - sets of $\mathcal{L}(x,y)$. Let E_X and E_Y denote the sets of all edges of $\mathcal{L}(x,y)$ being in the bounding cycles of a face of O_X or O_Y , respectively. By $\{E_X, E_Y\}$ a partition of the edge - set E of $\mathcal{L}(x,y)$ is obtained, i. e., $E_X \cup E_Y = E$ and $E_X \cap E_Y = \emptyset$ (see Fig. 3.2).
By clockwise orienting the edges of the bounding cycles of the faces of O_X and O_Y we obtain a directed quadrangulation. To each edge of E_X and E_Y the elements $x\,\mathfrak{z}(x,y)$ and $y\,\mathfrak{z}(x,y)$ are assigned, respectively. This directed 2 - cell decomposition of $S(x,y)$ with this assignment of group elements of $G(x,y)/\mathfrak{z}(x,y)$ is denoted by $\mathcal{L}(x\mathfrak{z}, y\mathfrak{z})$.
It turns out that $\mathcal{L}(x\mathfrak{z}, y\mathfrak{z})$ is a so called Cayley - colour - graph imbedded in $S(x,y)$.
Next we define this concept for finite groups.
Let G be a finite group with generators $h_1, h_2, \ldots, h_s$.

For G with these generators there is associated a <u>Cayley - colour - graph</u> ; the vertices correspond to the elements of the group; next imagine the generators of the group to be associated with different colours.

(*) If vertices V_1 and V_2 correspond to group elements g_1 and g_2 , respectively, then there is an arc (V_1, V_2) coloured by generator h iff $h\,g_1 = g_2$ (see Fig. 3.1) .

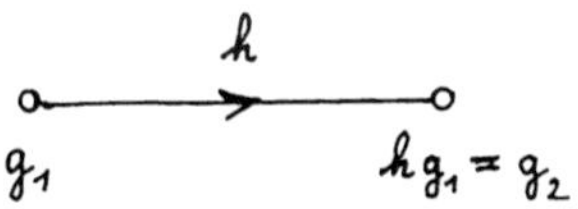

Fig. 3.1

Our definition slightly differs from that of A.T. White [15] : in Fig. 3.1 he requires $g_1 h = g_2$ instead of $h\,g_1 = g_2$.

<u>Theorem 3.2</u>. The 2 - cell decomposition $\mathcal{L}(x\mathfrak{z},y\mathfrak{z})$ with the orientation and the assignment as described above is the Cayley - colour - graph of $G(x,y)/\mathfrak{z}(x,y)$ with the generators $x\mathfrak{z}(x,y)$ and $y\mathfrak{z}(x,y)$ imbedded in $S(x,y)$ (details of $Q(X,Y)$ and $\mathcal{L}(x\mathfrak{z},y\mathfrak{z})$ are depicted in Fig. 3.2, where $F(X,Y)$ is shortly denoted by F).

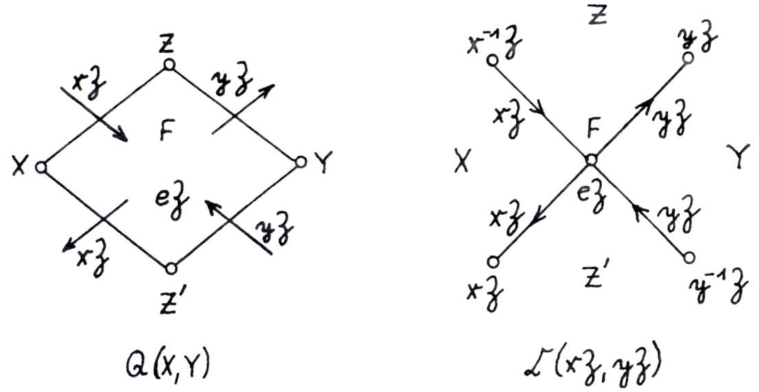

Fig. 3.2

<u>Proof</u>. Let (F,W) be the arc of $\mathcal{L}(x\mathfrak{z},y\mathfrak{z})$ such that the faces F, W of Q(X,Y) have a common O_X, O_Z - edge. Hence (F,W) is labelled by $x\mathfrak{z}$. In Q(X,Y) the automorphism induced by f_x maps F onto W , where f_x is the inner automorphism of $G(x,y)$ defined by $f_x(g) = x^{-1} g\, x$ for all $g \in G(x,y)$. By Theorem 3.1 the map f_x is the only automorphism mapping F onto W . Hence W is labelled by $x\mathfrak{z}$ and F, W, (F,W) satify condition (✱) .

Next let (F′,W′) be an arbitrary arc of $\mathcal{L}(x\mathfrak{z},y\mathfrak{z})$ such that faces F′,W′ of Q(X,Y) have a common O_X, O_Z - edge. Hence (F′,W′) is labelled by $x\mathfrak{z}$.

By Theorem 3.1 there is precisely one automorphism f_p mapping F onto F′. Hence F′ is labelled by $p\mathfrak{z}$. Obviously, $f_p(W') = W$. Then $f_x \circ f_p$ maps F onto W′ (Remember: $f_x \circ f_p(t) = f_p(f_x(t)))$. Hence the vertex W′ is labelled by $x\mathfrak{z}p\mathfrak{z}$. Thus F′,W′, (F′,W′) are labelled by $p\mathfrak{z}$, $x\mathfrak{z}p\mathfrak{z}$, $x\mathfrak{z}$, respectively. Consequently, F′,W′, (F′,W′) satisfy (✱).

Similarly, it can be shown: if (F′,V′) is an edge of $\mathcal{L}(x\mathfrak{z},y\mathfrak{z})$ such that the faces F′,V′ of Q(X,Y) have a common O_Y, O_Z - edge then F′,W′, (F′,W′) are labelled by $p\mathfrak{z}$, $y\mathfrak{z}p\mathfrak{z}$, $y\mathfrak{z}$, respectively. Consequently, F′,W′, (F′,W′)

satisfy condition (✱) .
This completes the proof of Theorem 3.2 . □

<u>Case II</u>. $O_X \neq O_Y = O_Z$.

This is true iff one of the following two conditions holds (see [7]):

(i) Precisely one of the valencies v_X, v_Y, v_Z is odd, namely v_X (an example is T((12),(23)) - Fig. 78 in [6]).

(ii) All valencies v_X, v_Y, v_Z are even, T has only one face orbit and $A^2(T)$ contains no rotation of a triangle through its centre (an example is T((1234)(5678), (15)(28)(49)(67)) - see Fig. 5 in [7]).

In Case II to each O_Y,O_Z-edge [U,V] there are precisely two automorphisms $\sigma_{U,V}$ and $\sigma_{V,U}$ of $A^2(T)$ mapping [Y,Z] onto [U,V] , where $\sigma_{U,V}$ maps Y onto U and Z onto V and $\sigma_{V,U}$ maps Y onto V and Z onto U .

<u>Theorem 3.3</u>. The triangle D(X,Y,Z) is the fundamental region of $A^2(T)$, i.e. ,

(i) each automorphism of $A^2(T)$ maps D(X,Y,Z) onto a triangle of T , and

(ii) to each triangle D of T there is precisely one automorphism $d \in A^2(T)$, which maps D(X,Y,Z) onto D . □

Again we assign to each face D′ of T(x,y) the coset of $\mathfrak{z}$ which represents the map of D(X,Y,Z) onto D′.
We remark that $A^2(T)$ contains the automorphism $\sigma_{Z,Y}$ which rotates the edge [Y,Z] through its middle, i. e. , $\sigma_{Z,Y} \in A^2(T)$ maps Y and Z onto Z and Y , respectively.
Let $s \in G(x,y)$ such that $f_s = \sigma_{Z,Y}$. Then to D(Z,Y,X′) the coset $s\mathfrak{z}$ of $\mathfrak{z}$ is assigned.
Let ψ_X denote the automorphism of $A^2(T)$ which clockwise rotates the basic disc K(X) of T(x,y) over one sector.
Let $r \in G(x,y)$ be such that $f_r = \psi_X$. Therefore, $r\mathfrak{z}$ is assigned to D(Y,X,Z′) (see Fig. 3.3).
Next we form the dual 2 - cell decomposition $\mathcal{L}(x,y)$ of T(x,y) .

If the vertex V of $\mathcal{L}(x,y)$ corresponds to the face L of T(x,y) then V is marked by the label of L, i. e. , by the coset of $\mathfrak{z}$ assigned to L .

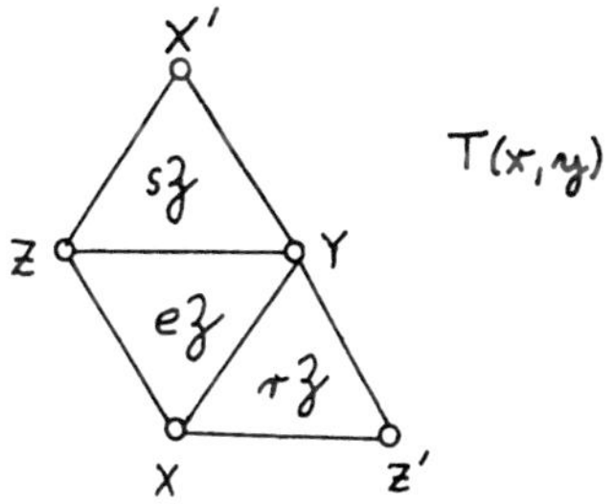

Fig. 3.3

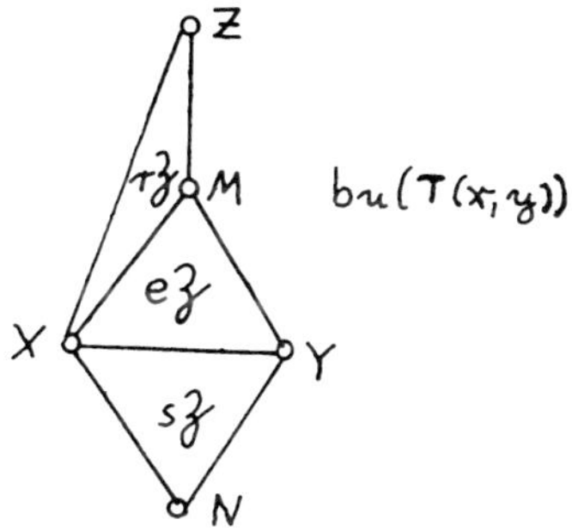

Fig. 3.7

The vertex - sets O_X, O_Y of T(x,y) are the face - sets of $\mathcal{L}(x,y)$.

Let E_X and E' denote the sets of all edges of $\mathcal{L}(x,y)$ being in a bounding cycle of a face of O_X or being in no such cycle, respectively. By $\{E_X, E'\}$ a partition of the edge set of $\mathcal{L}(x,y)$ is obtained.

Next the edges of the bounding cycles of the faces of O_X are clockwise oriented and labelled by $r\mathfrak{z}(x,y)$. Each edge k of E' is replaced by two opposite directed arcs joining the endvertices of k ; where each of the two new arcs is labelled by $s\mathfrak{z}(x,y)$.

This directed 2 - cell decomposition of S(x,y) with this assignment of group elements of $G(x,y)/\mathfrak{z}(x,y)$ is denoted by $\mathcal{L}(r\mathfrak{z},s\mathfrak{z})$.

It turns out that $\mathcal{L}(r\mathfrak{z},s\mathfrak{z})$ is a Cayley - colour - graph imbedded in S(x,y) .

<u>Theorem 3.4</u>. The 2 - cell decomposition $\mathcal{L}(r\mathfrak{z},s\mathfrak{z})$ of S(x,y) with the orientation and the assignment as described above is the Cayley - colour - graph of $G(x,y)/\mathfrak{z}(x,y)$ with generators $r\mathfrak{z}(x,y)$ and $s\mathfrak{z}(x,y)$ imbedded in S(x,y) (details of T(x,y) and $\mathcal{L}(r\mathfrak{z},s\mathfrak{z})$ are depicted in Fig. 3.4 , where D(X,Y,Z) is abbreviated by D). □

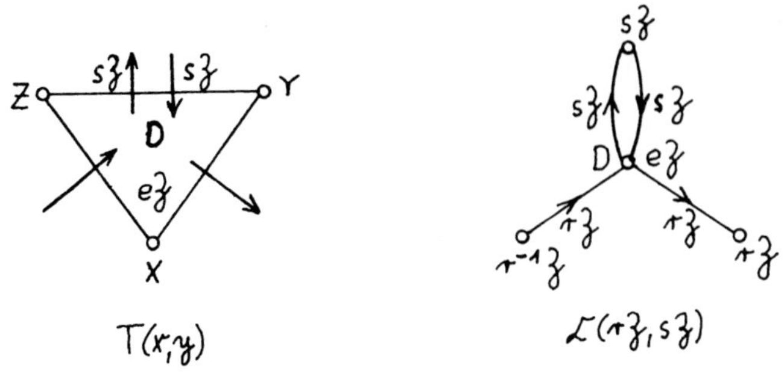

Fig. 3.4

Case III. $O_X = O_Y = O_Z$ and T has two face - orbits.

This condition holds iff $v_X = v_Y = v_Z$ is even, T has two face - orbits and no suitable 3 - colouring of its vertices ([7]) . An example is T((135)(246),(263)(457)) - see Fig. 94 in [6] .

In Case III to each edge [U,V] there is precisely one automorphism in $\sigma_{U,V}$ mapping [X,Y] onto [U,V] .

$A^2(T)$ contains the rotation of T through the centre of each triangle over two sectors.

A barycentric subdivision bu(T) is formed by introducing a new vertex in each face F and joining this new vertex with all vertices of the bounding cycle of F (the tetrahedron - triangulation T_4 and its barycentric subdivision $bu(T_4)$ is depicted in Fig. 3.5) . Obviously, $A^2(bu(T)) \supseteq A^2(T)$.

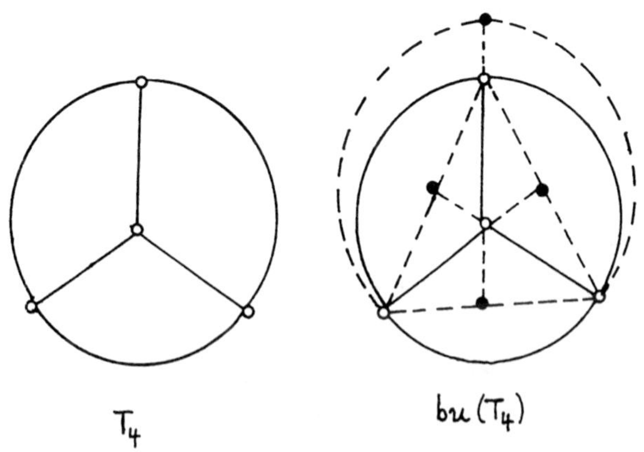

Fig. 3.5

In bu(T) the automorphism $\sigma_{U,V}$ maps $D(X,Y,M) \cup D(Y,X,N)$ onto $D(U,V,M') \cup D(V,U,N')$, where M, N, M', N' are the new vertices in the triangles in question, respectively.
We delete in bu(T) all edges of T and obtain a quadrangulation Q(X,Y). Let F(X,Y) denote the quadrangle of Q(X,Y) obtained by suppressing the edge [X,Y] in $D(X,Y,M) \cup D(Y,X,N)$.

<u>Theorem 3.5.</u> The quadrangle F(X,Y) is the fundamental region of $A^2(T)$, i. e.,

(i) each automorphism of $A^2(T)$ maps F(X,Y) onto a quadrangle of Q(X,Y), and

(ii) to each quadrangle R of Q(X,Y) there is precisely one automorphism $r \in A^2(T)$, which maps F(X,Y) onto R. □

In Q(X,Y) we assign to each quadrangle R the coset of $\mathfrak{z}$ representing the automorphism r.
Again we form the dual 2 - cell decomposition of Q(X,Y) and assign to each vertex of $\mathcal{L}(x,y)$ the corresponding coset of $\mathfrak{z}$.
Let $m, n \in G(x,y)$ such that $f_m = \psi_M$ and $f_n = \psi_N$, the rotations through M and N over one sector, respectively. Then $m\mathfrak{z}$ and $n\mathfrak{z}$ are assigned to faces neighboured to F = F(X,Y) and incident with the edges [X,M] and [Y,N] of Q(X,Y), respectively (see Fig. 3.6).
The vertex - sets O_M, O_N of Q(X,Y) are the face - sets of $\mathcal{L}(x,y)$.
Let E_M and E_N denote the set of all edges of $\mathcal{L}(x,y)$ being in the bounding cycle of a face of O_M or O_N, respectively. By $\{E_M, E_N\}$ a partition of the edge set E of $\mathcal{L}(x,y)$ is obtained.
Next the edges of the bounding cycles of the faces of O_M and O_N are clockwise oriented and labelled by $m\mathfrak{z}(x,y)$ and $n\mathfrak{z}(x,y)$, respectively. This directed 2 - cell decomposition with this assignment of group elements of $G(x,y)/\mathfrak{z}(x,y)$ is denoted by $\mathcal{L}(m\mathfrak{z}, n\mathfrak{z})$. It turns out that $\mathcal{L}(m\mathfrak{z}, n\mathfrak{z})$ is a Cayley - colour - graph imbedded in S(x,y).

<u>Theorem 3.6.</u> The 2 - cell decomposition $\mathcal{L}(m\mathfrak{z}, n\mathfrak{z})$ of S(x,y) with the orientation and the assignment as described above is the Cayley - colour - graph of $G(x,y)/\mathfrak{z}(x,y)$ with generators $m\mathfrak{z}(x,y)$ and $n\mathfrak{z}(x,y)$ imbedded in S(x,y) (details of T(x,y)

and $\mathcal{L}(m\mathfrak{z},n\mathfrak{z})$ are depicted in Fig. 3.6 , where $F(X,Y)$ is abbreviated by F).□

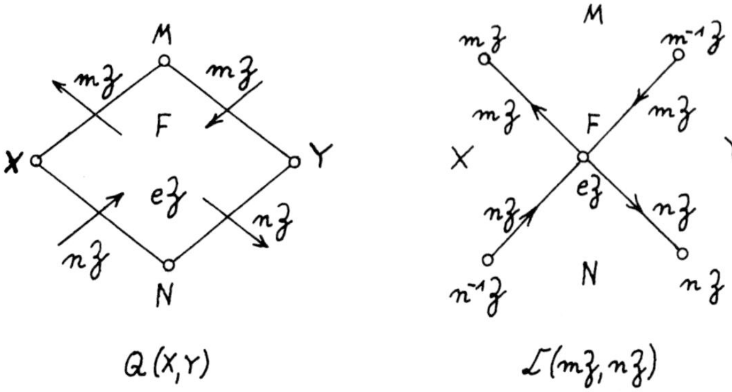

Fig. 3.6

<u>Case IV</u>. $O_X = O_Y = O_Z$ and T has precisely one face - orbit.

The triangulation is vertex - transitive and face - transitive. This is true iff one of the two conditions holds ([7]):

(i) $v_X = v_Y = v_Z$ is odd (an example is the tetrahedron - triangulation T((123),(134)) - see Fig. 79a in [6]), or

(ii) $v_X = v_Y = v_Z$ is even, T has only one face - orbit and $A^2(T)$ contains a rotation of a triangle (an example is not known til now).

In Case IV to <u>each</u> edge [U,V] there are precisely two automorphisms $\sigma_{U,V}$ and $\sigma_{V,U}$ mapping [X,Y] onto [U,V] , where $\sigma_{U,V}$ maps X onto U and Y onto V and $\sigma_{V,U}$ maps X onto V and Y onto U .

We form bu(T). Obviously, $A^2(bu(T)) \supseteq A^2(T)$.

In bu(T) the automorphism $\sigma_{U,V}$ maps D(X,Y,M) onto D(U,V,M′) and $\sigma_{V,U}$ maps D(X,Y,M) onto D(V,U,N′) , where M, M′, N′ are the new vertices in the triangles D(X,Y,Z) , D(U,V,W) , D(V,U,W′) of T , respectively.

<u>Theorem 3.7</u>. The triangle D(X,Y,M) is a fundamental region of $A^2(T)$, i. e. ,

(i) each automorphism of $A^2(T)$ maps D(X,Y,M) onto a triangle of bu(T) , and

(ii) to each triangle D of bu(T) there is precisely one automorphism $d \in A^2(T)$, which maps D(X,Y,M) onto D.

Again we assign to each face D′ of bu(T(x,y)) the coset of $\mathfrak{z}$ which represents the map D(X,Y,M) onto D′.

We remark that $A^2(T)$ contains the automorphism $\sigma_{Y,X}$ which rotates the edge [X,Y] through its middle, i. e. , $\sigma_{Y,X} \in A^2(T)$ maps X and Y onto Y and X , respectively. Let $s \in G(x,y)$ such that $f_s = \sigma_{Y,X}$. Then to D(Y,X,N) the coset $s\mathfrak{z}$ of $\mathfrak{z}$ is assigned.

Let ψ_M denote the automorphism of $A^2(T)$ which clockwise rotates the basic disc K(M) of bu(T(x,y)) over one sector. Let $m \in G(x,y)$ such that $f_m = \psi_M$. Therefore, $m\mathfrak{z}$ is assigned to D(Z,X,M) (see Fig. 3.7)

In bu(T(x,y)) we assign to each triangle D the coset of $\mathfrak{z}$ representing the unique automorphism $d \in A^2(T)$ which maps D(X,Y,M) onto D .

Again we form the dual 2 - cell decomposition of bu(T(x,y)) and assign to each vertex of $\mathcal{L}(x,y)$ the corresponding coset of $\mathfrak{z}$.

The vertex - set O_M of bu(T(x,y)) is the face - set of $\mathcal{L}(x,y)$ (and of T(x,y) , too) .

Let E_M and E′ denote the sets of all edges of $\mathcal{L}(x,y)$ being in a bounding cycle of a face of O_M or being in no such cycle, respectively. By $\{E_M, E'\}$ a partition of the edge set of $\mathcal{L}(x,y)$ is obtained.

Next the edges of the bounding cycles of the faces of O_M are clockwise oriented and labelled by $m\mathfrak{z}(x,y)$. Each edge k of E′ is replaced by two opposite directed arcs joining the endvertices of k ; where each of the two new arcs is labelled by $s\mathfrak{z}(x,y)$.

This directed 2 - cell decomposition of S(x,y) with this assignment of group elements of $G(x,y)/\mathfrak{z}(x,y)$ is denoted by $\mathcal{L}(m\mathfrak{z},s\mathfrak{z})$. It turns out that $\mathcal{L}(m\mathfrak{z},s\mathfrak{z})$ is a Cayley - colour - graph imbedded in S(x,y) .

<u>Theorem 3.8</u>. The 2 - cell decomposition $\mathcal{L}(m\mathfrak{z},s\mathfrak{z})$ of S(x,y) with the orientation and the assignment as described above is the Cayley - colour - graph of $G(x,y)/\mathfrak{z}(x,y)$ with generators

$m\mathfrak{z}(x,y)$ and $s\mathfrak{z}(x,y)$ imbedded in $S(x,y)$ (details of $T(x,y)$ and $\mathcal{L}(m\mathfrak{z},s\mathfrak{z})$ are depicted in Fig. 3.8 , where $D(X,Y,M)$ is abbreviated by D). □

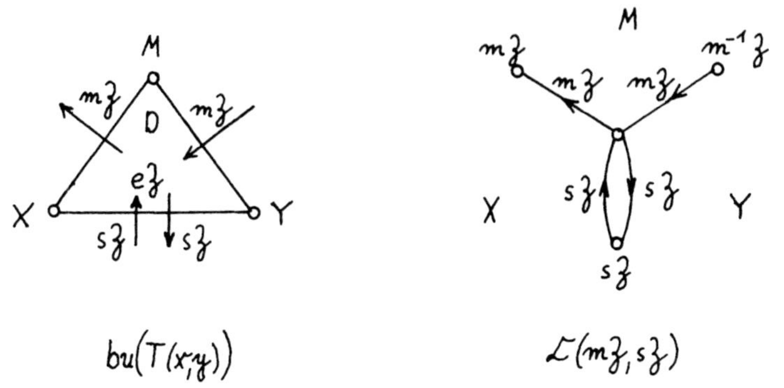

Fig. 3.8

4. The genus of group triangulations

The genus of a finite group G is the maximum genus of the triangulations of the set $\{T(x,y) / x,y \in G \wedge xy \neq yx\}$.
The genus of some groups and some infinite series of groups has been determined by W. Voss in [11] , [12] and [14] (see also [6]) .
The groups S_n , A_n , Z_n , D_n are respectively the symmetric and the alternating groups of degree n , the cyclic group of order n , and the dihedral group of order 2n .

Theorem 4.1. Let G be a finite non - commutative group. If the genus of G is 0 then for each pair of non - commutative elements $x, y \in G$ the subgroup $G(x,y)/\mathfrak{z}(x,y)$ is a dihedral group or is isomorphic to one of the three groups A_4 , or S_4 , or A_5 .

Sketch of a proof. In Theorem 5.64 of [6] I presented group-triangulations of the sphere. We could recently show that these triangulations are the only group triangulations on the sphere. Hence the proof can be accomplished by determining the subgroup $A^2(T)$ of the automorphism group of each possible group triangulation of the sphere. □

A graph is said to be planar if it can be imbedded in the plane (or, equivalently, in the sphere) . A graph imbedded in the plane (or sphere) is called a plane graph.
Let G be a finite group and $x, y \in G$, $xy \neq yx$.
Then $\mathcal{L}[x,y]$ denotes the Cayley - colour - graph imbedded in S(x,y) obtained from T(x,y) as described above in the Cases I - IV .
The imbeddings Γ_1, Γ_2 of two Cayley - colour - graphs in an oriented surface are called isomorphic if there exists an isomorphism of the underlying 2 - cell decompositions of Γ_1 and Γ_2 (preserving the orientation of the edges) such that two edges of Γ_1 have the same colour iff their images in Γ_2 have the same colour,too .
Let H be a Cayley - colour - graph on two generators of some finite group. Let Γ' denote the imbedding of H in an oriented surface of minimum genus.
We ask: Can H be chosen so that for all finite groups G and all elements $x, y \in G$, $xy \neq yx$, the Cayley - colour - graph $\mathcal{L}[x,y] \not\cong \Gamma'$?
The answer is: Yes ?
An example is $Z_2 \times Z_n$, $n \geq 5$. The corresponding plane Cayley - colour-graph Γ is depicted in Fig. 4.1 (Let a, b be the generating elements of Z_2 and Z_n , respectively. Then each thick line represents two opposite directed arcs labelled by (a,e) and each thin line represents an arc coloured by (e,b)).

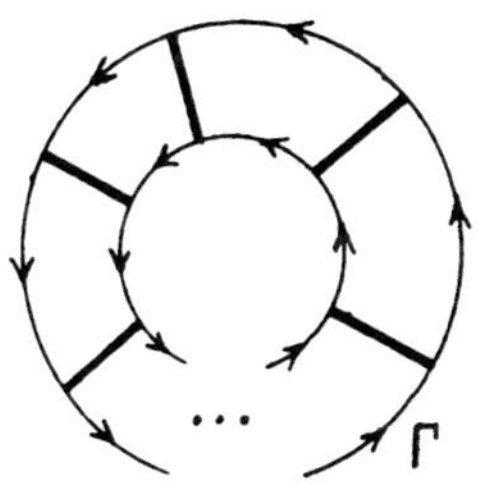

Fig. 4.1

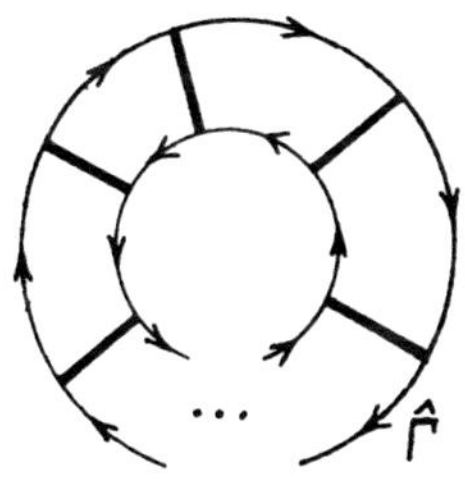

Fig. 4.2

The (undirected) dual 2 - cell decomposition of Γ is a triangulation T of the sphere, which is topologically equivalent to the n - gonal dipyramid.

If we take into consideration that the thick lines of Fig. 4.1 represent two arcs then we arrive at a quadrangulation Q of the sphere instead of T , where Q is obtained from T by subdividing each edge which joins endvertices of valency 4 by precisely one vertex of valency 2 .

The dipyramid has only one group A' of sense preserving automorphisms with a face as a fundamental region. This is the symmetry group D_n of the semiregular dipyramid. Hence if we form the Cayley - colour - graph from T or Q (as described in the Cases I - IV) then we arrive at a Cayley - colour - graph of $A' \cong D_n \not\cong Z_2 \times Z_n$.

Consequently, $\Gamma \not\cong \mathcal{L}[x,y]$ for all finite groups G and all elements $x, y \in G$, $xy \neq yx$.

We remark that a triangulation or quadrangulation isomorphic to T or Q , respectively, is obtained from the Cayley - colour - graph $\hat{\Gamma}$ of D_n depicted in Fig. 4.2 (The generating elements of D_n are the permutations $a = (1\ n)(2\ (n-1))(3\ (n-2) \ldots$ and $b = (123 \ldots n)$. Then each thick line represents two opposite directed arcs labelled by a and each thin line represents an arc coloured by b) . Really, $\hat{\Gamma} \cong \mathcal{L}[a,b]$.

All finite groups on two generators having a planar Cayley - colour - graph are well known (see A.T. White [15]). These are the dihedral groups D_n and the groups $Z_2 \times Z_n$, S_4, A_4, A_5 (the last three groups are the symmetry groups of regular polyhedra), and $Z_2 \times A_4$. It can be proved that $Z_2 \times Z_n$, $n \geqslant 3$ (see our remark above) and $Z_2 \times A_4$ are not isomorphic to $A^2(T)$ for all group - triangulations T of the sphere. Thus a new proof of Theorem 4.1 is obtained.

We conclude this paper with two open problems.

1.) Characterize all finite non-commutative groups of given genus g , $g \geqslant 1$.

2.) We ask: is the genus of $\mathcal{L}[x,y]$ equal to the genus of $S(x,y)$ for <u>all</u> finite groups G and <u>all</u> elements $x, y \in G$, $xy \neq yx$?

References

[1] H. Sachs, Problem 26 in: Theory of Graphs and its Applications. Proc. Symposium held at Smolenice in June 1963, Publ. House Czechoslovak Acad. Sci.; 162 - 163

[2] H. Sachs, Graphs and surfaces assigned to finite groups. Colloques internationaux C. N. R. S. 260 - Problèmes Combinatoires et Thérie des Graphes (1978); 371 - 375

[3] St. Ulbricht and H.-J. Voss, Groups and graphs. Tagungsberichte des 3. Koll. Geometrie u. Kombinatorik, Karl-Marx-Stadt, Sept. 9 - 11, 1987, in: Wiss. Tagungen TU Karl-Marx-Stadt 3 (1988); 65 - 66

[4] St. Ulbricht and H.-J. Voss, Gruppen und Graphen (engl.). 33. Internat. Wiss. Koll. TH Ilmenau 1988, Vortragsreihe C1/C2; 265 - 267

[5] St. Ulbricht, Direkte Produkte von Gruppen und Triangulationen. Wiss. Z. PH Dresden, Math.-Nat. Reihe 23(1989)

[6] H.-J. Voss, Beschreibung von Gruppen durch Triangulationen orientierbarer Flächen. Kapitel 5 in "Graphentheorie Band 1: Anwendungen auf Topologie, Gruppentheorie und Verbandstheorie" von K. Wagner and R. Bodendiek. BI - Wiss. Verlag Mannheim, Wien, Zürich, 1989; 92 - 156

[7] H.-J. Voss, Symmetries of groups and of triangulations of oriented surfaces. In: "Category Theory and Applications" (Potsdam, 1988, edited by K. Denecke and H.-J. Vogel), Potsdamer Forschungen, Schriftenreihe PH Potsdam, Nat. Reihe B62 (1989); 130 - 141

[8] H.-J. Voss, Groups and triangulations of oriented surfaces. In: Rostock. Math. Koll., Proc. 7 - th Fischland - Coll. Discrete Math. and Appl., Sept. 12 - 17, 1988, at Wustrow, GDR

[9] W. Voss, Über eine Möglichkeit, enlichen Gruppen orientierbare Flächen und diese triangulierende Graphen zuzuordnen. Dissertation, TH Ilmenau, 1977

[10] W. Voss, Gruppen und Graphen I. Wiss. Z. TH Ilmenau 24 (1978) 3; 63 - 78

[11] W. Voss, Gruppen und Graphen II. Wiss. Z. TH Ilmenau 24 (1978); 47 - 61

[12] W. Voss, Groups and graphs. In:"Algebraische Modelle, Kategorien und Gruppoide"(edited by H.-J. Hoehnke, 1979) Akad. - Verlag Berlin, 1979; 91 - 98

[13] W. Voss, Gruppen und Graphen. Tagungsberichte des 2. Koll. Geometrie u. Kombinatorik, Karl-Marx-Stadt, Oct. 13 - 14 , 1983; 207 - 210

[14] W. Voss, Gruppen und Graphen III. Wiss. Z. TH Ilmenau 30 (1984) 3; 61 - 69

[15] A. T. White, Graphs, groups and surfaces (book). North-Holland, Math. Studies 8. Amsterdam - London: North-Holland Publ. Company, New York (1973)

Experimental Mathematics - Tesselations of Convex Polygons in a Hexagonal Lattice

G. Walther, Kiel, FRG

1 Introduction

The following considerations can be seen in the context of a world wide demand put forward by mathematics educators to give more room in mathematics education to problem solving.
Teachers and teacher educators know very well that there are various obstacles, produced partly by the educational system itself, to acomplish this goal in the every-day mathematics lesson. Experience, however, has shown that almost in each mathematics class there is a couple of - not necessarily gifted - students who want to leave the more consumptive attitude in every-day lessons and want to become more deeply involved in doing mathematics. Doing mathematics does not mean for these students to work through additional routine problems which can be solved by applying in a more or less direct way the just learnt procedures.
Rather, those students like to work with challenging mathematical problems or situations where no previously prepared algorithm for the solution of the problem is available. Of course, problems of this kind must be "solvable" in the sence that the concepts and methods necessary to attack the problem are within the students' range of mathematical knowledge and metaknowledge.
About 5 years ago we started in some of the Gymnasien in Schleswig-Holstein the MATHEMA project with the aim to give mathematically interested students in the age range between 13 and 16 years the opportunity to do mathematics in the sense of problem solving outside the every-day mathematics lesson.
The students meet at least once in a month in a working group within their school to discuss their results and open questions of mathematical problems they had been given three of four weeks before. To do this work outside the normal mathematics lesson has the advantage to avoid constraints which make it difficult to do it at all. To do this work within the Gymnasium has among other things the advantage that the group can ask the supervising

R. Bodendiek, R. Henn (Eds.)
Topics in Combinatorics and Graph Theory

mathematics teacher's advice.
So in this open setting conditions are established that the indispensable individual phase in struggling with the mathematical problem is supplied by **cooperative** work within the group. None of the young problem solvers is left alone. This is a totally different situation compared with problem solving in the context of mathematical contests. More informations about MATHEMA can be found in Walther (1988a).

2 Doing mathematics - two extreme views

When students are practicing e. g. the addition of fractions they are certainly convinced - and perhaps their teacher, too - that they are doing mathematics. Many students even like it to work through series of such tasks, carefully arranged by the teacher or the textbook author according to increasing difficulty or other didactical principles. The underlying tacit contract between teacher, textbook author and students protect the latter (and the former) against "inconvenient" surprises. Other students, however, find this kind of work at least from a certain moment on boring but would like to do a task e. g. as "Find different unit fractions whose sum is 1".
In the first case the students have at hand a certain previously taught algorithm and a set of problems which can be immediately solved by a more or less direct application of the procedure in question. The main educational goal is to get routine in adding fractions. When the teacher is satisfied with the students' achievement he goes on to the next chapter.
In the second case it is presupposed that the students already know how to add fractions, but there is no prefabricated algorithm to solve the task. Rather, the students must experiment with unit fractions, observe patterns, make guesses, generalize, develop their own strategies for solution, generate their own subproblems etc.

For the sake of clearness the first scenario had probably been exaggerated. Nevertheless there is a distinct difference in the two scenarions between the quality of mathematical activity involved. Compared to the first task of adding fractions according to a specified algorithm the second task stimulates a much broader spectrum of mathematical, heuristic and creative activities. To put it briefly: doing mathematics in the latter sense is a complex system in which various levels of activity are interwoven. Such a dynamical conception of doing mathematics is no special pedagogical invention, rather it is a view which reflects parts of the methodology used within the mathematical science. Prominent mathematicians like Euler, Hermite, Poincaré Choquet and especially Polya, who not only have done mathematics but also have reported about their mathematical activity strongly support this view.

3 Mathematical problems in MATHEMA

Bearing in mind the background sketched in the preceding sections, conclusions can be drawn for the type of mathematical problems which might be appropriate for the MATHEMA working groups.

1. The problems should cause curiosity, they should be challenging and motivating.
2. The problems should stimulate and allow a broad spectrum of various mathematical-heuristic activities.
3. The problems should be productive and elastic, that means it should be not too hard for the students to get at least partial solutions and to develop - starting from the problems given - "new" problems related to the former. In this way starting from a problem germ, a field of related problems could be generated leading perhaps finally to some kind of mathematical micro-theory.
4. In order to avoid isolated puzzles, the problems should be - at least in an inituitive sense - mathematically substantial.
5. Group work in MATHEMA should be neither coaching for mathematically weak students nor an anticipation of subject matter of the normal mathematics course.

Experience has shown that these conditions are a good quideline for choosing suitable tasks but on the other hand it is not at all easy to take each of them into account with the same importance.
In the next section I'll give a more elaborate example of a sequence of problems appropriate for MATHEMA.

4 t-Tesselations of convex polygons in a hexagonal lattice

A couple of years ago I saw in a conference room strange looking tables. The table - tops were (congruent) trapezoids with the longest side twice the length e of the short sides, which were of equal length. These small tables could be grouped in order to make up larger tables of various shapes.

It is obvious, for example, that a regular hexagonal table could be composed using two small tables.
Furthermore, "long" tables shaped like a "stretched" parallelogram or a trapezoid could be constructed.
Observations and first considerations of this kind finally inspired a series of mathematical problems for MATHEMA. In Walther (1988) I discussed a first series of such problems, which in the meantime were to some extent already used as MATHEMA tasks. Only for a short time in the course of the problem introduction the "language of tables" is used. To avoid some kind of speudo-applications the situational context is changed as soon as possible.

Instead of designing and calculating tables from special trapezoidal tables with smallest side length e, certain lattice polygons in a hexagonal lattice with fundamental length e are studied.

Let H be a hexagonal lattice (grid) with fundamental length e.

Definition 1. A line segment s in K is an H-line segment iff the end points of s are lattice points and the length of s is a multiple of e.

Definition 2. A polygon P is an H-polygon iff the vertices of P are lattice points and the sides of P are H-line segments.

Examples for H-polygons are: equilateral triangles with side length a multiple of e (if the side length equals e the triangle is called fundamental triangle). Parallelograms with opposite angles equal to 60° resp. 120°.

Definition 3. A (symmetric) H-trapezoid with three side-lengths equal to e and the other side length equal to 2 e is called fundamental H-trapezoid.

Our initial context of table designs suggests to study tesselations of simple H-polygons with fundamental H-trapezoids.

Definition 4. Let P be a simple H-polygon. If there exists a tesselation (tiling) of P with fundamental trapezoids, we say that P can be t-tesselated (t-tiled).

From now on we confine ourselves to simple H-polygons; so we oftenly omit for brevity the "H".

Proposition 1. Each regular hexagon can be t-tesselated.

Experiments and heuristical reasoning are necessary to get an idea for proof. In Walther (1988) a constructive proof for this proposition is given. In a next step of generalization we ask for other polygons which can be t-tesselated. it is obvious to try in a first step convex polygons.

Proposition 2. There are exactly the following types of convex H-polygons: equilateral triangle, parallelogram, trapezoid, pentagon, hexagon.

Proof. We use the well-known formula for the sum of the interior angels of a convex polygon P with n vertices. Let p resp. q be the number of vertices of P with interior angle of 60° resp. 120°.

From $p+q=n$ and $60 \cdot p + 120 \cdot q = (n-2) \cdot 180$ we get $p = 6 - n$ and $q = 2n - 6$, hence $3 \leq n \leq 6$.

$n = 3$: $p = 3$, $q = 0$, P is an equilateral triangle.

$n = 4$: $p = 2$, $q = 2$, P is either a parallelogram or an isosceles trapezoid.

$n = 5$: $p = 1$, $q = 4$, P is a pentagon with exactly one interior angle of 60°, the other interior angles being 120°.

$n = 6$: $p = 0$, $q = 6$, P is a hexagon.

Remarks.

1. Each H-quadrangle is convex. A non-convex quadrangle must have an interior angle of 240°. Each of the three acute interior angles must be 60°. This is a contradiction to the sum of the interior angels, which is 360°.
2. There are non convex pentagons and hexagons (Figure 1, Figure 2).

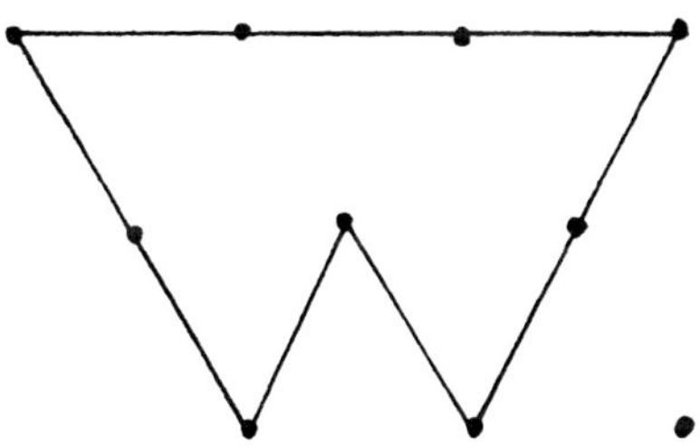

Figure 1

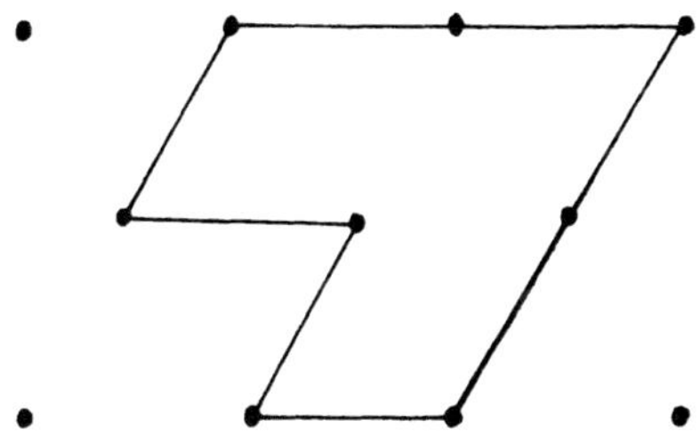

Figure 2

Since we are studying H-polygons it is obvious to measure the area F(P) of a simple H-polygon by the number of fundamental triangles in P.

For each fundamental trapezoid T, F(T) = 3; hence we have the next

Proposition 3. If P is a simple H-polygon, which can be t-tesselated, then $3|F(P)$.

The examples in Figure 1, 2 show that the converse of proposition 3 in general is not true. If we restrict ourselves to convex H-polygons, however, we have the

Proposition 4. If P is a convex H-polygon and $3|F(P)$, then P can be t-tesselated.

Combining, the last two propositions we have a characterization of convex H-polygons, which can be t-tesselated.

Proposition 5. A convex H-polygon P can be t-tesselated if $3|F(P)$.

To prove Proposition 4 we try to get informations about the side lengths of P via the condition $3|F(P)$. Then either we construct a t-tesselation for P or we dissect P into pieces each of which can be t-tesselated.

Lemma 1. Let P be a convex H-hexagon. There exists an H-parallelogram P' with sides x and y, two pairs of parallel sides of P coincide with the sides of P'. Then with the symbols of Figure 3 $F(P) = 2xy - (u^2 + v^2)$.

Proof. Each interior angle of P is 120°. Since P is an H-polygon there are three pairs of parallel sides of P.

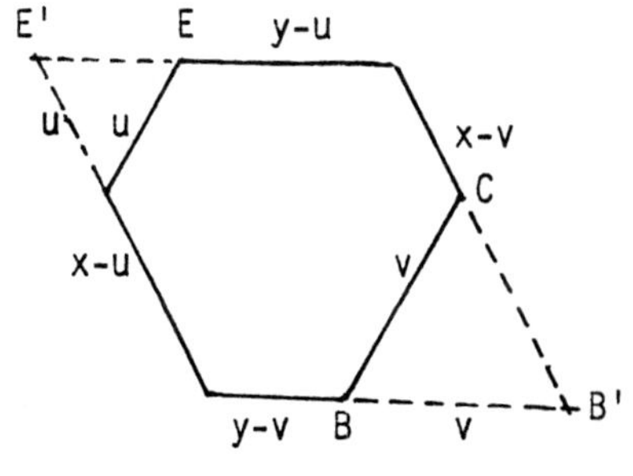

Figure 3

Without loss of generality we choose P' as in Figure 3. Triangles B B' C and E E' F are equilateral.

$F(P') = 2xy$

$F(BB'C) = v^2 \quad F(EE'F) = u^2$

Hence $F(P) = 2xy - (u^2 + v^2)$

The property of an H-polygon P to admit a t-tesselation resp. the area F(P) does not depend on the geometric position of P in the hexagonal lattice. Choosing special values for u,v,x,y in the formula of the last Lemma, we get F(P) also for the convex H-polygons different from the convex hexagon.

Lemma 2. Let P be a convex H-polygon and let us use the symbols of Figure 3.
If 1. P is a pentagon, then $F(P) = 2xy - v^2$
2. P is a trapezoid, then $F(P) = 2xy - x^2$
3. P is a parallelogram, then $F(P) = 2xy$
4. P is a triangle, then $F(P) = x^2$

Proof. For a convex H-hexagon P', $F(P') = 2xy - (u^2 + v^2)$ is true.
1. Each convex H-pentagon can be abtained from P' by setting $u = 0$.
2. Each H-trapezoid can be obtained from P' by setting $u = 0$, $v = x$.
3. Each H-parallelogram P can be obtained from P' by setting $u = 0$, $v = 0$.
4. Each H-triangle P can be obtained from P' by setting $x = y = v$, $u = 0$.

Remark. In 3. and 4. F(P) also could have been directly obtained from counting fundamental triangles in P.

The t-tesselation of convex regular hexagons P of side length x has shown that the type of t-tesselation does not depend on the factual length of x (in multiples of e) but on the concruence class mod 3 to which x belongs (Walther 1988). The same observation can be made in our generalized context. To get information about x,y,u,v mod 3 from the condition $3|F(P)$ we solve $F(P) = 0$ in Z_3.

Lemma 3. The equation $2xy - (u^2 + v^2) = 0$ which is symmetric in x,y resp. u,v has in Z_3 the following solutions:

i) $u = 0,\ v = 0,\ x = 0,\ y \in Z_3$

ii) $u = 0,\ v = 1,\ x = 1,\ y = 2$

$u = 0,\ v = 2,\ x = 1,\ y = 2$

iii) $u = 1,\ v = 1$

$u = 1,\ v = 2$ and $x = 1,\ y = 1$ resp. $x = 2,\ y = 2$

$u = 2,\ v = 2$

Proof. Obvious.

The last Lemma also gives the solutions mod 3 for the other equations in Lemma 2.
To prove our main proposition 4, we show that each of the 5 types of convex H-polygons P with $3|F(P)$ can be t-tesselated.

Lemma 4. Let P be a convex H-triangle with $3|F(P)$, then P can be t-tesselated.

Proof. Let x be the side length of P, then $F(P) = x^2$ and Lemma 3 shows that $x \equiv 0(3)$ (which, of course, can also be directly seen). If $x = 3$ then Figure 4 shows that P can be t-tesselated.
If $x = 3k$, $k > 1$, then P can be decomposed into k^2 Triangles of the preceding type.

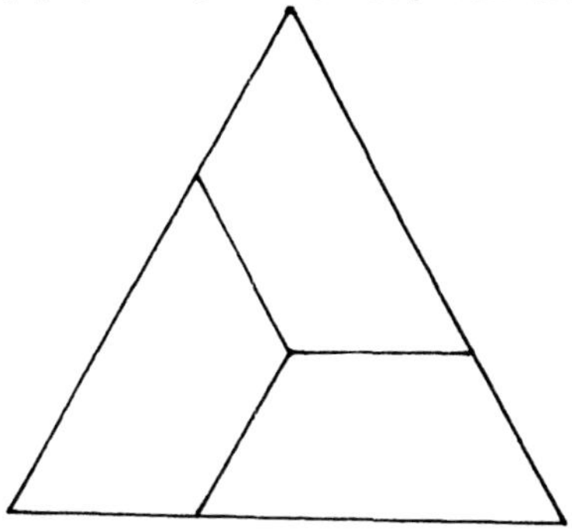

Figure 4

Lemma 5. Let P be a convex H-parallelogram with $3|F(P)$, then P can be t-tesselated.

Proof. Let x,y be the side lengths of P, then $F(P) = 2xy$.
From Lemma 3 follows that $x \equiv 0(3)$ or $y \equiv 0(3)$.
Let P' be a H-parallelogram with $x = 1$, $y \equiv 0(3)$ then P' (Figure 5) can be trivially t-tesselated.

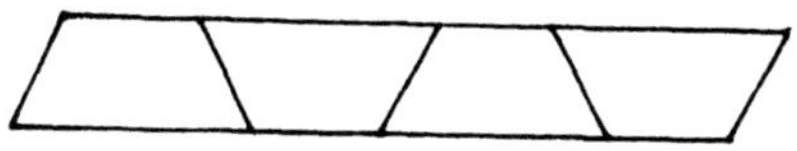

Figure 5

Each parallelogram with $x > 1$ can obviously be obtained in combining x parallelograms of the previous type.

Lemma 6. Let P be a convex H-trapezoid with $3|F(P)$ then P can be t-tesselated.

Proof. Let x,y,y-x be the side lengths of P, then from Lemma 2 $F(P) = 2xy - x^2$.

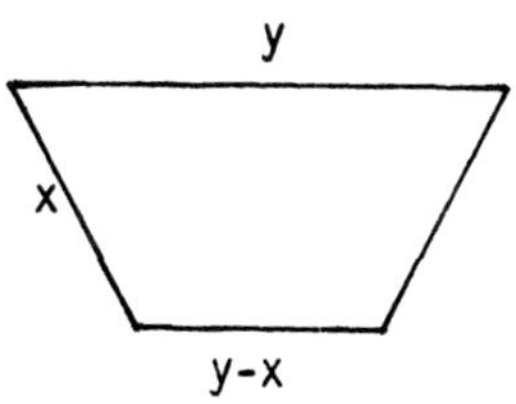

Figure 6

Lemma 3 shows that either

i) $x \equiv 0(3)$, $y \in \mathbb{N}$ or

ii) $x \not\equiv 0(3)$ and $x \equiv y-x \equiv 1$, $y \equiv 2$ or $x \equiv y-x \equiv 2$, $y \equiv 1$.

In i) P is cut up into a parallelogram and a triangle each with side length $x \equiv 0$ which can be t-tesselated according to Lemmas 4 and 5.
In ii) let $x \equiv y-x \equiv 1$, $y \equiv 2$. P can be decomposed into two trapezoids A,B. (Figure 7) The shortest side of B can be chosen to be 1, its longest side is $y \equiv 2(3)$, hence B can be obviously t-tesselated.

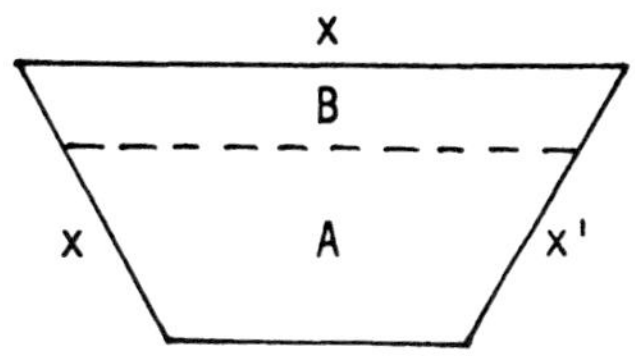

Figure 7

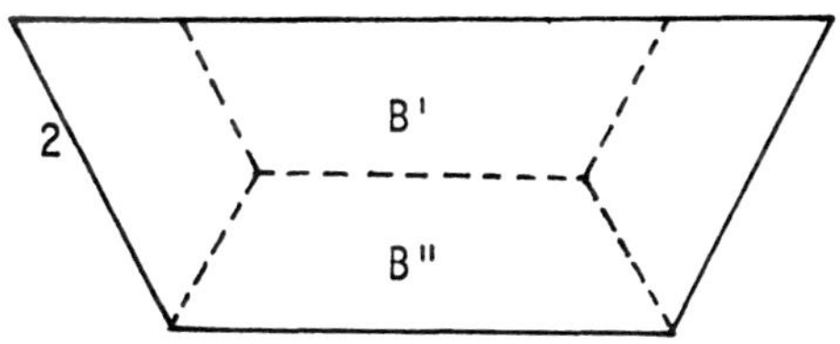

Figure 8

The side length x' of A is $\equiv 0(3)$, so A can be t-tesselated according to i).

If $x \equiv y - x \equiv 2$, $y \equiv 1$ we choose B with shortest side length 2, and longest side length $y \equiv 1(3)$.

The symmetric trapezoids B', B" with shortest side length 1 and longest side length $y' \equiv 2(3)$ can be t-tesselated according to Lemma 5. A can be t-tesselated according to i).

Lemma 7. Let P be a convex H-pentagon with $3|F(P)$, then P can be t-tesselated.

Proof. Let $x, y, v, x - v$, $y - v$ be the side lengths (Figure 9) of P suitably chosen (i.e $v < x$, $v < y$, then from Lemma 2: $F(P) = 2xy - v^2$.

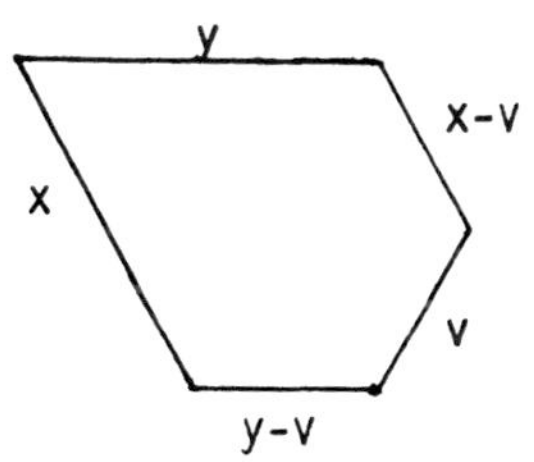

Figure 9

$3|F(P)$ yields according to Lemma 3:

i) $v \equiv 0(3)$, $x \equiv 0$, $y \in \mathbb{N}$

ii) $v \equiv 1(3)$, $x \equiv 1$, $y \equiv 2$

iii) $v \equiv 2(3)$, $x \equiv 1$, $y \equiv 2$.

If x and y are permutet and v is kept fixed we get a symmetric H-pentagon. So the problem of t-tesselation of H-pentagons can be reduced to those three cases. In each of these cases P can be dissected into a parallelogram P' one of its siden having length $\equiv 0(3)$, and a trapezoid T. P' can be t-tesselated according to Lemma 5. It is easily checked that T satisfies in i) the conditions of Lemma 6i) and in ii) those of Lemma 6ii), hence P can be t-tesselated.

Figure 10 shows case ii), the numbers next to the sides of the pentagon standing for the congruence class to which the respective side belongs. Numbers in parentheses are derived from the numbers given.

Lemma 8. Let P be a convex H-hexagon with $3|F(P)$, then P can be t-tesselated.

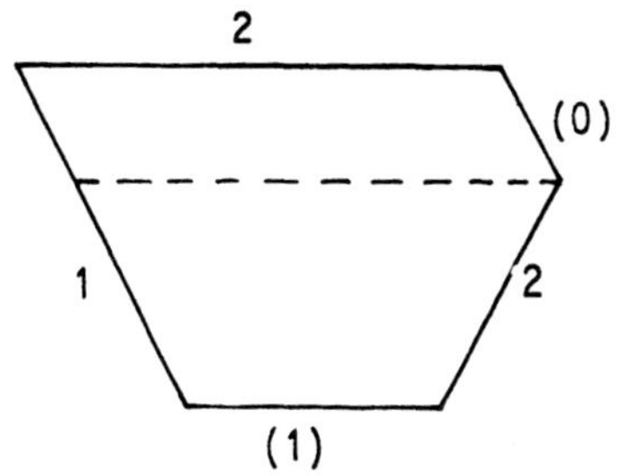

Figure 10

Proof (Sketch). Let $x-u$, $y-v$, v, $x-v$, $y-u$, u be the side lengths (Figure 11) of P suitably chosen (i.e $v < x$, $v < y$, $u < x$, $u < y$), then from Lemma 2: $F(P) = 2xy-(u^2+v^2)$.

According to Lemma 3 we have three main cases i), ii) and iii), the second resp. third of which splits up into two, resp. six subcases.

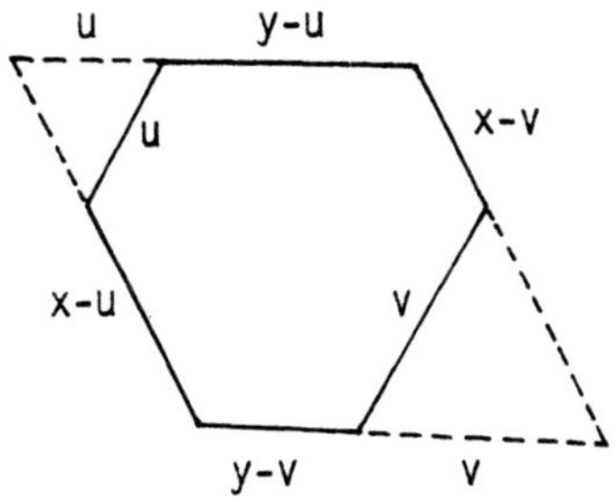

Figure 11

By means of an H-line h, which starts in a suitable vertex A of P, P can be dissected into a H-trapezoid T and a H-pentagon P', each of which can be t-tesselated. For P' the interior angle of 60° turns up at the vertex A.

It may happen, however, that P' degenerates to a trapezoid (if h meets another vertex of P). In Lemma 7 we have seen that the question whether a H-pentagon P' can be t-tesselated only depends on the congruence class to which the sides x', y' including the 60°-angle and the "non-parallel" side v' of P' belong. If P' in fact degenerates to a trapezoid P', then for example we have necessarily $x' = v'$ (resp. $y' = v'$). Hence conditions i) - iii) of Lemma 7 change into conditions i), ii) of Lemma 6 and therefore P' as a trapezoid can be t-tesselated. So, without loss of generality we can assume that P is cut up into a suitable H-trapezoid and a H-pentagon.

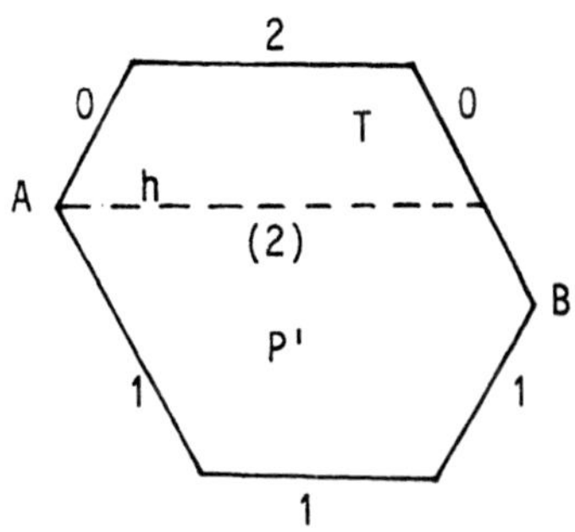

Figure 12

Let's come back to our nine cases, for each of which an appropriate dissection must be specified.

If we consider, for example, in ii) Lemma 3) $u \equiv 0$, $v \equiv 1$, $x \equiv 1$, $y \equiv 2$ then T can be t-tesselated according to Lemma 2i), and P' according to Lemma 7ii).

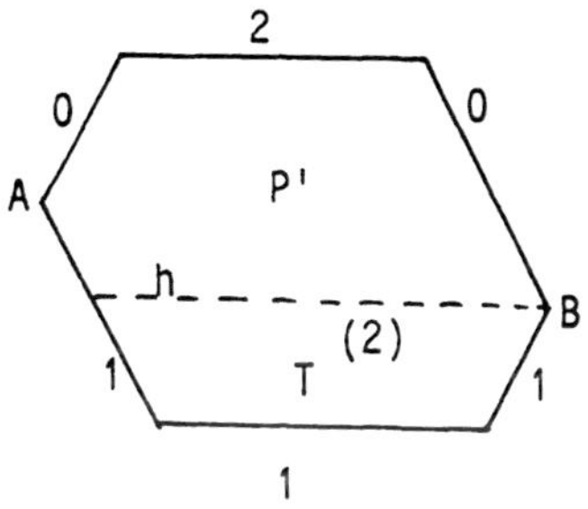

Figure 13

Since Figure 12 is only a diagram which does not reflect the actual side lengths of P, rather indicates **the congruence** classes to which its side lengths belong, it could further happen that T is "really" a H-pentagon.
This case can be taken into account if we consider also a parallel dissection line through vertex B (Figure 13).

Now, T can be t-tesselated according to Lemma 2ii) and P' according to Lemma 7i).
This procedure assures that we only use congruence classes of sides mod 3 and no actual side lengths of P.
The same produre can be used in the remaining 8 cases. For brevity the diagrams and proofs only for case i), the remaining case of ii), and one of the 6 cases iii) are given.

i) $u \equiv 0,\ v \equiv 0,\ x \equiv 0,\ y$ arbitrary

T : Lemma 2i)
P': Lemma 7i)

Figure 14

T : Lemma 2i)
P': Lemma 7i)

ii) $u \equiv 0,\ v \equiv 1,\ x \equiv 1,\ y \equiv 2$

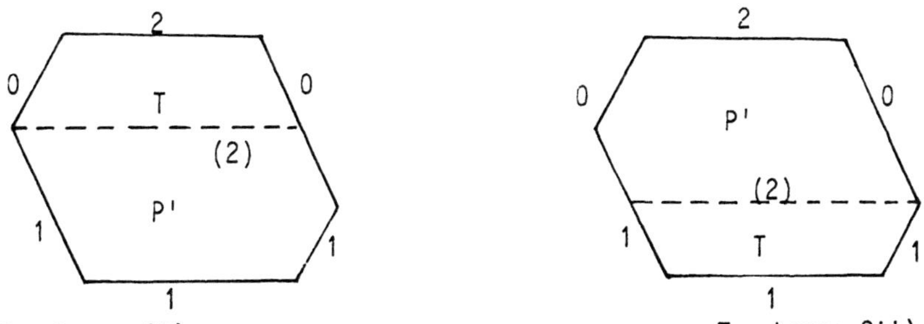

T : Lemma 2i)
P': Lemma 7ii)

Figure 15

T : Lemma 2ii)
P': Lemma 7i)

iii) $u \equiv 1,\ v \equiv 2,\ x \equiv 2,\ y \equiv 2$

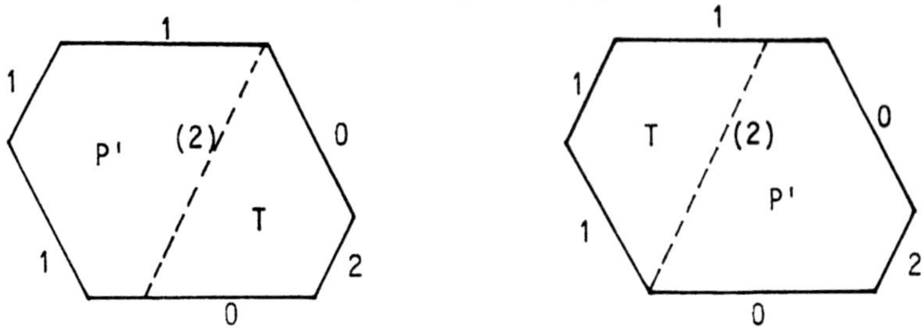

T : Lemma 2i)
P': Lemma 7ii)

Figure 16

T : Lemma 2ii)
P': Lemma 7i)

5 Summary

From the mathematician's point of view the proof of the main proposition 4 is perhaps not satisfactory, because so many cases had to be treated. Some kind of "unifying" argument is missing in order to reduce the number of cases. This, however, is no peculiar situation in discrete mathematics. From our educational point of view outlined in the first sections it is just this aspect - that the proof can be done in that way - which provides our students with rich material for experimentation and investigation.
If we look back to the task criteria given in section 2 its obvious that at least points 2 to 4 are satisfied. I do not aim at doing book-keeping, rather confine myself to give some catchwords. Various mathematical and heuristic activities can be stimulated by our topic: from drawing H-polygons in a given hexagonal lattice, constructing such a lattice, cutting out (with scissors) H-polygons, comparing, tesselating, dissecting H-polygons, counting (systematically) fundamental triangles in order to get F(P) etc. to classifying, investigating special cases, logical ordering, thinking out proof strategies etc.
Obviously our topic can also be used - perhaps as a module of some larger unit of "mathematics in the lattice plane" - in mathematics teacher education. For this purpose, however, something more should be done to elaborate the educational ideas in this module (c.f. Wittmann 1988).
Since our topic is a tiny part of the mathematical substantial domain of tesselation problems (Grünbaum, Shephard 1987) it is easy to use it as an entrance to this "problem area". But there are also elementary follow-up problems within our topic, one of which shall be finally mentioned: given n fundamental H-trapezoids, which types of different convex H-polygons can

be t-tesselated with exactly these n pieces, and what is the number of these H-polygons? For an investigation of this question it might be helpful to use the computer.

6 References

Grünbaum,B./Shephard,G.C.: Tilings and Patterns. New York 1987.

Walther,G.: MATHEMA in Schleswig-Holstein: Eine Konzeption zur Förderung mathematisch interessierter Schüler der gymnasialen Mittelstufe. In: Beiträge z. Mathematikunterricht. Bad Salzdetfurth 1988a, pp 307-310.

Walther,G.: Geometrie mit Tischen: Ein Unterrichtsvorschlag für geometrisch-kombinatorische Aktivitäten. In: Mathematikdidaktik: Theorie und Praxis. Festschrift für H.Winter. Bender,P. (Hrsg.). Berlin 1988.

Wittmann,E.Ch.: The mathematical training of teachers from the point of view of education. Survey lecture in Action Group A6, ICME-6 Budapest 1988, to appear.

Domination in Cubic Graphs

B. Zelinka, Liberec, CSSR

In this paper we shall study the domatic number, the total domatic number and the connected domatic number of cubic graphs, i. e. regular graphs of degree 3. We consider finite undirected graphs without loops and multiple edges.

A subset D of the vertex set $V(G)$ of a graph G is called dominating (or total dominating), if for each $x \in V(G) - D$ (or each $x \in V(G)$ respectively) there exists a vertex $y \in D$ adjacent to x. A dominating set in G is called connected, if it induces a connected subgraph of G.

A domatic (or total domatic, or connected domatic) partition of G is a partition of $V(G)$, all of whose classes are dominating (or total dominating, or connected dominating respectively) sets in G. The maximum number of classes of a domatic (or total domatic, or connected domatic) partition of G is called the domatic (or total domatic, or connected domatic respectively) number of G. The domatic number of G is denoted by $d(G)$, the total domatic number by $d_t(G)$, the connected domatic number by $d_c(G)$.

The domatic number was introduced by E. J. Cockayne and S. T. Hedetniemi in [3], the total domatic number by the same authors and R. M. Dawes in [4] and the connected domatic number by R. Laskar and S. T. Hedetniemi in [5].

For $d(G)$ and $d_t(G)$ the inequalities $d(G) \leq md(G) + 1$, $d_t(G) \leq md(G)$ hold, where $md(G)$ is the minimum degree of a vertex in G. A graph G for which $d(G) = md(G) + 1$ is called domatically full; a graph G for which $d_t(G) = md(G)$ is called totally domatically full. A cubic graph G has $md(G) = 3$; it

R. Bodendiek, R. Henn (Eds.)
Topics in Combinatorics and Graph Theory

is domatically full, if $d(G) = 4$, and totally domatically full, if $d_t(G) = 3$.

Before stating theorems, we shall introduce some auxiliary concepts.

Let G be a cubic graph, let Q be a quadratic factor of G. Let O(Q) be an orientation of Q such that all circuits of Q are cycles (directed circuits) in it. Now let C be an arbitrary circuit in G. Choose an orientation O(C) of C such that C is a cycle in it. Now we define f(C, O(C)) to be the difference between the number of common edges of Q and C which are equally directed in O(Q) and O(C) and the number of common edges of Q and C which are oppositely directed in Q and C. We see that there are two possibilities to choose O(C) at the given circuit C and the corresponding values of f differ only by the sign. Therefore we may define f(C) as the absolute value of f(C, O(C)) for an arbitrary orientation O(C) in which C is a cycle. Analogously we define f(P, O(P)), where P is a path and O(P) is an orientation of P in which P is a directed path.

Further by g(C) for a circuit C we denote the number of edges belonging to C and not to Q. Obviously the values of f and g depend on Q and O(Q). Analogously we define g(P) for a path P.

Now we state a theorem.

Theorem 1. A finite connected cubic graph G is domatically full if and only if there exists a quadratic factor Q of G and its orientation O(Q) in which all circuits of Q are cycles with the following properties:

(i) The lengths of all circuits of Q are divisible by 4.

(ii) If f, g are defined as above with respect to Q and O(Q), then

$$f(C) + 2g(C) \equiv 0 \pmod 4$$

for each circuit C of G.

Remark. In fact (i) follows from (ii), if we choose C to be a circuit of Q. But for a better imagination we write both (i) and (ii).

Proof. Suppose that G is domatically full. Therefore

$d(G) = 4$ and there exists a domatic partition $\underline{D}$ with the classes D_1, D_2, D_3, D_4 in G. The subscripts at D will be always taken modulo 4. As G is cubic, each vertex is adjacent to exactly one vertex from each class of $\underline{D}$ different from its own one and to no vertex of its own class. Let L be the spanning subgraph of G whose edge set is the set of all edges which join either a vertex from D_1 with a vertex of D_3, or a vertex from D_2 with a vertex of D_4. Evidently each vertex of G is incident with exactly one edge of L and therefore L is a linear factor of G. By deleting all edges of L from G we obtain a quadratic factor Q of G. All connected components of Q are circuits. Let Q_0 be a circuit in Q, let its vertices be $u_1, \ldots, u_k$ and edges $u_i u_{i+1}$ for $i = 1, \ldots, k - 1$ and $u_k u_1$. Without loss of generality we may assume that $u_1 \in D_1$ and $u_2 \in D_2$. Then necessarily $u_3 \in D_3$, because u_2 is adjacent to $u_1 \in D_1$ and to a vertex of D_4 joined with it by an edge of L. By induction we obtain that $u_i \quad D_h$, where $h \equiv i \pmod 4$, for each $i = 1, \ldots, k$. Going in the inverse direction, we obtain that $u_k \in D_4$ and this implies that k is divisible by 4. Now let $O(Q)$ be the orientation of Q in which each edge is directed from a vertex of D_i towards a vertex of D_{i+1} for each $i \in \{1, 2, 3, 4\}$, the sum $i + 1$ being taken modulo 4. Now let C be a circuit in G, let its vertices be $v_1, \ldots, v_m$ and edges $v_i v_{i+1}$ for $i = 1, \ldots, m - 1$ and $v_m v_1$. Let $O(C)$ be the orientation of C in which each edge is directed from v_i towards v_{i+1} for all $i = 1, \ldots, m$, the sum $i + 1$ being taken modulo m. Let $v_1 \in D_p$ for some $p \in \{1, 2, 3, 4\}$. If the edge $v_1 v_2$ belongs to Q and is directed equally in $O(Q)$ and $O(C)$, then $v_2 \in D_{p+1}$. If $v_1 v_2$ belongs to Q and is directed oppositely in $O(Q)$ and $O(C)$, then $v_2 \in D_{p-1}$. If $v_1 v_2$ belongs to L, then $v_2 \in D_{p+2}$. We proceed further in this way; at each step we determine uniquely the class which a vertex of C belongs to. At the end we investigate the edge $v_m v_1$ and we obtain that $v_1 \in D_q$, where $q \equiv p + f(C, O(C)) + 2g(C) \pmod 4$. But $v_1 \in D_p$ and thus $q \equiv p \pmod 4$ and $f(C, O(C)) + 2g(C) \equiv 0 \pmod 4$. We have either $f(C) = f(C, O(C))$, or $f(C) = -f(C, O(C))$. In the first case we have directly $f(C) + 2g(C) \equiv 0 \pmod 4$. In the second case we have $f(C) - 2G(C) \equiv 0 \pmod 4$. As $G(C)$ is an integer, always $4g(C) \equiv 0 \pmod 4$; by adding these congruences, we obtain the required result.

Now suppose that in G there exist Q and O(Q) satisfying the conditions (i) and (ii). Let u, v be two arbitrary vertices of G. The condition (ii) implies that if P_1, P_2 are two paths connecting u and v and $O(P_1)$, $O(P_2)$ are their orientations from u towards v, then $f(P_1, O(P_1)) + {} + 2g(P_1) \equiv f(P_2, O(P_2)) + 2g(P_2) \pmod 4$. Therefore we choose an arbitrary vertex u of G. For each $i \in \{1, 2, 3, 4\}$ let D_i be the set of all vertices x of G such that $f(P, O(P)) + 2g(P) \equiv i \pmod 4$, where P is a path connecting u and x and O(P) is its orientation from u towards x. Then $\{D_1, D_2, D_3, D_4\}$ is a partition of V(G), As each vertex x is incident with two edges of Q and one of L, it is adjacent to vertices of D_{i-1}, D_{i+1}, D_{i+2}, where i is such that $x \in D_i$. We have evidently $\{D_{i-1}, D_{i+1}, D_{i+2}, D_i\} = \{D_1, D_2, D_3, D_4\}$ and thus it is a domatic partition of G and $d(G) = 4$. The graph G is domatically full.

Corollary 1. Let G be a cubic graph whose number of vertices is divisible by 4. Let G have a Hamiltonian circuit Q and let all edges of G not belonging to Q join pairs of vertices whose distance in Q is congruent with 2 modulo 4. Then G is domatically full.

Another corollary will concern permutation graphs of circuits. Let C, C' be two circuits of the same length n, let p be a permutation of the number set $\{1,\ldots, n\}$. Let the vertices of C be $u_1,\ldots, u_n$ and the edges u_iu_{i+1} for $i = 1,\ldots, n-1$ and u_nu_1; let the vertices of C' be $u'_1,\ldots, u'_n$ and the edges $u'_iu'_{i+1}$ for $i = 1,\ldots, n-1$ and $u'_nu'_1$. From the circuits C, C' we construct the graph G(n, p) by joining the vertices u_i, u'_j by edges for all i and j such that $j = p(i)$. This graph is called the permutation graph of the circuit of length n with respect to the permutation p.

Permutation graphs (for graphs in general, not only for circuits) were introduced by G. Chartrand and F. Harary in [1] and further studied by G. Chartrand and J. B. Frechen in [2]. Especially for circuits they were studied in [6].

Corollary 2. Let n be a positive integer divisible by 4, let p be a permutation of the number set $\{1, \ldots, n\}$ such that $p(i) \equiv i \pmod 4$ for $i = 1, \ldots, n$. Then the graph $G(n, p)$ is domatically full.

If p is the identical permutation of the set $\{1, \ldots, n\}$, then $G(n, p)$ is the graph of the n-side prisma. We have a further corollary.

Corollary 3. Let n be a positive integer divisible by 4. Then the graph of the n-side prisma is domatically full.

Now we turn to the total domatic number.

Theorem 2. A finite connected cubic graph G is totally domatically full if and only if there exists a quadratic factor Q of G and its orientation O(Q) in which all circuits of Q are cycles with the following properties:
(i) The lengths of all circuits of Q are divisible by 3.
(ii) If f is defined as above with respect to Q and O(Q), then

$$f(C) \equiv 0 \pmod 3$$

for each circuit C of G.

Remark. Again (i) follows from (ii).

Proof. Suppose that G is totally domatically full. Therefore $d_t(G) = 3$ and there exists a total domatic partition $\underline{D} = \{D_1, D_2, D_3\}$ of G. The subscripts at D will be always taken modulo 3. As G is cubic, each vertex is adjacent to exactly one vertex from each class of $\underline{D}$. Now we proceed analogously as in the proof of Theorem 1. Let L be the spanning subgraph of G whose edge set is the set of all edges which join vertices of the same class of $\underline{D}$. Evidently each vertex of G is incident with exactly one edge of L and therefore L is a linear factor of G. By deleting all edges of L from G we obtain a quadratic factor Q. Analogously to the proof of Theorem 1 we prove that the lengths of all circuits of Q are divisible by 3. The orientations O(Q) and O(C) (where C is an arbitrary circuit of G) and the notation of vertices of C are analogous to the proof of Theorem 1. Let $v_1 \in D_p$ for some $p \in \{1, 2, 3\}$.

If the edge v_1v_2 belongs to Q and is directed equally in O(Q) and O(C), then $v_2 \in D_{p+1}$. If v_1v_2 belongs to Q and is directed oppositely in O(Q) and O(C), then $v_2 \in D_{p-1}$. If v_1v_2 belongs to L, then $v_2 \in D_p$. Analogously as in the proof of Theorem 1 we obtain that $f(C, O(C)) \equiv 0 \pmod 3$. As $f(C) = f(C, O(C))$ or $f(C) = -f(C, O(C))$, we have $f(C) \equiv 0 \pmod 3$.

Now suppose that there exist Q and O(Q) satisfying the conditions (i) and (ii). Again we choose an arbitrary vertex u. For each $i \in \{1, 2, 3\}$ let D_i be the set of all vertices x of G such that $f(P, O(P)) \equiv i \pmod 3$, where P is a path connecting u and x and O(P) is its orientation from u to x; the unicity of i can be proved analogously to the proof of Theorem 1. Also analogously we prove that $\{D_1, D_2, D_3\}$ is a total domatic partition of G and $d_t(G) = 3$. The graph G is totally domatically full.

<u>Corollary 4.</u> Let G be a cubic graph whose number of vertices is divisible by 3. Let G have a Hamiltonian circuit Q and let all edges not belonging to Q join pairs of vertices whose distance in Q is divisible by 3. Then G is totally domatically full.

<u>Corollary 5.</u> Let n be a positive integer divisible by 3, let p be a permutation of the number set $\{1, \dots, n\}$ such that $p(i) \equiv i \pmod 3$ for $i = 1, \dots, n$. Then the graph G(n, p) is totally domatically full.

<u>Corollary 6.</u> Let n be a positive integer divisible by 3. Then the graph of the n-side prisma is totally domatically full.

Note that a totally domatically full graph need not be domatically full.

Now we shall consider the connected domatic number.

A chord of a circuit is an edge which joins two vertices of that circuit and does not belong to it. Analogously a chord of a path is defined. If a circuit or a path has no chords, it is called chordless.

<u>Theorem 3.</u> Let G be a cubic graph with $n \geq 8$ vertices. Then $d_c(G) \leq 2$. There is $d_c(G) = 2$ if and only if one of the

following cases occurs:

(a) G contains two vertex-disjoint chordless circuits of length $n/2$.

(b) G contains two vertex-disjoint chordless paths of length $n/2 - 1$.

(c) G contains a chordless circuit of length $n/2 + 1$ and a chordless path of length $n/2 - 2$ vertex-disjoint with that circuit.

Proof. Suppose that $d_c(G) > 2$. Then there exists a connected domatic partition $\{D_1, D_2, D_3\}$ of G. As $n \geq 8$, at least one of the sets D_1, D_2, D_3 must contain at least three vertices; without loss of generality let $|D_1| \geq 3$. The subgraph G_1 of G induced by D_1, being connected, contains a vertex v of degree at least 2 in G. The vertex v is in D_1 and is adjacent to two vertices of D_1; as its degree in G is 3, it cannot be adjacent to vertices of both D_2 and D_3, which is a contradiction. We have $d_c(G) \leq 2$.

Now suppose that $d_c(G) = 2$ and let $\{D_1, D_2\}$ be a connected domatic partition of G.Let G_1 (or G_2) be the subgraph of G induced by D_1 (or D_2 respectively). The graph G_1 contains no vertex of degree 3; otherwise this vertex would belong to D_1 and be adjacent only to vertices of D_1, which would be a contradiction. Analogously the graph G_2 contains no vertex of degree 3. Each of the graphs G_1, G_2 is a connected graph in which the degree of a vertex is at most 2, i. e. a path or a circuit. Suppose that both G_1, G_2 are circuits. As they are induced subgraphs of G, they are chordless. As D_1, D_2 is a partition of V(G), the circuits G_1, G_2 are vertex-disjoint and form a quadratic factor of G. Each vertex of D_1 is adjacent to two vertices of D_1 and thus to exactly one vertex of D_2; similarly each vertex of D_2 is adjacent to exactly one vertex of D_1. Hence there is a one-to-one correspondence between D_1 and D_2 and $|D_1| = |D_2| = n/2$; the circuits G_1, G_2 have both the length $n/2$ and the case (a) occurs. If G_1, G_2 are paths, then they are also chordless and cover all vertices of G. Each inner vertex of the path G_1 is adjacent to exactly one vertex of G_2, each terminal vertex of G_1 is adjacent to exactly two vertices of G_2; analogously for G_2. Therefore again $|D_1| = |D_2| = n/2$ and each of the paths G_1, G_2 has $n/2$ vertices and its length is $n/2 - 1$; the case (b) occurs. If G_1 is a circuit and G_2 is a path, then each vertex of G_1 is adjacent to exactly one vertex of G_2, while each inner vertex of G_2 is adjacent to exactly one vertex of G_1 and each terminal vertex of G_2 is adjacent to exactly two vertices of G_1. This implies $|D_1| = |D_2| + 2$, i. e. $|D_1| = n/2 + 1$, $|D_2| = n/2 - 1$; the length of the circuit G_1 is $n/2 + 1$, the length of the path G_2 is $n/2 - 2$ and the case (c) occurs.

Now suppose that (a) occurs. Then D_1, D_2 are vertex sets of the mentioned circuits. As a circuit contains no vertex of degree 3, each vertex of one circuit is adjacent to a vertex of the other and $\{D_1, D_2\}$ is a connected domatic partition of G. The cases (b) and (c) are analogous.

Corollary 7. Let n be a positive integer, $n \geq 4$, let p be a permutation of the number set $\{1, \ldots, n\}$. Then the permutation graph G(n, p) has the connected domatic number equal to 2.

To describe the singular cases, we state a proposition whose proof is easy.

Proposition. The unique cubic graph G with $d_c(G) = 4$ is the complete graph K_4 with four vertices. The unique cubic graphs G with $d_c(G) = 3$ are the graph of the three-side prisma and the graph obtained from a circuit of length 6 by joining all pairs of opposite vertices by edges.

References:

[1] Chartrand G. - Harary F.: Planar permutation graphs. Ann. Inst. H. Poincaré Sect. B 3, 1967, 433-438.

[2] Chartrand G. - Frechen J. B.: On the chromatic number of permutation graphs. In: Proof Techniques in Graph Theory, New York 1969, 21-24.

[3] Cockayne E. J. - Hedetniemi S. T.: Towards a theory of domination in graphs. Networks 7, 1977, 247-261.

[4] Cockayne E. J. - Dawes R. M. - Hedetniemi S. T.: Total domination in graphs. Networks 10, 1980, 211-219.

[5] Laskar R. - Hedetniemi S. T.: Connected domination in graphs. Report No. 414, Clemson Univ., Clemson, South Carolina 1983.

[6] Zelinka B.: Edge-colourings of permutation graphs. Matem. časop. 23, 1973, 193-198.

A Generalization of the Bodendiek Conjecture About Graceful Graphs

C. Zhi-Zeng, Huhhot, China

ABSTRACT

R. Bodendiek conjectured that every graph consisting of a cycle plus a chord is graceful, and various proofs have been given. [1] Here we use GL-matrix as a tool in proving the following generalization: Apart from four exceptional cases, simple graphs consisting of three independent paths joining two vertices are graceful.

1. INTRODUCTION

A simpe graph $G=(V,E)$ is graceful if there is a labelling φ of its vertices with distinct integers from the set $\{0,1,2,\cdots,|E|\}$, So that the induced edge labelling L defined by $L(uv)=|\varphi(u)-\varphi(v)|$ assigns each edge a different label.

The labelling φ is called a graceful labelling, and designated as GL-φ.

We suppose that $G=(V,E)$ is a connected simpe graph and there exists a labelling φ of its vertices with distinct integers from the set $\{0,1,2,\cdots,|E|\}$. If rows of adjacency matrix of G are arranged according to the order $0,1,2,\cdots,|E|$, which are the labels of the corresponding vertices of the rows, then we get the matrix A (Because $|E|\geqslant|V|-1$ is tenable, then there may be $r\in\{0,1,2,\cdots,|E|\}$, which hasn't been used as label. We suppose that components of the row, corresponding r, are zeros, and the row is denoted by point $\star$).

Definition 1. The matrix A of order $|E|+1$ is called the labelling matrix of G, and designated as L-matrix A.

Obviously L-matrix $A=(a_{ij})$ is a symmetrical matrix. When $uv\in E$, there u, v corresponding respectively ith and jth rows, then $a_{ij}=a_{ji}=1$ (this 1 is called the edge index). Otherwise, it is equal to zero.

Definition 2. In the upper part of the main diagonal of L-matrix A of order m, there are altogether $m-1$ oblique lines, which parallel the main diagonal. If there is only an edge index 1 in each oblique line, then the L-matrix A is called "graceful L-matrix", usually expressed by

R. Bodendiek, R. Henn (Eds.)
Topics in Combinatorics and Graph Theory

GL-matrix or M.

Example 1. Let $\varphi(v_1)=0, \varphi(v_2)=1, \varphi(v_3)=3$ be a labelling φ of cycle $v_1v_2v_3v_1$, thus its L-matrix A is

$$A = \begin{array}{c} \\ v_1 \\ v_2 \\ \star \\ v_3 \end{array} \begin{array}{c} \begin{array}{cccc} v_1 & v_2 & \star & v_3 \end{array} \\ \begin{pmatrix} 0 & 1 & 0 & 1 \\ 1 & 0 & 0 & 1 \\ 0 & 0 & 0 & 0 \\ 1 & 1 & 0 & 0 \end{pmatrix} \end{array}.$$

the A is just its GL-matrix. For short, it is usually written as

$$M = \begin{pmatrix} v_1 & 0 & 1 & & 1 \\ v_2 & & 0 & & 1 \\ \star & & & 0 & \\ v_3 & & & & 0 \end{pmatrix} \triangleq (v_1v_2\star v_3).$$

Proposition 1. A graph G is graceful iff there is a GL-matrix of G.

proof. (We prove only the sufficiency.) We remember that GL-matrix is a L-matrix. Hence, $\varphi(v)=i-1$, in which the vertex v corresponds to ith row of GL-matrix of order $|E|+1$. Thus, there exists the labelling φ of vertices of G so that different vertex v has different labeled $\varphi(v)$, and each $\varphi(v)\in\{0,1,2,\cdots,|E|\}$.

On the other hand, we suppose that the GL-matrix of order $|E|+1$ is $M=(a_{ij})$. When the edge uv index 1 exists in oblique line i, which is counted from the main diagonal up to the upper part of M, then the edge uv index 1 is just $a_{t\,t+i}=1, t\in\{1,2,\cdots,1+|E|-i\}$. And we may as well suppose that v lies behind u in M. Hence $\varphi(u)=t-1, \varphi(v)=t+i-1$. Now we get $L(uv)=|\varphi(u)-\varphi(v)|=i$, i. e. different edge uv has different labelling $L(uv)$. Therefore the labelling φ is GL-φ. □

We quote symbol: $N(u,v)$—— the number of all vertices between u and v in the vertex sequence of L-matrix, including u, v and point $\star$. Hence we have arrived at the following result:

Corollary 1. Let $(\cdots u\cdots v\cdots)$ be L-matrix of G, $uv\in E$, then $L(uv)=N(u,v)-1$.

Definition 3. Suppose that the graceful graph G is a bipartite graph. Thus the vertices of G has been divided into vertex classes V_1 and V_2. If the vertices of V_1 are arranged in the front rows of the GL- matrix of G and the vertices of V_2 are arranged in the following rows, then this GL-matrix is called a standard matrix, and designated as S-matrix.

Example 2. When n is even integer, GL-matrix of the path $A_1A_2A_3\cdots A_{n-1}A_n$ is respectively $(A_1A_3\cdots A_{n-3}A_{n-1}A_nA_{n-2}\cdots A_4A_2)$、$(A_nA_{n-2}\cdots A_4A_2A_1A_3\cdots A_{n-3}A_{n-1})$ etc.. And when n is odd integer, GL-matrix of the path mentioned above is $(A_1A_3\cdots A_{n-2}A_nA_{n-1}A_{n-3}\cdots A_4A_2)$ etc.. So, these GL-matrices are just S-matrices.

Definition 4. The pendant vertex of the path, which is arranged in the centre of its S-matrix, is called the demarcation vertex.

In Example 2, the demarcation vertices of three S-matrices are A_n、$A_1$and $A_n$respectivelly.

We quote some symbols: N_1 —— the set of all odd integers; N_2 —— the set of all even integers; $\star \in (A\cdots)$—— point $\star$ is being arranged in a certain row behind A; meanwhile $\otimes\in(\cdots A)$ & $\star\in(A\cdots B)$ being the analogies to $\star\in(A\cdots)$; $|M|$—— the number of all vertices in L-matrix M.

Proposition 2. In S-matrix of the path, we suppose that its demarcation vertex v belongs to the vertex class, which is arranged in the front (following) rows of the S-matrix. Upon that, if we put a point $\star$ into the S-matrix, and $\star\in(v\cdots)$, then in the S-matrix is emptied out an oblique line, the corresponding edge labelling L of which is an odd (even) integer; if $\star\in(\cdots v)$, then relevant line emptied is the corresponding even (odd).

For instance, if we put point $\star$ into $(A_1A_3A_2)$, which is a S-matrix of the path $A_1A_2A_3$, and $\star\in(\cdots A_3)$, then we can obtain:

$$(A_1 * A_3A_2) = \begin{bmatrix} A_1 & 0 & & & 1 \\ \star & & 0 & & \\ A_3 & & & 0 & 1 \\ A_2 & & & & 0 \end{bmatrix}$$

Although that is not L-matrix of the path again, its second oblique line is emptied out. We maintain that $\varphi(A_1)=0, \varphi(A_3)=2, \varphi(A_2)=3$, thus the edge labelling 2 is emptied out.

Definition 5. When we first put the index 1 of a new edge e into a GL-matrix M and then put point $\star$ into it again so that the final result is still a GL-matrix, thus this operation is called "add edge operation", and designated as $M+\{e\}$.

About $M+\{v\}$, M_1+M_2, $M_1\oplus M_2$and their applications, see[2].

Example 3. The graph G in figure 1 is graceful.

proof. The S-matrix M of the path $B_1C_2C_1A_1A_2$ is $M=(B_1C_1A_2A_1C_2)$. We do $M+\{B_1C_1\}$ first, and then do $M+\{B_1C_1\}+\{A_2C_2\}$. So we obtain GL-matrix of G by Definition 5: $M+\{B_1C_1\}+\{A_2C_2\}=$

$$(B_1C_1\otimes A_2\star A_1C_2).$$

Thus GL-φ of G is $\varphi(B_1)=0, \varphi(C_1)=1, \varphi(A_2)=3, \varphi(A_1)=5, \varphi(C_2)=6$.

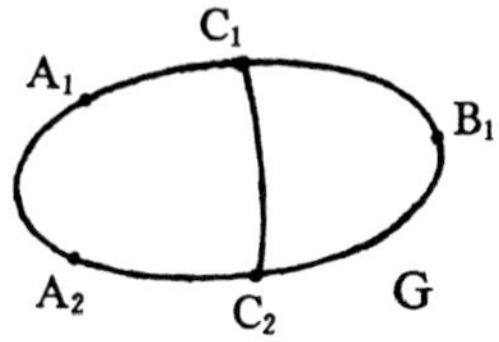

Fig. 1

2. A GENERALIZATION OF BODENDIEK'S CONJECTURE

Let G be a simple graph consisting of three independent paths joining two vertices (see figure 2). We may as well suppose $m\geqslant n\geqslant p-2$. Considering G is simple graph again, thus $mn\neq 0$, $p\geqslant 2$.

Theorem 1. The graph G in figure 2 is graceful except for the following four cases:

(a) $m\in N_2, n、p\in N_1, \frac{m}{2}+\frac{n+1}{2}\in N_1, \frac{n+1}{2}+\frac{p+1}{2}\in N_2,$

$m>n+2p-3$;

(b) $m\in N_1, n、p\in N_2, \frac{m+1}{2}+\frac{n}{2}\in N_2, \frac{n}{2}+\frac{p}{2}\in N_1,$

$n=p-2>0$;

(c) $m\in N_1, n、p\in N_2, \frac{m+1}{2}+\frac{n}{2}\in N_1, \frac{n}{2}+\frac{p}{2}\in N_1,$

$n=p-2>0$;

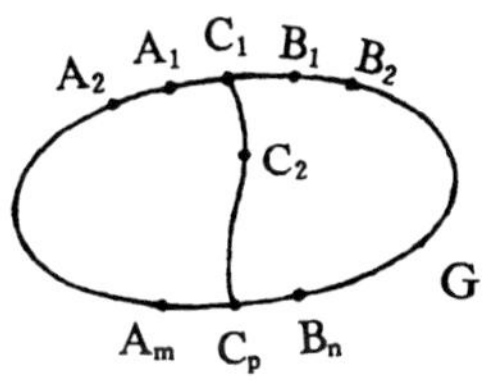

Fig. 2

(d) $m、n、p\in N_1, \frac{m+1}{2}+\frac{n+1}{2}\in N_2, \frac{n+1}{2}+\frac{p+1}{2}\in N_1, m=p-2>0.$

proof. (1) $m、n、p\in N_2$:

(1−1) When $\frac{m}{2}+\frac{n}{2}、\frac{n}{2}+\frac{p}{2}\in N_2$: S-matrix M of the path $A_1A_2\cdots A_mC_p\cdots C_2C_1B_1B_2\cdots B_n$ is

$$M=(A_1A_3\cdots A_{m-1}C_p\cdots C_4C_2B_1B_3\cdots B_{n-1}B_n\cdots B_4B_2C_1C_3\cdots C_{p-1}A_m\cdots A_4A_2).$$

Now we will do $M+\{B_nC_p\}$ first. In M, we should have $L(B_nC_p)=N(B_n, C_p)-1=\frac{n}{2}+\frac{p}{2}\in N_2$. As the demarcation vertex B_n belongs to the vertex class, which is arranged in the following rows of M, we put point $\star\in(B_n\cdots)$ by Proposition 2, in order to empty out even edge labelling $\frac{n}{2}+\frac{p}{2}$. When $B_n\star B_{n-2}$, edge labelling 2 is emptied out. When $B_{n-2k}\star B_{n-(2k+2)}$, edge labelling $2k+2$ is emptied out. We take

$$2k + 2 = \frac{n}{2} + \frac{p}{2},$$

thus we put point $\star$ at:

$$B_{\frac{n}{2}-\frac{p}{2}+2} \star B_{\frac{n}{2}-\frac{p}{2}}.$$

Hence the original edge labellings less than $\frac{n}{2}+\frac{p}{2}$ are constant, and 1 is added respectively to those no less than $\frac{n}{2}+\frac{p}{2}$, i. e. edge labelling $\frac{n}{2}+\frac{p}{2}$ is emptied out. There is exactly $L(B_nC_p)=\frac{n}{2}+\frac{p}{2}$ here, so we put the index 1 of B_nC_p into M again, we obtain GL-matrix $M+\{B_nC_p\}$.

And then we will do $M+\{B_nC_p\}+\{A_1C_1\}$. We get $\frac{n}{2}-\frac{p}{2}\geqslant 0$ since $n\geqslant p-2$, $\frac{n}{2}+\frac{p}{2}\in N_2$. Hence aforesaid $\star\in(\cdots C_1)$. Therefore, in $M+\{B_nC_p\}$, $L(A_1C_1)=N(A_1,C_1)-1=\frac{m}{2}+n+\frac{p}{2}+1\in N_1$, and $L(A_1C_1)>L(B_nC_p)$. thus we must put $\otimes\in(\star\cdots)$, in order to empty out the odd edge labelling. When $A_m\otimes A_{m-2}$, edge labelling $n+p+3$ is emptied out, since the degree of vertices behind B_n are 2, and $\star$ is considered; when $A_{m-2k}\otimes A_{m-(2k+2)}$, edge labelling $n+p+2k+3$ is emptied out. We take

$$n + p + 2k + 3 = \frac{m}{2} + n + \frac{p}{2} + 1,$$

thus we put point $\otimes$ at:

$$A_{\frac{m}{2}+\frac{p}{2}+2} \otimes A_{\frac{m}{2}+\frac{p}{2}}.$$

And it is obvious that, the point $\otimes\in(C_{p-1}\cdots)$. Hence, in $M+\{B_nC_p\}$, the original edge labellings less than $\frac{m}{2}+n+\frac{p}{2}+1$ (including $L(B_nC_p)$) are constant, and 1 is added respectively to those no less than $\frac{m}{2}+n+\frac{p}{2}+1$, i. e. edge labelling $\frac{m}{2}+n+\frac{p}{2}+1$ is emptied out. There is exactly $L(A_1C_1)=\frac{m}{2}+n+\frac{p}{2}+1$ here, so we put the index 1 of A_1C_1 into $M+\{B_nC_p\}$ again, we obtain GL-matrix of $G: M+\{B_nC_p\}+\{A_1C_1\}=$

$$(A_1A_3\cdots A_{m-1}C_p\cdots C_4C_2B_1B_3\cdots B_{n-1}B_n\cdots B_{\frac{n}{2}-\frac{p}{2}+2}\star B_{\frac{n}{2}-\frac{p}{2}}\cdots B_4B_2C_1C_3\cdots C_{p-1}A_m\cdots A_{\frac{m}{2}+\frac{p}{2}+2}\otimes A_{\frac{m}{2}+\frac{p}{2}}\cdots A_4A_2).$$ ①

and we can get $\star\in(B_n\cdots C_1)$, $\otimes\in(C_{p-1}\cdots A_2)$.

---

① When $m=n=p=2$, we get $B_2\star B_0$ and $A_4\otimes A_2$. In fact, the GL-matrix is $(A_1C_2B_1B_2\star C_1\otimes A_2)$, since there are no B_0 and A_4.

(1−2) when $\frac{m}{2}+\frac{n}{2}\in N_2, \frac{n}{2}+\frac{p}{2}\in N_1$: S-matrix M of the path $C_{p-1}\cdots C_2C_1A_1A_2\cdots A_mC_pB_n\cdots B_2B_1$ is $M=(C_{p-1}\cdots C_3C_1A_2A_4\cdots A_mB_n\cdots B_4B_2B_1B_3\cdots B_{n-1}C_pA_{m-1}\cdots A_3A_1C_2C_4\cdots C_{p-2})$.

Now we will do $M+\{B_1C_1\}$ first. In M, we put temporarily point 0 at C_10A_2, so $L(B_1C_1)=N(B_1,C_1)-1=\frac{m}{2}+\frac{n}{2}+2\in N_2$. As the demarcation vertex B_1belongs to the vertex class, which is arranged in the following rows of M, we put point $\star\in(B_1\cdots)$ by Proposition 2, in order to empty out the even edge labelling. When $A_{m-1}\star A_{m-3}$, edge labelling $n+4$ is emptied out. When $A_{m-(2k-1)}\star A_{m-(2k+1)}$, edge labelling $n+2k+2$ is emptied out. We take

$$n+2k+2=\frac{m}{2}+\frac{n}{2}+2,$$

thus we put point $\star$ at:

$$A_{\frac{m}{2}+\frac{n}{2}+1}\star A_{\frac{m}{2}+\frac{n}{2}-1}.$$

We can get $\star\in(C_p\cdots A_1)$, and the original edge labellings less than $\frac{m}{2}+\frac{n}{2}+2$ are constant, and 1 is added respectively to those no less than $\frac{m}{2}+\frac{n}{2}+2$, i. e. edge labelling $\frac{m}{2}+\frac{n}{2}+2$ is emptied out. There is exactly $L(B_1C_1)=\frac{m}{2}+\frac{n}{2}+2$, so we put the index 1 of B_1C_1into M again, we obtain matrix $M+\{B_1C_1\}$, which is not a GL-matrix.

And then we will do $M+\{B_1C_1\}+\{C_pC_{p-1}\}$. As C_1OA_2 and $\star\in(C_p\cdots)$, we should get $L(C_pC_{p-1})=N(C_p,C_{p-1})-1=\frac{m}{2}+n+\frac{p}{2}+1\in N_2$ and $L(C_pC_{p-1})>L(B_1C_1)$. If we put point $\otimes\in(A_{\frac{m}{2}+\frac{n}{2}}\cdots\star)$ (here $\frac{m}{2}+\frac{n}{2}$ is average value of footnumbers of two adjoining vertices of point $\star$), then the edge labelling emptied out by point $\star$ is increased by 1. If we put point $\otimes\in(\cdots A_{\frac{m}{2}+\frac{n}{2}})$, then the edge labelling emptied out by point $\star$ is constant, and the even edge labelling is emptied out. When $A_{m-2}\otimes A_m$ and $\otimes\in(\cdots A_{\frac{m}{2}+\frac{n}{2}})$, edge labelling $n+4$ is emptied out. When $A_{m-2k}\otimes A_{m-(2k-2)}$and $\otimes\in(\cdots A_{\frac{m}{2}+\frac{n}{2}})$, edge labelling $n+2k+2$ is emptied out. We take

$$n+2k+2=\frac{m}{2}+n+\frac{p}{2}+1$$

thus we put point $\otimes$ at:

$$A_{\frac{m}{2}-\frac{p}{2}+1}\otimes A_{\frac{m}{2}-\frac{p}{1}+3}.$$

We get $\frac{m}{2}-\frac{p}{2}+1\geqslant 0$ since $m\geqslant p-2$. And we get $\frac{m}{2}-\frac{p}{2}+3\leqslant\frac{m}{2}+\frac{n}{2}\leqslant m$ since $m\geqslant n>1$, $p\geqslant 2$, $\frac{n}{2}+\frac{p}{2}\in N_1$. Therefore $\otimes\in(C_1\cdots A_{\frac{m}{2}+\frac{n}{2}})$ and $\otimes\in(\cdots A_m)$. Hence we take point 0 away from C_10A_2, for its function is replaced by point $\otimes$. thus the original edge labellings less than

$\frac{m}{2}+n+\frac{p}{2}+1$ (including $L(B_1C_1)$) and the edge labelling emptied out by point $\star$ are constant, and 1 is added respectively to those no less than $\frac{m}{2}+n+\frac{p}{2}+1$, i. e. edge labelling $\frac{m}{2}+n+\frac{p}{2}+1$ is emptied out. There is exactly $L(C_pC_{p-1})=\frac{m}{2}+n+\frac{p}{2}+1$, so we put the index 1 of C_pC_{p-1} into $M+\{B_1C_1\}$ again, we get GL-matrix: $M+\{B_1C_1\}+\{C_pC_{p-1}\}=$

$$(C_{p-1}\cdots C_3C_1A_2A_4\cdots A_{\frac{m}{2}-\frac{p}{2}+1}\otimes A_{\frac{m}{2}-\frac{p}{2}+3}\cdots A_mB_n\cdots B_4B_2B_1B_3\cdots B_{n-1}C_pA_{m-1}\cdots$$
$$A_{\frac{m}{2}+\frac{n}{2}+1}\star A_{\frac{m}{2}+\frac{n}{2}-1}\cdots A_3A_1C_2C_4\cdots C_{p-2}).$$

Where $\star\in(C_p\cdots A_1)$ and $\otimes\in(C_1\cdots A_{\frac{m}{2}+\frac{n}{2}})$.

(1—3) When $\frac{m}{2}+\frac{n}{2}\in N_1, \frac{n}{2}+\frac{p}{2}\in N_2$: If we take the path $C_2C_3\cdots C_pA_m\cdots A_2A_1C_1B_1B_2\cdots B_n$, we obtain GL-matrix of G in the same method: $M+\{B_nC_p\}+\{C_1C_2\}=$

$$(C_2C_4\cdots C_pA_{m-1}\cdots A_3A_1B_1B_3\cdots B_{n-1}B_n\cdots B_4B_2C_1A_2A_4\cdots A_{\frac{m}{2}-\frac{n}{2}-1}\star A_{\frac{m}{2}-\frac{n}{2}+1}\cdots$$
$$A_{\frac{m}{2}+\frac{p}{2}-3}\otimes A_{\frac{m}{2}+\frac{p}{2}-1}\cdots A_mC_{p-1}\cdots C_5C_3).\ ①$$

Where $\star\in(C_1\cdots A_m)$ and $\otimes\in(\star\cdots A_m)$.

(1—4) When $\frac{m}{2}+\frac{n}{2}\in N_1, \frac{n}{2}+\frac{p}{2}\in N_1$: If we take the path $C_{p-1}\cdots C_2C_1A_1A_2\cdots A_mC_pB_n\cdots B_2B_1$, we obtain GL-matrix of G in the same method: $M+\{B_1C_1\}+\{C_pC_{p-1}\}=$

$$(C_{p-1}\cdots C_3C_1A_2A_4\cdots A_{\frac{m}{2}+\frac{n}{2}-1}\star A_{\frac{m}{2}+\frac{n}{2}+1}\cdots A_mB_n\cdots B_4B_2B_1B_3\cdots B_{n-1}C_pA_{m-1}\cdots$$
$$A_{\frac{m}{2}-\frac{p}{2}+3}\otimes A_{\frac{m}{2}-\frac{p}{2}+1}\cdots A_3A_1C_2C_4\cdots C_{p-2}).$$

Where $\star\in(A_2\cdots A_m)$ and $\otimes\in(A_{\frac{m}{2}+\frac{n}{2}}\cdots A_1)$.

(2) $m、n\in N_2, p\in N_1$:

(2—1) When $\frac{m}{2}+\frac{n}{2}、\frac{n}{2}+\frac{p+1}{2}\in N_2$: If we take the path $A_m\cdots A_2A_1C_1B_1B_2\cdots B_nC_p\cdots C_3C_2$, we obtain GL-matrix M_0 of G in the same method:

$$M_0=(A_m\cdots A_4A_2C_1B_2B_4\cdots B_nC_{p-1}\cdots C_4C_2C_3C_5\cdots C_pB_{n-1}\cdots B_{\frac{n}{2}+\frac{p+1}{2}+1}\star B_{\frac{n}{2}+\frac{p+1}{2}-1}\cdots B_3B_1A_1A_3\cdots$$
$$A_{\frac{m}{2}-\frac{n}{2}-3}\otimes A_{\frac{m}{2}-\frac{n}{2}-1}\cdots A_{m-1}).\ ②$$

Where $\star\in(C_p\cdots B_3)$ and $\otimes\in(B_3\cdots A_{m-3})$.

---

① When $m=4$ and $n=p=2$, we get $A_0\star A_2$ and $A_0\otimes A_2$. In fact, the GL-matrix is $(C_2A_3A_1B_1B_2C_1\star\otimes A_2A_4)$, since there is no A_0, and $\otimes\in(\star\cdots)$. It is the same with the similar place in the future.

② When $m=n$, we get $A_{-3}\otimes A_{-1}$. In fact, it's $B_3\otimes B_1$.

(2—2) $\frac{m}{2}+\frac{n}{2}\in N_2, \frac{n}{2}+\frac{p+1}{2}\in N_1$; (2—3) $\frac{m}{2}+\frac{n}{2}\in N_1, \frac{n}{2}+\frac{p+1}{2}\in N_2$; (2—4) $\frac{m}{2}+\frac{n}{2}$、$\frac{n}{2}+\frac{p+1}{2}\in N_1$. Since the proofs of these cases are like above, we leave them as exercises.

(3) m、$p\in N_2, n\in N_1$:

(3—1) When $\frac{m}{2}+\frac{n+1}{2}$、$\frac{n+1}{2}+\frac{p}{2}\in N_2$: $M_0=$

$(B_1B_3 \cdots B_nC_{p-1}\cdots C_3C_1A_2A_4 \cdots A_{\frac{m}{2}-\frac{p}{2}}\otimes A_{\frac{m}{2}-\frac{p}{2}+2}\cdots A_mA_{m-1}\cdots A_{m-\frac{n+1}{2}-\frac{p}{2}+3}\star A_{m-\frac{n+1}{2}-\frac{p}{2}+1}\cdots$ $A_3A_1C_2C_4\cdots C_pB_{n-1}\cdots B_4B_2)$. ①

Where $\star\in(A_m\cdots A_1)$ and $\otimes\in(C_1\cdots A_{m-\frac{n+1}{2}-\frac{p}{2}+2})$.

(3—2) $\frac{m}{2}+\frac{n+1}{2}\in N_2, \frac{n+1}{2}+\frac{p}{2}\in N_1$; (3—3) $\frac{m}{2}+\frac{n+1}{2}\in N_1, \frac{n+1}{2}+\frac{p}{2}\in N_2$;

(3—4) $\frac{m}{2}+\frac{n+1}{2}\in N_1; \frac{n+1}{2}+\frac{p}{2}\in N_1$. We leave them as exercises.

(4) $m\in N_2, n$、$p\in N_1$:

(4—1) $\frac{m}{2}+\frac{n+1}{2}\in N_2; \frac{n+1}{2}+\frac{p+1}{2}\in N_2$; (4—2) $\frac{m}{2}+\frac{n+1}{2}\in N_2, \frac{n+1}{2}+\frac{p+1}{2}\in N_1$; and (4—4) $\frac{m}{2}+\frac{n+1}{2}\in N_1, \frac{n+1}{2}+\frac{p+1}{2}\in N_1$. We leave them as exercises.

(4—3) When $\frac{m}{2}+\frac{n+1}{2}\in N_1, \frac{n+1}{2}+\frac{p+1}{2}\in N_2$: We have $m\leqslant n+2p-3$ due to the case (a) of Theorem 1. S-matrix M of the path $A_1A_2\cdots A_mC_pB_n\cdots B_2B_1C_1C_2\cdots C_{p-1}$ is

$$M=(A_1A_3\cdots A_{m-1}C_pB_{n-1}\cdots B_4B_2C_1C_3\cdots C_{p-2}C_{p-1}\cdots C_4C_2B_1B_3\cdots B_nA_m\cdots A_4A_2).$$

In M, we put temporarily point 0 at C_p0B_{n-1}, so we should have $L(C_pC_{p-1})=\frac{n+1}{2}+\frac{p+1}{2}\in N_2$. Therefore we put point $\star$ at:

$$B_{\frac{n+1}{2}-\frac{p+1}{2}+1}\star B_{\frac{n+1}{2}-\frac{p+1}{2}+3}.$$

We put the index 1 of C_pC_{p-1} into M again, so we get $M+\{C_pC_{p-1}\}$.

In $M+\{C_pC_{p-1}\}$, we should have $L(A_1C_1)=\frac{m}{2}+\frac{n+1}{2}+1\in N_2$. We proved that $m\geqslant p+3$ due to $m\geqslant n\geqslant p-2$ and $\frac{m}{2}+\frac{n+1}{2}\in N_1$ etc., so $L(A_1C_1)>L(C_pC_{p-1})$. If we put point $\otimes\in(\cdots B_{\frac{n+1}{2}-\frac{p+1}{2}+2})$, then the edge labelling emptied out by point $\star$ is constant, and the even edge labelling is emptied out. When $\otimes\in(\cdots B_{\frac{n+1}{2}-\frac{p+1}{2}+2})$ and $B_{2k}\otimes B_{2k-2}$, edge labelling $p+2k-1$ is emptied out. Therefore we put point $\otimes$ at:

$$B_{\frac{m}{2}+\frac{n+1}{2}-p+2}\otimes B_{\frac{m}{2}+\frac{n+1}{2}-p}.$$

---

① When $n=1, p=2$, we get $A_{m+1}\star A_{m-1}$. In fact, it's $A_m\star A_{m-1}$.

We get $\frac{m}{2}+\frac{n+1}{2}-p\geqslant\frac{n+1}{2}-\frac{p+1}{2}+2$ due to aforesaid $m\geqslant p+3$. and we get $\frac{m}{2}+\frac{n+1}{2}-p+2\leqslant n+1$ due to aforesaid $m\leqslant n+2p-3$. Hence $\otimes\in(C_p\cdots B_{\frac{n+1}{2}-\frac{p+1}{2}+2})$, and we have $\otimes\in(\cdots B_2)$. Therefore we take point 0 away from C_p0B_{n-1}, for its function is replaced by point $\otimes$. So we put the index 1 of A_1C_1 into $M+\{C_pC_{p-1}\}$ again, we get GL-matrix:

$M+\{C_pC_{p-1}\}+\{A_1C_1\}=$

$(A_1A_3\cdots A_{m-1}\ C_pB_{n-1}\cdots\ B_{\frac{m}{2}+\frac{n+1}{2}-p+2}\otimes\ B_{\frac{m}{2}+\frac{n+1}{2}-p}\cdots\ B_4B_2C_1C_3\cdots\ C_{p-2}\ C_{p-1}\cdots\ C_4C_2B_1B_3\cdots$
$B_{\frac{n+1}{2}-\frac{p+1}{2}+1} * B_{\frac{n+1}{2}-\frac{p+1}{2}+3}\cdots B_nA_m\cdots A_4A_2)$.

Where $\star\in(B_1\cdots B_n)$ and $\otimes\in(C_p\cdots B_{\frac{n+1}{2}-\frac{p+1}{2}+2})$.

(5) $m\in N_1, n, p\in N_2$:

(5−1) $\frac{m+1}{2}+\frac{n}{2}\in N_2, \frac{n}{2}+\frac{p}{2}\in N_2$; (5−3) $\frac{m+1}{2}+\frac{n}{2}\in N_1, \frac{n}{2}+\frac{p}{2}\in N_2$. We leave them as exercises.

(5−2) When $\frac{m+1}{2}+\frac{n}{2}\in N_2, \frac{n}{2}+\frac{p}{2}\in N_1$: We have $n\geqslant p$ due to the case (b) of Theorem 1. S-matrix M of the path $B_1B_2\cdots B_nC_pA_m\cdots A_2A_1C_1C_2\cdots C_{p-1}$ is

$M=(B_1B_3\cdots B_{n-1}C_pA_{m-1}\cdots A_4A_2C_1C_3\cdots C_{p-1}C_{p-2}\cdots C_4C_2A_1A_3\cdots A_mB_n\cdots B_4B_2)$.

In M, we put temporarily point 0 at C_p0A_{m-1}, so we should have $L(C_pC_{p-1})=\frac{m+1}{2}+\frac{p}{2}\in N_1$. We put point $\star$ at:

$$A_{\frac{m+1}{2}-\frac{p}{2}}\star A_{\frac{m+1}{2}-\frac{p}{2}+2}.$$

We put the index 1 of C_pC_{p-1} into M again, so we get $M+\{C_pC_{p-1}\}$.

In $M+\{C_pC_{p-1}\}$, we get $\frac{m+1}{2}-\frac{p}{2}\geqslant 1$ and $\frac{m+1}{2}-\frac{p}{2}+2\leqslant m$ since $m>n\geqslant p\geqslant 2$. Hence $\star\in(A_1\cdots A_m)$. So $L(B_1C_1)=\frac{m+1}{2}+\frac{n}{2}+1\in N_1$. And we have $L(B_1C_1)>L(C_pC_{p-1})$ due to aforesaid $n\geqslant p$. If we put point $\otimes\in(\cdots A_{\frac{m+1}{2}-\frac{p}{2}+1})$, then the edge labelling emptied out by point $\star$ is constant, and the odd edge labelling is emptied out. We put point $\otimes$ at:

$$A_{\frac{m+1}{2}+\frac{n}{2}-p+2}\otimes A_{\frac{m+1}{2}+\frac{n}{2}-p}.$$

We get $\frac{m+1}{2}+\frac{n}{2}-p+2\leqslant m-1$ since $m>n, p\geqslant 2, \frac{m+1}{2}+\frac{n}{2}\in N_2$. And we get $\frac{m+1}{2}+\frac{n}{2}-p\geqslant\frac{m+1}{2}-\frac{p}{2}+1\geqslant 2$ since $m>n\geqslant p, \frac{n}{2}+\frac{p}{2}\in N_1$. Hence $\otimes\in(A_{m-1}\cdots A_{\frac{m+1}{2}-\frac{p}{2}+1})$, and $\otimes\in(\cdots A_2)$. Therefore we take point 0 away from C_p0A_{m-1}, and we put the index 1 of B_1C_1 into $M+\{C_pC_{p-1}\}$ again, we get GL-matrix: $M+\{C_pC_{p-1}\}+\{B_1C_1\}=$

$(B_1B_3\cdots B_{n-1}C_pA_{m-1}\cdots A_{\frac{m+1}{2}+\frac{n}{2}-p+2}\otimes A_{\frac{m+1}{2}+\frac{n}{2}-p}\cdots A_4A_2C_1C_3\cdots C_{p-1}C_{p-2}\cdots C_4C_2A_1A_3\cdots$

$A_{\frac{m+1}{2}-\frac{p}{2}} \star A_{\frac{m+1}{2}-\frac{p}{2}+2} \cdots A_m B_n \cdots B_4 B_2)$.

Where $\star \in (A_1 \cdots A_m)$ and $\otimes \in (A_{m-1} \cdots A_{\frac{m+1}{2}-\frac{p}{2}+1})$.

(5—4) When $\frac{m+1}{2}+\frac{n}{2}$、$\frac{n}{2}+\frac{p}{2} \in N_1$: We have $n \geqslant p$ due to the case (c) of Theorem 1. If we take the path $B_n \cdots B_2 B_1 C_1 A_1 A_2 \cdots A_m C_p \cdots C_3 C_2$, we get GL-matrix M_0 of G in the same method as proof (5—2): $M_0=$

$(B_n \cdots B_4 B_2 C_1 A_2 A_4 \cdots A_{\frac{m+1}{2}+\frac{p}{2}-2} \star A_{\frac{m+1}{2}+\frac{p}{2}} \cdots A_{m-1} C_p \cdots C_4 C_2 C_3 C_5 \cdots C_{p-1} A_m \cdots$

$A_{\frac{m+1}{2}-\frac{n}{2}+p} \otimes A_{\frac{m+1}{2}-\frac{n}{2}+p-2} \cdots A_3 A_1 B_1 B_3 \cdots B_{n-1})$.

Where $\star \in (A_2 \cdots A_{m-1})$ and $\otimes \in (A_{\frac{m+1}{2}+\frac{p}{2}-1} \cdots A_1)$.

(6) $m \in N_1, n \in N_2, p \in N_1$; (7) m、$n \in N_1, p \in N_2$; (8) m、n、$p \in N_1$. We leave them as exercises.

Corollary 2. Bodendiek's conjecture is correct.

proof. If $p=2$ in proofs (1), (3), (5), (7) of Theorem 1, then Corollary 2 is proved.

Example 4. As to G in Theorem 1, when $m=3, n=2, p=3$, we get GL-matrix M_0 of G by proof (6—4) of Theorem 1: $M_0=(B_1 C_3 C_1 A_2 A_3 \star \otimes A_1 C_2 B_2)$.

Example 5. As to G in Theorem 1, when $m=18, n=11, p=3$, the case (a) of Theorem 1 is satisfied, we can not get GL-matrix of G by Theorem 1. But we have GL-matrix M_0 of G:

$M_0=(A_1 A_3 \cdots A_{17} C_3 C_1 B_2 \otimes B_4 \cdots B_{10} B_{11} \cdots B_5 \star B_3 B_1 C_2 A_{18} \cdots A_4 A_2)$.

Conjecture. As to G in Theorem 1, when four exceptional cases (a), (b), (c), (d) of Theorem 1 is satisfied respectively, G are graceful. (And we have proved them recently)

REFERENCES

[1] Chen, Zhi Zeng: A proof of Bodendiek Conjecture in Graceful Graphs. Journal of Inner Mongolia Teacher's University (Chinese) 3 (1986) 1—9.

[2] Chen, Zhi Zeng: A proof of the Gracefulness of a Class of Graphs. Journal of Mathematics (Chinese) (*Wu Han* University) 1989, will be published.

A Sparse Gallai-Witt Theorem

H.-J. Prömel, Bonn, FRG
B. Voigt, Bonn, FRG

ABSTRACT

Generalizing a result of the authors (Trans. Am. Math. Soc. **309**, 113–137, 1988) and simplifying the proof thereof a sparse Gallai-Witt theorem is proved.

1. Introduction

The celebrated theorem of van der Waerden [1927] on arithmetic progressions states that for every partition $\{1,\ldots,n\} = A_0 \cup \ldots \cup A_{r-1}$, where $n \geq n(k,r)$ is sufficiently large, there exists a k-term arithmetic progression A which totally belongs to one of the A_i's. In the language of combinatorial set theory van der Waerden's result may be formulated by saying that for fixed k and n tending to infinity the chromatic numbers of the set-systems $AP(n,k)$ tend to infinity. Hereby the vertices of $AP(n,k)$ are the positive integers $i \leq n$ and a set $A \subseteq \{1,\ldots,n\}$ is an edge of $AP(n,k)$ if and only if A is a k-term arithmetic progression.

As clearly the chromatic number of a set-system may only be large if a

R. Bodendiek, R. Henn (Eds.)
Topics in Combinatorics and Graph Theory

lot of cycles exist, the question arises to what extent cycles, in particular short cycles, may be excluded but still maintaining a large chromatic number. Apparently the first question in this direction is due to Tutte who (using the alias Blanche Descartes [1948]) asked whether there exist triangle-free graphs with arbitrary large chromatic numbers. Tutte showed that graphs without triangles can have arbitrary large chromatic numbers. Later on this result was rediscovered several times (e.g., Zykov [1949], Mycielski [1955]). Eventually Erdős [1959] gave a complete solution by showing that there exist graphs of arbitrary large chromatic number and, simultaneously, arbitrary high girth. As Erdős' proof uses the so-called probabilistic method the question remaining was to construct such graphs explicitly. For the general case of k-uniform set-systems a construction has been given by Lovász [1968]. A much simpler one has been found by Nešetřil and Rödl [1979]. The Nešetřil and Rödl construction is based on a certain amalgamation technique, a variation of which is also applied successfully in our present paper.

The set-system $AP(n,k)$ suggests two different approaches to exclude short cycles, both of which have been suggested by Spencer [1975] in his paper on restricted Ramsey theorems. Say, we exclude cycles of length at most g but still want a chromatic number larger than r. The weaker approach asks whether for an appropriate $n \geq n(k,g,r)$ there exists a set $\mathcal{A}$ of (some) k-term arithmetic progressions $A \subseteq \{1,\ldots,n\}$ such that $\text{girth}(\mathcal{A}) > g$ and $\chi(\mathcal{A}) > r$. Using a probabilistic argument, Spencer showed that the answer is positive. The second, stonger approach is motivated by the fact that arithmetic progressions carry an inherent arithmetic structure, only few k-element sets are arithmetic progressions. This suggests to define for any set S of positive integers the set-system $AP(S,k)$ of k-term arithmetic progressions in S. More precisely, vertices of $AP(S,k)$ are the integers in S and $A \subseteq S$ is an edge if A is a k-term arithmetic progression. Spencer [1975] then asked whether for any choice of k, g and r there exists a finite set S such that $\text{girth}(AP(S,k)) > g$ and $\chi(AP(S,k)) > r$. In Prömel and Voigt [1988] it is shown that in fact the answer to Spencer's conjecture is affirmative, viz.

Theorem A. *Let k, g and r be positive integers. Then there exists a finite set S of positive integers such that $AP(S,k)$ does not contain any cycle of length at most g but still $AP(S,k)$ has a chromatic number larger than r.*

The proof of Theorem A, as given in Prömel and Voigt [1988], requires a sparse version of Hales-Jewett's theorem and contains several delicate technical details.

In this paper we give a short proof of Theorem A which is somewhat more direct and conceptually easier. As it turns out, this new approach allows to prove a much more stronger result, namely a sparse multidimensional version of van der Waerden's theorem, to be explained in the next section.

2. The sparse Gallai-Witt theorem

Gallai, cf., Rado [1933], and Witt [1951] established a multidimensional version of van der Waerden's theorem. The crucial notion to be used here is that of a *homothetic mapping*. Let d be a positive integer. Recall that a mapping $f : \mathbb{R}^d \to \mathbb{R}^d$ is a *homothety* (homothetic mapping) if there exist a d-dimensional vector $\vec{a} \in \mathbb{R}^d$ as well as a positive real $\alpha \in \mathbb{R}_{>0}$ such that $f(\vec{x}) = \vec{a} + \alpha \cdot \vec{x}$ for all $\vec{x} \in \mathbb{R}^d$. If T and T' are (finite) subsets of $\mathbb{R}^d$ we say that T' is a *homothetic copy* of T if there exists a homothety $f : \mathbb{R}^d \to \mathbb{R}^d$ such that $f(T) = T'$. The Gallai-Witt extension of van der Waerden's theorem asserts the following:

Theorem B. *Let T be a finite subset of $\mathbb{R}^d$ and let r be a positive integer. Then there exists a finite subset $S \subseteq \mathbb{R}^d$ such that for every partition $S = A_0 \cup \ldots \cup A_{r-1}$ there exists a homothetic copy T' of T which is totally contained in one of the A_i's.*

Notice that the special case $d = 1$ and $T = \{1, \ldots, k\}$ reduces to van der Waerden's theorem on arithmetic progressions. Formally this requires the knowledge that one may assume that every $\vec{x} \in S$ is a rational linear combination of the $\vec{a} \in T$, which readily follows from any proof of Theorem B. But even for $d = 1$ Theorem B is stronger than van der Waerden's result, consider, e.g., the set $T = \{1, \sqrt{2}, \sqrt{3}, \pi\}$.

In the language of combinatorial set theory, Theorem B may be formulated by saying that for every finite set $T \subseteq \mathbf{R}^d$ and every positive integer r there exists a finite set $S \subseteq \mathbf{R}^d$ such that the set-system $AP(S,T)$ has a chromatic number larger than r. Hereby the set-system $AP(S,T)$ has vertex set S and $T' \subseteq S$ is an edge of $AP(S,T)$ if T' is a homothetic copy of T. Notice that $AP(\{1,\ldots,n\},\{1,\ldots,k\}) = AP(n,k)$ and $AP(S,\{1,\ldots,k\}) = AP(S,k)$ for any finite set S of positive integers.

The main result of this paper is a sparse version of Theorem B, where the notion of being sparse follows the stronger approach of Spencer.

Theorem C. *Let d be a positive integer, let $T \subseteq \mathbf{R}^d$ be a finite set containing at least three elements, and let g and r be positive integers. Then there exists a finite set $S \subseteq \mathbf{R}^d$ such that the set-system $AP(S,T)$ does not contain any cycles of length at most g but yet has chromatic number larger than r.*

The requirement that $|T| \geq 3$ is obviously necessary as for $|T| = 2$ any two-element set T' is a homothetic copy of T.

From what we said before it should be clear that the particular case $d = 1$ and $T = \{1,\ldots,k\}$ reduces to Theorem A, taking for granted that every $\vec{x} \in S$ may be assumed to be a rational linear combination of the $\vec{a} \in T$.

3. An auxiliary lemma

An *s-partite system* is a pair $(\Pi_{i<s} X_i, \mathcal{E})$, where the X_i's are nonempty sets, the sets of vertices, and $\mathcal{E} \subseteq \Pi_{i<s} X_i$ is the set of edges. If $(\Pi_{i<s} Y_i, \mathcal{F})$ is another s-partite system, an embedding $f : (\Pi_{i<s} X_i, \mathcal{E}) \to (\Pi_{i<s} Y_i, \mathcal{F})$ is given

by an injection $f : \bigcup_{i<s} X_i \to \bigcup_{i<s} Y_i$ such that $f(X_i) \subseteq Y_i$ and such that

$$(x_0, \dots, x_{s-1}) \in \mathcal{E} \text{ if and only if } (f(x_0), \dots, f(x_{s-1})) \in \mathcal{F}$$

holds for all $(x_0, \dots, x_{s-1}) \in \Pi_{i<s} X_i$. A sequence $(f^0, \dots, f^{c-1})$ of embeddings $f^j : (\Pi_{i<s} X_i, \mathcal{E}) \to (\Pi_{i<s} Y_i, \mathcal{F})$ forms a *cycle* of length c if there exist mutually distinct edges $F_0, \dots, F_{c-1} \in \mathcal{F}$ such that $F_0 \in \Pi_{i<s} f^0(X_i) \cap \Pi_{i<s} f^1(X_i), \dots,$ $F_{c-2} \in \Pi_{i<s} f^{c-2}(X_i) \cap \Pi_{i<s} f^{c-1}(X_i)$ and $F_{c-1} \in \Pi_{i<s} f^{c-1}(X_i) \cap \Pi_{i<s} f^0(X_i)$. A set $\{f^j \mid j < z\}$ of such embeddings is said to have *girth* larger than g, $\text{girth}(\{f^j \mid j < z\}) > g$, if no cycle of length $c \leq g$ may be formed from the $\{f^j \mid j < z\}$. $\{f^j \mid j < z\}$ has *chromatic number* larger than r if for every partition $\mathcal{F} = A_0 \cup \dots \cup A_{r-1}$ there exists some $j < z$ such that $f_j(\mathcal{E})$ is contained in one of the A_i's.

Lemma. *Let $z_0, \dots, z_s$ be positive integers, let $\mathcal{E} \subseteq \Pi_{i \leq s} \mathbf{z}_i$ be the edge-set of an $(s+1)$-partite system (where $\mathbf{z}_i = \{1, \dots, z_i\}$) and let g and r be positive integers. Then there exist positive integers $\hat{z}_0, \dots, \hat{z}_s$, there exists $\mathcal{F} \subseteq \Pi_{i \leq s} \hat{\mathbf{z}}_i$ and there exists a set $\{f_j \mid j < z\}$, for some z, of embeddings $f_j : (\Pi_{i \leq s} \mathbf{z}_i, \mathcal{E}) \to (\Pi_{i \leq s} \hat{\mathbf{z}}_i, \mathcal{F})$ such that girth $(\{f_j \mid j < z\}) > g$ and $\chi(\{f_j \mid j < z\}) > r$.* □

A proof of this Lemma can be found in Prömel and Voigt [1986].

4. Proof of Theorem C

4.1. Let $S' \subseteq \mathbb{R}^d$ be according to Theorem B, i.e., for every partition $S' = A_0 \cup \dots \cup A_{r-1}$ there exists a homothetic copy T' of T which is totally contained in one of the A_i's. Let $\{\vec{b}_s \mid s < t\}$ be some fixed enumeration of S'. Let q be a sufficiently large prime number. The precise meaning of "sufficiently large" will be clear throughout the argumentation. The reader should observe that at any stage q may be replaced by any larger prime without spoiling the proof.

4.2. Let $(T_j)_{j<z_0^0}$ be an enumeration of the homothetic copies T' of T in S'.

For each $j < z_0^0$ let $\hat{T}_j = q^j \cdot T_j$ and for $\vec{b} \in S'$ let

$$S_{\vec{b}}^0 = \{q^j \cdot \vec{b} \mid j < z_0^0,\ \vec{b} \in T_j\}.$$

Notice that the family $(S_{\vec{b}}^0 \mid \vec{b} \in S')$ has the following properties:

(4.2.1) $\mathrm{girth}(AP(\bigcup_{\vec{b}\in S'} S_{\vec{b}}^0, T)) > g$.

In fact, if q is sufficiently large, then the only homothetic copies of T in $\bigcup_{\vec{b}\in S'} S_{\vec{b}}^0$ are the $\hat{T}_j$'s, $j < z_0^0$, and these are mutually disjoint.

(4.2.2) For every partition $\bigcup_{\vec{b}\in S'} S_{\vec{b}}^0 = A_0 \cup \ldots \cup A_{r-1}$ such that each $S_{\vec{b}}^0$ totally belongs to one of the A_i's there exists some $j < z_0^0$ such that $\hat{T}_j$ is contained in one of the A_i's.

4.3. We first show how to construct $(S_{\vec{b}}^1 \mid \vec{b} \in S')$ and then, by induction on $s < t$, we construct $(S_{\vec{b}}^{s+1} \mid \vec{b} \in S')$. Consider the 1-partite system $(\mathbf{z}_0^0, \mathcal{E}_0)$ which is defined by $\mathcal{E}_0 = \{j < \mathbf{z}_0^0 \mid \vec{b}_0 \in T_j\}$. Thus $\mathcal{E}_0$ provides a copy of $S_{\vec{b}_0}^0$. According to the Lemma there exists a 1-partite system $(\mathbf{z}_0^1, \mathcal{E}_0')$ and there exists a set $\{f_j \mid j < z_1^1\}$ of embeddings $f_j : (\mathbf{z}_0^0, \mathcal{E}_0) \to (\mathbf{z}_0^1, \mathcal{E}_0')$, $j < z_1^1$, such that $\mathrm{girth}(\{f_j \mid j < z_1^1\}) > g$ and $\chi(\{f_j \mid j < z_1^1\}) > r$. For $j < z_1^1$ and $\vec{b} \in S'$ we define

$$S_{\vec{b},j}^1 = \{q^{f_j(\alpha)} \cdot \vec{b} + q^{z_0^1} \cdot (q^j \cdot \vec{b} + \sum_{i<z_0^1, i\neq j} q^i \cdot \vec{b}_0) \mid \alpha \in S_{\vec{b}}^0\}$$

and

$$S_{\vec{b}}^1 = \bigcup_{j<z_1^1} S_{\vec{b},j}^1.$$

Notice that the family $(S_{\vec{b}}^1 \mid \vec{b} \in S')$ satisfies the following properties:

(4.3.1) For every $j < z_1^1$ the family $(S_{\vec{b},j}^1 \mid \vec{b} \in S')$ is a copy of $(S_{\vec{b}}^0 \mid \vec{b} \in S')$.

(4.3.2) The sets $\{S_{\vec{b}_0,j}^1 \mid j < z_1^1\}$ do not form cycles of length at most g.

(4.3.3) For $\vec{b} \neq \vec{b}_0$ and $j \neq j'$ the sets $S_{\vec{b},j}^1$ and $S_{\vec{b},j'}^1$ are disjoint.

(4.3.4) If $T' \subseteq \bigcup\{S_{\vec{b},j}^1 \mid \vec{b} \in S',\ j < z_1^1\}$ is any homothetic copy of T then there exists some $j < z_1^1$ such that already $T \subseteq \bigcup_{\vec{b}\in S'} S_{\vec{b},j}^1$.

(4.3.5) girth($AP(\bigcup_{\vec{b}\in S'} S^1_{\vec{b}}, T)) > g$.

Properties (4.3.1)-(4.3.3) are obvious from the construction, (4.3.4) follows as the prime number q is chosen to be sufficiently large. (4.3.5) follows immediately from (4.3.1) and (4.3.2), recall that girth($AP(\bigcup_{\vec{b}\in S'} S^0_{\vec{b}}, T)) > g$.

4.4. Now we show how to construct $\{S^{s+1}_{\vec{b}} \mid \vec{b} \in S'\}$ in general. Assume by induction that for some $1 \leq s < s'$ a family $\{S^s_{\vec{b}} \mid \vec{b} \in S'\}$ has been constructed such that the following properties are satisfied:

(4.4.1) There exists a sequence $(z^s_j)_{j\leq s}$ of positive integers such that

$$\begin{aligned} S^s_{\vec{b}} \subseteq \{\vec{b}(j_0,\ldots,j_s) = {} & q^{j_0}\cdot\vec{b} + q^{z^s_0}\cdot(q^{j_1}\cdot\vec{b} + \sum_{i<z^s_1, i\neq j_1} q^i\cdot\vec{b}_0) \\ & + q^{z^s_0+z^s_1}\cdot(q^{j_2}\cdot\vec{b} + \sum_{i<z^s_2, i\neq j_1} q^i\cdot\vec{b}_1) + \ldots \\ & + q^{z^s_0+\ldots+z^s_{s-1}}\cdot(q^{j_s}\cdot\vec{b} + \sum_{i<z^s_s, i\neq j_{s-1}} q^i\cdot\vec{b}_{s-1}) \\ & \mid j_0 < z^s_0, j_1 < z^s_1, \ldots, j_s < z^s_s\}, \end{aligned}$$

moreover, for every fixed $j < z^s_s$ the family

$$(S^s_{\vec{b},j} = \{\vec{b}(j_0,\ldots,j_{s-1},j) \in S^s_{\vec{b}} \mid j_0 < z^s_0,\ldots,j_{s-1} < z^s_{s-1}\} \mid \vec{b} \in S')$$

is a copy of $(S^{s-1}_{\vec{b}} \mid \vec{b} \in S')$.

(4.4.2) girth($AP(\bigcup_{\vec{b}\in S'} S^s_{\vec{b}}, T)) > g$.

(4.4.3) For every partition $\bigcup_{\vec{b}\in S'} S^s_{\vec{b}} = A_0 \cup \ldots \cup A_{r-1}$ there exists some $j < z^s_s$ such that $S^s_{\vec{b}_{s-1}}$ is contained in one of the A_i's.

To construct $(S^{s+1}_{\vec{b}} | \vec{b} \in S')$ observe that according to (4.4.1) the set $S^s_{\vec{b}_s}$ defines an $(s+1)$-partite system on $\Pi_{i\leq s}\mathbf{z}^s_i$ by

$$\mathcal{E}_s = \{(j_0,\ldots,j_s) \in \Pi_{i\leq s}\mathbf{z}^s_i \mid \vec{b}_s(j_0,\ldots,j_s) \in S^s_{\vec{b}_s}\}.$$

By the Lemma there exists an $(s+1)$-partite system $(\Pi_{i\leq s}\mathbf{z}^{s+1}_i, \mathcal{E}'_s)$ and there exists a set $\{f_j \mid j < z^{s+1}_{s+1}\}$ of embeddings $f_j : (\Pi_{i\leq s}\mathbf{z}^s_i, \mathcal{E}_s) \to (\Pi_{i\leq s}\mathbf{z}^{s+1}_i, \mathcal{E}'_s)$

such that $\mathrm{girth}(\{f_j \mid j < z_{s+1}^{s+1}\}) > g$ and $\chi(\{f_j \mid j < z_{s+1}^{s+1}\}) > r$. For $j < z_{s+1}^{s+1}$ we define

$$\begin{aligned} S_{\vec{b},j}^{s+1} = & \{q^{f_j(\alpha_0)} \cdot \vec{b} + q^{z_0^{s+1}} \cdot (q^{f_j(\alpha_1)} \cdot \vec{b} + \sum_{i<z_1^{s+1}, i \neq j_1} q^i \cdot \vec{b}_0) + \ldots \\ & + q^{z_0^{s+1}+\ldots+z_{s-1}^{s+1}} \cdot (q^{f_j(\alpha_s)} \cdot \vec{b} + \sum_{i<z_s^{s+1}, i \neq j_s} q^i \cdot \vec{b}_{s-1}) \\ & + q^{z_0^{s+1}+\ldots+z_s^{s+1}} \cdot (q^j \cdot \vec{b} + \sum_{i<z_{s+1}^{s+1}, i \neq j} q^i \cdot \vec{b}_s) \\ & \mid \alpha_i < z_i^s,\ i \leq s, \text{ and } \vec{b}(\alpha_0, \ldots, \alpha_s) \in S_{\vec{b}}^s\}. \end{aligned}$$

Then, using essentially the same arguments as under 4.3 one readily shows that the family $(S_{\vec{b}}^{s+1} \mid \vec{b} \in S')$ again satisfies properties (4.4.1)-(4.4.3), where $S_{\vec{b}}^{s+1} = \bigcup_{j<z_{s+1}^{s+1}} S_{\vec{b},j}^{s+1}$.

4.5. Finally we consider the set $S = \bigcup_{\vec{b} \in S'} S_{\vec{b}}^t$. By (4.4.2) it follows that $\mathrm{girth}(AP(S,T)) > g$, thus it remains to show that $\chi(AP(S,T)) > r$. Let $S = A_0 \cup \ldots \cup A_{r-1}$ be any partition. Working backwards the construction of the families $(S_{\vec{b}}^s \mid \vec{b} \in S')$, $s \leq t$, using property (4.4.3) at each step, we see that there exists a copy of $(S_{\vec{b}}^0 \mid \vec{b} \in S')$ such that each $S_{\vec{b}}^0$, $\vec{b} \in S'$ (resp., its copy), totally belongs to one of the A_i's. According to the choice of S' (compare property (4.2.2)) then there exists a homothetic copy T' of T which is totally contained in one of the A_i's, thus completing the proof of Theorem C. □

References

Descartes, B. (1948): A three color problem. Eureka

Erdős, P. (1959): Graph theory and probability. Can. J. Math. **11**, 34–38

Lovász, L. (1968): On the chromatic number of finite set-systems. Acta Math. Acad. Sci. Hung. **19**, 59–67

Mycielski, J. (1955): Sur le coloriage des graphes. Colloq. Math. **3**, 161–162

Nešetřil, J., Rödl, V. (1979): A short proof of the existence of highly chromatic hypergraphs without short cycles. J. Comb. Theory, Ser. B **27**, 225–227

Prömel, H.J., Voigt, B. (1986): Ramsey theorems for finite graphs. To appear, cf. Report No. 86447-OR, Institut für Operations Research, Universität Bonn

Prömel, H.J., Voigt, B. (1988): A sparse Graham-Rothschild theorem. Trans. Am. Math. Soc. **309**, 113–137

Rado, R. (1933): Studien zur Kombinatorik. Math. Z. **36**, 424–480

Spencer, J. (1975): Restricted Ramsey configurations. J. Comb. Theory, Ser. A **19**, 278–286

van der Waerden, B.L. (1927): Beweis einer Baudetschen Vermutung. Nieuw Arch. Wiskd. **15**, 212–216

Witt, E. (1951): Ein kombinatorischer Satz der Elementargeometrie. Math. Nachr. **6**, 261–262

Zykov, A.A. (1949): On some properties of linear complexes. Mat. Sb., Nov. Ser. **24**, 163–188 (In Russian; English translation: Transl., II. Ser., Am. Math. Soc. **79** (1952))

Edges With at Most One Crossing in Drawings of the Complete Graph

H. Harborth, Braunschweig, FRG
I. Mengersen, Braunschweig, FRG

In 1963 G. Ringel ([3]) determined 2n-2 to be the maximum number of edges without crossings in drawings $D(K_n)$ of the complete graph K_n in the plane. A drawing D(G) is a realization of G in the plane with distinct points (also called vertices) for the vertices of G, and curves (also called edges) for the edges of G such that two edges have at most one point in common, either an endpoint or a crossing. As generalizations it may be asked for the maximum numbers $H_s(n)$ or the minimum numbers $h_s(n)$ of edges with at most s crossings in drawings $D(K_n)$. In [1] $h_0(n)$ is determined, and $h_s(n)=0$ for $n \geqq 4s+8$ holds in general ([2]). Here we will give an alternative proof of Ringel's result, $H_0(n)=2n-2$, and estimations of $H_1(n)$.

Theorem 1 (G. Ringel 1963): $H_0(n)=2n-2$ for $n \geqq 4$.

Proof. Let d be the minimum number of edges without a crossing incident to a vertex v in a $D(K_n)$ with $H_0(n)$ edges without crossings. If all edges not incident to v and with at least one crossing are deleted from $D(K_n)$ then a plane graph remains. Since this graph can have at most 3n-6 edges it follows

$$\frac{nd}{2} \leqq H_0(n) \leqq 3n-6-(n-1-d) = 2n-5+d.$$

This inequality yields $d \leqq 4-2/(n-2)$, and with $d \leqq 3$ it is proved $H_0(n) \leqq 2n-2$. Drawings as for n=10 in Figure 1 where all missing edges have to be added outside of the (n-1)-gon prove $H_0(n) \geqq 2n-2$.

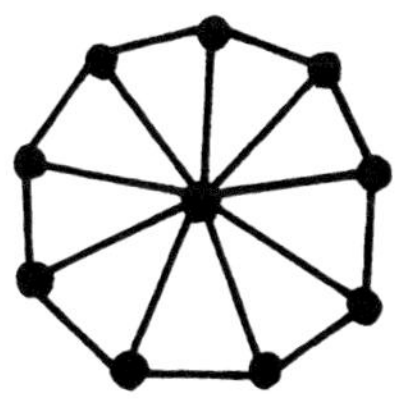

Figure 1.

R. Bodendiek, R. Henn (Eds.)
Topics in Combinatorics and Graph Theory

Theorem 2: $H_1(n) = \binom{n}{2}$ for $n \leqq 6$,

$H_1(n) = 18, 20, 22$ for $n=7,8,9$, and

$$2n-2+\left[\frac{n-1}{2}\right] \leqq H_1(n) \leqq 2n-2+\left[\frac{n-1}{2}\right]+9 \text{ for } n \geqq 8.$$

Proof. For $n \leqq 6$ one can easily find a $D(K_n)$ where each edge has at most one crossing. Let $D_0 = D_0(K_n)$ be a drawing of K_n with the maximum number $H_1(n)$ of edges with at most one crossing. If all edges with at least two crossings are deleted from D_0 then the remaining drawing is denoted by D_1. If in D_1 all edges incident to a vertex v are deleted then v lies in a region F_v with its boundary B_v. A cycle in D_1 which does not intersect itself and which has at least one vertex inside so as at least one vertex outside is called a separating cycle C. Let d(v) denote the degree of v in D_1. To prove the upper bound for $H_1(n)$ we first give three lemmas.

Lemma 1: In D_1 the four vertices of any pair of intersecting edges are connected by four edges without crossings, and these four edges do not enclose other vertices or edges.

Lemma 2: Each edge of B_v can be intersected in D_1 by at most one edge incident to v.

Lemma 3: In D_1 the length l of a separating cycle C fulfills

$$l \geqq \frac{mn - m^2}{m+1},$$

if at least m vertices are inside and at least m vertices are outside of C.

The proofs of Lemmas 1 and 2 are evident. Since in D_0 every edge connecting one of the x vertices inside of C to one of the n-l-x vertices outside of C intersects exactly one edge of C, Lemma 3 can be derived from

$$l \geqq x(n-l-x) \geqq m(n-l-m).$$

To prove the general upper bound of $H_1(n)$ in Theorem 2 we consider a vertex v in D_1 which has minimum degree $d=d(v)$. We discuss the three cases where no (I), one (II), or at least two (III) vertices besides v do not belong to B_v.

Case I: Vertex v can be connected without crossings to all

remaining n-1 vertices to give n-1 triangles without crossings (if necessary additional edges without crossings can be added). If one of two intersecting edges is deleted we obtain a plane graph. Since every plane graph with n vertices has at most 3n-6 edges and at most 2n-4 triangles we conclude

$$\frac{nd}{2} \leqq H_1(n) \leqq 3n-6-(n-1-d)+\left[\frac{2n-4-(n-1)}{2}\right] = 2n-6+d+\left[\frac{n-1}{2}\right]$$

assuming that every pair of 2n-4-(n-1) triangles could be completed by an edge with one crossing. With $d \leqq 4$ which follows from this inequality we have

$$H_1(n) \leqq 2n-2+\left[\frac{n-1}{2}\right].$$

Case II: Vertex v can be connected without crossings to all n-2 vertices of B_v. At least n-3 triangles without crossings arise (if necessary additional edges without crossings can be added). If one of two intersecting edges is deleted we obtain corresponding to Case I

$$\frac{nd}{2} \leqq H_1(n) \leqq 3n-6-(n-2-d)+\left[\frac{2n-4-(n-3)}{2}\right] = 2n-4+d\left[\frac{n-1}{2}\right].$$

This yields $d \leqq 5$, and thus

$$H_1(n) \leqq 2n+1+\left[\frac{n-1}{2}\right].$$

Case III: A separating cycle C is contained in B_v. We assume v to be outside of C. All vertices besides v and those of B_v are inside of C, since otherwise, outside of C, another separating cycle is contained in B_v which has at most one vertex in common with C, and with Lemma 3 both cycles together would have at least $2\frac{n-1}{2}-1=n-2$ vertices which is impossible since B_v has at most n-3 vertices.

Let C have length l with

$$l = \left\{\frac{n-1}{2}\right\}+r.$$

Because of Lemma 2 edges from any inner vertex of C to C are completely inside of C. Assume an inner vertex w is connected to the vertices $w_1, w_2, \ldots, w_k$ of C with $k \geqq r+4$. Then a second vertex within C lies for example in the sector determined by w_1, w, w_k. The boundary of this sector is a separating cycle of length l_1 with

$$l_1 = 3+l-k \leqq 3+l-r-4 < \{\frac{n-1}{2}\}$$

which contradicts Lemma 3. Thus every vertex inside of C is connected to at most r+3 vertices of C, and at most to vertex v outside of C because of Lemma 2. We can choose w inside of C such that w is connected to at most 4 vertices inside of C, since no cycle exists separating the vertices inside of C (contradiction to Lemma 3), and then Case I can be used. Altogether we obtain for the degree of w in D_1

$$d(w) \leqq r+3+1+4 = r+8.$$

Since C separates w and v there exists in B_w a separating cycle C_1 no parts of which are outside of C. The length of C_1 is at least $\{(n-1)/2\}$ because of Lemma 3. Let $d_i(v)$ and $d_a(w)$ be the numbers of edges in D_1 connecting v and w to vertices inside of C and outside of C_1, respectively. Now similar to Cases I and II we add edges without crossings incident to v or w, and delete one of two intersecting edges in D_1 to get a plane graph. It follows

$$\begin{aligned}\frac{nd}{2} \leqq H_1(n) &\leqq 3n-6-(\{\frac{n-1}{2}\}+r-(d-d_i(v)))-(\{\frac{n-1}{2}\}-(d(w)-d_a(w)))+ \\ &\quad +\frac{1}{2}[2n-4-(\{\frac{n-1}{2}\}+r-d_i(v))-(\{\frac{n-1}{2}\}-d_a(w))] \leqq \\ &\leqq 2n-5-r+d+d(w)+\frac{1}{2}(n-3-r-(d_i(v)+d_a(w))) \leqq \\ &\leqq \frac{1}{2}(5n+1)+d.\end{aligned}$$

The last step uses that either $d(w)<r+8$, or $d_i(v)\geqq 1$ so as $d_a(w)\geqq 1$. Now this inequality yields $d\leqq 6$ for $n\leqq 8$, and the upper bound in Theorem 2 is proved. For $n\geqq 14$ we even can prove +8 instead of +9.

The general lower bound of $H_1(n)$ in Theorem 2 follows as in the examples for n=10 and n=11 in Figure 2 where the missing edges are drawn outside of the (n-1)-gons.

To prove $H_1(n) \leqq 18$, 20, and 22 for n=7,8, and 9 we assume the existence of drawings $D(K_n)$ with more than 18, 20, or 22 edges with at most one crossing, and try to find a contradiction.

Let Q denote a subgraph K_4 in $D(K_n)$ with one pair of intersec-

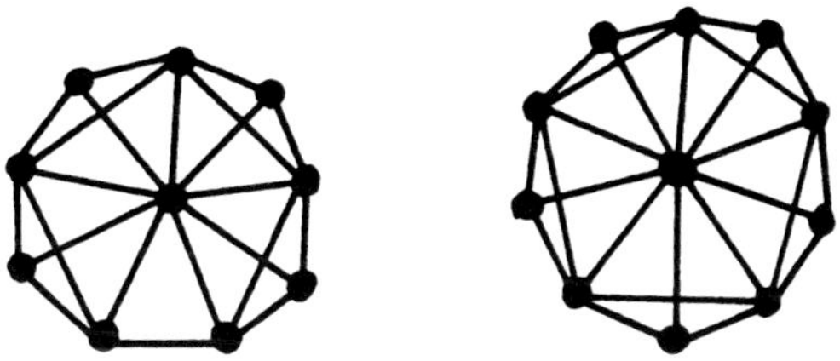

Figure 2.

ting edges and no other crossings. Let N(Q) denote the number of subgraphs Q in $D(K_n)$. Since at most 3n-6+N(Q) edges with at most one crossing occur in $D(K_n)$ the above assumption yields $N(Q)>2$ for n=7 and 8, and $N(Q)\geqq 2$ for n=9.

First we consider the case that two subgraphs Q have an edge in common. This is possible only as in Figure 3, where vertices 7,8, or 9, so as all missing edges are outside (Lemma 1).

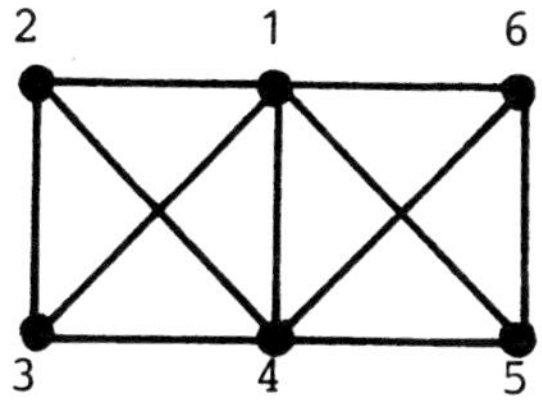

Figure 3.

If $n\geqq 8$ then all 4 edges (2,6),(3,5),(2,5),(3,6) have at least two crossings. Moreover, each vertex v=7,8, or 9 has at least 4 edges to 1,...,6 with at least two crossings since every triangle of v and (1,3),(1,5),(2,4), or (4,6), so as both quadrangles of v and (3,1,5) and (2,4,6) have at least one edge with at least two crossings if vertices from inside are connected to those outside. Thus we obtain

$$H_1(8) \leqq \binom{8}{2}-4-2\cdot 4 = 16 \quad \text{and} \quad H_1(9) \leqq \binom{9}{2}-4-3\cdot 4 = 20$$

which are contradictions to the above assumptions.

In the second case two subgraphs Q have at most one vertex in common. For $n\leqq 8$ this is impossible because of $N(Q)>2$. It remains n=9.

If two subgraphs Q with vertices 1,2,3,4 and 1,5,6,7 have (1,3) and (2,4), so as (1,6) and (5,7) as their intersecting edges, then at least 7 of the remaining edges, excluded possibly (2,7) and (4,5), have at least two crossings. At least 4 edges from each vertex v=8, or 9 have at least two crossings to vertices 1,...,7, since each of the triangles of v together with (1,3),(2,4),(1,6),(5,7), and the quadrangle (v,3,1,6) has at least one edge with at least two crossings. Thus

$$H_1(9) \leqq \binom{9}{2}-7-2\cdot 4 = 21$$

gives the contradiction.

If vertices 1,2,3,4 and 5,6,7,8 determine two subgraphs Q then N(Q)=2. Each of two vertex disjoint edges without crossings of one subgraph Q has at least 5 edges with at least two crossings to the secons subgraph Q if appropriate triangles and quadrangles are discussed as above. Also vertex 9 is incident to at least 4 vertices of 1,...,8 by edges with at least two crossings so that

$$H_1(9) \leqq \binom{9}{2}-4+2\cdot 5 = 22$$

finishes the proof of the upper bounds.

Finally, $H_1(n) \geqq 18, 20$, and 22 for n=7,8, and 9 is proved by Figures 4,5, and 6.

In general we conjecture that the lower bound of Theorem 2 for $n \geqq 8$ is the exact value of $H_1(n)$.

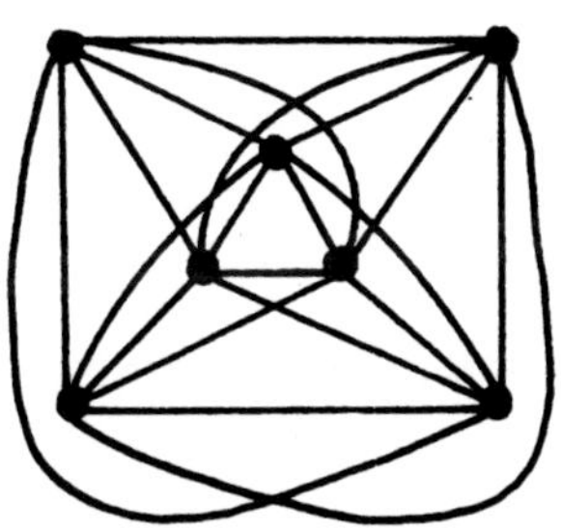

Figure 4.

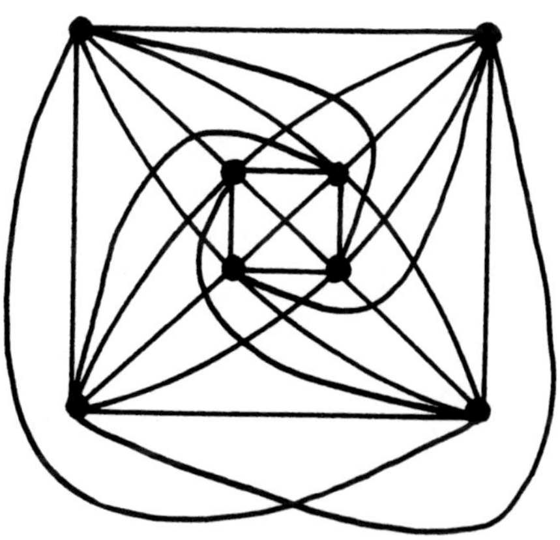

Figure 5.

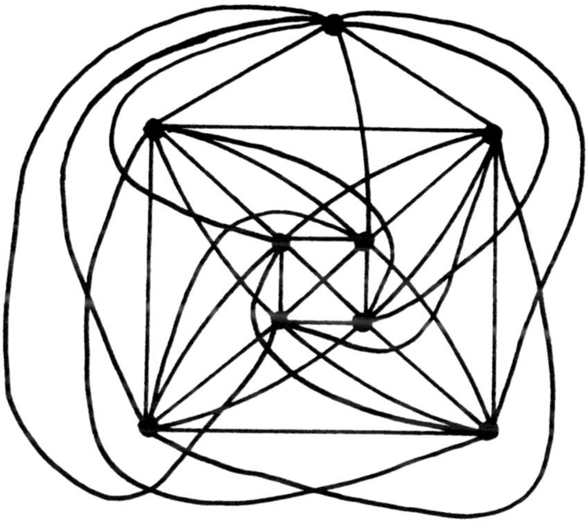

Figure 6.

REFERENCES

[1] H. Harborth and I. Mengersen: Edges without crossings in drawings of complete graphs. J. Combinatorial Theory (B) 17(1974), 299-311.

[2] H. Harborth: Darstellungen des vollständigen Graphen ohne Kanten mit wenigen Kreuzungen. 33. Intern. Wiss. Koll. TH Ilmenau 1988, Heft 4, C2, Graphen und Netzwerke - Theorie und Anwendungen, 1988, 179-181.

[3] G. Ringel: Extremal problems in the theory of graphs. Theory of Graphs and its Applications (Proc. Sympos. Smolenice, 1963). Publ. House Czechoslovak Acad. Sci., Prague 1964, 85-90.

Long Cycles in Graphs with Moderate Connectivity

H. A. Jung, Berlin, FRG

1. Introduction

In this paper we supply bounds for the length of a longest cycle C in G in terms of the structure of G-V(C).

We call a tree H a *star (double-star)* if all vertices of H with at most one (exactly two) exceptions have degree one. A *quasistar,* by definition, is a star, a double star or obtained from a star H with $|H|\geq 4$ by adding one edge. Let $\mathcal{K}_5$ denote the class of all stars and $\mathcal{K}_6$ the class of all quasistars, further $\mathcal{K}_k=\emptyset$ for k<5.

Theorem 1. *Let C be a longest cycle in the graph G, and let H be a component of G-V(C) such that $|H|\geq k-1$ (k=2,3,4,5,6). There exists a vertex v in H such that*

(i) $|C| \geq kd(v)-k(k-2)$ if G is k-connected and $H \notin \mathcal{K}_k$

(ii) $|C \cup H| > kd(v)-(k-1)(k-2)$ if G is (k-1)-connected and $H \notin \mathcal{K}_k$

Parts of Theorem 1 were proved by G.A. Dirac [3] (k=2), C.St.J.A. Nash-Williams [9] (k=3,(ii)), also by H.J. Voß [11] and the author [7] (k≤3) and by G. Fan [4] (k≤4). If H is not complete, the bounds can be improved by the additive constant min(4,k).

For related results we refer the reader to [1], [2], [4], [6] and [12], further to [5] and [10] .

2. Notations and preliminaries

For vertices a and b in a connected graph H let $D_H(a,b)$ denote the length of a longest ab-path in H . Considering a longest ab-path in H one easily verifies that $D_H(a,b) \leq D_H(a,c)+D_H(c,b)$ for any three vertices a,b and c in a connected graph H. For a complete graph H we set $D(H)=|H|-1$, and for a 2-connected graph H we set $D(H)= \max(D_H(v,v'):v,v' \in V(H)$ and $v\neq v')$.

R. Bodendiek, R. Henn (Eds.)
Topics in Combinatorics and Graph Theory

If H and K are subgraphs of a given graph G or $K \subseteq V(G)$ we denote by $N_H(K)$ the set of all vertices in H, which are adjacent in G to at least one vertex in K. For $K=\{v_1,v_2,...,v_m\} \subseteq V(G)$ we also write $N_H(v_1,v_2,...,v_m)$ for $N_H(K)$, also $d_H(v_1)=|N_H(v_1)|$ and in particular $d(v_1)=d_G(v_1)$. The number of edges joining disjoint subgraphs H and K of G will be denoted by e(H;K); analogously $e(v_1,...,v_n;K)$ and $e(v_1,v_2,...,v_m;w_1,w_2,...,w_n)$ are defined. Paths and cycles in a graph G are viewed as subgraphs of G. For a path P and vertices a and b on P let P [a,b] (P(a,b] and P(a,b)) denote the subpath with terminal vertices a and b (P[a,b]-a, P(a,b]-b). Similarly if C is a cycle, on which one of the two cyclic orientations is given, C[a,b], C(a,b] and C(a,b) are the subpaths in that direction from a to b.

For a path P in G and a component H of G-V(P) let P^H denote the maximum subpath P[a,b] of P such that $a,b \in N_P(H)$.

We will use the following result from [8] (Theorem 1).

Lemma 2.1. *Let a,b be distinct vertices in the 2-connected graph G, and let P be a longest ab-path in G. Each component H of G-V(P) contains a vertex v such that* $|P^H| \geq d_G(v)+1$.

Lemma 2.2. *Let H be a 2-connected graph. There exist distinct vertices v_1,v_2 and v_3 in H such that*

(i) $d_H(v_i) \leq D(H)$ $(i=1,2)$,

(ii) $d_H(v_3) \leq \min(D(H)+1,\ D_H(v_1,v_2))$,

(iii) $2D_H(v_1,v_2) \geq d_H(v_1)+d_H(v_2)+2$ *if G is not complete,*

(iv) $D_H(v_1,v_2) \geq 4$ *if* $|H| \geq 5$.

Proof. Abbreviate D=D(H) and $d_H(v)=d(v)$ for $v \in V(H)$. According to the definition there exist a,b in H and an ab-path P in H such that |P|-1=D.

If |P|=|G| then the claims trivially hold, since in that case D=|H|-1. Assuming |P|<|H| we list the components $H_1\ H_2,..., H_m$ of H-P so that $|P^{H_1}| \leq ... \leq |P^{H_m}|$.

For $1 \leq \mu \leq m$ we determine, by Lemma 2.1, a vertex v_μ in H_μ such that $|P^{H_\mu}| \geq d(v_\mu)+1$ .For w on P let w^+ and w^- denote respectively the successor and predecessor of w (if any) in the order from a to b. Abbreviate $P_\mu=P^{H_\mu}$ $(1 \leq \mu \leq m)$.

Case 1. $m \geq 3$.

If $|P_1|+|P_2| \leq 2|P|-2$ then $2D=2|P|-2 \geq d(v_1)+d(v_2)+2 \geq 6$ and $D \geq d(v_3)$. In that subcase also $D_H(v_1,v_2) \geq 4$ since |P|=4 implies $N_P(H_\mu) \cap \{a,b\} \neq \emptyset (\mu=1,2)$ hence $D_H(v_1,v_2) \geq 4$.

If $|P_1|+|P_2| \geq 2|P|-1$, then $P_2=P$ and $\{a,b\}\cap P_1 \neq \emptyset$. In that subcase $D_H(v_1,v_2)\geq D+2$ and $d(v_3)\leq D$.

Case 2. m=2.

As in Case 1 we obtain $D \geq d_H(v_i)$ (i=1,2), $2D_H(v_1,v_2)\geq d(v_1)+d(v_2)+2$ and $D_H(v_1,v_2))\geq 4$.

For any vertex v_3 in $P-N_p(H_1\cap H_2)$ clearly $D=|P|-1\geq d_P(v_3)=d(v_3)$. Such a vertex exists since otherwise say $a\in N_P(H_1)$, further successively $a^+\in N_P(H_2)$, $a^{++}\in N_P(H_1)$ and $a^{+++}\in N_P(H_2)$. But then P would not be a longest ab-path.

Case 3. m=1.

Let $P_1 = P[a_1,b_1]$. If $P \neq P_1$, say $a \neq a_1$, by the maximality of P, vertices , b_1^- and a_1^- are not adjacent hence $|P_1|1| > d_H(v)$ for all v in $\{v_1, b_1^- , a_1^-\}$. In that subcase (i),(ii) and (iii) hold with $\{b_1^-,a_1^-\} = \{v_2,v_3\}$. Further $D_H(v_1,v_2) \geq 4$, since $a_1 \neq a$ and $|P| = 4$ imply $N_P(H_1) = \{a^+,b\}$ and $N_H(a) = \{a^+,b\}$ hence $D_H(v_1,b^-) \geq 4$.

Now assume $P = P_1$. If $|P| \geq 4$ then the claims hold with $\{v_1,v_2,v_3\} = \{v_1,a^+,b^-\}$. Finally, let $|P| = 3$ hence $|H| = 4$. If d(a)=2, the claims hold with $v_2=a$ and $v_3=a^+$. If d(a)=3, the claims hold with $v_2=a^+$ and $v_3=a$. □

3. Proof of Theorem 1

In this section a finite graph G is given, and the notions of the previous section refer to G. For a non-empty subgraph H of G the *relative connectivity* of H in G is $\kappa_H = \min (|N_{G-K}(K)| : \emptyset \neq K \subseteq V(H)$ and $N_{G-K}(K) \neq V(G)-K)$ unless the minimum ranges over the empty set, in which case $\kappa_H=|G|-1$. Thus κ_G is the connectivity of G and $\kappa_G \leq \kappa_H \leq d(v)$ for all $v\in V(H)$.

In the sequel also a longest cycle C in G and a component H of G-V(C) are fixed. We set $\beta_H = 0$ if H is a complete graph, otherwise $\beta_H = 1$.

We first treat the case, when H is 2-connected or $|H|\leq 2$.

Theorem 3.1. *Let H be 2-connected or |H|=2. If $\kappa_H \geq 2$ there exists a vertex v in H such that $|C|\geq 2d(v)+2\beta_H+(\kappa_H-2)D(H)$ and $|C|+|H|\geq 3d(v)+3\beta_H-1+(\kappa_H-2)(D(H)-1)$, moreover $|C|\geq 3d(v)+3\beta_H-3+(\kappa_H-3)(D(H)-1)$, if $\kappa_H\geq 3$.*

Theorem 3.2. *Let H be 2-connected and $|H|\geq k-1 (k=4,5,6)$. There is a vertex v in H such that (i) $|C|\geq kd(v)+4\beta_H-k(k-2)$ if $\kappa_H \geq k$*

and

(ii) $|C\cup H| > d(v)+4\beta_H-(k-1)(k-2)$ if $\kappa_H\geq k-1$.

Proof of (3.1) and 3.2): We fix one of the two cyclic orientations of C and label $N_C(H)=\{x_1,x_2,...,x_s\}$ (indices mod s) accordingly. Let $X=\{x_i:d_H(x_i)\geq 2\}$ and $y=\{y\in V(H):d_{C-X}(y)\geq 2\}$. Since $N_C(H-Y)\subseteq X$ we have $\kappa_H\leq|X|+|Y|$ or $V(H)=Y$.

We call $C_i=C(x_i,x_{i+1}]$ a good segment if $|N_H(x_i,x_{i+1})|\geq 2$. Since $|H|\geq 2$ and $\kappa_H\geq 2$, for each $y\in Y$, we can determine a vertex $\varphi(y)=x_i$ in $N_C(H)$ such that $N_H(x_i)=\{y\}$ and $N_H(x_i,x_{i+1})\neq\{y\}$. Clearly $|X\cup\varphi(Y)|=|X|+|Y|$, and for any $x_i\in X\cup\varphi(Y)$ the segment C_i is good. Hence $\ell\geq\ell_1$, where $\ell_1=|X|+|Y|$ and ℓ is the number of good segments.

If $|H|=2$ let $V(H)=\{v_1,v_2\}$. If H is 2-connected we determine v_1,v_2 and v_3 according to Lemma 2.2. For $1\leq i\leq s$ and, if $|H|\geq 3$, for $h\in\{0,1,2\}$, we abbreviate $\alpha_i=|C_i|-1/2\ e(v_1,v_2;x_i,x_{i+1})$ and $\varepsilon_i^{(h)}=|C_i|-e(v_1,v_2;x_1,x_{i+1})-(h/2)\ e(v_3;x_i,x_{i+1})$, further $\alpha=\sum_{i=1}^{s}\alpha_i$ and $\varepsilon^{(h)}=\sum_{i=1}^{s}\varepsilon_i^{(h)}$. Since $|C|=\sum_{i=1}^{s}|C_i|$ and $\sum_{i=1}^{s}e(v;x_i,x_{i+1})=2d_C(v)$ for all $v\in V(H)$ we have

(1) $|C|=d_C(v_1)+d_C(v_2)+\alpha$ and
(2) $|C|=2d_C(v_1)+2d_C(v_2)+hd_C(v_3)+\varepsilon^{(h)}$, if $|H|\geq 3$.

Next we provide bounds for α_i and $\varepsilon_i^{(h)}$. Abbreviating $R_1=\min(D(H), D_H(v_1,v_2)-1)$ and $R_2=\min(D(H), D_H(v_1,v_2)-2)$ we claim that for each good segment C_i
(3) $\alpha_i\geq R_1+1$ and $\varepsilon_i^{(h)}\geq R_2-h$.

If C_i is good, then there exist distinct vertices $u\in N_H(x_i)$ and $w\in N_H(x_{i+1})$. Since C is a longest cycle, we must have $|C_i|\geq D_H(u,w)+2\geq D(H)+2$. Thus if in addition $e(v_1,v_2;x_i,x_{i+1})\leq 2$ then $\alpha_i\geq D(H)+1$. If $e(v_1,v_2,x_i,x_{i+1})\geq 3$ we can choose $\{u,w\}=\{v_1,v_2\}$ and so obtain $|C_i|\geq D_H(v_1,v_2)+2$ whence $\alpha_i\geq D_H(v_1,v_2)$ and $\varepsilon_i^{(h)}\geq D_H(v_1,v_2)-2-h$. This proves (3).

Now consider a segment C_i, which is not good. For such C_i we have $N_H(x_i,x_{i+1})=\{y_i\}$, where $y_i\in Y$. Thus $\alpha_i\geq|C_i-1|\geq 1$ and $\varepsilon_i^{(h)}\geq|C_i|-2\geq 0$. If in addition $y_i\notin\{v_1,v_2\}(y_i\notin\{v_1,v_2,v_3\}$, then $\alpha_i\geq 2$ and $\varepsilon_i^{(h)}\geq 2-h$, (respectively $\varepsilon^{(h)}\geq 2$). Let $\ell^{(h)}$ denote the number of segments C_i such that $N_H(x_i,x_{i+1})=\{y_i\}$ and $y_i\in V(H)-\{v_1,v_2\}(h=0)$, $y_i\in V(H)-\{v_1,v_2,v_3\}(h>0)$.

From (3) and the preceding observations we obtain

(4) $\alpha\geq\ell R_1+\ell^{(0)}+s$ and $\varepsilon^{(h)}\geq\ell(R_2-h)+2\ell^{(h)}$.
We claim

(5) $\alpha \geq \kappa_H R_1+s$ and

(6) $\varepsilon^{(h)} \geq \kappa_H(R_2-h)$ if $|H|\geq 3+h$.

(5) and (6) trivially hold if $\ell \geq \kappa_H$ or $|H|=2$ Now suppose $\ell<\kappa_H$ and $|H|>2$ .Since $\ell_1 \geq \ell$ we then have $Y=V(H)$.

For any $v\in V(H)$ and $x_i \in N_{C-X}(v)$ either C_i is good or $N_H(x_i, x_{i+1})=\{v\}$. Therefore

$\ell^{(0)} \geq \Sigma(d_{C-X}(v) : v\in V(H)-\{v_1, v_2\})-(\ell-2-|X|)$

$= \Sigma(d_{C-X}(v)-1 : v\in V(H)-\{v_1, v_2\})-(\ell-\ell_1)$.

As $\kappa_H \leq d(v) \leq |X|+|H|-1+d_{C-X}(v)$, we obtain

$\ell^{(0)} \geq (|H|-2)(\kappa_H-\ell_1)-(\ell-\ell_1) = (|H|-2)(\kappa_H-\ell)+(|H|-3)(\ell-\ell_1)$.

Since $|H|-2 \geq D_H(v_1, v_2)-1 \geq \max(R_1, R_2+1)$

we obtain (5) and (6) for h=0. If h>0 similarly

$\ell^{(h)} \geq (|H|-3)(\kappa_H-\ell_1)-(\ell-\ell_1)=(|H|-3)(\kappa_H-\ell)+(|H|-4)(\ell-\ell_1)$

hence

$\ell^{(h)} \geq R_2(\kappa_H-\ell)+(|H|-4)(\ell-\ell_1)$ and (6).

We may without loss of generality assume $d_H(v_1)\leq d_H(v_2)$. By the choice of v_1, v_2 and v_3 we have

(7) $\min(R_1+1, R_2+2 \geq \max((1/2)\, d_H(v_1)+(1/2)d_H(v_2)+\beta_H, d_H(v_3))$

and by (1), (5) and (7)

(8) $|C|\geq d(v_1)+d(v_2)+2\beta_H-2+(\kappa_H-2)R_1+s$.

From (8) and $s\geq \kappa_H$ we obtain $|C|\geq d(v_1)+d(v_2)+2\beta_H+(\kappa_H-2)(R_1+1)$. From (8), $s\geq d_C(v_1)$ and $R_1\geq d_H(v_1)+\beta_H-1$ we obtain $|C|\geq 2d(v_1)+d(v_2)+3\beta_H-3+(\kappa_H-3)R_1$.

As $|H|\geq d_H(v_1)+1+\beta_H$ and $s\leq d_C(v_1)$

also by (8)

$|C|+|H|\geq 2d(v_1)+d(v_2)+3\beta_H-1+(\kappa_H-2)R_1$.

As $R_1\geq D(H)-1$ we have proved Theorem 3.1.

Finally, we evaluate (6) and (7). Let $h\in\{0,1,2\}$ be given such that $|H|\geq 3+h$. By Lemma 2.2 we have $D_H(v_1,v_2)\geq 2+h$ hence $R_2\geq h$. By (2) and (6),

(9) $|C|\geq 2d_C(v_1)+2d_C(v_2)+hd_C(v_3)+\kappa_H(R_2-h)$.

From (9) and (7)

$|C|\geq 2d(v_1)+2d(v_2)+hd(v_3)+4\beta_H-(4+h)(2+h)+(\kappa_H-4-h)(R_2-h)$.

By (9) and (7) also

$|C| \geq d_C(v_1)+d(v_1)+2d(v_2)+hd(v_3)+3\beta_H-(3+h)(2+h)+(\kappa_H-3-h)(R_2-h)$.
That inequality and $|H| \geq d_H(v_1)+\beta_H+1$ yield
$|C|+|H| > 2d(v_1)+2d(v_2)+hd(v_3)+4\beta_H-(3+h)(2+h)+(\kappa_H-3-h)(R_2-h)$. □

Remark 3.3. If $D(H) \geq k$ the bounds (i) and (ii) in Theorem 3.2 can be sharpened by the additive constants respectively $(\kappa_H-k)(D(H)-k)$ and $(\kappa_H-k+1)(D(H)-k)$.

This follows from the above estimates and $R_2 \geq D(H)-2$.

If $V(H)=\{v\}$ then clearly $|C| \geq 2d(v)$. Thus for the proof of Theorem 1 it remains to consider the case when H has cut vertices.

Theorem 3.4. *Let H be a separabel component of G - V(C) and* $k \in \{4, 5, 6\}$. *There is a vertex v in H such that*

(i) $|C| \geq 2d(v) + 2$,

(ii) $|C| \geq kd(v) - 4k + 12$, *if* $\kappa_H \geq k - 1 \geq 3$ *and* $H \notin \mathcal{K}_k$,

(iii) $|C \cup H| \geq kd(v) - 3k + 11$, *if* $\kappa_H \geq k - 2 \geq 2$ *and* $H \notin \mathcal{K}_k$.

Proof: We fix a cyclic orientation of C and continue the meaning of $N_C(H) = \{x_1,...,x_s\}$ and $C_i = C(x_i, x_{i+1}]$ as introduced in the preceding proof. For an endblock B of H let c(B) denote the unique cut vertex of H in V(B). Among all pairs of distinct endblocks B and B' of H we consider those with maximum D(B) + D(B') and determine among them one with maximum $D_H(c(B), c(B'))$. We assume $D(B) \geq D(B')$ and abbreviate $c = c(B)$, $c' = c(B')$, $\Delta_0 = D_H(c, c')$, $\Delta' = D(B')$ and $\overline{\Delta} = \Delta' + \Delta_0$. If $|B| \neq 4$ we set $\Delta = D(B)$, otherwise $\Delta = 3$.

Next we determine suitable vertices v_1, v_2 and v_3 in H.

Let v_2 be a vertex in B' - c' with minimum $d_H(v_2)$. If $|B'| = 2$ then $D(B') = 1 = d_H(v_2)$, and if $B' \geq 3$, then by Lemma 2.2, $\Delta' \geq d_{B'}(v_2) = d_H(v_2)$. Further let v_1 be a vertex in V(B) with minimum $d_H(v_1)$; if B is the 4-cycle choose v_3 adjacent to c.

If $|B| \geq 3$ let v_3 be a vertex in B - c - v_1 with minimum $d_H(v_1)$; if B is the 4-cycle choose v_3 adjacent to v_1. By Lemma 2.2 we have $D(B) \geq d_H(v_3)$ unless $B = K_4^-$ and $d_B(c) = 2$. In the latter subcase $\Delta = 3 = D_H(v_1,v2) = D_H(v_1,c) = d_H(v_3)$. Hence if $|B| \geq 3$,

(10) $d_H(v_3) \leq \min (\Delta, D_H(v_1,v2), D_H(v_1,c))$.

Now assume $|B| = 2$. Then any endblock of H has exactly two vertices. If there is an endblock other than B and B' we can find v_3 in $H - v_1 - v_2$ such that $d_H(v_3) = 1$. Finally let v_1 and v_2 be the only vertices of degree 1 in H. We determine a cc'-path P in H of length $D_H(c,c')$. All cut vertices of H are on P. Let those be $c = c_1,...,c_m = c'$ in the order on P from c to c'. If $D_H(c_\mu, c_{\mu+1}) \geq 3$ for some μ, $1 \leq \mu < m$, then $P[c_\mu, c_{\mu+1}] \subseteq B''$ for a unique block B" of H and $|B''| \geq 3$. In that case we determine a vertex v_3 in $B'' - c_\mu - c_{\mu+1}$ with minimum $d_H(v_3)$. By Lemma 2.2 we have $D(B'') \geq d_H(v_3)$ unless $B = K_4^-$ and $d_{B''}(c_\mu) = d_{B''}(c_{\mu+1}) = 2$. In the latter subcase $D_H(c_\mu, c_{\mu+1}) \geq 3 \geq d_H(v_3)$. If $D_H(c_\mu, c_{\mu+1}) = 2$ for $1 \leq \mu \leq m$, then H is a path. In that case let v_3 be a centre vertex of H.

Thus in the case, when $\Delta = 1$, we have

(10') $\quad \Delta_0 \geq d_H(v_3)$, if $\Delta_0 \geq 1$ and $|H| \geq 5$.

Let us first treat the special case, when $|N_C(B - c) \cup N_C(B' - c')| \leq 1$, that is $N_C(B - c) = N_C(B' - c') = \{x_i\}$ for some i, $1 \leq i \leq s$. Then $\{c, x_i\}$ is a cut set of G hence $\kappa_H = 2$. Pick $w \in N_H(x_{i+1})$. Since $w \notin V(B - c)$ and $x_i \in N_C(B - c)$ we have $|C_i| \geq D(B) + D_H(c,w) + 2$ by the maximality of C. Similarly $|C(x_{i+1}, x_i]| \geq D(B') + D_H(w, c') + 2$.

Hence
$$\begin{aligned}|C| &\geq D(B) + D(B') + \Delta_0 + 4\\ &\geq d_H(v_1) + d_H(v_2) + \Delta_0 + 4\\ &\geq d(v_1) + d(v_2) + \Delta_0 + 2.\end{aligned}$$

Also
$$\begin{aligned}|C| + |H| &\geq 2D(B) + 2D(B') + 2\Delta_0 + 5\\ &\geq 2d_H(v_1) + 2d_H(v_2) + 2\Delta_0 + 5\\ &\geq 2d(v_1) + 2d(v_2) + 2\Delta_0 + 1.\end{aligned}$$

For the remainder of the proof we assume $|N_C(B - c) \cup N_C(B' - c')| \geq 2$. We call a segment C(y, z] of C a (B, B')-segment ((B', B)-segment) if $N_{C(y,z)}((B - c) \cup (B' - c')) = \emptyset$, further $y \in N_C(B - c)$ and $z \in N_C(B' - c')$ (respectively $y \in N_C(B' - c')$ and $z \in N_C(B - c)$). If C(y,z] is as (B, B')-segment or (B', B)-segment, we call it a {B, B'}-segment.

By the above assumption there exist distinct $x \in N_C(B - c)$ and $x' \in N_C(B' - c')$. Clearly C(x, x'] contains a (B, B')-segment and C(x', x] contains a (B', B)-segment . Let $C(y_1, z_1], C(y_2, z_2], ..., C(y_m, z_m]$ be all {B, B'}-segments of C, where $y_1, y_2,...,y_m$ occur in this order on C in the given orientation of C. Thus $m \geq 2$. Set $y_{m+1} = y_1$.

For $h \in \{0,1,2\}$ and $1 \leq i \leq s$ we abbreviate

$$\varepsilon_i^{(h)} = |C_i| - e(v_1, v_2\,; x_i, x_{i+1}) - \frac{h}{2}\, e(v_3\,; x_i, x_{i+1}).$$

Since $|C| = \sum_{i=1}^{s} |C_i|$ and $2d_C(v) = \sum_{i=1}^{s} e(v; x_i, x_{i+1})$ for all $v \in V(H)$ we have

(11) $|C| = 2d_C(v_1) + 2d_C(v_2) + hd_C(v_3) + \sum_{i=1} \varepsilon_i^{(h)}.$

For $h \in \{0,1,2\}$ and $1 \leq \mu \leq m$ let $\delta_\mu^{(h)}$ and $\sigma_\mu^{(h)}$ denote respectively the sum of all $\varepsilon_\mu^{(h)}$ with $C_i \subseteq C(y_\mu, z_\mu]$ and $C_i \subseteq C(z_\mu, y_{\mu+1}]$.

Abbreviating $\delta^{(h)} = \sum_{\mu=1}^{m} \delta_\mu^{(h)}$ and $\delta^{(h)} = \sum_{\mu=1}^{m} \sigma_\mu^{(h)}$ we can give (11) the form

(12) $|C| = 2d_c(v_1) + 2d_c(v_2) + hd_c(v_3) + \delta^{(h)} + \sigma^{(h)}.$

Next we supply bounds for $\sigma_\mu^{(h)}$ and $\delta_\mu^{(h)}$ ($h \in \{0,1,2\}$, $1 \leq \mu \leq m$). For any $C_i \subseteq C(y_\mu, y_{\mu+1}]$ such that $e(v_1, v_2, v_3; x_i, x_{i+1}) \leq 2$ we have $\varepsilon_i^{(h)} \geq |C_i| - 2$ hence $\varepsilon_i^{(h)} \geq 0$. Note that for a (B,B')-segment $C(y_\mu, z_\mu]$ with $z_\mu \neq y_{\mu+1}$ one must have $N_{C(y_\mu, y_{\mu+1}]}(B - c) = \varnothing$, since otherwise $C(z_\mu, y_{\mu+1}]$ would contain another {B, B'}-segment. Similarly $N_{C(y_\mu, y_{\mu+1}]}(B' - c') = \varnothing$, whenever $C(y_\mu, z_\mu]$ is a (B', B)-segment with $z_\mu \neq y_{\mu+1}$. For $C_i \subseteq C(z_\mu, y_{\mu+1}]$ with $e(v_1, v_2, v_3; x_i, x_{i+1}) \geq 3$ therefore holds $\varepsilon_i^{(h)} \geq |C_i| - 2 - h \geq 1 - h$.

We first show

(13) $\delta_\mu^{(h)} \geq \Delta + \overline{\Delta} - h - 2$, moreover

$\delta_\mu^{(h)} \geq \Delta + \overline{\Delta} - h - 1$, unless $e(v_1, v_2; y_\mu, y_{\mu+1}) = 4$.

Let $C(y_\mu, z_\mu] = C(x_j, x_k]$. If $x_{j+1} = x_k$ then $|C_j| \geq D(B) + D_k(c, c') + D(B') + 2$ since C is a longest cycle. If in addition $e(v_1, v_2; x_i, x_{i+1}) \geq 3$ then $|C_j| \geq D_H(v_1, c) + D_H(c,c') + D(B') + 2$. Hence (13) if $x_{j+1} = x_k$.

Now assume $x_{j+1} \neq x_k$, and let $C(y_\mu, z_\mu]$ be a (B, B')-segment. Choose $u \in N_H(x_{j+1})$ and $w \in N_H(x_{k-1})$. Then $|C_j| \geq D(B) + D_H(c, u) + 2$, moreover $|C_j| \geq \Delta + D_H(c, u) + 2$ if $v_1 \in N_H(x_j)$. Also $|C_{k-1}| \geq D_H(w, c') + 2$ hence

$$\varepsilon_j^{(h)} + \varepsilon_{k-1}^{(h)} \geq \Delta + \Delta' + D_H(c, u) + D_H(w, c') + 2 - 2h.$$

If $x_{j+1} = x_{k-1}$ we may choose $u = w$ to obtain $\delta_\mu^{(h)} = \varepsilon_j^{(h)} + \varepsilon_{k-1}^{(h)} \geq \Delta + D(B') + D_H(c, c') - h$.
If $x_{j+1} \neq x_{k-1}$ then $\varepsilon_{j+1}^{(h)} + \cdots + \varepsilon_{k-2}^{(h)} = |C_{j+1}| + \cdots + |C_{k-2}| = |C(x_{j+1}, x_{k-1}]| \geq D_H(u, w) + 2$.
In that case $D_H(c, u) + D_H(u, w) + D_H(w, c) \geq D_H(c, c')$ hence
$\delta_\mu^{(h)} = \varepsilon_j^{(h)} + \cdots + \varepsilon_{k-1}^{(h)} \geq \Delta + D(B') + D_H(c,c') + 2 - h$.

A symmetric argument yields (13), if $C(y_\mu, z_\mu]$ is a (B', B)-segment with $z_\mu \neq y_{\mu+1}$.

Case 1. $\Delta + \overline{\Delta} \leq 3$.

Then $D(B') = 1$ and $D_H(c, c') \leq 1$ hence all endblocks of H other than B have only two vertices, and H has no cut vertices other than c and c'. Also $\Delta \leq 2$, that is $|B| \leq 3$.

Case 1.1 $\Delta + \overline{\Delta} = 2$.

Then $\Delta = 1$ and $D_H(c, c') = 0$, hence H is a star. In that subcase by the preceding estimates $\sigma_\mu^{(0)} \geq 0$ and $\delta_\mu^{(0)} \geq 0$ hence by (12)
$|C| \geq 2d_C(v_1) + 2d_C(v_2) = 2d(v_1) + 2d(v_2) - 4$.

Case 1.2 $\Delta = 2$.

Then $D_H(c, c') = 0$, hence H is a quasistar. Let $h \in \{0, 1\}$. By the above estimates $\sigma_\mu^{(h)} \geq 0$ and $\delta_\mu^{(h)} \geq 1 - h$ for $1 \leq \mu \leq m$. If $e(v_1, v_2; y_\mu) = 1$ for some μ, then $\delta_{\mu-1}^{(h)} + \delta_\mu^{(h)} > 4 - 2h$ by (13), hence
$|C| \geq 2d_C(v_1) + 2d_C(v_2) + hd_C(v_3) + m(1 - h) + 2$
$\geq 2d(v_1) + 2d(v_2) + hd(v_3) - 4 - 2h + m(1 - h)$.

If $e(v_1, v_2; y_\mu) = 2$ for $1 \leq \mu \leq m$ then $d_C(v_1) = d_C(v_2)$ hence $d(v_1) > d(v_2)$ and
$|C| \geq 4d(v_2) + hd(v_3) - 4 - 2h + m(1 - h)$.

Therefore $|C| \geq (4 + h)\, d(v) - 2 - 4h$ for some $v \in \{v_1, v_2, v_3\}$.

Case 1.3 $D_H(c, c') = 1$.

Then H is a double star. As in the preceding subcase we obtain for $h \in \{0, 1\}$
$|C| \geq 2d_C(v_1) + 2d_C(v_2) + hd_C(v_3) + m(1 - h)$ hence
$|C| \geq 2d(v_1) + 2d(v_2) + hd(v_3) - 4 - 2h + m(1 - h)$. Thus also
$|C| \geq (4 + h)\, d(v) - 2 - 4h$ for some $v \in V(H)$.

For the rest of the proof we assume

(14) $\Delta + \overline{\Delta} \geq 4$.

Then $v_3 \in V(B - c)$ or $v_3 \in V(H - B - B')$.

Consider a segment $C_i \subseteq C(z_\mu, y_{\mu+1}]$. As shown above $e(v_1; C(y_\mu, y_{\mu+1}]) = 0$ or $e(v_2, C(y_\mu, y_{\mu+1}]) = 0$. If $e(v_1,v_2,v_3;x_i,x_{i+1}) \geq 3$ therefore $|C_i| \geq 2 + \min(D_H(v_1, v_3), D_H(v_2, v_3)) \geq 4$ and $\varepsilon_i^{(h)} \geq 2 - h$.

Therefore for any $h \in \{0, 1, 2\}$

(15) $\sigma^{(h)} \geq 0$.

Case 2. $\Delta = 1$.
Then $\Delta' = 1$ hence $d_H(v_1) = d_H(v_2) = 1$ and $d_H(v_3) \leq \Delta_0$. For $1 \leq \mu \leq m$ we have $\delta_\mu^{(h)} \geq \Delta_0 - h \geq 2 - h$ by (13) and (14).

Therefore $\delta^{(h)} \geq 2\Delta_0 - 2h + (m - 2)(2 - h) \geq hd_H(v_3) + (2 - h)\Delta_0 + (m - 2)(2 - h) - 2h$.
Hence by (12) for $h \in \{0, 1, 2\}$,
$|C| \geq 2d(v_1) + 2d(v_2) + hd(v_3) + (2 - h)(\Delta_0 + m) - 8$.

Case 3. $\Delta \geq 2$ and $m \geq 4$.
By (13) we have for any $h \in \{0, 1, 2\}$,

$$\begin{aligned}\delta^{(h)} &\geq m(\Delta + \overline{\Delta} - 2 - h)\\ &= (2 + h)\Delta + 2\Delta' + (m - 2 - h)(\Delta + \overline{\Delta}) + h\overline{\Delta} - (2 + h)m + 2\Delta_0\end{aligned}$$

hence, by (14) and (15),

$$\sigma^{(h)} + \delta^{(h)} \geq (2 + h)\Delta + 2\Delta' + 4(m - 2 - h) + h\overline{\Delta} - (2 + h)m + 2\Delta_0.$$

As $\Delta \geq d_H(v_p)$ $(p = 1, 3)$ and $\Delta' \geq d_H(v_2)$, we obtain by (12),
$$\begin{aligned}|C| &\geq 2d(v_1) + 2d(v_2) + hd(v_3) + (2 - h)(m + 4) + h\overline{\Delta} + 2\Delta_0 - 16\\ &\geq 2d(v_1) + 2d(v_2) + hd(v_3) + h(\overline{\Delta} - 8).\end{aligned}$$

If $\overline{\Delta} = 1$, then as in Case 1.2, either $\delta^{(h)} \geq 2 + m(\Delta - 1 - h)$ or $d(v_1) > d(v_2)$.

Hence $|C| \geq 4d(v) + hd(v_3) - 4(h + 1)$ for $v = v_1$ or $v = v_2$.

Case 4. $\Delta \geq 2$ and $m = 3$.

We first treat the subcase, when some (B', B)-segment $C(y_\lambda, z_\lambda]$ satisfies $z_\lambda \neq y_{\lambda+1}$ and $|N_{C[z_\lambda, y_{\lambda+1}]}(H)| \geq 2$, say $C(y_1, z_1]$.

For $\mu = 1$ and 2 we have by (13), $\delta_\mu^{(h)} \geq \Delta + \overline{\Delta} - 1 - h$ hence

(16) $\delta^{(h)} \geq 3\Delta + 3\overline{\Delta} - 4 - 3h.$

Let x_j be the first vertex on $C(z_1, y_2]$ in $N_C(H)$ such that $|N_{C[z_1, x_j]}(H)| \geq 2$.

Then $x_{j-1} \in N_C(B - c)$ hence $|C_{j-1}| \geq D(B) + 2$. If $e(v_1; x_{j-1}, x_j) = 2$, moreover $|C_{j-1}| \geq \Delta + 2$. Therefore $\varepsilon_{j-1}^{(h)} \geq \Delta - h$. As shown above $\varepsilon_i^{(h)} \geq 0$ for all $C_i \subseteq C(z_1, y_2]$. We infer $\sigma_1^{(h)} \geq \Delta - h$ and, using (16),

$$\begin{aligned}\sigma^{(h)} + \delta^{(h)} &\geq 4\Delta + 3\overline{\Delta} - 4 - 4h.\\ &= (2 + h)\,\Delta + 2\overline{\Delta} + (2 - h)\,(\Delta + 4) + \overline{\Delta} - 12.\end{aligned}$$

Thus by (12)

$|C| \geq 2d(v_1) + 2d(v_2) + hd(v_3) + (2 - h)(\Delta + 4) + \overline{\Delta} - 12.$

For the remainder of Case 4 we assume that for any (B', B)-segment $C(y_\mu, z_\mu]$ either $z_\mu = y_{\mu+1}$ or $N_{C[z_\mu, y_{\mu+1}]}(H) = \{u_{\mu+1}\}$.

If $C(y_\mu, z_\mu]$ is a (B, B')-segment or $z_\mu = y_{\mu+1}$, set $u_{\mu+1} = y_{\mu+1}$. For those segments we have $N_{C[y_\mu, y_{\mu+1}]}(B - c) \subseteq \{u_{\mu+1}\}$. For every other segment $C(y_\mu, z_\mu]$ we have $z_\mu \neq y_{\mu+1}$ and $N_{C[y_\mu, y_{\mu+1}]}(H) = \{u_{\mu+1}\} \subseteq V(B - c)$.

If there were two segments of the latter type, say $C(y_1, z_1]$ and $C(y_2, z_2]$ then $N_{C(y_1, y_3]}(B' - c') = \emptyset$ and $C(y_2, z_2]$ would not be a {B, B'}-segment. Therefore $\{c, u_1, u_2, u_3\}$ is a cut set of G and $\kappa_H \leq 4$.

For $h \in \{0, 1, 2\}$ we have by (13) and (15)

$$\begin{aligned}\delta^{(h)} + \sigma^{(h)} &\geq 3\Delta + 3\overline{\Delta} - 6 - 3h\\ &= (2 + h)\,\Delta + 2\Delta' + (1 - h)\,(\Delta + \overline{\Delta} + 3) + h\overline{\Delta} + 2\Delta_0 - 9.\end{aligned}$$

Hence by (12)
$|C| \geq 2d(v_1) + 2d(v_2) + hd(v_3) + (1-h)(\Delta + \overline{\Delta} + 3) + h\overline{\Delta} - 9.$

Thus for $h \in \{0, 1\}$ by (13),
$|C| \geq 2d(v_1) + 2d(v_2) + hd(v_3) + h(\overline{\Delta} - 7) - 2$ and, as in Case 3,
$|C| \geq 4d(v) + hd(v_3) - 5h - 2$ for $v = v_1$ or $v = v_2$.

Since $|H| \geq \Delta + \overline{\Delta} + 1$ for $h \in \{0, 1, 2\}$ also
$|C \cup H| \geq 2d(v_1) + 2d(v_2) + hd(v_3) + (2-h)(\Delta + \overline{\Delta} + 3) + h\overline{\Delta} - 11$
$\geq 2d(v_1) + 2d(v_2) + hd(v_3) + h(\overline{\Delta} - 7) + 3$ and
$|C \cup H| \geq 4d(v) + hd(v_3) - 5h + 3$ for $v = v_1$ or $v = v_2$.

Case 5. $\Delta \geq 2$ and $m = 2$
Without loss of generality let us assume that $C(y_1, z_1]$ is a (B', B)-segment while $C(y_2, z_2]$ is a (B, B')-segment.

Then $N_{C[y_1, y_2]}(B' - c') \subseteq \{y_2\}$ and $N_{C[y_2, y_1]}(B - c) \subseteq \{y_1\}$. By (13) and (14) we have for $h \in \{0, 1, 2\}$, $\varepsilon^{(h)} \geq 2\Delta + 2\overline{\Delta} - 4 - 2h$. Hence by (12) with $h = 0$
$|C| \geq 2d(v_1) + 2d(v_2) + 2\Delta_0 - 4.$
As $|H| \geq \Delta + \overline{\Delta} + 1$ also

$$\begin{aligned} \varepsilon^{(h)} + |H| &\geq 3\Delta + 3\overline{\Delta} - 3 - 2h \\ &= (2 + h)\Delta + 2\overline{\Delta} + (1 - h)(\Delta + \overline{\Delta} + 2) + h\overline{\Delta} - 5. \end{aligned}$$

Hence by (12) and (14), for $h \in \{0, 1\}$,
$|C \cup H| \geq 2d(v_1) + 2d(v_2) + hd(v_3) + h(\overline{\Delta} - 6) + 1$ and
$|C \cup H| \geq 4d(v) + hd(v_3) - 4h + 1$ for $v = v_1$ or $v = v_2$.

For the rest of Case 5 assume that $\kappa_H \geq 4$. Then $z_1 \neq y_2$ since otherwise $N_C(B - c) \subseteq \{y_1, y_2\}$ and $\{c, y_1, y_2\}$ would be a cut set. Similarly we deduce $z_2 \neq y_1$. Moreover $|N_{C[z_1, y_2]}(B - c)| \geq 2$ since otherwise $N_{C[z_1, y_2]}(B - c) = \{u\}$ and $\{u, c\}$ would be a cut set of G. Let x_j be the first vertex on $C(z_2, y_1]$ such that $|N_{C[z_2, x_j]}(H)| \geq 2$, and let x_k be the last vertex on $C[z_1, y_2)$ such that $|N_{C[x_k, y_2]}(H)| \geq 2$. If $x_{j-1} \neq z_1$, then $N_H(C[z_1, x_{j-1}]) = \{u\} \subseteq V(B-c)$. If $x_{j-1} = z_1$, set $u = z_1$. Similarly we have $x_{k+1} = y_2 = w$ or $N_H(C[x_{k+1}, y_2]) = \{w\} \subseteq V(B-c)$.

Case 5.1 $C_{j-1} = C_k$

Then $x_k = x_{j-1}$ hence $x_{j-1} \neq z_1$, since otherwise $\{c, z_1, w\}$would be a cut set of G. Symmetrically $x_{k+1} \neq y_2$ hence $\{u, w\} \subseteq V(B-c)$. Further $N_{C[z_1, x_{j-1}]}(B - c) = \{u\}$ and $N_{C[x_j, y_2]}(B - c) = \{w\}$. Hence $V(B - c) = \{u, w\}$ since otherwise $\{c, u, w\}$ would be a cut set of G.

Now $\delta^{(h)} \geq 2\Delta + 2\overline{\Delta} - 2 - 2h$ by (13) and $\sigma^{(h)} \geq \varepsilon_k^{(h)} \geq D(B) + 2 - 1 - h = 2\Delta - 1 - h$. Hence for any $h \in \{0, 1, 2\}$,

$$\begin{aligned}\sigma^{(h)} + \delta^{(h)} &\geq 4\Delta + 2\overline{\Delta} - 3 - 3h \\ &=(2 + h)\Delta + 2\overline{\Delta} + (2 - h)(\Delta + \overline{\Delta} + 3) + h\overline{\Delta} - 9.\end{aligned}$$

Consequently, by (12) and (14),
$|C| \geq 2d(v_1) + 2d(v_2) + hd(v_3) + h(\overline{\Delta} - 7) + 5.$

Case 5.2 $C_{j-1} \neq C_k$

As $x_{j-1} \in N_C(B - c)$ we clearly have $\varepsilon_{j-1}^{(h)} \geq \Delta - h$ (cf. Case 4). Also $\varepsilon_k^{(h)} \geq \Delta - h$ hence $\sigma^{(h)} \geq 2\Delta - 2h$. Therefore $\sigma^{(h)} + \delta^{(h)} \geq 4\Delta + 2\overline{\Delta} - 2 - 4h$.

Hence for any $h \in \{0, 1, 2\}$,
$|C| \geq 2d(v_1) + 2d(v_3) + hd(v_2) + h(\overline{\Delta} - 8) + 6.$

The proof of Theorem 3.4 is complete. ❑

Clearly (3.1), (3.2) and (3.4) imply Theorem 1.

References

[1] Min Aung, Longest cycles in triangle-free graphs, J. Combin. Theory B, 47 (1989), 171-186.

[2] J.A. Bondy, Longest paths and cycles in graphs of high degree, University of Waterloo. Preprint CORR 80-76.

[3] G.A. Dirac, Some theorems on abstract graphs, Proc. London Math. Soc. 2 (1952), 69-81.

[4] G. Fan, Extremal theorems on paths and cycles in graphs and weighted graphs, Thesis, Waterloo, Ontario 1987.

[5] P. Fraisse, D_λ-cycles and their applications for hamiltonian graphs, preprint (1987).

[6] P. Fraisse and H.A. Jung, Longest cycles and independent sets in k-connected graphs, in: V.R. Kulli, ed., Recent Studies in Graph Theory (Vishwa Intern. Publ. Gulbarga, India, 1989), 114-139.

[7] H.A. Jung, Longest circuits in 3-connected graphs, Coll. Math. Soc. J. Bolyai, 37. Finite and infinite sets, Eger (1981), 403-438.

[8] H.A. Jung, Longest paths joining given vertices in a graph, Abh. Math. Sem. Univ. Hamburg 56 (1986), 127-137.

[9] C. St. J.A. Nash-Williams, Edge-disjoint hamiltonian circuits in graphs with vertices of large degrees, in: L. Mirsky, ed., Studies in Pure Mathematics (Acad. Press, London 1971), 157-183.

[10] H.J. Veldman, Existence of D_λ-cycles and D_λ-paths, Discr. Math. 44 (1983), 309-316.

[11] H.-J. Voss, Maximal circuits and paths in graphs. Extreme cases, Coll Math. Soc. J. Bolay, 18. Combinatorics, Keszthely (1976), 1099-1122.

[12] H.-J. Voss - C. Zuluaga, Maximale gerade und ungerade Kreise in Graphen I, Wiss. Z Techn. Hochschule Ilmenau, 4 (1977), 57-70.

Independent Covers in Plane Graphs

M. M. Syslo, Wroclaw, Poland

ABSTRACT

A subset W of vertices of a plane graph G is said to be a face-independent vertex cover (FIVC) if each face-boundary contains exactly one vertex of W. A FIVC naturally corresponds to a vertex-independent face cover (VIFC) in the dual graph G^* of G. In this paper we review the results on the existence of such covers in arbitrary planar graphs and in outerplanar graphs, in particular.

1. INTRODUCTION

Let $G=(V,E)$ be a 2-connected planar graph embedded in the plane. In what follows we call G a *plane* graph. A subset $W \subset V$ of vertices is called a *face-independent vertex cover* (simply, a *FIVC*) of faces of G if:

1. no two vertices of W belong to the same face of G,
2. every face of G has a vertex in W.

Equivalently, W is a FIVC if every face of G has exactly one vertex in W. If W satisfies only condition 1, then it is called a *face-independent* set and if W satisfies only condition 2 - it is called a *vertex cover of faces*. Note that if G is a maximal planar graph, then the face-independence of vertices is equivalent to the independence in the usual sense and a FIVC is an independent vertex set of G which covers all its faces.

A FIVC in G corresponds in the geometric dual G^* of G to a set of faces $\mathscr{E}$ which we call a *vertex-independent face cover* (simply, *VIFC*) of vertices of G^* since no two faces of $\mathscr{E}$ share a vertex and every vertex of G^* is contained in exactly one

R. Bodendiek, R. Henn (Eds.)
Topics in Combinatorics and Graph Theory

face of $\mathcal{E}$. Evidently, VIFC's in G are in one-to-one correspondence with 2-factors of G consisting of facial cycles. A set of faces $\mathcal{E}$ in G^* which corresponds to a vertex cover of faces in G is called a *face cover* of vertices of G.

Every 2-connected plane graph admits a cycle cover of vertices and a vertex cover of faces. In contrast to this and to almost all other variants of covering sets (of cycles and/or vertices), a graph may not have a FIVC and /or VIFC (see [5] for an infinite family of plane graphs with no FIVC and VIFC).

The *face cover problem* is to determine, for an input triple $(G, V' \subseteq V(G), k\text{-integer})$, if G can be embedded in the plane so that at most k faces cover V'. One can similarly define the FIVC and VIFC problems. It was independently proved in [1] and [2], that the face cover problem (G, V, k) is NP-complete in general. Several other versions of this problem are also NP-complete as well as determining if a plane graph has a VIFC [3].

The purpose of this note is to review some known results related to the existence of FIVC and VIFC. In the next section we consider arbitrary planar graphs and Section 3 deals with outerplanar graphs. Some other special classes of planar graphs have been also considered, see [4] and [5].

2. INDEPENDENT COVERS IN PLANAR GRAPHS

Let G be a maximal planar graph. Hence, its dual graph G^* is a 3-valent and 3-connected plane graph. Moreover, both graphs G and G^* have a unique embedding in the sphere (or equivalently, a unique set of faces). If G is Eulerian, that is, every vertex in G is of even degree, then it has a unique 3-coloring of its vertices. Each color class contains exactly one vertex from each face of G. Hence we obtain

THEOREM 1.[5] Every Eulerian maximal plane graph G has exactly three FIVC of faces which correspond to exactly three VIFC of vertices in its dual graph G^*.

Note that from the well-known theorem about the existence of 1-factors in bipartite regular graphs, it follows only that G^* contains a cover of vertices by vertex-disjoint cycles, which however are not necessarily facial.

We now characterize maximal planar graphs with odd degree vertices which admit a FIVC.

Let G be a maximal planar graph. Every vertex v in G together with its neighbors generate a wheel in G. If v is of even degree, then there are two ways to cover the faces containing v: either by taking v or by taking every second neighbor of v on the circle. For an odd degree vertex v, the latter set is not independent and hence v must belong to every FIVC of faces in G. Thus, all odd degree vertices of G belong to every FIVC of G, provided G admits such a set. It follows that no two odd degree vertices in G are adjacent, but this is not sufficient, as one can easily see.

A pair of triangles sharing an edge is called a *diamond*. (Here and in what follows, a *triangle* is understood as a triangular face.) The vertices in a diamond which belong to only one traingle are called *terminal*. It is easy to see that if a maximaximal plane graph G has a FIVC W then for every diamond H in G, either both terminal vertices of H or none of them belongs to W. This observation applied iteratively to every t-path (to be defined) will result in a global condition for G to have a FIVC. A *t-path* in a maximal plane graph G is a sequence of distinct triangles $\rho=(T_1,T_2,\dots,T_k)$ such that:

(i) T_i and T_{i+1} form a diamond in G for $i=1,3,\dots,2\lfloor k/2\rfloor-1$, that is T_i shares an edge with T_{i+1}. Let H_i denote the diamond consisting of T_{2i-1} and T_{2i} for $i=1,2,\dots,\lfloor k/2\rfloor$.

(ii) The terminal vertex of H_1 which belongs to T_1 does not belong to any other triangle of ρ and the other terminal vertex of H_1 is a terminal vertex of H_2. For every $i=2,\dots,\lfloor k/2\rfloor-1$, the diamond H_i shares its terminal vertices with those of H_{i-1} and H_{i+1}. For $l=\lfloor k/2\rfloor$, one terminal vertex of H_l is also a terminal vertex of H_{l-1}. Moreover, if k is odd then the other terminal vertex of H_l belongs to T_k.

(iii) H_i and H_{i+1} may share at most one edge, for $i=1,2,\dots,\lfloor k/2\rfloor-1$.

(iv) No terminal vertex of H_i is a non-terminal vertex of any other diamond in ρ.

The *length* of ρ is equal to k. The terminal vertices in H_i's are called *even* with respect to ρ and the other ones are called *odd*. Property (iv) ensures that no vertex contained in ρ is simultaneously even and odd. The terminal vertex of H_1 which belongs to T_1 is called an *initial* vertex of ρ. If k is even then the terminal vertex of H_l which belongs to T_k is called an *end-vertex* of ρ, and when k is odd - each of the two vertices of T_k which is not a terminal vertex of H_l is an *end-vertex* of ρ. For the sake of simplicity, if u is the initial vertex of ρ and v is an end-vertex of ρ then we say that ρ *connects* u and v.

One can easily see that if the initial vertex v of a t-path ρ in a maximal plane graph G belongs to a FIVC W of G then every even vertex of ρ has to belong to W and no odd vertex of ρ can be in W. Hence, if v is of odd degree then no odd vertex of ρ is of odd degree. As a main result of this section we state that the converse also holds.

THEOREM 2.[5] A maximal plane graph G has a FIVC if and only if no two odd degree vertices in G are connected by an odd-length t-path.

COROLLARY 1. A maximal plane graph G has exactly three or at most one FIVC of faces. The latter may occure when G contains odd degree vertices.

Transforming the result of Theorem 2 to the dual graphs we obtain

COROLLARY 2. A 3-connected cubic plane graph has a 2-factor consisting of facial cycles if and only if no two odd-length faces are connected by an even sequence of faces.

Theorem 2 can be also applied to some special cases of arbitrary plane graphs. Observe, that a FIVC W in a plane graph G corresponds to that in a maximal plane graph G' which satisfies the conditions of Theorem 2. The graph G' is obtained from G by adding to it all edges between w and every vertex which lies in the same face as w and is not adjacent to w, for $w \in W$. For instance, if a plane graph G has an odd-length face C with all vertices of odd degree and C is incident with only triangular faces of G then G has no FIVC since every triangulation of C results in an odd-length t-path between two odd-degree vertices.

3. INDEPENDENT COVERS IN OUTERPLANAR GRAPHS

A 2-connected planar graph G is *outerplanar* if it can be embedded in the plane with all vertices on the exterior face. Such an embedding is called an *outerplane graph*. A *maximal outerplanar graph* (*mop*) has all interior faces triangular.

If G is a plane graph then G^w denotes the *weak dual* of G which can be obtained from the dual graph G^* of G by removing the vertex corresponding to the exterior face of G. The weak dual of an outerplane graph G is a tree and is denoted by T^w, or simply by T.

It is easy to provide an infinite family of mops, each of which does not admit a FIVC for any of its plane embeddings. If all nodes of a weak dual T of a mop G have degree 1 or 3, then both T and G are called *full*. If all nodes of T have degree 1 or 2, then both T and G are called *linear*.

LEMMA 1.[6] If a mop is full, then none of its plane embeddings has a FIVC.

LEMMA 2.[6] Every linear mop has a FIVC.

In the rest of this section we present a method for enumerating all possible plane embeddings of a mop, discuss a dynamic programming algorithm for finding a smallest FIVC in such graphs and characterize mops possesing a FIVC. These results can be easily generalized to arbitrary 2-connected outerplanar graphs.

Let G be a mop and T be the weak dual of its outerplane embedding. Any partially directed tree obtained from T by:

- selecting an arbitrary $t \in V(T)$ as a root, and
- directing some of the branches of T toward t,

is called an *in-tree rooted at* t, and will be denoted by $\vec{T}$. We have

LEMMA 3.[7] Every in-tree $\vec{T}$ specifies a plane embedding of G.

LEMMA 4.[7] Every plane embedding of G can be specified by an appropriate in-tree $\vec{T}$ rooted at an arbitrarily chosen node t.

The proofs of these lemmas suggest a very simple linear-time (in the number of veritices) algorithm for identifying all boundaries of the embedding of G corresponding to a given in-tree $\vec{T}$ tooted at a node t.

Let f denote the number of faces in a mop G, i.e., $f=m-n+1$, where m is the number of edges and n is the number of vertices in G. Note that the weak dual T of G has $f-1$ branches. If we fix a root t of T, then evidently there are exactly 2^{f-1} different in-trees rooted at t. One can easily show that a transformation $\Phi(\vec{T})$ of $\vec{T}$ obtained by replacing:

- each directed branch incident to t by an undirected one,
- each undirected branch incident to t by a directed (toward t) one,

preserves the set of boundaries and no two other in-trees rooted at t generate the same set of faces. Hence, all in-trees of T rooted at any node t produce 2^{f-2} different embeddings of G. Moreover, there are no other embeddings of G, since every in-tree rooted at a node s, $s\neq t$, can be transformed to an in-tree rooted at t and generating the same set of faces. Thus,

THEOREM 3. [7] A mop G has exactly 2^{f-2} different embeddings which can be generated by the maximal family of in-trees rooted at any node t of T which is complement-free (i.e., if $\vec{T}$ belongs to the family then $\Phi(\vec{T})$ does not belong).

Theorem 3 shows that the number of different plane embeddings of a mop grows exponentially with the size of a mop. Thus, the space to be searched when looking for a FIVC of a mop is non-polynomial. However, the tree-like structure of mops can be used to obtain a polynomial-time algorithm for finding an embedding admitting minimum cardinality FIVC and determining this FIVC. Such an algorithm based on dynamic programming technique is presented in detail in [8].

To characterize mops admitting a FIVC for at least one of their plane embeddings we need another tree associated with a mop. A *semidual* graph G^s of a plane graph G is obtained from the dual G^* of G by splitting the node corresponding to the exterior face of G into as many copies as there are edges on the exterior face. It is obvious that the semidual of a mop is a tree. Let T^s denote the semidual of a maximal outerplane graph G. A subtree of T^s is called *one-pendant* if it contains exactly one node which is pendant in T^s. Let W denote an arbitrary subset of vertices in G, and let B_w, $w \in W$, denote the branches of T^s intersected by the edges of G incident with w. We refer to B_w as a *branch boundary* of w. Moreover, we denote by T_w the forest obtained from T^s by deleting all branches in $\langle B_w : w \in W \rangle$.

THEOREM 4. [8] A mop G admits W as a FIVC if and only if T_W is a forest consisting of disjoint one-pendant subtrees.

This theorem provides a method of verifying whether a given subset of vertices of a mop G is a FIVC without the need of investigating embeddings of G. It is unfortunately not true that every mop whose semidual can be covered by one-pendant subtrees has a FIVC; they must be generated by branch boundaries.

The results obtained for mops have been recently generalized to arbitrary 2-trees, see [4].

REFERENCES

[**1**] D. Bienstock, C.L. Monma, On the complexity of covering vertices by faces in a planar graph, *SIAM J. on Comput.* **17**(1988), 53-76.

[2] M. Fellows, F. Hickling, M.M. Syslo, Topological parameterization and hard graph problems (extended abstract), *Congressus Numerantium* 59(1987), 69-78.

[3] M. Fellows, Personal communication, 1986.

[**4**] A. Proskurowski, M.M. Syslo, P. Winter, Face-independent vertex covers in 2-trees, in preparation.

[5] M.M. Syslo, Independent face and vertrex covers in plane graphs, in: M. Borowiecki, Z. Skupien (eds.), *Proceedings of the Banach Center*, PWN, Warszawa (in press).

[6] M.M. Syslo, P. Winter, Independent covers in outerplanar graphs, in: R. Karlsson, A. Lingas (eds.), *SWAT 88: Proceedings of the 1-st Scandinavian Workshop on Algorithm Theory* LNinCS, Vol. **318**, Springer-Verlag, Berlin-Heidelberg, 1988, 243-254.

[7] M.M. Syslo, P. Winter, In-trees and plane embeddings of outerplanar graphs, *BIT*, 1989 (in press).

[8] P. Winter, M.M. Syslo, Outerplanar graphs with perfect face-independent vertex covers, manuscript, 1989.

List of Contributors

AHLSWEDE,R., Prof. Dr., Universität Bielefeld, Fakultät für Mathematik, Universitätsstraße 1, D-4800 Bielefeld 1, FRG

AIGNER,M., Prof. Dr., FU Berlin, II. Math. Institut, Arnimallee 3, D-1000 Berlin 33, FRG

ALSPACH;B., Prof. Dr., Simon Fraser University, Dept. of Mathematics and Statistics, Burnaby, B.C. V5A 1S6, Canada

ANDRE,J., Prof. Dr., Universität des Saarlandes, Fachbereich Mathematik, Bau 36, D-6600 Saarbrücken, FRG

ARCHDEACON,D., Prof. Dr., The University of Vermont, Dept. of Mathematics and Statistics, 16, Colchester Ave., Burlington, VT 05405, USA

BÄR,G., Prof. Dr., Ernst-Moritz-Arndt-Universität, Sektion Mathematik, Jahnstraße 15a, DDR-2200 Greifswald, GDR

BANDELT,H.-J., Vakgroep Kwantitatieve Economie, Rijksuniversiteit Limburg, Postbus 616, NL-6200 MD Maastricht, The Netherlands

BAUER,D., Prof. Dr., Dept. of Pure and Applied Mathematics, Stevens Institute of Technology, Hoboken, NJ 07030, USA

BAUMANN,U., Dr., Päd. Hochschule K.F.W. Wander, Sektion Mathematik, Wigardstraße 17, DDR-8060 Dresden, GDR

BEZDEK,A., Prof. Dr., Hungarian Academy of Sciences, Math. Institute, P.O.B. 127, H-1364 Budapest, Hungary

BEZDEK,K., Prof. Dr., Eötvös Lorand University, Dept. of Geometry, Râkôczi ut 5, H-1088 Budapest, Hungary

BIGGS,N., Prof. Dr., London School of Economics, Dept. of Statistical and Mathematical Sciences, GB-London WC2A 2AE, Great Britain

BODENDIEK,R., Prof. Dr., Pädagogische Hochschule Kiel, Institut für Mathematik und ihre Didaktik, Olshausenstraße 75, D-2300 Kiel 1, FRG

BÖHME,T., Dr., TH Ilmenau, Sektion Mathematik, Rechentechnik und Ökonomische Kybernetik, Postfach 327, DDR-6300 Ilmenau, GDR

BORODIN,O.V., Prof. Dr., Academy of Sciences of USSR, Universitetski pr 4, SU-630090 Novosibirsk, USSR

BOUCHET,A., Prof. Dr., Dépt. de Mathématiques, Université du Maine, F-72017 Le Mans Cedex, France

BRANDSTÄDT,A., Prof. Dr., Friedrich-Schiller-Universität, Sektion Mathematik, UHH, DDR-6900 Jena, GDR

BREHM,U., Prof. Dr., Institut für Mathematik, TU Berlin, Straße des 17. Juni 135, D-1000 Berlin 12, FRG

BROERSMA,H.J., Prof. Dr., University of Twente, Dept. of Applied Mathematics, P.O.B. 217, NL-7500 AE Enschede, The Netherlands

BUROSCH,G., Prof. Dr., Wilhelm-Pieck-Universität, Sektion Mathematik, Universitätsplatz 1, DDR-2500 Rostock, GDR

CAHIT,I., Prof. Dr., Turkish Telecommunications, Box 446, Lefkosa Mersin 10, Turkey

CAI,N., Prof. Dr., Universität Bielefeld, Fakultät Mathematik, Universitätsstraße 1, D-4800 Bielefeld, FRG

CATLIN,P.A., Prof. Dr., Dept. of Mathematics, Wayne State University, Detroit, MI 48202, USA

CHARTRAND,G., Prof. Dr., Dept. of Mathematics, Western Michigan University, Kalamazoo, MI 49008-3899, USA

CIESLIK,D., Prof. Dr., Ernst-Moritz-Arndt-Universität, Sektion Mathematik, Jahnstraße 15a, DDR-2200 Greifwald, GDR

CZYZOWICZ,J., Prof. Dr., Dépt. d'Informatique, Université du Québec à Hull, C.P. 1250, Succ. "B",Hull, Québec J8X 3X7, Canada

DAMASCHKE,P., Prof. Dr., Friedrich-Schiller-Universität, Sektion Mathematik, UHH, DDR-6900 Jena, GDR

DEMARIA,D.C., Prof. Dr., Instituto di Matematica, Università di Torino, Via Principe Amedeo 8, I-10123-Torino, Italy

FAUDREE,R.J., Prof. Dr., Dept. of Mathematics Sciences, Memphis State University, Memphis, Tennessee 38152, USA

FISCHER,T., Prof. Dr., Akademie der Wissenschaften der DDR, Karl-Weierstraß-Institut für Mathematik, Mohrenstraße 30, DDR-1086 Berlin, GDR

FLEISCHNER,H., Prof. Dr., Institut für Informationsverarbeitung, Österreichische Akademie der Wissenschaften, Sonnenfelsgasse 19, A-1010 Wien, Austria

GEORGE,J.C., Prof. Dr., University of Illinois, Dept. of Mathematics, Urbana, IL 61801, USA

GIANELLA,G.M., Prof. Dr., Instituto di Matematica, Università di Torino, Via Principe Amedeo 8, I-10123-Torino, Italy

GIUDJENOV,I.O., Prof. Dr., University of Blagoevgrad, Dept. of Mathematics, Blagoevgrad -2700, Bulgaria

GOULD,R.J., Prof. Dr., Emory University, Dept. of Mathematical Sciences, Memphis, Tennessee 38152, USA

GRÖTSCHEL,M., Prof. Dr., Lehrstuhl f. Angew. Math. II., Universität Augsburg, Memminger Straße 6, D-8900 Augsburg, FRG

GRÜNWALD,N., Prof. Dr., Ing.-Hochschule für Seefahrt, Sektion Grundlagenausbildung, DDR-2530 Warnemünde, GDR

GUTMAN,I., Prof. Dr., University of Kragujevac, Faculty of Science, P.O.B. 60, YU-34000 Kragujevac, Yugoslavia

GUY,R.K., Prof. Dr., The University of Calgory, Dept. of Mathematics & Statistics, 2500 University Drive N.W., Calgary, Alberta T2N 1N4, Canada

HARANT,J., Prof. Dr., Technische Hochschule Ilmenau, Sektion Mathematik, Postfach 327, DDR-6300 Ilmenau, GDR

HARBORTH,H., Prof. Dr., Technische Universität Braunschweig, Institut für Geometrie, Pockelsstraße 14, D-3300 Braunschweig, FRG

HAMIDOUNE,Y.O., Prof. Dr., Université Paris, E. Combinatoire, 4 Place Jussieu, F-75230 Paris, France

HARNAU,W., Prof. Dr., Päd. Hochschule K.F.W. Wander, Sektion Mathematik, Wigardstraße 17, DDR-8060 Dresden, GDR

HARTSFIELD,N., Prof. Dr., Western Washington University, Dept. of Mathematics, Bellingham, WA 98225, USA

HARZHEIM,E., Prof. Dr., Universität Düsseldorf, Math. Institut, Universitätsstraße 1, D-4000 Düsseldorf 1, FRG

HAVEL,I., Prof. Dr., CSAV Mathematicky, ústav Zitna 25, CS-11567 Praha 1, CSSR

HILTON,A.J.W., Prof. Dr., University of Reading, Dept. of Mathematics, GB-Reading RG6 2AX, Great-Britain

HOEDE,C., Prof. Dr., Twente University of Technology, Dept. of Applied Mathematics, P.O.B. 217, NL-7500 AE Enschede, The Netherlands

HORAK,P., Prof. Dr., Dept. of Mathematics, SvF SVST, Radlinského 11, CS-813 68 Bratislava, CSSR

JACOBSON,M.S., Prof. Dr., Dept. of Mathematics Sciences, Memphis State University, Memphis, Tennessee 38 152, USA

JINGZHONG,M., Prof. Dr., Huazhong Normal University, Dept. of Mathematics, Wuhan, Hubei, P.R.C., China

JOHNS,G., Prof. Dr., Dept. of Mathematics, Western Michigan University, Kalamazoo, MI 49008-3899, USA

JOHNSON,P.D., Prof. Dr., ACA, Auburn University, Auburn, Alabama 36849, USA

JORDAN,D., Prof. Dr., Päd. Hochschule K.F.W. Wander, Sektion Mathematik, Wigardstraße 17, DDR-8060 Dresden, GDR

JUNG,H.A., Prof. Dr., Technische Universität Berlin, Fachbereich 3 - Mathematik, Straße des 17. Juni 135, D-1000 Berlin 12, FRG

KADEN,F., Prof. Dr., Akademie der Wissenschaften der DDR, ZKI, Kurstraße 33, DDR-1086 Berlin, GDR

KALHOFF,F.B., Prof. Dr., Universität Dortmund, Fachbereich Mathematik, Postfach 50 05 00, D-4600 Dortmund 50, FRG

KATONA,G.O.H., Prof. Dr., Magyar Tudományos Akadémia, Matematikai Kutató Intézete, Budapest, Reáltanoda u. 13-15, P.F. 127, 1364 Hungary

KEMNITZ,A., Prof. Dr., Universität Braunschweig, Institut für Geometrie, Pockelsstraße 14, D-3300 Braunschweig, FRG

KLOTZ,W., Prof. Dr., Technische Universität Clausthal, Institut für Mathematik, Erzstraße 1, D-3392 Clausthal-Zellerfeld, FRG

KOEBE,M., Prof. Dr., Ernst-Moritz-Arndt-Universität, Sektion Mathematik, Friedrich-Ludwig-Jahnstraße 15a, DDR-2200 Greifswald, GDR

KOPPITZ,J., Prof. Dr., Pädagogische Hochschule "N.K. Krupskaja", Sektion Math./Physik, Kröllwitzer Straße 44, DDR-4050 Halle/S., GDR

KRASIKOV,I., Prof. Dr., School of Mathematical Sciences, Tel-Aviv University, Tel-Aviv, Israel

KUPERBERG,W., Prof. Dr., Mathematics-F.A.T. Auburn University, Auburn AL 36849-5310, USA

LABAHN,R., Prof. Dr., Wilhelm-Pieck-Universität, Sektion Mathematik, Universitätsplatz 1, DDR-2500 Rostock, GDR

LABORDE,J.-M., Prof. Dr., Laboratoire de Structures Discrêtes et de Didactique (IMAG), BP 53x, F-38041-Grenoble cedex, France

LANG,R., Dr., Universität Hamburg, Fachbereich Informatik, Bundesstraße 55, D-2000 Hamburg 13, FRG

LASSMANN,G., Prof. Dr., Research Institute of the German PTT, Ringbahnstraße 130, D-1000 Berlin 42, FRG

LESCH,M., Prof. Dr., Pädagogische Hochschule K.F.W. Wander, Sektion Mathematik, Wigardstraße 17, DDR-8060 Dresden, GDR

LITTLE,C.H.C., Prof. Dr., Massey University, Private Bag, Palmerston North, New Zealand

LLOYD,K.E., Prof. Dr., University of Southampton, Faculty of Mathematical Studies, GB-Southampton SO9 5NH, Great Britain

LOEBL,M., Prof. Dr., Universität Bonn, Forschungsinstitut für Diskrete Mathematik, Nassestraße 2, D-5300 Bonn 1, FRG

MAHRHOLD,K., Prof. Dr., Ing.-Hochschule für Seefahrt, Sektion Grundlagenausbildung, DDR-2530 Warnemünde, GDR

MARCU,D., Prof. Dr., Str. Pasului 3, Sect. 2, R-70241, Bucharest, Romania

MENGERSEN,I., Prof. Dr., Technische Universität Braunschweig, Institut für Geometrie, Pockelsstraße 14, D-3300 Braunschweig, FRG

MOHAR,B., Prof. Dr., University of Ljubljana, Dept. of Mathematics, Jadranska 19, YU-61111 Ljubljana, Yugoslavia

MULDER,H.M., Prof. Dr., Erasmus Universiteit Rotterdam, Econometrisch Instituut,Postbus 1738, NL-3000 DR Rotterdam, The Netherlands

NEDELA,R., Prof. Dr., Dept. of Mathematics, College of Education, CS-975 49 Banskâ Bystrica, CSSR

NOVAK, J., Prof. Dr., Halkova 1, CS-46001 Liberec I, CSSR

NOWAKOWSKI,R.J., Prof. Dr., The University of Calgory, Dept. of Mathematics & Statistics, 2500 University Drive N.W., Calgary, Alberta T2N 1N4, Canada

OELLERMANN,O.R., Prof. Dr., University of Natal, Dept. of Computer Science, Durban, 4001, South Africa

OKAMURA,H., Prof. Dr., Osaka City University, Faculty of Engineering, Sugimoto, Osaka 558, Japan

PACH,J., Prof. Dr., Courant Institute of Mathematical Sciences, New York University, 251 Mercer Street, New York, N.Y. 10012, USA

PELC,A.R., Prof. Dr., Dépt. d'Informatique, Université du Québec à Hull, C.P. 1250, Succ. "B", Hull, Québec J8X 3X7, Canada

POLLACK,R., Prof. Dr., Courant Institute of Mathematical Sciences, New York University, 251 Mercer Street, New York, N.Y. 10012, USA

POLJAK,S., Prof. Dr., Charles University, Dept. of Applied Mathematics, Malostranské n. 25, CS-11800 Prague 1, CSSR

PRÖMEL,H.J., Prof. Dr., Universität Bonn, Forschungsinstitut für Diskrete Mathematik, Nassestraße 2, D-5300 Bonn 1, FRG

QUILLIOT,A., Prof. Dr., Université Clermont Blaise Pascal, BP 45, F-63177 Aubiere, France

RAUSCHE,A., Prof. Dr., Ernst-Moritz-Arndt-Universität, Sektion Mathematik, Jahnstraße 15a, DDR-2200 Greifswald, GDR

RIVAL,I., Prof. Dr., Dept. of Computer Science, University of Ottawa, Ontario K1N 6N5, Canada

RODITTY,Y.R., Prof. Dr., School of Mathematical Sciences, Tel-Aviv University, Tel-Aviv, Israel

SACHS,H., Prof. Dr., Technische Hochschule Ilmenau, Abt. Mathematik, Rechentechnik, Ökonomische Kybernetik, Postfach 327, DDR-6300 Ilmenau, GDR

SCHEFFLER,P., Prof. Dr., Akademie der Wissenschaften der DDR, Karl-Weierstraß-Institut für Mathematik, Mohrenstraße 39, DDR-1086 Berlin, FRG

SCHELP,R.H., Prof. Dr., Memphis State University, Dept. of Mathematical Sciences, Memphis, Tennessee 38152, USA

SCHILD,G., Dr., Technische Hochschule Ilmenau, Abt. Mathematik, Rechentechnik, Ökonomische Kybernetik, Postfach 327, DDR-6300 Ilmenau, GRD

SCHMEICHEL,J., Prof. Dr., Päd. Hochschule K.F.W. Wander, Sektion Mathematik, Wigardstraße 17, DDR-8060 Dresden, GDR

SCHMITT,P., Prof. Dr. , Universität Wien, Institut für Mathematik, Strudlhofgasse 4, A-1090 Wien, Austria

SCHNABEL,R., Dr., Universität Kiel, Seminar für Mathematik, Olshausenstraße 40, D-2300 Kiel 1, FRG

SCHÖNE,W., Dr., Technische Universität Karl-Marx-Stadt, Sektion Mathematik, PSF 964, DDR-9010 Karl-Marx-Stadt, GDR

SCHUMACHER,H., Prof. Dr., Pädagogische Hochschule Kiel, Institut für Mathematik und ihre Didaktik, Olshausenstraße 75, D-2300 Kiel 1, FRG

ŠIRAŇ,J., Prof. Dr., Dept. of Mathematics, SvF SVST, Radlinskêho 11, CS-81368 Bratislava, CSSR

SKOVIERA,M., Prof. Dr., Dept. of Theoretical Cybernetics, MFF UK, Mlynskâ dolina, CS-842 15 Bratislava, CSSR

SKUPIEN,Z., Prof. Dr., Institute of Mathematics, Academy of Mining and Metallurgy, PL-30-059 Krakow, Poland

SMIT,H.J., Prof. Dr., University of Twente, Dept. of Applied Mathematics, P.O.B. 217, NL-7500 AE Enschede, The Netherlands

SPENCER,J., Prof. Dr., Courant Institute of Mathematical Sciences, New York University, 251 Mercer Street, New York, N.Y. 10012, USA

STIEFEL,B., Prof. Dr., Ing.-Hochschule Mittweida, Sektion Informationselektronik, Platz des DSF 17, DDR-9250 Mittweida, GDR

SWART,H.C., Prof. Dr., University of Natal, Dept. of Mathematics, Durban, 4001, South Africa

SYSLO,M.M., Prof. Dr., University of Wroclaw, Institute of Computer Science, ul. Przesmyckiego 20, PL-51151 Wroclaw, Poland

SZE-CHIN,S., Prof. Dr., National University of Singapore, Dept. of Mathematics, Singapore 0511

TEUMER,G.W., Prof. Dr., Ernst-Moritz-Arndt-Universität, Sektion Mathematik, Jahnstraße 15a, DDR-2200 Greifswald, GDR

THIES,C., Dr., Pädagogische Hochschule Kiel, Institut für Mathematik und ihre Didaktik, Olshausenstraße 75, D-2300 Kiel 1, FRG

THRON,R., Prof. Dr.,Päd. Hochschule "N.K. Krupskaja", Sektion Math./ Physik, Kröllwitzer Straße 4, DDR-4050 Halle/S., GDR

TRIESCH,E., Prof. Dr., RWTH Aachen, Lehrstuhl C für Mathematik, Templergraben 55, D-5100 Aachen, FRG

VELDMAN,H.J., Prof. Dr., Dept. of Applied Mathematics, Twente University of Technology, P.O.B. 217, NL-7500 AE Enschede, The Netherlands

VINCE,A., Prof. Dr., University of Florida, Gainesville, Florida 32611, USA

VOIGT,B., Prof. Dr., Universität Bonn, Forschungsinstitut für Diskrete Mathematik, Nassestraße 2, D-5300 Bonn 1, FRG

VOSS,H.-J., Prof. Dr., Päd. Hochschule K.F.W. Wander, Sektion Mathematik, Wigardstraße 17, DDR-8060 Dresden, GDR

WAGNER,K., Prof. Dr., Wodanstraße 57, D-5000 Köln 91, FRG

WAKABAYASHI,Y., Prof. Dr., Universidade de Sao Paulo, Instituto de Matemâtica e Estatistica, Caixa Postal 20.570 (Agência Iguatemi), 01498 Sao Paulo SP, Brazil

WALTHER,G., Prof. Dr., Pädagogische Hochschule Kiel, Institut für Mathematik und ihre Didaktik, Olshausenstraße 75, D-2300 Kiel 1, FRG

WALTHER,H.-J., Prof. Dr., Technische Hochschule Ilmenau, Abt. Mathematik, Rechentechnik, Ökonomische Kybernetik, PF 327, DDR-6300 Ilmenau, GDR

WARNKE,I., Prof. Dr., Wilhelm-Pieck-Universität, Sektion Mathematik, Universitätsplatz 1, DDR-2500 Rostock, GDR

WEBER,K., Prof. Dr., Ing. Hochschule für Seefahrt, Sektion Grundlagenausbildung, DDR-2530 Warnemünde, GDR

WENGER,E., Prof. Dr., Institut für Informationsverarbeitung, Österreichische Akademie der Wissenschaften, Sonnenfelsgasse 19/2, A-1010 Wien, Austria

WERTH,J.S., Prof. Dr., Western Washington University, Bellingham, WA 98225, USA

WILSON,R.J., Prof. Dr., The Open University, Faculty of Mathematics, GB Milton Keynes MK7 6AA, Great Britain

ZELINKA,B., Prof. Dr., Dept. of Metal Forming and Plastics, Institute of Mechanical and Textile Technology, Studentskâ 1292, CS-46117 Liberec 1,CSSR

ZHANG-Z., Prof. Dr., Universität Bielefeld, Fakultät für Mathematik, Universitätsstraße 1, D-4800 Bielefeld 1, FRG

ZHI-ZENG,C., Prof. Dr., Dept. of Mathematics, The Inner Mongolia Teacher's, University Huhhot, China

Kolibri 360 GmbH
Ferdinand-Jühlke-Straße 7
99095 Erfurt, Deutschland
produktsicherheit@kolibri360.de